Algorithms for Intelligent Systems

This book series publishes research on the analysis and development of algorithms for intelligent systems with their applications to various real world problems. It covers research related to autonomous agents, multi-agent systems, behavioral modeling, reinforcement learning, game theory, mechanism design, machine learning, meta-heuristic search, optimization, planning and scheduling, artificial neural networks, evolutionary computation, swarm intelligence and other algorithms for intelligent systems.

The book series includes recent advancements, modification and applications of the artificial neural networks, evolutionary computation, swarm intelligence, artificial immune systems, fuzzy system, autonomous and multi agent systems, machine learning and other intelligent systems related areas. The material will be beneficial for the graduate students, post-graduate students as well as the researchers who want a broader view of advances in algorithms for intelligent systems. The contents will also be useful to the researchers from other fields who have no knowledge of the power of intelligent systems, e.g. the researchers in the field of bioinformatics, biochemists, mechanical and chemical engineers, economists, musicians and medical practitioners.

The series publishes monographs, edited volumes, advanced textbooks and selected proceedings.

More information about this series at http://www.springer.com/series/16171

Srikanta Patnaik · Xin-She Yang ·
Ishwar K. Sethi
Editors

Advances in Machine Learning and Computational Intelligence

Proceedings of ICMLCI 2019

Springer

Editors
Srikanta Patnaik
School of Computer
Science and Engineering
SOA University
Bhubaneswar, Odisha, India

Xin-She Yang
Simulation and Modelling, School
of Science and Technology
Middlesex University
London, UK

Ishwar K. Sethi
Department of Computer
Science and Engineering
Oakland University
Rochester, MI, USA

ISSN 2524-7565 ISSN 2524-7573 (electronic)
Algorithms for Intelligent Systems
ISBN 978-981-15-5242-7 ISBN 978-981-15-5243-4 (eBook)
https://doi.org/10.1007/978-981-15-5243-4

This Springer imprint is published by the registered company Springer Nature Singapore Pte Ltd.
The registered company address is: 152 Beach Road, #21-01/04 Gateway East, Singapore 189721, Singapore

Preface

Advances in Machine Learning and Computational Intelligence is the outcome of the researchers who have been working in the area of Artificial Intelligence, Soft Computing and Machine Learning. It has provided an outlet for researchers, academicians, software developers and industry experts to share their research works, findings and new research directions with each other. The volume covers research and developments in the field of machine learning and computational intelligence along with innovative applications in various areas. This volume has been divided into five major parts i.e., (i) Modeling and Optimization, (ii) Machine Learning Techniques, (iii) Computational Intelligence, (iv) IOT and Cloud Computing and (v) Applications.

The first part comprises the papers that considered optimization problems exist in the real-world. It is often difficult to find an optimal solution and that demands highly efficient techniques. In this part, many of the papers proposed and modeled new innovative techniques to solve these challenging and practically relevant optimization problems while others presented improved methods compared with existing ones.

The second part consists of contributions showcasing works from machine learning techniques. It provides a strong combination of machine learning techniques with mathematical models. The underlying techniques are capable of extracting large amount of data from different sources and working with huge unstructured datasets and apply efficient techniques to extract meaningful, effective and accurate insights using data mining, neural networks and core machine learning techniques. These insights further help in identifying hidden patterns, future trends and making actionable decisions for solving significant problems and create impact. Today, machine learning techniques are being employed to solve some of the most crucial problems in the sectors ranging from health sector to business sector to aviation sectors. Many of the submissions fall into this part of the proceeding.

The third part deals with papers in the area of computational intelligence. Computational Intelligence is a multidisciplinary area that involves existing as well as emerging fields such as computer science, information technology, evolutionary computing, metaheuristics, and swarm intelligence etc., for solving several crucial problems of real world. These intelligent techniques are being used to process data collected from heterogeneous sources for extracting information and provide significant insights. These insights support the users by enhancing their understanding about important problems in hand and make crucial decisions in several uncertain and critical situations. It thus adds value to various problem-solving approaches when integrated into the systems. We have received a good number of submissions in this part.

The fourth part includes papers related to IoT and cloud computing which is an extension of Information Technology and involves all the infrastructure and components that enable modern computing including computer systems, mobile devices, networking components, and applications. It allows organizations (both private and public) and individuals to interact with each other in the digital world. This part of the conference has also received lots of attention.

The last part considers the cutting-edge applications in various sectors and applications areas like IoT, Cloud Computing and Blockchain approach that utilize both machine learning and computational intelligence indirectly for solving problems while scrutinizing the relevant papers. These papers are categorized into the applications part.

Bhubaneswar, India Prof. Srikanta Patnaik
London, UK Prof. Xin-She Yang
Rochester, USA Prof. Ishwar K. Sethi

Acknowledgements

The contributions covered in this proceeding are the outcome of the contributions from more than one hundred researchers. We are thankful to the authors, paper contributors of this volume and the departments which supported the event a lot.

We are thankful to the series editors of the Springer Book series on *Algorithms for Intelligent Systems* Prof. Dr. Jagdish Chand Bansal, Prof. Kusum Deep and Prof. Atulya K. Nagar for their support to bring out this volume.

We are also very much thankful to Aninda Bose, Senior Editor–Hard Sciences, SpringerNature Publishing, and his team for his constant encouragement support and time-to-time monitoring.

We are thankful to all the reviewers, who have given their time for reviewing the papers. Lastly but not the least, we are thankful to all the authors of ICMLCI-2019 and MOSICOM-2020, without whom the volume was impossible. We are sure the readers shall be benefited immensely, by referring this volume.

Prof. Srikanta Patnaik
Prof. Xin-She Yang
Prof. Ishwar K. Sethi

Contents

About the Editors

Dr. Srikanta Patnaik is a Professor at the Department of Computer Science and Engineering, Faculty of Engineering and Technology, SOA University, Bhubaneswar, India. He received his Ph.D. (Engineering) on Computational Intelligence from Jadavpur University, India, in 1999 and has since supervised 27 Ph.D. theses and more than 60 M.Tech. theses in the areas of Computational Intelligence, Soft Computing Applications and Re-Engineering. Dr. Patnaik has published over 100 research papers in international journals and conference proceedings. He is the author of 2 textbooks and editor of 42 books, published by leading international publishers like Springer-Verlag and Kluwer Academic. He is Editor-in-Chief of the International Journal of Information and Communication Technology and the International Journal of Computational Vision and Robotics, published by Inderscience, UK, and of the book series "Modeling and Optimization in Science and Technology", published by Springer, Germany.

Xin-She Yang obtained his D.Phil. in Applied Mathematics from the University of Oxford and then worked at Cambridge University and UK's National Physical Laboratory as a Senior Research Scientist. Now, he is a Reader at Middlesex University London. He is also the IEEE CIS task force chair for business intelligence and knowledge management. With more than 250 research publications and 25 books, he has been a highly cited researcher for consecutive four years (2016–2019), according to Web of Science.

Ishwar K. Sethi is currently a Professor in the Department of Computer Science and Engineering at Oakland University in Rochester, Michigan, where he served as the chair of the department from 1999 to 2010. From 1982 to 1999, he was with the Department of Computer Science at Wayne State University, Detroit, Michigan. Before that, he was a faculty member at Indian Institute of Technology, Kharagpur, India, where he received his Ph.D. degree in 1978.

His current research interests are in data mining, pattern classification, multimedia information indexing and retrieval and deep learning and its applications. He has authored or co-authored over 180 journal and conference articles. He has served on the editorial boards of several prominent journals including IEEE Transactions on Pattern Analysis and Machine Intelligence, and IEEE MultiMedia. He was elected IEEE Fellow in 2001 for his contributions in artificial neural networks and statistical pattern recognition and achieved the status of Life Fellow in 2012.

Modeling and Optimization

Intrusion Detection Using a Hybrid Sequential Model

Abhishek Sinha, Aditya Pandey, and P. S. Aishwarya

1 Introduction

1.1 Context

A network intrusion could be any unauthorized activity on a computer network. Virus attacks, unauthorized access, theft of information, and denial-of-service attacks were the greatest contributors to computer crime. Detecting an intrusion depends on the defenders having a clear understanding of how attacks work. Detecting an intrusion is the first step to create any sort of counteractive security measure. Hence, it is very important to accurately determine whether a connection is an intrusion or not.

1.2 Categories

There are four broad categories of intrusions in a network of systems:

1.2.1 Dos

In computing, a denial-of-service attack is a cyber-attack in which the perpetrator seeks to make a machine or network resource unavailable to its intended users by

A. Sinha · A. Pandey · P. S. Aishwarya (✉)
PES University, 100 Feet Ring Road, Banashankari III Stage, Bangalore 560085, India
e-mail: aishwarya.ps31@gmail.com

A. Sinha
e-mail: sinha.abhish3k@gmail.com

A. Pandey
e-mail: pandeyan98@gmail.com

S. Patnaik et al. (eds.), *Advances in Machine Learning and Computational Intelligence*, Algorithms for Intelligent Systems, https://doi.org/10.1007/978-981-15-5243-4_1

temporarily or indefinitely disrupting services of a host connected to the Internet, e.g., back, land, and Neptune (Tables 1, 2, 3, 4).

1.2.2 U2r

These attacks are exploitations in which the hacker starts off on the system with a normal user account and attempts to abuse vulnerabilities in the system in order to gain superuser privileges, e.g., perl, xterm.

1.2.3 R2l

Remote to local is an unauthorized access from a remote machine, by maybe guessing the password of the local system and accessing files within the local system, e.g., ftp write, guess passwd, imap.

1.2.4 Probe

Probing is an attack in which the hacker scans a machine or a networking device in order to determine weaknesses or vulnerabilities that may later be exploited so as to compromise the system. This technique is commonly used in data mining, e.g., saint, portsweep, mscan, and nmap.

1.3 Importance of Intrusion Detection

Intrusion detection is important for both military as well as commercial sectors for the sake of their information security, which is the most important topic of research for future networks. It is critical to maintaining a high level of security to ensure safe and trusted communication of information between various organizations.

Intrusion detection system is a new safeguard technology for system security after traditional technologies, such as firewalls and message encryption. An intrusion detection system (IDS) is a device or software application that monitors network system activities for malicious activities or policy violations and produces reports to a management station.

2 Previous Work

The two approaches to an intrusion detection system are misuse detection and anomaly detection. A key advantage of misuse detection techniques is its high degree of accuracy in detecting known attacks and their variations. Their obvious drawback is the inability to detect attacks whose instances have not yet been observed. Anomaly detection schemes, on the other hand, suffer from a high rate of false alarms. This occurs primarily because previously unseen (yet legitimate) system behaviors are also recognized as anomalies and are hence flagged as potential intrusions. Researchers have used various algorithms to solve this specific problem.

A data mining algorithm such as random forest can be applied for misuse, anomaly, and hybrid network-based intrusion detection systems. [7] uses a hybrid detection technique where they employ misuse detection followed by anomaly detection. This approach, however, has multiple limitations:

- Intrusions need to be much lesser than the normal data. Outlier detection will only work when the majority of the data is normal.
- A novel intrusion producing a large number of connections that are not filtered out by the misuse detection could decrease the performance of the anomaly detection or even the hybrid system as a whole.
- Some of the intrusions with a high degree of similarity cannot be detected by this anomaly detection approach.

The paper 'Artificial Neural Networks for Misuse Detection' [8] talks about the advantages of using an artificial neural network approach over a rule-based expert system approach. It also tells us about the various approaches with which these neural networks are applied to get high accuracy. A review paper on misuse detections [9] sheds light on the most common techniques to implement misuse detection. These are:

- Expert systems, which code knowledge about attacks as 'if-then' implication rules.
- Model-based reasoning systems, which combine models of misuse with evidential reasoning to support conclusions about the occurrence of a misuse.
- State transition analysis, which represents attacks as a sequence of state transitions of the monitored system.
- Keystroke Monitoring, which uses user keystrokes to determine the occurrence of an attack.

Landge and Wadh introduces a pattern matching model along with an artificial neural network model for their implementation of a misuse detection system. Further [9] performed intrusion detection using data mining techniques along with fuzzy logic and basic genetic algorithms. [1–5] are varied approaches, but their accuracy is not as good as the approach in [7]. The approach specified in [6] has good accuracy and also supports unsupervised learning algorithms, but it requires a very complicated state machine specification model.

3 Problem Statement

Here, our goal is to detect network intrusions and to create a predictive model that is able to differentiate between 'good' or 'bad' connections (intrusions) and classify those intrusions into known categories. To that end, we used the KDD cup dataset from 1999 which was created by simulating various intrusions over a network over the course of several weeks in a military environment.

The dataset itself consists of 42 attributes that are used in the various models according to relevance and importance. It contains approximately 1,500,000 data points. For the purposes of this project, we use a 10% representative dataset, since our machines did not have the processing power to handle the larger dataset. The dataset, as is, is unbalanced across the various result categories, and hence, we balance it by upsampling and downsampling. The dataset is also cleaned. Certain categorical variables like the 'protocol type' column are one-hot encoded. We found out that the 'flag' and 'services' attributes are not of much value as they are nominal variables, and hence, they are dropped.

4 Approach

Due to the various drawbacks of the individual anomaly detection models as well as the individual misuse detection models, we use a combined approach, i.e., a hybrid of the two. We combine the models serially such that the anomaly detection is followed by the misuse detection. This approach provides us with multiple advantages:

- The misuse detection acts as a verification model where it verifies whether an anomaly is actually an intrusion or not. This helps us reduce the false positives coming from the anomaly detection (primary drawback).
- The misuse detection also helps us classify the intrusions detected into various categories based on the intrusions 'signature' (centroid of sample data from the training dataset).

Figure 1 shows the approach for the intrusion detection system.

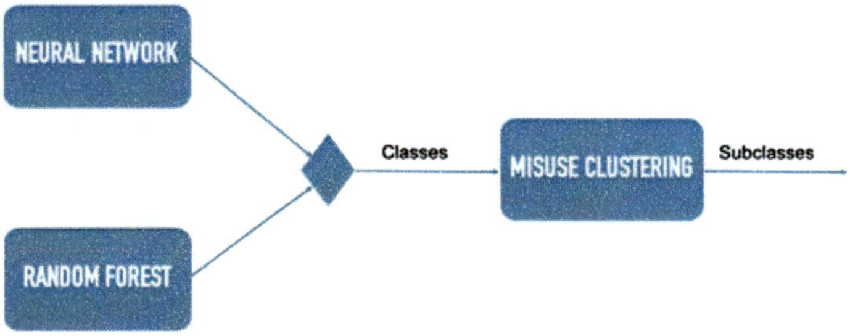

Fig. 1 Block diagram of approach

Table 1 Number of data points

Label:	Dos	Normal	Probe	Rtl	U2r
Before sampling	54,572	87,832	2131	999	52
After sampling	27,285	39,524	2131	999	86

4.1 Modeling

For anomaly detection, we use two models:

- Random forest model.
- Neural network model.

After running both these classification models parallelly, we take the combined output of the anomalies detected by either of the two. Doing so helps us reduce the rate of false negatives. The connections which are identified as anomalies by either of the two are then passed on to the misuse detection model. A false positive from the anomaly detection model classified as 'normal' in the misuse detection model reduces the number of false positives.

For misuse detection, we use a K-means-based clustering model. Therefore, the use of the misuse detection model on the anomalies helps trim down the number of false positives. The clustering model can also be used to further classify the 5 classes of attacks into 24 subclasses which will help in fighting or preventing the occurrence of the attack.

5 Components of the Intrusion Detection System

Our intrusion detection system consists of 3 components which have been combined to create one complete robust system. The components of the anomaly detection system are as follows.

5.1 Neural Network

A neural network is a system of hardware and/or software patterned after the operations of neurons in the human brain.

For the neural network, we created a sequential neural network with two hidden layers. The neural network consists of 41 input nodes and 5 output nodes. The 41 input nodes are the numerical attributes that are obtained after preprocessing, and the 5 output nodes are used to classify the data point into one of the 5 classes of connections that we have (u2r, rtl, normal, dos, probe).

The model is validated by a K-fold cross-validation with k as 2, and an average accuracy of 99.57% was obtained. In an attempt to reduce processing time without much drop in accuracy, we reduced the number of splits in the k-fold cross-validation as well as used a lower number of hidden layers in the network with minimal loss of accuracy.

5.2 Random Forest

A random forest is essentially a multitude of decision trees. The output obtained from a random forest model is a combination of the outputs obtained from all the decision trees.

In our case, we use the mode of all the decision tree outputs as the final output from the model. Based on trial and error, the number of decision trees in our random forest was taken as 100. We use this random forest classifier to classify the connection among the 5 main classes.

Based on the importance value of variables, all the non-essential variables are dropped before the model is retrained. This helps in removing all the redundant and useless attributes in the model. Using the random forest model, an average accuracy of 99.78% is obtained. In order to reduce the processing time, the number of decision trees was reduced from an initial value of 1000–100 without resulting in any significant loss in accuracy of the model. Figure 2 shows a part of the large random forest model created (Fig. 2).

5.3 Misuse Clustering

We used a K-means-based clustering algorithm as the misuse detection model. This clustering algorithm is used to further classify the five classes of connections into 24 classes. During training, clusters are made for each known class and their centroids are stored. When the model receives a data point during testing, the minimum distance to any of the centroids is calculated to assign the point to the relevant class. If a

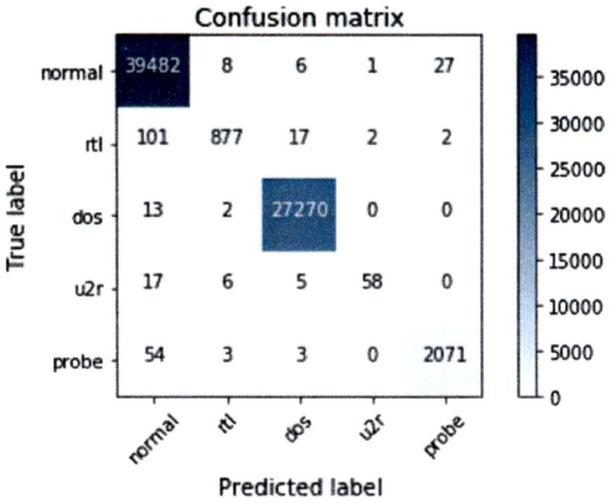

Fig. 2 Confusion matrix for 5 classes

connection arrives during testing that is falsely tagged as an attack, then the cluster it is assigned to should theoretically be 'normal,' thereby reducing false positives.

5.4 Combination of Models

The output from both the models in the anomaly detection stage is verified with each other, and the connections labeled as some kind of attack or connections not having the same labels are passed on to the misuse clustering model which will further classify the attack into its subclass. Another version of our final model is to use the misuse clustering model to reduce the number of false positives, thereby increasing precision and accuracy.

6 Results

6.1 Discussion of Results

Since we are using a synthetic dataset, we obtain accuracies that are quite high. However, the precision and recall for u2r in specific, in both the models are very low compared to the rest of our numbers. This can be attributed to the fact that the number of connections causing u2r attacks itself is very low, making up only around 70 of the 70,000 connections.

Table 2 Neural network results

Label:	Normal	Rtl	Dos	U2r	Probe
Precision	98.391	97.560	99.955	79.710	97.468
Recall	99.817	80.080	99.146	63.953	90.333
Accuracy	98.976	99.687	99.650	99.935	99.634

Table 3 Random forest results

Label:	Normal	Rtl	Dos	U2r	Probe
Precision	99.820	99.382	99.985	87.500	99.669
Recall	99.949	96.596	99.978	81.395	98.967
Accuracy	99.870	99.942	99.985	99.962	99.958

Table 4 Misuse clustering results

Type of classification:	5 class	24 class
Accuracy	99.651	91.31

As a result, the models have not been trained enough to be able to detect these attacks. Hence, a good proportion of u2r attacks have been misclassified. This problem can be rectified if a more balanced dataset is provided.

From the tabular columns, we can see that most of the attacks have not only been detected but also correctly classified in the anomaly detection model. In an attempt to further classify the attacks into its subcategories, we lose a bit of accuracy but it is still at an acceptable 91.3%. A further classification is necessary from the perspective of dealing with these intrusions. For example, A and B might both belong to 'dos' but they might have different security requirements to stop these attacks. We can also use the misuse clustering to cut down on false positives and that will only make our accuracy and precision better if in case the drop in accuracy is not preferable.

From the confusion matrix, we can also see that the number of 'normal' and 'dos' connections are significantly higher than the number of all the other connections even after downsampling these categories. Therefore, in order to achieve a balance that is acceptable, we downsampled these 2 categories and upsampled the 'u2r' category.

What can also be seen is that the number of misclassifications is fairly less. The only issue which can be seen is that a fraction of the misclassification is classifying one of the kinds of attacks as normal connections. The most probable reason for this is due to the fact that the number of 'normal' connections make up more than 55% of the total dataset. However, considering the number of false negatives versus the total number of connections, this error is almost insignificant.

Another important aspect of an intrusion detection system is its applicability to be used in real-time systems. We believe that our solution is easily deployable as once our model is trained (on previous data), and the classification happens almost instantaneously. The clustering algorithm is also able to identify new anomalies by comparing the incoming traffic to the cluster centroids that are already there using a cutoff to determine the novelty of that particular packet.

7 Conclusion

Even though technology has advanced leaps and bounds over the past few decades, intrusion detection continues to be an active research field. This paper outlines our approach to solving this problem by using a combination of anomaly and misuse detection models. The techniques used to perform the same have been explained and illustrated. We believe that this approach is quite successful in classifying connections into their respective categories.

8 Future Work

The current approach is able to spot and classify new intrusions as such. However, the number of false positives for this is high. Furthermore, the connections detected as new intrusions could be further categorized into subclasses based on similarity (clustering) and dealt with accordingly. Different approaches could also be tried in order to obtain more accurate signatures for the different categories of intrusions which would improve the accuracy of the misuse detection model.

References

1. A. Lazarevic, L. Ertoz, V. Kumar, A comparative study of anomaly detection schemes in network intrusion detection. in *Proceedings of the 2003 SIAM International Conference on Data Mining* (2002)
2. D. Barbara, N. Wu, S. Jajodia, Detecting novel network intrusions using bayes estimators. in *Proceedings of the 2001 SIAM International Conference on Data Mining* (2001)
3. S.A. Hofmeyr, S. Forrest, A. Somayaji, Intrusion detection using sequences of system calls. J. Comput. Security. **6**(3), 151–180 (1998)
4. A. Ghosh, A. Schwartzbard, A study in using neural networks for anomaly and misuse detection. in *Proceedings of the 8th USENIX Security Symposium*, August 23–36,(1999), pp. 141–152
5. E. Eskin, W. Lee, S.J. Stolfo, Modeling system calls for intrusion detection with dynamic window sizes. in *Proceedings DARPA Information Survivability Conference and Exposition II*, (DISCEX'01, 2001)
6. R. Sekar, A. Gupta, J. Frullo. Specification-based anomaly detection: A new approach for detecting network intrusions. in *CCS '02: Proceedings of the 9th ACM conference on Computer and communications security*, November 2002, (2002), pp. 265–274
7. J. Zhang, M. Zulkernine, A. Haque, Random-forests-based network intrusion detection systems. in *IEEE Transactions on Systems, Man, and Cybernetics—part C: Applications and Reviews*, vol. 38, No. 5 (2008)

8. J. Cannady, Artificial neural networks for misuse detection. in *National Information Systems Security Conference*, (1998)
9. R.S. Landge, A.P. Wadh, Misuse detection system using various techniques: A review. Int. J. Adv. Res. Comput. Sci., Udaipur **4**(6) (2013)

Simulation and Analysis of the PV Arrays Connected to Buck–Boost Converters Using MPPT Technique by Implementing Incremental Conductance Algorithm and Integral Controller

D. S. Sumedha, R. Shreyas, Juthik B. V., and Melisa Miranda

1 Introduction

Photovoltaic cells are device modules that convert directly solar energy into electrical energy. As solar energy is the renewable source of energy, it is most advantageous to drive as a power source. Once installed, these produce no pollution to the environment. Efficiency of PV panels is extensively affected by certain factors such as climatic shade, orientation, the intensity of illumination, and temperature. Unless an ideal solar cell, it is impossible to achieve 100% efficiency. About 55% efficiency can be achieved when a sun tracking system is implemented. There are ongoing efforts and research made in regard to counteract the inefficiency of PV arrays [3]. MPPT technique enhances the efficiency of the PV arrays by 30% which can be further cascaded with solar modules [1]. Results reveal that a significant amount of power is wasted when a non-MPPT scheme is introduced. According to measured data, 81.56% average power is achieved compared to the non-MPPT technique which is 11.4%. Moreover, the result shows that MPPT is more efficient than non-MPPT which is about 86.02% according to average output power [2]. In [2], the comparison of systems with and without MPPT is proposed extensively. Here, we are going to implement the MPPT technique, which can be added in combination with PV modules.

Different algorithms have been developed to implement MPPT techniques [2]. Detailed work has been proposed to compare and analyze these algorithms. Out of all the algorithms developed, perturb and observe and incremental conductance algorithms are widely used [4]. P&O algorithm considers the swift change in irradiation level as a change in MPP due to perturbation and correspondingly changes the MPPT which eventually results in calculation error of the system. To overcome the anomalous calculation, an incremental conductance algorithm is much preferred.

D. S. Sumedha · R. Shreyas · B. V. Juthik · M. Miranda (✉)
Department of E. C. E, PES University Bengaluru, Karnataka, India
e-mail: melisamiranda@pes.edu

S. Patnaik et al. (eds.), *Advances in Machine Learning and Computational Intelligence*, Algorithms for Intelligent Systems, https://doi.org/10.1007/978-981-15-5243-4_2

A proposal has been made to stimulate and provide a detailed analysis of MPPT techniques implemented through an incremental conductance algorithm [5]. The efficiency of the system is further increased by connecting an integral controller. In this paper, a proposal has been made to simulate efficient MPPT technique using integral controller and incremental conductance algorithm which can be employed in various applications like storage and transmission of power in gridlines. All the results are supported using MATLAB and SIMULINK. In further sections, there are simulations and analyses of our proposed approach toward the implementation of MPPT techniques for PV arrays. In the following section, the detailed analysis is mentioned about the design of the complete photovoltaic system.

2 Photovoltaic System

The proposed photovoltaic system is designed to get the maximum output power efficiency from the PV arrays before supplying it to the load. Figure 1 shows the block diagram of the system consists of important components like photovoltaic (PV) arrays, buck–boost converters, incremental conductance algorithm block, integral controller, and PWM generator.

The PV arrays change their output characteristics when there are variations in temperature and irradiation values [2]. The *V* and *I* blocks act as voltage and current sensors which calculate and feed the output to the incremental conductance block. Incremental conductance being an MPPT algorithm tracks the maximum power point. The output of this block is fed to integral controllers [13], which calculates the error in output voltage and incremental conductance output. The PWM generator is further controlled by the integral controllers. The output of the PWM is given to the buck–boost converters by varying the duty cycle of the PWM generated. Finally, the buck–boost converter increases or decreases the voltage of the PV array accordingly to maximum power output.

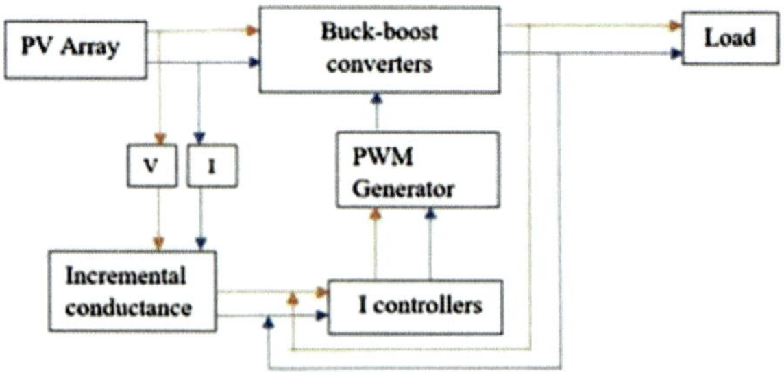

Fig. 1 Block diagram of a proposed PV system

2.1 Photovoltaic Arrays

PV arrays are composed of smaller photovoltaic cells arranged in different configurations to achieve better voltages or better current output. These configurations are either series or parallel connections of the cells. A basic PV cell is generally a photodiode whose output current is dependent on both radiation levels and temperature changes. The temperature dependence of the photovoltaic cell on the current is given by the formula:

$$I_O = \frac{eADn_i^2}{LN_D} \tag{1}$$

where e is the electronic charge (C), A is the area of the channel, D is the diffusivity of minority carriers, L is the diffusion length, and N_D is the doping level. The intrinsic carrier (n_i) is further dependent on temperature which results in corresponding change in the output current. Open-circuit voltage is also dependent on temperature and is given by the formula:

$$V_{OC} = \frac{kT}{e}\left[(\ln I_{SC}) - (\ln B) - \gamma \ln T + \frac{eV_{GO}}{kT}\right] \tag{2}$$

where k is Boltzmann's constant, V_{GO} is the band gap voltage, B and γ re constants. Differentiating the above equation proves that the voltage of the PV array is inversely proportional to the temperature of the PV array system. For silicon-based photovoltaic cells, the rate of change of voltage is -2.2 mV per °C. In Fig. 2, temperature dependence of current and voltage is plotted in the top plot and the power

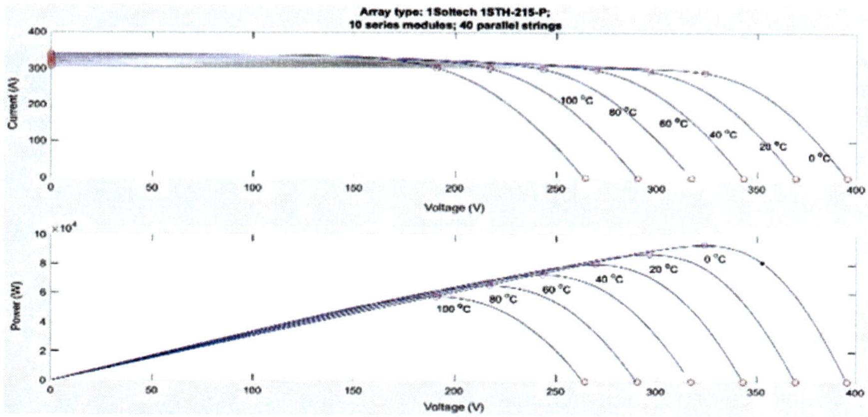

Fig. 2 Output characteristics of PV arrays for different temperatures. At the top, current versus voltage plot and power versus voltage plot at the bottom

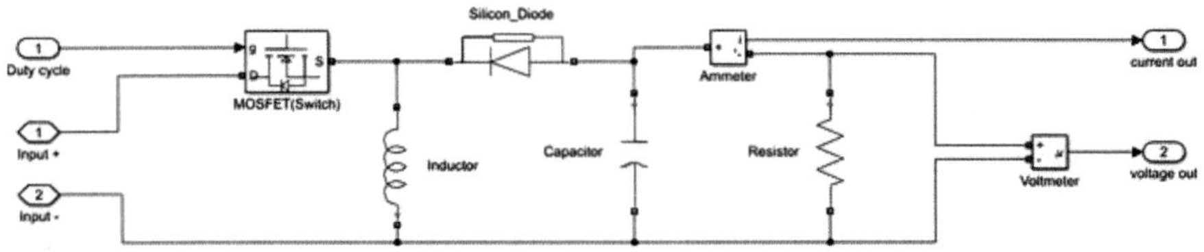

Fig. 3 Circuit of inverting buck–boost converter

voltage characteristics in the bottom plot. These graphs are plotted on SIMULINK using the PV array block with the model 1SOLTECH 1STH-215-P.

2.2 *Buck–Boost Converters*

Buck–boost converters widely come with two topologies, namely inverting and non-inverting buck–boost converters [12, 13]. These converters contain a semiconductor switch, inductor, capacitor, and diode. Diode is usually reverse biased, and this makes the output voltage to be inverted with respect to the input voltage. The switching frequency used in simulation is 50 kHz. Therefore, such converters are called inverting buck–boost converters. Figure 3 shows the circuit of inverting buck–boost converters.

2.3 *Integral Controller*

Integral controller produces an output which is proportional to the error of the system. Here, integral controller is fed with input which is the combination of the output from the buck–boost converter and incremental conductance. It compares the output current and voltage from the buck–boost converter and minimizes the steady-state error of the voltage and the current [6, 8]. The transfer function of the integral controller system is given by:

$$TransferFunction = \frac{K_I}{s} \tag{3}$$

where K_I is called the integral constant. Integral controllers are employed in this design so that the steady-state error of the system becomes very negligible and can improve the accuracy of the system on long runs. The integral constant used in the simulation is 200,000. Integral controllers when compared with other controllers provide better results with more accuracy and make the system more stable. The rise time of the controller is also very less. The output from the integral controller is fed to the PWM generator. Eventually, when duty cycle is varied, the buck–boost converters work as buck only or boost only converters to step down or step up the voltage, respectively.

2.4 *Incremental Conductance Algorithm*

Incremental conductance is the most widely used technique for maximum power point tracking (MPPT). This algorithm employs measurement of voltage and current coming out of the PV arrays [9, 10]. Implementation of incremental conductance algorithm requires a memory element since it uses the previously measured voltage and current. Figure 4 shows the flowchart of incremental conductance algorithm. The change in the voltage and current values is determined. These variations are used to further change the duty cycle of the PWM generated [7, 11]. If there is no change in current and voltage, the previous values are retained. If the change in current is positive, the duty cycle should be decreased [5]. When the change in current is negative, duty cycle must be increased. If the instantaneous change in conductance is equal to the conductance of the system, it is not necessary to change the duty cycle and the maximum power point is retained. Duty cycle is decreased if the instantaneous change in conductance is greater than the conductance of the

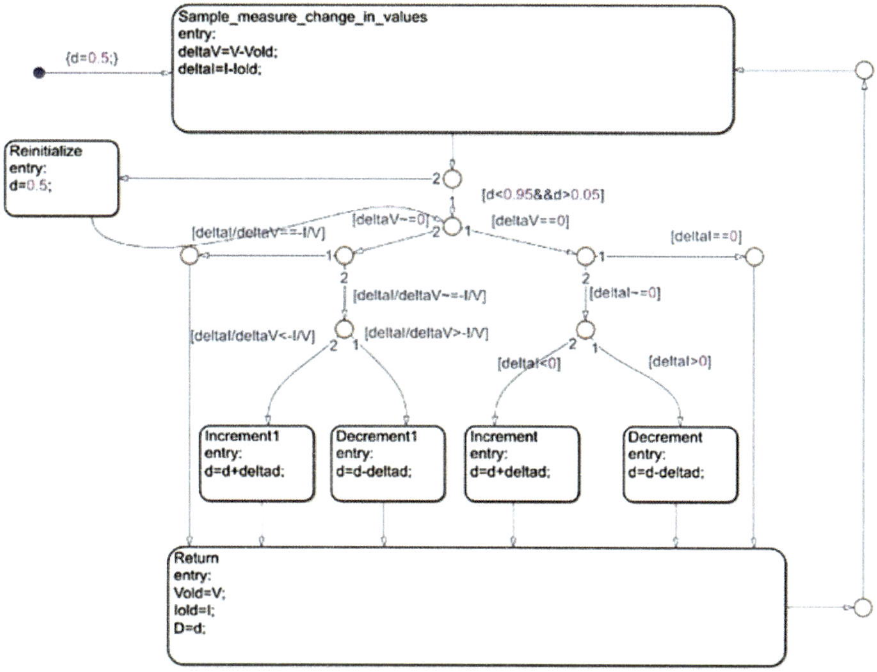

Fig. 4 Flowchart of incremental conductance algorithm in SIMULINK state-flow model

system. Further, duty cycle is increased if the instantaneous change in conductance is lesser than the conductance of the system.

3 System Performance And Analysis

Extensive simulations have been performed to implement the MPPT using incremental conductance algorithm. Various PV array modules are present in the SIMULINK. Out of these 1SOLTECH 1STH-215-P is used to design the system, and the specifications of this model are given in Table 1.

The input values to the PV array are irradiation values and the temperature values. To make the system nearer to the real-world application, irradiation and temperature values are considered in such a way that it depicts a day. Initially, the values are moderate which depicts the morning irradiation and temperate values. Later, it is increased to the maximum value which depicts the noon part of the day. Further, values are taken to a lower level which depicts the evening and dusk. Figs. 5 and 6 represent the irradiation and temperature values considered in the simulation of the system. The output voltage, current, and power of the PV array are graphically represented in Figs. 7, 8 and 9. All the graphs depicted have the time axis in seconds(s). Figure 5 has the y-axis illumination measured in lux(lx). Fig. 6 has the y-axis temperature measured in (°C). Figs. 7 and 11 have the y-axis voltage measured in Volts(V). Figs. 8 and 12 have the y-axis current measured in Ampere(A). Figs. 9 and 13 have the y-axis power measured in Watts(W).

Figs. 7, 8 and 9 depict the output parameters of the PV array without employing MPPT techniques for various values of irradiation and temperature. The different values of irradiance and temperature used in the simulation are given in Table 2.

The complete system of PV array with incremental conductance algorithm employed for MPPT technique shows the results tracking the maximum power at the output stage of the system. Fig. 10 shows the complete system designed in the SIMULINK. Further, the incremental conductance block is designed using a state flow model in SIMULINK. The output of the system after including incremental conductance algorithm and integral controllers has been depicted in Figs. 11 and 12. Fig. 11 represents the output voltage of the PV array while tracking the maximum

Table 1 Specification of PV array (ISOLTECH 1STH-215-P)

Parameter	Value
Maximum power	213.15 W
Voltage at maximum power point	29 V
Current at maximum power point	7.35A
Number of cells in the module	60
Temperature coefficient of open-circuit voltage	−0.36099%/°C

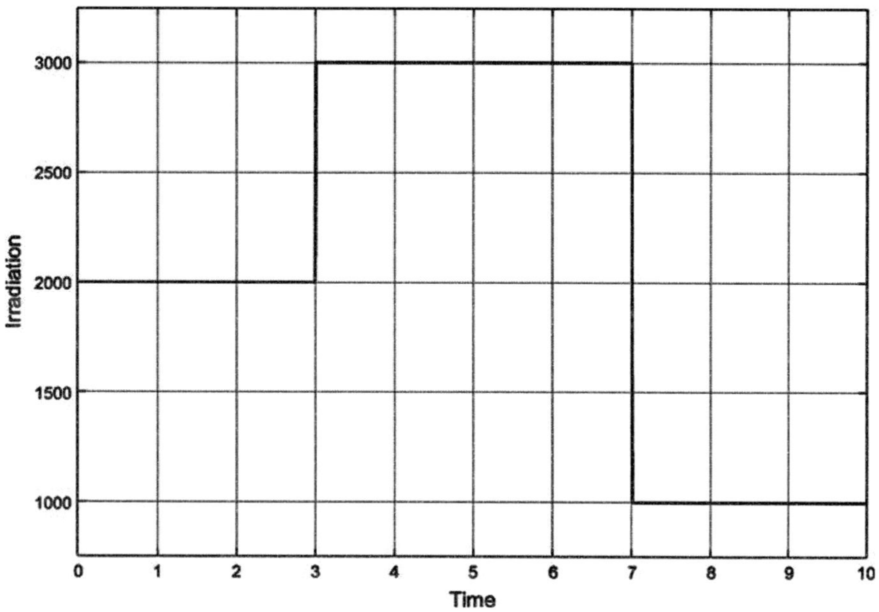

Fig. 5 Irradiation values given to the PV Array

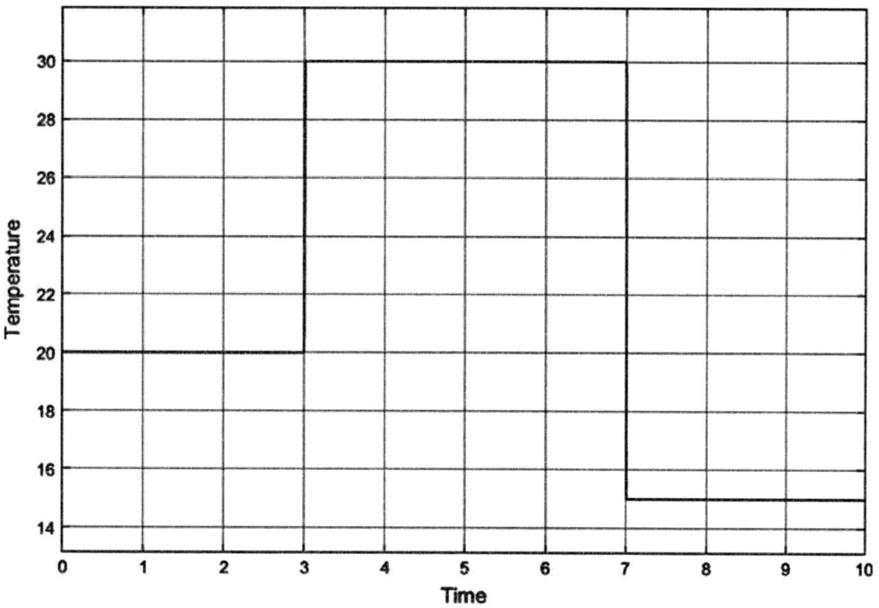

Fig. 6 Temperature values given to the PV array

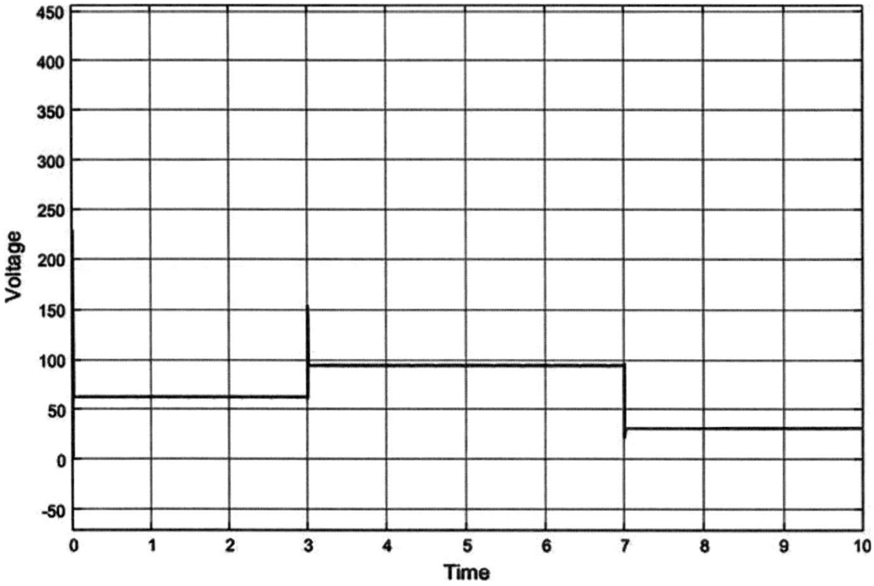

Fig. 7 Output voltage of PV array without MPPT

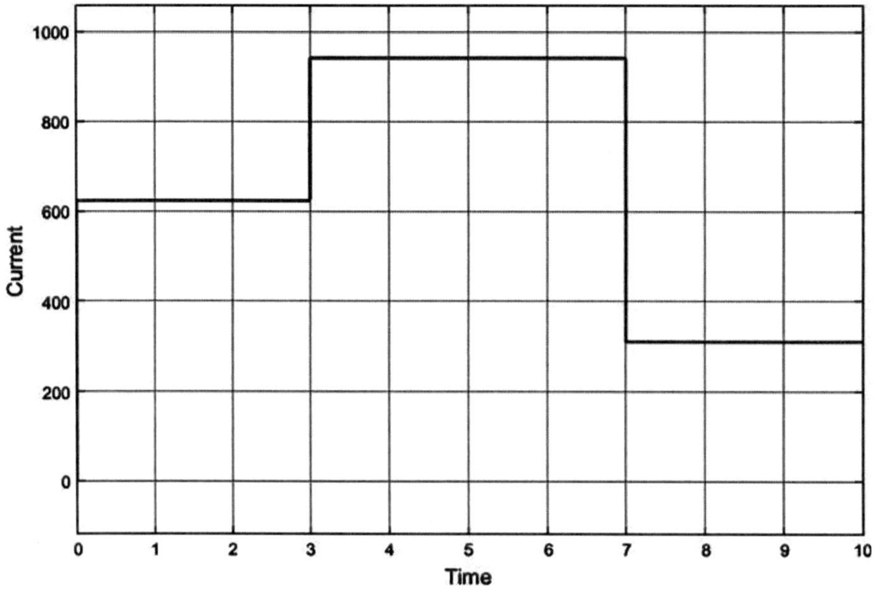

Fig. 8 Output current of PV array without MPPT

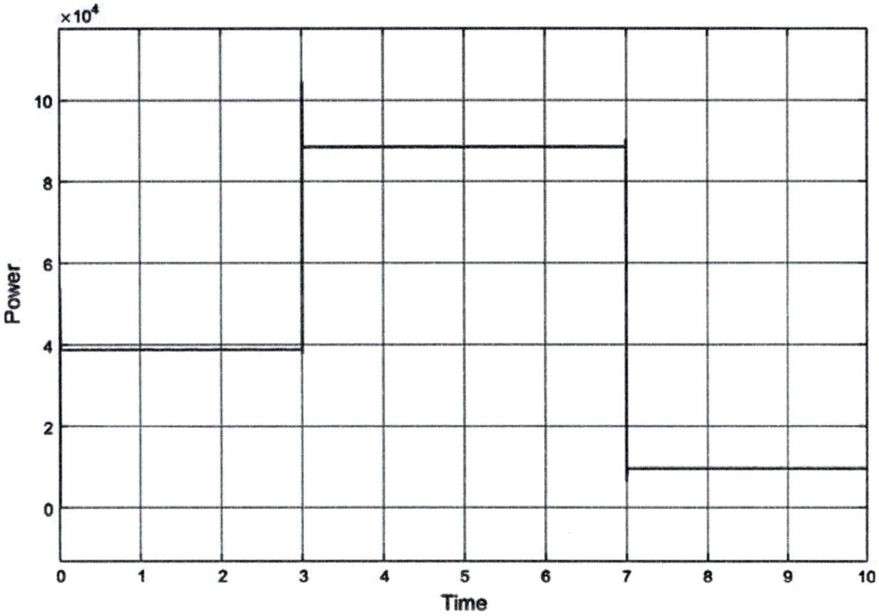

Fig. 9 Output power of PV array without MPPT

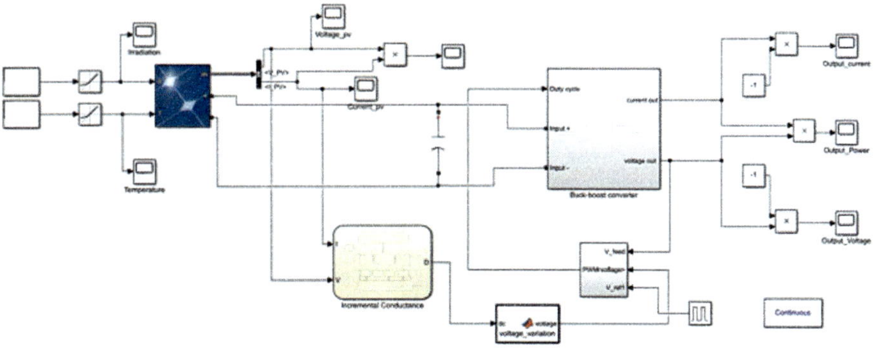

Fig. 10 SIMULINK model designed for MPPT technique

power point through the different values of irradiance and temperature. Fig. 12 represents the output current of the PV array. Fig. 13 depicts the output power of the PV array.

A nominal voltage of 60 V is chosen so that the output voltage varies around this mark. Most of the applications like charging a battery or in gridlines require a constant voltage with a variable current. In case of charging a battery, the battery is rated to a voltage. Anything more than this voltage will lead to loss of power. In this paper, the output voltage is fixed, and the current can be varied. The assumed

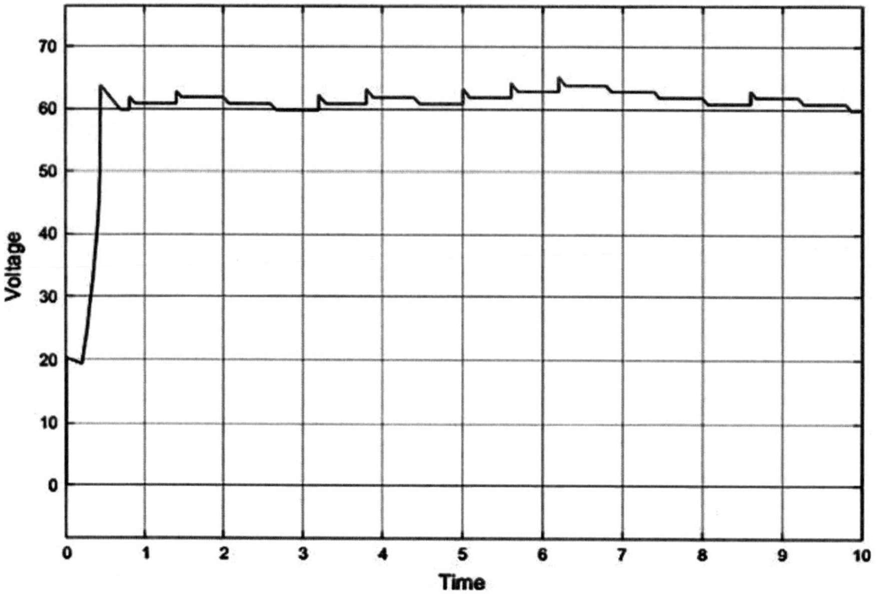

Fig. 11 Output voltage after incremental conductance

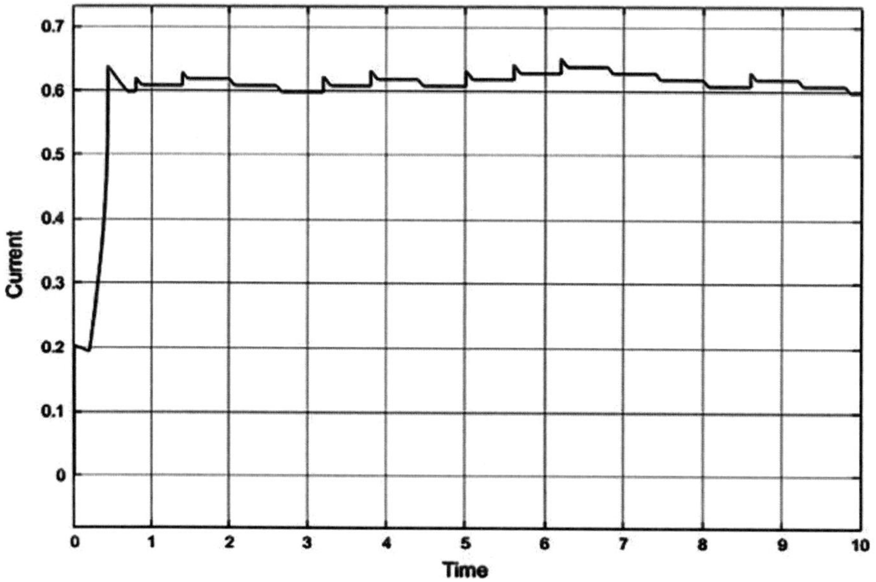

Fig. 12 Output current after incremental conductance

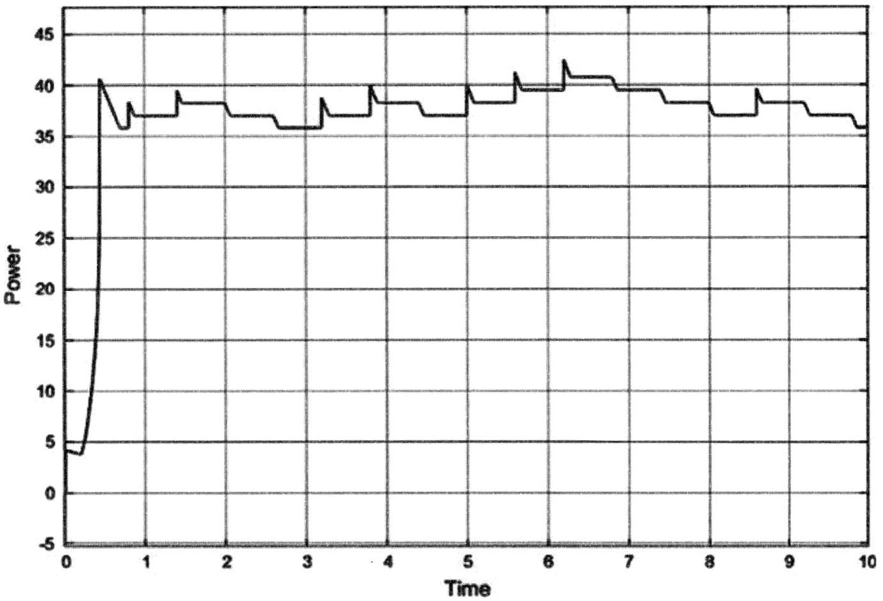

Fig. 13 Output power after incremental conductance

Table 2 Different values of irradiation and temperature

Parameter	Morning	Afternoon	Evening
Irradiance (W/m^2)	2000	3000	1000
Temperature (°C)	20	30	15

load is resistive in nature with resistance equal to 100 Ω. The output current can be furthered varied with different loads. The average power of the PV array before and after applying MPPT technique is same, whereas the instantaneous power values are different representing the tracking of maximum power point. The output voltage remains almost constant making it feasible for various applications. The maximum output power obtained with the resistive load is 42 W.

4 Conclusion

The complete simulation results of the PV system employing the incremental conductance algorithm and integral controllers prove that the system is more power efficient when compared with the normal output of the PV system. The output voltage is boosted, while the current is decreased making it useful for transmission of power through gridlines. The simulations are considered with some real-time values making it more efficient to be practically implemented. The output voltage remains almost constant with variations of atmospheric conditions making it feasible for various applications. However, accurate numerical values are not possible with the graphical simulations obtained. As a measure to this, a hardware can be developed to obtain accurate numerical values. These simulations are based on a resistive load at the output. Certain parameters will change based on the different loads used. A more robust model can be designed as further work.

References

1. Y. Dong, J. Ding, J. Huang, L. Xu, W. Dong, Investigation of PV inverter MPPT efficiency test platform. in *International Conference on Renewable Power Generation (RPG 2015)*, (Beijing, 2015), pp. 1–4
2. D.K. Chy, M. Khaliluzzaman, Experimental assessment of PV arrays connected to buck-boost converter using MPPT and Non-MPPT technique by implementing in real time hardware. in *2015 International Conference on Advances in Electrical Engineering (ICAEE)*, (Dhaka, 2015), pp. 306–309
3. A.H.M. Nordin, A.M. Omar, Modeling and simulation of PHOTOVOLTAIC (PV) array and maximum power point tracker (MPPT) for grid-connected PV system. in *2011 3rd International Symposium & Exhibition in Sustainable Energy & Environment (ISESEE)*, (Melaka, 2011), pp. 114–119
4. D. Ryu, Y. Kim, H. Kim, Optimum MPPT control period for actual insolation condition. in *2018 IEEE International Telecommunications Energy Conference (INTELEC)*, (Turin, 2018), pp. 1–4
5. A. Saleh, K.S. Faiqotul Azmi, T. Hardianto, W. Hadi, Comparison of MPPT fuzzy logic controller based on perturb and observe (P&O) and incremental conductance (InC) algorithm on buck-boost converter. in *2018 2nd International Conference on Electrical Engineering and Informatics (ICon EEI)*, (Batam, Indonesia, 2018), pp. 154–158
6. J. Zhang, L. Li, D.G. Dorrell, Y. Guo, Modified PI controller with improved steady-state performance and comparison with PR controller on direct matrix converters. Chin. J. Electr. Eng. **5**(1), 53–66 (2019)
7. K.S. Tey, S. Mekhilef, Modified incremental conductance MPPT algorithm to mitigate inaccurate responses under fast-changing solar irradiation level. Sol. Energy. 101, 333–342 (2014) https://doi.org/10.1016/j.solener.2014.01.003
8. S. Khatoon, Ibraheem, M.F. Jalil, Analysis of solar photovoltaic array under partial shading conditions for different array configurations, in *2014 Innovative Applications of Computational Intelligence on Power, Energy and Controls with their impact on Humanity (CIPECH)*, (Ghaziabad, 2014), pp. 452–456
9. T.M. Chung, H. Daniyal, M. Sulaiman, M. Bakar, Comparative study of P&O and modified incremental conductance algorithm in solar maximum power point tracking. in *4th IET Clean Energy and Technology Conference (CEAT 2016)*, (Kuala Lumpur, 2016), pp. 1–6
10. M.J. Hossain, B. Tiwari, I. Bhattacharya, An adaptive step size incremental conductance method for faster maximum power point tracking. in *2016 IEEE 43rd Photovoltaic Specialists Conference (PVSC)*, (Portland, OR, 2016), pp. 3230–3233
11. M.H. Anowar,, P. Roy, A modified incremental conductance based photovoltaic MPPT charge controller. in *2019 International Conference on Electrical, Computer and Communication Engineering (ECCE)*, (Cox'sBazar, Bangladesh, 2019), pp. 1–5

12. D. Lakshmi, M.R. Rashmi, A modified incremental conductance algorithm for partially shaded PV array, in *2017 International Conference on Technological Advancements in Power and Energy (TAP Energy)*, (Kollam, 2017), pp. 1–6
13. S.N. Soheli, G. Sarowar, M.A. Hoque, M.S. Hasan, Design and analysis of a DC -DC buck boost converter to achieve high efficiency and low voltage gain by using buck boost topology into buck topology. in *2018 International Conference on Advancement in Electrical and Electronic Engineering (ICAEEE)*, (Gazipur, Bangladesh), 2018, pp. 1–4

A New Congestion Control Algorithm for SCTP

S Syam Kumar and T. A. Sumesh

1 Introduction

SCTP [1] is a transport layer protocol which combines specific properties of TCP and UDP. SCTP is designed for sending control signals, but nowadays, it is also used for transmitting data packets. SCTP is called as next generation TCP because its design is apt for wireless and multimedia communication [2]. SCTP inherits the same congestion control strategy from TCP [1]. Most of the congestion control algorithm uses the packet drop as an intimation to congestion. In a wired media, the main reason for packet drop is congestion. In a wireless network, the devices are allowed to move within the range. The mobility of devices sometimes causes transmission errors, which will result in packet drops [3]. TCP/SCTP congestion control algorithm does not have a facility to distinguish between packet drops due to congestion with packet drop due to transmission errors. As soon as a packet drop is detected, the sender will reduce the packet delivery rate, thereby decreasing the end-to-end throughput and results in performance loss [4]. Several modifications to TCP are suggested, addressing the issue of how to separate the packet loss due to transmission errors. One of the most successful protocols proposed in this area is TCP Westwood. TCP Westwood is conceived as a sender-side modification. TCP Westwood estimates the end-to-end bandwidth and using the calculated bandwidth for setting slow start threshold (*ssthresh*) and congestion window (*cwin*) after a congestion episode [5]. The principal objective of this paper is to suggest a Westwood-based congestion control algorithm for SCTP. The idea is to modify the SCTP SACK so that we can use bandwidth estimate to distinguish between transmission loss and congestion loss.

S. Syam Kumar · T. A. Sumesh (✉)
Department of Computer Science and Engineering, National Institute of Technology Calicut, Calicut, Kerala, India
e-mail: sumesh@nitc.ac.in

S. Syam Kumar
e-mail: syamkumarcek@gmail.com

S. Patnaik et al. (eds.), *Advances in Machine Learning and Computational Intelligence*, Algorithms for Intelligent Systems, https://doi.org/10.1007/978-981-15-5243-4_3

2 Related Works

The effective way to control the congestion is to reduce the number of packets that transport layer sends to network. Some of the popular congestion control algorithms are TCP Tahoe, TCP Reno, TCP SACK [6], and TCP Westwood [5]. SCTP inherits the same congestion control strategy from TCP Reno.

One of the challenges in wireless networking scenario is to distinguish between congestion loss and transmission loss. Data loss notification [7] is one of the solutions. There were some attempts [8–12] in the literature which are related to our work. They aimed to improve performance in the wireless network and not to distinguish between transmission loss and congestion loss.

3 SCTP Westwood—New Congestion Control Algorithm for SCTP

This section describes the new congestion control algorithm for SCTP. The first part of this section describes the bandwidth estimation procedure. The second part describes how the estimated bandwidth is used for congestion control.

3.1 Bandwidth Estimation

SCTP Westwood estimates the end-to-end bandwidth by monitoring the arrival of SACK segments. The calculated bandwidth is used for setting the parameters *ssthresh* and *win* after a congestion episode. The bandwidth estimation procedure of SCTP Westwood is conceived as a modification to SCTP SACK. SCTP SACK gives a clear snapshot of the receiver side, i.e., which all segments are delivered and outgoing so far. We use this information to calculate the bandwidth more accurately. The proposed algorithm will also handle both the multi-streaming and multi-homing cases.

The algorithm uses the following symbols. All the below variables are suffixed with symbol '*k*' which denotes the acknowledgment number. All the below variables are computed on the arrival of a new acknowledgment.

t_k	Time stamp at which the ACK (or DUPACK) is received at the sender.
con_no	Stands for connection number, which denotes a connection within an SCTP association.
Δ_k	Denotes the gap between two successive ACKs, $\Delta_k = t_k - t_{k-1}$.
$d_k[con\_no]$	This value indicates the amount of bytes that are received at the SCTP receiver through a particular connection.
$b_k[con\ no]$	Denotes the sample bandwidth of the connection.

$$b_k[con\_no] = d_k[con\_no]/\Delta_k \tag{1}$$

$B_k[con\_no]$ The actual value of bandwidth for a particular connection estimated at time t_k.
$\alpha_k[con\_no]$ The filter value for a connection.
$\beta_k[con\_no]$ $\beta k[con\_no] = (1 - \alpha k\ [con\_no])/2$.

The bandwidth estimation algorithm (see Algorithm-1) initiates when a new ACK or a duplicate ACK arrives at the sender. First, we calculate the amount of data that are delivered successfully at the receiver side. The cumulative TSN ACK denotes the segments which are delivered in order. The first part of the algorithm counts the data segments which are delivered in order and calculates the volume of data. The gaps in the SACK segments denote the segments which are not delivered in the expected order. These segments are also counted for bandwidth estimation. The cumulative volume of successfully delivered data (both ordered and unordered) is computed for each connection and store in the list d_k. Then, sample bandwidth of each channel is computed using the expression $b_k[con\ no] = d_k[con\ no]/\Delta_k$. Once the sample bandwidth is computed, it is applied in the following expression to compute the average value of bandwidth,

$$B_k[con\_no] = w1 + w2 \tag{2}$$

where
$w1 = \alpha_k\ [con\_no]B_{k-1}[con\_no]$ and $w2 = \beta_k\ [con\_no](b_k[con\_no] + b_{k-1}[con\ no])$.

This estimated value of bandwidth is used to set the *ssthresh* and *cwin* after a congestion episode.

3.1.1 Bandwidth Estimation Using IP Time Stamping

The proposed bandwidth estimation algorithm computes the value of Δ_k using the arrival time of ACK. The ACK can be affected by several delays in the network like propagation delay, buffering delay, etc. These delays can be excluded while computing the bandwidth between peers. The SCTP sender requests the timestamp value of the client, and client includes the timestamp value in the acknowledgment packet. The sender uses the timestamp value reported by the client to estimate the value of Δ_k by the expression $\Delta_k = T_k - T_{k-1}$, where T_{k-1} is the value of timestamp reported last time. T_k is the current value of timestamp reported by the client.

3.2 Congestion Control

This section describes how the estimated bandwidth is used in congestion control. The slow start and congestion avoidance are kept unchanged. When a packet loss is reported, then SCTP Westwood behaves differently. The key idea is to use the estimated bandwidth to set the parameters *cwin* and *ssthresh* after a congestion episode. The following section describes the algorithm for setting the *cwin* and *ssthresh* after n duplicate (generally $n = 3$) ACKs and after a coarse-grained time-out. Each connection has separate value for *cwin* and *ssthresh.* The below algorithms will be executed for each connection within a SCTP association separately.

```
for each SACK Chunk do
   /* Count the packets which are arrived in order at the receiver side */
   for seg_id = last_acked_seqno to current_acked_seqno do
   /* Find the connection through which the segment is sent */
      con_no = get_connection(seg id) ;
      dk[con_no] = dk[con_no] + size(seg id);
   end
   /* Count the packets which are not arrived in order at the receiver side */
   for each gap do
      for seg_id = first seg to last_seg do
         con_no = get connection(seg id) ;
         dk[con_no] = dk[con_no] + size(seg_id);
      end
   end
   /* Calculate the sample bandwidth */
   for con_no = 0 to number of connections do
      bk[con_no] = dk[con_no] / Δk;
   end
   /* Calculate the actual bandwidth */
   for con_no = 0 to number of connections do
         w1 = αk [con_no]Bk-1[con_no]
         w2 = βk [con_no](bk[con_no] + bk-1[con no])
         Bk[con no] = w1 + w2 ;
   end
end
```

Algorithm 1 Bandwidth estimation algorithm.

```
if timer expires then
   for con_no = 0 to maximum number of connections do
      ssthresh = max (Bk[con_no],2);
      cwin = 1;
   end
end
```

Algorithm 2 Algorithm after n duplicate ACKs.

```
if n duplicate ACKs are received then
   for con_no = 0 to maximum number of connections do
      ssthresh = B_k[con_no] ;
      if cwin > ssthresh then
         cwin = ssthresh;
      end
   end
end
```

Algorithm 3 Algorithm after time-out.

4 Performance Evaluation

In this section, the improved performance of SCTP Westwood over basic SCTP is illustrated with several comparison studies.

4.1 NS2 Implementation

The following scenarios are simulated in ns-2 [13] for comparing the performance of SCTP Westwood with basic SCTP congestion control algorithm.

Scenario 1 Mobile Ad hoc Network (MANET): To demonstrate the improved performance, a MANET is simulated in ns-2 with SCTP Westwood, and basic SCTP congestion control algorithms and then fundamental features are compared. The variation in values of *ssthresh/cwin* and outstanding bytes are carefully compared in both cases. Basic SCTP is equally sensitive to transmission loss and congestion loss, and in both cases, SCTP sender will reduce the traffic rate that will result in significant performance loss. The unnecessary reduction in the values of *ssthresh* and *cwin* is not present in SCTP Westwood, and hence, SCTP Westwood outperforms basic SCTP in a wireless network (see Figs. 1 and 2).

Another parameter which is considered for comparison study is outstanding data (see Fig. 3). Outstanding data represent the volume of data which is pushed into the network by the SCTP sender. SCTP Westwood estimates *ssthresh* and *cwin* based on the end-to-end bandwidth available so that sender can pump more data into the network.

Scenario 2 Wired Network: A fully wired network is simulated in ns-2 for the purpose of comparison. This simulation shows that the modified congestion control algorithm in SCTP Westwood maintains the quality of communication in the wired network. The parameters *ssthresh* and *cwin* are compared (see Figs. 4 and 5) to verify

Fig. 1 ssthresh versus time[MANET]

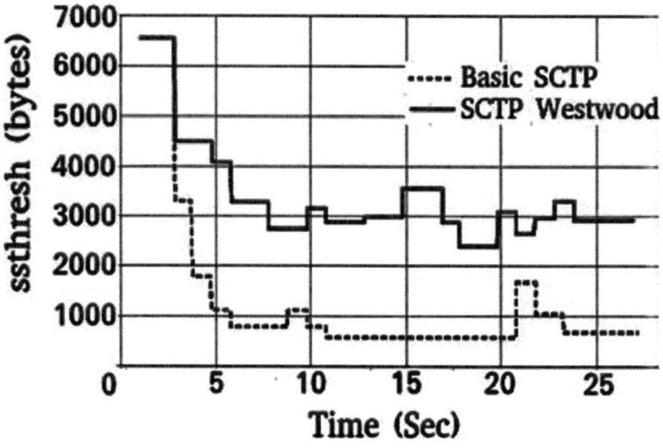

Fig. 2 cwin versus time[MANET]

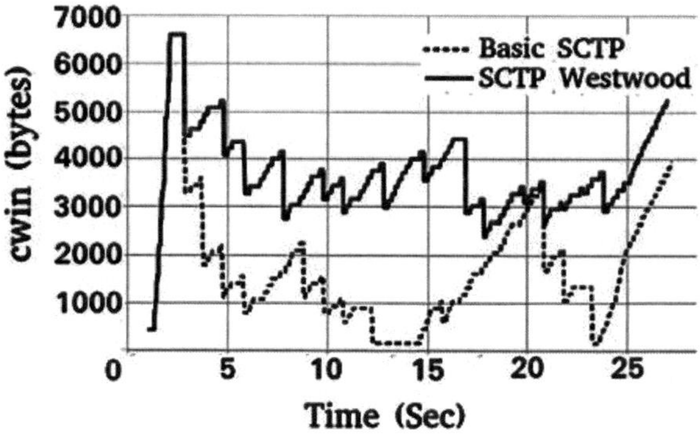

Fig. 3 Outstanding data versus time [MANET]

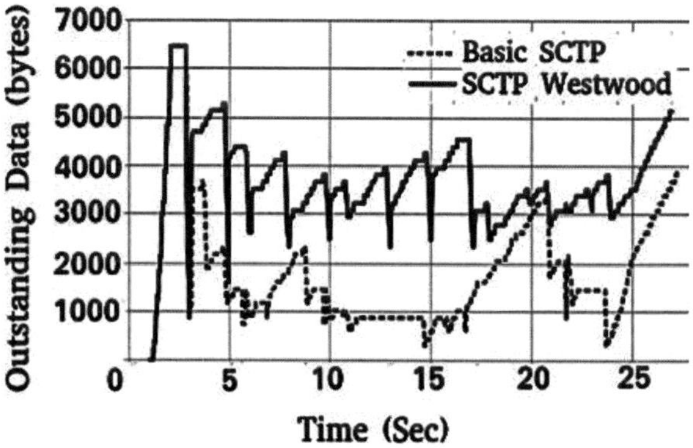

this. The simulation results tell that SCTP Westwood outperforms basic SCTP in a fixed network also.

Scenario 3 Mixed Network: This section aims at comparing the performance of basic SCTP and SCTP Westwood in a mixed network. For this purpose, a mixed

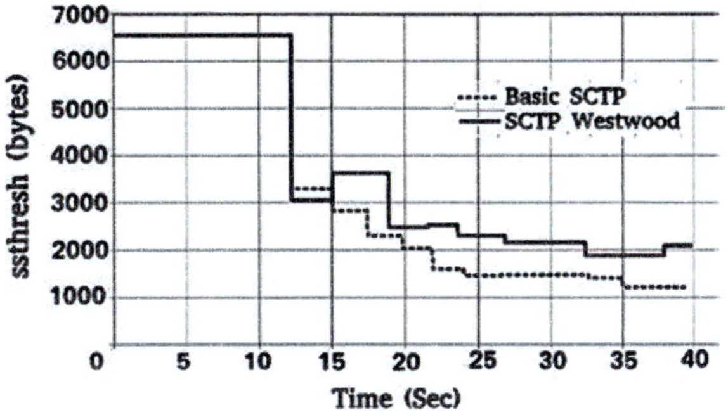

Fig. 4 *ssthresh* versus time [Wired Network]

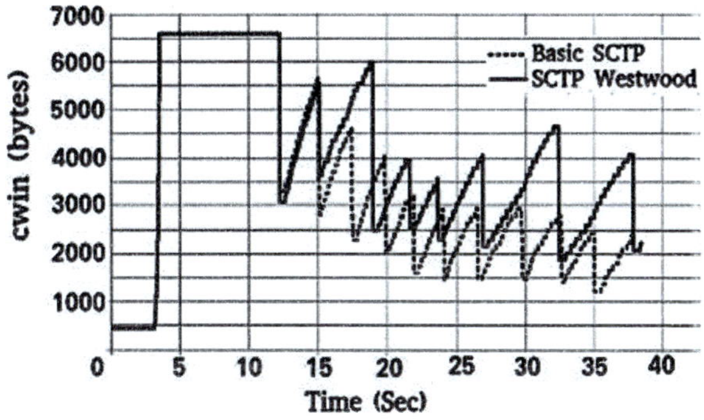

Fig. 5 *cwin* versus time [Wired Network]

network, as shown in Fig. 6, is simulated in ns-2. The comparison results are shown in Fig. 7 and Fig. 8. In this case, also SCTP Westwood outperforms basic SCTP.

Scenario 4 Bottleneck Link: This section aims at comparing the effective bandwidth utilization on a bottleneck link by basic SCTP and SCTP Westwood. The simulation scenario is depicted in Fig. 9 in which three SCTP and three UDP connections share a 3Mbps bottleneck link. The simulation result is presented in Fig. 10, and it is evident that the bandwidth of a bottleneck link is effectively utilized with SCTP Westwood. This is a testimony for the effectiveness of bandwidth estimation procedure in SCTP Westwood algorithm.

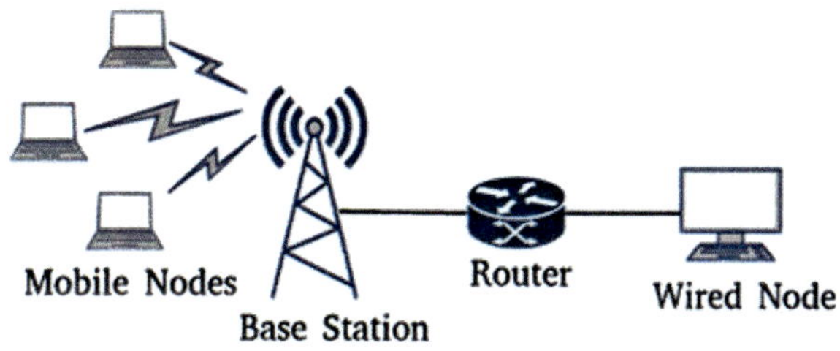

Fig. 6 A mixed network

Fig. 7 ssthresh versus time [Mixed Network]

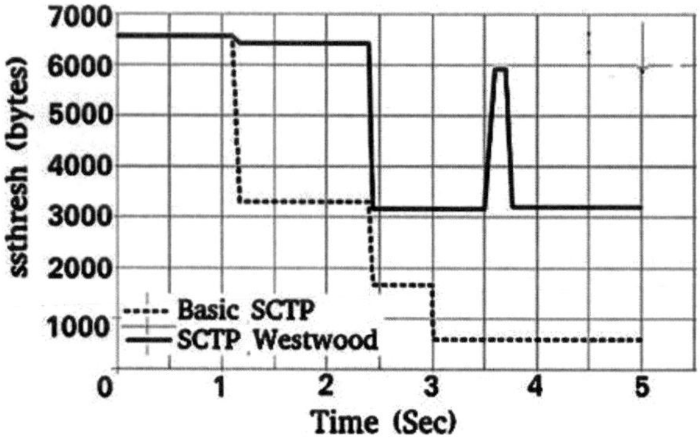

Fig. 8 cwin versus time [Mixed Network]

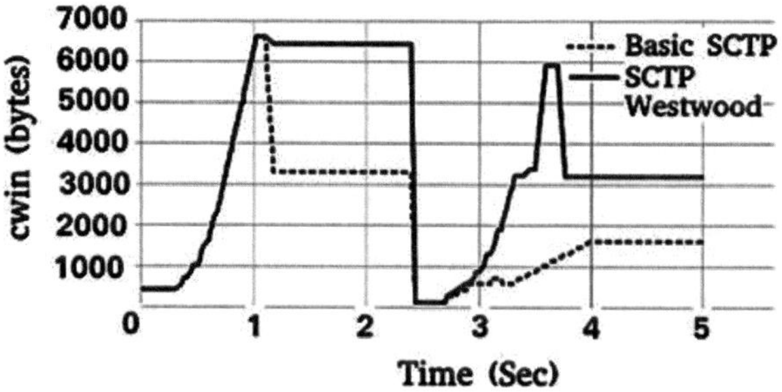

Fig. 9 A network with bottleneck link

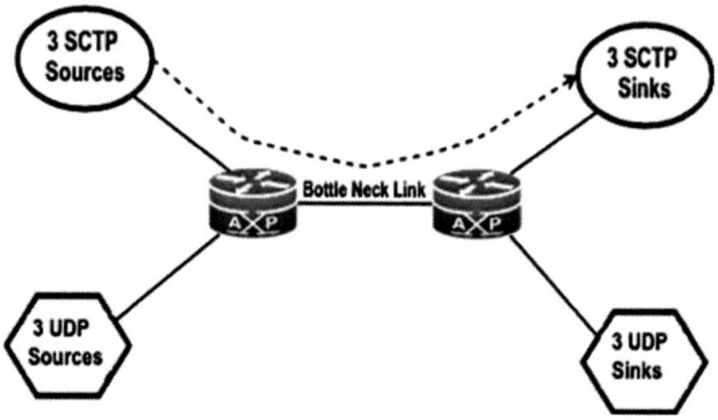

Fig. 10 Bottleneck link utilization

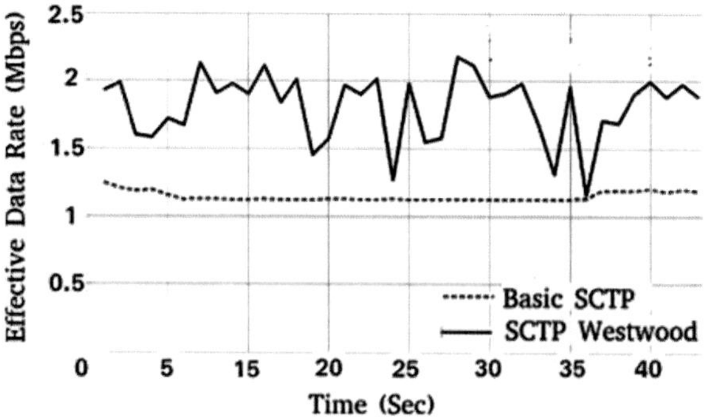

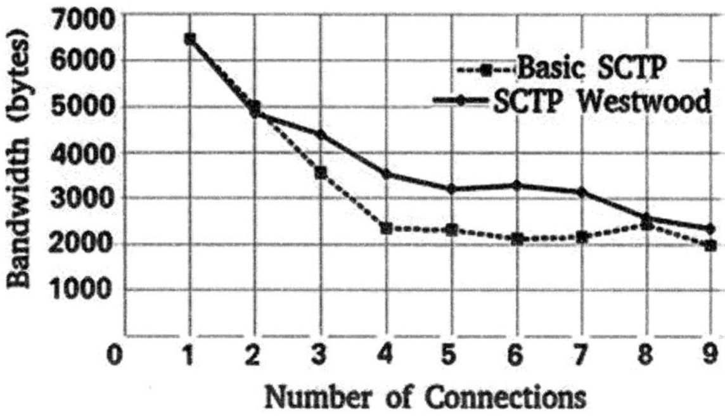

Fig. 11 Sharing of bandwidth

Scenario 5 Fairness: This section aims to analyze the sharing of bandwidth among SCTP connections. Here, a wireless network is simulated and with the n-SCTP connection. Initially, all SCTP connections are idle. At 1 s, one connection is started, at 2 s, one more connection is started, and this process is repeated until all n connections are active. Then, bandwidth share of a single connection is studied and found that more bandwidth is conferred with SCTP Westwood (see Fig. 11). This experiment reiterates the effectiveness of bandwidth estimation procedure in SCTP Westwood.

4.2 *Linux Implementation*

This section describes the details of SCTP Westwood implementation in Linux kernel. A modified Linux kernel is used to compare basic SCTP with SCTP Westwood in a LAN. This experiment is scheduled during peak time so that transmission errors are maximum. In the experiment, a large file is transferred using basic SCTP and SCTP Westwood. The experiment is repeated several times, and the average values of throughput are tabulated (see Table 1). In this case, also SCTP Westwood performs better than basic SCTP.

Table 1 Internet Throughput Measurements

	Basic SCTP	SCTP Westwood
File size	2.6 GB	2.6 GB
Time	6.3828 min	5.8847 min
Throughput	51.32 Mbps	55.66 Mbps

5 Conclusion

This paper suggests a new congestion control algorithm for SCTP, SCTP Westwood. The main soul of SCTP Westwood is the bandwidth estimation procedure, in which the end-to-end bandwidth is computed on the arrival of SCTP SACK message. All the simulation results tell that SCTP Westwood outperforms basic SCTP.

References

1. R. Stewart, Q. Xie, K. Morneault, C. Sharp, H. Schwarzbauer, T. Taylor, I. Rytina, M. Kalla, L. Zhang, V. Paxson, RFC 4960, Stream control transmission protocol (SCTP). Internet Eng. Task Force (IETF) (2007)
2. A. Elhabib Elhabib, B. Amin, H. Abdel Nabi, A.J. Alzubaidi, Performance of SCTP congestion control over network. IJAR, **2**(2), 470–473 (2016)
3. J-C. Lin, ed. Recent advances in wireless communications and networks. BoD–Books on Demand, (2011)
4. Y. Tian, K. Xu, N. Ansari, TCP in wireless environments: problems and solutions. IEEE Commun. Mag. **43**(3), S27–S32 (2005)
5. S. Mascolo, C. Casetti, M. Gerla, M.Y. Sanadidi, R. Wang, TCP westwood: Bandwidth estimation for enhanced transport over wireless links. in *Proceedings of the 7th annual international conference on Mobile computing and networking*, (ACM, 2001), pp. 287–297
6. J.F. Kurose, K.W. Ross, Computer networking. Fifth edition. (Pearson India Education Services Pvt Ltd, India, 2012)
7. S. Saravanan, E. Karthikeyan, A protocol to improve the data communication over wireless network. Int. J. Wireless Mobile Networks **3**(5), 95 (2011)
8. G. Ye, T. Saadawi, M. Lee, SCTP congestion control performance in wireless multi-hop networks. in *MILCOM 2002. Proceedings*, vol 2 (IEEE, 2002), pp. 934–939
9. L. Ma, F.R. Yu, V.C Leung, Performance improvements of mobile SCTP in integrated heterogeneous wireless networks. IEEE Trans. Wireless Commun. **6**(10), 3567–3577 (2007)
10. T.D. Wallace, A. Shami, Concurrent multipath transfer using SCTP: Modelling and congestion window management. IEEE Trans. Mobile Comput. **13**(11), 2510–2523 (2014)
11. I.A. Najm, M. Ismail, J. Lloret, K.Z. Ghafoor, B.B. Zaidan, A.A.R.T. Rahem, Improvement of SCTP congestion control in the LTE-A network. J. Network Comput. Appl. 58, 119–129 (2015)
12. C. Xu, J. Zhao, G-M. Muntean, Congestion control design for multipath transport protocols: A survey. IEEE commun. surveys tutorials **18**(4), 2948–2969 (2016)
13. T. Issariyakul, E. Hossain, Introduction to network simulator NS2. 26–90 (X, Springer US 2009)

RGNet: The Novel Framework to Model Linked ResearchGate Information into Network Using Hierarchical Data Rendering

Mitali Desai, Rupa G. Mehta, and Dipti P. Rana

1 Introduction

Modelling social media information as a network of linked data has its roots in many profound applications such as influence finding, community detection, expert or topic identification, clustering and recommendation systems [1]. As the social media platforms provide means of social or personal information sharing and social communication [2], likewise academic social network sites (ASNS) are profusely used for various research activities such as research sharing, academic collaborations, scientific communications, feedbacks collection and technical information dissemination [3]. The emergence of ASNS brings the benefits of social network sites to academics and research communities [4]. ASNS provides a flexible medium to manage the explosive growth in research activities compared to traditional scientific outlets such as journals and conferences [5]. Widespread utilization of ASNS such as ResearchGate (RG), Google Scholar, Academia.edu, Mendeley has open a novel paradigm of research in the field of scholarly literature analysis, inspired from social network analysis [5, 6].

M. Desai (✉) · R. G. Mehta · D. P. Rana
Sardar Vallabhbhai National Institute of Technology, Surat, Gujarat 395007, India
e-mail: mitalidesai17@gmail.com

R. G. Mehta
e-mail: rgm@coed.svnit.ac.in

D. P. Rana
e-mail: dpr@coed.svnit.ac.in

S. Patnaik et al. (eds.), *Advances in Machine Learning and Computational Intelligence*, Algorithms for Intelligent Systems, https://doi.org/10.1007/978-981-15-5243-4_4

1.1 RG Growth Statistics

A study was carried out in 2015 to analyse the distribution of profiles of researchers from Spanish National Research Council, i.e. CSIC [7]. Three ASNS namely Academia.edu, Google Scholar Citations and RG were used for the purpose. The study discovered the higher utilization of RG among all in terms of number of registered profiles. While among all present ASNS, three main platforms, i.e. Academia.edu, Mendeley and RG that are currently leading the market, a survey conducted in August 2018 reveals RG receiving the greatest attention [8] (see Fig. 1). Another survey was conducted in 2018 to learn the usage frequency of various ASNS by researches who indicated to be familiar with a particular platform (denoted by *n*). The results show substantially more frequent usage of RG among all. 26% of RG users were found to be daily users, 41% were weekly users, and 18% were using RG at least monthly. In total, 85% of RG users specified using RG least monthly [9] (see Fig. 2).

The recent year surveys demonstrate the extreme active participation on RG among other well-known ASNS which motivates our research focus mainly on RG.

1.2 Identified Research Openings and Research Objectives

Information shared on RG is categorized into user demographics: research items, citations, reads, recommendations, affiliation, etc., and user associations: followers and followings. Existing research on RG is based upon the user demographics and user profile statistics without any consideration of user-to-user connections [10–14]. Various engrained social theories illustrate that the impact (influence) of one entity propagates to other connected entity. Such influence can be effectively analysed using

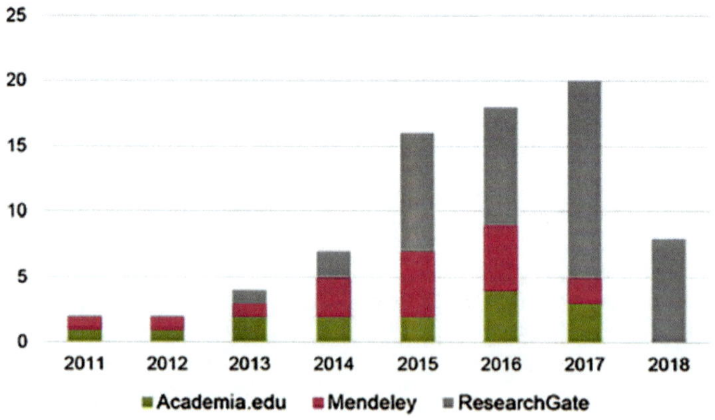

Fig. 1 Yearly publications count per platform

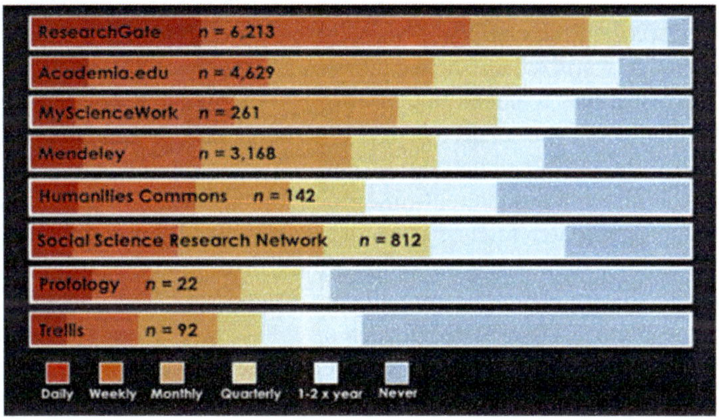

Fig. 2 Usage frequency of ASNS

network models [15]. Once a network of linked RG information is formed, it enables various network-based analysis to be carried out on RG information more accurately. The linked RG information can be precisely represented and explored leveraging the concepts of network modelling and network theory. The novel methodology is proposed to generate and utilize network-based model of RG information considering user demographics and user associations information present on RG.

This research aims to

- propose a novel hierarchical data rendering process to collect the linked RG information.
- propose a novel framework, RGNet; to construct the RG network from collected linked RG information, leveraging the user demographics and user associations information present on RG.

In Sect. 2, the recent research status of the studies using RG information in various fields is reviewed. The concept of networks, network terminologies and network storage mechanisms is explicated in Sect. 3. The proposed methodology and the modelling framework are elaborated along with the implementation details and testing of the proposed RGNet framework in Sect. 4. Finally, the future aspects and possible applications of the constructed RG network are described in Sect. 5.

2 Literature Survey

In research community, the ASNS are gaining popularity as they provide flexible and easily accessible medium to the prominent researchers of various disciplines and geographically disperse locations to share theirs as well as access others' ad hoc, pre-print or fill fledge research, collaborate, communicate and collect feedbacks from scientific experts [2]. Due to which, ASNS such as RG, Mendeley or Academia.edu boast to have several million registered users [2, 5]. Among much popular ASNS, RG has been extensively utilized and hence labelled as 'Facebook for scientists' [6]. RG facilitates the researches to generate their profiles, share their research and research interests, find the peers with expertise in similar or specific domain to communicate or work in corporation [7] and getting insights into trending research [8, 9].

Hammook et al. explored the real-time RG publications authored or co-authored by the Canadian computer science researchers to analyse the correlation between various RG features such as views, publications, downloads, citations, questions, answers and followers [10]. The research investigations show that the Canadian researchers are highly active by the means of collaborations and knowledge sharing, while the statistics illustrate that number of views is around $2\frac{1}{2}$ times of the number of downloads and number of citations is around $\frac{2}{3}$ times of the number of downloads for the collected data. Thelwall and Kousha inspected the effect of RG statics on institute ranking [11], in which, the results exhibit moderate correlation. Research collaboration-based RG network of Macedonian University was explored by Kadriu for specific research areas, and then, centrality measures were used to find the important users having knowledge of multiple research areas [12].

Fu et al. have carried out cluster analysis on RG information based on various RG features such as projects, publications, followers, followings, skills, questions, articles and full texts [13]. The data are clustered into Active users, Representors and Lurkers. The results disclose that in collected data, the 8% of total users are active in communication, whereas 4% of total users are only representors of their scientific work. The activities of the remaining users seemed to be not significant. Priem et al. have uncovered that the researchers of the USA, Japan, Sweden, and to a smaller degree, of China, Iran and Russia effectively utilize the open doors given by ASNS to present their own research [14]. The study also illustrates that there is an immediate reliance between the quantity of downloads and the 'freshness' of the paper.

RG has been highly utilized among all ASNS, and it is yet to be explored in various applications similar to network-based analysis of connected social media data.

3 Technical Background

A network N is represented as N (V, E), where set V represents entities (nodes) and set E represents relationships (edges) [15, 16]. Networks naturally model the relationships among data and hence are proven to be highly efficient in storing, retrieving and analysing exponentially growing complex relationships [17, 18].

In literature, various network models have been illustrated [19, 20]. Simple networks have single edges between a pair of nodes, whereas multi-network may have multiple edges. Single networks display a single unit of entity, whereas multiple networks consist of many disjoint networks. Finite network has finite number of relations and finite number of nodes; infinite network has either infinite number of nodes or infinite number of edges. In directed network, each edge has predefined direction, whereas edges in undirected network have no directions. Weighted networks are special purpose networks where weights are assigned to the edges; unweighted networks do not have weighted edges. In connected network, there present an edge from any node to any other node.

This research exploits the multi, single, finite, directed, unweighted and connected network to model RG information.

Amidst all recent network databases such as Neo4j, ArangoDB, OrientDB, MarkLogic, AllegroGraph, Flock, Titan, GraphDB, HyperGraphDB, Neo4j has been emerged as a leading network database [1]. Prominent features of Neo4j like application diversity, ACID property, interactive GUI, simple query language, provision for property network models and extensive primitive indexing make Neo4j highly utilized and explored network database. Neo4j organizes network as a collection of nodes and relations along with their properties [21]. Additionally, to incorporate heterogeneous networks, nodes and relations have distinguishable parameters, i.e. nodes labels and relations types. It utilizes a declarative and flexible Cypher Query Language (CQL) for traversal, data and property queries [22]. Additionally, Neo4j provides an automatic schema index for fast data retrieval on node properties.

This research utilizes the highly efficient network database storage platform, i.e. Neo4j to store the RG network.

4 RGNet: The Proposed Framework

In this section, the work flow of the proposed RGNet framework (see Fig. 3) with details of data rendering process is explained.

In the proposed novel RGNet framework, the followings' and followers' relations present on RG are explored systematically along with user meta data, as a whole to get insight into how the information propagates in the connected network.

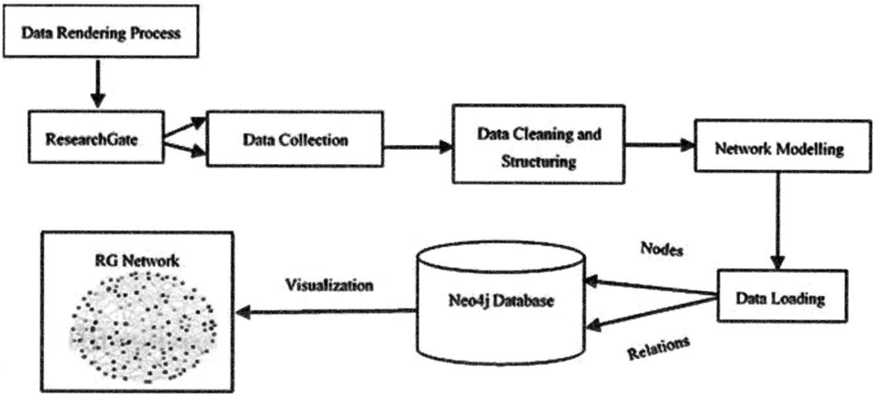

Fig. 3 Proposed RGNet framework

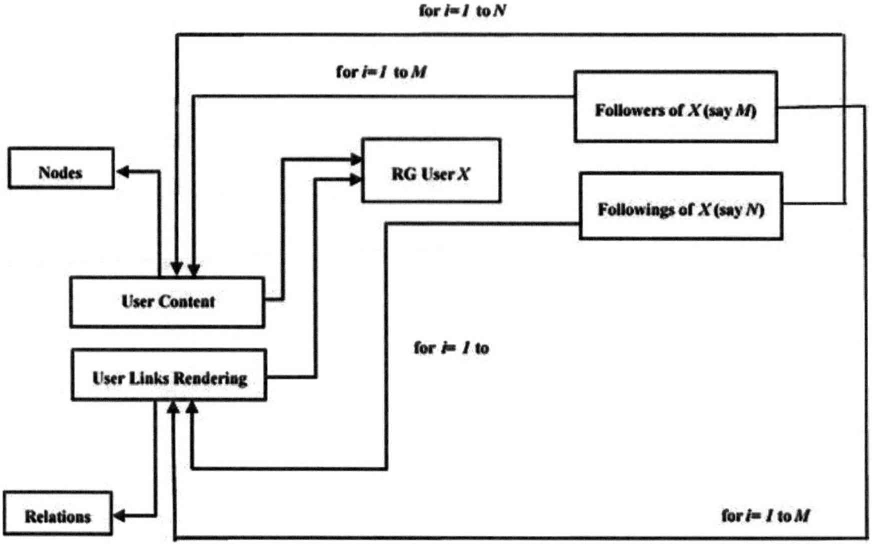

Fig. 4 Proposed hierarchical data rendering process

To collect the real-time RG user demographics and user associations, we propose a hierarchical data rendering process (see Fig. 4).

The process has two functions: demographics rendering and associations rendering. Initially, one RG user's registered profile (say *X*) is rendered with all user demographics. Assuming the user *X* has *M* followers and *N* followings, the user demographics rendering and user associations rendering processes are repeated for total $(M + N)$ times to capture user *X*'s demographics as well as user associations. User demographics are stored along with user nodes, while user associations help in connecting users in the network.

In initial stage, the real-time RG data in terms of user demographics and user associations information are collected implementing the proposed hierarchical data rendering process. In the second stage, the collected data are pre-processed to remove outliers and handle missing value glitches. In the third stage, data model (see Fig. 5) to transform collected data into network structure is prepared, based upon the following and follower relations among users. Each node in the network represents an RG user,

Fig. 5 RG network model

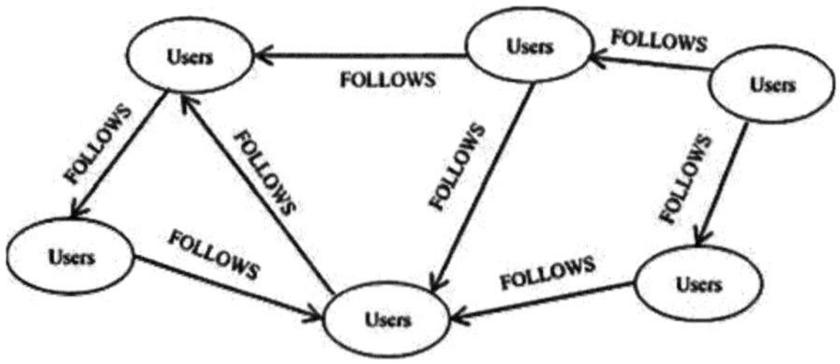

while their correlations are shown by the directed edge ‘FOLLOWS’. The nodes are imported in Neo4j, and relations are formed among users based upon collected user associations information. Further, the constructed RG network is visualized using network visualization tool.

The proposed hierarchical data rendering process and the proposed RGNet framework are tested with Ubuntu 18.04 LTS (64-bit), 8 GB RAM and Intel Core i7-7700 processor. For data storage, network database Neo4j (version 3.5.8), and for visualization, Gephi (version 0.9.2) are used.

Total 1544 RG user profiles which are linked in terms of followers’ and followings from various research branches of ‘Economics’ are collected along with their user demographics. Total 1646 associations among 1544 user profiles are found and rendered. Employing the RG network model (see Fig. 5), the collected information is transformed into nodes and relations forming the connected RG network.

The constructed network is visualized in Gephi (see Fig. 6). Total number of nodes and edges are denoted by $|v|$ and $|E|$.

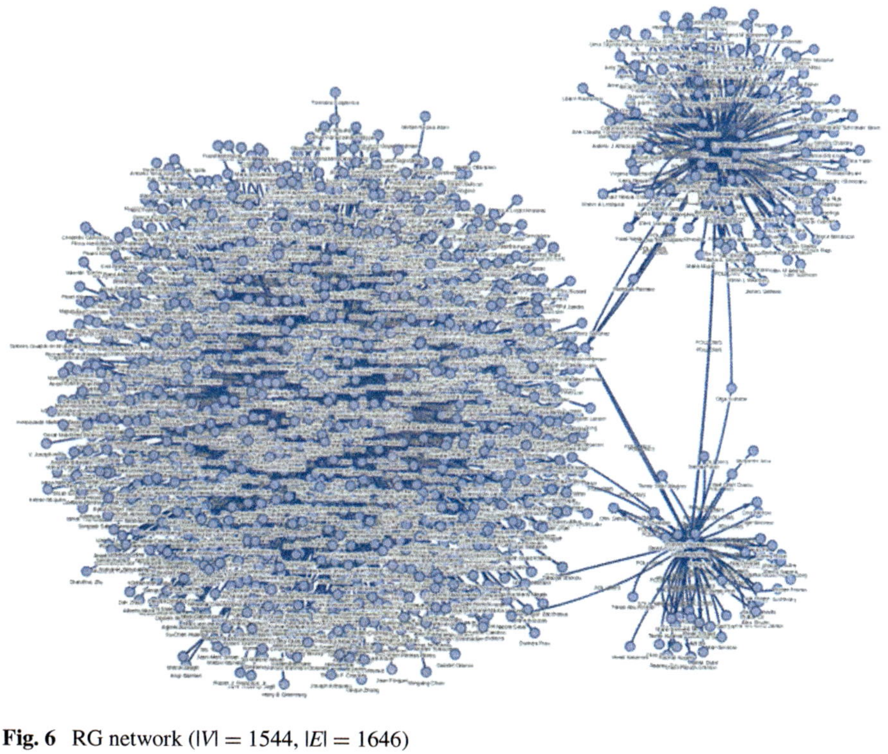

Fig. 6 RG network ($|V| = 1544$, $|E| = 1646$)

5 Summary and Future Directions

The emerging Web-based platforms in form of ASNS facilitate the increasing numbers of research-related activities in scholarly literature domain. Among other well-known ASNS platforms, according to recent years surveys, RG is widely exploited in terms of number of active users, communications, collaborations and research sharing. Various research conducted in exploration of RG are user demographics-oriented research, without any consideration of user-to-user relations and the impact of one user's statistics on his other connected users. To incorporate user demographics and user associations information on RG, this research suggests to transform RG information into connected network so that the RG platform can be explored and analysed for various social network similar applications such as influence finding, community detection, expert finding, topic discovery, opinion leader detection and recommendation systems. In order to construct RG network, this research proposes a novel RGNet framework along with hierarchical data rendering process to collect linked information from RG. Implementing the proposed hierarchical data rendering process, first the linked RG information is rendered and then transformed into the connected network utilizing the data modelling concepts. Later, the constructed network is visualized by means of graph visualization techniques.

The research outcomes disclose the new possibilities to utilize the network analytics concepts in various domain of scholarly literature inspired from social media data analysis. Various efforts can be made to carry out well-established social network application on the constructed network to deliver more concise outcomes leveraging the notions of social theories in scholarly community.

References

1. M. Desai, R.G. Mehta, D.P. Rana, An empirical analysis to identify the effect of indexing on influence detection using graph databases. Int. J. Innov. Technol. Exploring Eng. **8**(9s), 414–421 (2019)
2. X. Wu, C. Zhang, Finding high-impact interdisciplinary users based on friend discipline distribution in academic social networking sites. Scientometrics **119**(2), 1017–1035 (2019)
3. A. Mas-Bleda, M. Thelwall, K. Kousha, I.F. Aguillo, Do highly cited researchers success-fully use the social web? Scientometrics **101**(1), 337–356 (2014)
4. W. Yan, Y. Zhang, Research universities on the ResearchGate social networking site: an examination of institutional differences, research activity level, and social networks formed. J. Inform. **12**(1), 385–400 (2018)
5. K. Mangan, Social networks for academics proliferate, despite some doubts. Chron. Higher Educ. **58**(35), 1–7 (2012)
6. A.M. Elsayed, The use of academic social networks among Arab researchers: a survey. Social Sci. Comput. Rev. **34**(3), 378–391 (2016)
7. J.L. Ortega, Toward a homogenization of academic social sites: a longitudinal study of profiles in Academia edu, google scholar citations and researchgate. Online Inform. Rev. **41**(6), 812–825 (2017)
8. K. Jordan, From social networks to publishing platforms: a review of the history and scholarship of academic social network sites, in *Frontiers in Education 2019* (In-Press)

9. C. Rapple, Understanding and supporting researchers' choices in sharing their publications: the launch of the FairShare Network and Shareable PDF. Insights 31 (2018)
10. Z. Hammook, J. Misic, V.B. Misic, Crawling ResearchGate. net to measure student/supervisor collaboration, in *2015 IEEE Global Communications Conference (GLOBECOM)* (2015)
11. M. Thelwall, K. Kousha, Researchgate: disseminating, communicating, and measuring scholarship? J. Assoc. Inform. Sci. Technol. **66**(5), 876–889 (2014)
12. A. Kadriu, Discovering value in academic social networks: a case studying researchgate, in 35th international conference on information technology interfaces (ITI2013) (Cavtat, Croatia, 2013), pp. 57–62
13. T.Z.J. Fu, Q. Song, D.M. Chiu, The academic social network. Scientometrics **101**(1), 203–239 (2014)
14. J. Priem, D. Taraborelli, P. Groth, C. Neylon, Altmetrics: A manifesto, http://altmetrics.org/manifesto. Last accessed 11/5/2018
15. C.C. Aggarwal, An introduction to social network data analytics, in Social network data analytics (Springer, Boston, MA, 2011), pp. 1–15.
16. M. Desai, R.G. Mehta, D.P. Rana, Issues and challenges in big graph modelling for Smart City: An extensive survey. Int. J. Comput. Intell. IoT **1**(1) (2018)
17. M. Hunger, From relational to network: a developers' guide, https://dzone.com/refcardz/from-relational-to-network-a-developers-guide?chapter=1. Last accessed 18/06/2018
18. T. Petkova, Why network databases make a better home for interconnected data than the relational databases, https://ontotext.com/network-databases-interconected-data-relational-databases/. Last accessed 11/06/2018
19. A. Patel, J. Dharwa, Network data: the next frontier in big data modeling for various domains. Ind. J. Sci. Technol. **10**(21), 1–7 (2017)
20. S. Srinivasa, Data, storage and index models for network databases. Netw. Data Manage. 4770–4796 (2011)
21. P. Jadhav, R. Oberoi, Comparative analysis of different graph databases. Int. J. Eng. Res. Technol. **3**(9), 820–824 (2014)
22. S. Agrawal, A. Patel, A study on graph storage database of NoSQL. Int. J. Soft Comput. Artif. Intell. Appl. **5**(1), 33–39 (2016)

A New Approach for Momentum Particle Swarm Optimization

Rohan Mohapatra, Rohan R. Talesara, Saloni Govil, Snehanshu Saha, Soma S. Dhavala, and TSB Sudarshan

1 Introduction

Real-world optimization problems are mostly, multidimensional, and multi-objective. Compared to unconstrained optimization, constrained optimization problems pose many challenges. Constraints are difficult to model in the problem and are generally represented by equality or non-equality. Also, the constraints modify the search space. Consequently, any solution to optimize the problem must do so in the bounds of the constraints. A traditional way to approach constrained optimization problems is to use mathematical programming. However, it is restrictive in the sense that it is heavily dependent on the problem being optimized and how it is modelled. If the problem is not modelled well, mathematical programming may

R. Mohapatra (✉) · R. R. Talesara · S. Govil
PES University, Bangalore, India
e-mail: rohanmohapatra@pesu.pes.edu

R. R. Talesara
e-mail: rohantalesara@pesu.pes.edu

S. Govil
e-mail: salonigovil@pesu.pes.edu

S. Saha
Department of Computer Science and Information Systems and APPCAIR, BITS Pilani, K K Birla, Goa, Sancoale, India
e-mail: snehanshu.saha@ieee.org

S. S. Dhavala
Center for Astroinformatics, Modeling and Simulation (CAMS), Department of Computer Science, PES University, Bengalore, India
e-mail: soma.dhavala@gmail.com

T. Sudarshan
Department of Computer Science, PES University, Bengalore, India
e-mail: sudarshan@pes.edu

S. Patnaik et al. (eds.), *Advances in Machine Learning and Computational Intelligence*, Algorithms for Intelligent Systems, https://doi.org/10.1007/978-981-15-5243-4_5

not yield good results. Another widely used approach is to calculate the gradient and use it to optimize the problem, as in the gradient descent algorithm. However, this approach requires one to find the derivative of the function, which may be difficult in real-world problems which are generally constrained, multidimensional, non-continuous, and non-convex. As a result of the shortcomings of traditional optimization algorithms, there has been significant interest in using meta-heuristic algorithms for optimization. In particular, there has been significant interest in nature-inspired and evolutionary algorithms such as genetic algorithm [1], ant colony algorithm, simulated annealing, and particle swarm optimization. It is important to note that meta-heuristic algorithms generally give approximate solutions and not exact solutions. The particle swarm optimization algorithm was proposed in 1995 as a swarm intelligence-based evolutionary algorithm for optimization problems. It is inspired by the flocking behavior of birds and is characterized by two equations: velocity update and position update. The velocity update is controlled by both a global and local component. PSO, however, has its own shortcomings. It may get stuck in the local-extremum, and is prone to pre-mature convergence. Various modified versions of PSO [2] have been proposed to overcome its shortcomings and make it more robust and fast. Garg [3] presented a hybrid of genetic algorithm and PSO, where the PSO is used for improving the velocity vector and the GA has been used for modifying the decision vectors using genetic operators. Tripathi et al. [4] proposed the time-variant PSO where the coefficients of the PSO velocity update equation (inertia weight and acceleration coefficients) are allowed to change every iteration. This has been said to help the algorithm explore the search space more efficiently. Sun et al. [5] proposed a quantum-behaved PSO, wherein the particles follow quantum mechanics as opposed to Newtonian mechanics. In quantum mechanics, it is not possible to determine both the position and velocity of a particle simultaneously, which leads to radically different PSO equations. Nebro et al. [6] present speed constrained multi-objective PSO (SMPSO) which provides a mechanism to control the velocity of the particles when it becomes too high. It also includes a turbulence factor and an external archive to store non-dominated solutions found during the search. Xiang et al. [7] presented the momentum PSO algorithm which uses a momentum term in the velocity update equation. However, the momentum term is defined such that the effect of the previous velocity terms is diminishingly small, and hence, the momentum term does not provide great acceleration compared to the original PSO algorithm. To this end, in this article, we propose a new momentum PSO, wherein the momentum term is influenced by the previous velocities to a greater extent. The proposed algorithm has been tested on 20 benchmark test optimization functions, and results show that the proposed algorithm performs better than both weighted PSO and momentum PSO in almost all cases. Performance here is measured in terms of how fast the optimal value is reached. Further, the proposed PSO is used to optimize two habitability scores, namely, Cobb–Douglas habitability score [8, 9] and constant elasticity earth similarity approach [10] by maximizing their respective production functions. These scores are used to assess extra-solar planets and assign them scores based on their habitability and earth similarity. CDHS considers the planet's radius, mass, escape velocity, and surface temperature, while CEESA includes a fifth parameter, the orbital eccentricity of the planet. Ultimately, we also show that the proposed PSO is equivalent to gradient descent with momentum. The contents of the paper are organized as follows: Sect. 2 explains the weighted PSO and momentum PSO, Sect. 3 explains the proposed PSO algorithm and its equivalence to gradient descent with momentum. Section 4 explains the two habitability scores, namely CDHS and CEESA. Section 5 presents the results of running the proposed algorithm against 20 benchmark test optimization functions, CDHS and CEESA, with graphs and tables comparing it to both weighted PSO and momentum PSO.

2 Particle Swarm Optimization with Its Variants

2.1 Particle Swarm Optimization Algorithm with Inertial Weight

The particle swarm optimization algorithm [11] is an optimization algorithm inspired by the flocking behaviour of birds. It is characterized by a population of particles in space, which aim to converge to an optimal point. The movement of the particles in space is characterized by two equations, namely velocity and position update equations, which are as follows:

$$v_i^{t+1} = \omega v_i^t + c_1 r_1 (p_i^{best} - x_i^t) + c_2 r_2 (g^{best} - x_i^t) \tag{1}$$

$$x_i^{t+1} = x_i^t + v_i^{t+1} \tag{2}$$

where ω, c_1, $c_2 \geq 0$. Here, x_i^t refers to the position of particle i in the search space at time t, v_i^t refers to the velocity of particle i at time t, p_i^{best} is the personal best position of particle i, and g_i^{best} is the best position amongst all the particles of the population. The movement of the particles is decided by updating the current position of the particles with the velocity as shown in (3).

2.2 Particle Swarm Optimization Algorithm with Momentum

The back-propagation algorithm is one of the most frequently used algorithms for training multilayer feed-forward neural networks. It uses gradient descent method to minimize the error between actual output and expected output, but it tends to stagnate or oscillate in the superficial local minima and fail to converge to global minimum. The momentum term was introduced to deal with this issue. It helps by acting as a low pass filter to smoothen.

Inspired by this, a momentum is introduced in the velocity updating equation of *PSO*. Thus, the new equation along with the momentum term was introduced by the following equation,

$$v_i^{t+1} = (1-\lambda)(v_i^t + c_1 r_1 (p_i^{best} - x_i^t) + c_2 r_2 (g^{best} - x_i^t)) + \lambda v_i^{t-1}$$

where c_1, c_2, x_i^t, v_i^t, p_i^{best}, g_i^{best} mean the same as described in the previous section. The momentum factor is indicated by λ.

3 New Approach to Momentum Particle Swarm Optimization

In this section, we propose a rather new approach to momentum particle swarm optimization. The problem with the currently available *m-PSO* that weighted average which is computed takes care of both exploration and exploitation simultaneously. Since PSO tries to search the space by exploring, it makes more sense to give more weight to the exploration part of the equation. Another problem with the above term is that it takes longer iteration to reduce error and reach the minimum.

To counter the above-said problems, we mathematically formulate a new particle swarm optimization with momentum as follows:

$$v_i^{t+1} = M_i^{t+1} + c_1 r_1 (p_i^{best} - x_i^t) + c_2 r_2 (g^{best} - x_i^t) \tag{3}$$

where

$$M_i^{t+1} = \beta M_i^t + (1-\beta) v_i^t \tag{4}$$

Here β is the momentum factor, and M_i^{t+1} indicates the effect of the momentum. The above equation can be rewritten as the following by combining (4) and (5),

$$v_i^{t+1} = \beta M_i^t + (1-\beta) v_i^t + c_1 r_1 (p_i^{best} - x_i^t) + c_2 r_2 (g^{best} - x_i^t) \tag{5}$$

We understand that PSO is composed of two phases: the exploration phase and the exploitation phase.

$$\underbrace{v_i^t}_{\text{Exploration}} \underbrace{+ c_1 r_1 (p_i^{best} - x_i^t) + c_2 r_2 (g^{best} - x_i^t)}_{\text{Exploitation}}$$

With above proposed approach, the exploration phase is determined by the **exponential weighted average of the previous velocities seen so far** only. The negligible weights applied in the momentum PSO do not help much in providing us that acceleration required.

3.1 Exponentially Weighted Average

In the previous section, we defined the momentum (5) and we also mentioned that it was an exponentially weighted average of the previous velocities seen so far. We prove it by expanding M_i^t(5); we get the following derivation.

$$M_i^t = \beta M_i^{t-1} + (1-\beta)v_i^{t-1} \tag{6}$$

$$M_i^t = \beta[\beta M_i^{t-2} + (1-\beta)v_i^{t-2}] + (1-\beta)v_i^{t-1} \tag{7}$$

$$M_i^t = \beta^2 M_i^{t-2} + \beta(1-\beta)v_i^{t-2} + (1-\beta)v_i^{t-1} \tag{8}$$

$$M_i^t = \beta^2[\beta M_i^{t-3} + (1-\beta)v_i^{t-3}] + \beta(1-\beta)v_i^{t-2} + (1-\beta)v_i^{t-1} \tag{9}$$

$$M_i^t = \beta^3 M_i^{t-3} + \beta^2(1-\beta)v_i^{t-3} + \beta(1-\beta)v_i^{t-2} + (1-\beta)v_i^{t-1} \tag{10}$$

Generalizing M_i^t, it can be written as the follows,

$$\begin{aligned} M_i^t =& \beta^n M_i^{t-n} + \beta^{(n-1)}(1-\beta)v_i^{t-n} + \beta^{(n-2}(1-\beta)v_i^{t-(n-1)} + \ldots \\ &+ \beta(1-\beta)v_i^{t-2} + (1-\beta)v_i^{t-1} \end{aligned}$$

3.2 Equivalence to Stochastic Gradient Descent Rule

For functions of multiple variables, the Taylor expansion with remainder is f(x) = f(a) + f'(a)(x - a) + $E_n(x)$.

The gradient descent weight update with momentum is given by,

$$w^{(t)} = w^{(t-1)} + \eta V_{dw}^t \tag{11}$$

$$V_{dw}^t = \beta V_{dw}^{t-1} + (1-\beta)\frac{\partial f}{\partial w} \tag{12}$$

Combining the equations and dividing (13) by $(1-\beta)$, we get

$$w^{(t)} = w^{(t-1)} + \alpha V_{dw}^{t-1} + \eta\frac{\partial f}{\partial w} \tag{13}$$

Here, η is the learning rate and V_{dw}^{t-1} is the momentum applied to the weight update.

Let us apply Taylor expansion to the above to get,

$$w^{(t)} = w^{(t-1)} + \eta f(w\prime) + \eta \sum_{j=1}^{n} \frac{\partial f}{\partial w_j}(w\prime)(w_j^{t-1} - w_j') + E_n(x) + \alpha V_{dw}^{t-1} \tag{14}$$

at some optimum value $w\prime$. The equation of the exponentially weighted momentum particle swarm optimization is,

$$v_i^t = M_i^t + c_1 r_1 (p_i^{best} - x_i^{t-1}) + c_2 r_2 (g^{best} - x_i^{t-1}) \tag{15}$$

$$x_i^t = x_i^{t-1} + v_i^t \tag{16}$$

Combining (16) and (17) and expanding, we get

$$x_i^t = x_i^{t-1} + \beta M_i^{t-1} + (1-\beta) v_i^{t-1} + c_1 r_1 (p_i^{best} - x_i^{t-1}) + c_2 r_2 (g^{best} - x_i^{t-1}) \tag{17}$$

The term $c_2 r_2 (g^{best} - x_i^{t-1})$ in PSO is the social factor. We equate this social factor with the error term in the Taylor expansion. We get

$$k_1 (x-a)^{n+1} \leq E_n(x) \leq k_2 (x-a)^{n+1} \tag{18}$$

$$E_n(x) = k(x-a)^{n+1} \tag{19}$$

Expanding the above at optimum value w', and equating to the social factor we get,

$$k(w-w')(w-w')^n = c_2 r_2 (g^{best} - x_i^{t-1}) \tag{20}$$

We interpret as follows, and k plays the role or c_2, and if we consider that $w\prime$ is the optimum value, then the corresponding value in PSO is g^{best}

Comparing the weight update rule (15) and the PSO Eq. (18), considering a single particle we find,

$$\begin{aligned}
&\eta \equiv (1-\beta) \\
&f(w\prime) \equiv v^{t-1} \\
&\eta \sum_{j=1}^{n} \frac{\partial f}{\partial w_j}(w\prime)(w_j^{t-1} - w_j') \equiv -c_1 r_1 (p_i^{best} - x_i^{t-1}) \\
&\alpha \equiv \beta
\end{aligned}$$

By using the above equations for mathematical convenience, we find that M_i^{t-1} works the same way as the momentum term in the (14) i.e. V_{dw} which helps in smoothing and faster convergence to the minimum. Additionally, $c_1 r_1$ serves the same purpose of gradient of the loss function.

$$\eta = (1 - \beta) = (1 - \alpha) \tag{21}$$

This implies the learning rate is tuned by another parameter β. Therefore,

$$\frac{\partial f}{\partial w_j} = \frac{-c_1 r_1}{\eta} = \frac{-c_1 r_1}{1 - \alpha} \tag{22}$$

So, whenever the gradient of the function needs to be computed, the above equivalence can be extended to functions which are non-differentiable!

This can be experimentally proved for various equations. Taking one such example,

$$f(x) = x^2 - 10x + 17 \tag{23}$$

Differentiating (24), we obtain

$$f\prime(x) = 2x - 10 \tag{24}$$

Setting the initial parameters for m-PSO as $c_1 = 0.8,\ c_2 = 0.9,\ \beta = 0.7$ and solving for *30* particles, the algorithm converges towards the global minima in *29* iterations.

4 Representing the Problem

We apply the proposed momentum PSO on both benchmark test optimization functions and habitability optimization problems in astronomy. The benchmark functions considered are standard optimization functions described in Sect. 4.1. The optimization problems tested against, namely CDHS and CEESA, have been described in Sects. 4.2, 4.3. The results of applying the proposed momentum PSO on the above test functions and problems have been discussed in Sect. 5, with supporting values, tables, and graphs.

Usually, a problem may be constrained or unconstrained depending on the search space that has been tackled with. An unconstrained problem's space is the full search space for the particle swarm. The difficulty arises only when it's constrained.

Theophilus, Saha et al. [10] describe a way to handle constrained optimization. We use the same method to represent the test functions and represent few standard optimization problems.

We also consider two habitability scores that are the Cobb–Douglas habitability (CDH) score and the constant elasticity earth similarity approach (CEESA) score. Estimating these scores involves maximizing a production function while observing a set of constraints on the input variables.

4.1 Standard Test Optimization Problems

In this section, we briefly describe the benchmark optimization functions chosen to evaluate our proposed algorithm and compare its performance to that of the weighted PSO and momentum PSO described in Sect. 2. For the purpose of assessing the performance of the proposed algorithm, we have considered single-objective unconstrained optimization functions Rastrigin, Ackley, Sphere, Rosenbrock, Beale, Goldstein-Price, Booth, Bukin N.6, Matyas, Levi N.13, Himmelblau's, Three-hump camel, Easom, Cross-in-tray, Eggholder, Holder table, McCormick, Schaffer N.2, Schaffer N.4, Styblinski–Tang as well as constrained optimization functions Rosenbrock constrained with a cubic line, Rosenbrock constrained to a disc and Mishra's bird.

Rosenbrock function has a long, closed, parabolic-shaped valley where the global minima are present. Finding that valley is an easy task, but converging to the global minima is challenging. Goldstein-Price, Cross-in-Tray, Holder table, and Himmelblau's are 2-dimensional continuous non-convex multimodal functions with four global minima. Three-hump camel is a 2-dimensional continuous non-convex multimodal function with single global minima and three local ones. Easom function has 2 dimensions and is unimodal consisting of multiple local minima but a single global minima where the global minima has a small area relative to the search space. Mishra's Bird is a constrained optimization function which is 2-dimensional and non-convex with two global minima. It is one of the most challenging constraint optimization problem introduced so far.

Rest all the benchmark optimization functions considered except Matyas have multiple local minima and only a single global one. Matyas has no local minima, except the global minima. The results for the above-mentioned benchmark optimization functions are summarized in Tables 1 and 2.

4.2 Representing CDHS

The Cobb–Douglas habitability score is based on the Cobb–Douglas production function. The Cobb–Douglas [12] is a production function very popularly used in economics. It represents the relationship between the values of two or more inputs (particularly physical capital and labour) and the amount of output that can be produced by those inputs.

Table 1 Unconstrained test optimization functions formulas

Name	Formula
Ackley function	$-20\exp[-0.2\sqrt{0.5(x^2+y^2)}]-$ $\exp[0.5(\cos(2x\pi)+\cos(2y\pi))]+e+20$
Rosenbrock 2D function	$(1-x)^2+100(y-x^2)^2$
Beale function	$(1.5-x+xy)^2+(2.25-x+xy^2)^2+(2.625-x+xy^3)^2$
Goldstein–Price function	$[1+(x+y+1)^2(19-14x+3x^2-14y+6xy+3y^2)]$ $+[30+(2x-3y)^2(18-32x+12x^2+48y-36xy+27y^2)]$
Booth function	$(x+2y-7)^2+(2x+y-5)^2$
Bukin function N.6	$100\sqrt{\lvert y-0.01x^2\rvert}+0.01\lvert x+10\rvert$
Matyas function	$0.26(x^2+y^2)-0.48xy$
Lévi function N.13	$\sin^2(3x\pi)+(x-1)^2(1+\sin^2(3y\pi))+(y-1)^2(1+\sin^2(2y\pi))$
Himmelblau's function	$(x^2+y-11)^2+(x+y^2-7)^2$
Three-hump camel function	$2x^2-1.05x^4+\frac{x^6}{6}+xy+y^2$
Easom function	$-\cos(x)\cos(y)\exp(-((x-\pi)^2+(y-\pi)^2))$
Cross-in-tray function	$-0.0001\left[\lvert\sin(x)\sin(y)\exp\left(\lvert 100-\frac{\sqrt{x^2+y^2}}{\pi}\rvert\right)\rvert+1\right]^{0.1}$

The general form of Cobb–Douglas production function is:

$$Q = A\prod_{i=1}^{L} x_i^{\lambda_i} \quad x = (x_1, \ldots, x_L) \tag{25}$$

where

- Q = quantity of output
- A = efficiency parameter
- L = total number of goods
- $x_1, \ldots, x_L$ = (non-negative) quantities of good consumed, produced, etc.
- λ_i = elasticity parameter for good i

The Cobb–Douglas habitability score can be constituted from two components, *the interior score* (Y_i) and *the surface score* (Y_s). Both have to maximized, so the objective function can be represented as follows:

$$Y_i = R^{\alpha}.D^{\beta} \tag{26}$$

$$Y_s = V_e^{\gamma}.T_s^{\delta} \tag{27}$$

Table 2 Unconstrained test optimization functions

Name	Global minimum	m-PSO optimized value	Iterations	Proposed momentum PSO optimized value	Iterations
Ackley function	0	0.001	59	0.001	47
Rosenbrock 2D function	0	$2*10^{-8}$	215	67.14	124
Beale function	0	$4.38*10^{-7}$	68	0	136
Goldstein–Price function	3	3	61	3	53
Booth function	0	$1.07*10^{-7}$	90	$4.09*10^{-7}$	49
Bukin function N.6	0	0.05	503	0.047	191
Matyas function	0	0	55	0	30
Lévi function N.13	0	0	83	0	56
Himmelblau's function	0	$2.51*10^{-7}$	95	0	48
Three-hump camel function	0	$2*10^{-8}$	59	0	36
Easom function	−1	$-3*10^{-10}$	45	−1	40
Cross-in-tray function	−2.062	−2.06	43	−2.064	30

where $R,\ D,\ V_e$ and T_s are density, radius, escape velocity, and surface temperature for a particular exoplanet, respectively, and $\alpha,\ \beta,\ \gamma$, and δ are elasticity coefficients.

The final CDH score can be represented as weighted sum of the interior score and surface score as follows,

$$\begin{aligned} &\underset{\alpha,\ \beta,\ \gamma,\ \delta}{maximize} \quad Y = w_i Y_i + w_s Y_s \\ &subject\ to\ \ 0 < \phi < 1, \forall \phi \in \{\alpha,\ \beta,\ \gamma,\ \delta\}, \end{aligned}$$

$$\begin{aligned} \alpha + \beta - 1 - \tau &\le 0 \\ 1 - \alpha - \beta - \tau &\le 0 \\ \gamma + \delta - 1 - \tau &\le 0 \\ 1 - \gamma - \delta - \tau &\le 0 \end{aligned}$$

It can be subjected to two scales of production: constant return to scale (CRS) and decreasing return to scale (DRS). The above two Eqs. (26 and 27) are concave under constant returns to scale (CRS), when $\alpha + \beta = 1$ and $\gamma + \delta = 1$, and also under decreasing returns to scale (DRS), $\alpha + \beta < 1$ and $\gamma + \delta < 1$.

4.3 Representing CEESA

The constant elasticity earth similarity approach score is based on the CES production function. The general form of CES production function is:

$$Q = F.\left[\sum_{i=1}^{n} a_i X_i^r\right]^{\frac{1}{r}}$$

where

- Q = quantity of output
- F = factor of productivity
- a_i = share parameter of input *i*, $\sum_{i=1}^{n} a_i = 1$
- X_i = quantities of factors of production (i = 1, 2... n)
- $s = \frac{1}{1-r}$ = elasticity of substitution

The objective function for CEESA to estimate the habitability score of an exoplanet is:

$$\underset{r,d,t,v,e,\rho,\eta}{\text{maximize}} \quad Y = (r.R^\rho + d.D^\rho + t.T^\rho + v.V^\rho + e.E^\rho)^{\frac{\eta}{\rho}}$$

$$\text{Subject to } 0 < \phi < 1, \forall\phi \in \{r, d, t, v, e\},$$
$$0 < \rho \leq 1,$$
$$0 < \eta < 1,$$
$$(r + d + t + v + e) - 1 - \tau \leq 0,$$
$$1 - (r + d + t + v + e) - \tau \leq 0$$

where *E* represents orbital eccentricity, and τ is tolerance. Two scales of production are used: constant return to scale (CRS) and decreasing return to scale (DRS). Under DRS, $0 < \eta \leq 1$. Under CRS, $\eta = 1$; hence, the objective function reduces to:

$$\underset{r,d,t,v,e,\rho,\eta}{\text{maximize}} \quad Y = (r.R^\rho + d.D^\rho + t.T^\rho + v.V^\rho + e.E^\rho)^{\frac{1}{\rho}}$$

5 Experiments and Discussions

5.1 Experimental Setup

The confirmed exoplanets catalogue was used for dataset maintained by the Planetary Habitability Laboratory (PHL). We use the parameters described in Tables 3, 4, 5. Surface temperature and eccentricity are not recorded in Earth Units; we normalized these values by dividing them with earth's surface temperature (288 K) and eccentricity (0.017). The PHL-EC records are empty for those exoplanets whose surface temperature is not known. We drop these records from the experiment.

We conveniently first test the **proposed swarm algorithm** on test optimization functions mentioned below. We used $n = 1000$ with a *target error* $= 1 * 10^-6$ and *50* particles. Then, the algorithm was used to optimize the CDHS and CEESA objective functions.

Table 3 Constrained test optimization functions

Name	Formula	Global minimum	mPSO Optimized Value	Iterations	Proposed momentum PSO Optimized Value	Iterations
Mishra bird function	$\sin(y)e^{\|[(1-\cos x)^2]\|} + \cos(x)e^{\|[(1-\sin y)^2]\|} + (x-y)^2$ *subjected to* $(x+5)^2 + (y+5)^2 < 25$	−106.76	−106.76	121	−106.76	55
Rosenbrock function constrained with a cubic and a line function	$(1-x)^2 + 100(y-x^2)^2$ *subjected to* $(x-1)^3 - y + 1 \leq 0$ *and* $x + y - 2 \leq 0$	0	0.99	109	0.99	64
Rosenbrock function constrained to a disc	$(1-x)^2 + 100(y-x^2)^2$ *subjected to* $x^2 + y^2 \leq 2$	0	0	69	0	39

Table 4 Estimated Cobb–Douglas habitability scores under CRS

Name	Algorithm	α	β	Y_i	γ	δ	Y_s	Iterations	CDHS
TRAPPIST-1 b	Momentum PSO	0.99	0.01	1.09	0.01	0.99	1.38	130	1.234
	Proposed momentum PSO	0.99	0.01	1.09	0.01	0.99	1.38	75	1.234
TRAPPIST-1 c	Momentum PSO	0.99	0.01	1.17	0.01	0.99	1.21	65	1.19
	Proposed momentum PSO	0.99	0.01	1.17	0.01	0.99	1.21	80	1.19
TRAPPIST-1 d	Momentum PSO	0.01	0.99	0.9	0.01	0.99	1.02	94	0.96
	Proposed momentum PSO	0.01	0.99	0.9	0.01	0.99	1.02	64	0.96
TRAPPIST-1 e	Momentum PSO	0.99	0.01	0.92	0.01	0.99	0.88	30	0.9096
	Proposed momentum PSO	0.99	0.01	0.92	0.2	0.8	0.88	69	0.9096
TRAPPIST-1 f	Momentum PSO	0.99	0.01	1.04	0.95	0.05	0.8	93	0.92
	Proposed momentum PSO	0.99	0.01	1.04	0.7	0.3	0.8	59	0.92
TRAPPIST-1 g	Momentum PSO	0.99	0.01	1.13	0.99	0.01	1.09	117	0.92
	Proposed momentum PSO	0.99	0.01	1.13	0.99	0.01	1.09	66	0.92
TRAPPIST-1 h	Momentum PSO	0.01	0.99	0.81	0.99	0.01	0.68	88	0.7449
	Proposed momentum PSO	0.036	0.963	0.807	0.99	0.01	0.68	86	0.7438

5.2 *Test Optimization Functions*

The particle swarm optimization mentioned in Sect. 2 was tested on the unconstrained test optimization functions defined in Table 2 and constrained test optimization functions in Table 3.

Table 5 Estimated CEESA scores under CRS

Name	Algorithm	r	d	t	v	e	ρ	Iterations	CEESA Score
TRAPPIST-1 b	Momentum PSO	0.556	0.000	0.398	0.045	0.000	0.629	76	1.193
	Proposed Momentum PSO	0.107	0.314	0.578	0.001	3.704	0.999	92	1.126
TRAPPIST-1 c	Momentum PSO	0.117	0.384	0.273	0.225	0.000	0.999	54	1.161
	Proposed Momentum PSO	0.053	0.348	0.212	0.386	0.000	0.999	60	1.161
TRAPPIST-1 d	Momentum PSO	0.127	0.306	0.566	0.000	0.000	0.167	71	0.948
	Proposed Momentum PSO	0.413	0.283	0.304	0.000	0.000	0.820	38	0.882
TRAPPIST-1 e	Momentum PSO	0.455	0.486	0.033	0.027	5.182	0.504	82	0.868
	Proposed Momentum PSO	0.264	0.004	0.626	0.105	0.000	0.936	7	0.897
TRAPPIST-1 f	Momentum PSO	0.718	0.000	0.276	0.006	0.000	0.969	77	0.972
	Proposed Momentum PSO	0.382	0240	0.093	0.284	5.392	0.719	46	0.836
TRAPPIST-1 g	Momentum PSO	0.256	0.232	0.009	0.501	0.002	0.106	4	1.038
	Proposed Momentum PSO	0.337	0.251	0.000	0.412	0.000	0.986	66	1.066
TRAPPIST-1 h	Momentum PSO	0.434	0.259	0.001	0.306	0.000	0.538	73	0.743
	Proposed Momentum PSO	0.295	0.181	0.294	0.230	0.000	0.999	56	0.709

Table 6 Equivalence of new approach to stochastic gradient descent: The table is proof that using m-PSO, we are able to compute the derivative of functions in a non-classical, iterative manner. This could be easily be applied to functions whose derivatives are difficult to find, analytically. We call this "derivative-free" optimization

Parameter Set 1 : $\eta = 0.1,\ \beta = 0.9,\ c_1 = 0.8,\ c_2 = 0.9$

Parameter Set 2 : $\eta = 0.3,\ \beta = 0.7,\ c_1 = 0.8,\ c_2 = 0.9$

Function	Derivative	Number of particles	SGD Iterations		Proposed momentum PSO Iterations		Weighted PSO Iterations	
			Set 1	Set 2	Set 1	Set 2	Set 1	Set 2
$\sin x + \cos^2 x$	$\cos x - 2\sin x \cos x$	15	16	5	162	36	42	45
		30			91	27	42	37
		60			87	31	40	33
x^2	$2x$	15	68	35	65	27	24	32
		30			62	26	24	26
		60			53	23	24	25
$x^2-10x+17$	$2x - 10$	15	37	26	66	29	25	29
		30			66	29	26	23
		60			55	23	25	18
$\sqrt{x^2+2}$	$\frac{x}{\sqrt{x^2+2}}$	15	535	209	64	29	23	28
		30			57	25	22	25
		60			54	22	22	22
$\lvert x \rvert$	$\frac{x}{\lvert x \rvert}$	15	3472	1600	105	41	52	39
		30			107	42	48	44
		60			89	41	43	40

5.3 *A Comparative Study of PSO Variants on CDHS Scores*

The particle swarm optimization mentioned in Sect. 2 was used to optimize the CDHS score to find if the planets are habitable or not. The 3 variants of PSO mentioned in 2 were compared, the Table 4 gives an overview that the proposed momentum particle swarm optimization converges faster to the global minima compared to the other variants. A representational graph attached below gives a clearer picture. The three particle swarm optimizers mentioned in Sect. 2 have been tested with *maximum-iterations* = *1000*, *number-of-particles* = *50* and *threshold-error to be at* 10^{-6} (Fig. 1).

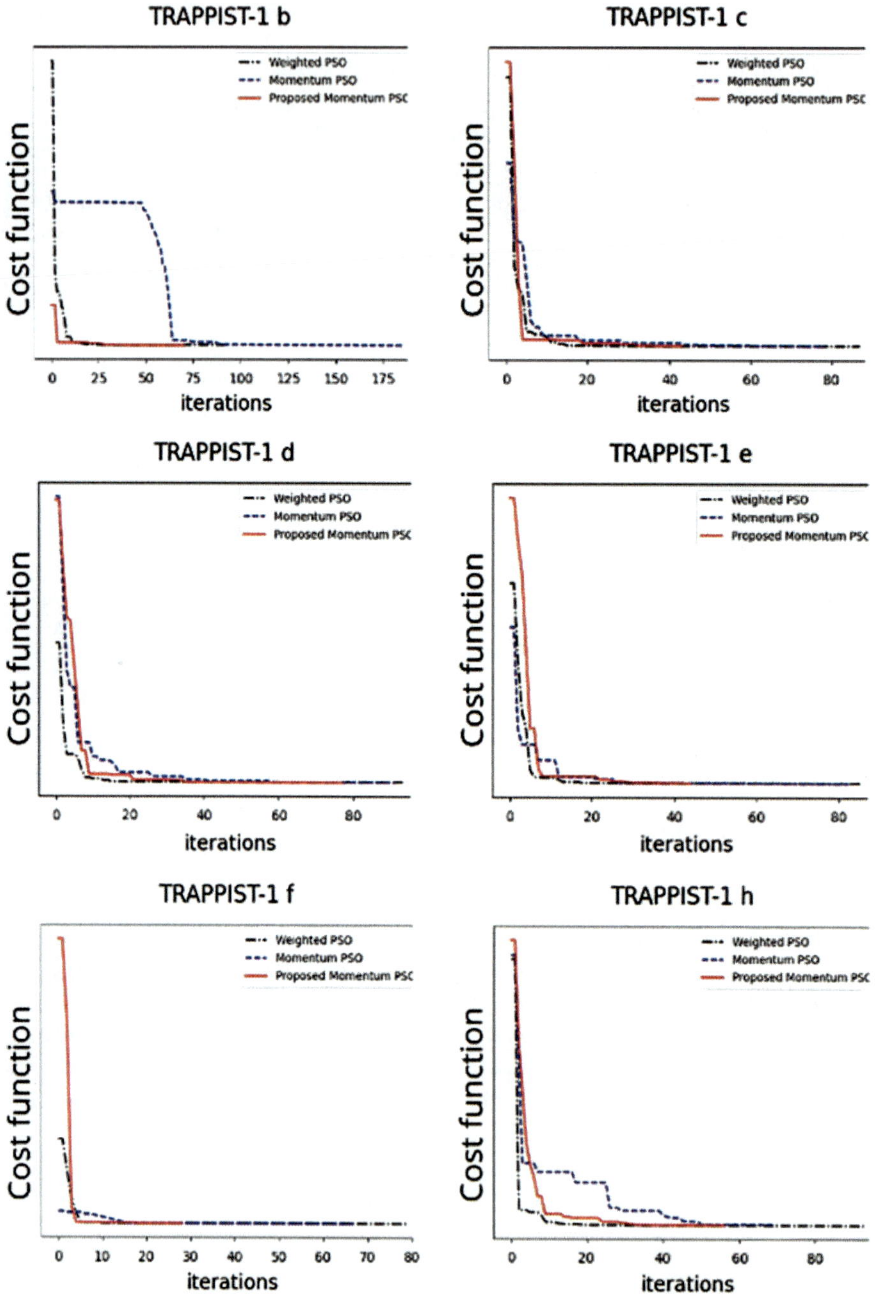

Fig. 1 CDHS: Comparison between 3 variants

6 Conclusion

The paper presents a momentum-enhanced particle swarm approach to solve unconstrained and constrained optimization problems. We also establish the equivalence between the classical stochastic gradient descent/ascent type approaches and the new approach proposed here. This throws some interesting insights to derivative-free optimization, apart from demonstrating empirical evidence of our method surpassing standard PSO in terms of speed of convergence. We conclude by noting that the proposed method can be extended to energy Hamiltonian approach to Momentum PSO as energy can be computed from momentum easily. We also note that acceleration from momentum can also be derived, and therefore, controlling acceleration becomes easier. This is handy when excessive acceleration may push the algorithm further away from minima/maxima. Approximation of derivatives is thus a subtask that can also be accomplished by our approach.

References

1. D.E. Goldberg, *Genetic Algorithms in Search* (Addison-Wesley Longman Publishing, Optimization and Machine Learning, 1989)
2. S. Chen, J. Montgomery, Particle swarm optimization with threshold convergence. in *2013 IEEE Congress on Evolutionary Computation, CEC 2013*, (2013)
3. H. Garg, A hybrid pso-ga algorithm for constrained optimization problems. Appl. math. comput. **274**, 292–305 (2015)
4. P.K. Tripathi, S. Bandyopadhyay, S.K. Pal, Multi-objective particle swarm optimization with time variant inertia and acceleration coefficients. Inf. Sci. **177**, 5033–5049 (2007)
5. J. Sun, B. Feng, W. Xu, Particle swarm optimization with particles having quantum behavior. in *Proceedings of the 2004 Congress on Evolutionary Computation (IEEE Cat. No.04TH8753)*, vol. 1, (2004), pp. 325–331
6. A.J. Nebro, J.J. Durillo, J. Garcia-Nieto, C.C. Coello, F. Luna, E. Alba, Smpso: A new pso-based metaheuristic for multi-objective optimization. in *2009 IEEE Symposium on Computational Intelligence in Multi-Criteria Decision-Making, MCDM 2009—Proceedings*, (2009), pp. 66–73
7. T. Xiang, J. Wang, X. Liao, An improved particle swarm optimizer with momentum. in *2007 IEEE Congress on Evolutionary Computation*, (2007), pp. 3341– 3345
8. K. Bora, S. Saha, S. Agrawal, M. Safonova, S. Routh, A. Narasimhamurthy, Cd-hpf: New habitability score via data analytic modeling. Astron. Comput. **17**, 129–143 (2016)
9. S. Saha, S. Basak, K. Bora, M. Safonova, S. Agrawal, P. Sarkar, J. Murthy, Theoretical validation of potential habitability via analytical and boosted tree methods: An optimistic study on recently discovered exoplanets. Astronomy and Comput., **23**, 141–150 (2018)
10. A. Theophilus, S. Saha, S. Basak, J. Murthy, A novel exoplanetary habitability score via particle swarm optimization of ces production functions. (2018)
11. J. Kennedy, R. Eberhart, Particle swarm optimization. in *Proceedings of ICNN'95—International Conference on Neural Networks*, vol. 4, (1995), pp.1942–1948
12. C.W. Cobb, P.H. Douglas, A theory of production. The American Economic Review, vol. 18, No. 1, Supplement, Papers and Proceedings of the Fortieth Annual Meeting of the American Economic Association (1928), pp. 139–165

Neural Networks Modeling Based on Recent Global Optimization Techniques

Anwar Jarndal, Sadeque Hamdan, Sanaa Muhaureq, and Maamar Bettayeb

1 Introduction

Global optimization techniques can be classified into evolutionary and swarm intelligence algorithms. The evolutionary algorithms such as genetic algorithm (GA) are stochastic optimizations inspired by nature evolution and heredity mechanisms [1, 2]. Swarm intelligence algorithms such as PSO, ant colony optimization and artificial bee colony are inspired by collective intelligence, which appears through the interaction of animals and social insects [3–6]. These optimization techniques are used in wide range of engineering problems. Recently, some works started to combine global optimization techniques with artificial neural networks (ANNs). The key element of the ANNs is the neurons (processing units), which are connected together into a network and arranged in layers to perform the required computing. Each neuron within the network takes one or more inputs and produces an output. At each neuron, all inputs are scaled by weights to modify their strengths and processed by proper base function to produce an output, which will be considered as an input for the next neuron (in the next layer). During training, the ANN processes the inputs and compares its resulting outputs with the desired outputs to calculate the errors, which are then propagated back through the system to adjust the weights for best fitting. This typically used back-propagation technique works well for simple model of lower number of weights (variables) [7]. However, for larger scale ANN model of

A. Jarndal (✉) · S. Muhaureq · M. Bettayeb
Electrical Engineering Department, University of Sharjah, Sharjah, United Arab Emirates
e-mail: ajarndal@sharjah.ac.ae

S. Hamdan
Sustainable Engineering Asset Management (SEAM) Research Group, University of Sharjah, Sharjah, United Arab Emirates

M. Bettayeb
Center of Excellence in Intelligent Engineering Systems (CEIES), King Abdulaziz University, Jeddah, Saudi Arabia

S. Patnaik et al. (eds.), *Advances in Machine Learning and Computational Intelligence*, Algorithms for Intelligent Systems, https://doi.org/10.1007/978-981-15-5243-4_6

larger number of weights, the final solution will have higher dependency on the initial guess and may be trapped in a local minimum. In this case, the user needs more effort to re-initiate the training process many times to get the best fitting, which is not practical for some application such as automatic controlling. For that reason, training the ANN using global optimization methods could provide an efficient alternative and avoids drawbacks of local minimization methods. Various optimization techniques in the literature have been used to optimize ANN weight. For instance Whitley et al. [8] discussed using GA for optimizing neural network weights and architectures in the form of connectivity patterns. Yamazaki et al. [9] used simulated annealing to optimize neural network architectures and weights which resulted in good generalization performance and low complexity. Karaboga et al. [10] used artificial bee colony optimization to optimize neural network weights and overcome local optimal solutions. Mirjalili et al. [11] implemented a hybrid technique of PSO and gravitational search algorithm. On the other hand, some research has been conducted to compare the back-propagation for training neural networks with other global optimization techniques. For example, Gudise and Venayagamoorthy [12] showed that ANN weight convergence is faster using PSO, in learning feed-forward ANN, with respect to back-propagation. Mohaghegi et al. [13] compared PSO with back-propagation in optimizing neural network weights. The comparison revealed that PSO is efficient, requires less effort and is more efficient in finding optimal weights compared with back-propagation.

To the best knowledge of the authors, the research on application of global optimization for ANN is still limited and more extensive researches are needed. Also, many recently developed swarm intelligence optimization techniques have not been used, except for grey wolf optimization (GWO). Mirjalili [14] used GWO for training multi-layer perception and found that it outperforms some other global optimization techniques. As a result, this paper presents combined techniques of ANN modeling with different recent global optimizations and investigates their performance (in terms of the efficiency and effectiveness) in solving practical modeling problems.

The contribution of this paper can be summarized as follows:

1. This work investigates the performance of recent global optimization techniques such as dragonfly algorithm (DA), grasshopper optimization algorithm (GOA), whale optimization algorithm (WOA) and GWO in training the ANNs.
2. The developed techniques are demonstrated by modeling the GaN transistor and compared their performance with the measured data.
3. This paper compares the performance of the widely used optimization technique GA with the recently developed swarm optimization techniques in terms of speed and accuracy.

Figure 1 illustrates the proposed methodology used in this work. In the first phase, measured data are used as inputs for the ANN model, which consists of two hidden layers (see details in Sect. 2). In the second phase, the ANN model weights are optimized using five different optimization techniques. In the last phase, the output of each combined ANN model is compared with the actual data. Finally, the results of all considered optimization techniques are compared and discussed. The organization

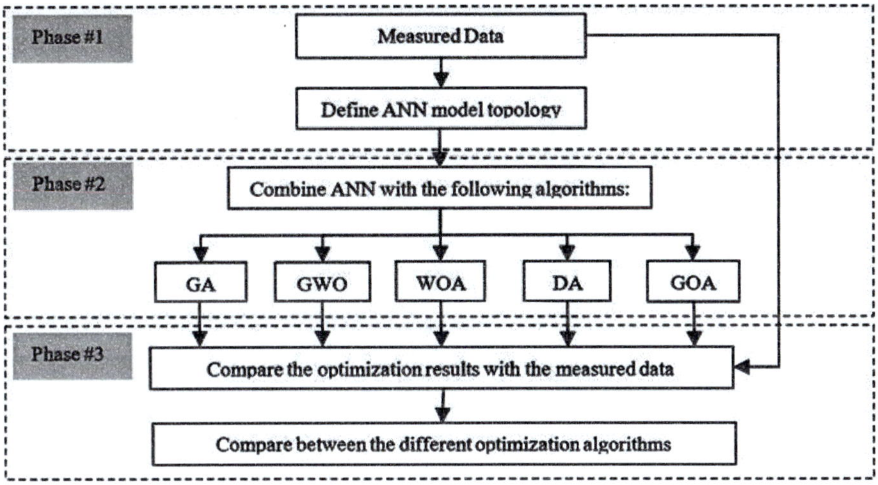

Fig. 1 Methodological framework

of this paper can be summarized as follows: Sect. 2 describes the ANN model used in this work. Section 3 describes the ANN-GA model and Sect. 4 describes ANN-GWO model. Sections 5, 6, 7 provide details on ANN-WOA, ANN-DA and ANN-GOA models. Section 8 describes the case study of modeling the GaN transistor. Section 9 provides analyses, results, discussion and insights. The last section provides a conclusion.

2 ANN Model and Experiment Design

ANN is used to capture the relationship between inputs and outputs. In Fig. 2, we show a simple ANN model consisting of two hidden layers, two inputs (X_1, X_2) and one output (Y). The output can be calculated using the following equation:

$$Y = \sum_{i=1}^{3} W_i \tan h\left(W_{i1} + \sum_{j=2}^{4} W_{ij} \tanh\left(W_{1j}X_1 + W_{2j}X_2 + W_{3j}\right)\right). \quad (1)$$

In Eq. (1) the input weights are W_{1j}, W_{2j} and W_{3j}, the intermediate weights are expressed by W_{ij} and the output weight is expressed by W_i. The activation function used in this model is *tanh*(.). In this paper, we optimize the weights of the ANN model using five global optimization techniques. The five global optimization techniques are to be tested over five numbers of solutions (50, 100, 200, 300 and 500) and over six different maximum numbers of iterations (50, 100, 200, 400, 600 and 800). Each configuration (number of solution and number of iteration) is to be run 30 times and

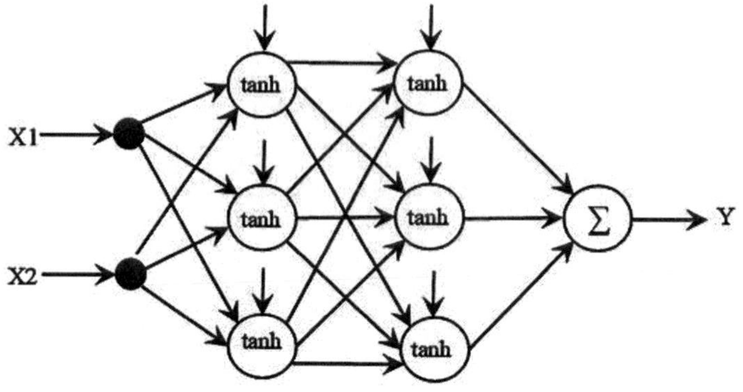

Fig. 2 A simple illustrative ANN model

the average performance is measured. Moreover, all the techniques will be fed with the same initial population.

3 Training ANN Using Genetic Algorithm Optimization

GA is inspired by the evolutionary biology, and it mimics the natural evolution processes such as reproduction, mutation, selection and crossover [1]. GA is classified as an evolutionary technique and was found to provide optimal or near-optimal solutions with good quality. Moreover, GA has been used to find optimal values for the weight of the ANN model in Fig. 2. The optimization approach starts by generating a set of random population (24 weights) within the range from -1 to 1. Next, the fitness of all solutions is evaluated, which is the total error among the measured and simulated values in Eq. (2). The objective is to minimize the difference between the measured data and the simulated results obtained from Eq. (1) in ANN-GA model.

$$\text{Error} = \frac{1}{M}\sum_{i=1}^{M}(Y_{\text{measured}} - Y_{\text{simulated}})^2 \tag{2}$$

where M is the total number of the measured data. $Y_{measured}$ and $Y_{simulated}$ represent the measured and the simulated data, respectively. After that, the parents are selected based on their fitness. Next, the parents are recombined using the crossover operator. Then, the offspring is mutated randomly with low probably. It is worth to note that the crossover and mutation ensure the exploration in GA algorithm and avoid getting stuck in local optima. Furthermore, the new offspring fitness is evaluated and the next generation population is selected. The solution with best fitness (least error) is selected and returned as the optimal weights for the ANN-GA model.

4 Training ANN Using Grey Wolf Optimization

GWO algorithm is inspired by grey wolves, and it mimics their strict leadership hierarchy and hunting mechanisms [15]. The grey wolves' hierarchy contains of four types: alpha, beta, delta and omega. The alpha wolf is the leader and responsible for the decision making in the pack. In the following types (beta, delta and omega) the dominance decreases in each level. The grey wolves hunting behavior for preys and selecting the best prey (optimal solution) among these multiple preys can be described by: tracking, encircling and attack the prey [15]. In order to mathematically model the GWO algorithm the alpha wolf (α) is considered to be the optimal solution, the beta (β) and delta (δ) wolves are considered second and third best solutions [15]. The remaining candidate solutions are assumed to be the omega wolves (ω). First the mathematical model inserts the measured ANN weights. Next the grey wolves' population positions are initialized (24 weights) and the corresponding coefficients are calculated using the equations in [15]. The fitness of grey wolves is calculated, and alpha, beta and delta wolves are selected. Subsequently, GWO starts optimizing positions of packs as described in [15]. It is worth mentioning that the adaptive values of GWO coefficients allow a good balance between explorations, exploitation and avoiding stagnation in local solutions [15]. Note that in the mathematical model the alpha, beta and delta positions are saved and all search agents update their positions accordingly [15]. It can be observed that the search agents' new position will be directed by the alpha, beta and delta wolves with some randomness as a result of the coefficients. In other words, the alpha, beta and delta wolves estimate the preys' position and the search agents update their positions around the prey. Furthermore, after obtaining the packs' new positions, the fitness is calculated and the new positions for alpha, beta and delta are updated. Finally, alpha is returned as the optimal solution for the ANN model.

5 Training ANN Using Humpback Whales Optimization

WOA is inspired by the social behavior of humpback whales. The WOA mathematical model mimics the bubble-net hunting strategy to find optimal solutions for optimization problems. The bubble-net hunting method is basically when the humpback whale circulates around the prey (optimal solution) going up in a helix-shaped path while releasing bubbles along the path. The humpback whale will get closer and closer to the prey while circulating and then reaches the optimal solution [16]. The mathematical model for encircling the prey for WOA is similar to the GWO algorithm. However, there are differences because the WOA depends on the bubble-net feeding method [16]. Similar to other models, initially, the measured data are inserted and then population is initialized. To mathematically model the bubble-net attacking method of the humpback whales, two approaches are used: Shrinking encircling mechanism and spiral updating position. The humpback whales use the shrinking encircling and spiral updating position approaches simultaneously. To model this behavior, there is a probability of 50% the whales will choose either of two approaches to update the position and reach optimal solution.

6 Training ANN Using Dragonfly Optimization (ANN-DA)

Another swarm intelligence optimization technique was developed by Mirjalili [3]. The DA imitates the social behavior of dragonflies in avoiding enemies and searching for food sources. Dragonflies have mainly two types of movement, static and dynamic movements, which correspond to exploitation and exploration needed for global optimization. Static movement is used for hunting and represents the exploitation phase of the optimization. On the other hand, dynamic movement is when dragonflies move in bigger swarms to migrate in one direction for long distances, this represents the exploration phase. This optimization technique is similar to the conventional particles swarm optimization, however; instead of being based on the velocity of PSO, DA updates the position of the individuals by using the step vector. The DA-ANN model starts by inserting the measured data. Next, the dragonflies' positions and step vectors are initialized. Then the fitness values for all dragonflies' positions are evaluated using Eq. (2). It is worth mentioning that the food source represents the best (optimal) value and enemy represents the worst value, and it is updated in each iteration. After that, DA updates social factors such as separation, alignment, cohesion, attraction to food and distraction from the enemy as described in [3]. In addition to the five factors that control the updated position of each individual, the step vector takes into account the inertia weight factor ω of the current position. Finally, check the boundaries and modify the new position if the boundaries are violated. Continue until stopping criterion is met and return best optimal solution (food source) [3].

7 Training ANN Using Grasshopper Optimization Algorithm

GOA is inspired by the behavior of grasshopper swarms in nature. Saremi et al. [4] developed a global optimization algorithm by modeling grasshoppers in searching for food and mimicking their exploration and exploitation movements. GOA, as all the other swarm techniques, is based on a set of individuals with random position, where the position of each agent is updated until it converges to an optimal solution. GOA updates the position of each search agents taking into account social interaction, gravity force and wind advection. However, this mathematical model cannot be used directly to solve optimization problems. The main reason is because grasshoppers quickly reach their comfort zone and don't converge to optimal solution. Therefore, a modified version of the position equation is proposed in [5]. It is assumed there is no gravity force and that the wind direction is always toward target. First, measured data for ANN-GOA is inserted. Next, the grasshopper population is initialized (24 weights) and parameters. Then Eq. (2) is used to evaluate the fitness of each search agent and best search agent is selected. Subsequently, the GOA starts its iteration with updating the coefficient that shrinks the comfort zone, attraction zone and repulsion zone. After that each search agent normalizes the distance between grasshoppers [4]. Then a check is performed on all search agents if the new position goes beyond boundaries. After that all solutions are evaluated again, and the best optimal solution is updated until the maximum iteration is reached. It is worth mentioning that GOA updates the position of search agents based on the current position, global-best position and position of all other grasshoppers in population. This is different from the well-known PSO where the position is updated using current position, self-best position and global-best position.

8 ANN Modeling of GaN Transistor (Case Study)

ANN can be used to solve a wide range of modeling problems. One of these problems is transistor modeling, which is needed for circuit design purposes, especially, for the new technology such as Gallium Nitride High Electron Mobility Transistor (GaN HEMT). The current-voltage characteristics of this power transistor affected mainly by power dissipation induced self-heating which results in current collapsing in higher power dissipation operating conditions (Fig. 3a) [17]. This regenerative process, in addition to the strong nonlinear behavior of the current with respect to

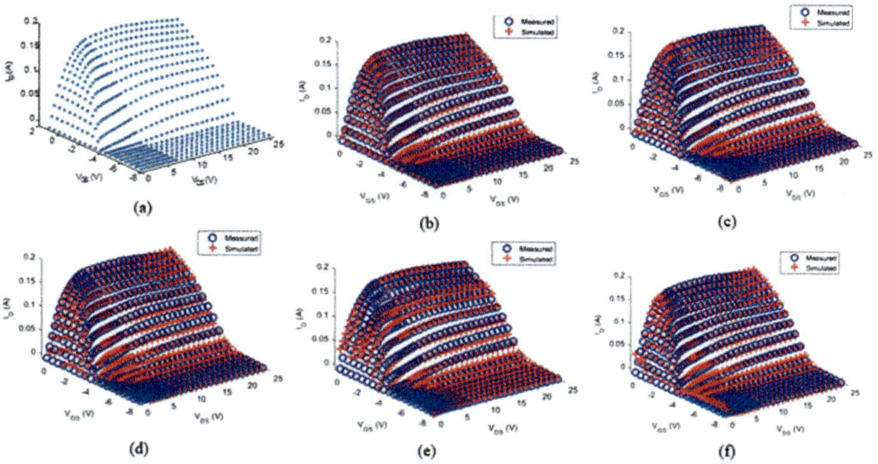

Fig. 3 **a** Measured current–voltage characteristics of GaN HEMT. **b** ANN-GA best fitting. **c** ANN-GWO best fitting. **d** ANN-WOA best fitting. **e** ANN-DA best fitting. **f** ANN-GOA best fitting

applied drain and gate voltages (V_{GS} and V_{DS}), put more effort on the implemented model. Analytical modeling is one of the commonly used techniques to simulate this device. However, this approach needs higher effort to adopt an efficient closed formula and also to find the corresponding fitting parameters [18]. ANN modeling can provide an optimal solution with lower effort and higher accuracy [19]. Of course, to obtain an accurate simulation for the nonlinear behavior of the current, a higher ANN order model with a higher number of weights is required. In this case, the modeling process will be reduced to an optimization problem and global optimization techniques should be used. The model, illustrated in Fig. 2, has been used efficiently to simulate the current–voltage characteristics of the GaN HEMT. The model inputs X_1 and X_2 represent the gate and drain voltages, V_{GS} and V_{DS}, respectively. The output Y represents the drain current I_D. In Fig. 3(b–f) we can observe the obtained results for the best fitting for ANN-GA, ANN-GWO, ANN-WOA, ANN-DA and ANN-GOA. It can be noticed the ANN-GA and ANN-GWO give the best fitting performance.

9 Results and Discussion

GaN transistor modeling data, shown in Fig. 3, was used in the ANN model combined with five global optimizations, one technique at a time. The analysis was performed by changing the number of iterations and number of individuals. Since all the optimization techniques mentioned in this paper share the same starting process, which consists in imitating a set of individuals that represent the 24 weights of the ANNs model, the same population set was used to test all the techniques. Thus, before starting any optimization technique, the population was generated and stored then used as initial guess for model optimization. This was done in order to guarantee equal conditions for all optimization techniques during the comparison. All optimization techniques were used with their default parameters. Each ANNs optimization model was tested with five different numbers of solutions (50, 100, 200, 300 and 500). While changing the number of iterations (50, 100, 200, 400, 600 and 800). The performance of each model was evaluated with respect to two criteria: efficiency in terms of time and effectiveness in terms of solution's quality (error value). Figure 4 shows the optimization value (error) of different ANN-optimization models. It can be seen that ANN-GWO outperforms all the other optimization techniques under small number of iterations and solutions (see Fig. 4, solutions = 50, iteration 50 and 100). It is noted that ANN-GA consistently had a stable robust performance that didn't fluctuate while increasing the number of solutions or iterations unlike the ANN-DA and ANN-GOA.

It was noticed that the best fitting using ANN-GA was obtained for 500 solutions and 800 iterations with an error equal to 0.02022 (see Fig. 4).The best result for ANN-GWO was obtained for 300 solutions and 800 iterations with an error equal to 0.026 (see Fig. 4).Moreover, for ANN-WOA, ANN-DA and ANN-GOA, the best fitting was obtained for 100, 100 and 100 solutions with 800, 400 and 400 iterations, respectively. The obtained errors, 0.04992, 0.169 and 0.03375, respectively. To sum

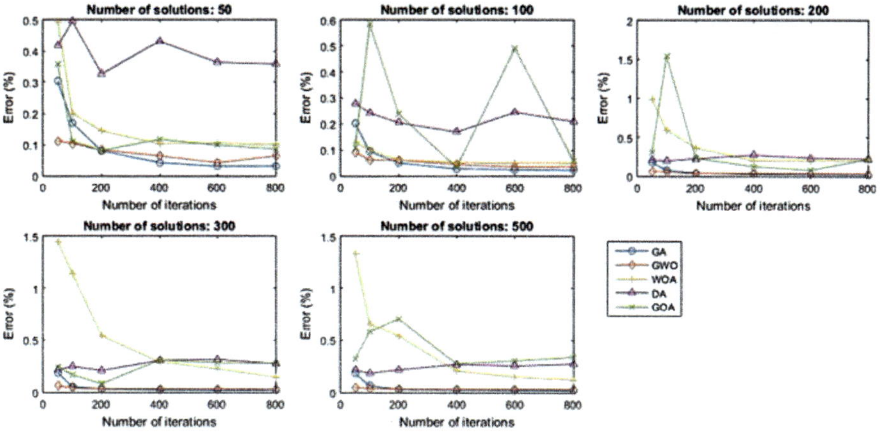

Fig. 4 Average error values of different ANN-optimization models as the number of iterations increases for different number of solutions

up, ANN-GA resulted to be the most accurate fitting; however, ANN-GWO gave a very competitive result under reasonable number of iterations, solution and time. Both ANN-GA, ANN-GWO and ANN-WOA are efficient and fast enough with a minimum computation time of 3.42 s for 50 solutions and 50 iterations. Furthermore, 471.27 s for 500 solutions and 800 iterations. On the other hand, ANN-DA and ANN-GOA have the highest computation time range (5.77–3314 s for ANN-DA and 20.38–27070 s for ANN-GOA) (Fig. 5). It is worth mentioning that all the ANNs optimization models have been run on a computer equipped with an Intel(R) Core (TM) i5-4590, CPU @ 3.3 GHz 3.3 GHZ, 8.00 GB RAM and Windows 7 64-bit operating system.

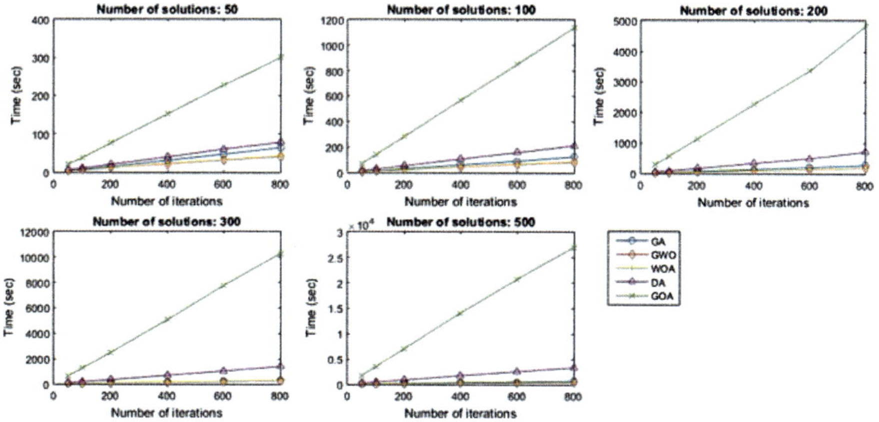

Fig. 5 Average CPU time of different ANN-optimization models as the number of iterations increases for different number of solutions

10 Conclusion

In this paper, five different global optimization techniques were used to optimize the weights of an ANN model. The performance and computational time of these techniques were compared. It has been found that ANN-GA and ANN-GWO show competitive performance in most of the considered cases. Furthermore, ANN-GA shows robustness and improved performance with increasing number of iterations and solutions. The ANN-GWO obtained competitive results with high speed of convergence. In general, ANN-GA provided the most accurate fitting (lowest error value) compared to others, but the convergence speed was less than that of AAN-GWO. On the other hand, ANN-DA and ANN-GOA are considered extremely slow with respect to the other techniques and didn't achieve best fitting values. As a further step, it is suggested to study the enhanced version of the developed global optimization algorithms and to improve the algorithms to account for parallel computing in order to enhance the speed of the algorithms. In addition, studying the effect of the parameters of each algorithm in feeding neural networks might be of interest as well.

Acknowledgements The authors gratefully acknowledge the support of University of Sharjah (Sharjah, United Arab Emirates).

References

1. W. Zhong, J. Liu, M. Xue, L. Jiao, A multiagent genetic algorithm for global numerical optimization, IEEE Trans. Syst. Man. Cybern. B. Cybern. **34**(2), 1128–1141 (2004)
2. S. Hamdan, A. Jarndal, A two stage green supplier selection and order allocation using AHP and multi-objective genetic algorithm optimization, in *2017 7th International Conference on Modeling, Simulation, and Applied Optimization, ICMSAO* (2017). https://doi.org/10.1109/icmsao.2017.7934843
3. S. Mirjalili, Dragonfly algorithm: a new meta-heuristic optimization technique for solving single-objective, discrete, and multi-objective problems. Neural Comput. Appl. **27**(4), 1053–1073 (2016). https://doi.org/10.1007/s00521-015-1920-1
4. S. Saremi, S. Mirjalili, A. Lewis, Grasshopper optimisation algorithm: theory and application. Adv. Eng. Softw. **105**, 30–47 (2017). https://doi.org/10.1016/j.advengsoft.2017.01.004
5. N. Nawayseh, A. Jarndal, S. Hamdan, Optimizing the parameters of a biodynamic responses to vibration model using particle swarm and genetic algorithms, in *2017 7th International Conference on Modeling, Simulation, and Applied Optimization, ICMSAO* (2017). https://doi.org/10.1109/icmsao.2017.7934851
6. A. Jarndal, S. Hamdan, Forecasting of peak electricity demand using ANNGA and ANN-PSO approaches, in *2017 7th International Conference on Modeling, Simulation, and Applied Optimization, ICMSAO* (2017). https://doi.org/10.1109/icmsao.2017.7934842
7. J F. Kolen, J.B. Pollack, Back propagation is sensitive to initial conditions, in *Proceedings of the 1990 Conference on Advances in Neural Information Processing Systems*, pp. 860–867 (1990)
8. D. Whitley, T. Starkweather, C. Bogart, Genetic algorithms and neural networks: optimizing connections and connectivity. Parallel Comput. **14**(3), 347–361 (1990). https://doi.org/10.1016/0167-8191(90)90086-O

9. A. Yamazaki, M.C.P. de Souto, T.B. Ludermir, Optimization of neural network weights and architectures for odor recognition using simulated annealing, in *Proc. 2002 Int. Jt. Conf. Neural Networks. IJCNN'02 (Cat. No.02CH37290)*, vol. 1, no. 3, pp. 547–552, (2002). https://doi.org/10.1109/ijcnn.2002.1005531
10. D. Karaboga, B. Akay, C. Ozturk, Artificial Bee Colony (ABC) optimization algorithm for training feed-forward neural networks, in *Proceedings of the 4th International Conference on Modeling Decisions for Artificial Intelligence*, pp. 318–329 (2007). https://doi.org/10.1007/978-3-540-73729-2_30
11. S. Mirjalili, S.Z. Mohd Hashim, H. Moradian Sardroudi, Training feedforward neural networks using hybrid particle swarm optimization and gravitational search algorithm. Appl. Math. Comput. **218**(22), 11125–11137 (2012). https://doi.org/10.1016/j.amc.2012.04.069
12. V.G. Gudise,G.K. Venayagamoorthy, Comparison of particle swarm optimization and back-propagation as training algorithms for neural networks, in *Swarm Intelligence Symposium, 2003. SIS'03. Proceedings of the 2003 IEEE*, pp. 110–117 (2003)
13. S. Mohaghegi, Y. Del-Valle, G.K. Venayagamoorthy, A comparison of PSO and backpropagation for training RBF neural networks for identification of a power system with STATCOM, in *Swarm Intelligence Symposium,. SIS 2005*, pp. 381–384, (2005). https://doi.org/10.1109/SIS.2005.1501646
14. S. Mirjalili, How effective is the Grey Wolf optimizer in training multi-layer perceptrons. Appl. Intell. **43**(1), 150–161 (2015). https://doi.org/10.1007/s10489-014-0645-7
15. S. Mirjalili, S.M. Mirjalili, A. Lewis, Grey wolf optimizer. Adv. Eng. Softw. **69**, 46–61 (2014). https://doi.org/10.1016/j.advengsoft.2013.12.007
16. S. Mirjalili, A. Lewis, The whale optimization algorithm. Adv. Eng. Softw. **95**, 51–67 (2016). https://doi.org/10.1016/j.advengsoft.2016.01.008
17. I. Saidi, Y. Cordier, M. Chmielowska, H. Mejri, H. Maaref, Thermal effects in AlGaN/GaN/Si high electron mobility transistors. Solid-State Electron. **61**(1), 1–6 (2011). https://doi.org/10.1016/j.sse.2011.02.008
18. A. Jarndal, F.M. Ghannouchi, Improved modeling of GaN HEMTs for predicting thermal and trapping-induced-kink effects. Solid State Electron. **123**, 19–25 (2016). https://doi.org/10.1016/j.sse.2016.05.015
19. A. Jarndal, Genetic algorithm-based neural-network modeling approach applied to AlGaN/GaN devices. Int. J. RF Microw. Comput. Eng. **23**(2), 149–156 (2013). https://doi.org/10.1002/mmce.20660

Machine Learning Techniques

Network Intrusion Detection Model Using One-Class Support Vector Machine

Ahmed M. Mahfouz, Abdullah Abuhussein, Deepak Venugopal, and Sajjan G. Shiva

1 Introduction

Cyber-attacks have become more widespread as intruders employ system vulnerabilities for theft of intellectual properties, financial gain, or even destruction of the whole network infrastructure [1]. In many cases, a security breach is inevitable which makes early detection and mitigates the best defense for surviving an attack. To reduce the risk of security breaches, security professionals use different prevention and detection techniques. Prevention techniques such as applying complex configurations and establishing a strong security policy try to make attacks more difficult. Detection techniques are either signature-based or anomaly-based. Classical security solutions such as virus scanners, intrusion detection systems, and firewalls depend on the 'misuse detection' also known as 'signature-based' approaches that compare a hash of the payload to a set of known malicious signatures [2].

One of the most strenuous efforts in the fields of virus and intrusion detection is the continuous need for up-to-date definitions of different attacks. This is due to the use of the signature-based methods which work on defining the anomaly traffic rather than defining the normal traffic. Although those approaches have been widely effective in nearly all the recent antivirus and intrusion detection tools as they precisely identify new instances of attacks, they are useless when facing an unknown

A. M. Mahfouz (✉) · D. Venugopal · S. G. Shiva
Department of Computer Science, University of Memphis, Memphis, TN, USA
e-mail: amahfouz@memphis.edu

D. Venugopal
e-mail: dvngopal@memphis.edu

S. G. Shiva
e-mail: sshiva@memphis.edu

A. Abuhussein
Department of Information Systems, St. Cloud State University, St. Cloud, MN, USA
e-mail: aabuhussein@stcloudstate.edu

S. Patnaik et al. (eds.), *Advances in Machine Learning and Computational Intelligence*, Algorithms for Intelligent Systems, https://doi.org/10.1007/978-981-15-5243-4_7

attack. This problem was somehow successfully addressed in the area of antivirus with a 24-hour response team updating the database of malicious signatures. But, in the area of intrusion detection, maintaining such an up-to-date knowledge base is ultimately a lost battle.

This problem is not only because of the massive number of daily discovered vulnerabilities but also because of the unknown number of the hidden vulnerabilities that can be exposed and exploited. Furthermore, a skilled attacker can right away study some forms of attacks, just to hit a single or a few systems. Moreover, computer attacks have a polymorph nature, as an attacker can exploit the same vulnerability in different ways. Consequently, it is not easy to develop adequate signatures. Going back to the basics by implementing anomaly detection methods that model what is normal instead of what is anomalous would be an obvious solution [3].

In this paper, we propose a new network intrusion detection model that trains on normal network traffic data and searches for anomalous behaviors that deviate from the normal model. We apply one-class support vector machine (OCSVM) algorithm to detect the anomalous activities in network traffic. Our approach models the regions where normal data has a high probability density in n-dimensional feature space and considers anomalous data as those that do not occur in these regions [4]. Thus, our approach is capable of detecting threats to the network without using any labeled data. Further, by using kernel functions with OSVMs, our approach can capture complex, nonlinear regions in the feature space where the normal data is most likely to reside as compared to anomaly detectors that assume a specific shape/form for the normal class. For example, methods that assume that the normal class follows a normal distribution and detects deviation from this distribution [5, 6].

The rest of the paper is organized as follows: In Sect. 2, we provide a brief introduction to OCSVM algorithm. In Sect. 3, we describe the dataset used in the paper and how it is collected. The related work is presented in Sect. 4. The experimental results are given in Sect. 5. Conclusions and further work are finally discussed in Sect. 6.

2 One-Class Support Vector Machine (OCSVM)

One-class classification approaches are essentially helpful in solving two-class learning problems, whereby the first class which is mostly well-sampled is known as 'target' class, and the other class which severely under-sampled is known as the 'outlier' class [7]. The goal is to build a decision surface around the samples in the target class with a view to distinguish the target objects from the outliers (all the other possible objects) [8].

Support vector machine (SVM) [9] is a supervised learning model that can be used in the process of analyzing data and recognizing patterns, which is the core of any classification task. The SVM algorithm takes training dataset with labeled samples that belong to one of two classes and divides the samples into separate groups through a wide gap while penalizing all samples that appear on the wrong side of the gap.

Then, the SVM model produces his predictions by assigning points to one side of the gap or the other. Occasionally, to increase the existing samples to be able to build the two-class model, over-sampling is used; however, it is impossible to predict all new patterns of a network intrusion detection system from limited examples. Furthermore, a collection of even limited examples can be expensive.

Adapting the one-class classification from a two-class SVM problem was proposed in [10]. The basic idea was to use an appropriate kernel function to map the input data to a high dimensional feature space. Doing this, it was possible to create a decision function that best separates up one-class samples from the second class samples with the maximum margin.

3 The Dataset

Finding a comprehensive and valid dataset to train and evaluate proposed intrusion detection techniques is a significant challenge to many researchers [11]. Although there exist a number of available datasets that researchers have used in evaluating their proposed approaches performance, most of them are unreliable and out of date and do not reflect the current trends. On the other hand, most of the adequate and suitable datasets are not publicly available due to privacy issues. In this paper, we produce a new reliable IDS dataset that contains benign and different common attack network flows that meets the real-world criteria. In this section, we describe the dataset and how it was collected.

To collect the data, we have implemented the modern honey network (MHN) [12], which is a centralized server to manage and collect data from honeypots. MHN has an easy to use Web interface that helps in quickly deploying the sensors and immediately collecting a viewable data. MHN deploys scripts which include several common honeypot technologies, like Snort, Cowrie, Dionaea, and glastopf. MHN can aid in developing stronger network security by analyzing the data from honeynets. Unfortunately, no tool currently exists to aggregate data from the sensors implemented with MHN environments. So, we created a dataset tool using Excel, which aggregates data from separate network monitors into a single spreadsheet.

We used Google Cloud to create four instances of Ubuntu 16.05 LTS servers, where we had one MHN server, and three sensor servers. The first sensor (Sensor-1) was set up with Conpot, p0f, and Snort. The second sensor (Sensor-3) was set up with Shockpot, Elastic honey, p0f, and Snort. The third sensor (Sensor-4) was set up with Wordpot, p0f, and Snort. Using this architecture, we were able to collect a large amount of data through the sensors. We also were able to sort through the data and create a better format and similar data structure for all the collected data. Figure 1 shows the MHN implementation.

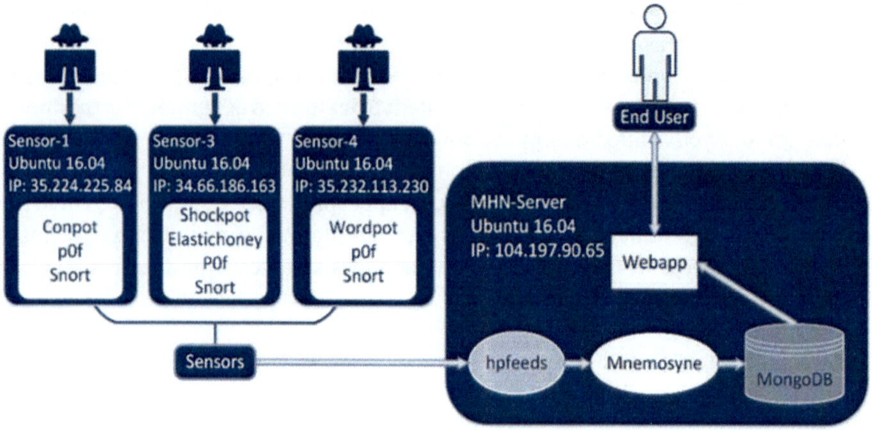

Fig. 1 Modern honey network implementation

4 Related Work

The topic of the anomaly detection has been the subject of many surveys as well as review articles. Chandola et al. [13] presented a structured survey of research on anomaly detection in different research areas and application fields, including network intrusion detection systems. Also, researchers heavily depend on ML algorithms as a primary focus for designing different network intrusion detection systems [14]. Likewise, different one-class classification techniques have been utilized in solving the intrusion detection problem. Giacinto et al. [15] proposed an intrusion detection model based on an ensemble of one-class classifiers. In their model, they used v-SVM, k-means, and Parzen density estimation to construct a modular approach where each single module models a group of similar network protocols and services. They evaluated their model using the KDD 99 dataset and concluded that a high detection rate and a lower false alarm rate can be achieved by dividing the problem into different modules. In Kang et al. [16], utilized the one-class classification method to propose a differential support vector data descriptor-based intrusion detection model. Their experiments which were performed on a simulated dataset and DARPA 1998 dataset showed that their model performs better than existing techniques in detecting specific types of attacks. In ‘ [17] have employed two different one-class classification methods on SCADA networks that monitor and control different industrial and public service processes. Their results show that both methods can tightly surround the normal flow behavior in the SVM hypersphere and also detect intrusions.

Furthermore, the research of using OCSVM in intrusion detection for traditional TCP/IP networks includes the following: In [18], Ghorbel has associated the coherence parameter with the least-squares optimization problem to examine a new one-class classification model. He applied his model to a wireless sensor network and obtained a good detection rate. Using the OCSVM, Kaplantzis [19] has provided new centralized IDS that can detect black hole attacks with high accuracy. In order to improve the detection accuracy of OCSVM models, Xiao [20] proposed the DFN and the DTL methods to select the model Gauss Kernel function. Similarly, Amer [21] tried to reduce the effect of the outliers on the normal data boundary decision. He proposed two boosted OCSVM models for the unsupervised-based anomaly detection systems. By hierarchically integrating both misuse and anomaly detection methods in a decomposition structure, Kim [22] designed a hybrid intrusion detection model. In his model, Kim used multiple one-class methods. Winter [23] used the OCSVM algorithm to propose an IDS model derived from inductive learning.

Considering all these previous works, we have implemented a new network intrusion detection model based on integrating the OCSVM and the MHN. Our model was able to get an overall accuracy of more than 97%.

5 Experimental Results

In this section, we discuss the experimental details, which include the experimental setup, the performance evaluation metrics, and the experimental results.

5.1 The Experimental Setup

Data Source

Our dataset consists of 41,770 entries with 25 feature columns which are a mix of numeric and categorical types plus one label column. In order to train our anomaly detector, we will be using only the samples with a 'normal' label while ignoring those with 'anomaly' label. However, we will be using both categories for evaluating the anomaly detector. In our experiment, we used Azure Machine Learning (AML) [24] which is a cloud-based environment from Microsoft to preprocess data, train, test, deploy, manage, and track machine learning models.

Preprocessing

The first step in the experiment is the dataset preprocessing where we use the AML Metadata Editor module to mark the target column 'Class' as type 'label'. Next, we split the data into two sets, one (70%) for training and the second (30%) for testing. To make sure that the training set contains only normal traffic, we use the Split module with a regex filter on the label column to remove rows with 'anomaly' label. This process is being repeated on the data that is used for parameter sweep including a further splitting of training data into train and validation sets.

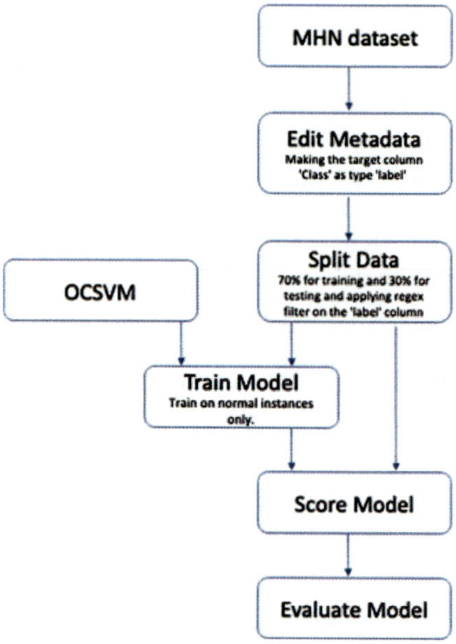

Fig. 2 OCSVM anomaly detector model using AML

Model Training

The second step is the model training process using the OCSVM. In this process, the model works on separating the training data collection from the origin using maximum margin. By default, a radial basis kernel is used. To specify the parameters of the model, we use the Create Trainer Mode option from the module properties and select the parameter sweep mode which is used in conjunction with the sweep parameters module. This module produces a learner with the best settings.

Model Evaluation

The next step is using the generic score model module to obtain the predictions from the OCSVM anomaly detector, and finally, evaluate the proposed model. Figure 2 shows the whole process as applied in the AML.

5.2 *Performance Evaluation Metrics*

To evaluate the performance of the proposed model, we use precision, recall, and F1 score which are the most common measures to evaluate the performance of anomaly detection models. Precision refers to the portion of the relevant instances among the retrieved instances. Recall refers to the portion of relevant retrieved instances from the total number of the relevant instances. F1 score is harmonic mean of the precision and the recall.

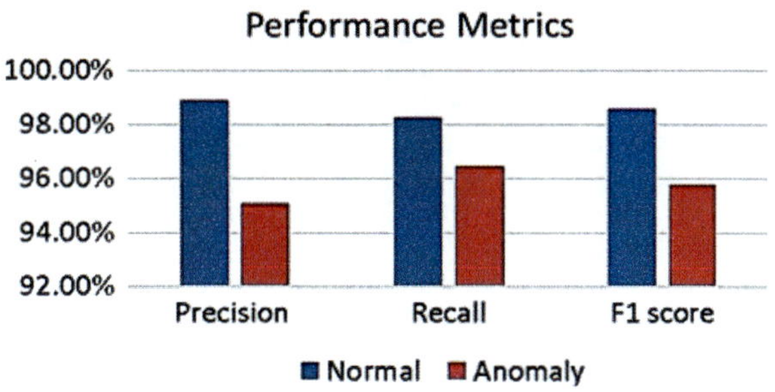

Fig. 3 Per-class comparison based on precision, recall, and F1 score

5.3 The Experimental Result

Since our dataset was randomly divided into 70% for training and 30% for testing, in the preprocessing phase, we tried to run the experiment several times to compute the average and variance of the results on the test data. The AML evaluation module shows that there is no big variance in the results, and the average accuracy of the proposed anomaly detection model was 97.61%. Figure 3 shows the Per-class comparison for precision, recall, and F1 score.

6 Conclusion and Future Work

This paper proposed a new network intrusion detection model that applied the OCSVM algorithm. The proposed model method worked by modeling what is normal instead of what is anomalous and was able to detect the anomalies with a high detection rate. The model has been experimented on a new dataset collected from real network traffic using the modern honey network. For future work, we plan to modify the MHN by adding more sensors and generating more network traffic. Also, we will work on applying the model to real-time network traffic to have an online network intrusion detection system with high accuracy and a low false alarm rate.

References

1. Mohamed Abomhara, Cyber security and the internet of things: vulnerabilities, threats, intruders and attacks. J. Cyber. Secur. Mobility. **4**(1), 65–88 (2015)
2. R. Singh et al., Internet attacks and intrusion detection system: a review of the literature. Online. Inf. Rev. **41**(2), 171–184 (2017)
3. S. Zanero, S.M. Savaresi, Unsupervised learning techniques for an intrusion detection system. in *Proceedings of the 2004 ACM Symposium on Applied Computing* (ACM, 2004)
4. B. Schölkopf et al., Estimating the support of a high-dimensional distribution. Neural. comput. **13**(7), 1443–1471 (2001)

5. K. Yamanishi et al., On-line unsupervised outlier detection using finite mixtures with discounting learning algorithms. Data Min. Knowl. Disc. **8**(3), 275–300 (2004)
6. K. Yamanishi, J. Takeuchi Discovering outlier filtering rules from unlabeled data: combining a supervised learner with an unsupervised learner. in *Proceedings of the Seventh ACM SIGKDD International Conference on Knowledge Discovery and Data Mining* (ACM, 2001)
7. L.M. Manevitz, Y. Malik, OCSVMs for document classification. J. Mach. Learn. Res. **2**, 139–154 (2001)
8. D. Tax, One-class classification; Concept-learning in the absence of counterexamples. Ph. D. thesis. Delft University of Technology, ASCI Dissertation Series. p. 146 (2001)
9. C.J.C. Burger, A tutorial on support vector machines for pattern recognition, data mining and knowledge discovery. Workshop on data mining and knowledge discovery (1998)
10. JC Platt et al, Estimating the support of a high-dimensional distribution. Technical Report MSR-T R-99–87, Microsoft Research (MSR) (1999)
11. R. Koch, G. Mario, G.D. Rodosek, Towards comparability of intrusion detection systems: new data sets. in: *TERENA Networking Conference*, vol. 7 (2014)
12. https://www.anomali.com/blog/mhn-modern-honey-network
13. Varun Chandola, Arindam Banerjee, Vipin Kumar, Anomaly detection: a survey. ACM Comput. Surv. (CSUR) **41**(3), 15 (2009)
14. H.J. Liao et al., Intrusion detection system: a comprehensive review. J. Netw. Comput. Appl. **36**(1), 16–24 (2013)
15. G. Giacinto, Intrusion detection in computer networks by a modular ensemble of one-class classifiers. Inf. Fusion **9**(1), 69–82 (2008)
16. Inho Kang, Myong K. Jeong, Dongjoon Kong, A differentiated one-class classification method with applications to intrusion detection. Expert Syst. Appl. **39**(4), 3899–3905 (2012)
17. P. Nader, P. Honeine, P. Beauseroy, Intrusion detection in SCADA systems using one-class classification. in *21st European Signal Processing Conference* (*EUSIPCO 2013*) (IEEE, 2013)
18. O. U. S. S. A. M. A. Ghorbel, H. I. C. H. E. M. Snoussi, M. O. H. A. M. E. D. Abid, Online OCSVM for outlier detection based on the Coherence Criterion in Wireless Sensor Networks. in *Proc International Conference*. vol. 12 (2013)
19. S. Kaplantzis et al, Detecting selective forwarding attacks in wireless sensor networks using support vector machines. in *2007 3rd International Conference on Intelligent Sensors, Sensor Networks and Information* (IEEE, 2007)
20. Y. Xiao et al., Two methods of selecting Gaussian kernel parameters for OCSVM and their application to fault detection. Knowl.-Based Syst. **59**, 75–84 (2014)
21. M. Amer, M. Goldstein, S. Abdennadher, Enhancing one-class support vector machines for unsupervised anomaly detection. in *Proceedings of the ACM SIGKDD Workshop on Outlier Detection and Description* (ACM, 2013)
22. Gisung Kim, Seungmin Lee, Sehun Kim, A novel hybrid intrusion detection method integrating anomaly detection with misuse detection. Expert Syst. Appl. **41**(4), 1690–1700 (2014)
23. P. Winter, E. Hermann, M. Zeilinger, Inductive intrusion detection in flow-based network data using one-class support vector machines. in *2011 4th IFIP international conference on new technologies, mobility and security* (IEEE, 2011)
24. https://azure.microsoft.com/en-us/

Query Performance Analysis Tool for Distributed Systems

Madhu Bhan and K. Rajanikanth

1 Introduction

With the growing size of Distributed Database Systems, and complex query workload, the task of assessing the performance of these systems is a difficult task. Since Distributed Systems are complex systems with many interacting components, it is essential to have a framework that supports early assessment of such systems. Software Performance Engineering (SPE) is an approach that proposes building and solving quantitative performance models to assess the performance of the system early in the software development life cycle [1]. The most common and popular approach for solving the performance models developed using the SPE approach is to use simulation. Two such models are proposed in this approach. The first one, "Software Execution Model", accepts the performance characteristics of all the resources used in building the system as inputs and builds a model that can be solved to obtain the performance characteristics of the software system. This model ignores issues like the presence of other service requests, the consequent contention for resources, the need for buffering requests waiting for service, service policies, etc. All such issues ignored by the first model are considered by the second performance model, the "System Execution Model". This second model makes use of the results produced by the first model. Additionally, it requires data regarding the workload characteristics and execution environment including all the resources for which contention is possible [2].

The DSPAT tool can solve both the Software Execution Model and System Execution Model and provide appropriate performance metrics like response ratio, average

M. Bhan (✉)
Ramaiah Institute of Technology, Bangalore, India
e-mail: madhoobhan@yahoo.co.in.com

K. Rajanikanth
Visvesvaraya Technological University, Belgavi, India

S. Patnaik et al. (eds.), *Advances in Machine Learning and Computational Intelligence*, Algorithms for Intelligent Systems, https://doi.org/10.1007/978-981-15-5243-4_8

waiting time, average idle time, and server utilization. We can analyze how the system will perform in its "as-is" configuration and under myriad possible "to-be" alternatives, so that we can analyze better ways to run analytical queries.

2 Design Strategy

The tool is based on two important design strategies of databases to improve performance of queries. These two design techniques used are materialized views and indexing. The approach of materializing views refers to the evaluation of frequent queries beforehand and storing of the results as materialized views for later use by the Proxy Server to provide fast response to those queries [3, 4]. A popular indexing scheme in the data warehousing environments is "BitMap Indexes". We have measured the performance under BitMap Indexing, Binary Indexing, new multi-level (NML) and new multi-component (NMC) Indexing techniques [5, 6].

2.1 Features

The salient features of the DSPAT are that it simulates the Web-based Distributed System multi-tier architecture with open workload. The workload for the tool is represented by query/query classes. Each query is realized as a sequence of requests to be processed by respective servers. The tool computes the processing times of queries based on hardware/software configuration and environment of servers deployed at different layers of the architecture. DSPAT considers exponential and normal distribution for query inter-arrival times. The simulation can be run for any number of queries that arrive for service. Performance metrics like response ratio, server utilization, average service time, etc., can be computed. It supports simulation of the system with different indexing schemes at the Database Server level/Proxy Server level. The tool can determine the views to be materialized from a set of possible views.

2.2 Working

The simulation tool DSPAT models the Distributed System architecture as shown in Fig. 1. The sequence of operations involved in processing queries as given in [7] starts when the client submits the query through the client application. The request reaches the Web Server over the Internet/Intranet. Web Server submits the received query to the Proxy Server. If the query can be satisfied with the data available at the Proxy Server level, the Proxy Server will generate the results for the query and return them to the Web Server. Otherwise, the Proxy Server forwards the query up one more layer, i.e., to the Database Server. The Database Server processes the query by

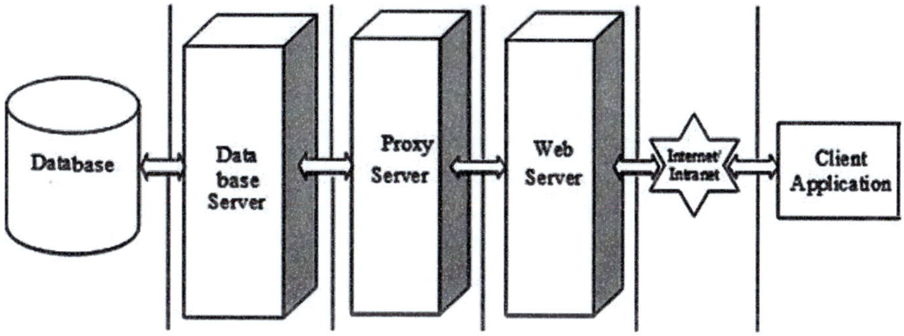

Fig. 1 Distributed system architecture

fetching the required data from the Database and returns the result of the query back to the Proxy Server. The Proxy Server then returns the results to the Web Server. The Web Server returns the results over the Internet/Intranet to the client application. The main sub-models of this tool are resource model, workload model, and query model. The aim of resource model is to define the resources and their characteristics. In our simulation model, the servers are assumed to have resources such as the CPUs, disks, memory, and network interfaces. The workload model is a stream of requests arriving into the system. Workload in a system may be open workload or closed workload. We have considered an open workload in our simulation model. The inter arrival time between the queries is generated using exponential or normal distribution. The Query Execution Model specifies whether the query is to be served by the Proxy Server itself or it must be served by Database Server. In either case, user provides an estimate of size of the data to be scanned and processed.

3 Implementation

The tool is built in the NetBeans IDE using Java language with Swings for developing GUIs and JFreeChart for generating graphs. NetBeans IDE is a free, open-source integrated development environment that provides a rich set of features. The NetBeans platform is a generic framework for Swing applications. Swings is a primary Java GUI widget toolkit. Swing provides a more sophisticated set of GUI components than the earlier Abstract Window Toolkit. JFreeChart is a free open source Java chart library that can be used to generate professional quality charts [8, 9].

4 Usage

The first step in the simulation is to specify the hardware configuration of resources deployed at different layers of the Distributed System architecture—Client, Proxy Server, and Database Server. In each case, we need to specify the number of Servers, CPU speed in MIPS (million instructions per second), disk speed in pages per millisecond, page size in KB, and network speed in KB per second. Accordingly, the tool presents three screens, one for the client side configuration, one for the Proxy Server configuration, and one for Database side configuration. For example, the Proxy Server side configuration screen is as shown in Fig. 2.

4.1 Main Menu Options

After the specification of the hardware configuration and workload, the user is presented with the Main Menu. The Menu bar has options of “Software”, “System”, “View Materialization”, “Analysis”, and “Configure”.

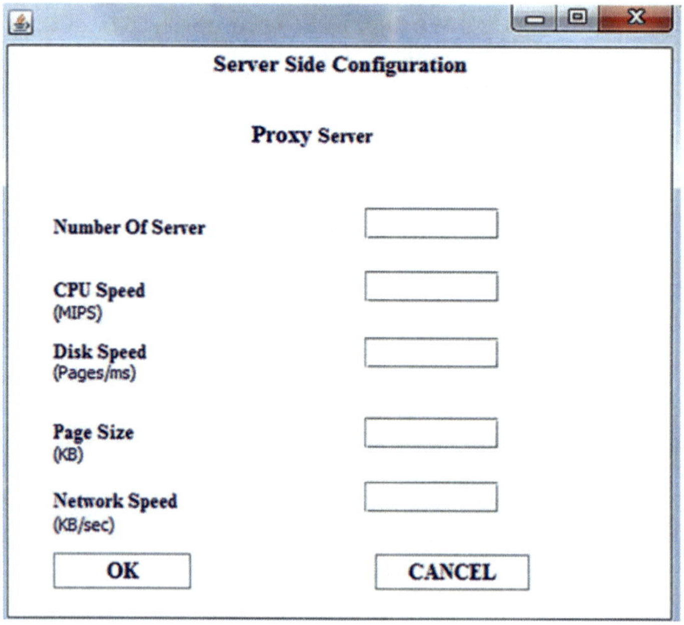

Fig. 2 Input screen for proxy server configuration

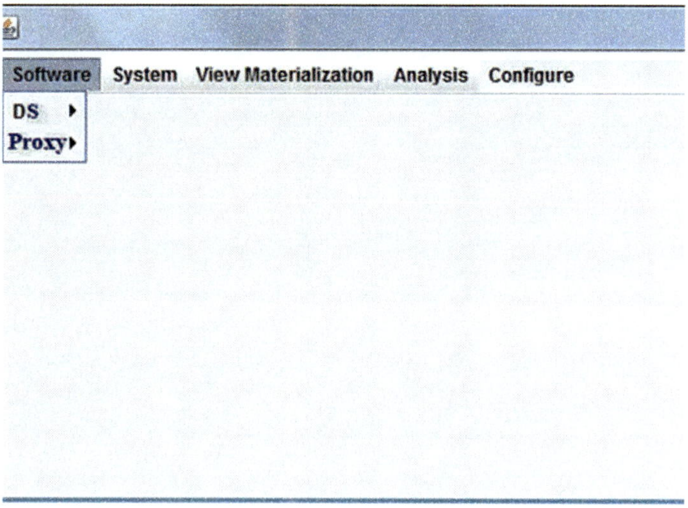

Fig. 3 Options under software model

4.2 *Software*

This option directs the tool to solve the Software Execution Model. Solving this model provides a static analysis of the response times. On clicking the "Software", the user gets a pull-down menu with two options of Database Server (DS) and "Proxy" as shown in Fig. 3.

DS option indicates that the queries in the workload are run by the Database Server. Proxy indicates that the queries in the workload are run by the Proxy Server, i.e., the queries must get executed based on the views present in the Proxy Server. The following screen shots illustrate how the tool DSPAT solves the Software Model. For the below input/output screen shots, we consider a Database application which analyzes trends in sales and supply [7]. User can estimate the data size using any suitable approach like linear search (for select) and nested loop (for join) [10]. When the "Next" button is clicked, the processing time of each query/query class is displayed. The panel down in the window of Fig. 4 displays a corresponding bar chart of query/query class.

4.3 *System*

This main option indicates that a System Execution Model is to be solved. Such a model is a dynamic one that characterizes the software performance in the presence of factors which could cause contention for resources [2]. When the System Execution Model is solved, we get the performance metrics of response ratio of queries, waiting

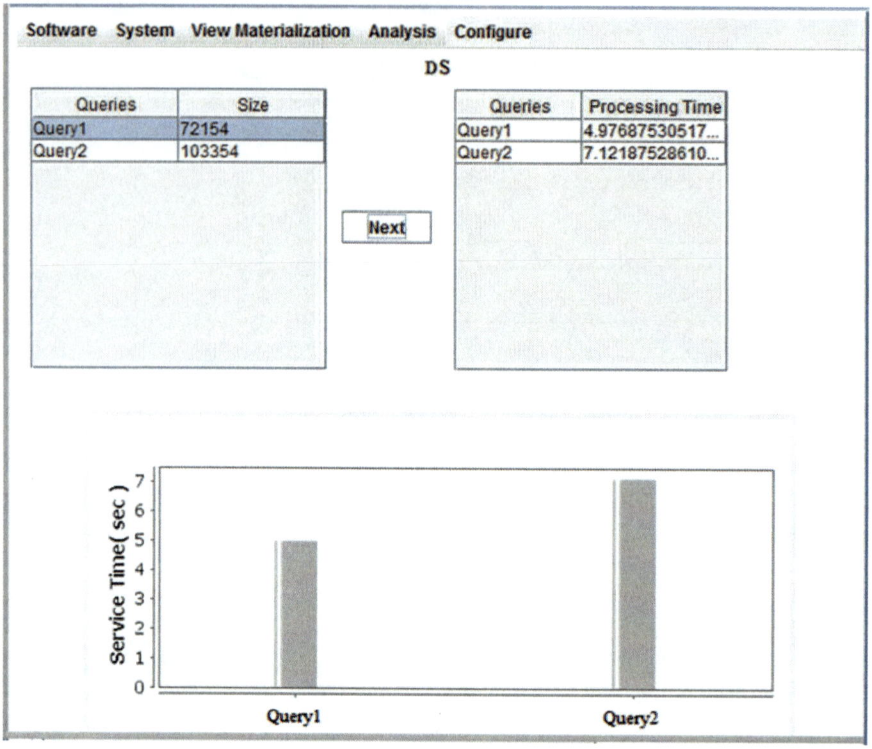

Fig. 4 Input/output screen

time of queries, idle time of server, total service time, and server utilization for a specific arrival pattern of the queries. Generally, we are interested in the variations of these performance metrics with different arrival rates. All these arrival processes are assumed to follow either the exponential distribution or the normal distribution. Further, user must specify the data sizes associated with each query/query class. User also must specify the number of customers and queue size/buffer size based on the expected system characteristics. All these parameters are entered in the top portion of an input/output window as shown in Fig. 5.

The user can click on the Next button on the top panel to get the results. The system outputs performance metrics including the response ratios of queries, waiting time of queries, idle time of server, total service time, and server utilization for each value of the mean inter arrival time specified by the user earlier. This output appears as shown in bottom panel of the window in Fig. 5. Also shown in the same Figure is the option for selecting the performance metric that is to be displayed graphically. As can be seen from Fig. 5, the idle time increases with increasing mean inter arrival time.

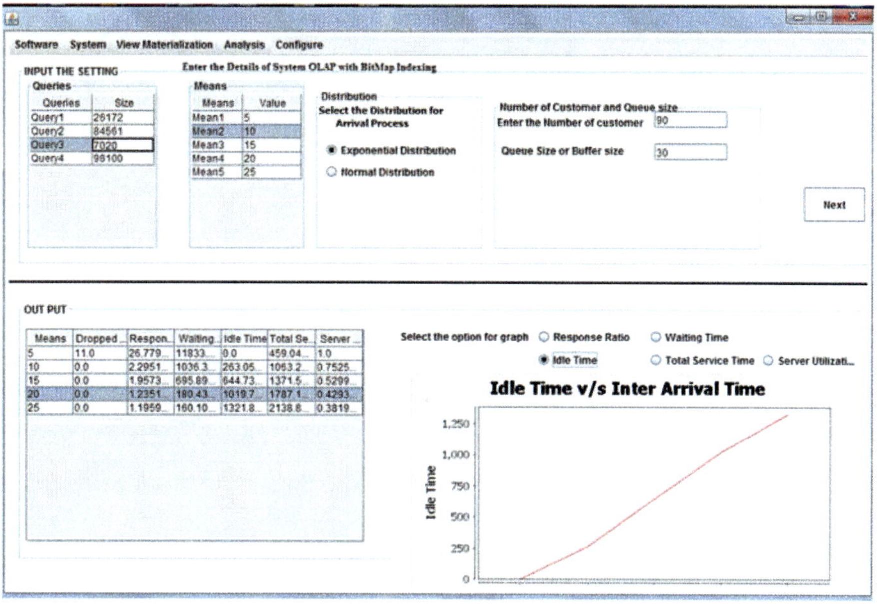

Fig. 5 Input/output screen for system execution model

4.4 *View Materialization*

Selecting the best possible views for materialization out of all possible views in a lattice structure given a constraint on the total number of views to be materialized is very important for the performance of Proxy Server. The tool provides an option for generating the views for materialization using a Greedy approach [3]. When the View Materialization choice is clicked from the Main Menu, the system presents a screen for getting the relevant parameters. User must specify the maximum number of views available, the details of the view lattice (number of nodes, root nodes, leaf nodes, links between nodes, and the weights associated with the nodes). Then, user has to specify the number of views to be materialized. This input screen in Fig. 6 shows the screen shot for View Materialization. After entering all the parameters, user clicks the Get Views button. The system then displays the specified number of views to be materialized based on a Greedy algorithm.

4.5 *Analysis*

This option allows the user to analyze a System Execution Model under different indexing schemes at the Database level/Proxy level. User can specify the performance metric of interest, the indexing schemes to be used at data warehouse level/OLAP

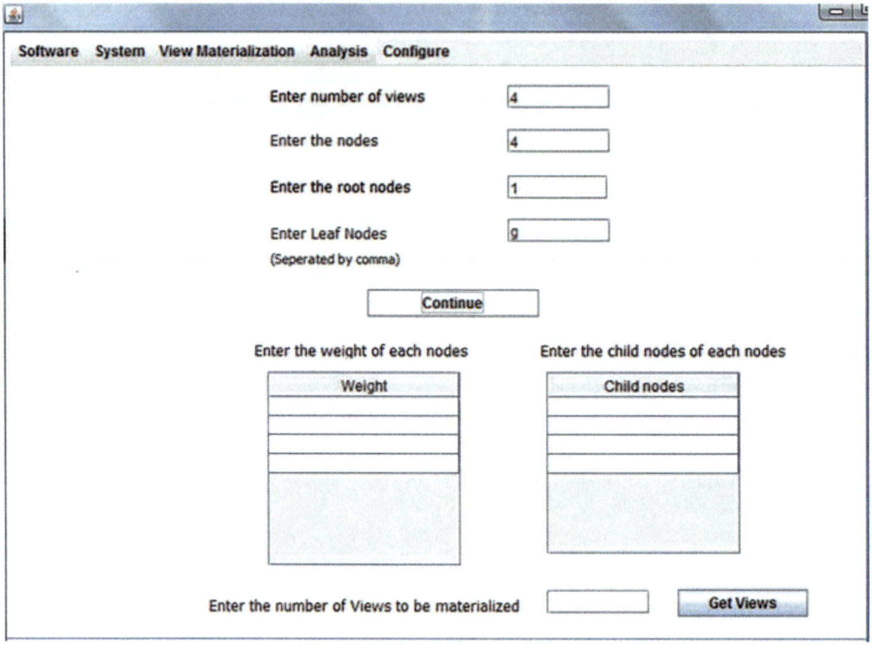

Fig. 6 View materialization

level to get the comparative analysis performed by the tool. An example is shown in Fig. 7 which shows the variation of the response ratio (*y*-axis) against inter arrival times of queries (*x*-axis) with BitMap Indexing and with Multi-level (NML) Indexing [6].

4.6 Configure

This option allows the user to change the hardware configurations set up earlier. User can change the configuration of the Client Server, Proxy Server, as well as Database Server. Clicking on Software or System options will result in solutions based on changed configuration.

5 Conclusion

"Distributed System Performance Analysis Tool (DSPAT)" is developed to support Software Performance Engineering concept of estimating and analyzing the performance in the early design phases of Distributed System development. The tool

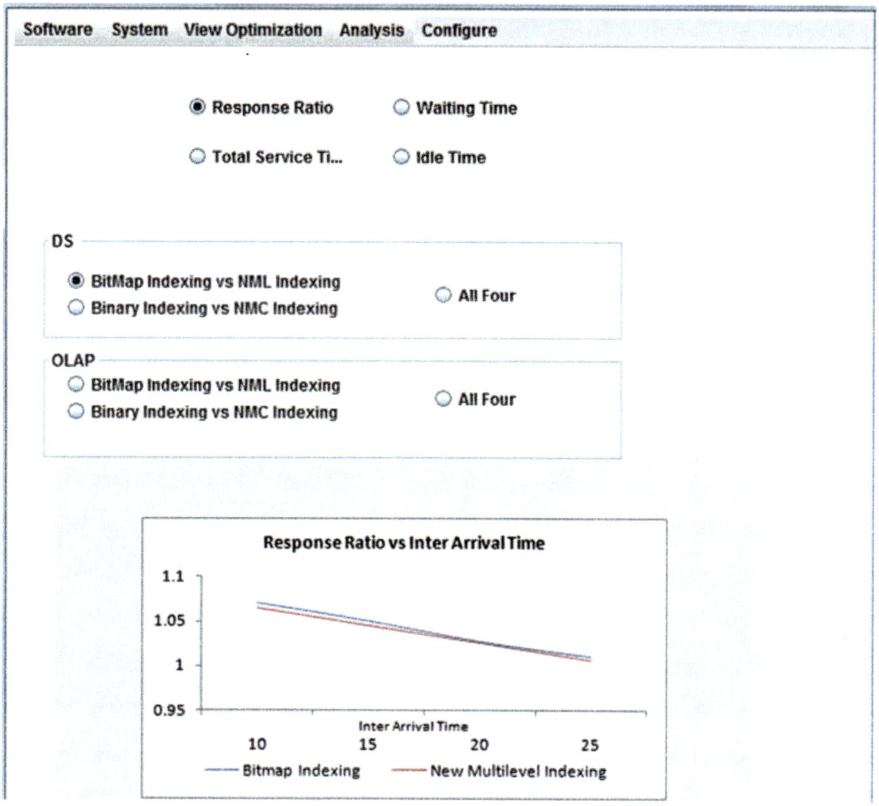

Fig. 7 Analysis of response ratio versus inter arrival times for BitMap and NML indexing

provides appropriate performance metrics like response ratio, average waiting time, average idle time, and server utilization etc. The tool is used to investigate the impact of a variety of operating environments on performance of Database/Proxy systems. The tool can be enhanced to cover the effects of concurrency to provide more realistic performance estimates.

References

1. C.U. Smith, L.G. Williams, *Performance solutions: a practical guide to creating responsive* (Addison Wesley, Scalable Software, 2001)
2. C.U. Smith, L.G. Williams, Performance engineering evaluation of object oriented systems with SPE-ED. Comput. Perform. Eval. Model. Tech. Tools LNCS **1245**, 135–154 (1997)
3. V. Harinarayan, A. Rajaraman, J.D. Ullman, Implementing datacubes efficiently, in *International Conference on Management of Data*, Canada (1996), pp. 205–216
4. R.N. Jogekar, A. Mohod, Design and implementation of algorithms for materialized view selection and maintenance in data warehousing environment. Int. J. Emerg. Technol. Adv. Eng.

3(9), 464–470 (2013)
5. P. O'Neil, D. Quass, Improved query performance with variant indexes, in *International Conference on Management of Data, Tucson*, Arizona, USA (ACM Press, 1997), pp. 38–47
6. K. Stockinger, K. Wu, A. Shoshani, Analyses of multi-level and multi-component compressed bitmap indexes. ACM Trans. Database Syst. **35**(1), 1–52 (2010). Article 2
7. C. Ju, M. Han, Effectiveness of OLAP based sales analysis in retail enterprises, in *The Proc. of ISECS International Colloquium on Computing, Communication, Control and Management*, vol. 3, (2008) pp. 240–244
8. H. Schildt, in *Java: The Complete Reference*, 7th edn. (Mc Graw Hill, 2007)
9. J. D'Anjou et al, in *The Java Developer's Guide to Eclipse*, 2nd edn. (Addison-Wesley Professional, 2005)
10. Viktor Leis et al., Query optimization through the looking glass, and what we found running the join order benchmark. Int. J. Very Large Databases **27**(5), 643–668 (2018)

A Robust Multiple Moving Vehicle Tracking for Intelligent Transportation System

N. Kavitha and D. N. Chandrappa

1 Introduction

Vehicular traffic management in our daily routine is of greater significance due to increased traffic on roads. Automatic detection of vehicles in busy roads helps to improve traffic management in heavy traffic scenes [1] by extracting the information from detected vehicles in busy roads. Vehicular traffic data such as vehicle speed [2, 3], flow of traffic, and number of vehicles are collected using image processing techniques [4], which leads to better controlling of traffic flow. Vehicular data such as number of vehicles, type of vehicle [5, 6], and speed and license plate detection of vehicles play a major role in monitoring the traffic flow in highway and toll collection. Automatic vehicle trajectory counting on highways roads from CCTV video footage [7, 8] is a very challenging domain in image processing and its applications [9, 10]. Hence, adaptive Gaussian distribution is one of the efficient techniques to detect congestions and to monitor the traffic at intersections. In [11, 12], manually collecting a large amount of vehicular data is often impractical. Hence, this paper aims in the automatic collection of vehicular traffic data for effective management of heavy traffic scenes. Our proposed methodology detects the vehicles using adaptive background subtraction technique [13, 14] followed by shadow and noise removal using morphological operations. After detected vehicles are being counted, it is followed by color classification of vehicles [15, 16] which is a major attribute to determine the external parameter of a vehicle for intelligent transportation system

N. Kavitha (✉)
Department of Electronics and Communication Engineering, Dayananda Sagar Academy of Technology and Management, Bengaluru 560082, India
e-mail: kavithaprashanthn@gmail.com

D. N. Chandrappa
Department of Electronics and Communication Engineering, SJB Institute of Technology, Bengaluru 560060, India
e-mail: chandrappa.dn@gmail.com

S. Patnaik et al. (eds.), *Advances in Machine Learning and Computational Intelligence*, Algorithms for Intelligent Systems, https://doi.org/10.1007/978-981-15-5243-4_9

[17]. Vehicle color is one of the significant feature to identify specific vehicle. Hence it is been written as vehicle color is significant feature [18, 19].

2 Related Works

Kavitha et al. [1] present a stationary wavelet transform (SWT) technique for removal of dynamic shadow using wavelet coefficients threshold. The threshold statistical parameter "skewness" is proposed for the detection and removal of shadow. The estimation of threshold value depends on the wavelet coefficients without the intervention of any supervised learning or manual calibration. Usually, the statistical parameter [20, 21], such as mean, variance, and standard deviation, does not show much variation in complex environments. Skewness shows a unique variation between the shadow and non-shadow pixels [22] under complex environmental conditions.

Naveen et al. [7] present Harris-Stephen corner detection algorithm for vehicle detection. The algorithm detects vehicles and estimates vehicle counts and speeds at arterial roadways and freeways. The performance of the proposed system is mainly used for intelligent transportation system for alerting the commuters in advance to reduce the speed, further vehicle counts allow to reduce the congestions at work zones and also at special events [23].

Asaidi et al. [11] have proposed a technique for tracking and classifying the vehicle for a given video input. Here contrast model is proposed for the elimination of dynamic shadow and also for changing the contrast, further Hidden Markov Model (HMM) is used to reduce the perspective effects and classify the vehicles into bus, cars, scooter, etc., on the knowledge basis.

Crouzil et al. [8] present a robust segmentation algorithm for detection of moving vehicles by extracting the foreground pixels. Initially, each pixel of the background is modeled with an adaptive Gaussian distribution. Further motion detection procedure is combined with adaptive Gaussian distribution model for accurate identification of moving vehicle location in space and time model.

Kishor Kumar et al. [4] have proposed methodology for vehicle counting and vehicle classification by SVM classifiers. Initially, background subtraction is carried out for identifying the vehicles and adaptive background mixture model with morphological transformation is implemented for preserving the edges for improvising the performance of vehicle detection, further detected vehicles are being counted by the mathematical operation to generate vehicle flow report. Here the proposed algorithm uses SVM classifiers for vehicle classification.

Gupta et al. [9] have proposed a technique for vehicle color identification. In this paper, the images are taken from CCTV installed at crossing roads where vehicles are being automatically cropped from the traffic image, and later the cropped vehicles are individually considered on the basis of probability frequency of each pixel RGB intensity from which predominant color from that RGB intensity will be projected as the vehicle color output.

Javadzadeh et al. [12] have proposed robust technique for tracking and counting of vehicles in highway. Here Prewitt filter is used for preprocessing followed by background subtraction and morphological operations for identifying the vehicles on the highways roads. Initially, moving vehicle objects are extracted from the video frame using adaptive background subtraction, and further irrelevant objects such as noise are eliminated by morphological operations. Experimental results of the proposed systems show high accuracy.

3 Our Approach

In this section, we propose a methodology for implementing vehicle tracking and color classification of vehicles using several standard databases such as CAVIAR database and Italy highway database, and it is also performed on online real-world data using OpenCV library. Here we first detect the vehicles by extracting moving foreground objects from a background frame in a video. Further detected vehicles are counted followed by vehicle color classification. It is predominantly carried out by following steps as shown in Fig. 1.

Initially, traffic video is captured from still camera having 48 mm focal length generating 25 frames per second with resolution of 720 × 576 pixels. The input video is taken from surveillance camera and is converted into subsequent frames to which preprocessing is carried out by Gaussian smoothing of the image to eliminate noise present in video. The region of interest (ROI) in each frame is convoluted using a 2D circular Gaussian function and its discrete approximation shown below:

$$G(x, y) = \frac{1}{(2\pi\sigma^2)} e^{\left(\frac{x^2+y^2}{2}\right)} \quad (1)$$

In addition to Gaussian smoothing median filter is used to preserve the edges of the vehicles in both background model and the current frame. After preprocessing, adaptive background subtraction technique with KNN classifiers is used to classify

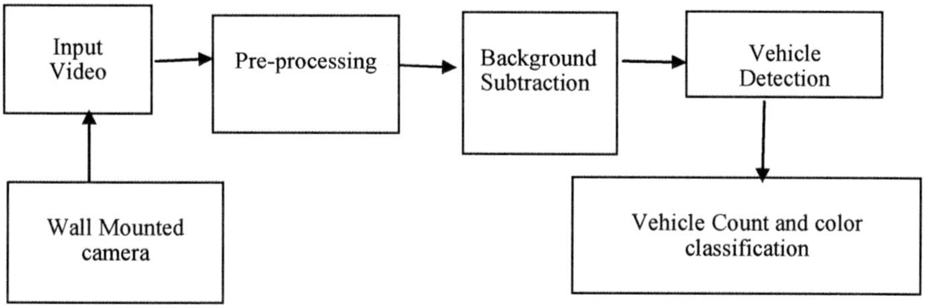

Fig. 1 Block diagram of proposed method

the input frame into vehicles and non-vehicles, further binary mask eliminates shadow present along with ROI, in second step morphological operations and thresholding are carried out for detecting the vehicles and removal of white noises in the scene. In third step, the mathematical operation is performed to count the detected vehicles by assigning unique Id. Finally, vehicle color classification is carried out by cropping a part of detected vehicles and training it using R, G, B color histogram to extract the features for predominant vehicle color identification.

4 Moving Vehicle Detection and Counting

The proposed system automatically detects and counts vehicles. It consists of five main functionalities such as motion detection, morphological operation, vehicle detection, and vehicle counting followed by vehicle color classification. The input to the system is a video recording, whereas the final displayed output will be total count of vehicles and color of an individual vehicle in particular video input. The following sections will describe the different processing steps involved in detecting and counting of vehicles.

4.1 Moving Vehicle Detection by Background Subtraction Using KNN Classifiers

Motion detection classifies the pixels into either foreground or background frame. Usually a moving object will be detected by background subtraction which is a major preprocessing step in many vision-based applications. The vital role of background subtraction algorithm is to extract the image's foreground for further processing. Foreground objects will be the image's region of interest. Foreground objects might be text humans or cars. As shown in Fig. 2, here sub-pixel edge map is used for background modeling, wherein it shows background model performs background and foreground segmentation. When a background model is initialized, the foreground can be segmented out by detecting differences between the current frame and the background model, the one which comes above threshold level is considered as a foreground object which also includes dynamic shadow for which we apply binary mask to efficiently eliminate dynamic shadow yielding in accurate detection of moving vehicles. The KNN classifiers are used to find disparity between the model and the current frame, by estimating the Euclidean distance, which classifies the input as vehicles and non-vehicles.

Exterior environment conditions such as occlusions, casted shadows, and illumination variations increase false detection rate and erroneous counting results. There are several algorithms have been introduced for background subtraction such as BackgroundSubtractorMOG(), and it creates the model of the background in an image

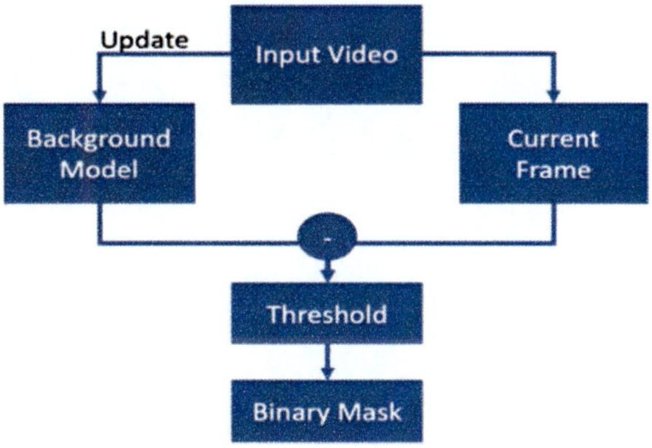

Fig. 2 Flow chart of background model

using 3–5 Gaussian distributions for each pixel. Another type of background subtraction implemented in OpenCV is called BackgroundSubtractorGMG(), a which combines the background image estimation technique with Bayesian segmentation. It is Gaussian mixture model being used on the pixel and background subtraction carried out using Mixture of Gaussian (MoG2). The above-discussed techniques fail to yield good results in the complex environment such as occlusions, casted shadows, and illumination changes in the scene. Hence, we propose an enhanced algorithm called background subtraction using KNN classifiers with binary mask, which outperforms in handling complex environmental changes compared to other state-of-the-art methods discussed. Henceforth, K-nearest neighbor (KNN) technique is used for background subtraction, which finds the nearest neighboring value by estimating the Euclidean distance between each segment to every training region. In K-nearest neighbor classifier, the K-nearest neighbors will be assigned a weight of 1\\$k$ rest all others will be assigned 0 weight. This is generalized as weighted nearest neighbor classifiers. That is, the ith nearest neighbor is assigned a weight $\mathbf{w_{ni}}$ with $\sum_1^{\mathbf{n}} \mathbf{w_{ni}} = 1$.

KNN Algorithm Working:

1. Let y denote the region of interest (ROI) which we need to estimate and y_i denote feature values.
2. Initially, we will find Euclidean distance between the points. "$d(y, y_i)$" $i = 1, 2, \ldots, \mathbf{n}$;
3. After estimating the Euclidean distance, it is arranged in the increasing order.
4. Let **k** represent a +ve integer, which will be taken from first **k** distances from the increasing order list.
5. Further, we find those **k**-points relevant to these **k**-distances and label it as ROI.
6. Let $\mathbf{k}_\mathrm{i}$ represent the amount of points corresponds to the ith class among **k** points, i.e., $k \geq 0$
7. If $k_i > k_j \; \forall \; i \neq j$, then call x as vehicles.

Fig. 3 Background subtraction of video

This method is implemented in Python using the OpenCV library, wherein background subtraction library with default parameters and classification is implemented. Here $\phi(\mathrm{x}^*)$ is a training set with a group of k-objects are identified, and further it is compared with closest set of the test object $\phi(x_i)$. On the bases of comparison of the label, we will be able to classify it as vehicle and non-vehicle object. It is given by

$$\left\|\phi(x^*) - \phi(x_i)\right\|^2 = (\phi(x^*) - \phi(x_i))^T\phi(x^*) - \phi(x_i))$$
$$= \phi(x_i)^T\phi(x^*) - 2\phi(x_i)^T\phi(x^*) + \phi(x_i)^T\phi(x^*) \quad (2)$$

As shown in Fig. 3, a correspondence metric is worn to figure out the value of k, distance between the objects, the numbers of nearest neighbors are then used to determine the vehicle and non-vehicle object by estimating the Euclidean distance between the image using the equation

$$E(X, Y) = \sum_{(k=0)}^{n} (x_i - y_i) \quad (3)$$

The segmented objects and centroids are extracted from each frame after segmenting a video. Instinctively, the adjacent frames will be connected if two segments are spatially the closest, and distance between their centroids will be estimated by Euclidean distance. This estimation helps us to classify the input frame as vehicle and non-vehicle objects.

4.1.1 Morphological Operation

After extracting vehicle region from background model, binary mask will be processed. Parts that are incorrectly classified as foreground, such as shadows or noise, will be removed from the foreground as shown in Fig. 4. It is given by

$$\text{dst}(x, y) = \begin{cases} \text{maxval}\,0 & \text{if src}(x, y) > \text{thresh} \\ 0 & \text{otherwise} \end{cases} \tag{4}$$

Morphology deals with shape and structure of all the vehicles. This morphological transformation will structure the shape of an object from the image we go for. It finds the holes in the image by overlapping, which includes basic operations in morphology such as dilation and erosion. In this section, a morphological operation is carried out to match the patterns related to the vehicles, whereas the output image includes only edges related to the vehicles. Parts that belong to the foreground but were incorrectly classified as background, such as vehicles, were added to the foreground. The resulting binary masks will be processed. To remove noise in the mask, we initially perform opening (erosion followed by dilation), subsequently to fill holes in the binary mask we perform closing (dilation followed by erosion). For each recordings, the kernels size differs in opening and closing as shown in Fig. 5.

Dilation is a convolution of an image with the matrix which is having the maximum values. It is given by

$$A\varphi B = \{Z\varepsilon E|(B^{s})_{z}\Pi\, A\} \tag{5}$$

Here overlapped values are estimated, and anchor point is replaced with the maximum value. The pixel value will be set to 1 if at least one of the pixels under the kernel has value 1. Similarly, erosion estimates a local minimum over the area of

Fig. 4 Binarization of cars

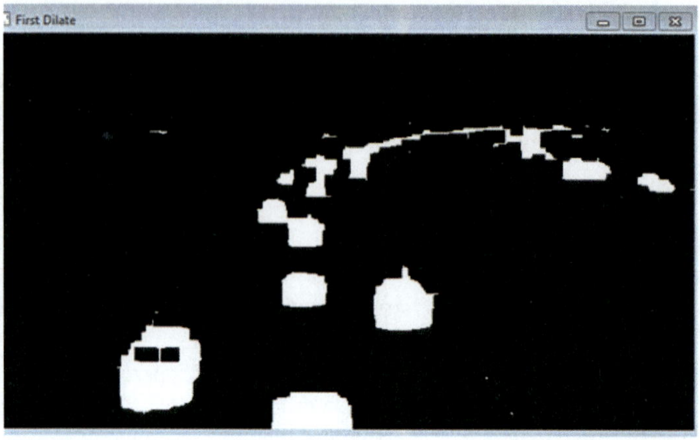

Fig. 5 Dilation

Fig. 6 Erosion

the kernel and when a kernel 1 slides over the binary mask and a pixel in the mask is set to 1 only if all pixels under the kernel are 1 as shown in Fig. 6; otherwise it is eroded to 0. It is given by

$$\{Z\varepsilon\ E\ E|(B^{s})_{z}\ c\ A\} \tag{6}$$

4.1.2 Vehicle Detection by Contour Extraction

After morphological operations, vehicles are detected by contour extraction, and contours are boundaries of a shape which are used for the shape detection and

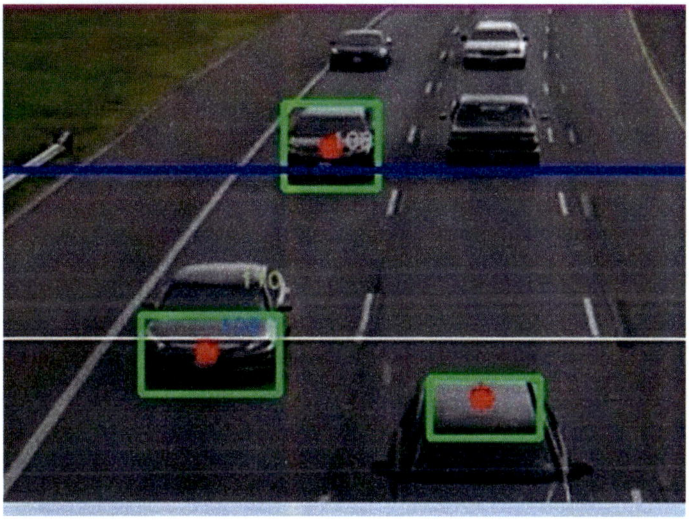

Fig. 7 Detected vehicles

recognition. The canny edge detector accurately finds the contours; and canny edge detection is also applied on a binary image. In OpenCV library cv2.findContours() instruction find the contours in an image. In contour detection, each frame object has been identified by extracting and stored in the contour vector. Initially, we draw a bounding box over the contour and locate its centroid when it intersects with an imaginary line, based on which vehicles are being detected. The features of an object are extracted by contour vector, attributes such as position and numbers of pixels of an object are stored in vector. Further, numbers of pixels are used to estimate the area. Similarly, other features are extracted in this step for detecting the vehicles as shown in Fig. 7. Rectangle is drawn around the detected vehicle by finding length and width of vehicle. The ratio of width and length creates importance in presence of occlusion of vehicles.

4.2 Vehicle Counting

After vehicles are detected, we assign a unique Id to all detected vehicles, further counting zone delimited by four virtual lines to count the vehicles moving upside and downside. The virtual zone has been created in such a way that it avoids too many false positives and counts every vehicle that passes through that zone. A vehicle which crosses the counting zone will be counted, i.e., if its trajectory begins before the entry line and continues after the exit line. Here assigning unique Id to detected vehicles avoids repetitive counting of same vehicles.

4.3 Vehicle Color Classification

Vehicle color classification is one of the major attributes in the application of intelligent transportation system. Here, the vehicle color is been extracted by cropping the detected vehicles for a traffic video frame acquired by a CCTV cameras. Initially, we convert RGB color space to the rgi space for a given cropped input. It is given by the following formula:

$$r = R/(R + G + B,\ g = G/(R + G + B),\ i = (R + G + B)/3 \quad (7)$$

In the next step, we define pixel-pairwise graph, where $w_{i,j}$ is the weight on the edge which is connected by node *i* and node *j*. It is estimated by following equation:

$$W_{i,j} = \|f_i - f_j\|^2 \quad (8)$$

where $f_i = (r, g, i)$ which is the intensity value of pixel *i*, and f_j which is the intensity value of pixel *j* in the rgi color space, and Euclidean distance between the corresponding pixels is used to estimate weight of arc. Further, dominant color is been extracted from pixel-pairwise graph by training the color recognition module and gradually dominant color modes are obtained. Here $C = \{c_1\}_{l=1}^{k}$. A predefined threshold is set based on which if left pixels are less than that threshold, then training process ends, the noisy image pixels are filtered out, and the main color modes are extracted. For a k-bin histogram $U = \{u_1\}_{l=1}^{k}$, a ratio matrix Ar is used to define the cross bin relationship. Each element in the matrix is represented in the form of (u_i/u_j) which estimates the relation between bin u_i and u_j. Henceforth, after training the R, G, B color histogram and extracting the color histogram features, the testing input is been compared with the trained database by considering weight of the corresponding color mode, and the spatial similarity measure is given by following formula:

$$S(UT, UC) = \sum_{(l=1)}^{k} w_{(1)}\exp\left\{-\frac{1}{2}(\mu_l^T - \mu_l^C)^T\left(\sum_l^T - \sum_l^C\right)^{-1}(\mu_l^T - \mu_l^C)\right\} \quad (9)$$

where μ_1 is the spatial mean and $\sum_l$ is the covariance. Based on this similarity measure, the color classification is carried out by k-Nearest Neighbors machine learning classifier, each cropped vehicle is individually analyzed from top to bottom, and pixelwise RGB intensity of cropped vehicle image is been estimated. Here, RGB intensity of each pixel is compared and updated when intensity matches with previous intensity. RGB intensity having maximum frequency of occurrence is reflected as the color of vehicle as shown in Fig. 8.

Fig. 8 Cropped detected vehicles

5 Discussions

The proposed methodology for counting and classifying vehicle color has been implemented using KNN classifiers and is been tested under several standard databases such as Taiwan highway, Italy highway, and CAVIAR database. The database has 2000, 3339, 2792 frames, respectively, and it is also performed on online real-world data. The methodology is been implemented using OpenCV library in Python. Here initially, foreground objects such as vehicles are being extracted from video sequence input through background subtraction technique but the background is inclusive of both static and dynamic background movements such as people movement, illumination changes, waving tree phenomena, and various environmental conditions in various scenes. Henceforth, we implemented background subtraction using KNN classifiers and binary mask which outperformed in accurately extracting the foreground object such as vehicles from background by efficiently eliminating the shadow under various traffic scenes. The dataset chosen is challenging due to various illumination changes and complexity of background and weather conditions which is summarized in Table 1. Further counting zone delimited by four virtual lines to count the vehicles moving upside and downside the traffic intensity is erroneously estimated by assigning unique Id to each detected vehicles and counts every vehicle that passes through that counting zone which eliminates repeated counting of detected vehicles. Finally, the part of detected vehicles is being cropped to extract dominant color by training the R, G, B color histogram for extracting color histogram features and classifying it based on the dominant color extracted. Hence, Fig. 9 shows the final output which displays count and color of the detected vehicles passing through the counting zone. The proposed methodology tracks the vehicles and displays count of the vehicles passing upside and downside along with vehicle color and shows overall accuracy of 94% for various case studies tabulated in Table 1 (Fig. 10).

Table 1 Result of the tests carried out for several traffic videos

Case studies	Number of vehicle in video	Number of vehicles detected and counted in video	Detection and counting rate (%)
Vehicles movement in sunny weather condition	7	7	100
Vehicles movement with dynamic changes in the background	13	12	92.3
Vehicles movement in light and illumination	24	24	100
Vehicle movement in underpass	11	9	81.8
Vehicle movement in junction crossing	10	9	90
Vehicles movement in highway	12	12	100
Vehicles movement with too slow/fast vehicles	32	29	90.6
Vehicle movement in rainy weather condition	10	10	100
Vehicle movement under different video noises	45	43	95.5
Vehicle movement in congested traffic conditions	27	24	88.8

Fig. 9 Count and color of detected vehicles

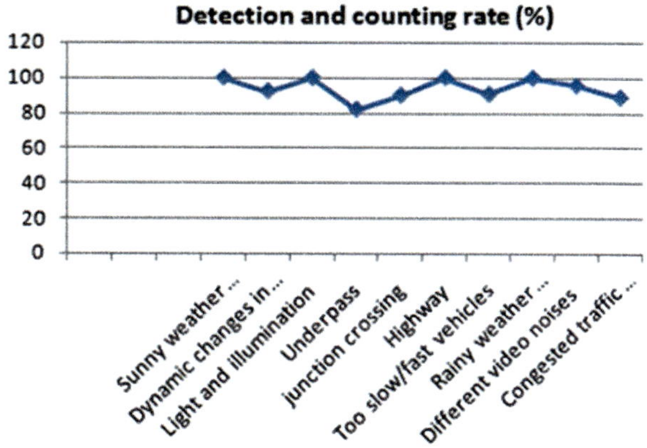

Fig. 10 Graph for the vehicle tracking under various traffic scenarios

6 Conclusion

In this paper, proficient system is implemented for estimating traffic intensity and determines external parameter such as color of a vehicle in a video to intelligent transportation system. In real time, vehicle monitoring system can erroneously count the detected vehicles by eliminating the dynamic shadow region using the coefficient of linearity between shadow and light region by utilizing contrast information. The proposed framework with background subtraction using KNN classifier and morphological operation will reconstruct the actual background robustly for a real-time video compared to other background subtraction techniques discussed previously.

The main strength of this method is it can handle occluded vehicle images and also it can distinguish and number the moving foreground objects (vehicles) from a background image. Furthermore, the mathematical operation is carried out for counting the detected vehicles by assigning unique Id to the detected vehicles and counts the vehicles that pass through that counting zone to avoid repetitive counting of the same vehicle. Finally, in addition to counting of the vehicles, color of the vehicles also been identified by training R, G, B color histogram and extracting color histogram features, and based on the predominant color output vehicles, color is been identified using the classification of k-Nearest Neighbors machine learning classifier. The proposed methodology outperforms with overall efficiency of 94%, and it also adds robustness and increased efficiency to the traditional background subtraction and addresses several instability issues. As a future work, it can be implemented for moving camera and also accurate counting of vehicles in heavy congested night traffic conditions.

References

1. N. Kavitha, R.S. Kathavarayan, Moving shadow detection based on stationary wavelet transform. EURASIP J. Image Video Process. **49** (2017). https://doi.org/10.1186/s13640-017-0198-x
2. S. Taghvaeeyan, R. Rajamani, Portable roadside sensors for vehicle counting classification and speed measurement. IEEE Trans. Intell. Transp. Syst. **15**(1), 73–83 (2014)
3. S. Ali, B. George, L. Vanajakshi, A multiple inductive loop vehicle detection system for heterogeneous and lane-less traffic. IEEE Trans. Instrum. Meas. **61**(5), 1353–1360 (2012)
4. S.R., Kishor Kumar, Sampath, Vehicle detection, classification and counting. IJIRT **3**(10) (2017). ISSN: 2349-6002
5. N. Seenouvong, U. Watchareeruetai, C. Nuthong, K. Khongsomboon, N. Ohnishi, A computer vision based vehicle detection and counting system, in *2016 8th International Conference on Knowledge and Smart Technology (KST)* (Chiangmai, 2016), pp. 234–238
6. Z. Zhang, K. Liu, F. Gao, X. Li, G. Wang, Vision-based vehicle detecting and counting for traffic flow analysis, in *2016 International Joint Conference on Neural Networks (IJCNN)* (Vancouver, BC, 2016), pp. 2267–2273
7. C. Naveen, M. Venkatesan, Video based vehicle detection and its application in intelligent transportation systems. Scrip J. Transp. Technol. **2**, 305–314 (2012)
8. A. Crouzil, Khoudour, Automatic vehicle counting system for traffic monitoring. J. Electron. Imag. **25**(5), 1–12 (2016). ISSN 1017-9909
9. P. Gupta, G.N. Purohit, Vehicle colour recognition system using CCTV cameras. Int. J. Adv. Res. Comput. Sci. Softw. Eng. **4**(1) (2014)
10. M.B. Subaweh, E.P. Wibowo, Implementation of pixel based adaptive segmenter method for tracking and counting vehicles in visual surveillance, in *2016 International Conference on Informatics and Computing (ICIC)* (Mataram, Indonesia, 2016), pp. 1–5
11. H. Asaidi, A. Aarab, Shadow elimination and vehicle classification approaches in traffic video surveillance context. J. Vis. Lang. Comput. (Elsevier) (2014). https://doi.org/10.1016/j.jvlc.2014.02.0011045-926x
12. R. Javadzadeh, E. Banihashemi, Fast vehicle detection and counting using background subtraction technique and prewitt edge detection. Int. J. Comput. Sci. Telecomm. **6**(10) (2015)
13. P. Prommool, S. Auephanwiriyakul, N. Theera-Umpon, Vision-based automatic vehicle counting system using motion estimation with Taylor series approximation, in *2016 6th IEEE International Conference on Control System, Computing and Engineering (ICCSCE)* (Penang, Malaysia, 2016), pp. 485–489
14. C.M. Bautista, C.A. Dy, M.I. Manalac, R.A. Orbe, M. Cordel, Convolutional neural network for vehicle detection in low resolution traffic videos, in *IEEE Region 10 Symposium (TENSYMP)* (2016), pp. 277–281
15. L. Xie, G. Zhu, Y. Wang, H. Xu, Z. Zhang, Real-time vehicles tracking based on Kalman filter in a video-based ITS, in *Proceedings of IEEE Conference on Communications, Circuits and Systems*, vol. 2 (2005), p. 886
16. N. Goyette, P. Jodoin, F. Porikli, Change detection .net: a new change detection benchmark dataset, in *Proceedings of the Computer Vision and Pattern Recognition Workshops (CVPRW), IEEE Computer Society Conference* (2012)
17. L. Unzueta, Adaptive multicue background subtraction for robust vehicle counting and classification. IEEE Trans. Intell. Transp. Syst. **13**, 527–540 (2012)
18. Z. Dong, M. Pei, Y. He, T. Liu, Y. Dong, Y. Jia, Vehicle type classification using unsupervised convolutional neural network, in *2014 22nd International Conference on Pattern Recognition (ICPR)* (IEEE, 2014), pp. 172–177
19. Y.T. Wu, J.H. Kao, M.Y. Shih, A vehicle color classification method for video surveillance system concerning model-based background subtraction, in *Advances in Multimedia Information Processing—PCM*, vol. 6297 (Springer, Berlin, Heidelberg, 2010), pp. 369–380

20. J.W. Son, S.B. Park, K.J. Kim, A convolution kernel method for color recognition, in *Proceedings of the Sixth International Conference on Advanced Language Processing and Web Information Technology* (2007), pp. 242–247
21. Z. Kim, J. Malik, High-quality vehicle trajectory generation from video data based on vehicle detection and description, in *Proceedings of IEEE Conference on Intelligent Transportation Systems* (2003), pp. 176–182
22. H. Tao, H.S. Sawhney, Object tracking with Bayesian estimation of dynamic layer representations. IEEE Trans. Pattern Anal. Mach. Intell. **24**(1), 75–89 (2002)
23. J.-W. Hsieh, S.-H. Yung-Sheng Chen, W.-F. Hu, Automatic traffic surveillance system for vehicle tracking and classification. IEEE Trans. Intell. Transp. Syst. **7**, 175–187 (2006)

Bug Priority Assessment in Cross-Project Context Using Entropy-Based Measure

Meera Sharma, Madhu Kumari, and V. B. Singh

1 Introduction

A large number of bugs are reported on bug tracking systems by different users, developers, and staff members located at different geographical locations. Bug priority (P1, the most important, to P5, the least important) is an important attribute which determines the importance and the order of fixing of the bugs in the presence of other bugs. To automate the bug priority prediction, we need historical data to train the classifiers. In reality, this data is not available easily in all software projects, especially in new projects. Cross-project priority prediction works well in such situation where we train the classifiers with historical data of projects other than the testing projects [1, 2].

The bug reports are reported by users having different levels of knowledge about the software which results in uncertainty and noise in bug reports data. "Without proper handling of these uncertainties and noise, the performance of learning strategies can be significantly reduced" [22]. The entropy-based measure has been used to calculate the uncertainty in bug summary reported by different users. In literature, researchers [1, 2] have made attempts for cross-project bug summary-based priority prediction. No attempt has been made to handle uncertainty in bug summary in cross-project context for bug priority prediction. We have proposed summary entropy-based cross-project priority prediction models using Support Vector Machine (SVM), k-Nearest Neighbors (k-NN), Naïve Bayes (NB), and Neural Network (NNET). In addition to the summary entropy, we have also considered bug severity and the derived bug summary weight attribute. Results show improvement in performance over summary-based cross-project priority prediction models [2].

M. Sharma
Swami Shraddhanand College, University of Delhi, Delhi, India

M. Kumari · V. B. Singh (✉)
Delhi College of Arts and Commerce, University of Delhi, Delhi, India
e-mail: vbsingh@dcac.du.ac.in

S. Patnaik et al. (eds.), *Advances in Machine Learning and Computational Intelligence*, Algorithms for Intelligent Systems, https://doi.org/10.1007/978-981-15-5243-4_10

The rest of the paper is organized as follows: Sect. 2 deals with related work. Section 3 describes the data description, bug attributes, and model building required to perform the analysis. Results have been discussed in Sect. 4. Finally, the paper is concluded in Sect. 5.

2 Related Work

Bug priority assessment helps in correct resource allocation and bug fix scheduling. A bug priority recommender has been proposed by Kanwal and Maqbool [3] by using SVM classification technique. The study was further extended for comparison of SVM and NB performance with different feature sets by Kanwal and Maqbool [4]. An attempt for bug priority prediction has been made by Alenezi and Banitaan [5] using NB, Decision Tree (DT), and Random Forest (RF) for Firefox and Eclipse datasets. Lian Yu et al. [6] proposed defect priority prediction using Artificial Neural Network (ANN) technique. Results show that ANN performs better than Bayes algorithm. Tian et al. [7] proposed a new framework called DRONE (PreDicting PRiority via Multi-Faceted FactOr ANalysEs) for Eclipse projects and compared it with Severis^{Prio} and Severis^{Prio+} [8].

In literature, several studies have been conducted in cross-project context [9–16].

Bug summary-based cross-project priority prediction models have been proposed by [1, 2] using SVM, NB, k-NN, and NNET. Results show that cross-project bug priority prediction works well. Another attempt has been made by authors to propose bug summary-based cross-project severity prediction models [17].

Software are evolved through source code changes done in it to fix different issues, namely bugs, new features, and feature improvements reported by different users. These source code changes result in uncertainty and randomness in the system. In literature, researchers have used entropy-based measures to quantify the code change process for defects prediction [18]. Researchers have used entropy-based measures to predict the potential code change complexity [19]. A software reliability uncertainty analysis method has been proposed by Mierswa et al. [20].

To our knowledge, no work has been done for measuring trustworthiness of bug summary data in bug repositories. The uncertainty/noise present in bug summary data can affect the performance of prediction models. In this paper, we have measured the uncertainty in bug summary by using entropy-based measures. In addition to summary entropy, bug severity and summary weight for bug priority prediction in cross-project context have been considered. We have compared our proposed summary entropy-based bug priority prediction models with Sharma et al. [2] and found improvement in performance of the classifiers.

Table 1 Priority-wise number of bug reports of different projects

Project	Product	Priority-wise number of bug reports					
		P1	P2	P3	P4	P5	Total
Eclipse	V2	923	1416	8609	370	229	11,547
Eclipse	V3	361	963	26,667	320	136	28,447
OpenOffice	DB	76	472	2834	243	38	3663
OpenOffice	SST	82	518	4210	316	114	5240
OpenOffice	PPT	62	553	2688	90	37	3430

3 Description of Datasets, Bug Attributes, and Model Building

In this section, description of datasets and bug attributes used for validation and the model building have been discussed.

3.1 *Description of Datasets*

We have taken different products, namely Platform Version 2 (V2), Platform Version 3 (V3) of Eclipse project (http://bugs.eclipse.org/bugs/) and Database Access (DB), Spreadsheet (SST), Presentation (PPT) of OpenOffice project (http://bz.apache.org/000/). We have considered the bug report for status "verified," "resolved," and "closed." Table 1 shows the distribution of bug reports of different priority levels.

3.2 *Bug Attributes*

To predict bug priority in cross-project context, we considered three attributes, namely severity, summary weight, and entropy of summary. Severity is a nominal attribute, whereas summary weight and entropy are continuous attributes. Bug severity gives the impact of bug on the functionality of software or its components. It is divided into seven levels, namely "Blocker, Critical, Major, Normal, Minor, Trivial, and Enhancement." Blocker is the highest level, and Enhancement is lowest level. Bug priority determines the importance of a bug in the presence of others. Bugs are prioritized by P1 level, i.e., the most important, to P5 level, i.e., the least important. The bug summary gives the textual description of the bug. Summary weight is extracted from the bug summary attribute, entered by the users.

The bug summary has been preprocessed with the RapidMiner tool [21] to calculate the summary weight of a reported bug [2].

Different users are reported bug on bug tracking system. The size of software repositories is also increasing by an enormous rate that enhances the noise and uncertainty in the bug priority prediction. If these uncertainties are not handled properly, the performance of the learning strategy can be significantly reduced [22]. We have proposed entropy-based measure to build the classifier for bug priority prediction to handle uncertainties in cross-project context. We have used Shannon entropy to build the classifier model.

Shannon entropy, S is defined as

$$S = -p_i \log_2 p_i$$

where $p_i = \frac{\text{Total number of occurrences of terms in } i\text{th bug report}}{\text{Total number of terms}}$.

The top 200 terms have been taken from all terms based on their weight. To rationalize the effect of the priority, we multiplied the entropy by 10 for P1 and P2 priority level bugs, 3 for the P3 priority level bug, and 1 for P4 and P5 priority level bugs [23].

3.3 Model Building

We have proposed summary entropy-based classifiers based on SVM, k-NN, NNET, and NB for bug priority prediction in cross-project context by taking bug attributes severity and summary weight. We have taken the bug reports of two products of Eclipse and three products of OpenOffice projects. To get the significant amount of performance, we have used the appropriate parameters values. "For SVM, we have taken polynomial kernel with degree 3, the value of k as 5 in case of k-NN and for NNET the training cycle as 100" [2]. Number of validations is taken as 10 and sampling types as stratified sampling for different classification techniques. The performance of the proposed models has been validated using different performance measures, namely Accuracy, Precision, Recall, and F-measure.

Figure 1 shows the main process of cross-project priority prediction.

4 Results and Discussion

We have validated the entropy-based classifier of different machine learning techniques, namely SVM, k-Nearest Neighbors, Naive Bayes, and Neural Network using 10 fold cross-validations for predicting the bug priority. We have compared the proposed entropy-based approach to Sharma et al. [2]. We have taken the same datasets and techniques as taken by Sharma et al. [2] to predict the bug priority in cross-project

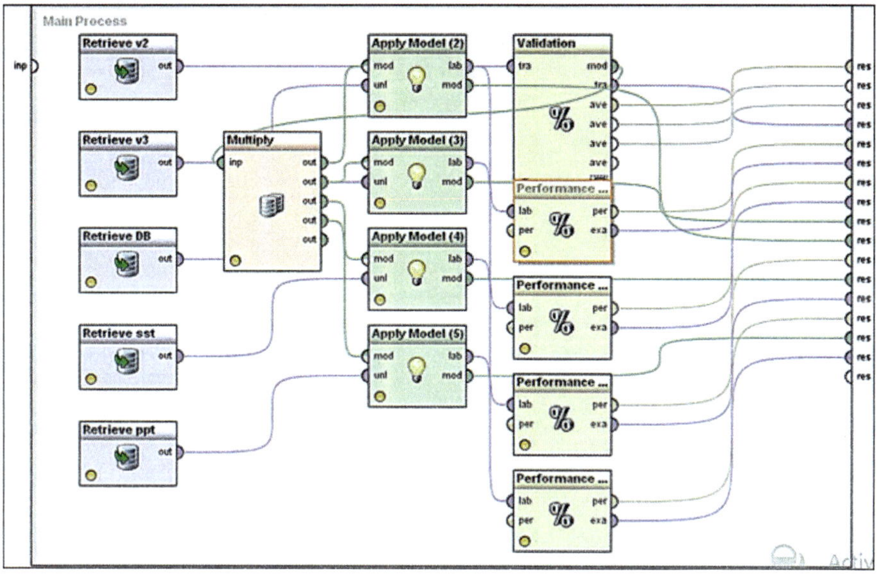

Fig. 1 RapidMiner process for bug priority prediction in cross-project context

context. Table 2 shows the Accuracy of different machine learning techniques to predict the priority of cross-validated projects.

Accuracy for Training Dataset V2

For testing dataset V3, our entropy-based approach improves the Accuracy by 3.46% and 91.93% for SVM and NB, respectively. Our entropy-based approach improves the Accuracy by 7.86%, 10.21%, 2.81%, and 82.85% for SVM, k-NN, NNET, and NB, respectively, for testing dataset DB. For testing dataset SST, our approach improves the Accuracy by 6.66%, 8.42%, 2.96%, and 82.08% for SVM, k-NN, NNET, and NB, respectively. Our entropy-based approach improves the Accuracy by 11.69%, 10.99%, 13.00%, and 85.19% for SVM, k-NN, NNET, and NB, respectively, for testing dataset PPT.

Accuracy for Training Dataset V3

Our entropy-based approach improves the Accuracy by 6.34%, 6.57%, 6.40%, and 82.10% for SVM, k-NN, NNET, and NB, respectively, for testing dataset DB. For testing dataset SST, our approach improves the Accuracy by 3.46% and 91.93% for SVM and NB, respectively. Our entropy-based approach improves the Accuracy by 9.39%, 8.16%, 7.41%, and 76.44% for SVM, k-NN, NNET, and NB, respectively, for testing dataset PPT.

Accuracy for Training Dataset DB

For testing dataset V2, our entropy-based approach improves the Accuracy by

Table 2 Accuracy (%) of cross-validated projects

Training versus testing dataset	Accuracy (%)			
	SVM	k-NN	NNET	NB
V2 versus V3	95.51	89.26	91.13	95.59
V2 versus DB	84.93	86.24	80.04	86.40
V2 versus SST	86.43	87.69	83.19	87.02
V2 versus PPT	89.53	88.13	91.02	89.48
V3 versus DB	86.13	86.27	86.21	86.32
V3 versus SST	86.66	86.89	86.74	86.95
V3 versus PPT	87.67	86.50	87.64	81.95
DB versus V2	77.73	83.29	83.14	68.21
DB versus V3	94.48	91.05	96.07	85.55
DB versus SST	85.10	92.18	96.53	86.34
DB versus PPT	86.73	83.97	86.82	69.30
SST versus V2	77.61	74.00	50.40	58.68
SST versus V3	94.47	88.15	81.09	80.72
SST versus DB	84.93	88.15	81.33	82.25
SST versus PPT	86.85	83.76	82.30	61.25
PPT versus V2	78.58	79.13	83.02	79.02
PPT versus V3	94.87	93.34	93.94	92.06
PPT versus DB	86.24	82.72	73.87	86.35
PPT versus SST	86.98	77.96	75.82	86.97

3.43%, 10.29%, 9.09%, and 60.21% for SVM, k-NN, NNET, and NB, respectively. Our entropy-based approach improves the Accuracy by 1.04%, 0.11%, 2.70%, and 79.27% for SVM, k-NN, NNET, and NB, respectively, for testing dataset V3. For testing dataset SST, our approach improves the Accuracy by 5.46%, 13.35%, 16.66%, and 76.19% for SVM, k-NN, NNET, and NB, respectively. Our entropy-based approach improves the Accuracy by 8.77%, 5.66%, 11.59%, and 60.12% for SVM, k-NN, NNET, and NB, respectively, for testing dataset PPT.

Accuracy for Training Dataset SST

For testing dataset V2, our entropy-based approach improves the Accuracy by 3.14%, 1.08%, and 49.11% for SVM, k-NN, and NB, respectively. Our entropy-based approach improves the Accuracy by 0.92% and 75.93% for SVM and NB, respectively, for testing dataset V3. For testing dataset DB, our approach improves the Accuracy by 7.89%, 11.11%, 4.18%, and 75.83% for SVM, k-NN, NNET, and NB, respectively. Our entropy-based approach improves the Accuracy by 8.89%, 5.25%, 4.02%, and 52.07% for SVM, k-NN, NNET, and NB, respectively, for testing dataset PPT.

Accuracy for Training Dataset PPT

For testing dataset V2, our entropy-based approach improves the Accuracy by 4.33%, 5.85%, 8.59%, and 70.95% for SVM, k-NN, NNET, and NB, respectively. Our entropy-based approach improves the Accuracy by 1.42%, 1.64%, 0.38%, and 84.93% for SVM, k-NN, NNET, and NB, respectively, for testing dataset V3. For testing dataset DB, our approach improves the Accuracy by 9.23%, 5.62%, and 72.92% for SVM, k-NN, and NB, respectively. Our entropy-based approach improves the Accuracy by 7.46% and 75.73% for SVM and NB, respectively, for testing dataset SST.

Out of 19 combination cases, SVM, k-NN, NNET, and NB outperform in 19, 16, 14, and 19 cases, respectively, in comparison with Sharma et al. [2]. Our approach improves the Accuracy 0.92–11.69% for SVM, 0.11–13.35% for k-NN, 0.38–16.66% for NNET, and 49.11–91.93% for NB across all the 19 combinations for bug priority prediction in cross-project context. SVM and NB outperforms for bug priority prediction across all the 19 combinations.

Table 3 shows the best training dataset with highest Accuracy for different machine learning techniques. Across all the machine learning techniques, on the basis of Accuracy, DB is the best training dataset for V2 testing dataset, DB is the best training dataset for V3 testing dataset, SST is best training dataset for DB testing dataset, DB is the best training dataset for SST testing dataset, and V2 is the best training dataset for PPT testing dataset.

Avg. F-Measure for Training Dataset V2

From Table 4, we observed that the value of F-measure (avg.) lies between 34.32%–48.49%, 30.69%–40.52%, 31.63%–40.04%, and 35.13%–39.44% for training candidates V3, DB, SST, and PPT, respectively, across all the machine learning techniques.

Avg. F-Measure for Training Dataset V3

We obtained the value of F-measure (avg.) that lies between 33.94%–35.22%,

Table 3 Classifier-wise best training candidate with highest accuracy

Best training dataset (Accuracy %)				
Testing datasets	SVM	k-NN	NNET	NB
V2	PPT (78.58)	DB (83.29)	DB (83.14)	PPT (79.02)
V3	V2 (95.51)	DB (91.05)	DB (96.07)	V2 (95.59)
DB	PPT (86.24)	SST (88.15)	V3 (86.21)	V2 (86.40)
SST	PPT (86.98)	DB (92.18)	DB (96.53)	PPT (86.97)
PPT	V2 (89.53)	V2 (88.13)	V2 (91.02)	V2 (89.48)

Table 4 Average precision (P), recall (R), and F-measure (F) for training dataset (V2 product)

Testing datasets	SVM			k-NN			NNET			NB		
	P (%)	R (%)	F (%)	P (%)	R (%)	F (%)	P (%)	R (%)	F (%)	P (%)	R (%)	F (%)
V3	40.75	33.47	36.19	52.29	53.91	48.49	31.53	41.44	34.32	76.83	40.15	42.58
DB	32.28	34.46	33.30	45.61	37.99	40.52	28.05	34.71	30.69	48.87	35.78	35.83
SST	33.94	33.53	33.52	48.91	36.81	40.04	28.78	35.84	31.63	54.93	34.10	34.76
PPT	39.90	34.14	35.13	65.55	35.33	38.56	36.27	35.72	35.88	45.58	37.45	39.44

33.53%–35.98%, and 32.04%–35.96% for training candidates DB, SST, and PPT, respectively, across all the machine learning techniques as given in Table 5.

Avg. F-Measure for Training Dataset DB
Table 6 shows the value of F-measure (avg.) that lies between of 25.23%–41.53%, 26.33%–38.09%, 30.93%–51.12, and 32.30%–42.75% for training candidates V2, V3, SST, and PPT, respectively, across all the machine learning techniques for bug priority prediction.

Avg. F-Measure for Training Dataset SST
From Table 7, we observed that the value of F-measure (avg.) lies between 25.00%–39.50%, 26.27%–38.80%, 32.46%–51.34%, and 31.84%–46.00% for training candidates V2, V3, DB, and PPT, respectively, across all the machine learning techniques.

Avg. F-Measure for Training Dataset PPT
Table 8 shows the value of F-measure (avg.) that lies between 26.71%–40.29%, 29.10%–39.88%, 27.88%–40.53%, and 27.64%–37.54% for training candidates V2, V3, DB, and SST, respectively, across all the machine learning techniques.

Table 9 shows the best training dataset with highest F-measure (avg.) for different machine learning techniques. Across all the machine learning techniques, on the basis of F-measure, DB is the best training candidate for V2 testing dataset, V2 is the best training candidate for V3 testing dataset, SST is best training candidate for DB testing dataset, DB is the best training candidate for SST testing dataset, and SST is the best training candidate for PPT testing dataset.

Figure 2 shows the Accuracy comparison using SVM machine learning technique for cross-project bug priority prediction.

Figure 3 shows the Accuracy comparison using k-NN machine learning technique for cross-project priority prediction.

Figure 4 shows the Accuracy comparison using NNET machine learning technique for cross-project priority prediction.

Figure 5 shows the Accuracy comparison using NB machine learning technique for cross-project priority prediction.

5 Conclusion

In the absence of data for building a classifier, cross-project study provides a solution. In this paper, we have proposed an approach for cross-project bug priority prediction using three attributes, bug severity, summary weight, and summary entropy. By considering learning from the uncertainty, we have derived an attribute termed as summary entropy using Shannon entropy. To build the classifier, we have used machine learning techniques, namely Support Vector Machine (SVM), k-Nearest Neighbors (k-NN), Naïve Bayes (NB), and Neural Network (NNET). The built-in classifiers

Table 5 Average precision (P), recall (R), and F-measure (F) for training dataset (V3 product)

Testing datasets	SVM			k-NN			NNET			NB		
	P (%)	R (%)	F (%)	P (%)	R (%)	F (%)	P (%)	R (%)	F (%)	P (%)	R (%)	F (%)
DB	34.96	33.67	33.94	52.33	34.55	35.22	34.92	33.87	34.06	33.82	34.99	34.31
SST	35.41	32.78	33.53	49.80	33.91	35.98	35.45	32.93	33.66	35.14	33.73	34.11
PPT	35.64	31.54	32.76	45.24	30.38	32.04	35.67	31.50	32.74	33.30	41.22	35.96

Table 6 Average precision (P), recall (R), and F-measure (F) for training dataset (DB product)

Testing datasets	SVM			k-NN			NNET			NB		
	P (%)	R (%)	F (%)	P (%)	R (%)	F (%)	P (%)	R (%)	F (%)	P (%)	R (%)	F (%)
V2	30.67	25.18	25.23	43.13	43.52	41.53	33.65	34.75	32.78	34.92	39.27	35.58
V3	35.26	24.36	26.33	40.30	43.85	38.09	38.57	34.30	34.36	29.05	42.27	31.58
SST	35.13	29.61	30.93	51.29	51.14	51.12	35.10	32.68	33.39	69.66	33.61	34.78
PPT	45.27	30.67	32.30	43.24	44.59	42.75	37.22	34.36	34.91	43.69	38.23	36.63

Table 7 Average precision (P), recall (R), and F-measure (F) for training dataset (SST product)

Testing datasets	SVM			k-NN			NNET			NB		
	P (%)	R (%)	F (%)	P (%)	R (%)	F (%)	P (%)	R (%)	F (%)	P (%)	R (%)	F (%)
V2	30.75	24.99	25.00	39.67	43.93	39.50	29.76	33.31	27.32	33.61	39.94	32.89
V3	35.17	24.32	26.27	41.63	42.40	38.80	34.33	35.96	32.87	31.54	42.31	31.46
DB	34.67	31.74	32.46	50.63	53.05	51.34	40.63	44.39	41.20	38.63	36.04	34.85
PPT	35.32	30.52	31.84	48.30	51.56	46.00	40.10	48.67	39.34	40.85	40.76	32.92

Table 8 Average precision (P), recall (R), and F-measure (F) for training dataset (PPT product)

Testing datasets	SVM			k-NN			NNET			NB		
	P (%)	R (%)	F (%)	P (%)	R (%)	F (%)	P (%)	R (%)	F (%)	P (%)	R (%)	F (%)
V2	30.45	26.57	26.71	49.55	32.43	35.41	29.40	34.77	31.84	41.53	39.68	40.29
V3	35.18	26.83	29.10	44.82	29.06	32.71	28.20	34.07	30.43	38.98	43.41	39.88
DB	34.13	34.65	34.24	47.71	41.70	40.53	25.48	33.72	27.88	33.76	35.04	34.30
SST	35.27	33.61	34.06	42.69	41.41	37.54	25.00	34.35	27.64	54.88	33.92	34.30

Table 9 Classifier-wise best training candidate with highest F-measure (average)

Best training dataset (average F-measure)				
Testing datasets	SVM	k-NN	NNET	NB
V2	PPT (26.71)	DB (41.53)	DB (32.78)	PPT (40.29)
V3	V2 (36.191)	V2(48.49)	DB (34.36)	V2 (42.58)
DB	PPT (34.24)	SST (51.34)	SST(41.20)	V2 (35.83)
SST	PPT (34.06)	DB (51.12)	V3 (33.36)	DB (34.78)
PPT	V2 (35.13)	SST (46.00)	SST (39.34)	V2 (39.44)

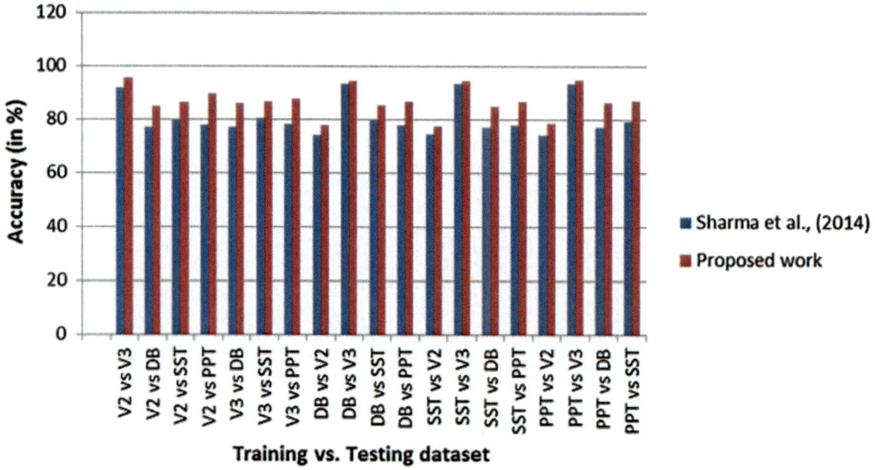

Fig. 2 SVM accuracy comparison (proposed work vs. Sharma et al., 2014 [2])

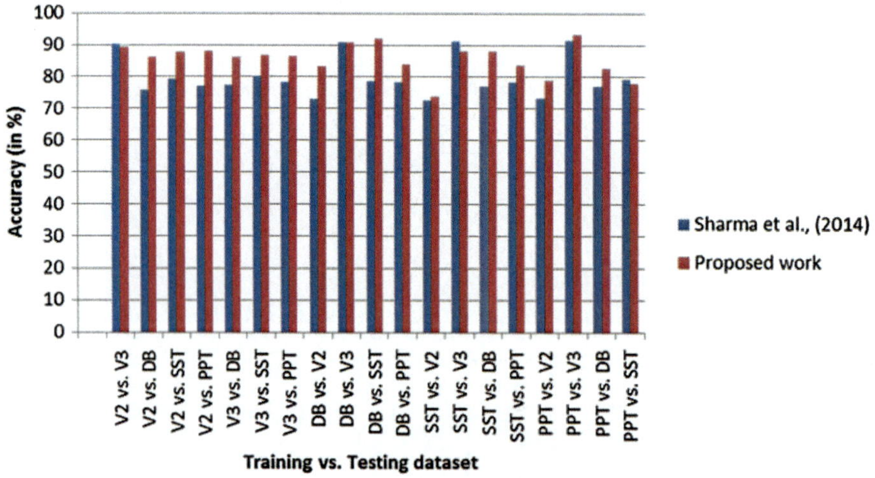

Fig. 3 k-NN accuracy comparison (proposed work vs. Sharma et al., 2014 [2])

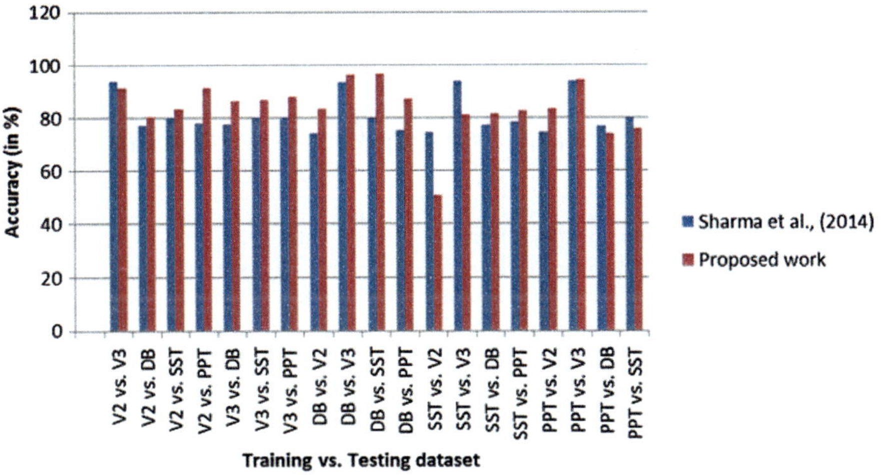

Fig. 4 NNET accuracy comparison (proposed work vs. Sharma et al., 2014 [2])

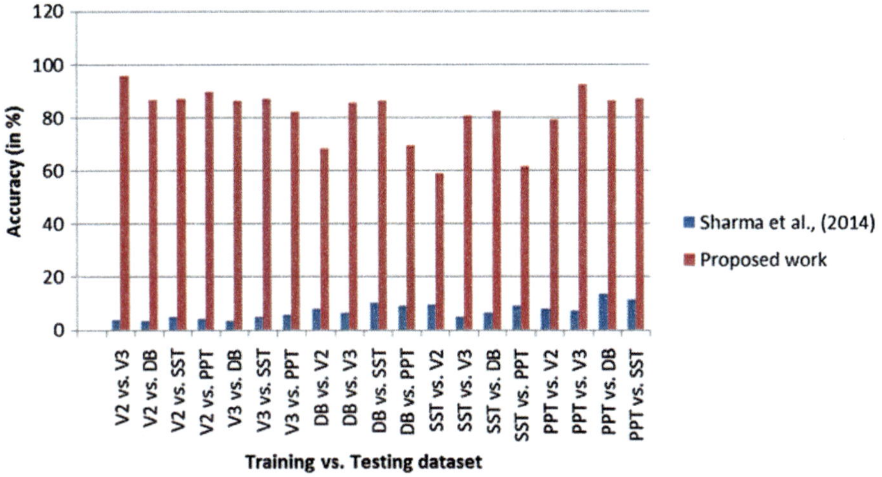

Fig. 5 NB accuracy comparison (proposed work vs. Sharma et al., 2014 [2])

based on these techniques predict the priority of a reported bug in cross-project context very accurately and outperform with the work available in the literature.

References

1. M. Sharma, P. Bedi, K.K. Chaturvedi, V.B. Singh, Predicting the priority of a reported bug using machine learning techniques and cross project validation, in *Proceedings of the 12th*

International Conference on Intelligent Systems Design and Applications (ISDA) (Kochi, India, 2012), pp. 539–545
2. M. Sharma, P. Bedi, V.B. Singh, An empirical evaluation of cross project priority prediction. Int. J. Syst. Assur. Eng. Manage. **5**(4), 651–663 (2014)
3. J. Kanwal, O. Maqbool, Managing open bug repositories through bug report prioritization using SVMs, in *Proceedings of the International Conference on Open-Source Systems and Technologies* (Lahore, Pakistan, 2010)
4. J. Kanwal, O. Maqbool, Bug prioritization to facilitate bug report triage. J. Comput. Sci. Technol. **27**(2), 397–412 (2012)
5. M. Alenezi, S. Banitaan, Bug reports prioritization: which features and classifier to use, in *12th International Conference on Machine Learning and Applications* (IEEE, 2013), pp. 112–116
6. L. Yu, W. Tsai, W. Zhao, F. Wu, Predicting defect priority based on neural networks, in *Proceedings of the 6th International Conference on Advanced Data Mining and Applications* (Wuhan, China, 2010), pp. 356–367
7. Y. Tian, D. Lo, C. Sun, DRONE: predicting priority of reported bugs by multi-factor analysis, in *IEEE International Conference on Software Maintenance* (2013), pp. 200–209
8. T. Menzies, A. Marcus, Automated severity assessment of software defect reports, in *Proceedings of International Conference on Software Maintenance* (IEEE, New York, 2008), pp. 346–355
9. T. Zimmermann, N. Nagappan, H. Gall, Cross-project defect prediction: a large scale experiment on data vs. domain vs. process, in *Proceedings of the 7th Joint Meeting of the European Software Engineering Conference and the ACM SIGSOFT Symposium on The Foundations of Software Engineering* (2009), pp. 91–100
10. B. Turhan, T. Menzies, A.B. Bener, J.D. Stefano, On the relative value of cross-company and within-company data for defect prediction. Empir. Softw. Eng. (2009). https://doi.org/10.1007/s10664-008-9103-7
11. Y. Ma, G. Luo, X. Zeng, A. Chen, Transfer learning for cross-company software defect prediction. Inf. Softw. Technol. **54**, 248–256 (2011)
12. Z. He, F. Shu, Y. Yang, M. Li, Q. Wang, An investigation on the feasibility of cross-project defect prediction, in *Automated Software Engineering* (2012), pp. 167–199
13. F. Peters, T. Menzies, A. Marcus, Better cross company defect prediction, in *10th IEEE Working Conference on Mining Software Repositories (MSR)* (IEEE, New York, 2013), pp. 409–418
14. G. Canfora, A. De Lucia, M. Di Penta, R. Oliveto, A. Panichella, S. Panichella, Multiobjective cross-project defect prediction, in *IEEE 6th International Conference on Software Testing, Verification and Validation (ICST)* (IEEE, New York, 2013), pp. 252–261
15. J. Nam, S.J. Pan, S. Kim, Transfer defect learning, in *Proceedings of International Conference on Software Engineering* (IEEE, New York, 2013), pp. 382–391
16. D. Ryu, O. Choi, J. Baik, Value-cognitive boosting with a support vector machine for cross-project defect prediction. Empir. Softw. Eng. **21**(1), 43–71 (2016)
17. V.B. Singh, S. Misra, M. Sharma, Bug severity assessment in cross project context and identifying training candidates. J. Inf. Knowl. Manage. **16**(01), 1750005 (2017)
18. A.E. Hassan, Predicting faults based on complexity of code change, in *Proceedings of International Conference on Software engineering (ICSE 09)* (2009), pp. 78–88
19. K.K. Chaturvedi, P.K. Kapur, S. Anand, V.B. Singh, Predicting the complexity of code changes using entropy-based measures. Int. J. Syst. Assur. Eng. Manage. Spr. **5**, 155–164 (2014)
20. S. Kamavaram, K. Goseva-Popstojanova, Entropy as a measure of uncertainty in software reliability, in *13th Int'l Symposium Software Reliability Engineering* (2002), pp. 209–210
21. I. Mierswa, M. Wurst, R. Klinkenberg, M. Scholz, T. Euler, YALE: rapid prototyping for complex data mining tasks, in *Proceedings of the 12th ACM SIGKDD International Conference on Knowledge Discovery and Data Mining (KDD-06)* (2006)
22. X. Wang, Y. He, Learning from uncertainty for big data. IEEE Syst. Man Cybern. Mag. 26–32 (2016)
23. [IEEE88], IEEE standard dictionary of measures to produce reliable software, IEEE Std 982.1-1988, Institute of Electrical and Electronics Engineers (1989). http://www.rapid-i.com

Internet of Things Security Using Machine Learning

Bhabendu Kumar Mohanta and Debasish Jena

1 Introduction

The IoT is one of the main research topics in both academic and in industry. A lot of smart devices are manufactured by the industry to sense and act in intelligent way. IoT devices are connected in wireless or through the wire to the network layer and next to the application layer. Basically, it follows a three-layer architecture. The security vulnerability of smart devices and future challenges are explained in papers [1, 2]. The IoT application needs a security issue to be addressed like the confidentiality and integrity of the user must be maintained which is described by authors in [3]. The basis of any security system is addressing the confidentiality, integrity, and availability. Before developing the security solution of the IoT system, it is essential to know the difference between the traditional security framework and the current system. To ensure the security in IoT framework design and challenges, the potential threads are explained by the authors in [4]. The IoT needs some enabling technologies like fog computing, software-defined network, to integrate with the model to address the security issue [5, 6]. As shown in Fig. 1, IoT has many applications. In a smart home system, a lot of home appliances are embedded with smart technology. Those devices are connected to the home network and communicate with the home user through mobile devices. The real-time monitoring, security of the system, even it is easy to monitor the fire in case of home from the kitchen or any unusual activity, can be detected. Similarly in the case of smart farming, using IoT makes the farmer's job easy by sensing the environment and processing this information. The IoT-based smart farming makes the system more efficient and reduces the excess use of material. Because of resource-constraint smart devices, different security and privacy issues

B. K. Mohanta (✉) · D. Jena
Information Security Laboratory, IIIT Bhubaneswar, Bhubaneswar, Odisha 751003, India
e-mail: C116004@iiit-bh.ac.in

D. Jena
e-mail: debasish@iiit-bh.ac.in

S. Patnaik et al. (eds.), *Advances in Machine Learning and Computational Intelligence*, Algorithms for Intelligent Systems, https://doi.org/10.1007/978-981-15-5243-4_11

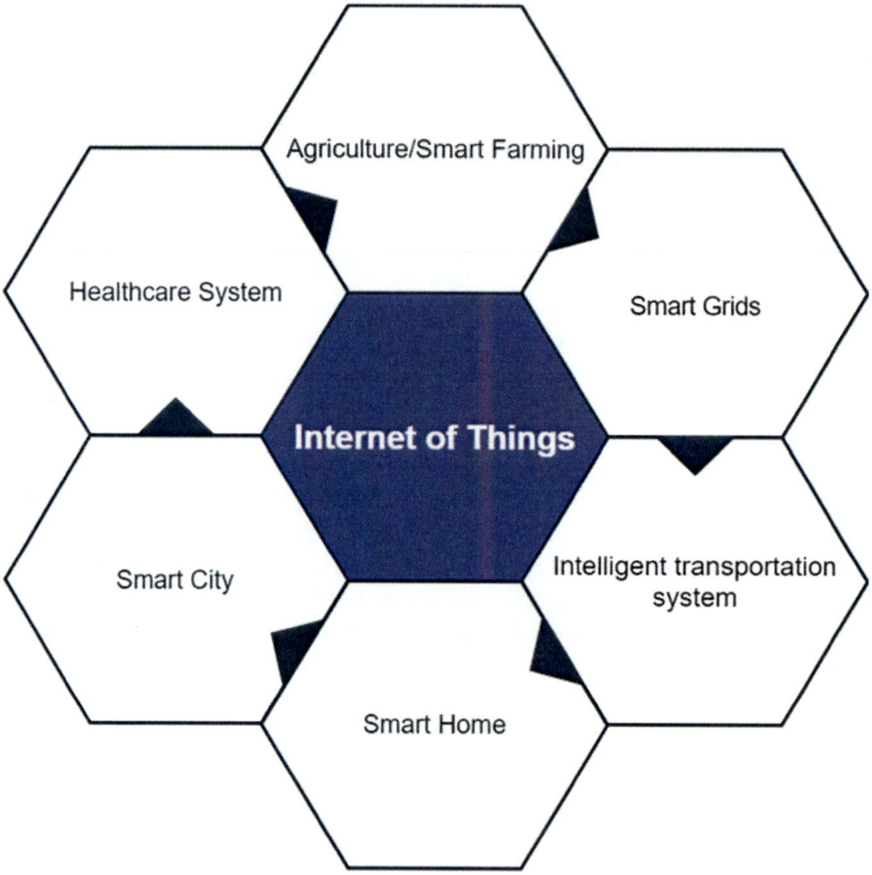

Fig. 1 Internet of things' different application areas

are existing in IoT applications. Some of the issues are already addressed by the research group. Still, a lot needs to be done to make the system more efficient and more trustful to the end-user.

1.1 Organization of the Paper

The rest of the paper is organized in this way. In Sect. 2, recent work on IoT security and machine learning is described. In Sect. 3, IoT architecture and its associated security issues are explained. In Sect. 4, machine learning technique and security issue addressed by the machine learning identified with their implementation details are discussed. In Sect. 5, the paper concludes with the future work.

2 Related Work

IoT is one of the promising technologies in the recent past. The grown of IoT is exponential, and as estimated, 50 billion devices are connected to the smart IoT network. It generated a huge volume of data for processing and computing. So some related techniques like fog computing and cloud computing are used to process and store the data generated from the IoT applications. Fog computing provides computation at the edge of the network. The security issues are the most challenging part to implement IoT applications. Machine learning techniques like supervised learning, unsupervised learning, and reinforcement learning are widely used for many domains for classification and regression purposes. The techniques such as machine learning and artificial neural network can address the security issue in IoT

Table 1 Related work on security issue in IoT and machine learning

Papers	Year	Contribution
Janice Canedo et al. [7]	2016	Machine learning is used in IoT gateway to secure the devices. The artificial neural network (ANN) is used to detect the anomalies present in the edge of the IoT application. The paper explained the IoT scenario where the temperature sensor was used to collect data and ANN technique is used to predict the anomalies present in the network. By doing so some of the security issues of IoT applications are solved
Xiao et al. [8]	2018	In this article, machine learning techniques, such as supervised, unsupervised, and reinforcement learning, are used to address the security issues in the IoT system. The privacy issues like secure authentication, data protection, malware detection, and secure offloading are investigated by the authors in this work. Finally, the paper also explained the challenges of implementing machine learning in IoT applications
Molanes et al. [9]	2018	The safety and security of IoT are explained in this paper using deep learning and neural network. The integration of machine learning and artificial network addressed the security issue in the IoT platform
Hussain et al. [10]	2019	In this paper, in-depth layer-wise (physical, network, and application) security analysis is done by the authors. The different techniques and algorithms used for IoT security are explained in detail in this paper
Zantalis et al. [11]	2019	The IoT and machine learning are integral to solve many security issues in IoT application. A lot of algorithms are designed and implemented to address different security issues in the IoT system. In this paper, an intelligent transportation system is considered as the application and the security issue associated with this infrastructure explained how machine learning can solve these issues

application. As shown in Table 1, related research done on IoT security and machine learning is explained.

3 Architecture Based on IoT-Fog Computing

The IoT architecture is of three layers which consist of the physical layer, network layer, and application layer. In the physical layer, all sensors or smart devices are deployed in the environment to monitor the system. The physical layer devices are connected to the network layer through wired or wireless. There is some protocol used for communication like Zigbee, 6LoWPAN, Bluetooth, Z-Wave, Wi-Fi, and near-field communications (NFCs). The smart devices are connected to the network layer, and from the network layer, it was connected to the application layer. As shown in Fig. 2, a three-layer consists of a physical layer where all devices are connected. In the second layer, processing and analysis are done using a machine learning technique. Similarly, in the third layer, all the notification or events are executed and broadcast to all the registered users.

3.1 Security and Privacy Issue in IoT

The Internet of Things provides a lot of services to the different applications. The used resource-constraint smart devices in applications make the system more vulnerable to the outside world. Most of the IoT devices are having less storage and low processing power. Because of resource-constraint devices, it is not feasible to use the existing high computation security protocol to the IoT system. The challenges are to develop a lightweight protocol that will suit the IoT devices. In paper [12], authors have

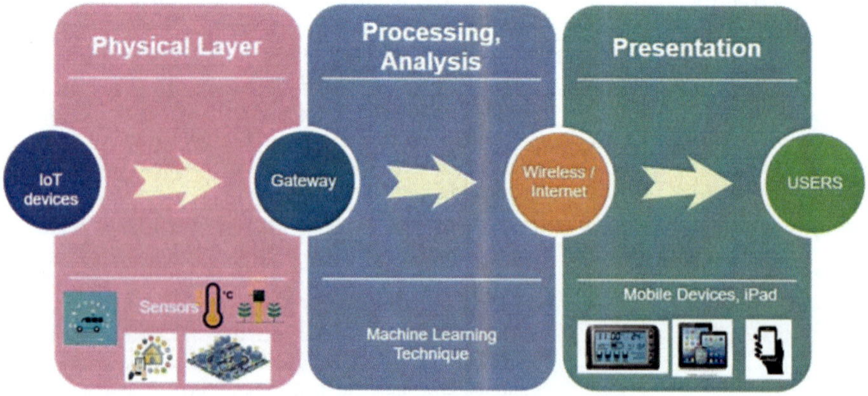

Fig. 2 Machine learning-based IoT application overview

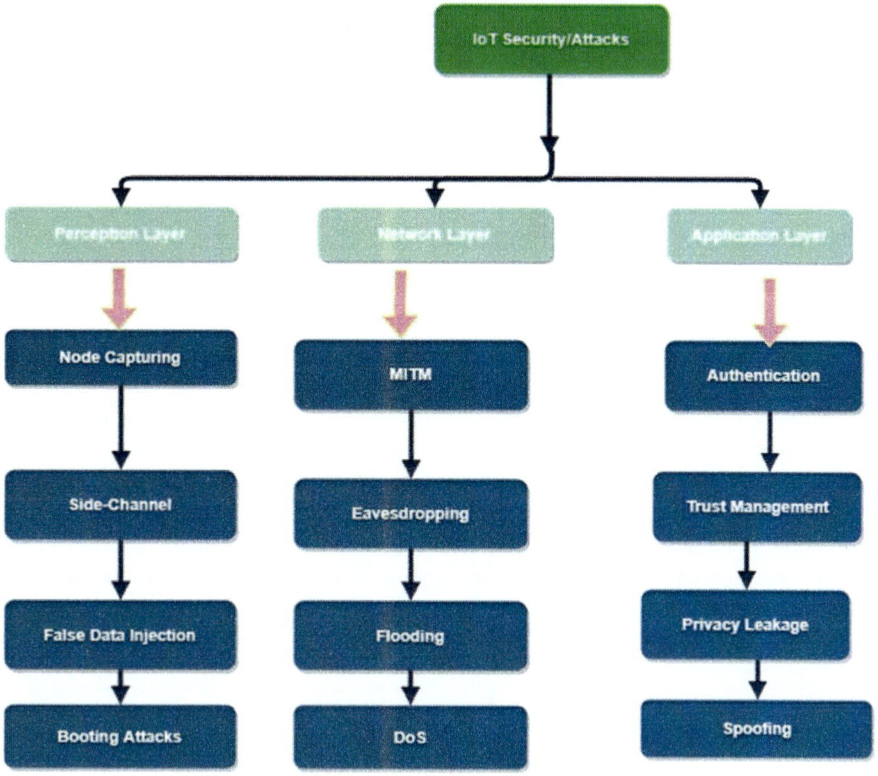

Fig. 3 IoT security and attacks on different layers

proposed an authentication technique based on neural networks. Figure 3 shows Internet of Things security attack types in different layers. The details of application area, security issues, and architecture

related to basic IoT are explained in paper [13]. Some of the security issues and their challenges are already addressed by the research community. The recent development of fog computing, blockchain technology, and machine learning provides a more solution approach to solved security and privacy issue in IoT. In this paper, we focus only on machine learning techniques which help in solving the security and privacy issues in IoT.

4 Machine Learning

In this section, various machine learning algorithms used in IoT security are explained. Machine learning is referred to an intelligent system that can optimize the system performance using the past data and applying some of the classification and

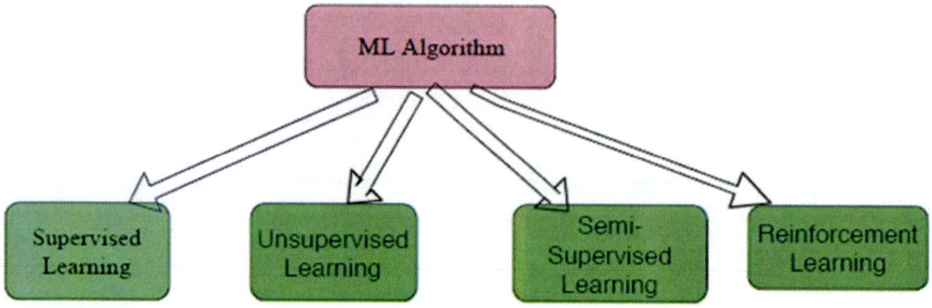

Fig. 4 Fundamental types of machine learning

clustering techniques. The basic working principle of machine learning is a model to build/train using the past dataset. Machine learning can help in developing intelligent IoT-based smart devices. Once the model is learned from the huge dataset, it can process and compute automatically. In IoT application systems like healthcare systems, smart homes, and smart transportation, deciding in real time is a challenging task. So machine learning can help in making intelligent decisions if a model trains properly using the past dataset.

4.1 Basic Machine Learning Algorithm

The machine learning types are basically four types—supervised, semi-supervised, unsupervised, and reinforcement learning—as shown in Fig. 4. In paper [14], authors address the machine learning techniques used in IoT and fog-based applications. Similarly in [15, 16], authors described security, privacy issue, healthcare system, and intrusion detection system identifications using hybrid machine learning system.

4.2 Security Issue Address Using Machine Learning

Machine learning helps in solving many security issues in IoT application as shown in Table 2. The machine learning and IoT integration makes the system more secure and efficient in decision-making process.

5 Conclusion

The number of interconnected smart devices has already crossed the total world population. The data generated by these devices are huge. The IoT used in different applications like smart monitoring, healthcare system, and smart home where sensitive information is shared among users. Security and privacy are some of the major

Table 2 Security issue in IoT and machine learning

Papers	Contribution	Application area
[17, 18]	Device identification done in secure way based on network traffic analysis and software define network	IoT
[19, 20]	Classifying unsolicited IoT devices and integration of machine learning in IoT	IoT
[21, 22]	Authentication using extreme learning machine algorithm and war soldiers technique for security in healthcare system	IoT-based healthcare
[23, 24]	Network vulnerability analysis and security issue discussion using machine learning	IoT-based smart grid
[25, 26]	Trust computational model and malware detection using machine learning technique for secure computation	IoT

challenges in IoT applications. Some traditional security and cryptographic techniques are tried to address some of the issues exist in the IoT system. In this paper, the authors tried to identify the existing security issue in IoT. The machine learning technique is used to solve the security issue in the IoT system. In this paper, the authors identified research already done in the integration of machine learning with IoT. In Sect. 4, machine learning and its associated technique are explained. In summary, this paper concludes with some of the security issues which are shown in tabular format. The integration of blockchain technology [27] with IoT can solve some of the security issues. In the future, we would like to implement an IoT application and analyze security properties.

References

1. X. Liu, M. Zhao, S. Li, F. Zhang, W. Trappe, A security framework for the internet of things in the future internet architecture. Future Internet **9**(3), 27 (2017)
2. I. Andrea, C. Chrysostomou, G. Hadjichristofi, Internet of Things: security vulnerabilities and challenges, in *2015 IEEE Symposium on Computers and Communication (ISCC)* (IEEE, 2015), pp. 180–187
3. W.H. Hassan, Current research on internet of things (IoT) security: a survey. Comput. Netw. **148**, 283–294 (2019)
4. M. Ammar, G. Russello, B. Crispo, Internet of things: a survey on the security of IoT frameworks. J. Inf. Sec. Appl. **38**, 8–27 (2018)
5. O. Salman, I. Elhajj, A. Chehab, A. Kayssi, IoT survey: an SDN and fog computing perspective. Comput. Netw. **143**, 221–246 (2018)
6. A. Olakovi, M. Hadiali, Internet of things (IoT): a review of enabling technologies, challenges, and open research issues. Comput. Netw. **144**, 17–39 (2018)
7. J. Canedo, A. Skjellum, Using machine learning to secure IoT systems, in *2016 14th Annual Conference on Privacy, Security and Trust (PST)* (IEEE, 2016), pp. 219–222
8. L. Xiao, X. Wan, X. Lu, Y. Zhang, D. Wu, IoT security techniques based on machine learning (2018). arXiv:1801.06275
9. R. Fernandez Molanes, K. Amarasinghe, J. Rodriguez-Andina, M. Manic, Deep learning and reconfigurable platforms in the internet of things: challenges and opportunities in algorithms and hardware. IEEE Ind. Electron. Mag. **12**(2) (2018)

10. F. Hussain, R. Hussain, S.A. Hassan, E. Hossain, Machine learning in IoT security: current solutions and future challenges (2019). arXiv:1904.05735
11. F. Zantalis, G. Koulouras, S. Karabetsos, D. Kandris, A review of machine learning and IoT in smart transportation. Future Internet **11**(4), 94 (2019)
12. McGinthy, J. M., Wong, L. J., Michaels, A. J. (2019). Groundwork for Neural Network-Based Specific Emitter Identification Authentication for IoT. IEEE Inter- net of Things Journal
13. V. Hassija, V. Chamola, V. Saxena, D. Jain, P. Goyal, B. Sikdar, A sur- vey on IoT security: Application areas, security threats, and solution architectures. IEEE Access **7**, 82721–82743 (2019)
14. M. Moh, R. Raju, Machine learning techniques for security of internet of things (IoT) and fog computing systems, in *2018 International Conference on High Performance Computing Simulation (HPCS)* (IEEE, 2018), pp. 709–715
15. P.M. Shakeel, S. Baskar, V.S. Dhulipala, S. Mishra, M.M. Jaber, Maintaining security and privacy in health care system using learning based deep-Q-networks. J. Med. Syst. **42**(10), 186 (2018)
16. M. Nivaashini, P. Thangaraj, A framework of novel feature set extraction based intrusion detection system for internet of things using hybrid machine learning algorithms, in *2018 International Conference on Computing, Power and Communication Technologies (GUCON)* (IEEE, 2018), pp. 44–49
17. Y. Meidan, M. Bohadana, A. Shabtai, J.D. Guarnizo, M. Ochoa, N.O. Tippenhauer, Y. Elovici, ProfilIoT: a machine learning approach for IoT device identification based on network traffic analysis, in *Proceedings of the Symposium on Applied Computing* (ACM, 2017), pp. 506–509
18. F. Restuccia, S. DOro, T. Melodia, Securing the internet of things in the age of machine learning and software-defined networking. IEEE IoT J. **5**(6), 4829–4842 (2018)
19. F. Shaikh, E. Bou-Harb, J. Crichigno, N. Ghani, A machine learning model for classifying unsolicited IoT devices by observing network telescopes, in *2018 14th International Wireless Communications Mobile Computing Conference (IWCMC)* (IEEE, 2018), pp. 938–943
20. L. Xiao, X. Wan, X. Lu, Y. Zhang, D. Wu, IoT security techniques based on machine learning: How do IoT devices use AI to enhance security? IEEE Signal Process. Mag. **35**(5), 41–49 (2018)
21. A. Gondalia, D. Dixit, S. Parashar, V. Raghava, A. Sengupta, V.R. Sarobin, IoT-based healthcare monitoring system for war soldiers using machine learning. Proc. Comput. Sci. **133**, 1005–1013 (2018)
22. N. Wang, T. Jiang, S. Lv, L. Xiao, Physical-layer authentication based on extreme learning machine. IEEE Commun. Lett. **21**(7), 1557–1560 (2017)
23. M. Zolanvari, M.A. Teixeira, L. Gupta, K.M. Khan, R. Jain, Machine learning based network vulnerability analysis of industrial internet of things. IEEE IoT J. (2019)
24. E. Hossain, I. Khan, F. Un-Noor, S.S. Sikander, M.S.H. Sunny, Application of big data and machine learning in smart grid, and associated security concerns: a review. IEEE Access **7**, 13960–13988 (2019)
25. U. Jayasinghe, G.M. Lee, T.W. Um, Q. Shi, Machine learning based trust computational model for iot services. IEEE Trans. Sustain. Comput. **4**(1), 39–52 (2018)
26. L. Wei, W. Luo, J. Weng, Y. Zhong, X. Zhang, Z. Yan, Machine learning-based malicious application detection of android. IEEE Access **5**, 25591–25601 (2017)
27. B.K. Mohanta, D. Jena, S.S. Panda, S. Sobhanayak, Blockchain technology: a survey on applications and security privacy challenges. IoT, 100107 (2019)

Churn Prediction and Retention in Banking, Telecom and IT Sectors Using Machine Learning Techniques

Himani Jain, Garima Yadav, and R. Manoov

1 Introduction

Customer churn is when customers stop using a company's product or service. Also called customer attrition, customer churn is a very important metric as it is significantly less expensive to retain the existing customers than acquiring new customers. Churn prediction plays a vital role in customer retention because it predicts customers who are at risk of leaving the firm.

According to various studies and models, churn is one of the major factors that decrease the enterprise value of an industry. As compared to the average revenue per user (ARPU) which is one of the most significant key performance indicators, churn has a greater impact on the customer lifetime value (CLTV).

Hence, the motive of our paper is to successfully detect the early signs of churn so that we can take preventive actions. In order to understand early signs of customer churn, we should have a detailed information about our customers and their interactions across various channels like calls to customer services, social media interactions, mobile and Web-based transactions. If we can successfully detect these early signs, we can take preventive actions like giving the customers a hefty discount, special offers, perks, etc., to prevent churn. Also, if we are familiar with the behaviour and preferences of the churners and we understand how the loyal customers with similar preferences reacted to the various offers for retention, we can drastically improve the "accept rate" of the retention offers to the potential churners. The major objective of our paper is:

H. Jain · G. Yadav · R. Manoov (✉)
VIT University, Vellore, Tamil Nadu, India
e-mail: manoov.r@vit.ac.in

S. Patnaik et al. (eds.), *Advances in Machine Learning and Computational Intelligence*, Algorithms for Intelligent Systems, https://doi.org/10.1007/978-981-15-5243-4_12

- To compare the performances of various algorithms and build the best model for churn prediction in telecom, banking and IT sectors, respectively.
- Based on the results and exploratory data analysis, we will derive which attributes contribute more towards the customer churn and accordingly develop various retention strategies for the respective domains.

2 Previous Work

For the major time, the focus of churn prediction-based research was on applying one customer churn prediction algorithm only on telecom Industry, because telecom industry was facing a huge impact and loss due to customer churn but with the emergence of online banking and IT sector, customer churn is no longer a domain-oriented concern, rather it has become a threat to all the domains.

Focusing on research is done in the telecom industry. Can and Albey [1] in their paper researched about the increasing inclination of customers towards prepaid lines rather than post-paid lines because it is more convenient for the customers to handle the cost and also to end the subscription at their own convenience without having to contact the vendor. The mechanisms used are divided into two models, one is logistic regression-based model, and other is Pareto/NBD model—out of them the latter proves out to be more efficient. One major drawback of this research is that the dataset used is not real time; hence, the results concluded cannot be implemented on large datasets.

Many researchers have focused on features focusing on customer churn rather than domain orientation. For example, Hopner et al. [2] in their research paper have focused on the profit as a major feature to describe customer churn. They have proposed a new algorithm called ProfTree which is based on the decision tree and is a profit-inducing tree algorithm. Although there are no domain constraints in this algorithm, no implementation has been done so we cannot say for sure whether the algorithm is actually feasible or not.

Faris [3] has suggested a new neural network-based model to identify influencing factors and features which can be responsible for customer churn. This has been done by assigning feature weights to assist in classification. But the drawback is that the results are not satisfying enough because the accuracy rate which is measured by F-score is pretty low and not much impressive.

Cotter et al. have tried to optimize some of the features which are acting as non-differentiable constraints [4] using various mathematical equations and Lagrange's theorem along with neural networks.

Spanoudes et al. [5] used unsupervised feature learning which is against company-independent feature. Firstly, the data is taken from the company and is gone through feature engineering where important and relevant features are extracted after dataset is preprocessed and is made ready for the algorithm. After this, the dataset is worked upon by the random forest machine learning algorithm and after this, the final results are obtained. Since the implementation is not company specific, it cannot be revealed how the algorithm will work on large and real-time datasets. Other than telecom industry, Yang et al. [6] have applied churn prediction to real-time physical situation of urban migrants. The authors examine the integration process of migrants in the first weeks after moving to a new city and the disintegration process for some migrants who left early. In the case study, the city of Shanghai is taken into consideration using clustering and statistics.

Other than telecom industry, customer churn is also seen on social media which is done by spatial diffusion by Lengyel et al. [7] which is done by various chats and distribution of customers. Other than customer prediction, one more area of focus is customer retention. Zhang et al. [8] in their paper focus on three major factors of customer churn problem which are taken into consideration, namely prediction performance, churn rate and retention capability. In this paper, the analysis of profit maximization of a retention campaign is done taking into account both prediction performance and retention capacity. The algorithm used for this was exponential retention model. The conclusion derived from this research is increased capability of retention will lead to increased profit. And on the other hand, when both prediction performance and retention capacity are not good enough, it is advisable to not take any actions. Coming to a real-life domain of gym membership Semrl et al. [9] worked on their paper which is mainly focused on predicting customer behaviour to enhance customer retention. A comparison is done between two machine learning platforms—Azure ML and Big ML—which can comfortably be used by non-ML experts as well to conclude that both platforms perform well and can be used to improve their services and to gain profit in the business. Benoit et al. [10] in their paper focus on the foundation that the mass marketing approach where the marketing strategy is same for all the customers cannot succeed in the pool of diversified customers. Therefore, companies are now focusing on the marketing strategy called Customer Relationship Management which includes detailed information about the potential customers in order to understand and satisfy their needs. This paper illustrates how important it is to include social network variables in the prediction model. Banking sector is also facing an increase in the amount of customer churn; this is majorly because of customer awareness and their inclination towards a better service quality.

Nie et al. [11] in their research paper use a logistic regression-based algorithm to find out customer churn for credit card holders in China. Their data includes 5000 credit card holders. The research is done by splitting time window into two parts—one is observation and other is performance period. The accuracy measured is highly incredible (91.9%), but it just focuses on understanding one algorithm and no new algorithm has been suggested. Jing et al. [12] have researched on a commercial bank's VIP customers' database, and on applying various machine learning algorithms, it has been found out that SVM is the best algorithm suitable for the case, which is concluded by considering accuracy rate, hit rate, covering rate and lift coefficient.

This thesis has been used to improve customer relationship management. Szmydt [13] has researched how customer churn in electronic banking is more intense than in regular banking. Also why is such a situation and happening and what all behavioural and psychological aspects should be considered to increase customer retention. Chen et al. [14] in their research paper have researched in the field of retail banking and how customer churn can be prevented in this domain. They have utilized five different algorithms in which neural networks outshined, and hence, they have concluded that deep ensemble classifiers are best for this job. Anil Kumar et al. [15] have used various data mining techniques and sampling techniques to predict customer churn and have been successful with impressive accuracy rates. They have also mentioned why there is a high level of churn rate in banking sector. E-commerce industry has also been a major victim of customer churn. One such solution to combat this problem is given by Abbet et al. [16] where they suggest the use of multilingual chat bots to predict customer churn. Firstly, customers chat through a bot, and if a churning behaviour is seen, then they are redirected to a human in charge. But the problem is not all the customers will actually use chat bots before leaving the service.

Hence, from the above survey we can conclude that no work has been done till now in which a comparison of various algorithms on multiple domains has been done. Therefore, our work will majorly focus on applying four machine algorithms, i.e. logistic regression, random forest, SVM and XGBoost, in three major domains which are banking, telecom and IT sectors.

Predicting customer churn only is not enough if relevant marketing offers such as discounts, vouchers and customized schemes are not developed which can satisfy the customers who are showing churn. Usually, generic marketing offers lead to less retention rate and very low success to prevent customer churn.

There is a need to implement efficient machine learning algorithm in other fields with ultimate accuracy which can lead to a visible decrease in the percentage of customer churn. From the literature survey, it can be noticed that most of the authors have not used more than one algorithm although it is proven that different algorithms have different performance metrics in different scenarios so through our project, we are carrying out a comparative analysis of the performances of various algorithms in multiple domains and then proceeding with the algorithm with the highest performance score for developing retention strategies.

3 Theoretical Background

Let us see about the domains, algorithms and data description in detail.

3.1 Domains

3.1.1 Banking Sector

The main reason behind choosing banking domain is that according to various studies and models, churn is one of the major factors that decrease the enterprise value of banks. As compared to the average revenue per user (ARPU) which is one of the most significant key performance indicators, churn has a greater impact on the customer lifetime value (CLTV). The banking dataset for predicting customer churn consists of 10,000 entries and 14 attributes. The following are the attributes for the banking dataset—credit score, geographical location, gender, age, tenure, balance, number of products, has credit card or not, is active member or not.

3.1.2 Telecom Sector

The reason for choosing telecom is because churn is significantly higher in telecom as compared to other domains. The telecom dataset consists of 7,043 rows and 21 attributes. The following are the attributes for the telecom dataset—gender, senior citizen, partner, dependence, tenure, phone service, multiple lines, Internet service, online security, online backup, device protection, technical support, streaming TV, streaming movies, contract, paperless billing, payment method, monthly charges and total charges.

3.1.3 IT Sector

The reason behind including IT domain is because employee churn is predominant in IT sector, with so many IT firms providing competitive wages, perks and career growth. The IT dataset for employee churn prediction consists of 1,470 entries and 36 attributes. The following are the attributes for the IT dataset—age, business travel, daily rate, department, distance from home, education, educational field, employee count, employee number, environmental satisfaction, gender, hourly rate, job involvement, job level, job role, job satisfaction, marital status, monthly income, monthly rate, number of companies worked, overtime, percentage salary hike, performance rating, relationship satisfaction, standard hours, stock option level, total working years, training time, work–life balance, years at company, years in current role, years since last promotion and years with current manager.

3.2 *Algorithms*

3.2.1 Logistic Regression

It is a classification algorithm which is majorly used for binary classification; in our case, it will be customer churned and not churned. It is one of the most basic and easily comprehensible algorithms. It is based on the sigmoid function or the logit function whose value lies between 0 and 1.

$$\partial(z) = \frac{1}{(1 + e - z)} \tag{1}$$

Here,

∂ is sigmoid function.
z is function value.

Graphically, we can represent the logistic function as follows:
The advantages of using logistic regression are that:

- It can be easily changed and reshaped such that we can avoid overfitting of the data (Figs. 1 and 2).
- It comes in extremely handy while updating the data, i.e. adding new data (in our case when we will get live stream of data) as we can use stochastic gradient descent.

The one weakness when it comes to logistic regression is that it might underperform in some cases.

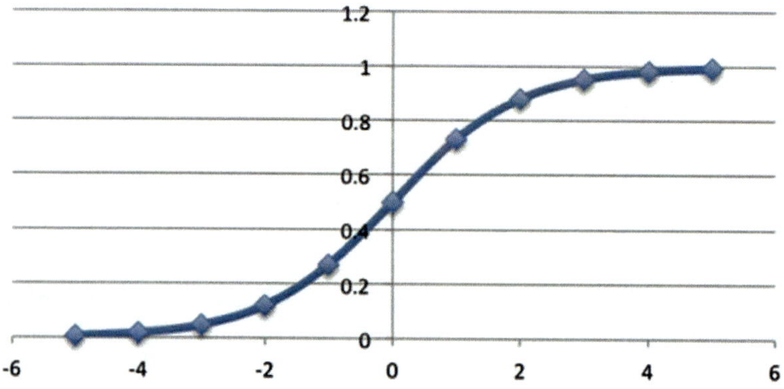

Fig. 1 Logistic function

3.2.2 Random Forest

Random forest is one of the most powerful machine learning algorithms. It is a type of ensemble algorithm called bagging (bootstrap aggregating). Random forests are an improved version of decision tress because it creates multiple decision trees and then merges them to get more stable and accurate results. The disadvantage with decision trees is that there is high correlation in the predictions. Random forest is enhanced because there is less correlation between the resulting predictions from all the sub-trees. In CART algorithm, the algorithm looks through all variables and their values to select an optimal split-point, but in random forest, the algorithm is restricted to search from a randomly selected sample of features. The number of features to be searched for a split-point (m) can be passed as an argument to the algorithm.

For classification m can be calculated as:

$$m = \sqrt{p} \tag{2}$$

For regression m can be calculated as:

$$m = p/3 \tag{3}$$

where p is total no. of input variables.

The advantages of random forest are as follows:

- The main advantage of random forest is its application in banking due to its tendency to predict stable and accurate results and its powerful nature. It is used to distinguish between loyal customers and customers who are about to churn.
- Other advantage of random forest is that it reduces overfitting. Because it averages several decision trees, the chances of overfitting are significantly reduced.

The main limitation of random forest is that it is complex and difficult to visualize the model and understand the reason of its prediction.

3.2.3 SVM

Support vector machine (SVM) is a classification-based algorithm which makes use of kernels to basically find out the distance between two values. It then creates a boundary or a hyperplane whose major function is to maximize the distance between the nearest value points of two different classes, in some sense kind of separating them. In our case, the two classes are customers churned and not churned. The name of the hyper plane is maximum-margin hyperplane. To find out the distance between two points, dot product of vectors is used and the function can be defined as:

$$S(x) = B_0 + \sum_{i=1}^{n} (a_i \times (x \times x_i)) \tag{4}$$

where

$S(x)$ is SVM distance function
B_0, a_i are coefficients defined while training of data
x is new input vector
x_i is previous support vectors.

The main advantages of using SVM are:

- In SVM, we have multiple options of kernel to choose from; in our case, we took polynomial kernel as it suited the dataset best.
- Also, SVM is immune to overfitting especially with the high-density dataset.

One weakness when it comes to SVM is that it depends on the previous memory of the support vectors; hence when it comes to larger datasets, the situation might get trickier.

3.2.4 XGBoost

XGBoost is an implementation of gradient boosting algorithm, but its uniqueness lies in using an advanced model to control overfitting that helps in giving better performance.

We have used XGBoost for classification into "churn" or no churn" but it can be used for regression and ranking as well. XGBoost is capable of performing feature selection automatically and includes a parameter for randomization which is helpful in reducing the correlation between each tree.

Gradient boosting machine (GBM) uses two steps to solve the optimization problem:

Step-1: To determine the direction of step
Step-2: To optimize the step length

whereas XGBoost determines the step directly using only one equation.
Advantages:

- Easy to interpret and can be constructed quickly.
- Can easily deal with categorical as well as numerical data and it can also deal with outliers and missing data.

Limitations:

- Tendency to select attributes that have a greater no. of distinct values.
- Can lead to over fitting in case of attributes with numerous categories.

3.3 Data Description

3.3.1 Banking Sector

From Fig. 3, we can see that in the chosen banking dataset, it has 20.4% of customers who showed churning as compared to 70.6% customers who did not churn.

From Fig. 4, we can see the comparative analysis of customer churn rate with respect to their geographical location, gender has credit card and active membership. 0 means the customer has not churned and 1 stands for customers showing churn (Fig. 5).

```
df.dtypes
```

```
CreditScore          int64
Geography           object
Gender              object
Age                  int64
Tenure               int64
Balance            float64
NumOfProducts        int64
HasCrCard            int64
IsActiveMember       int64
EstimatedSalary    float64
Exited               int64
dtype: object
```

Fig. 2 Data types for banking sector

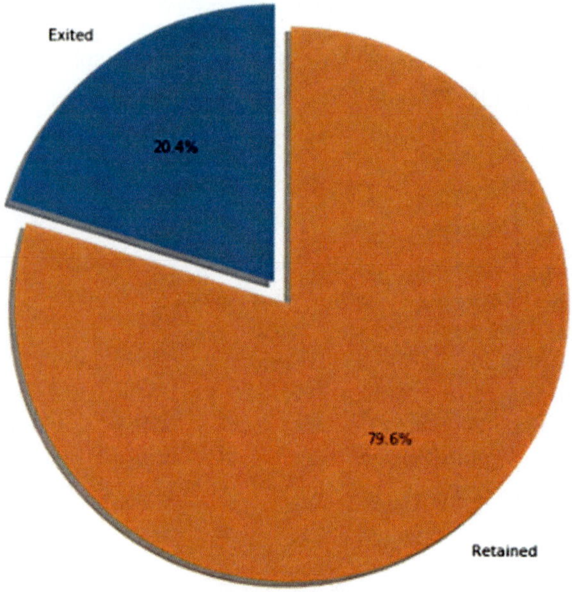

Fig. 3 Proportion of customers churned in banking dataset

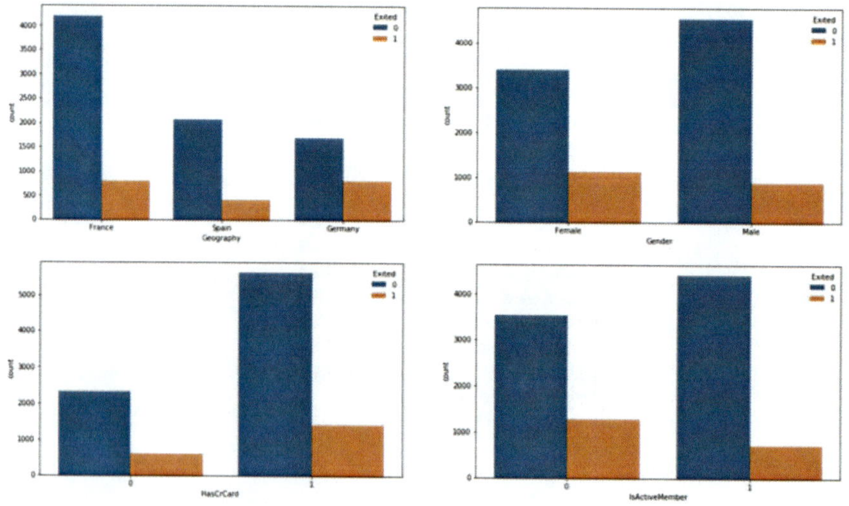

Fig. 4 Bivariate analysis with respect to churn for banking dataset

3.3.2 Telecom Sector

From Fig.6, we can see that in the chosen telecom dataset we have 26.6% of customers have churned as compared to 73.4% customers who did not churn.

Figure 7 gives the correlation between churn and other attributes of the dataset such as month to month contract, whether the customer is a senior citizen, streaming services provided, security, Internet speed and all the other attributes leading to customer churn (Figs. 8, 9 and 10).

Fig. 5 Data types for telecom sector

```
telecom_cust.dtypes

customerID          object
gender              object
SeniorCitizen        int64
Partner             object
Dependents          object
tenure               int64
PhoneService        object
MultipleLines       object
InternetService     object
OnlineSecurity      object
OnlineBackup        object
DeviceProtection    object
TechSupport         object
StreamingTV         object
StreamingMovies     object
Contract            object
PaperlessBilling    object
PaymentMethod       object
MonthlyCharges     float64
TotalCharges        object
Churn               object
dtype: object
```

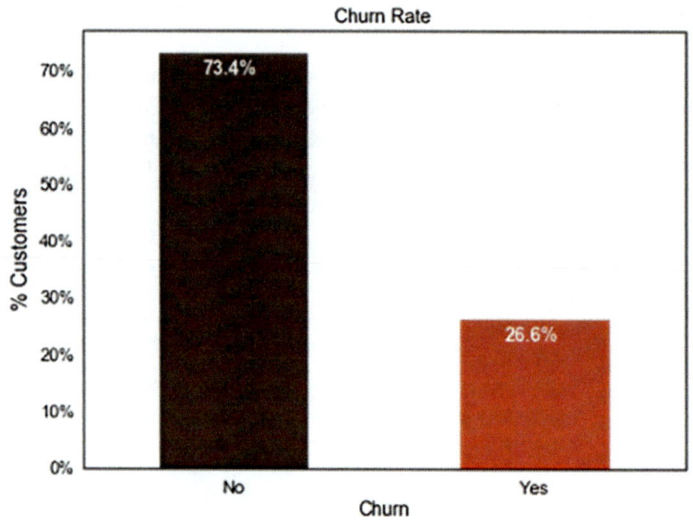

Fig. 6 Proportion of customers churned

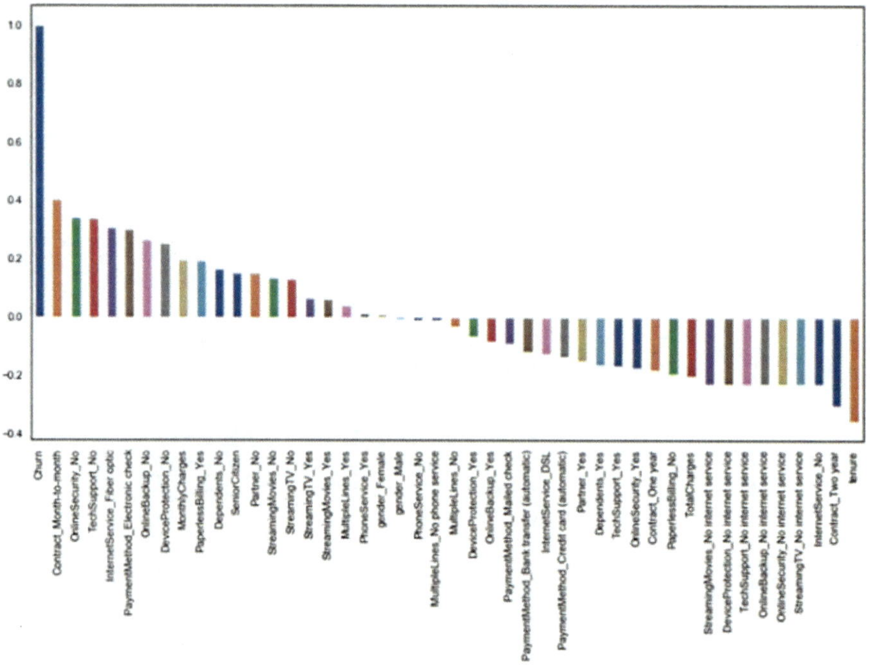

Fig. 7 Correlation between churn and the attributes for telecom dataset

3.3.3 IT Sector

In Fig. 10, we can see that first metric gives us the average monthly income according to the job role and the second metric gives us the churn count in each of the job role. We can see a rough trend that people in roles having high income are less likely to churn.

Fig. 8 Data types for IT sector

```
df.dtypes

Age                          int64
Attrition                   object
BusinessTravel              object
DailyRate                    int64
Department                  object
DistanceFromHome             int64
Education                    int64
EducationField              object
EmployeeCount                int64
EmployeeNumber               int64
EnvironmentSatisfaction      int64
Gender                      object
HourlyRate                   int64
JobInvolvement               int64
JobLevel                     int64
JobRole                     object
JobSatisfaction              int64
MaritalStatus               object
MonthlyIncome                int64
MonthlyRate                  int64
NumCompaniesWorked           int64
Over18                      object
OverTime                    object
PercentSalaryHike            int64
PerformanceRating            int64
RelationshipSatisfaction     int64
StandardHours                int64
StockOptionLevel             int64
TotalWorkingYears            int64
TrainingTimesLastYear        int64
WorkLifeBalance              int64
YearsAtCompany               int64
YearsInCurrentRole           int64
YearsSinceLastPromotion      int64
YearsWithCurrManager         int64
dtype: object
```

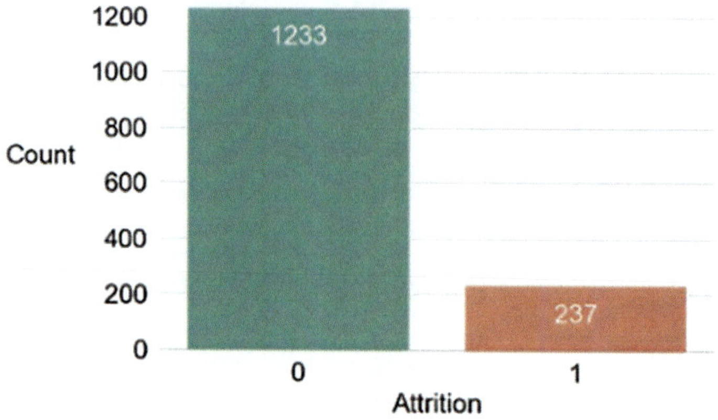

Fig. 9 Proportion of customers churned in IT dataset

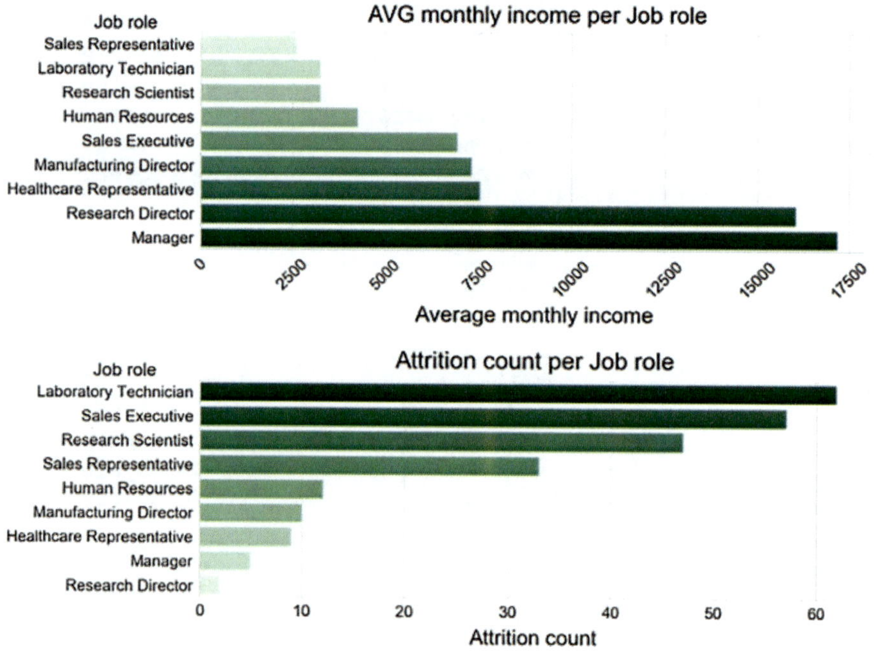

Fig. 10 Attrition count for various job roles with respect to monthly income in IT dataset

4 Methodology

The following steps were used during the implementation of the work:

1. Begin
2. Collection of datasets for banking, telecom and IT sectors.
3. Data cleansing and pre-processing
4. Exploratory Data Analysis
For each dataset do the following steps:
4.1 Find the proportion of customer churned and retained.
4.2 Find correlation of other attributes with the target
4.3 Box plot to visualize outliers
5. Feature Engineering
6. Data Preparation for model fitting
7. Model Fitting and Selection
Apply following models on each dataset:
7.1 Logistic Regression
7.2 Random Forest
7.3 SVM
7.4 XG Boost
8. Compare all the algorithms using confusion matrix
9. Analyze ROC curve and bar graphs to compare algorithms.
10. Identify the best algorithm for each domain.
11. Formulate retention strategies for each domain.
12. End

Pseudo-code 1 Pseudo-code for churn prediction and retention

Figure 11 is the step-by-step guide for implementation. We can see all the three datasets used and all four algorithms applied. Also, the metrices are used to get the results and the final outcome.

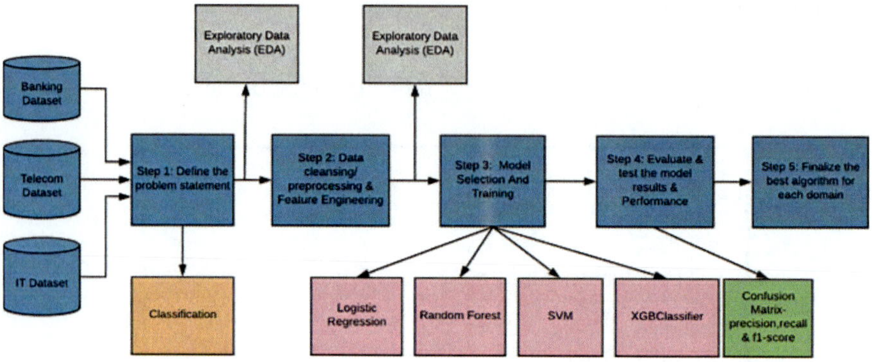

Fig. 11 Workflow diagram

5 Result and Discussions

From our study and implementation, we obtained the following results.

5.1 Churn Prediction

Figure 12 is the ROC curve for the banking sector dataset. In this, we can see that random forest algorithm has the highest area under the curve and hence is the most appropriate algorithm for this sector.

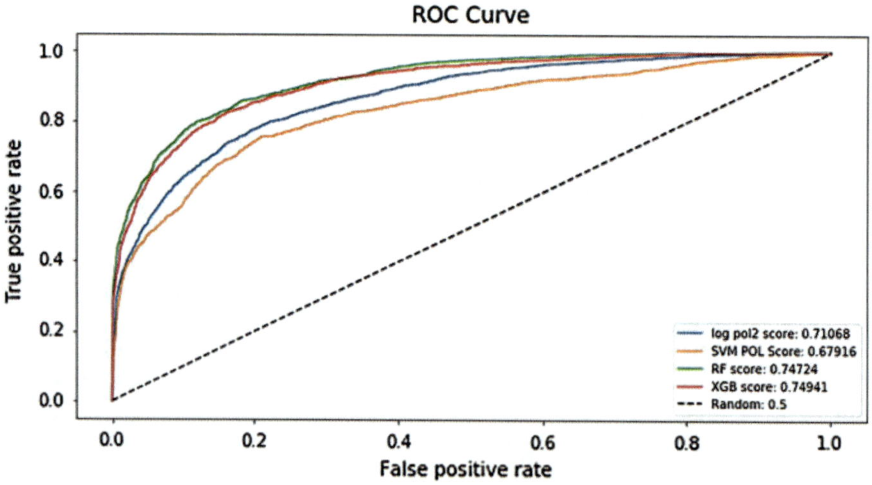

Fig. 12 ROC curve for banking sector

Figure 13 is the bar graph showing the accuracy percentage of all the algorithms. We can see that logistic regression has the most accuracy, and hence, it is most suitable for telecom sector.

Figure 14 is the bar graph showing the accuracy percentage of all the algorithms. We can see that XGBoost has the most accuracy, and hence, it is most suitable for IT sector.

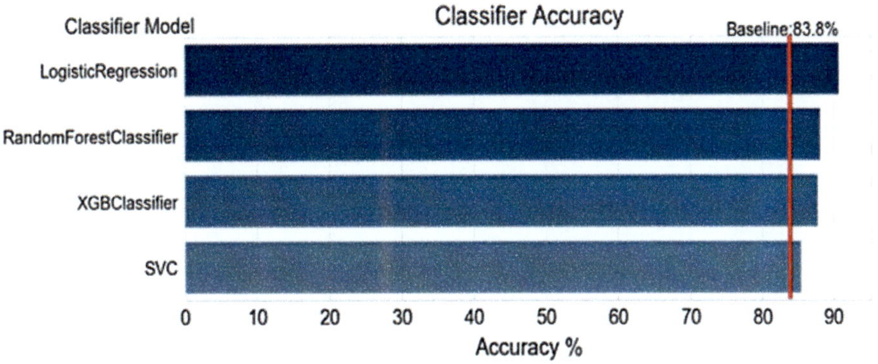

Fig. 13 Accuracy bar graph for telecom sector

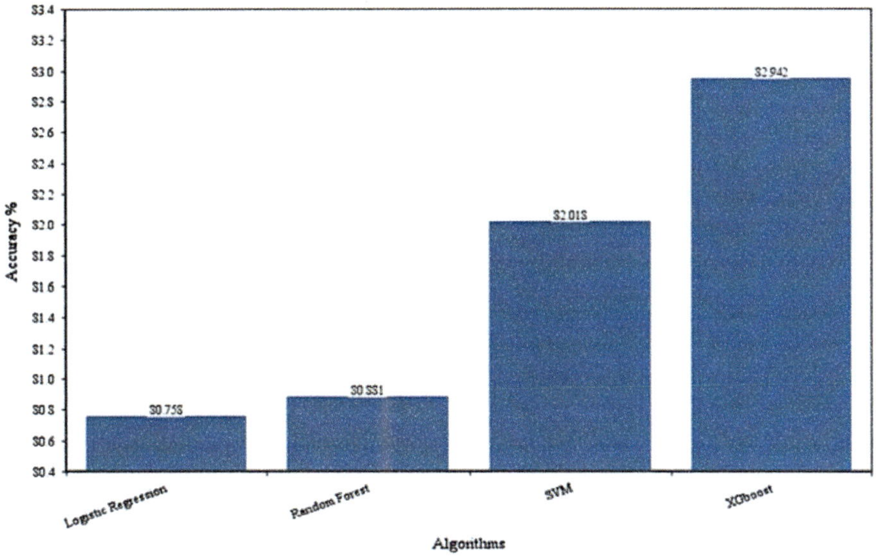

Fig. 14 Accuracy bar graph for IT sector

Table 1 Algorithmic accuracies for all the domains

Domain	Logistic regression	Random forest	SVM	XGBoost
Banking sector	85.562	86.312	85.450	86.305
Telecom sector	80.758	80.881	82.018	82.942
IT sector	90.136	85.714	85.374	87.755

From the Table 1, we can see that the best models for the respective domains are:

1. Banking sector: random forest with accuracy 86.312%
2. Telecom sector: XGBoost with accuracy 82.942%
3. IT sector: logistic regression with accuracy 90.136%.

5.2 *Churn Retention*

5.2.1 Banking Sector

1. Giving customer a competitive and low loan rate, which will encourage them to open an account in the bank.
2. Free credit card offers which will encourage the customers to use the credit card more often and hence increase the activity of their account.
3. Distributing free goodies on opening an account, in this way the customers will be attracted towards the free gifts and open an account, also they will encourage their friends and family to do the same.
4. If a customer has only one asset with the bank, per say only a bank account then it is more likely that they might leave the bank. Hence, we can encourage them to invest in more assets such as mutual funds, fixed deposits so that it will be more difficult for them to leave the bank and hence they will not churn.
5. Making the customer call services hassle-free and solving their problems fast will help the banks in gaining the patience and trust of the banks.

5.2.2 Telecom Sector

1. The prepaid and post-paid plans can be tailored in such a way that the total charges for 2–3 months are available at a discounted price thus luring the subscribers to choose a long-term plan.
2. The correlation analysis shows that there is greater churn percentage in customers who have month-to-month subscription, whereas customers with yearly subscription tend to stay so the strategy will be to successfully upgrade customer's subscription from "month to month" to "year to year". This can be done

by giving them a decent discount at the yearly subscription to convince them to take this leap.
3. Internet service-fast-speed Internet at a reasonable price with better offers than the competitors.
4. Offer providing free message pack on specific talk-time recharges.
5. Encouraging customers to go for post-paid accounts instead of prepaid accounts because in case of prepaid, the customers can churn easily, whereas in post-paid, the customer has to go to the office to unsubscribe from the service.

5.2.3 IT Sector

1. Support for higher education studies.
2. Better wages as compared to the competitors.
3. Giving perks like free transport and food.
4. Making the office more accommodating with vibrant colours and indoor activities for relaxation like gym, games, etc.
5. Making location transfers hassle-free in case of family emergencies.

6 Conclusion

Customer churn prevention is one of the most important factors when it comes to maximizing the profits of any organization. Also known as customer attrition, it occurs when customers stop using products or services of a company.

Through our project, we predicted customer churn beforehand in three major domains, namely banking, telecom and IT. After applying the four machine learning algorithms which are logistic regression, random forest, SVM and XGBoost, we found out that:

- Random forest is the best algorithm for banking sector with accuracy 86.312%
- Logistic regression is the best algorithm for IT sector with accuracy 90.136%
- XGBoost is the best algorithm for telecom sector with accuracy 82.942%.

After predicting the customer churn accurately, the next step is to stop the attrition by various retention strategies which were developed by extensive use of exploratory data analysis and hence will be used for increasing the company's profit.

In future, we can collaborate with various organizations of different domains to help them increase their profit by reducing their customer churn rate, as we now know which algorithm to use for which domain and give retention strategy according to the live stream of data. Also, we can improvise our program according to the feedbacks given by the industry.

References

1. Z. Can, E. Albey, Churn prediction for mobile prepaid subscribers, in *DATA* (2017), pp. 67–74
2. S. Höppner, E. Stripling, B. Baesens, S. vanden Broucke, T. Verdonck, Profit driven decision trees for churn prediction. Eur. J. Oper. Res. (2018)
3. H. Faris, A hybrid swarm intelligent neural network model for customer churn prediction and identifying the influencing factors. Information **9**(11), 288 (2018)
4. A. Cotter, H. Jiang, S. Wang, T. Narayan, M. Gupta, S. You, K. Sridharan, Optimization with non-differentiable constraints with applications to fairness, recall, churn, and other goals (2018). arXiv:1809.04198
5. P. Spanoudes, T. Nguyen, Deep learning in customer churn prediction: unsupervised feature learning on abstract company independent feature vectors (2017). arXiv:1703.03869
6. Y. Yang, Z. Liu, C. Tan, F. Wu, Y. Zhuang, Y. Li, To stay or to leave: churn prediction for urban migrants in the initial period, in *Proceedings of the 2018 World Wide Web Conference on World Wide Web* (International World Wide Web Conferences Steering Committee, 2018), pp. 967–976
7. B. Lengyel, R. Di Clemente, J. Kertész, M.C. González, Spatial diffusion and churn of social media (2018). arXiv:1804.01349
8. Z. Zhang, R. Wang, W. Zheng, S. Lan, D. Liang, H. Jin, Profit maximization analysis based on data mining and the exponential retention model assumption with respect to customer churn problems, in *2015 IEEE International Conference on Data Mining Workshop (ICDMW)* (IEEE, 2015), pp. 1093–1097
9. J. Semrl, A. Matei, Churn prediction model for effective gym customer retention, in *2017 International Conference on Behavioral, Economic, Socio-cultural Computing (BESC)* (IEEE, 2017), pp. 1–3
10. D.F. Benoit, D. Van den Poel, Improving customer retention in financial services using kinship network information. Expert Syst. Appl. **39**(13), 11435–11442 (2012)
11. G. Nie, G. Wang, P. Zhang, Y. Tian, Y. Shi, Finding the hidden pattern of credit card holder's churn: a case of china, in *International Conference on Computational Science* (Springer, Berlin, Heidelberg, 2009), pp. 561–569
12. J. Zhao, X.H. Dang, Bank customer churn prediction based on support vector machine: taking a commercial bank's VIP customer churn as the example, in *2008 4th International Conference on Wireless Communications, Networking and Mobile Computing* (IEEE, 2008), pp. 1–4
13. M. Szmydt, Predicting customer churn in electronic banking, in *International Conference on Business Information Systems* (Springer, Cham, 2018), pp. 687–696)
14. Y. Chen, Y.R. Gel, V. Lyubchich, T. Winship, Deep ensemble classifiers and peer effects analysis for churn forecasting in retail banking, in *Pacific-Asia Conference on Knowledge Discovery and Data Mining* (Springer, Cham, 2018), pp. 373–385)
15. D.A. Kumar, V. Ravi, Predicting credit card customer churn in banks using data mining. Int. J. Data Anal. Tech. Strat. **1**(1), 4–28 (2008)
16. C. Abbet, M. M'hamdi, A. Giannakopoulos, R. West, A. Hossmann, M. Baeriswyl, C. Musat, Churn intent detection in multilingual chatbot conversations and social media (2018). arXiv:1808.08432

Reinforcement Learning-Based Resource Allocation for Adaptive Transmission and Retransmission Scheme for URLLC in 5G

Annapurna Pradhan and Susmita Das

1 Introduction

The fifth-generation (5G) wireless network is the backbone of applications like Internet of things (IoT), intelligent transport system, robotic surgery, augmented reality (AR), virtual reality (VR), industrial automation and control. These applications demand for quality of service like high data rate, reliability, latency and coverage. Therefore, the new service classes of 5G provide a way to achieve above mentioned QoS. Among all the services provided by 5G, ultra-reliable low-latency communication (URLLC) is unarguably innovative with a stringent requirement of low latency (less than 1 ms) and high reliability (in order of 99.999%) for mission-critical applications [1]. Ensuring reliability requires successful data packet transmission. If the packet is lost during transmission, the retransmission of same packet should restore reliability within the limited predefined low-latency bound for URLLC. In these data (re)transmission, it requires allocation of certain amount of radio resources to packets for collision-free reliable transmissions.

The URLLC use cases rely on time-sensitive reliable data transmission. These data packets are generally affected by variable channel conditions. Due to this, there are chances of packet loss or delay in packet arrival. A packet is dropped by the receiver if it does not arrive within the URLLC time limit. Hence, for ensuring reliability, the lost packets are retransmitted. These data transmissions are scheduled with available radio resources. A grant-free reservation-based radio resource allocation [2] for sporadic URLLC traffic is considered here. In case of increased URLLC traffic, a contention-based resource allocation can be useful rather than classical HARQ method [3] because of low-latency criteria (Fig. 1).

A. Pradhan (✉) · S. Das
Department of Electrical Engineering, National Institute of Technology, Rourkela, Rourkela, India
e-mail: pradhan.annapurna37@gmail.com

S. Das
e-mail: sdas@nitrkl.ac.in

S. Patnaik et al. (eds.), *Advances in Machine Learning and Computational Intelligence*, Algorithms for Intelligent Systems, https://doi.org/10.1007/978-981-15-5243-4_13

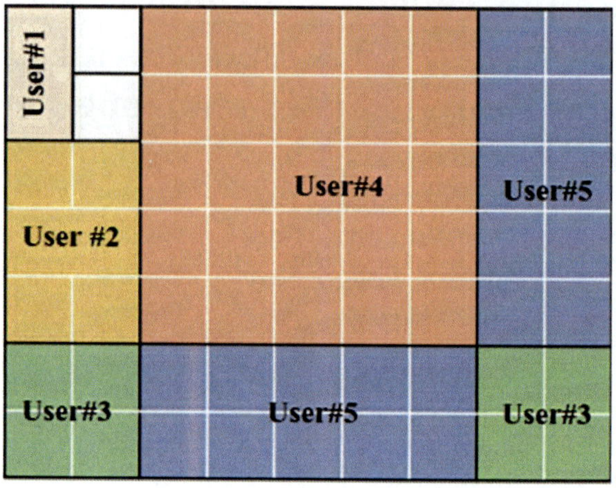

Fig. 1 Resource allocation in 5G

Minimizing the number of radio resource reservation for data retransmission schemes is a complex optimization task because of sporadic random nature of URLLC packet generation. Therefore, an efficient reservation policy based on environment should be adopted. The issues of resource allocation in multiuser scenario have been investigated by several research works including [4, 5]. The authors in [4] discussed the trade-off between feedback less and with feedback transmission frameworks. Resource allocation challenges for cyber-physical systems are handled using dynamic programming by the authors in [6]. Upper layer network management for URLLC traffic and resource allocation is discussed for improving network capacity in [5]. With the advancement of computing technologies, many complex tasks can be handled by machine learning approaches [7]. The capability of using ML for solving various wireless communication challenges has been explained in [8, 9]. Hence, we propose a reinforcement learning (RL)-based algorithm to solve the resource usage optimization problem. The state, action and reward define the learning process in reinforcement learning. Hence, for any given state and corresponding action, there is a reward. The RL agent interacts with the environment and is trained in such a manner that the corresponding action maximizes the average system reward (Fig. 2).

The rest of the paper is organized as follows. Section 2 provides a brief idea about the scope of machine learning in the field of wireless communication. In Sect. 3, the system model and problem formulation for resource allocation are described. A reinforcement learning-based approach for solving the resource allocation problem is provided in Sect. 4. In Sect. 5, the proposed model implementation with the related results is presented. And finally, we concluded the paper in Sect. 6.

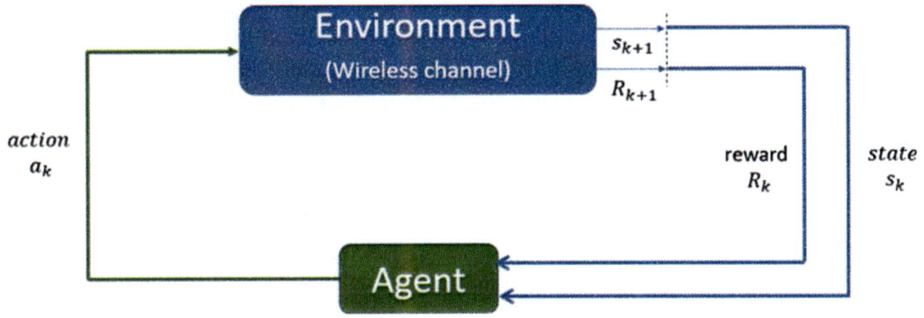

Fig. 2 Reinforcement learning model

2 Role of Machine Learning in Wireless Communication

Machine learning is the key enabler of the future technological revolution. It is capable of solving many complex engineering problems in the field of wireless communication. With the introduction of fifth-generation wireless systems, various applications require the use of machine learning techniques to achieve desired quality of service. The successful application of machine learning in the areas like computer vision, recommendation system and many more attracted both the academia and industry to apply these ML techniques in wireless communication field.

The application of machine learning in resource management, spectral management, mobility management [10], networking, beam-forming methods, localization [11] and routing can achieve significant performance gain in comparison to the traditional algorithm-based solutions and can adaptively operate under dynamic environmental conditions. ML techniques like supervised, unsupervised and reinforcement learning, deep learning [12] are suitable for cognitive radio (CR) applications [13], decision making under uncertainty, classification task [14], self-organizing networks and many more. In all these areas, the machine learning tools can speed up the computation without explicit programming, reduce complexity of design and provide architectural enhancement for future generation of wireless systems.

3 System Model and Problem Formulation

The system model considered here contains a network of base station and user equipment (UE) generating URLLC data traffic. Maximum U URLLC users can be accommodated in the system. We consider a reservation-based resource allocation to a UE for initial transmission and subsequent retransmissions. The number of resource blocks required to transmit a data packet [2] can be determined as, $R = \frac{d}{\eta t b}$, where d is the URLLC packet size in bits, η is the spectral efficiency of used modulation and coding scheme, t is the total time allotted to each resource block and b is the fraction of bandwidth assigned to a RB out of total available bandwidth B.

A single resource unit consists of R number of RBs. If the initial transmission fails, then upon receiving ACK signal, retransmission is initiated. The reserved resources for first transmission should be equal to the number of UEs in the system, as explained in [2]. Then the retransmitted data is scheduled with the corresponding reserved resources. This resource allocation for initial transmission and retransmissions should satisfy the desired reliability target and latency con-strains of URLLC use cases [15]. Therefore, the minimization of total resource blocks required for both initial transmission and retransmissions is defined as per [2]. This complex minimization problem is formulated subject to targeted reliability and latency bound of URLLC, which can be expressed as:

$$\text{Minimization: } \min(U + P)\frac{d}{\eta t b}$$

where the minimization is considered for a total selected modulation and coding scheme. The total reserved resources for initial transmission are U, which is equal to the number of active URLLC UEs. Similarly, P is the number of reserved resources for retransmissions.

4 Reinforcement Learning-Based Solution

A reinforcement learning-based approach is adopted to optimize the resource allocation policy for the mentioned URLLC data packet transmissions and retransmissions (in case of packet loss). The RL agent interacts with the environment, observes state S_k, and performs corresponding action a_k. Then the state transits to the state S_{k+1} providing a reward R_k, to the RL agent.

This learning process leads to the maximization of expected reward based on the trajectory of actions taken (i.e., resource reservation and minimization of drop probability). The neural network is trained to obtain the resource reservation policy. The state action and reward for this learning-based approach are defined as:

- State: State S, at Kth retransmission, depends on packet drop probability subject to URLLC reliability constrain and latency budget T.
- Action: It is finite set of reserved resources P = $(P_1,\ldots, P_k)$, satisfying equation [1], where P_k is the number of resources reserved for Kth retransmission.
- Reward: The reward R_k is defined for the state S_k, while choosing action a_k as the total expected number of successful URLLC data packet transmission.

$$R_k = E\left[\sum_{k=1}^{K} \gamma^k \{(1 - e_k) I\{n \leq P/2\}\right]$$

where e_k is the packet loss probability for Kth retransmission as per [2], n is the number of lost packet after initial transmission and I is the indicator function which is equal to 1 if the condition satisfies and 0 otherwise.

The RL-based system is trained using state-of-the-art Actor-Critic algorithm [16]. At a given state S_k, the agent selects a policy $\pi \leftarrow [0, 1]$, which is the probability of selecting an action ak. Performance evaluation of actions taken is observed from rewards obtained following the above-described policy. The Actor-Critic algorithm uses a policy gradient method to estimate the expected total reward observed from rewards obtained following the above-described policy. The Actor-Critic algorithm uses a policy gradient method to estimate the expected total reward.

5 Performance Evaluation

The reinforcement learning agent is trained using neural network-based action critic model. The architecture of this proposed method is implemented using TensorFlow. The learning rate is set to 0.001. The network assumes the URLLC packets from variable users at different time. A total of 100 RBs are available. The resource block reservation policy depends on current active URLLC users, packet drop probability of initial transmission, reliability target and latency bound. Maximum four retransmissions allowed for maintaining low packet drop probability due to hard latency requirement of URLLC.

We finally evaluate the performance of the proposed RL-based optimal resource reservation policy with the baseline using random resource allocation in retransmission of packets. As shown in Fig. 3, the packet drop probability decreases with increase in number of retransmissions. For first retransmission, there is a small improvement in probability of loss for proposed RL-based solution. But, a significant reduction of packet drop probability is observed for proposed method with respect to random resource allocation, as the number of retransmissions increases

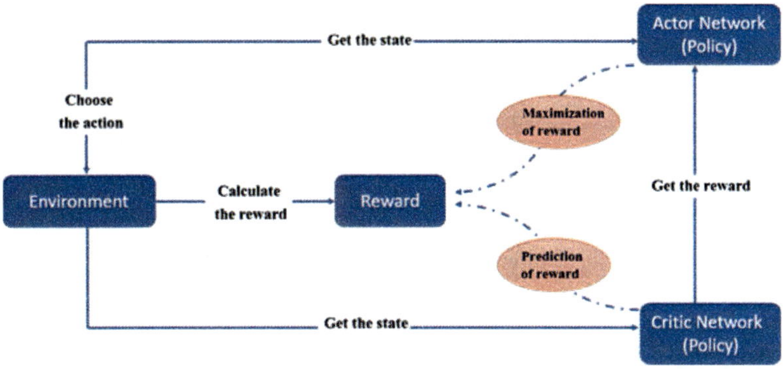

Fig. 3 Actor-critic algorithm-based reinforcement learning model

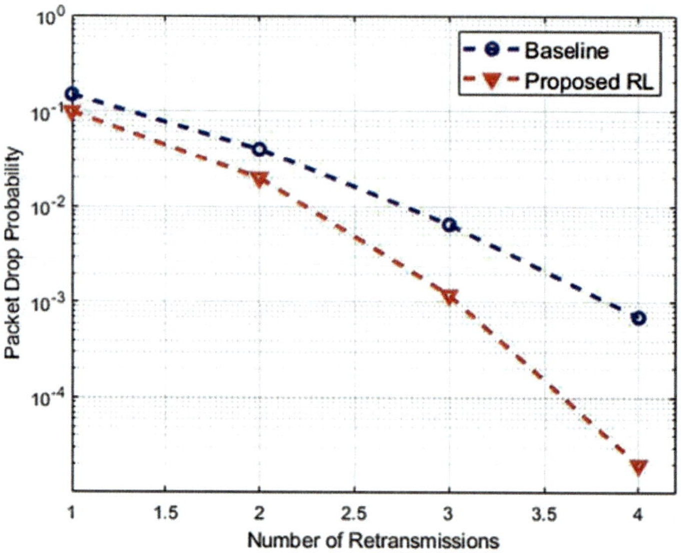

Fig. 4 Comparison of packet drop probability under different retransmissions

to four. Again, the number of retransmissions should be chosen optimally depending on the latency budget of URLLC users. The percentage of resource block usage under variable URLLC traffic load is presented in Fig. 4. The plot shows a significant reduction in the RB usage percentage compared to the baseline policy without reservation (Fig. 5).

6 Conclusion

In this work, we investigated the problem of radio resource reservation policy for URLLC data packet (re)transmission under variable load conditions. A novel reinforcement learning-based algorithm is proposed to solve this complex problem of resource reservation. A state-of-the-art action-critic algorithm is used to train the proposed learning algorithm. The simulation result shows a significant performance gain over the baseline method in terms of packet drop probability and resource usage. In future, we would like to optimize the resource allocation and scheduling of resource data packet (re)transmission for coexistence of multiple services like eMBB, URLLC and mMTC in 5G wireless systems.

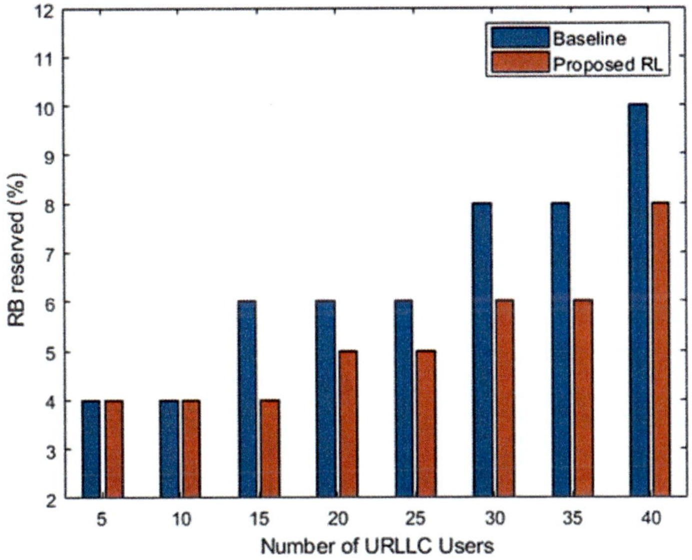

Fig. 5 Percentage of RB reservation versus number of URLLC users

References

1. G. Pocovi, H. Shariatmadari, G. Berardinelli, K. Pedersen, J. Steiner, Z. Li, Achieving ultra-reliable low-latency communications: challenges and envisioned system enhancements. IEEE Netw. **32**(2), 8–15 (2018)
2. S.E. Elayoubi, P. Brown, M. Deghel, A. Galindo-Serrano, Radio resource allocation and retransmission schemes for URLLC over 5G networks. IEEE J. Sel. Areas Commun. **37**(4), 896–904 (2019)
3. S.R. Khosravirad, G. Berardinelli, K.I. Pedersen, F. Frederiksen, Enhanced HARQ design for 5G wide-area technology, in *2016 IEEE 83rd Vehicular Technology Conference (VTC Spring)* (2014), pp. 1–5
4. S.-Y. Lien, S.-C. Hung, D.-J. Deng, Y.J. Wang, Optimum ultra-reliable and low latency communications in 5G new radio. Mob. Netw. Appl. **23**(4), 1020–1027 (2018)
5. E. Khorov, A. Krasilov, A. Malyshev, Radio resource and traffic management for ultra-reliable low latency communications, *2018 IEEE Wireless Communications and Networking Conference(WCNC)* (2018), pp. 1–6
6. A. Vora, K.-D. Kang, Effective 5G wireless downlink scheduling and resource allocation in cyber-physical systems. Technol. **6**(4), 105 (2018)
7. A. Azari, M. Ozger, C. Cavdar, Risk-aware resource allocation for URLLC: challenges and strategies with machine learning. IEEE Commun. Mag. **57**(3), 42–48 (2019)
8. O. Simeone, A very brief introduction to machine learning with applications to communication systems. IEEE Trans. Cogn. Commun. Netw. **4**(4), 648–664 (2018)
9. Y. Sun, M. Peng, Y. Zhou, Y. Huang, S. Mao, Application of machine learning in wireless networks: key techniques and open issues. IEEE Commun. Surv. Tutor. (2019)
10. M. Simsek, M. Bennis, I. Guvenc, Context-aware mobility management in hetnets: a reinforcement learning approach, in *2015 IEEE Wireless Communications and Networking Conference (WCNC)* (2015), pp. 1536–1541
11. D. Mascharka, E. Manley, Machine learning for indoor localization using mobile phone-based sensors (2015). arXiv:1505.06125

12. Z. Qin, H. Ye, G. Y. Li, B.-H. F. Juang, Deep learning in physical layer communications. IEEE Wirel. Commun. **26**(2), 93–99 (2019)
13. K.-L.A. Yau, P. Komisarczuk, P.D. Teal, Applications of reinforcement learning to cognitive radio networks, in *2010 IEEE International Conference on Communications Workshops* (2010), pp. 1–6
14. N. Strodthoff, B. Goktepe, T. Schierl, C. Hellge, W. Samek, Enhanced machine learning techniques for early HARQ feedback prediction in 5G (2018). arXiv:1807.10495
15. A. Anand, G. de Veciana, Resource allocation and HARQ optimization for URLLC traffic in 5G wireless networks. IEEE J. Sel. Areas Commun. **36**(11), 2411–2421 (2018)
16. A. Elgabli, H. Khan, M. Krouka, M. Bennis, Reinforcement learning based scheduling algorithm for optimizing age of information in ultra reliable low latency networks. arXiv:1811.06776 (2018)

Deep Learning-Based Ship Detection in Remote Sensing Imagery Using TensorFlow

Atithee Apoorva, Gopal Krishna Mishra, Rashmi Ranjan Sahoo, Sourav Kumar Bhoi, and Chittaranjan Mallick

1 Introduction

Images have been that unique way of sharing information and knowledge from a long time and unlike human beings, computers are not expected to treat the images and understand it in the same way as humans do. One plausible reason for this could be the way computers accept the images, i.e., they acquire it as a numeric matrix. Hence, the introduction of the concept of computer vision through image processing came into existence through the institutes that had been trying to excel in the field of artificial intelligence [1, 17]. Identifying or detecting small images from the satellite images, and especially doing it through a computer was a tedious task which was still possible in a homogenous environment [2]. However, as already mentioned, this was quite complex, and one possible approach to simplify the whole thing was to apply deep learning.

Deep learning is the popularly emerging technology that helps provide a very precise result in case of problems dealt with by the computer vision [1]. This technology with the aid of its models not only helps to simplify the complexity but also helps do the same in the heterogeneous environment. By the heterogeneous environment, we

A. Apoorva · G. K. Mishra · R. R. Sahoo · S. K. Bhoi
Computer Science and Engineering, Parala Maharaja Engineering College, Luhajhara, India
e-mail: itsatithee@gmail.com

G. K. Mishra
e-mail: gopal.mishra25798@gmail.com

R. R. Sahoo
e-mail: rashmiranjan.cse@pmec.ac.in

S. K. Bhoi
e-mail: souravbhoi@gmail.com

C. Mallick (✉)
Department of Basic Science, Parala Maharaja Engineering College, Luhajhara, India
e-mail: cmallick75@gmail.com

S. Patnaik et al. (eds.), *Advances in Machine Learning and Computational Intelligence*, Algorithms for Intelligent Systems, https://doi.org/10.1007/978-981-15-5243-4_14

mean an environment that not only consists of the homogenous background and the soul object to be detected but also several other objects along with it that can affect the accuracy of the result that is going to be produced.

Ship detection using remote sensing imagery is a very important concept with much significance in today's world. Remote sensing imagery is the two-dimensional representation of the perceived images of the part of the earth's surface visualized from the space. The concept of remote sensing plays a very significant role in solving the segmentation and detection problem of ships [2]. Ship's detection has substantial advantages which promote our idea of enhancing the ship detection methods through our proposal. Ship detection plays a major role in protecting ocean security, nautical transportation, traffic surveillance of ships, fishery management and so on. It also helps in detecting illegal smuggling and in protecting the coasts and seas from the military perspective [3–5].

Ships are large watercraft that helps to travel through the water bodies like ocean and sea to carry passengers, goods, etc. Ships can also be used for protection against illegal fisheries, oil discharge control and the severe pollution levels in the sea. Ship detection has been made possible through several ways almost all ways including the remote sensing [18, 19]. Here in this proposed approach, we have tried a solution through the application of deep learning concepts. We have extensively trained our data using the convolution neural network, to process the SAR images. SAR stands for synthesized aperture radar images which are taken with the help of satellite and have quite a high resolution. Extraction of information from the SAR images which are also known to be panchromatic images [2] helps to gain information regarding the detection of ships.

2 Related Works

Ship detection is a class of problems that covers object detection as well as segmentation. These are studied from the remote images sensed from satellites and are present in sea-based images. This was an easy task if there were only a few or similar types of ships that continued to exist within the waterways. But there exists a huge variety of ships that makes the whole process complex. Not only the types of ships but also the type of background renders its role toward the complexity of the case. The whole scenario was still tangible with homogenous backgrounds where patterns were easily recognizable. But with the heterogeneous background which had other ship-like structures around, in the form of coral reefs, islands and other structures increases the complexity. Thus, due to the huge importance of the detection of ships, there is extensive research being carried out by the researchers. Along with our work, several other related works helped us in many different ways.

In [6], the researchers have used a digital terrain model which is a mathematical representation of the ground surface to mask the land areas. Then the ships were detected by training with data called ERS-1 SAR images, and the detection was done on the basis of human visual interpretation. This was compulsive because they had no other information regarding the ship. The search was made through a wake search, and there was also a homogeneity and wake behaviour test to reduce the falseness of information. In [7], an innovative method known as MSER method (maximally stable extremal region) is followed to pre-screen the image. This is followed by the adoption of the proposed approach which is local contrast variance weighted information entropy (LCVWIE) that aids in evaluating complexity and dissimilarity of the candidate regions with the neighbourhood.

There are many other research works that are being carried out in the recent times with the use of deep learning. One such research was made using Res-Net in [2, 8, 9], choosing which made their proposal at least semiautomatic. They implemented Res-Net using transfer learning which renders quite a high efficiency and accuracy [19, 24, 25]. In a very recent research [10], a method that is based upon deep learning known as the deep neural method is proposed that helps in providing faster and accurate results.

3 Literature Review

3.1 Deep Learning

Deep learning is a huge accumulation of several neural networks, hierarchical probabilistic models and a huge variety of learning algorithms [11, 17]. It happens to be a subspace of the huge area of artificial intelligence that successfully imitates the processing happening in a human brain to perceive something. Deep learning has efficaciously increased and advanced the state of the art in vision speech, etc. [12]. Its evolution has happened hand in hand along with the digital era and deals with a vast amount of data. These data are unstructured, and deep learning helps in learning the relevant information from this cluttered unstructured data. The normal human takes decades in understanding and processing the same amount of data, which deep learning takes manifold lesser time. This efficiently applies the hierarchical artificial neural network to process the data collected, and the whole process is done just like a brain working to solve the issue. This paradigm of artificial intelligence not only works by finding the patterns but also knows where the pattern is implying to and what actions need to follow the detection. Use of deep learning in computer vision has been a huge advantage as it helped simplify the working with the huge amount of data [26]. Earlier, there existed several traditional approaches that did work well like K-nearest neighbour and linear classifiers but the introduction of deep learning has boosted the performance like no other.

3.2 Computer Vision

Computer vision is one of the interdisciplinary fields that play a vital role in dealing with the process of how computers are used to extract the information from the captured high-resolution digital images. This is a separate topic from the normal domain as computers treat the images to be number matrix. Computer vision is that the emergent technology that we find embedded with the different devices we use in our day-to-day life [17, 26]. It aids in object segmentation, in the different games we play, in access control systems, etc. It deals with understanding or making sense from the image and video content automatically. It is something which human brains perform instantaneously, but it is a big deal for computers to do the same. This helps in movement identification, action identification or 3D reconstruction. This is nothing but simply teaching the computer to see and to perceive and extract information from the seen things as we as humans do.

3.3 Automatic Identification System

A tracking system that works automatically with the aid of transponders planted on the ships. The transponders are the devices that receive incoming radio signals and then retransmit some different signals carrying the required information [21, 23]. All of this is carried out in a particular frequency range, and this completely helps in telecommunication. In case of satellites, transponders are interconnected series of communicating channels which helps to establish communication between the sender and the receiver.

3.4 Convolutional Neural Network

The convolutional neural network is an algorithm of deep learning that extensively helps in dealing with the images or high-resolution satellite images. It is a type of artificial neural network that takes out some form of pattern to make sense of those patterns. They work by taking in an image as input, assigning importance like weights and bias to the objects of the images, and thus, detect them differently from one another. Convolutional neural networks have brought revolutionary changes in the computer vision, and when compared to the other deep neural networks, these provide more efficient and effective results [13]. CNN is used in semantic or object segmentation efficiently as a very powerful feature extractor with the ability to learn feature representation in an end to end manner [14].

CNN is an algorithm that constitutes three significant layers that help in achieving its goals to solve the issues. To state these layers, they are (i) convolutional layer (ii) pooling layers and (iii) fully connected layers. Each layer has its role to render with its significance and altogether helps CNN attain its efficiency in working for the solution. The convolution layers help to convolve the whole image and with the help of numerous kernels and feature maps to attain a very fast learning rate that has rendered it to substitute many fully connected layers [11].

The pooling layers render its role in diminishing the dimensions of the input image of spatial origin for preparing the resulting image to be the input for the next convolutional layer. This layer is said to be performing the down-sampling as the diminishing size refers to the loss of information. The last but not the least is the fully connected layers which are layers to the activations of the previous layers and renders its roles in converting the two-dimensional feature maps into one-dimensional feature vector which could be used to be fed into the inputs for classification or further processing [11].

3.5 Remote Sensing Imagery

Remote sensing imagery is the generalized concept of colour imagery [15], and this whole technology has evolved due to the evolution of three interrelated factors [16, 20, 21]. These factors are (i) advancements in sensor technology (ii) improvement and standardization of remote sensing methods and iii) applications of the same in research fields. This, in simple words is a two-dimensional representation of the objects present in the real scenario. These remotely sensed images are nothing but images perceived by the satellites about the surface of the earth form the space. Satellites which cover land mass as well as water bodies. Our paper deals with remotely sensed images covering seas as we are focusing on the ship detection to aid for the ocean security and all the related issues.

4 Proposed Steps

In this section, the paper has proposed steps to undergo to design the model's code to facilitate the detection of the ship which is something that is of huge importance in carrying out maritime surveillance. The whole process has been worked out in the exact flow which has been depicted (See Fig. 1). The individual steps are followed in the rest of this section elaborating the process followed in each step.

There are eight steps stated and have been followed to give results irrespective of the customized input. (See Fig. 1). In step 1, for the image to be used as input and be processed, there is a need to import all the required modules, libraries and packages that aid in doing the needful. Here, in this step, the paper has focused on importing all the modules like Keras, random, NumPy, pandas, matplotlib, PIL, os, sklearn, etc. This helps in providing predefined functions that help in training the machine as well as processing the image to recognize the pattern.

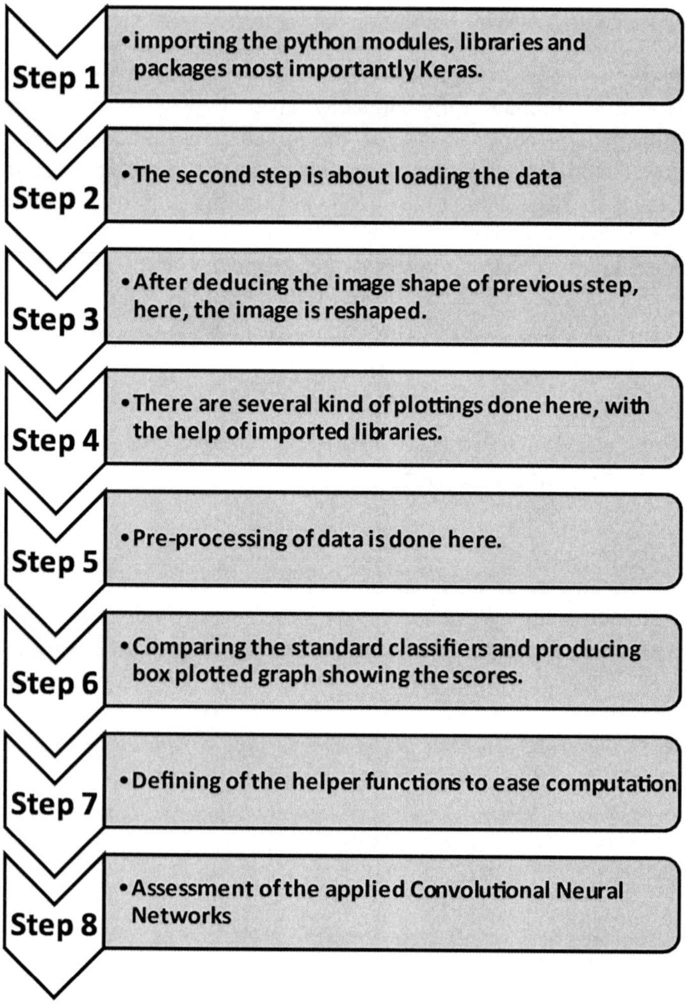

Fig. 1 Steps involved in detection of ship through the Keras model

In step 2; the dataset that is taken here for evaluating the model is loaded, and this is achieved by the help of load() function that helps in loading the JSON file. After loading the data, it is properly described by defining a function that prints the number of images and relevant information regarding the initial shape and ship or no ship condition. This is followed by step 3 which reshapes the data. Reshaping is sometimes required to manipulate the image, and thus, easing the complexity (Fig. 2).

After this, there is the next step which is step 4 that concentrates on several types of plotting, these plotting are aided by the matplotlib import and here in this phase there occurs three kinds of plotting.

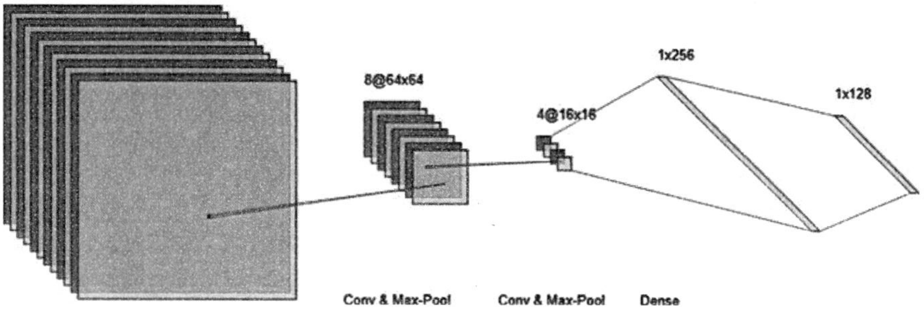

Fig. 2 CNN model

1. Plotting of a single NumPy array denoting the instance of the labels which were either ship conditioned label or no ship conditioned label.
2. Plotting of several NumPy arrays sorted according to their labels.
3. Plotting a histogram graph contemplating the RGB pixel intensities.

This was followed by preprocessing of the input image in step 5 where after the plotting of the data, train_test_split is imported from the sklearn. Cross_validation and then the sample size are reduced that aids n debugging. After this, there follows the next step which is step 6 which aids in comparing all the standard classifiers. In this process, a list of standard classifiers is compared, and a comparison is box plotted to show the level of accuracy of each classifier. The list of standard classifiers is as follows:

1. Logistic regression
2. Random forest classifier
3. K-neighbor classifier
4. Support vector machine
5. Linear SVC
6. Gaussian NB
7. Decision tree classifier
8. XGB classifier.

This produces a graph which is a box plotted with standard classifiers against the cross-validation accuracy score in the x- and y-axis, respectively. The next step is step 7 that is all about defining helper functions. These are the functions that aid in computing for some other function and make the complexity easier by promoting easier readability of the programs. These computations which are performed by the helper functions can be reused too. In this step, the first phase is about plotting a confusion matrix with either normalization = false or normalization = true. This is completely at your discretion to apply normalization or not. After plotting the graph, this method also helps in plotting the Keras learning curve which is a tool to diagnose the learning performance over time. Along with the Keras learning curve, it also targets in plotting and summarizing the loss of the model.

By the completion of the previous step, commences the next and the final stage which is step 8 that corresponds to the evaluation of the applied convolutional model. In this step, there occurs running of the CNN augments, max pooling, dropping out, flattening and other operations that help in generating graphs of model accuracy and model loss after getting calculated in each epoch. Also, there appears a plotted confusion matrix at the end of this step.

5 Training and Results

After the above work -flow to be followed, during designing the Python codes, let us now properly say how the model was trained and the resulting values. The model was trained for the dataset which consisted of image chips. Each image chip was of size 2800 80 * 80 RGB images and was labeled. The images had clear labels of "Ship" or "No Ship" for getting sorted later after getting represented as numpy arrays.

The entire dataset is zipped and contains.png format images and is made available through Planet's open California dataset that is completely open licensed. This model with the given dataset was trained for 12 epochs, and the package sklearn was aiding to realize each applied algorithm. This process was done without batch normalization, and certain effects appeared due to feature engineering.

Effects on accuracy score with differing standard classifiers:
The comparison among several standard classifiers gave in as a result several accuracy scores that is represented in Table 1.

After the pltshow (), the above accuracy scores were produced on the basis of estimating skill of the model. After this, a histogram graph is produced for the same scores which is shown in Fig. 3.

This application of cross-validation to produce the accuracy score gives the solution to the overfitting problem. The training then continues to be done through the CNN model, and for all a total of 12 epochs and the observed results are stated in the table below names to be Table 2.

Table 1 The relative accuracy score of the respective compared standard classifier

Algorithm	Cross-validation accuracy
Logistic regression (LR)	0.866
Random forest classifier (RF)	0.928
K-nearest neighbor classifier (KNN)	0.916
Support vector machine (SVM)	0.745
Linear SVM (LSVM)	0.870
Gaussian NB (GNB)	0.642
Decision tree classifier (DTC)	0.892
XGB classifier (XGB)	0.946

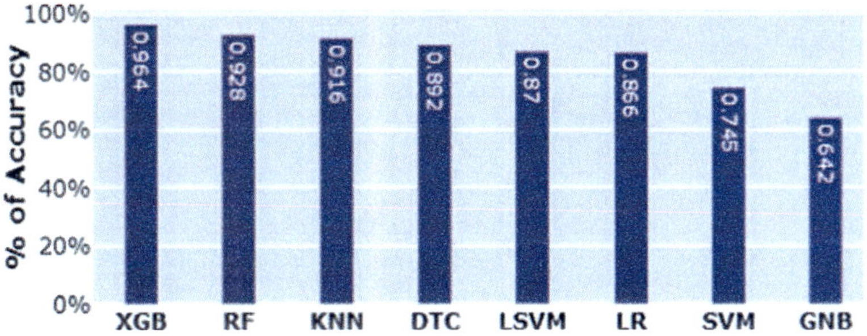

Fig. 3 Chart of histogram type showing the trend of accuracy score of each standard classifier

Table 2 Accuracy and loss before and after validation for each epoch

Epochs	Loss	Acc	Val_loss	Val-acc
I	0.6601	0.7835	0.3115	0.8839
II	0.3416	0.8647	0.3376	0.875
III	0.3082	0.8665	0.2004	0.9
IV	0.2627	0.8839	0.1788	0.9125
V	0.2554	0.8902	0.1669	0.9196
VI	0.2266	0.9054	0.1802	0.9125
VII	0.2146	0.908	0.1558	0.9214
VIII	0.1869	0.9096	0.1559	0.9196
IX	0.2006	0.9161	0.1232	0.95
X	0.1703	0.925	0.1305	0.9411
XI	0.1842	0.9214	0.1446	0.9321
XII	0.1542	0.9348	0.1493	0.9357

The line graph in Fig. 4 shows the trend that was observed in the variation of the scores of accuracy and loss before and after the validation.

6 Confusion Matrix

To describe a given classification model, the confusion matrix is the best table that on a given dataset gives the known values. In this model, a confusion matrix was generated in step 7 of the previous section, and for the applied two-dimensional model, the generated confusion matrix looks like something in Fig. 5.

In Fig. 5, a confusion matrix is plotted by the help of the plot_confusion_matrix() for two labeled classes, Ship and No ship. The applied model made a total of 560 predictions where ship class denoted the presence of the ship in the image chip and no Ship depicted the absence of ship. Out of the 560 prediction, 158 predictions

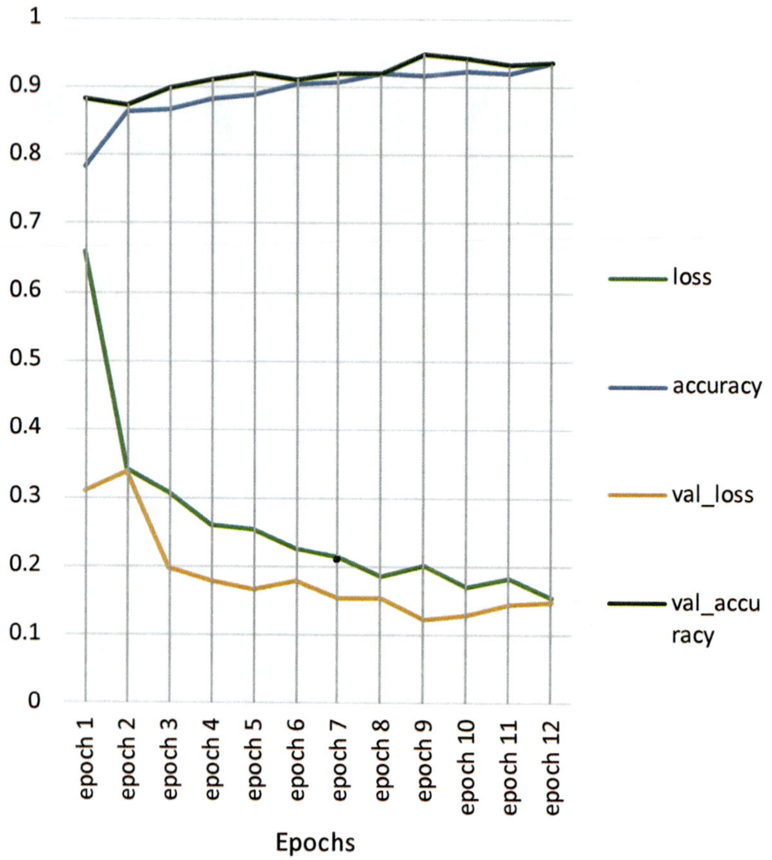

Fig. 4 Variation of accuracy and loss before and after validation in each epoch based upon the data mentioned in Table 2

predicted ship condition denoting the presence of ship and 402 predictions about the absence. However, in reality, out of the 560 predictions, 430 predictions were no ship depiction and only 130 were ship condition.

7 Disadvantages and Advantages of the Proposed Scheme

This whole experiment could help us in many ways but could produce only an accuracy slightly more than 90%, and hence, it would be a slight demerit of this model. Apart from this, there exists every other advantage that is given to us by this model. To summarize the advantages, it produces a result with 90% accuracy and a graph showing the model accuracy, loss functions and produces a confusion matrix that helps us diagnose the error in our model. This model not just helps in applying a

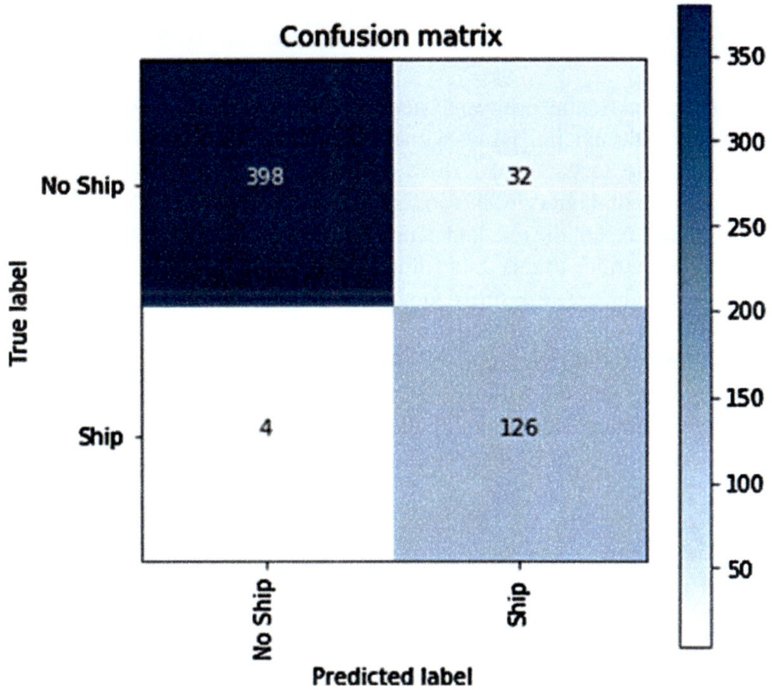

Fig. 5 The confusion matrix for the predictions made

given convolutional neural network but also helps in comparing standard classifiers for any given dataset which is a huge plus point of this model.

8 Conclusion

Ship detection is a very important aspect that requires advanced detecting technologies from time and then. Waterways are that means through which maximum of import–export trafficking is carried out, and also, in recent years, there is a huge increase in ocean trafficking that requires ship detection techniques from high-resolution satellite images. This would help in a manifold amount to overcome many issues endangering the ocean security as well as help in maritime surveillance. Through this paper, a step flow of a program that is flexible with your own devised CNN model with the number of layers at your discretion has been proposed. Also, proper plotted graphs relating to accuracy and loss before and after validation have been shown basing upon the data mentioned in the tables. The flexibility of having any CNN model as per your defined levels of accuracy would be working here and that would completely help the sole motive of this paper, which is ship detection.

9 Future Scope

As mentioned already, the proposed steps and the results obtained have tried to overcome many of the existing issues with the use of deep learning in computer vision. As mentioned in the Sects. 1 and 2, there were several traditional approaches which deep learning has combatted in due course of time along with its newly developed algorithms. Also, CNN or the residual neural network applications with several layers are not the only solutions to this. Our future scope would involve a statistical review and comparison of the existing ship detection techniques through the algorithms and would focus on devising customized filters that would increase the overall efficiency. Our prime motive would be to render reduced complexity and increased efficiency and directly recognizing the patterns to categorize them into the respective class of problems. For example, recognizing a pattern and thus, categorizing the problem into commercial, economical or military importance would help for faster prevention of the incoming problem.

References

1. J.S. Thanakumar, R.S. Edwin, J. Sasikala, J.D. Sujitha, Smart Vessel detection using deep convolutional neural network, in *2018 Fifth HCT Information Technology Trends (ITT)* (2018)
2. R.J. Richa, *Smart Ship Detection using Transfer Learning with ResNet* (2019)
3. Z. Shao, W. Wu, Z. Wang, W. Du, C. Li, SeaShips: a large-scale precisely annotated dataset for ship detection. IEEE Trans. Multimed. (2018)
4. C. Zhu, H. Zhou, R. Wang, J. Guo, A novel hierarchical method of ship detection from spaceborne optical image based on shape and texture features. IEEE Geosci. Remote Sens.Lett. **48**(9), 3446–3456 (2010)
5. S. Sheng, J. Ma, J. Lin, et al., Unsupervised ship detection based on saliency and S-HOG descriptor from optical satellite images. IEEE Geosci. Remote Sens. Lett. **11**(3), 641–645 (2014)
6. K. Eldhuset, An automatic ship and ship wake detection system for spaceborne SAR images in coastal regions. IEEE Trans. Geosci. Remote Sens. (1996)
7. W. Huo, Y. Huang, J. Pei, Q. Zhang, Q. Gu, J. Yang, Ship detection from ocean sar image based on local contrast variance weighted information entropy. Sensors (2018)
8. A Gentle Introduction to Transfer Learning for Deep Learning by Jason Brownlee on December 20, 2017
9. An Overview of ResNet and its Variants by Vincent Fung
10. S.I. Joseph, J. Sasikala, D.S. Juliet, Detection of ship from satellite images using deep convolutional neural networks with improved median filter, in *Artificial Intelligence Techniques for Satellite Image Analysis* (2020)
11. A. Voulodimos, N. Doulamis, A. Doulamis, E. Protopapadakis, Deep learning for computer vision: a brief review. Comput. Intell. Neurosci. (2018)
12. S. Ioffe, C. Szegedy, Batch normalization: accelerating deep network training by reducing internal covariate shift (2015)
13. Z. Zuo, B. Shuai, G. Wang, X. Liu, X. Wang, B. Wang, Y. Chen, Convolutional recurrent neural networks: learning spatial dependencies for image representation, in *Proceedings of the IEEE Conference on Computer Vision and Pattern Recognition Workshops* (2015)
14. V. Iglovikov, S. Mushinskiy, V. Osin, Satellite imagery feature detection using deep convolutional neural network: a kaggle competition (2017)

15. B. Kartikeyan, A. Sarkar, K.L. Majumder, A segmentation approach to classification of remote sensing imagery. Int. J. Remote Sens. **19**(9), 1695–1709 (1998)
16. J. Rogan, D. Chen, Remote sensing technology for mapping and monitoring land-cover and land-use change. Progr. Plan. **61** (2004)
17. K. Simonyan, A. Zisserman, Very deep convolutional networks for large-scale image recognition (2014)
18. T.N. Hannevik, Ø. Olsen, A.N. Skauen, R. Olsen, Ship detection using high resolution satellite imagery and space-based AIS, in *2010 International WaterSide Security Conference* (2010)
19. A. Bauer, J. Ball, J. Colmer, S. Orford, S. Griffiths, J. Zhou, Combining computer vision and deep learning for high-throughput aerial phenotypic analysis in wheat pre-breeding
20. J.A. Carballo, J. Bonilla, M. Berenguel, J. Fernández-Reche, G. García, New approach for solar tracking systems based on computer vision, low cost hardware and deep learning. Renew. Energy (2019)
21. F. Hu, G.S. Xia, J. Hu, L. Zhang, Transferring deep convolutional neural networks for the scene classification of high-resolution remote sensing imagery. Remote Sens. (2015)
22. Y. Chen, J. Zheng, Z. Zhou, Airbus ship detection-traditional versus convolutional neural network approach
23. K. Kang, W. Ouyang, H. Li, X. Wang, Object detection from video tubelets with convolutional neural networks, in *Proceedings IEEE Conference in Computers Vision and Pattern Recognition* (2016), pp. 817–825
24. X. Wang, A. Shrivastava, A. Gupta, A-fast-RCNN: hard positive generation via adversary for object detection, in *Proceedings IEEE Conerence in Computers Vision and Pattern Recognitio*n (2017), pp. 3039–3048
25. S. Ren, K. He, R. Girshick, J. Sun, Faster R-CNN: towards real-time object detection with region proposal networks, IEEE Trans. Pattern Anal. Mach. Intell. **39**(6), 1137–1149 (2017)
26. K. Alex, et al., ImageNet classification with deep convolutional neural networks. Neural Inf. Process. Syst. 1097–1105 (2012)

Modern Approach for Loan Sanctioning in Banks Using Machine Learning

Golak Bihari Rath, Debasish Das, and BiswaRanjan Acharya

1 Introduction

Loan analysis is a process adopted by banks used to check the credibility of loan applicants who can pay back the sanction loan amount within regulations and loan amount term mentioned by the bank. Most banks use their common recommended procedure of credit scoring and background check techniques to analyze the loan application and to make decisions on loan approval. This is overall a risk-oriented and a time-consuming process. In some cases, people suffer through financial problems while some intentionally try to fraud. As a result, such delay and default in payment by the loan applicants can lead to loss of capital of the banks. Hence to overcome this, banks need to adopt a better procedure to find the trustworthy applicants for granting loan from the list of all applicants applied for the loan, who can pay can their loan amount in stipulated time [1].

In the modern-day age and advance of technology, we adopt a machine learning approach to reduce the risk factor and human errors in the loan sanction process and determine where an applicant is eligible for loan approval or not. Here, we examine various features such as applicant income, credit history, education from past records of loan applicants irrespective of their loan sanction, and the best features are determined and selected which have a direct impact on the outcome for loan approval [2].

G. B. Rath (✉) · D. Das
Department of Computer Science & Engineering, Centurion University of Technology & Management, Bhubaneswar, Odisha, India
e-mail: golakk.2009@gmail.com

D. Das
e-mail: debasish.das@cutm.ac.in

B. Acharya
School of Computer Engineering, KIIT Deemed to University, Bhubaneswar, Odisha, India
e-mail: acharya.biswa85@gmail.com

S. Patnaik et al. (eds.), *Advances in Machine Learning and Computational Intelligence*, Algorithms for Intelligent Systems, https://doi.org/10.1007/978-981-15-5243-4_15

Various machine learning algorithms such as logistic regression, decision tree, and SVM have been tested, and their results have been compared. The performance of logistic regression was found more than other models, and hence, it was assumed that it could be used as a predictive model which could predict future payment behaviors of the loan applicants. Thus, the bank could adopt this model for loan sanction process whenever new applicants apply for a loan, and the loan can be processed instantly with minimum time and reduced risk.

2 Literature Review

Zurada J. (2002) found ensemble model performs better in comparison with other data mining techniques by exploring the application details of both paid and defaulters [1]. Turkson et al. (2016) analyzed the performance of 15 different classification methods with 23 different features of applicant data and found that linear regression was used to formulate as the final model [2]. Vaidya, A. (2017) found out taking more number of attributes will result in the model learning better using logistic regression techniques by statistically analyzing the distribution of features and prediction of the model [3]. The work proposed by Hamid and Ahmed (2016) depicts the data mining process for loan classification using Weka application. Algorithms such as j48, Bayes Net, and Naive Bayes were used for model creation. Results showed j48 had the highest accuracy and low mean absolute error so considered the best suited for prediction [4]. Sivasree and Rekha Sunny (2015) used the Weka explorer tool for data exploration and implementation of using decision tree induction algorithm to find relevant attributes in decision making. ASP.NET-MVC5 was used as the platform for implementing the model into the application for use [5]. The authors Arun et al. (2016) explained the application of six algorithms applying parameter setting using R open-source software [6]. Shin et al. (2005) presented the prediction of the bankruptcy prediction model using support vector machine and good classifier to attain correct prediction performance from smaller sets [7]. Salmon et al. (2015) measured the performance scores of different classifiers using the evaluation technique of the confusion matrix [8]. Panigrahi and Palkar (2018) used various feature selection techniques for various models to determine fraud claims and found random forest model has the best accuracy and precision, whereas decision tree has the best recall using various feature selection methods applied on the dataset [9]. Soni and Paul (2019) compared the results of the model in R and Weka and found Weka results were better for making an optimized random forest classifier [10].

Arutjothi and Senthamarai (2017) proposed a K-NN credit scoring system using where the error rate is recorded minimum when the iteration level is increased [11]. Priyanka and Baby (2013) suggested a Naive Bayesian algorithm for classifying a customer according to posterior probability using records of customers in banks [12]. Sudhamathy G. (2016) implemented R package for visualization using data mining techniques. For the prediction of labels, the tree model was used [13]. Eweoya et al.

(2019) found that incorrect predictions can be reduced with using stratified cross-validation on a decision tree model [14]. Jency et al. (2019) preferred using an exploratory data analysis approach for graphical representation and understanding of different features of the people. It was found that short-term loans were preferred more and people having a home as mortgage apply more for a loan [15]. Rawate and Tijare (2019) used a Naive Bayes approach along with a combination of various algorithms such as K-NN, binning for consistency in dataset and improvement in the accuracy rate [16]. Priya et al. (2018) concluded that loan proposals of people with good credit history with high income have a better chance of approval [17]. Torvekar and Game (2019) experimented with various classifiers on two different datasets, and classifiers performance is reduced when operating with a large number of features [18]. Singh and Chug (2017) suggested that the calculation of the error rate metric is needed to find the best algorithms whenever various algorithms have the same accuracy rate. The linear classifier was found best for determining a software defect using tenfold cross-validation techniques [19].

In our paper, we try to use a similar machine learning approach exploring 12 different features of a banking dataset which affect the loan approval directly in some manner. We then apply feature selection techniques on the existing dataset to manually select those features based on weighting scores which contribute most to predicting a category or outcome using certain techniques of feature selection. We use the scikit-learn library which provides the SelectKBest class using the chi-squared (chi^2) test to select a specific number of features [20]. We select seven relevant features from 11 different attributes of previous loan applicants. Finally, the selected best features are fed into classification models for the classifying the outcome.

3 Proposed Model

To achieve our objective, we train the system with two parameters. First parameter is the predictive features referred as independent variables, and the other referred as categorical class for those variables. This system is a predictive model to determine the class as yes for approve and no for disapproval of loan dataset of loan applicants. Data is collected from a banking dataset containing recent records of applicants whose loan application has been either approved or disapproved. Supervised machine learning techniques are performed on that historical data to create a proposed model which can forecast the identification of loan applicant's repayment. We used scikit-learn library which supports Python functions in this process to visualize the distribution of data features and creating a classification model [20]. As this is classification problem of supervised learning, so we use algorithms such as logistic regression, decision tree, and SVM which can be best effective to solve the classification problem by categorizing the class for loan applied by the applicant is either risky or safe.

We implement a two-step process: model learning and model prediction. In model learning, a model is developed by using different classifier functions for the construction of a classification based on given training data. In model prediction, the model is used to predict the response for given data. Initially, we train a dataset with a set of features as input or predictor variables, and with an outcome variable Loan Status. All models are constructed and trained for classification with the help of classification algorithms. The logistic regression model is finally used as the best model for classification after the evaluation results of the prediction.

3.1 Logistic Regression

It is a statistical-based algorithm that determines the probability of an event as a function of variables using classification function. The classification function calculates the statistics for a logistic response function. It shows the relationship between the dependent variable and independent variables [1].

$$P(y) = 1/1 + \mathrm{e}^{-z} \tag{1}$$

Here $P(y)$ is our result which is determined with the help of dependent variable y, where z represents the function of independent variables used in the dataset. The range of values that $P(y)$ predicts is from 0 to 1 which helps us to identify our category as no or yes as results [1].

3.2 Decision Tree

Decision tree uses the tree representation to produce a model with the most relevant predictor attributes. Attribute with the highest-ranking attribute of the dataset is placed as the root node, and other attributes are placed to the leaf node. At each node, a decision is made where leaf nodes give us the final result. The tree building is continued until we get an outcome in the internal nodes. The overall results are calculated in the form of decisions, and the final decisions constitute our category [5].

3.3 SVM

Support vector machine is a classifier which is represented by a line that splits the data between the two differently classified groups of data representing the training data into two-dimensional planes as data points. We plot the features of the dataset in either side of the plane forming two different categories. When the testing data lands on either side of the line, we can classify the data for approval as yes or no [7].

4 Experiments and Results

Data is collected in the form of banking credit dataset containing records of past loan applicants from the UCI machine repository. There is a definite set of inputs defined as independent variables and a corresponding output referred as dependent variable or class. As the class is binary in nature, so we find a solution to the binary classification problem by adopting various classification techniques of machine learning such as decision tree, logistic regression, and SVM with the features of the dataset, and the best performance is calculated among them to create a predictive model. We use Python with the help of reading the data, exploring with visualizing different features of an existing dataset, and implementation of the model for prediction.

We implement our methodology with the processes of data exploration, data preprocessing, feature selection, model training, and model evaluation. Initially, we extract the data from CSV dataset file to pandas data frames using pandas class with read_csv() function. After reading the records and having a general understanding about various roles of different variables in the dataset, we prepare the data for building a model and prediction of results using various functions supported by scikit-learn [20].

4.1 Data Exploration

Before making a prediction model, we try to have a basic idea about both categorical and numeric data of the current dataset containing a vast number of data. The data in tabular format is displayed in graphical format by exploring the variables for calculation of frequency. Frequency of categorical variables is represented in Fig. 1 as a bar chart.

Here, we observe that the maximum number of applicants applied for the loan are male, graduates, employed, and married. Visualization of numerical variables like loan amount and applicant income is displayed in Figs. 2 and 3.

We can see that there is a considerable amount of outliers in the numerical variables along with missing values in categorical values which can be resolved in the next phase of preprocessing.

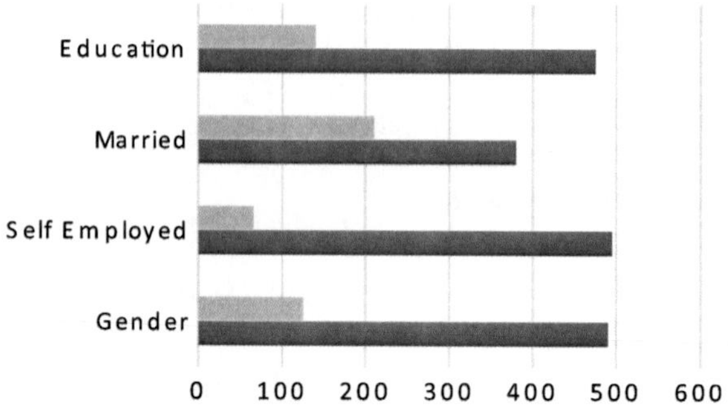

Fig. 1 Distribution of categorical values in chart

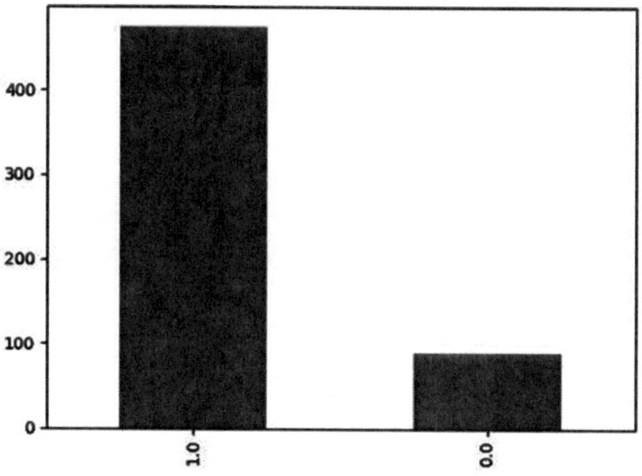

Fig. 2 Bar chart displaying credit history with 1 as yes and 0 for no

4.2 Data Preprocessing

The preparation and cleansing of the dataset are performed in this processing phase. It includes data cleansing which involves detecting and correcting (or removing) corrupt or inaccurate records from the current dataset. We check the missing data and fill them with mean or median values. Next, we perform numeric conversion of all categorical data types. The scikit-learn library supports only numerical variables during processing, hence we convert all categorical variables into numerical data types by encoding. We check that our data is now complete and numeric in nature which is suitable for other processing phases [20].

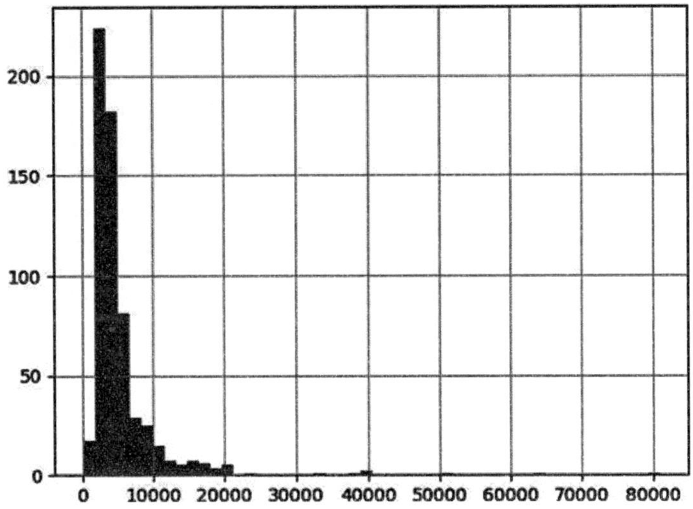

Fig. 3 Histograms displaying annual income of applicants

4.3 Feature Selection

When all features are not related to the outcome of a class, hence we select the relevant features with the help of technique such as feature selection. It helps in increasing the interpretability of the model and reducing the complexity and the training time of the model. Feature selection is performed by a class called SelectKBest provided with the scikit-learn library to identify the most important features which impact the loan approval results. This class uses a chi-squared (chi^2) test to determine the relative importance of independent features with the outcome variable [20].

In Table 1, we can see the output of the test and find seven dominant features scores of variables related to target variable 'Loan_Status'. So accordingly we choose some predictors in order of their weight to fit into the model.

Table 1 Feature score of variables

Feature	Score
Applicant income	5342.194844
Co-applicant income	4988.307182
Loan amount	110.4371
Credit history	19.617746
Married	2.132101
Education	1.793838
Dependents	0.806228

4.4 Training a Model

Once all predictive variables are selected, we need a test dataset for evaluating the performance. Hence, we split the current dataset into two sets, one for training and the other for testing. We randomly split the data with the help of scikit library supported train_test_split function. Thus, random sampling is performed using train_test_split function with three parameters such as predictors, outcome variable, and test size [20]. Our existing dataset contains 615 applicants. We split this dataset into the training and the testing datasets in certain proportions with the test size parameter. We split this dataset into the training and the testing datasets in certain proportions like 70% of the data for training the model and 30% of the data as testing dataset. We store predictors in x and the target variable in y separately in an array. We try with a different set of instances and predictors using different classification algorithms. First, we create an object of these classifiers to build the model by using its classification function for the algorithm and then fit the model with our parameters. Now the model starts to learn from the training data and is ready to predict for the testing dataset. Using the predict function, we predict and compare the results.

4.5 Model Performance Evaluation

We used a confusion matrix as a model evaluation metric to find the best and effective model among all models we tested. Here confusion matrix is shown as a 2×2 matrix, having $N = 2$ where two classes are being predicted as approve or disapprove. It results in showing the number of predicted values with the actual values. By continuous test on the datasets, the best model is selected in terms of the score of its classification report.

Performance of each classification algorithm is tested by the measure of its classification report function supported by sklearn [20]. Accuracy is the correct predictions made by the predictive model, whereas precision is determined on the numerical proportion of all predictions that we made with our predictive model are true in nature. Recall gives us an idea about total predictions correctly identified, while F1-score represents the harmonic mean of precision and recall [8]. All actual and predicted values are calculated, and thus metric scores are depicted for each model. These evaluation metrics scores help to determine our final model which is reliable for making predictions.

On the basis of the confusion matrix scores for each model, the evaluation metrics are calculated. Table 2 describes scores of confusion matrix for models.

$$\text{Accuracy} = (\text{TP} + \text{TN})/(\text{TP} + \text{TN} + \text{FP} + \text{FN}) \tag{2}$$

$$\text{Recall} = \text{TP}/(\text{TP} + \text{FN}) \tag{3}$$

Table 2 Results of confusion matrix for each model

Algorithm	True positive	True negative	False positive	False negative
Logistic regression	21	93	1	28
Decision tree	30	74	20	20
SVM	2	90	4	48

Table 3 Evaluation scores of model

Algorithm	Accuracy score	Precision	Recall	F1-score
Logistic regression	0.79	0.83	0.80	0.77
Decision tree	0.72	0.72	0.72	0.72
SVM	0.64	0.54	0.64	0.63

$$\text{Precision} = \text{TP}/(\text{TP} + \text{FP}) \tag{4}$$

$$\text{Score} = (2 * \text{Recall} * \text{Precision})/(\text{Recall} + \text{Precision}) \tag{5}$$

Table 3 shows the resultant values of metrics for each of the algorithms. Based on metrics score, the best scoring model is selected among them to be used as predictive model.

After continuous test using various combinations of predictors, we found the performance of the logistic regression model has accuracy over 79% with a recall of 80%. After the final evaluation of all the models, we see that the accuracy of the logistic regression model is higher compared to other algorithms used. Hence, we can use the logistic regression model as the final predictive model.

5 Conclusion and Future Work

We find that the accuracy of the logistic regression model has a prediction accuracy of nearly 80% which is more than the performance of other models. Further, we can use this model as a prediction model to test with the data of the applicants. The model can provide fast, reliable approach in decision making which can be an alternative to the current procedures adopted by banks for processing loan approval of an applicant.

We can change the prediction variables used in the training of the model for gaining better accuracy and performance of the model. Further, maximum performance in predictability can be achieved through machine learning tools by tuning the variables and implementation with other classification algorithms.

References

1. J. Zurada, Data mining techniques in predicting default rates on customer loans. Databases Inf. Syst. II, 285–296 (2002). https://doi.org/10.1007/978-94-015-9978-8_22
2. R.E. Turkson, E.Y. Baagyere, G.E. Wenya, A machine learning approach for predicting bank credit worthiness, in *2016 Third International Conference on Artificial Intelligence and Pattern Recognition (AIPR)*. (2016). https://doi.org/10.1109/icaipr.2016.7585216
3. A. Vaidya, Predictive and probabilistic approach using logistic regression: application to prediction of loan approval, in *2017 8th International Conference on Computing, Communication and Networking Technologies (ICCCNT)* (2017). https://doi.org/10.1109/icccnt.2017.8203946
4. A.J. Hamid, T.M. Ahmed, Developing prediction model of loan risk in banks using data mining. Mach. Learn. Appl. Int. J. **3**(1), 1–9 (2016). https://doi.org/10.5121/mlaij.2016.3101
5. M.S. Sivasree, T. Rekha Sunny, Loan credibility prediction system based on decision tree algorithm. Int. J. Eng. Res. Technol. **V4**(09) (2015). https://doi.org/10.17577/ijertv4is090708
6. K. Arun, G. Ishan, K. Sanmeet, Loan approval prediction based on machine learning approach. IOSR J. Comput. Eng. **18**(3), 18–21 (2016)
7. K.-S. Shin, T.S. Lee, H.-J. Kim, An application of support vector machines in bankruptcy prediction model. Expert Syst. Appl. **28**(1), 127–135 (2005). https://doi.org/10.1016/j.eswa.2004.08.009
8. B.P. Salmon, W. Kleynhans, C.P. Schwegmann, J.C. Olivier, Proper comparison among methods using a confusion matrix, in *2015 IEEE International Geoscience and Remote Sensing Symposium (IGARSS)* (2015). https://doi.org/10.1109/igarss.2015.7326461
9. S. Panigrahi, B. Palkar, Comparative analysis on classification algorithms of auto-insurance fraud detection based on feature selection algorithms. Int. J. Comput. Sci. Eng. **6**(9), 72–77 (2018). https://doi.org/10.26438/ijcse/v6i9.7277
10. P.M. Soni, V. Paul, A novel optimized classifier for the loan repayment capability prediction system, in *2019 3rd International Conference on Computing Methodologies and Communication (ICCMC)* (2019). https://doi.org/10.1109/iccmc.2019.8819772
11. G. Arutjothi, C. Senthamarai, Prediction of loan status in commercial bank using machine learning classifier, in *2017 International Conference on Intelligent Sustainable Systems (ICISS)* (2017). https://doi.org/10.1109/iss1.2017.8389442
12. L.T. Priyanka, N. Baby, Classification approach based customer prediction analysis for loan preferences of customers. Int. J. Comput. Appl. **67**(8), 27–31 (2013). https://doi.org/10.5120/11416-6752
13. G. Sudhamathy, Credit risk analysis and prediction modelling of bank loans using R. Int. J. Eng. Technol. **8**(5), 1954–1966 (2016). https://doi.org/10.21817/ijet/2016/v8i5/160805414
14. I.O. Eweoya, A.A. Adebiyi, A.A. Azeta, A.E. Azeta, Fraud prediction in bank loan administration using decision tree. J. Phys: Conf. Ser. **1299**, 012037 (2019). https://doi.org/10.1088/1742-6596/1299/1/012037
15. X. Jency, V.P. Sumathi, J. Sri, An exploratory data analysis for loan prediction based on nature of the clients. Int. J. Recent Technol. Eng. **7**, 176–179 (2019)
16. K.R. Rawate, P.A. Tijare, Review on prediction system for bank loan credibility. Int. J. Adv. Eng. Res. Dev. **4**(12), 860–867 (2017)
17. K.U. Priya, S. Pushpa, K. Kalaivani, A. Sartiha, Exploratory analysis on prediction of loan privilege for customers using random forest. Int. J. Eng. Technol. **7**(2.21), 339 (2018). https://doi.org/10.14419/ijet.v7i2.21.12399
18. N. Torvekar, P.S. Game, Predictive analysis of credit score for credit card defaulters. Int. J. Recent Technol. Eng. **7**, 283–286 (2019)
19. P.D. Singh, A. Chug, Software defect prediction analysis using machine learning algorithms, in *2017 7th International Conference on Cloud Computing, Data Science & Engineering—Confluence* (2017). https://doi.org/10.1109/confluence.2017.7943255
20. Scikit-learn Machine Learning in Python. https://scikit-learn.org/stable

Machine Learning for Customer Segmentation Through Bibliometric Approach

Lopamudra Behera, Pragyan Nanda, Bhagyashree Mohanta, Rojalin Behera, and Srikanta Patnaik

1 Introduction

With the emergence of new businesses every day, survival of organizations in the cutthroat competition has become a challenge. To survive, the organizations and firms have to upgrade themselves with the flow of change or be prepared to face the respective consequences. Again, with the wide adoption of digital transformation in almost all sectors, e-commerce has become a media for conducting business successfully. Recently, this wide adoption of digitalization has given customer segmentation huge attention as one of the most relevant tools for e-commerce as well as other sectors dealing with customers. Further, while planning upscaling of an organization, instead of focusing on a wider group of customer, organizations need to focus on a specific group of customers. This brings in the essentiality of customer segmentation.

Customer segmentation plays a significant role in understanding both local and global market characteristics along with customer behaviors and actions. Customer segmentation is also important to recommend customized products to customers according to their preferences. However, customer segmentation involves grouping of customers into various categories on the basis of various characteristics. Some of the most commonly considered attributes include location, age, annual income, spending habit, purchase capability, buying frequency, and so on. Again, customer segmentation projects a major contribution to customer satisfaction and retention which again contributes to the overall profit maximization. Further, customer segmentation contributes in managing customer relationship management (CRM) systems. It also

L. Behera · P. Nanda (✉)
Interscience Institute of Management and Technology, Bhubaneswar, India
e-mail: n.pragyan@gmail.com

B. Mohanta · S. Patnaik
SOA University, Bhubaneswar, India

R. Behera
Sambalpur University, Sambalpur, India

S. Patnaik et al. (eds.), *Advances in Machine Learning and Computational Intelligence*, Algorithms for Intelligent Systems, https://doi.org/10.1007/978-981-15-5243-4_16

helps firms in visualizing targeted customer needs and preferences to provide better service and customized products. Typically, customer segmentation is performed by studying and analyzing discrete information about customers by focusing on geographical, behavioral, and demographical data. Accurate and appropriate segmentation may affect marketing strategies and lead to higher yield of profits. Machine learning methods have been proved to be useful in solving some of the most complex real-time problems. Also machine learning techniques include a huge collection of classification and clustering algorithms which make it more suitable depending on their performance for customer segmentation-related problems. Machine learning techniques assist in identifying hidden patterns from customer behaviors and other characteristics, thus grouping them into homogeneous segments.

Since the amount of literature available on customer segmentation is quite huge, definitely there is a chance that naïve researchers will be lost while searching for directions. This paper attempts to present a bibliometric analysis of the relevant literature available on customer segmentation and its relationship and impact with machine learning techniques. A bibliometric study employs both mathematical as well as statistical tools to analyze the relevancy of research works collected from scholarly databases such as Scopus, Web of Science, and DBLP. Bibliometric analysis further provides a quantitative approach to derive several subjective categories on the basis of citations, co-citations, keywords, authors, and also bibliographic coupling. In addition to this, the frequency of occurrence of keywords, number of publications of particular authors, citation scores of journals, and geographical distribution of publications are several other categories to be mentioned. A detailed bibliometric study can assist in identifying prominent research areas to focus on along with significant journals in the research areas.

This paper focuses on exploring significant research works done in the area of customer segmentation and its relationship with machine learning techniques along with the amount of work done in the area in terms of modeling, experimentations, comparative analysis, and implementation of novel models from a bibliometric perspective to provide insights into new researchers and make them familiar with the work done till now. The first section introduces various components; the second section discusses several works related to the components; the third section elaborates the data collected, while section four discusses the methodology adopted; the next section analyzes the findings, and finally section six concludes the work.

2 Related Works

Although a vast range of work has been done on customer segmentation, some works that provide base knowledge and helps in building a strong foundation for naïve researchers are discussed here. As discussed earlier, customer segmentation involves grouping of customers into several segments on the basis of various attributes such as geographic, demographic, behavioral, and psychographic data generated from customer's choices and lifestyle. The outcome of such segmentation can provide better view regarding customers' brand loyalty, current status, willingness to pay, readiness to buy, etc., thus helping in developing new marketing tactics [1–4]. Linden et al. [5] compared Amazon's own recommendation algorithm, i.e., item-to-item collaborative filtering with three frequently used methods like traditional collaborative filtering, cluster models, and search-based methods. As the item-to-item collaborative filtering algorithm works on a huge data set of approximately more than 29 million customers and several million catalog items, it shows better results compared to others and produces excellent recommendations in real time. This algorithm aims to find and recommend similar items with respect to the items viewed by the customers. Rao et al. [6] analyzed the utility of big data mining with ML to strengthen the profit via sale services. They had taken the data of tea device enterprise to evaluate persistent item sets by using FP grow algorithm, to estimate feature vector for user classification association rule data mining was considered, and finally to perform clustering learning for transparent trading and customized e-recommendation facilities the Naive Bayesian algorithm was evaluated. Fang et al. [7] also studied the online recommender system by providing customized product list to the customers to increase sale services. They followed three steps to achieve the target such as association rule data mining to nullify unrelated association of items, customer segmentation by K-means clustering algorithm to estimate each trade segment, and recommendation technique corresponding to these cluster segments. In an article by Pondel and Korczak [8], they discussed the significance of recommendation system in the e-commerce industry. The customer segmentation is predicted by different clustering algorithms like K-means algorithm, bisecting K-means algorithm, Gaussian mixture model (GMM), and DBSCAN identifying clusters algorithms influenced by RFM (recency, frequency, monetary) marketing technique. Tavakoli et al. [9] proposed customer segmentation and strategy development based on user behavior analysis, R + FM model which set up the customer segmentation and clustering or grouping of customer carried out by K-means clustering algorithm, and data mining techniques for better recommendation system, and applied the framework on Digikala company. They analyzed the research work by running a SMS campaign and derived that their proposed customer segmentation framework boosted the buying quantity. Yang et al. [10] presented a method that merge genetic algorithm (GA) and K-nearest neighbor (KNN) algorithm to evaluate customer's inclination toward particular products from their respective account profile and recommend most relevant items to reach their expectations. Chen et al. [11] studied six clustering algorithms on ten standard data sets and compared with the proposed two-level subspace weighting spectral clustering (TSW) algorithm for customer segmentation, and the exploratory outcome favors the new proposed TSW algorithm. Hossain [12] applied one of the density-based algorithms, i.e., density-based spatial clustering of applications with Noise (DBSCAN) algorithm and the centroid-based algorithms, i.e., K-means for customer segmentation. These algorithms are applicable in product recommender system that is beneficial to know the local and global wholesale or retail market. Chen et al. [13] presented a personalized product tree named as purchase tree model to maintain the transaction details of a customer, a partitional clustering algorithm, named PurTreeClust algorithm to efficiently cluster the purchase trees, an advanced distance metric to successfully calculate the distance between two purchase trees, and further to estimate the number of clusters they presented a method based on gap statistic. They led a sequence of experiments by taking ten live transaction data sets which out turned efficiently. Catal and Guldan [14] presented a framework to distinguished illusive remarks of discouraged customer by applying various classifier systems. They justified their model by collecting data set of rating of hotels from TripAdvisor. Five classifiers are taken in the suggested framework by considering some majority voting combination rule, namely libLinear, libSVM, sequential minimal optimization, random forest, and J48. The first two majority voting combination rule represented various applications of support vector machines (SVMs). Zhao et al. [15] analyzed different mobile users' attitude toward different online services, and based on these differences they evaluated a segmentation procedure which comprised four steps, such as (a) classify the customer response pattern into cycles, (b) specify customers' cyclical responses by considering the probability density distributions from the temporal dimension and frequency dimension, (c) evaluate the customer homogeneity by calculating the difference of the distributions, (d) apply the k-medoid clustering algorithm to classify the customers on the basis of the homogeneity pattern. Xi [16] proposed grouping of Chinese review spam and extracted three types of characteristics such as unigrams, attribute features, and semantic attribute by considering the data sets of mobile phone, digital camera, and laptop. With these attributes, he then classified the reviews by using LIBSVM to learn an SVM classifier. This paper focused on the classification of Chinese review spam by using machine learning procedure with the above stated three attributes and studied the effect of various attributes on classification implementation. Gallego and Huecas [17] proposed a context-aware mobile recommender system by considering factual banking data like customer profiles, credit card transactions, and information about their customers' previous investment areas, and the data are collected from a reputed Spanish bank and a mobile prototype was successfully deployed in the bank to estimate the outcome. To minimize the difficulties of the banking data mining process, canopy clustering algorithm was implemented, the Manhattan distance was used to estimate the likeness among client profiles, and K-means clustering algorithm was then applied over the canopies for clustering and named as Social Clusters comprised of banking customers of similar kind of profiles.

A quantitative application of bibliometric study was applied by adopting statistical and mathematical tools to review the information administration by scholars using various channels of information in scholarly communication [18]. A clear and concise prospective for quantitative approach for various subjectively derived categories in the related areas of research methods used in management and organizational perspective has been explored for future analysis [19]. The initiation and establishment of research aspects by using bibliometric study in the area of credit risk and bankruptcy and related data were available in the Web of Science was being studied during the period of 1968 and 2014 while being applicable in other knowledge disciplines [20]. Similarly, the bibliometric analysis was adopted to contribute in the field of medical science and health care such as analysis of related publications on spinal tuberculosis from the year of 1994–2015 and assessment of advancement and expansion in the area of neuroscience pertaining to changing scenario over a decade, i.e., from 2006 to 2015 related study by extracting the data from Web of Science [21, 22]. All these works emphasize on specific key terms that have been frequently used in publication numbers, authors cluster, citation scores, and regional contribution to productive endeavor in these particular areas. A comprehensive narration of the published papers employs qualitative comparative analysis tools and methods are prominent to discern the future applicability [23]. Recent research trends fueling the exploration in applications of AI in many prominent areas of science are to enhance the practice and applicability of intelligent interaction system such as natural language processing and functional magnetic resonance imaging to concede the contribution in terms of investigation, evaluation, analysis, and implication [24, 25]. To represent throughout research directions in the areas of ethics and entrepreneurship has reflected the insights into overall dimension for further scopes to be explored over time [26]. To complement the bibliometric study, an identifiable computer software tool *VOSviewer* is used to formulate and develop bibliometric networks pertaining to given research papers, articles, journals, and chapters of particular area of work. It is an open access software that has developed to visualize by exploring bibliometric maps using network data depending on authors clustering, co-authorship, citations, co-citation, and bibliographic coupling. Extracted data from bibliographic database files like Web of Science, Scopus Index, PubMed files, Journal Citation Reports, DBLP, etc., can be used as functional categorization of specific subject or area by relatedness and co-occurrences in VOSviewer tool to construct graphical representation of huge data easily and *quickly* compiling the technical implication as required [27].

3 Data

The data considered in this paper have been extracted from bibliographic database file of Scopus. The Scopus is Elsevier's database comprises nearly 36,377 of titles from approximately 11,678 publishers active since 2004 [28]. The database covers sources like Book Series, Journals, and Conference proceedings by to perceive qualitative and accurate data. In this paper, the authors tried to demonstrate a bibliometric study for the work has been done in the area of customer segmentation using machine learning techniques with the help of bibliographic database file for customer segmentation separately and machine learning tools applied to administrate customer segmentation during the time frame of 2009–2019. Originally, the data set of 1369 publications resultant has generated in the area of customer segmentation, and later by categorizing it into machine learning applications the result has brought down to 71 papers. Both of these data sets are being used to analyze the evolution of customer segmentation and machine learning tools applied to administrate customer segmentation by representing the number of contribution for the same during the last decade. The paper demonstrated major 29 sources of contribution as influential publications explained as seven interlinked clusters and four individual clusters by network visualization as a basis of at least one document and one citation to meet the threshold by VOSviewer. Again most 21 relevant keywords are being highlighted on the basic of citation and occurrence meeting the minimum occurrence requirement of five times, respectively.

4 Methods

The authors have used VOSviewer as a tool to reflect a bibliometric study on applied machine learning approaches in the area of customer segmentation during the last decade, i.e., 2009–2019. The study of related works and fields of bibliometric of this area is analyzed by constructing network and visualization diagrams to establish linkage between the relevant keywords and document sources based on database files generated from Scopus. The bibliographic information has been used to showcase the relatedness of major aspects of on extensive literature review based on various counting and analysis method such as co-occurrence of most relevant keyword and bibliographic coupling of major sources. The database is used as a unit of analysis to study the impact of most influential terms and sources based on the overall weightage of contribution in this field by demonstrating visualization map.

5 Analyzing the Framework of Machine Learning Approaches in Customer Segmentation Practices

5.1 Genesis of Applied Machine Learning Tools in Customer Segmentation

Figure 1 shows the extended significance of paper published on customer segmentation and the implementation of machine learning for the incremental transactions of customer segmentation over the period of 2009–2019. This figure is a showcase of Scopus database for the mentioned years. Customer segmentation being the inseparable area of management discipline holds a high degree of research prerequisites almost in every sector. The chart is a clear reflection of the dynamic evolution of this area reflects notable trend over last decade.

A total number of 1440 papers have been published in the area of customer segmentation over the designated periods, but only 71 articles fall under the category of customer segmentation with machine learning applications. The chart reveals boot starting year for customer segmentation and machine learning twosome in 2009. And for the subsequent three years, the trends get off the board but again picked up the trend with six publications and maintain the consistency. The last two years are the growth period for customer segmentation, and machine learning coupling as a

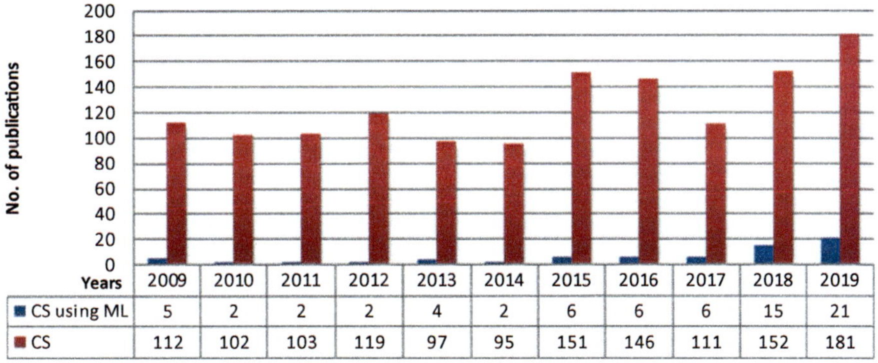

Fig. 1 Evolution of customer segmentation

divulge result of the figure published article for the year of 2018 and 2019 is 15 and 21, respectively.

Although the variation of customer segmentation with machine learning tool publication is very low for the nominated period, the overall development of the topic interest across the time period is impressive. It can be assumed the popularity of the topic will increase in the future research.

5.2 Connection of Most Relevant Keywords of Applied Machine Learning Tools in the Context of Customer Segmentation

In this section, keywords connected with machine learning and customer segmentation coupling research in specific context have been discussed.

Figure 2 represents the network map of most occurred keywords relevant to machine learning applications in the area of customer segmentation coupling research over the specified period of the last ten years. The authors identified 21 most relevant keywords meeting the threshold set as default criteria of minimum occurrence of 5 per each keyword.

The cluster along with the segregated keywords with the most frequent occurrence has been represented in Table 1. The result reveals that the most dominant keywords used are 'machine learning' and 'customer segmentation' with most occurrence assessment in total as 30 and 29, respectively. The visualization suggested the links between the most occurred keywords and strength of connection [29–31].

Cluster-1 (red) summarizes the highlighted terms of research articles are customer segmentation, clustering algorithms, image segmentation, artificial intelligence, learning systems, smart meters, and data mining. Carnein and Nilashiet al. [31, 32] highlighted systematic segmentations of customer are an extraction of information, discovery of knowledge from raw data easily by using data mining. Hung,

Fig. 2 Co-occurrence of keywords in machine learning and customer segmentation coupling research

P.D. focused on new learning algorithms, machine learning tools like clustering algorithm to grouping similar customer values and market demands. This cluster also highlights the researchers' interest toward new trends like image segmentation adapted by David and Averbuch [33] which provides the clarity of representation of an image into something meaningful that can lead to a systematic learning and developmental variations of customer segmentation.

Cluster-2 (green) drags the attention for exchange relationship and majors taken to adapt the new trends. The chronic change in the digital era leads the focus of the researcher Ozan, S into e-commerce. The online service is the hub point for the exchange of goods and services in large scale. This cluster mainly focused on business process, commercial activities, and market segmentation. Hung, P.D. pointed out segmentation in market mainly focuses on wide environmental trends to cover all aspects of socioeconomical scenario such as legal, economics, political, social, cultural, and technical system that varies from one place to another. Miller, C. gives insight into the applications of machine learning provides the ability to the marketer to learn for themselves about the customer process relationship and the process of making important decisions.

Cluster-3 (blue) finally consists of a series of articles completely focused on the customer satisfaction and sales growth. The process of customer relation management brings the fundamental thoughts of researcher Carnein, M. that every organization always concerns with optimum revenue generation and healthy customer relation

Table 1 Co-occurrence of keywords in machine learning and customer segmentation coupling research

	Keyword	Occurrences	Total link strength
Cluster 1	Cluster segmentation	29	114
	Learning systems	24	110
	Data mining	14	56
	Artificial intelligence	12	52
	Clustering algorithms	8	41
	Image segmentation	7	25
	Classification (of information)	6	22
	Cluster analysis	6	29
	Smart meters	5	17
Cluster 2	Machine learning	30	113
	Commerce	10	41
	Market segmentation	8	30
	Marketing	7	31
	Clustering	5	23
	Decision making	5	23
	Electronic commerce	5	24
Cluster-3	Sales	24	107
	Learning algorithms	7	33
	Customer relationship management	6	36
	Customer satisfaction	6	17
	Public relations	6	42

for suitability. In the words of Nilashi, M. et al., by using machine leanings application organizations can meet the expectation of the customers along with the sales acquisitions, and the algorithms of learning develop an effective business process of relationship with its customers.

5.3 Major Sources Contributed: Cluster Analysis

Cluster 1: Linking Machine Learning approaches and market segmentation context

Cluster 1 emulates the context of uncertainty and dynamism in aspect of data administration ranging from allocation to sustainability. This section particularly vindicates the preconditions of the machine learning approaches in field of market and customer segmentation to penchant the absolute dynamism in the composition of market demand and other organizational variables. The key prospective of segmentation lies between maintenance and development of vast data sets to arouse frequent consistent procedures [34]. The applications to pursuit accuracy in theft detection are to avoid unseen risk by developing customer segmentation model with definite forecasting tools [35]. Demand for flexible customer segmentation by an effective administration of after sales-service data has been ensured by reduce errors in the production process to get assurance of qualitative products, smooth manufacturing, accuracy in inspection, and transparency data management by connecting machine algorithms with engines [36]. Smart error detection techniques, effective risk measurement policies, flexible demand-supply forecasting, customer-sensitive pricing schemes, cost reduction strategies, etc., are indispensible attributes for optimal revenue generation and that only can be achieved by efficient market segmentation, whether in banking sector or energy sector even in agriculture [37–39]. Ghahari et al. [40] streamlined the application of artificial computational algorithms with factual prediction on weather complexity and dynamic climate in order to retain potential market suitability. An interesting approach toward validating the impacts of machine algorithms on customer loyalty and brand attractiveness to achieve competitive advantage as well as value creation analyzed by Stormi et al. [41] illustrated a clear framework for potential market segmentation (Fig. 3).

Cluster 2: Integration of Customer Segmentation Attributes and Machine Learning Tools

Cluster 2 emphasizes applicability of machine learning tools on marketing strategies to enhance the customer segmentation decisions whether for online or offline services. Further, this investigation aims to identify attributes responsible for the implementation of innovative marketing strategies. Particularly, strategic climate prevailing in the framework of segmentation by studying the behavioral aspect and

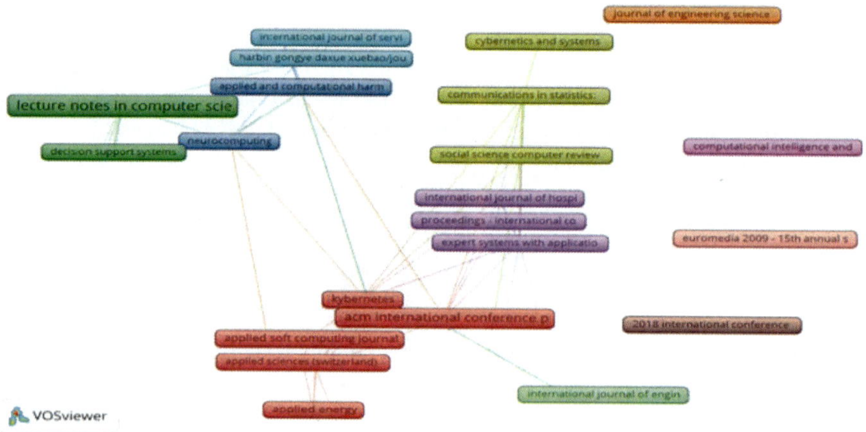

Fig. 3 Network visualization of 29 sources

comparing it with levels of consumption has been explained by case study analysis associated with machine learning techniques for better classification and prediction. The combined perquisites like identification of potential customers, predication of market preferences, and improvisation of marketing contents have developed a special requirement of optimization of marketing initiatives [42]. Thus, organizations need to streamline the customer segmentation aspects with artificial intelligent applications like meter data analysis, highly comparative time series analysis (HCTSA), clustering algorithms, and artificial neural network [43] to enhance the core competency. A related context of research addressing the marketing interdisciplinary approaches in a connection with marketing intelligent system, broadly pipelined the sub-systems like databases, data warehouse, procedures, software, and expert systems based on crisp and fuzzy set clustering for flexible and accurate decision making [44].

Cluster 3: Intelligent Interaction in Potential Customer Identification and Retention

Taking reference of data volatility, high-dimensional data sets are essential to be managed in order to eliminate the gap between real data and data affinities focusing on consistent organizational hierarchical system [45]. This basically relates programming languages like R language in association with natural language processing, operational and statistical methods, RFM model, and K-means clustering as an impact factor for brand attractiveness for potential customers and product categorization [46]. Another important contribution has justified the impact of cognitive decision-making tools in determining preferred customer zones [30].

Cluster 4: Customer Relationship Management as an Impact Factor of Automation

The document sources summarized in cluster 4 concentrated in distinct concept of customer segmentation with definite applications for a certain prospective. Two specific topics have identified to complement notable factors like profitable business opportunity, value creation, core competency achievement, brand identity creation, data structure administration, prevailing the drives for customer segmentation. Albuquerque et al. [47] purposed the predominant requirement of one of the machine learning techniques, support vector clustering to improve manager–customer relationship. Moreover, it has analyzed the influence of computational application on knowledge mapping, which is of intensifying relevance cognitive learning methods and automated system examining [48]. The integration of customer profitability accounting into customer segmentation is also assumed to have an upper hand in facilitating in global cost–benefit analysis by improvising the problem-solving practices [49].

Cluster 5: Targeted Market Strategy and Web Big Data

On the contrary of automation and machine learning applications on availing online data administration practices implicated for market analysis, it has become much bigger approach in target market segmentation and categorization. The fact imposed on transparency in market data sources is strongly driven by implementation of appropriate techniques, algorithms, clustering, etc., for analyzing customer behavior [50]. Another research related to the online data base in terms of customer satisfaction based on experience and information [51] to create brand attractiveness and enhance overall performance has outlined the accurate segmentation algorithm with flexible segmentation. Ahani et al. [52] have discussed the relevance of Web site data via reviews, blog, rating, and comments as one of the growing aspects of brand effectiveness using social media contexts through applied machine learning methods.

Cluster 6: Cost-Sensitive Learning Existing Versus Applied

A comparative analysis of original support vector machine and cost-sensitive support vector machine to study the optimum value creation of segmentation practices based on demographic attributes has supported the heterogamous cost analysis systems [53]. Referring to the risk and error occurrences, intensity to tolerance has geared us the ability to detect error beforehand to minimize uncertainty associated with misclassification of value [54].

Cluster 7: Customer Psychological Aspects

This cluster pipelined the customer psychology and buying behavior in the context of customer segmentation. A review on application of long-term and short-term memory networks addressing the behavioral practices and psychological analysis is [55] based on product reviews by customers. Additionally, sentimental attributes have great impact on physical attributes, and which affect the customers' biasness relating to product can be understood by categorizing accordingly using machine learning applications [56].

Other Clusters

These clusters are not interlinked by the threshold of bibliometric coupling stressed on studying the influencing factors, emerging issues, employed methods, and basic attributes by case study method [29] to tackle critical challenges [57] by forecasting methods (Table 2).

6 Conclusion

The study in this paper has been centered on the growing establishment of machine learning techniques as a decisive determinant of perspective marketing analysis with the objective to review the significant contributions. Findings emphasize the increasing trend by justifying the relevant attributes of machine learning in customer segmentation prospectus as well as objectively study the major influential contribution which can further emerge as baseline mapping source for research aspirants. According to the findings, North American Actuarial Journal has made a significant contribution as measured in terms of the highest citations received. Document source coupling wise analysis revealed that lecture notes in computer science have

Table 2 Major sources of contribution

Cluster	Source	Document contribution	Citations	Link strength	Author/s
1	North American Actuarial Journal	1	100	10	Ghahari A., Newlands N.K., Lyubchich V., Gel Y.R.
	Applied Energy	2	12	4	Razavi R., Gharipour A., Fleury M., Akpan I.J.
	Applied Sciences (Switzerland)	1	10	4	Koolen D., Sadat-Razavi N., Ketter W.
	Applied Soft Computing Journal	2	6	2	Arevalillo J.M.
	European Conference on Information Systems: Beyond Digitization—Facets of Socio-Technical Change	2	6	10	Stormi K., Laine T., Elomaa T.
	Industrial Management and Data Systems	1	5	1	Ko T., Hyuk Lee J., Cho H., Cho S., Lee W., Lee M.
	ACM International Conference Proceeding Series	4	4	10	Phridviraj M.S.B., Guru Rao C.V.
	Journal of Business Economics and Management	1	1	3	Smeureanu I., Ruxanda G., Badea L.M.
2	Journal of Computational and Theoretical Nanoscience	1	15	6	Tan K.S., Subramanian P.
	Lecture Notes in Computer Science (Including Subseries Lecture Notes in Artificial Intelligence and Lecture Notes in Bioinformatics)	6	4	15	Carnein M., Trautmann H.
	Studies in Fuzziness and Soft Computing	1	3	1	Markic B., Tomic D.
	Energy and Buildings	1	1	1	Miller C.
3	ACM International Conference Proceeding Series	1	36	23	Hung P.D., Ngoc N.D., Hanh T.D.

(continued)

Table 2 (continued)

Cluster	Source	Document contribution	Citations	Link strength	Author/s
	International Journal of Recent Technology and Engineering	1	34	5	Shetty P.P., Varsha C.M., Vadone V.D., Sarode S., Pradeep Kumar D.
	Applied and Computational Harmonic Analysis	1	22	19	David G., Averbuch A.
4	Cybernetics and Systems	1	6	1	Chen L.-S., Hsu C.-C., Chen M.-C.
	Social Science Computer Review	1	6	22	Florez-Lopez R., Ramon-Jeronimo J.M.
	Communications in Statistics: Simulation and Computation	1	2	18	Albuquerque P., Alfinito S., Torres C.V.
5	Proceedings—International Conference on Machine Learning and Data Engineering	1	25	29	Yoseph F., Heikkila M.
	International Journal of Hospitality Management	1	8	15	Ahani A., Nilashi M., Ibrahim O., Sanzogni L., Weaven S.
	Proceedings—International Conference on Machine Learning and Data Engineering	1	2	5	Tan K.S., Subramanian P.
6	Harbin Gongye Daxue Xuebao/Journal of Harbin Institute of Technology	1	4	4	Zou P., Yu B., Wang X.-Q.
	International Journal of Services, Technology and Management	1	4	4	Zou P., Hao Y., Li Y.
7	Journal of Engineering Science and Technology Review	1	6	1	Chen H., Li S., Wu P., Yi N., Li S., Huang X.
	Proceedings—International Conference on Industrial Informatics—Computing Technology, Intelligent Technology, Industrial Information Integration	1	1	1	Sun L.

(continued)

Table 2 (continued)

Cluster	Source	Document contribution	Citations	Link strength	Author/s
8	2018 International Conference on Artificial Intelligence and Data Processing	1	1	0	Ozan Åž.
9	Computational Intelligence and Neuroscience	1	2	0	Long H.V., Son L.H., Khari M., Arora K., Chopra S., Kumar R., Le T., Baik S.W.
10	Euromedia 2009—15th Annual Scientific Conference on Web Technology, New Media Communications and Telematics Theory Methods, Tools and Application	1	1	0	Sammour G., Schreurs J., Vanhoof K.
11	International Journal of Engineering and Advanced Technology	1	12	0	Mathew R.M., Suguna R., Shyamala Devi M.

major contribution with six documents. Thus, it can be concluded that customer segmentation, being the dominant precedent of marketing discipline, has occupied a new dimensional base for addressing its potential customers by managing the core competency. As a result, a prerequisite of machine learning applications has been fueling up to remove the barrier in market categorization to balance between customer relationship and organizational efficacy.

References

1. S. Goyat, The basis of market segmentation: a critical review of literature. Eur. J. Bus. Manag. **3**(9), 45–54 (2011)
2. J. Tikmani, S. Tiwari, S. Khedkar, An approach to consumer classification using K-Means. IJIRCCE **3**(11), 10542–10549 (2015)
3. C.P. Ezenkwu, S. Ozuomba, C. Kalu, Application of K-Means algorithm for efficient customer segmentation: a strategy for targeted customer services (2015)
4. V.R. Patel, R.G. Mehta, Impact of outlier removal and normalization approach in modified k-means clustering algorithm. Int. J. Comput. Sci. Issues (IJCSI) **8**(5), 331 (2011)
5. G. Linden, B. Smith, J. York, Amazon.com recommendations: item-to-item collaborative filtering. IEEE Internet Comput. **7**(1), 76–80 (2003)
6. H.K. Rao, Z. Zeng, A.P. Liu, Research on personalized referral service and big data mining for e-Commerce with machine learning, in *2018 4th International Conference on Computer and Technology Applications (ICCTA)* (IEEE, 2018, May), pp. 35–38

7. Y. Fang, X. Xiao, X. Wang, H. Lan, Customized bundle recommendation by association rules of product categories for online supermarkets, in *2018 IEEE Third International Conference on Data Science in Cyberspace (DSC)* (IEEE, 2018, June), pp. 472–475
8. M. Pondel, J. Korczak, Collective clustering of marketing data-recommendation system Upsaily, in *2018 Federated Conference on Computer Science and Information Systems (FedCSIS)* (IEEE, 2018, September), pp. 801–810
9. M. Tavakoli, M. Molavi, V. Masoumi, M. Mobini, S. Etemad, R. Rahmani, Customer segmentation and strategy development based on user behavior analysis, RFM model and data mining techniques: a case study, in *2018 IEEE 15th International Conference on e-Business Engineering (ICEBE)* (IEEE, 2018, October), pp. 119–126
10. H.W. Yang, Z.G. Pan, X.Z. Wang, B. Xu, A personalized products selection assistance based on e-commerce machine learning, in *Proceedings of 2004 International Conference on Machine Learning and Cybernetics (IEEE Cat. No. 04EX826)*, vol. 4 (IEEE, 2004, August), pp. 2629–2633
11. X. Chen, W. Sun, B. Wang, Z. Li, X. Wang, Y. Ye, Spectral clustering of customer transaction data with a two-level subspace weighting method. IEEE Trans. Cybern. **49**(9), 3230–3241 (2018)
12. A.S. Hossain, Customer segmentation using centroid based and density based clustering algorithms, in *2017 3rd International Conference on Electrical Information and Communication Technology (EICT)* (IEEE, 2017, December), pp. 1–6
13. X. Chen, Y. Fang, M. Yang, F. Nie, Z. Zhao, J.Z. Huang, Purtreeclust: a clustering algorithm for customer segmentation from massive customer transaction data. IEEE Trans. Knowl. Data Eng. **30**(3), 559–572 (2017)
14. C. Catal, S. Guldan, Product review management software based on multiple classifiers. IET Softw. **11**(3), 89–92 (2017)
15. H. Zhao, X.H. Zhang, Q. Wang, Z.C. Zhang, C.Y. Wang, Customer segmentation on mobile online behavior, in *2014 International Conference on Management Science & Engineering 21th Annual Conference Proceedings* (IEEE, 2014, August), pp. 103–109
16. Y. Xi, Chinese review spam classification using machine learning method, in *2012 International Conference on Control Engineering and Communication Technology* (IEEE, 2012, December), pp. 669–672
17. D. Gallego, G. Huecas, An empirical case of a context-aware mobile recommender system in a banking environment, in *2012 Third FTRA International Conference on Mobile, Ubiquitous, and Intelligent Computing* (IEEE, 2012, June), pp. 13–20
18. C.L. Borgman, Bibliometrics and scholarly communication: editor's introduction. Commun. Res. **16**(5), 583–599 (1989)
19. I. Zupic, T. Čater, Bibliometric methods in management and organization. Organ. Res. Methods **18**(3), 429–472 (2015)
20. J.W. Prado, V. Castro Alcântara, F. Melo Carvalho, K.C. Vieira, L.K. Machado, D.F. Tonelli, Multivariate analysis of credit risk and bankruptcy research data: a bibliometric study involving different knowledge fields (1968–2014). Scientometrics **106**(3), 1007–1029 (2016)
21. Y. Wang, Q. Wang, R. Zhu, C. Yang, Z. Chen, Y. Bai, Trends of spinal tuberculosis research (1994–2015): a bibliometric study. Medicine **95**(38) (2016)
22. A.W.K. Yeung, T.K. Goto, W.K. Leung, The changing landscape of neuroscience research, 2006–2015: a bibliometric study. Front. Neurosci. **11**, 120 (2017)
23. N. Roig-Tierno, T.F. Gonzalez-Cruz, J. Llopis-Martinez, An overview of qualitative comparative analysis: a bibliometric analysis. J. Innovation Knowl. **2**(1), 15–23 (2017)
24. X. Chen, Z. Liu, L. Wei, J. Yan, T. Hao, R. Ding, A comparative quantitative study of utilizing artificial intelligence on electronic health records in the USA and China during 2008–2017. BMC Med. Inform. Decis. Mak. **18**(5), 117 (2018)
25. A.W.K. Yeung, Bibliometric study on functional magnetic resonance imaging literature (1995–2017) concerning chemosensory perception. Chemosens. Percept. **11**(1), 42–50 (2018)
26. C. Vallaster, S. Kraus, J.M.M. Lindahl, A. Nielsen, Ethics and entrepreneurship: a bibliometric study and literature review. J. Bus. Res. **99**, 226–237 (2019)

27. N. Van Eck, L. Waltman, Software survey: VOSviewer, a computer program for bibliometric mapping. Scientometrics **84**(2), 523–538 (2010)
28. J.F. Burnham, Scopus database: a review. Biomed. Digital Libr. **3**(1), 1 (2006)
29. Ş. Ozan, A case study on customer segmentation by using machine learning methods, in *2018 International Conference on Artificial Intelligence and Data Processing (IDAP)* (IEEE, 2018, September), pp. 1–6
30. P.D. Hung, N.D. Ngoc, T.D. Hanh, K-means clustering using RA case study of market segmentation, in *Proceedings of the 2019 5th International Conference on E-Business and Applications* (2019, February), pp. 100–104
31. C. Miller, What's in the box?! Towards explainable machine learning applied to non-residential building smart meter classification. Energy Build. **199**, 523–536 (2019)
32. M. Nilashi, A. Mardani, H. Liao, H. Ahmadi, A.A. Manaf, W. Almukadi, A hybrid method with TOPSIS and machine learning techniques for sustainable development of green hotels considering online reviews. Sustainability **11**(21), 6013 (2019)
33. G. David, A. Averbuch, SpectralCAT: categorical spectral clustering of numerical and nominal data. Pattern Recogn. **45**(1), 416–433 (2012)
34. M.S.B. Phridviraj, C.G. Rao, A novel approach for unsupervised learning of transaction data, in *Proceedings of the 5th International Conference on Engineering and MIS* (2019, June), pp. 1–5
35. R. Razavi, A. Gharipour, M. Fleury, I.J. Akpan, A practical feature-engineering framework for electricity theft detection in smart grids. Appl. Energy **238**, 481–494 (2019)
36. T. Ko, J.H. Lee, H. Cho, S. Cho, W. Lee, M. Lee, Machine learning-based anomaly detection via integration of manufacturing, inspection and after-sales service data. Ind. Manag. Data Syst. (2017)
37. D. Koolen, N. Sadat-Razavi, W. Ketter, Machine learning for identifying demand patterns of home energy management systems with dynamic electricity pricing. Appl. Sci. **7**(11), 1160 (2017)
38. I. Smeureanu, G. Ruxanda, L.M. Badea, Customer segmentation in private banking sector using machine learning techniques. J. Bus. Econ. Manag. **14**(5), 923–939 (2013)
39. J.M. Arevalillo, A machine learning approach to assess price sensitivity with application to automobile loan segmentation. Appl. Soft Comput. **76**, 390–399 (2019)
40. A. Ghahari, N.K. Newlands, V. Lyubchich, Y.R. Gel, Deep learning at the interface of agricultural insurance risk and spatio-temporal uncertainty in weather extremes. North Am. Actuarial J. **23**(4), 535–550 (2019)
41. K. Stormi, T. Laine, T. Elomaa, Feasibility of B2C customer relationship analytics in the B2B industrial context (2018)
42. K.S. Tan, P. Subramanian, Proposition of machine learning driven personalized marketing approach for E-commerce. J. Comput. Theor. Nanosci. **16**(8), 3532–3537 (2019)
43. M. Carnein, H. Trautmann, Customer segmentation based on transactional data using stream clustering, in *Pacific-Asia Conference on Knowledge Discovery and Data Mining* (Springer, Cham, 2019, April), pp. 280–292
44. B. Markic, D. Tomic, Marketing intelligent system for customer segmentation. *Marketing Intelligent Systems Using Soft Computing* (Springer, Berlin, Heidelberg, 2010), pp. 79–111
45. G. David, A. Averbuch, Hierarchical data organization, clustering and denoising via localized diffusion folders. Appl. Comput. Harmonic Anal. **33**(1), 1–23 (2012)
46. P.P. Shetty, C.M. Varsha, V.D. Vadone, S. Sarode, D. Pradeep Kumar, Customers churn prediction with RFM model and building a recommendation system using semi-supervised learning in retail sector. Int. J. Recent Technol. Eng. **8**(1), 3353–3358 (2019)
47. P. Albuquerque, S. Alfinito, C.V. Torres, Support vector clustering for customer segmentation on mobile TV service. Commun. Stat.-Simul. Comput. **44**(6), 1453–1464 (2015)
48. L.S. Chen, C.C. Hsu, M.C. Chen, Customer segmentation and classification from blogs by using data mining: an example of VOIP phone. Cybern. Syst. Int. J. **40**(7), 608–632 (2009)
49. R. Florez-Lopez, J.M. Ramon-Jeronimo, Marketing segmentation through machine learning models: an approach based on customer relationship management and customer profitability accounting. Soc. Sci. Comput. Rev. **27**(1), 96–117 (2009)

50. F. Yoseph, M. Heikkila, Segmenting retail customers with an enhanced RFM and a hybrid regression/clustering method, in *2018 International Conference on Machine Learning and Data Engineering (iCMLDE)* (IEEE, 2018, December), pp. 108–116
51. J. Cuzzola, J. Jovanović, E. Bagheri, D. Gašević, Automated classification and localization of daily deal content from the Web. Appl. Soft Comput. **31**, 241–256 (2015)
52. A. Ahani, M. Nilashi, O. Ibrahim, L. Sanzogni, S. Weaven, Market segmentation and travel choice prediction in Spa hotels through TripAdvisor's online reviews. Int. J. Hosp. Manag. **80**, 52–77 (2019)
53. P. Zou, B. Yu, X.Q. Wang, Cost-sensitive learning method with data drift in customer segmentation. Harbin Gongye Daxue Xuebao (J. Harbin Inst. Technol. **43**(1), 119–124 (2011)
54. Z. Peng, H. Yuanyuan, L. Yijun, Customer value segmentation based on cost-sensitive learning support vector machine. Int. J. Serv. Technol. Manage. **14**(1), 126–137 (2010)
55. H. Chen, S. Li, P. Wu, N. Yi, S. Li, X. Huang, Fine-grained sentiment analysis of chinese reviews using LSTM network. J. Eng. Sci. Technol. Rev. **11**(1) (2018)
56. L. Sun, Research on product attribute extraction and classification method for online review, in *2017 International Conference on Industrial Informatics-Computing Technology, Intelligent Technology, Industrial Information Integration (ICIICII)* (IEEE, 2017, December), pp. 117–121
57. H.V. Long, L.H. Son, M. Khari, K. Arora, S. Chopra, R. Kumar et al., A new approach for construction of geodemographic segmentation model and prediction analysis. Comput. Intell. Neurosci. **2019** (2019)

Online Hostel Management System Using Hybridized Techniques of Random Forest Algorithm and Long Short-Term Memory

S. Suriya, G. Meenakshi Sundaram, R. Abhishek, and A. B. Ajay Vignesh

1 Introduction

Throughout the world, educational sector has seen a rapid increase. Many established institutions utilize old manual procedures for record-keeping activities in all departments. This can damage the institute's efficiency. Our proposed approach explains how records are maintained in a better way. Graphical UI-oriented which is the proposed framework here overcomes the disadvantages of traditional approach. We used HTML and CSS for the front end and PHP for the back end. We used three languages for developing the Web site: HTML—stands for HyperText Markup Language. It is used for creating a webpage and displaying all the contents in it, CSS—stands for Cascading Style Sheets. It is used to add style elements to the webpage such as layout, fonts, colors and PHP—stands for Hypertext Pre-processor. It is a server-side scripting used as a general-purpose programming language and also for Web development.

S. Suriya · G. Meenakshi Sundaram (✉) · R. Abhishek · A. B. Ajay Vignesh
Department of Computer Science and Engineering, PSG College of Technology, Coimbatore, India
e-mail: meenakshisundaram5698@gmail.com

S. Suriya
e-mail: suriyas84@gmail.com; ss.cse@psgetch.ac.in

R. Abhishek
e-mail: abhishekramji10@gmail.com

A. B. Ajay Vignesh
e-mail: ajayvigneshab@gmail.com

S. Patnaik et al. (eds.), *Advances in Machine Learning and Computational Intelligence*, Algorithms for Intelligent Systems, https://doi.org/10.1007/978-981-15-5243-4_17

2 Literature Review

An Optimized Random Forest Classifier for Diabetes Mellitus published in 2019. The main aim of the paper [1] is to develop an optimized random forest classifier by genetic algorithm for diabetes mellitus. Random forest algorithm and genetic algorithm were used for the same. The merit of using the algorithm was genetic algorithm-optimized random forest classifier gave accuracy higher than before. Demerits were oversampling and undersampling of the data. The main purpose of the algorithm was extracting the valuable information from the symptoms and providing an appropriate medication in a lesser amount of time. Crop phenology retrieval via polarimetric SAR decomposition and random forest algorithm published in 2019 [2], whose main aim was to monitor crop phenology and suggest the right time to harvest. SAR decomposition and random forest algorithm were used. Random forest algorithm outperformed all other algorithms. The demerit of the algorithm was the overfitting of data. Random forest algorithm takes advantage of inputs of multiple parameters, improving the ability to track the phenology. Automatic detection of Alzheimer disease based on histogram and random forest published in 2019 [3], whose main purpose was on-time detection of Alzheimer disease using random forest algorithm. Even though there was overfitting, accuracy using random forest algorithms was better than other ML algorithms. The functionality of the algorithm was getting information from dataset and predicting the disease with foremost accuracy. The main aim of geographical variations in the fatty acids of Zanthoxylum seed oils based on random forest classifier, published in 2019 [4], was to distinguish Zanthoxylum seed oils. The advantage of using random forest algorithm in the project is that the built classification model showed 100% accuracy despite the overfitting of data. FA composition together with chemometrics based on RF algorithm has been applied to distinguish the crude seed oils of Z. This chemometric approach seems effective over a wide geographical range. Gene pathogenicity prediction (GPP) of Mendelian diseases via the random forest algorithm in 2019 predicted Mendelian diseases based on RF algorithm [5]. The GPP score of random forest algorithm was significantly better than other algorithms. Random forest algorithm obtained an accuracy of 80%, recall of 93%, FPR of 34% and FNR of 7%. Effect of photic simulation for migraine detection using random forest and discrete wavelet transform in 2019 used random forest algorithm for migraine identification [6]. Even though the algorithm performed poorly on rare outcomes, due to its efficiency and reliability, it outperformed other algorithms. This is because RF algorithm takes advantage of inputs of multiple parameters. Estimating grassland aboveground biomass on the Tibetan Plateau using a random forest algorithm estimated an AGB score using the algorithm. RF model was successful than SVM, which was considered the most successful model [7]. RF model performed well in AGB estimation, which can explain 86% of the observational data. A generalized space-time OBIA classification scheme to map sugarcane areas at regional scale using Landsat images time series and the random forest algorithm (2019) used random forest algorithm for rapid and accurate monitoring of the sugarcane area at a regional scale, reaching accuracies similar to the currently used national statistics [8]. The merit is that RF classifier provides accurate classifications for all validation sites. The local RF models Dice Coefficient (DC) accuracies were around 90%. Applying a nonparametric random forest algorithm to assess the credit risk of the energy industry in China measured the credit risk of credit cards in China's energy industry and to lay a foundation for comprehensive credit risk management using random forest algorithm [9]. RF has a high prediction accuracy rate with a good tolerance of outliers. But the disadvantage is that its application remains limited by China's commercial banks when constructing credit risk assessment models. The overall prediction accuracy of the models cannot fully reflect the prediction accuracy of each sample category. The aim of a brief survey on random forest ensembles in classification model showed that the ultimate aim of ensemble methods is to find high accuracy with greater performance [10]. It efficiently handled thousands of input variables. The disadvantage is that classification of large amount of data with a single classifier will give less accuracy. Estimating leaf area index and light extinction coefficient using random forest regression algorithm in a tropical moist deciduous forest estimated LAI and k by integrating RS and in situ data using random forest regression algorithm. RF can be effectively applied to predict the spatial distribution of LAI and k [11]. Random forest (RF) algorithm was used to predict the spatial distribution of LAI and k using the best predictor variables. Predicting Blast-Induced Air Overpressure: A Robust Artificial Intelligence System Based on Artificial Neural Networks and Random Forest used ANN and Random Forest Algorithm [12]. The main aim of the project was to control the undesirable effects of blast-induced AOp. The RF-ANN models provided high performance for predicting blast-induced AOp in Nui Beo open-pit coal mine, Vietnam. The number of hidden layers is also one of the factors affecting the training time of the model. The best ANN model yielded an RMSE of 1.184, R2 of 0.960 and MAE of 0.813. ANN and RF algorithms were Land-Subsidence Spatial Modeling Using the Random Forest Data-Mining Technique [13]. The purpose of this study was spatial modeling of land subsidence using an RF data-mining model. The results indicated that the accuracy of the built map using an ROC curve was excellent (96.06%). Implementation of ANN-RF is complex and difficult. The results of the evaluation of the model indicate an AUC value of 94%. Crop type mapping without field-level labels, random forest transfer and unsupervised clustering techniques, used random forest algorithm and Gaussian mixture model (GMM), whose main aim is to grow demand for food from an increasing global population that necessitates close monitoring of agricultural activities [14]. The crop type maps may be generated by either by applying a supervised model trained elsewhere or using regional statistics. Lack of field-level crop labels for training supervised classification models. Random forests transfer with high accuracy to neighboring geographies with regional crop compositions and growing conditions. Random forest as one-class classifier and infrared spectroscopy for food adulteration detection used RF algorithm for adulteration detection purposes [15]. Random forest algorithm and chemometric algorithm [partial least squares for discriminant analysis (PLS-DA)] were used. The advantage of using RF algorithm is the high classification accuracy. PLS-DA algorithm presented a rate of false positive of 12.22%. The functionality to generate artificial outliers from the target samples was needed. So, random forest algorithm was used. A New Application of Random Forest Algorithm to Estimate Coverage of Moss-Dominated Biological Soil Crusts in Semi-Arid Mu Us Sandy Land, China, used random forest algorithm. This paper [16] aims at optimizing the extraction of BSCs using RF. The RF algorithm has high capability in detecting BSC information from multispectral channels. The disadvantage is that the performance reduced when the models were applied to remote sensing data. The use of the RF model with multispectral remote sensing data to detect BSCs is more convenient. Assessment of advanced random forest and decision tree algorithms for modeling rainfall-induced landslide susceptibility in the Izu-Oshima Volcanic Island, Japan, used RF algorithm and decision trees to model the massive rainfall-triggered landslide occurrences in the Izu-Oshima Volcanic Island, Japan, at a regional scale [17]. The RF model could provide an extension to the DT model with the target of increasing the robustness of the classification tool. An application of ensemble random forest classifier for detecting financial statement manipulation of Indian listed companies used many ML algorithms to create an efficient and effective framework for detection of financial fraud [18]. RF proved to be the best model with highest accuracy. The disadvantage is the selection of hyperparameters. PCA and RF algorithms were used in human activity recognition from smart watch sensor data using a hybrid of principal component analysis and random forest algorithm [19]. Its main purpose was the detection of human movements using mobile applications. The disadvantage of the project is that it was difficult to carry out the experiment with a large number of values and also difficult to determine parameter values that will give better results.

3 Existing Approach

In the recent times, there has been an exponential rise in the number of colleges and universities all over the world. Most of the newly established educational institutions, however, are using the old conventional techniques for managing their assets, especially hostel facilities. It in turn causes a lot of strain on the group of people who are responsible for running the hostel. The incumbent management of the hostel administration is done manually by many colleges and universities. The important details of the residents of the hostel are kept as a stack of papers. The hostel rooms are allotted to the students based on their request which may result in corruption in the allocation process. It may also cause redundancy of data. Hostel offices and other data are altogether kept in a record note or displayed on the notice board. Any complaints or requests must also be entered manually which may be misplaced. The existing approach does not deal with mess token distribution and mess fee calculation. There are many chances that there might be discrepancies in the fee calculation. Furthermore, the mess tokens are distributed manually. There are high chances that the tokens might be misplaced. The limitations of the existing approaches are the existing approach stands no chance against the natural calamities which results in loss of data, information management, searching and retrieval is very difficult because of the involvement of huge volumes of data, security and integration of data is a major concern in the existing approach. It is a tedious process which involves a huge amount of time and material assets.

4 Proposed Approach

After a thorough analysis of the existing approach, if there exists a need to automate the existing process, the proposed system should be adopted to overhaul the existing system which is characterized by data inconsistency, concurrency problem and a possibility of losing students' information. To overcome all the limitations of the existing approach, we propose a Web application called hostel management. The approach that we propose involves the automation of all the manual processes done by the hostel staffs. While automating, data redundancy is removed. There will be no need of standing in long queues to get tokens, room allocation, mess change or leave forms. Hostelers can do all these in their mobiles/laptops. Room allocation and token booking are based on **first come first serve policy**. It also shows the fee structure of all rooms to the users, and the hostelers can pay their fee as well. Hostelers can change their mess before the beginning of every month. Once a student books a token, they will get a token id, which can be seen as a notification. They should show it to the mess supervisor. Similarly, leave form also generates an id. The student should show it to the watchman before leaving and entering the hostel. They can view their fee balance. Students can send their feedback which will be sent to the admin. The profile module shows all the information about the user. There, the user can set profile photo, username, etc. The students can use their accounts only after admin verification. If admin rejects their request, they can use their account. Similarly, if the admin rejects their requests for leave form/room allocation, they will not receive any id for room allocation or leaving the hostel. Confirmation of the student's accounts and room allocation are done by the admin and sent to the user as a notification. Admin also sets the count of all tokens. Admin can also view the number of tokens booked. Admin also has the authority to accept/reject leave forms and room allocation. He can also view the feedback given by various users. The proposed approach reduces all the costs related to labor-intensive activities. Backup of data is easy and efficient. Data consistency, data integration and data security are the advantages of the system. This will also remove all the errors that occur while allocation. This system ensures that all records are stored correctly which cannot be ensured in the existing system. The proposed system being a Web application runs on robust databases eliminating the problem of data redundancy ensuring data integrity since there will be no duplicate entries. Username and password are incorporated to ensure data security. The proposed system is cost efficient since there is no need of printing documents, duplicating documents, etc. Machine learning is used for the automatic allocation of rooms to the students based on the training data, i.e., the hostel dataset. The machine learns how rooms are allocated in the previous years and allocates rooms accordingly (Fig. 1).

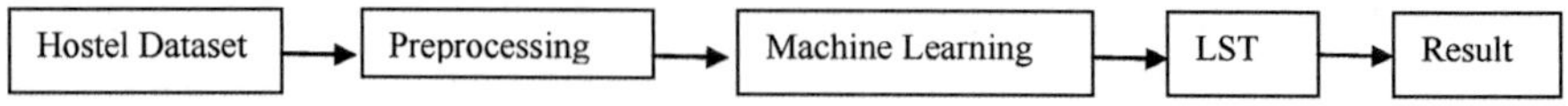

Fig. 1 Block diagram of the proposed system

LSTM is abbreviated as long short-term memory. LSTM is an artificial recurrent neural network (RNN). It is a general-purpose computer since it has feedback connections. They are well suited to classify process and make predictions based on data. They overcome the problems of traditional RNN. They can selectively remember patterns for a long duration of time. This is because information passes through a variety of cells in LSTM. This helps LSTM to selectively remember patterns. Machine learning is an application of artificial intelligence (AI) which without being explicitly programmed gives systems the ability to automatically learn and improve from experience. Machine learning focuses on the development of computer programs that can access data and use it learn for themselves. There are four types of machine learning algorithms, namely supervised, unsupervised, semi-supervised and reinforcement learning. Supervised learning is used in this application. The most common form of machine learning is supervised learning. In supervised learning, a set of examples and the training set are given as input to the system during the training phase. Each input is labeled with a desired output value so that the system knows what the output is, for that corresponding input. Let the training set be $\{(x_1, y_1), (x_2, y_2) \ldots (x_n, y_n)\}$ with x_i being the input and y_i being the class of the corresponding output. Here, labeled dataset is used which has both input and output parameters. While training the model, data is split in the ratio of 80:20 where 80% of the data is used for training and the rest is used for testing. The model learns from the training data and builds a model. Then, the testing data is passed through the model, and the derived output is compared to the original output, and accuracy is determined (Fig. 2).

Supervised learning is of two types—classification and regression. Classification is used when the output has a discrete value (defined labels). It can be a binary classification (where the output is either 0 or 1) or a multi-class classification (where there is more than one class). The goal here is to predict discrete values belonging to a particular class and evaluate on the basis of accuracy. When the output has a continuous value, regression is used. The goal here is to predict a value much closer

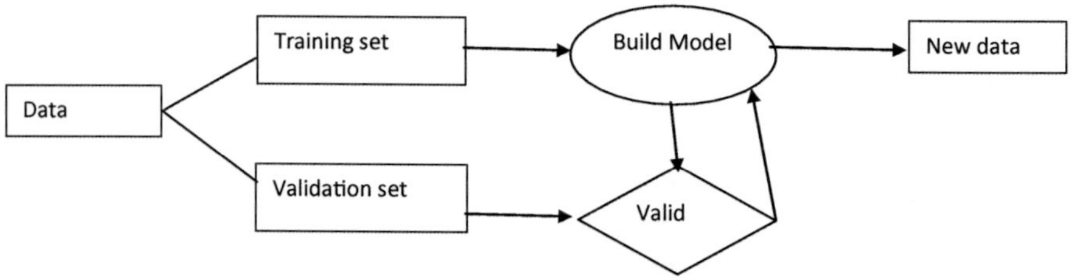

Fig. 2 Plan of execution

to the actual output. Random forest algorithm and support vector machine algorithms can be used for classification. Linear regression is used for regression. In random forest classifier, higher the number of trees in the forest, higher is the accuracy of the model. Decision tree concept is used to build the trees. Decision tree concept is based on rule-based system. Given a training dataset with a set of features and targets, the decision tree algorithm will give a set of rules which can be used for prediction on the test dataset. Calculating the nodes and forming the rules will happen using information gain and Gini index calculations. Random forest algorithm will handle the missing values. When we have more trees, random forest algorithm will not overfit the model. Categorical values can also be modeled and take the **test features** and use the rules of each randomly created decision tree to predict the outcome, and we calculate the votes for each predicted target, and then, we consider the high voted predicted target as the final prediction. The random forest is a classification algorithm consisting of many decision trees. **It uses bagging and feature randomness when building each individual tree to try to create an uncorrelated forest of trees** whose prediction by committee is more accurate than that of any individual tree.

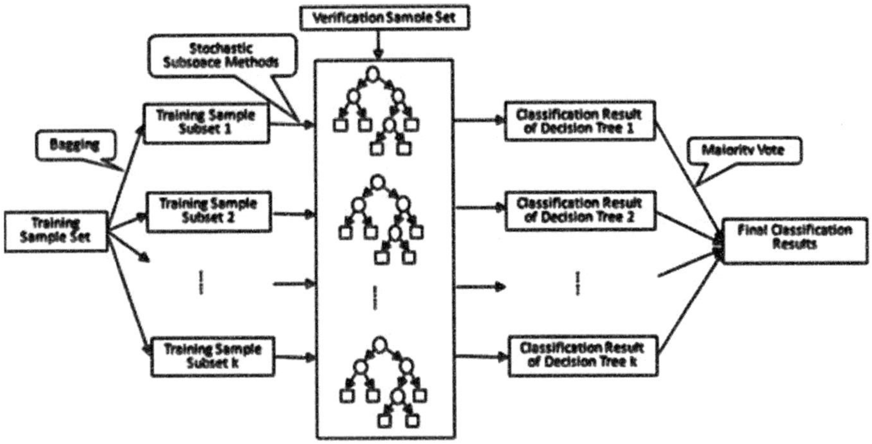

Overfitting to a particular data sample is one of the foremost problems observed in the literature especially in regression tasks. Random forests have trouble dealing with attributes that are multi-valued and multi-dimensional. They favor categorical variables with many levels. To classify a large amount of data with a single classifier is not worthy and may lead to less accuracy in the result. Hence, it is used on a dataset that has multiple classifiers.

Pseudocodes:
Random Forest:

1. From the total m features, randomly select k features such that $k \ll m$
2. Among the k features, calculate the root node using the best split approach
3. Split the node into daughter nodes using the best split approach
4. Repeat steps 1–3 until all the nodes have been reached

5. Repeat the steps 1–4 for *n* number of times to build n trees and hence, building the forest

Random Forest Prediction:

1. Using the rules of each randomly created decision tree and the test features, it predicts the outcome and stores it
2. Calculates the number of votes for each predicted target
3. From the random forest algorithm, it considers the highest voted predicted target as its final outcome.

5 Experimental Setup and Results

The hostel management system is a Web application developed for managing different activities in the hostel. It is very easy to use with a user-friendly GUI to automate, arrange and handle all the procedures of managing hostel offices. This is an online site which is developed using HTML and CSS for front end and PHP for back end. The site will be a great help to the workers. As the number of records is high, it is very helpful especially in large institutional organizations with a huge number of hostels.

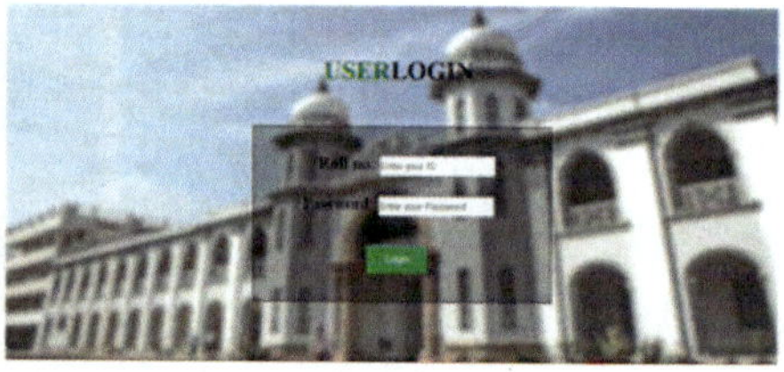

For registration, students should give all their details like name, roll number, password, phone number, email, etc. All these details will be stored in a database. Students cannot log in before admin validation. After admin validation, they can log in by giving their roll number and password. If both roll number and password match with a row in a database, the user will be logged in.

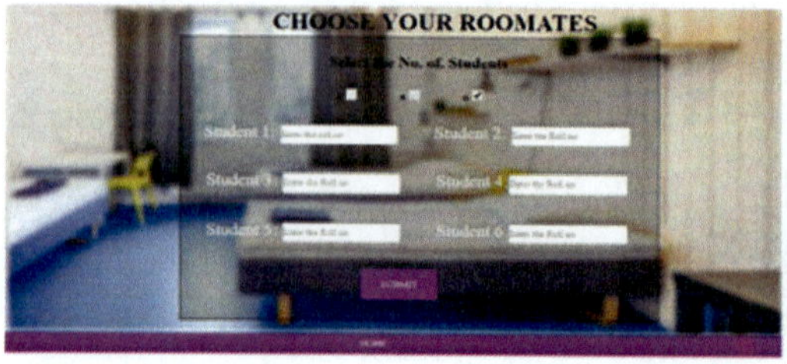

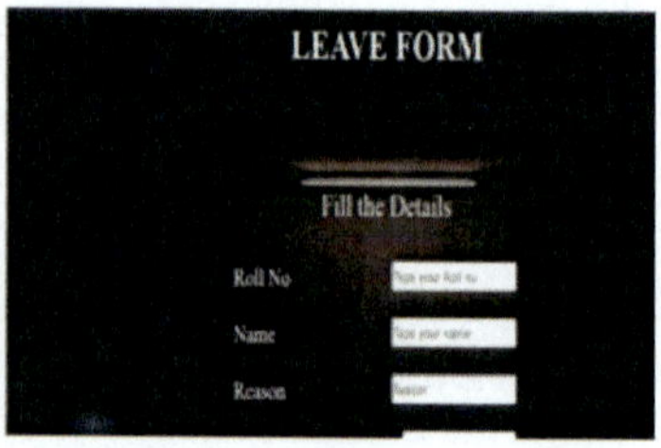

There are three types of rooms available to the student. Here, he should select the number of roommates and their roll numbers. A request will be sent to all the roommates. Once they accept that request, the room request will be sent to the admin. Room will be allocated after his verification. Request will not be sent to a student if he is already allocated a room. Here, the student should enter his roll number, reason for leave, entry time and exit time. Then, a request will be sent to the admin. The students will be permitted to leave the hostel only after admin's validation.

The student's current mess will be displayed. If he wishes to change the mess, he should select his preferred mess. However, the student can request for a mess change only before the beginning of each month. If he has sufficient balance in his account, mess will be changed. The students can book their tokens here. However, they can book tokens only during the specific days of the week when these items are available. A person belonging to south-Indian mess cannot request for tokens in north-Indian mess and vice versa. Once they book the tokens, they will get an id which should be shown to the mess supervisor. The count will be updated in the database which can be viewed by the admin.

This module displays which kind of room is available to which student and their price details. From this module, students can also pay their hostel fee. Here, the student should give all his card details and can pay their hostel fee.

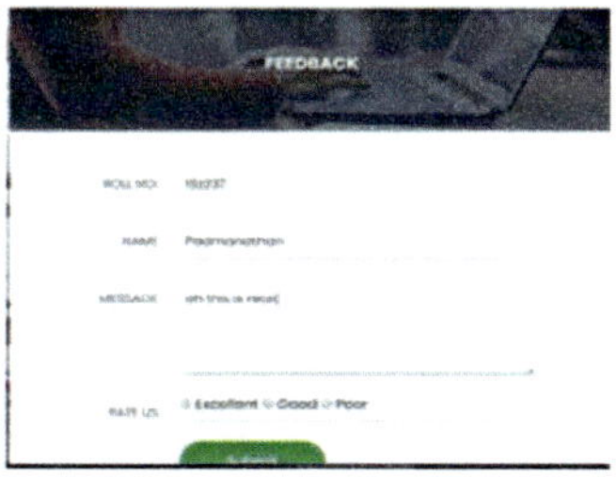

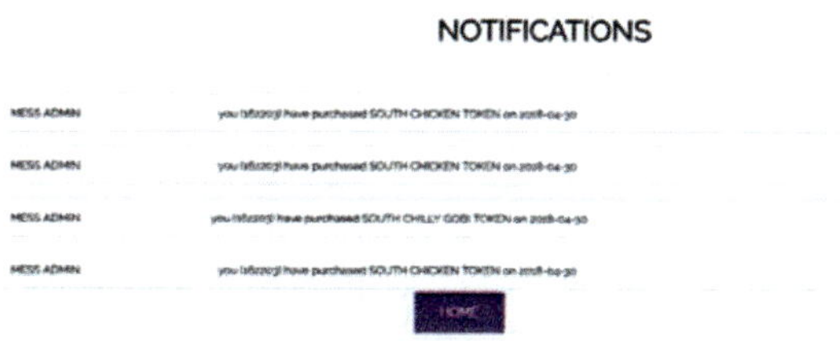

In this module, the student's roll number and name will be automatically displayed since they are logged into their account. He can send a message and also rate the application. The feedback will be sent to the admin. Here, all notifications such as token booking information, mess change information, room allocation info and leave request info will be displayed.

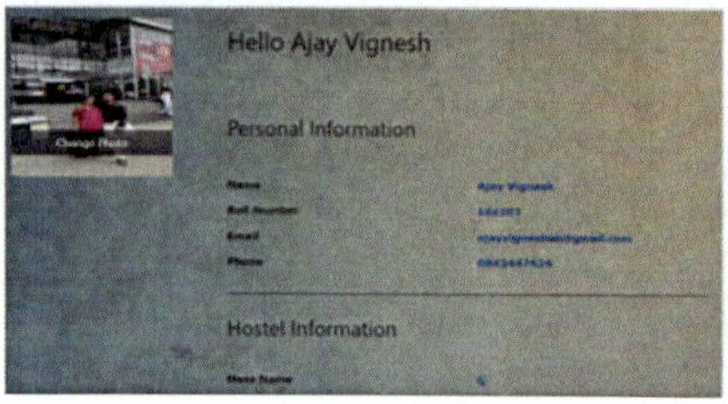

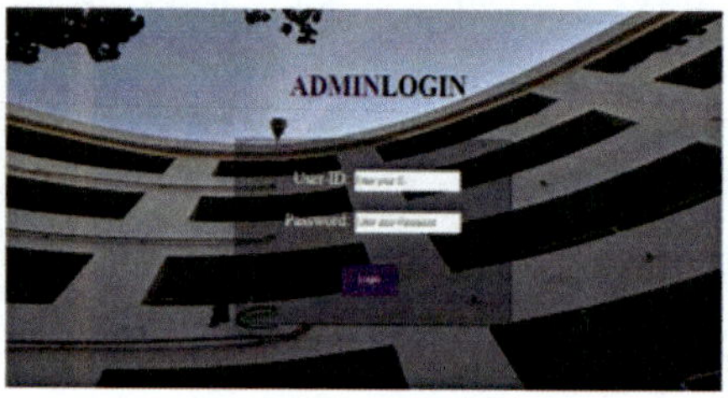

This is the student's profile module. Here, all his details such as name, roll number, email, phone, mess name and fee balance will be displayed. He can also change his display picture. In this module, the admin can log onto his account by specifying his user id and password. If the user id and password match with a record in the database, he can log into his account.

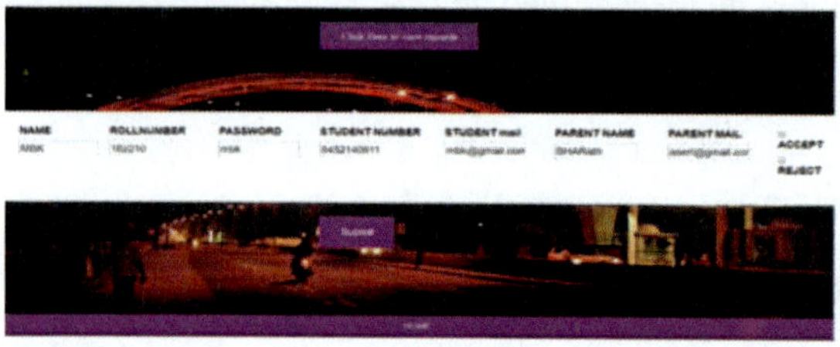

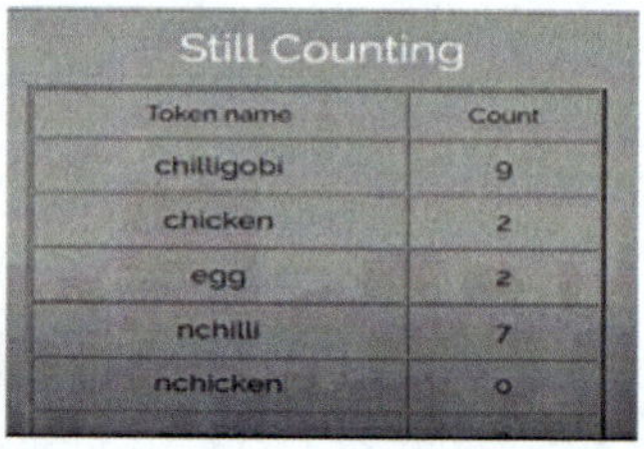
Still Counting

Token name	Count
chilligobi	9
chicken	2
egg	2
nchilli	7
nchicken	0

The admin should verify details from the college list and accept the student's account. Students can log into their account only after admin's verification. The count of all the tokens booked will be available only to the admin. He informs the mess supervisor the number of dishes to make only after the deadline to book tokens is over.

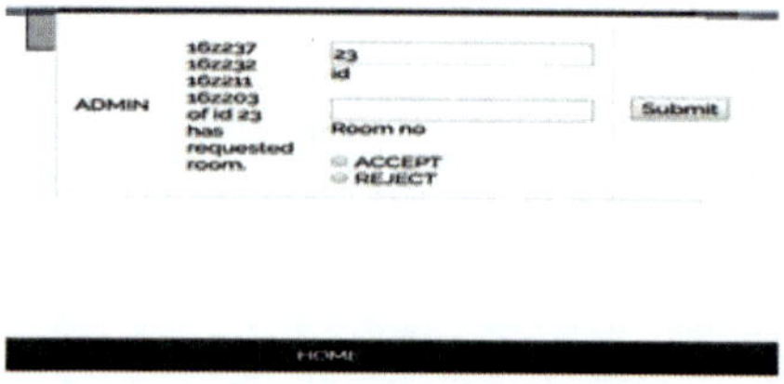

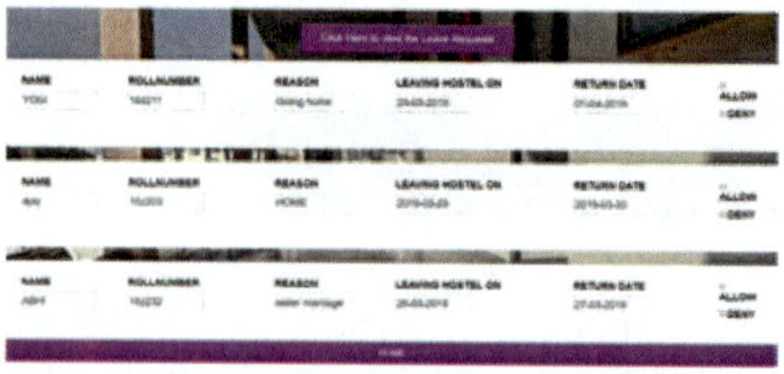

This is the room request notification. The admin can accept/reject their request. He should allocate a room based on the availability. Once the request is accepted, a notification is sent to the students. Admin can allow/deny the student requests to

Table 1 Modules comparison with and without machine learning support

Modules	Without machine learning	With machine learning
Registration	At a time, only five members can register in the application	It supports registration for up to 30 members at a time, which is a significant improvement
Room allotment	Supports requests from three students at a time	With machine learning, requests from 20 students are generated at a time
Leave form	Supports requests from four students at a time	No such issues in machine learning. Requests from 20 students are generated
Token booking	Since all students try to book tokens once the tokens are available, without machine learning only three students can book tokens at a time. So the time taken for each student to book a token is high	With machine learning, 30 students can book tokens at a time
Mess change	Just like token booking, only three students can request for mess change at a time	30 students can apply for mess change at a time

leave the hostel. Students cannot leave the hostel without admin validation. Admin can see all the feedbacks sent by the students.

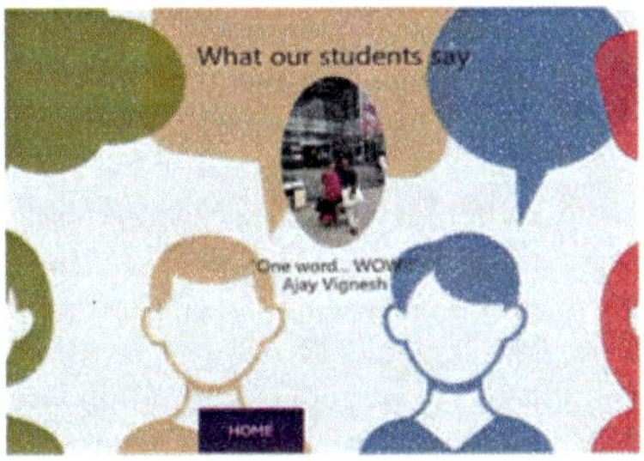

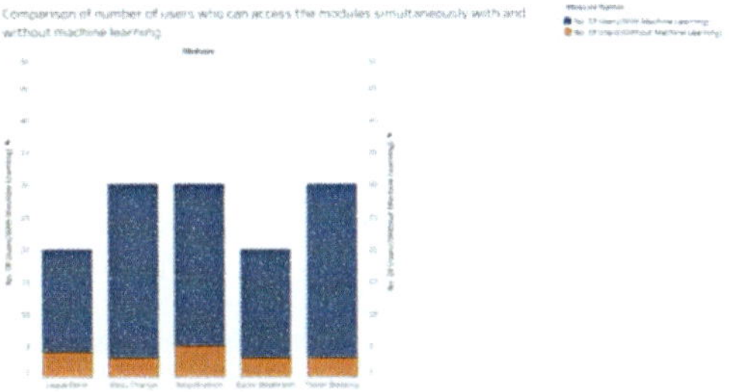

Since the dataset is a large one, we used random forest algorithm since it does not overfit the data no matter how big the dataset is. It handles all the missing data, and even categorical data can be modeled. Because of the random forest (Table 1).

References

1. N.K. Kumar, D. Vigneswari, M. Vamsi Krishna, G.V. Phanindra Reddy, An optimized random forest classifier for diabetes mellitus, in *Emerging Technologies in Data Mining and Information Security* (Springer, Singapore, 2019), pp. 765–773
2. H. Wang, R. Magagi, K. Goïta, M. Trudel, H. McNairn, J. Powers, Crop phenology retrieval via polarimetric SAR decomposition and Random Forest algorithm. Remote Sens. Environ. **231**, 111234 (2019)

3. E. Alickovic, A. Subasi, Alzheimer's Disease Neuroimaging Initiative, Automatic detection of Alzheimer Disease based on histogram and random forest, in *International Conference on Medical and Biological Engineering* (Springer, Cham, 2019), pp. 91–96
4. L. Hou, Y. Liu, A. Wei, Geographical variations in the fatty acids of Zanthoxylum seed oils: a chemometric classification based on the random forest algorithm. Ind. Crops Prod. **134**, 146–153 (2019)
5. S. He, W. Chen, H. Liu, S. Li, D. Lei, X. Dang, et al., Gene pathogenicity prediction of Mendelian diseases via the random forest algorithm. Hum. Genet. **138**(6), 673–679 (2019)
6. A. Subasi, A. Ahmed, E. Aličković, A.R. Hassan, Effect of photic stimulation for migraine detection using random forest and discrete wavelet transform. Biomed Sig Process Control **49**, 231–239 (2019)
7. N. Zeng, X. Ren, H. He, L. Zhang, D. Zhao, R. Ge, et al., Estimating grassland aboveground biomass on the Tibetan Plateau using a random forest algorithm. Ecol. Ind. **102**, 479–487 (2019)
8. A.C. dos Santos Luciano, M.C.A. Picoli, J.V. Rocha, D.G. Duft, R.A.C. Lamparelli, M.R.L.V. Leal, et al., A generalized space-time OBIA classification scheme to map sugarcane areas at regional scale, using Landsat images time-series and the random forest algorithm. Int. J. Appl. Earth Obs. Geoinf. **80**, 127–136 (2019)
9. L. Tang, F. Cai, Y. Ouyang, Applying a nonparametric random forest algorithm to assess the credit risk of the energy industry in China. Technol. Forecast. Soc. Chang. **144**, 563–572 (2019)
10. A.B. Shaik, S. Srinivasan, A brief survey on random forest ensembles in classification model, in *International Conference on Innovative Computing and Communications* (Springer, Singapore, 2019), pp. 253–260
11. R. Srinet, S. Nandy, N.R. Patel, Estimating leaf area index and light extinction coefficient using Random Forest regression algorithm in a tropical moist deciduous forest, India. Ecol. Inf. **52**, 94–102 (2019)
12. H. Nguyen, X.-N. Bui, Predicting blast-induced air overpressure: a robust artificial intelligence system based on artificial neural networks and random forest. Nat. Resour. Res. **28**(3), 893–907 (2019)
13. H.R. Pourghasemi, M.M. Saravi, Land-subsidence spatial modeling using the random forest data-mining technique, in *Spatial Modeling in GIS and R for Earth and Environmental Sciences* (Elsevier, 2019), pp. 147–159
14. S. Wang, G. Azzari, D.B. Lobell, Crop type mapping without field-level labels: random forest transfer and unsupervised clustering techniques. Remote Sens. Environ. **222**, 303–317 (2019)
15. F.B. de Santana, W.B. Neto, R.J. Poppi, Random forest as one-class classifier and infrared spectroscopy for food adulteration detection. Food Chem. **293**, 323–332 (2019)
16. X. Chen, T. Wang, S. Liu, F. Peng, A. Tsunekawa, W. Kang, et al., A new application of Random Forest Algorithm to estimate coverage of moss-dominated biological soil crusts in Semi-Arid Mu Us Sandy Land, China. Remote Sens. **11**(11), 1286 (2019)
17. J. Dou, A.P. Yunus, D.T. Bui, A. Merghadi, M. Sahana, Z. Zhu, et al., Assessment of advanced random forest and decision tree algorithms for modeling rainfall-induced landslide susceptibility in the Izu-Oshima Volcanic Island, Japan. Sci. Total Environ. **662**, 332–346 (2019)
18. H. Patel, S. Parikh, A. Patel, A. Parikh, An application of ensemble random forest classifier for detecting financial statement manipulation of Indian listed companies, in *Recent Developments in Machine Learning and Data Analytics* (Springer, Singapore, 2019), pp. 349–360
19. S. Balli, E.A. Sağbaş, M. Peker, Human activity recognition from smart watch sensor data using a hybrid of principal component analysis and random forest algorithm. Measure. Control **52**(1–2), 37–45 (2019)

Improving Accuracy of Software Estimation Using Stacking Ensemble Method

P. Sampath Kumar and R. Venkatesan

1 Introduction

Software managers have to do the estimation accurately in order to complete the project successfully. Software estimation is done along with planning to estimate the amount of resources, effort and time required to complete the project. Accuracy of software estimation is difficult during the early phases of software planning as there is no clear picture or the complete details of the entire software project [1]. But accurate estimation at early stages is required for successful completion of projects as overestimation leads to make the resources remain idle and underestimation leads to incomplete features or defective products and eventual failure in their customer delivery. The benefits of accurate estimation are higher quality, better coordination between project teams, improved status visibility and better budgeting [2]. Accurate software project estimates lead to better project planning, scheduling, effective utilization of resources, contract commitments, happy and successful project teams.

Software effort estimation traditionally was done using expert judgment, analogy, poker playing for agile, algorithmic approaches, and their advantages and disadvantages are discussed in Table 1. Traditional estimation methods tend to be inaccurate because of human bias and subjectivity. For tackling the subjectivity, bias and uncertainty in its estimation [3], data mining and machine learning algorithms were employed in software estimation which can adapt to the changes during the project. To still improve further in their estimation accuracy and stability, stacking ensemble of multiple machine learning algorithms is chosen, which will decrease the errors due to bias and variance. Estimation generally depends on the type of project, type of organization, experience and skill level of personnel.

Software effort estimation done using single machine learning model was found to be not so powerful in all conditions, and their performance varied from one dataset to

P. Sampath Kumar (✉) · R. Venkatesan
Department of CSE, PSG College of Technology, Coimbatore, India
e-mail: psk.cse@psgtech.ac.in

S. Patnaik et al. (eds.), *Advances in Machine Learning and Computational Intelligence*,
Algorithms for Intelligent Systems, https://doi.org/10.1007/978-981-15-5243-4_18

Table 1 Comparison of software estimation models

S. No.	Estimation method	Pros	Cons
1	Estimation by analogy	Estimation will be near accurate if analogous project is available. Best choice when you have very less data	Very difficult to find a near analogous project
2	Estimation by experts group	Group estimates improve the accuracy than individual estimates	Estimates vary with expertise, experience of the group
3	Parametric methods	Uses independent variables for estimation and tend to be quick and accurate if these independent variables of the project are captured by any process quickly	Cannot quantitatively estimate all project events and features
4	Decomposition and recomposition	Decomposition will be good for preliminary estimate, and recomposition will have better accuracy only when you have all details	

another. In order to overcome this issue, we are combining multiple machine learning models using stacking ensemble method to improve accuracy [4]. To effectively reduce the errors due to variance and bias, stacking ensemble has been adopted, where the first level has diverse base machine learning algorithms and the second level is a meta-learner which takes the predictions from the first level as the input and combines them to produce the final prediction. Stacking uses a technique where the base learners are different and heterogeneous; each learner will have a different learning bias and different knowledge representation, and so, their errors will not be correlated, and the combination of them will perform better than the individual base learners [5].

International Software Benchmarking Standards Group (ISBSG) [6] dataset has heterogeneous, diverse software project engineering data, and this dataset is better than older datasets from NASA and Promise repository in terms of recency and diversity. ISBSG dataset contains 252 variables, and out of that 20 variables are frequently used in projects for software estimation. By using data preprocessing techniques, the data has been transformed to a format where effective machine learning algorithms can be applied to extract knowledge and make accurate predictions in software estimation. The ISBSG dataset has 252 variables and those have been reduced to 12 relevant independent variables using data preprocessing and feature selection.

2 Related Work

Over the years, software estimation has been traditionally done using techniques like expert judgment, analogy-based estimation and parametric approaches [7]. When these methods are used, accuracy has always been a challenge due to various limitations found in these methods. In order to improve the accuracy of the algorithms, machine learning algorithms have been employed which extracted knowledge from the previous history. Wen et al. have done a comprehensive study on various machine learning algorithms for effort estimation [8], and in this study, it was found that neural networks outperformed others in getting more accuracy than the other machine learning models like case-based reasoning, decision tree and Bayesian networks.

Machine learning algorithms are basically sensitive to noises in the dataset, and this sensitivity will affect the prediction accuracy. In order to overcome this sensitivity issue, ensemble methods should be used which combines multiple machine learning algorithms, and this increases prediction accuracy [9]. Different kinds of ensemble methods like bagging, boosting and stacking can be tried to improve the accuracy. Bagging and boosting use homogenous ensemble technique, whereas stacking uses heterogeneous ensemble technique [10]. Also, accuracy of the ensemble methods will improve if the base learners are diverse [11].

The machine learning algorithm accuracy will improve when data is preprocessed and relevant features are selected. The ISBSG data contains missing values and outliers, and each feature has different ranges. To deal with these issues, imputation, listwise deletion [12] has been used to deal for missing values, outlier data is removed and normal normalization and standardization procedures are adopted. Apart from these preprocessing methods, feature selection has been implemented where unwanted dependent variables which are strongly correlated among themselves and weakly correlated with the target variable are removed [13].

Regression problems are evaluated using mean square error (MSE), mean absolute error (MAE) and mean magnitude of relative error (MMRE), and percentage relative error deviation PRED(x) is selected as evaluation metrics [14]. When the heterogeneous and the latest ISBSG dataset [15] is used, the benefits are (i) data about projects is public (ii) data can be filtered on quality rating (iii) size (functional) measurement is based on accepted international standards. In addition, the projects have been selected based on application type, organization type, development platform and development type.

Linear regression is one of the simplest models used in supervised machine learning for prediction purposes. But since the correlation between the dependent variables (effort) in software effort estimation does not have exact linear relationship with explanatory variables, the accuracy will be lower in that case. In order to overcome this issue in linear regression, ensemble technique is used with linear regression as one of the base learners [16].

Random forest is an ensemble method, and they operate by constructing many decision trees and output the result by calculating the mean prediction of the individual trees. Random forest produces better accuracy than linear regression and decision trees, but they are sensitive to the number of trees and the number of chosen attributes. This sensitivity can be reduced by choosing the best values for these two parameters [17].

Neural networks are nonlinear mapping from inputs to outputs, and they perform much better than the traditional approaches and simpler machine learning models when the data is nonlinear. But the disadvantage of neural network is that it converges to the determined local minima than global minima [18].

This paper aims to improve the accuracy in software estimation considering the above limitations and to implement an effective approach using latest datasets with required data preprocessing and using stacking ensemble method of diverse base machine learning models.

3 Theoretical Background

3.1 Software Estimation Techniques

The success of any software project depends on how close the effort estimation is close to the actual efforts. But it is difficult to predict the software effort estimate at the beginning of the software life cycle with so many factors that can vary during the execution of the project. The inaccurate estimates can affect the software project schedule and quality of the project.

The sources of software estimation error primarily can come from uncertain requirements, subjectivity, bias, skill mismatch of the personnel and unfamiliar business/technology areas. The traditional software estimation techniques used so far were expert judgment, estimation by analogy, parametric techniques like COCOMO and use case-based techniques. All the above mentioned techniques are inaccurate and have limitations when applied to generic project effort estimation due to subjectivity and human bias factors encountered. In order to overcome the limitations, machine learning techniques have been employed which will extract knowledge from software engineering data collected from various organizations and make near accurate predictions [19]. The machine learning algorithms make use of the heterogeneous data collected from various organizations and can easily adapt to the changing environments.

3.2 Using Ensemble of Machine Learning Algorithms

Ensemble Method is a very powerful technique to increase accuracy of target result when using with machine learning algorithms [20–22]. Different machine learning algorithms will have different levels of bias and variance on the software engineering data which is used to predict the effort estimate. To overcome the errors due to bias and variance of machine learning algorithms, outputs of multiple machine learning algorithms are combined through ensemble methods. Stacking uses meta-learner which combines multiple base machine learning models to improve the accuracy [23]. In this paper, a stacking ensemble model of three base machine learning models, namely linear regression, random forest regression and neural networks and a support vector regressor as a meta-learner, has been attempted, which combines these three models with a factored weightage. For regression-based estimation, MAE, MSE, MMRE and PRED(x) are used as evaluation metrics for accuracy measurement.

4 Proposed Approach

The proposed approach deals with the application of stacking ensemble of machine learning methods, to increase the prediction of software accuracy. All the machine learning algorithms' effectiveness depends on the quality of the dataset. The heterogeneous ISBSG dataset contains data about the projects collected over 35 countries and spanning different kinds of platforms and applications. The ISBSG dataset needs to be sent through the process of data preprocessing before applying the ensemble of machine learning algorithms for accurate effort prediction.

4.1 ISBSG Data Preprocessing

ISBSG dataset is a repository that contains heterogeneous data (8261 rows and 252 columns) with varying degree of quality. ISBSG dataset contains a lot of missing values, duplicate values, correlated features and categorical data, and this needs to be preprocessed into a mathematically acceptable numerical format before machine learning algorithms are applied. Missing data treatment (MDT), data reduction through feature selection and correlated features, outlier removal and data transformation through scaling and standardization are used in dataset preprocessing.

ISBSG dataset has many missing values, and these missing values affect the software effort estimation accuracies. List-wise deletion is done to remove the rows with null values from the dataset. The accuracy of the machine learning algorithm will improve when the correlation between dependent and independent variables is high and when the correlation among the independent variables is low. Outliers which are 3-sigma away from the mean value are removed. Highly correlated columns were removed, and only highly relevant features for the software effort estimation were retained. After these steps, the total number of features retained is 12.

Reduced features which will be highly relevant with the target variable 'Summarized Effort' are 'Application Group,' 'Development Type,' 'Development Platform,' 'Language Type,' 'Functional Size,' 'Relative Size,' 'Resource Level,' 'Team Size Group,' 'Development Methodologies,' 'Development Techniques,' 'Architecture,' 'Type of Server' and 'Target Platform.'

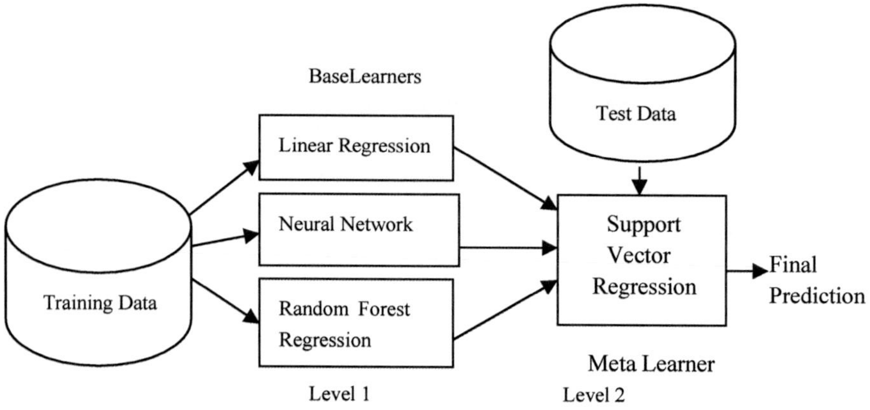

Fig. 1 Stacking with three base learners and one meta-learner

4.2 Stacking

Stacking regression is an ensemble technique to combine multiple machine learning regression models using a meta-learner. Each regression model chosen is trained on the complete training set, and then, the outputs from these models are fed into another meta-learner which combines each model with a factored weight to increase the accuracy.

Stacking is an ensemble technique which uses two levels as shown in Fig. 1. The first level or base level has many base level predictors, and the second-level predictor combines the outputs from the first level. Mostly, the stacked model (also called second-level model) will outperform each of the individual models due to its smoothing nature and ability to highlight each base model where it performs best and discredit each base model where it performs poorly. For this reason, stacking is the most effective when the base learners are significantly diverse.

4.3 Evaluation Criteria

For software estimation which is basically a regression measure, the most appropriate criteria for error measurement which were widely used are mean magnitude of relative error (MMRE) and percentage relative error deviation (PRED) apart from the standard error measures such as mean absolute error and mean square error. MMRE and PRED are both based on magnitude of relative error which is the difference between actual value and estimated value. The metrics used for evaluation are defined as follows:

(a) Magnitude of Relative Error (MRE) = |Actual Value − Estimated Value|/Actual Value.

Table 2 Results comparison for individual ML methods and stacking in effort estimation

Metrics	Linear_Regression	Random_Forest	Neural network	Stacking
MAE	0.15	0.12	0.18	0.11
MMRE	0.14	0.11	0.17	0.10
PRED (0.25)	0.86	0.92	0.82	0.925

(b) Mean Magnitude of Relative Error (MMRE) $= 1/n \sum_{(i=1)}^{n}$ MRE, where n is number of samples

(c) Percentage Relative Error Deviation PRED(x) $= 1/n \sum_{(t=1)}^{n} \begin{cases} 1 \text{ if MRE} < x \\ 0 \text{ Otherwise} \end{cases}$, where n is number of samples.

A software model estimate is considered accurate if MMRE < 0.25 and PRED (0.25) > 0.75 [24].

5 Results and Discussion

This section discusses the comparison of software effort estimation results obtained by stacking ensemble method and the results obtained by individual machine learning methods. The stacking ensemble model uses random forest regression, linear regression and neural networks as base learners and support vector regressor as the meta-learner.

The accuracy comparison is done initially with the estimates of the learners individually, and then, they are compared with the estimates of the stacking ensemble. The results are given in Table 2 which shows that the stacking model performs better than the individual machine learning models.

In the above experiment, the software effort estimation is done using the linear regression, neural networks and random forest regression individually and found that the random forest regression had better accuracy than the other two. Then, the same data is passed to the stacking ensemble, and the accuracy prediction improvement is observed to have error values of mean absolute error (0.11), MMRE (0.10) and PRED (0.925), which is slightly better than random forest regressor.

6 Conclusion

The traditional methods used for software estimation, in spite of advancement in software development methodologies, still remained inaccurate due to bias and subjectivity. All these techniques had some limitations and could not adapt to changing demands in software development. Knowledge is extracted from software engineering datasets using machine learning techniques, to improve accuracy in software estimation. To improve accuracy further, the stacking ensemble with two levels is employed where the first base level contains three base learners (linear regression, neural network, random forest regression) and support vector regression as the meta-learner. From the results, it is observed that the ensemble techniques have shown improvement in PRED (0.25) values and MMRE error scores.

The effort estimation accuracy scores can be improved further if we can get more software engineering data from the Information Technology industry and add more diverse level learners in the base level of stacking.

Acknowledgements The authors would like to extend gratitude to the Principal and Management of PSG College of Technology, India, for all their support rendered in carrying out this project. We would also like to thank ISBSG for the Software Engineering Dataset which was highly helpful in completion of this project.

References

1. B.W. Boehm, C. Abts, A.W. Brown, S. Chulani, B.K. Clark, E. Horowitz, R. Madachy, D.J. Reifer, B. Steece, *Software Cost Estimation with Cocomo II* (Prentice Hall PTR, 2000)
2. S. McConnell, *Software Estimation—Demystifying the Black Art* (Microsoft Press, 2006)
3. R.K. Wysocki, *Effective Project Management: Traditional, Agile, Extreme, Industry Week* (Wiley, 2014)
4. R. Hidaya, M.K. Benhachmi, *Smart Data and Computational Intelligence*, vol. 66 (Springer International Publishing, 2019)
5. A. Ledezma, R. Aler, A. Sanchis, D. Borrajo, GA-stacking: evolutionary stacked generalization. Intell. Data Anal. **14**(1), 89–119 (2010)
6. International Software Benchmarking Standards Group, ISBSG Repository Data Release 12—Field Descriptions (2018)
7. Z. Mansor, Z. Mohd Kasirun, S. Yahya, N.H. Hj Arshad, Current practices of software cost estimation technique in Malaysia context. Commun. Comput. Inf. Sci. **251** (CCIS, no. PART 1), 566–574 (2011)
8. J. Wen, S. Li, Z. Lin, Y. Hu, C. Huang, Systematic literature review of machine learning based software development effort estimation models. Inf. Softw. Technol. **54**(1), 41–59 (2012)
9. L.L. Minku, X. Yao, Ensembles and locality: insight on improving software effort estimation. Inf. Softw. Technol. **55**(8), 1512–1528 (2013)
10. S. Wan, H. Yang, Comparison among methods of ensemble learning, in *Proceedings—2013 International Symposium on Biometrics Security Technologies. ISBAST 2013* (2013), pp. 286–290
11. S. Ali, S. Tirumala, Ensemble methods for decision making, in *International Conference on Machine Learning and Cybernetics* (2015)
12. K. Strike, K. El Emam, N. Madhavji, Software cost estimation with incomplete data. IEEE Trans. Softw. Eng. **27**(10), 890–908 (2001)
13. J. Huang, Y.F. Li, M. Xie, An empirical analysis of data preprocessing for machine learning-based software cost estimation. Inf. Softw. Technol. **67** (2015)
14. S. Berlin, T. Raz, C. Glezer, M. Zviran, Comparison of estimation methods of cost and duration in IT projects. Inf. Softw. Technol. **51**(4), 738–748 (2009)
15. C. López-Martín, A. Abran, Neural networks for predicting the duration of new software projects. J. Syst. Softw. **101**, 127–135 (2015)
16. S.B. Kotsiantis, D. Kanellopoulos, I.D. Zaharakis, Bagged averaging of regression models. IFIP Int. Fed. Inf. Process. **204**, 53–60 (2006)

17. Z. Abdelali, H. Mustapha, N. Abdelwahed, Investigating the use of random forest in software effort estimation. Procedia Comput. Sci. **148**, 343–352 (2019)
18. P. Rijwani, S. Jain, Enhanced software effort estimation using multi layered feed forward artificial neural network technique. Procedia Comput. Sci. **89**, 307–312 (2016)
19. T.M. Mitchell, *Machine Learning* (McGraw-Hill, 1997)
20. D. Azhar, P. Riddle, E. Mendes, N. Mittas, L. Angelis, Using ensembles for web effort estimation, in *ACM/IEEE International Symposium on Empirical Software Engineering and Measurement* (IEEE, 2013)
21. A. Idri, M. Hosni, A. Abran, Improved estimation of software development effort using classical and fuzzy analogy ensembles. Appl. Soft Comput. **49**, 990–1019 (2016)
22. A. Krogh, J. Vedelsby, Neural network ensembles, cross validation, and active learning, in *Advances in Neural Information Processing Systems*, vol. 7 (1995), pp. 231–238
23. S. Wan, H. Yang, Comparison among methods of ensemble learning, in *Proceedings of International Symposium on Biometrics Security Technologies. ISBAST* (2013), pp. 286–290
24. S.D. Conte, H.E. Dunsmore, V.Y. Shen, *Software Engineering Metrics and Models* (Benjamin/Cummings Pub. Co., 1986)

EEG-Based Automated Detection of Schizophrenia Using Long Short-Term Memory (LSTM) Network

A. Nikhil Chandran, Karthik Sreekumar, and D. P. Subha

1 Introduction

Schizophrenia is a mental disorder usually found in teenagers or in early adults. Schizophrenia affects more than 21 million people worldwide [1] and represents a serious public health problem. Due to the absence of biological markers, diagnosis of schizophrenia largely depends upon the examination of the mental state of the patient through hours of counseling and interviewing over a period of time, usually extending up to months [2]. Common symptoms include lack of insight, auditory hallucination, suspiciousness and delusional mood [3].

EEG is a noninvasive physiological method for detecting electrical brain activity by mounting electrodes on the scalp surface. The signals are broadly divided into five different frequency bands, namely Delta (1–4 Hz), Theta (4–7 Hz), Alpha (7–12 Hz), Beta (12–30 Hz) and Gamma (>30 Hz). Neurological conditions reflected in EEG aid in identifying people with mental disorder/illness from healthy controls. This is possible as the brain behavior is different for both the groups [4]. Features are extracted from the raw EEG data to analyze the electrical brain activity. These features can be either time-domain features like mean, variance, skewness or several other nonlinear parameters like fractal dimension [5], approximate entropy [6], Hurst exponent [7], largest Lyapunov exponent [8], etc. Earlier studies have successfully employed these nonlinear features to characterize various mental disorders.

Artificial neural networks (ANNs) are biologically inspired (from the human brain) computation models which outperform the previous machine learning models.

A. Nikhil Chandran
School of Biotechnology, National Institute of Technology, Calicut, India

K. Sreekumar
Department of Physics, National Institute of Technology, Calicut, India

D. P. Subha (✉)
Department of Electrical Engineering, National Institute of Technology, Calicut, India
e-mail: subhadp@nitc.ac.in

S. Patnaik et al. (eds.), *Advances in Machine Learning and Computational Intelligence*, Algorithms for Intelligent Systems, https://doi.org/10.1007/978-981-15-5243-4_19

Long short-term memory (LSTM) is an artificial recurrent neural network (RNN) architecture in the realm of deep learning [9]. LSTM is prominent by its feedback connections unlike observed in standard neural networks and is therefore preferred when working with time series data. LSTM can handle a large data set and seems to overcome long-term dependencies like vanishing and exploding gradients [9]. These benefits motivated us to use LSTM network. Till date, not much work has been done on EEG signals using LSTM except for the prediction of depression [10] and epileptic seizures [11].

2 Data Recording

The data set used was published by RepOD in the year 2017 [12] and is comprised of 14 patients with paranoid schizophrenia (mean age: 28.1 $\pm$ 3.7 years) and 14 healthy controls (mean age: 27.8 $\pm$ 3.2 years). Data was recorded from 19 EEG channels using the standard 10–20 electrode placement system: Fp1, Fp2, F7, F3, Fz, F4, F8, T3, C3, Cz, C4, T4, T5, P3, Pz, P4, T6, O1, O2 with a sampling frequency of 250 Hz. The reference electrode was placed between electrodes Fz and Cz. Olejarczyk E and Jernajczyk W used this data set for graph-based analysis of brain connectivity in schizophrenia [13]. The EEG signals were acquired under eyes-closed resting state. Artifacts were removed from the signal using total variation denoising (TVD) method. In order to remove power line interference, a notch filter of 50 Hz was used.

3 Methodology

Nonlinear features such as Katz fractal dimension, approximate entropy and the time-domain feature of variance are extracted for the analysis of EEG signals acquired from schizophrenia patients and healthy controls. Feature extraction was carried out in MATLAB platform, and the LSTM network was modeled using Keras in Python. An i7 processor with clock speed of 2.8 GHz was used for computation.

3.1 Katz Fractal Dimension

Fractal dimension is a measure of complexity of a pattern that shows how the detail in a pattern changes with the scale at which it is measured. In this paper, the fractal dimension is calculated by a method proposed by Katz [14].

Consider a curve with points $S = \{s_1, s_2,\ldots,s_N\}$. The coordinates of the sequence s_i are represented as (x_i, y_i) [15]. The Euclidean distance between the points is:

$$\mathrm{dist}(s_i, s_j) = \sqrt{(x_i - x_j)^2 + (y_i - y_j)^2} \tag{1}$$

Let L be the total length of the curve (the sum of Euclidean distances between successive points) and d be the diameter (or planar extent) of the curve estimated as the distance between the first point of the sequence and the point of the sequence that provides the farthest distance.

$$d = \max\{\mathrm{dist}(s_i, s_j)\} \tag{2}$$

The fractal dimension D of the curve is:

$$D = \frac{\log(L)}{\log(d)} \tag{3}$$

Let 'a' be the average distance between two successive points in the curve. If n is the number of steps in the curve,

$$a = L/n \tag{4}$$

So, Eq. (3) becomes:

$$D = \frac{\log(n)}{\log\left(\frac{d}{L}\right) + \log(n)} \tag{5}$$

The above steps enumerate the calculation for the Katz fractal dimension of a waveform.

3.2 Approximate Entropy

Approximate entropy (ApEn) is a quantification method for identifying the regularity and complexity present in a time series data.

Let, $U(1), U(2), U(3), \ldots, U(N)$ form a time series data, with N raw data equally spaced in time [16].

Let $X(i)$ be a subsequence of U such that

$$X(i) = [U(i), U(i+1), \ldots, U(i+m-1)] \tag{6}$$

where m is the length of compared run of data.

Consider $\{x(j)\}$, a subsequence of X. Each element in $x(j)$ is compared to $x(i)$, and two parameters $C_i^m(r)$ and $C_i^{m+1}(r)$ are defined as:

$$C_i^m(r) = \frac{\sum_{j=1}^{N-m} k_j}{N-m} \tag{7}$$

where

$$k = \begin{cases} 1, & \text{if } |x(i) - x(j)| \text{ for } 1 \le j \le N-m \\ 0, & \text{otherwise} \end{cases} \tag{8}$$

and r represents the noise filter level defined as

$$r = k \times \mathrm{SD} \tag{9}$$

where SD is the standard deviation of X.

Define $\phi^m(r)$ and $\phi^{m+1}(r)$ as:

$$\phi^m(r) = (N-m) - 1 \sum_{i=1}^{N-m} \ln\left(C_i^m(r)\right) \tag{10}$$

$$\phi^{m+1}(r) = (N-m)^{-1} \sum_{i=1}^{N-m} \ln\left(C_i^m(r)\right) \tag{11}$$

ApEn(m, r, N) is calculated as:

$$\mathrm{ApEn} = \phi^m(r) - \phi^{m+1}(r) \tag{12}$$

In this work, value of m is taken as 1 and r is 0.5.

3.3 Variance

Variance measures the spread of a set of numbers around its mean value. Variance of a set of values is defined as:

$$\mathrm{Var}(X) = \frac{1}{n} \sum_{i=1}^{n} (x_i - \mu)^2 \tag{13}$$

where n is the sample frequency, μ is the mean of the sample and x_i is the ith sample.

4 LSTM Architecture and Modeling

LSTM, a type of recurrent neural network (RNN), resolves the problem of vanishing gradient seen in conventional RNN. LSTM uses sigmoidal gates for remembering important information and forgetting irrelevant data. This makes it more efficient in classifying or clustering groups of data or detecting anomalies. Aside from the regular recurrent neural network, LSTM has longer-term dependency. Figure 1 shows the LSTM architecture.

The EEG data from 14 schizophrenia patients and 14 healthy controls was split into 4 s windows, each of 1000 epochs. 6790 feature vectors of FD, ApEn and variance are calculated from each electrode of the EEG signals.

Of the input feature vectors, 6000 vectors are used for training, and the rest is used as test data set. The model has four LSTM layers with 32 hidden neurons each. A 30% dropout is added to each LSTM layer. The output layer is a single neuron dense layer. The structure of this LSTM network is shown in Fig. 2.

The model was trained for 250 epochs. Adam optimizer was used to fit the model with a learning rate of 0.0001, and binary cross-entropy was used as loss function.

Binary cross-entropy measures the difference between the true value and predicted value for each class [18]. The individual error is averaged and depicted as the final model loss.

$$L(y, \hat{y}) = -\sum_{i=0}^{N}\left(y_i \log(\hat{y}_i) + (1 - y_i)\log(1 - \hat{y}_i)\right) \quad (14)$$

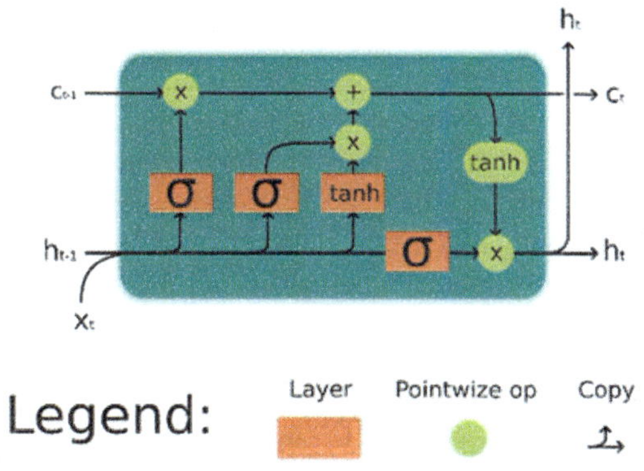

Fig. 1 LSTM architecture [17]. 'X_t' is the tth input, 'h_t' is the tth output, 'C_t' is the cell state, 'σ' represents a sigmoid function, 'tanh' represents the hyperbolic tangent function, 'X' represents pointwise multiplication operator and '+' is the pointwise addition operator

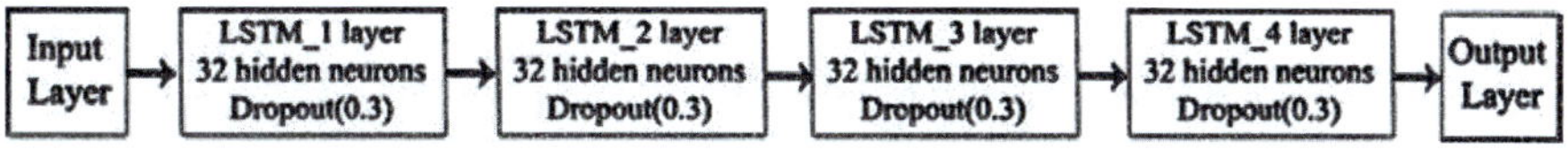

Fig. 2 Structure of the LSTM network

where $\hat{y}$ is the predicted value and y is the actual value.

The model is evaluated using four parameters: accuracy, precision, recall and f-score.

Accuracy is the ratio of number of correct predictions to total number of predictions

$$\text{Accuracy} = \frac{\text{TP} + \text{FN}}{\text{Total}} \tag{15}$$

Precision is the proportion of positive identifications that are correct.

$$\text{Precision} = \frac{\text{TP}}{\text{TP} + \text{FP}} \tag{16}$$

Recall is the proportion of positives which were identified correctly.

$$\text{Recall} = \frac{\text{TP}}{\text{TP} + \text{FN}} \tag{17}$$

F-score is the harmonic mean between precision and recall.

$$F\text{-score} = 2 \times \frac{\text{precision} \times \text{recall}}{\text{precision} + \text{recall}} \tag{18}$$

5 Results

In this work, classification of EEG signals of schizophrenic patients and healthy subjects is carried out using the long short-term memory (LSTM) technique. The data is obtained from 14 schizophrenic patients and 14 healthy subjects. The network is trained using 89% of the data set and tested using the remaining 11%.

The plot of model accuracy vs number of epochs is represented in Fig. 3, and the plot of model loss vs number of epochs is represented in Fig. 4. A classification accuracy of 99.0% is obtained while classifying schizophrenia patients and normal controls. The evaluation metrics are shown in Table 1. 20% of the training set was used for validation. It is observed that the proposed LSTM works very well in classifying schizophrenic patients from healthy subjects. This model is compared with the previous works on classification of EEG signals of schizophrenia patients and healthy controls using machine learning methods. Thilkavathi et al. [6] reported an accuracy of 81.5% using a feedforward neural network and an accuracy of 88.5% using support vector machine (SVM).

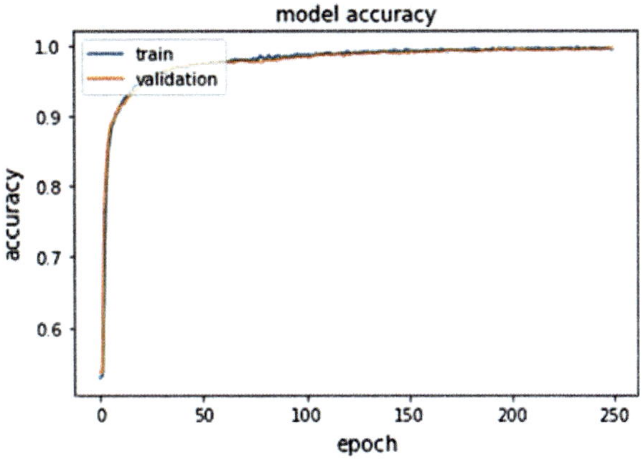

Fig. 3 Model accuracy versus number of epochs

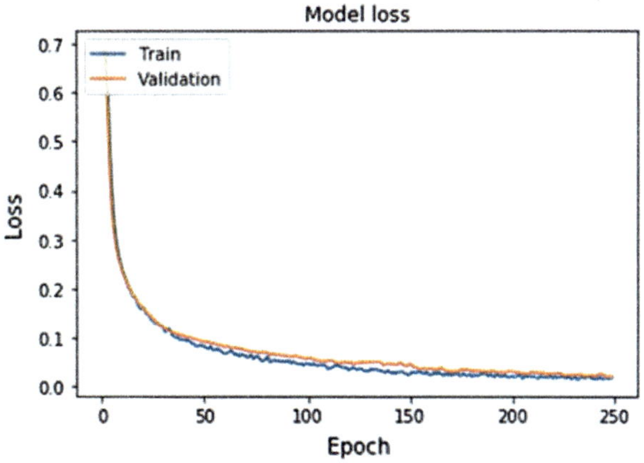

Fig. 4 Model loss versus number of epochs

Table 1 Evaluation metrics

	Accuracy (%)	Precision (%)	Recall (%)	*F*-score (%)
Test data	99.0	99.2	98.9	99.0
Holdout cross-validation	99.1	99.3	99.0	99.2
Fivefold cross-validation	99.1 ± 0.15	99.4 ± 0.3	99.0 ± 0.31	99.2 ± 0.15

6 Conclusion

In this paper, the classification capability of LSTM is evaluated. The nonlinear time-domain feature vectors used for classification are the fractal dimension and approximate entropy, whereas the time-domain feature vector used is variance. The EEG signals were acquired under eyes-closed resting state. The model was trained for 6000 feature vectors and tested on 790 feature vectors. It was found that the LSTM network could classify the patients and healthy controls more accurately than the machine learning techniques of FFNN and SVM with a classification accuracy of 99.0%. Hence, LSTM proves to be a better classifier for EEG signals. This computer-aided technique of LSTM could easily detect schizophrenia from EEG signals with a high accuracy which could revolutionize the healthcare sector.

References

1. Schizophrenia. https://www.who.int/mental_health/management/schizophrenia/en/. Last accessed 11 Oct 2019
2. A. Barbato, *Schizophrenia and public health* (Division of Mental Health and Prevention of Substance Abuse, World Health Organization, Geneva, 1998)
3. M.M. Picchioni, Robin M. Murray, Schizophrenia. BMJ (Clinical Research ed.) **335**(7610), 91–95 (2007)
4. S.D. Puthankattil, P.K. Joseph, Classification of EEG signals in normal and depression conditions by ANN using RWE and signal entropy. J. Mech. Med. Biol. **12**(4) (2012)
5. M. Sabeti, S.D. Katebi, R. Boostani, G.W. Price, A new approach for EEG signal classification of schizophrenic and control participants. Expert Syst. Appl. **38**, 2063–2071 (2011)
6. B. Thilakvathi, S. Shenbaga Devi, K. Bhanu, M. Malaippan, EEG signal complexity analysis for schizophrenia during rest and mental activity. Biomed. Res. **28**(1), 1–9 (2017)
7. O. Tan, S. Aydin, G.H. Sayar, D. Gürsoy, EEG complexity and frequency in chronic residual schizophrenia. Anatolian J. Psychiatry **17**(5), 385–392 (2016)
8. J. Röschke, J. Fell, P. Beckmann, Nonlinear analysis of sleep EEG data in schizophrenia: calculation of the principal Lyapunov exponent. Psychiatry Res. **56**(3), 257–269 (1995)
9. S. Hochreiter, J. Schmidhuber, Long short-term memory. Neural Comput. **9**(8), 1735–1780 (1997)
10. S. Dhananjay Kumar, D.P. Subha, Prediction of depression from EEG signal using Long Short Term Memory (LSTM), in *3rd International Conference on Trends in Electronics and informatics* (ICOEI-IEEE 2019), pp. 1248–1253 (23–25 April 2019)
11. K.M. Tsiouris, V.C. Pezoulas, M. Zervakis, S. Konitsiotis, D.I. Fotiadis, A long short-term memory deep learning network for the prediction of epileptic seizures using EEG signals. J. Comput. Biol. Med. 24–37 (2018)
12. E. Olejarczyk, W. Jernajczyk, EEG in schizophrenia. RepOD http://dx.doi.org/10.18150/repod.0107441 (2017). Last accessed 14 Oct 2019
13. E. Olejarczyk, W. Jernajczyk, Graph-based analysis of brain connectivity in schizophrenia. PLoS ONE **12**(11), e0188629 (2017)
14. J. Michael, Katz: fractals and the analysis of waveforms. Comput. Biol. Med. **18**(3), 145–156 (1988)
15. R. Esteller, G. Vachtsevanos, J. Echauz, B. Litt, A comparison of waveform fractal dimension algorithms. IEEE Trans. Circ. Syst. I: Fundam. Theory Appl. **48**(2), 117–183 (2001)
16. V. Srinivasan, C. Eswaran, N. Sriraam, Approximate entropy-based epileptic EEG detection using artificial neural networks. IEEE Trans. Inf Technol. Biomed. **11**(3), 288–295 (2007)
17. Wikimedia commons: The LSTM cell.png. https://commons.wikimedia.org/wiki/File:The_LSTM_cell.png. Last accessed 14 Oct 2019
18. Binary Crossentropy. https://peltarion.com/knowledge-center/documentation/modeling-view/build-an-ai-model/loss-functions/binary-crossentropy. Last accessed 18 Oct 2019

Prediction of University Examination Results with Machine Learning Approach

S. Karthik Viswanath, P. B. Mohanram, Krijeshan Gowthaman, and Chamundeswari Arumugam

1 Introduction

Universities and colleges usually find the dilemma of setting a tough question paper or a student-friendly easy question paper. However, they cannot visualize the marks the students may obtain beforehand. Furthermore, the average marks are calculated only after the exams are completed. If the exams were too tough, the average marks drop too drastically. To overcome this, if there is a system that predicts the average marks given the complexity level of the questions, then the question paper can be redesigned without compromising the standard to be met for the exams.

The objective of this work is to develop a machine learning (ML) model for teachers to predict the average marks of the question paper before setting it. The actual and predicted average marks are compared to determine the accuracy of the proposed ML model. The various sections discussed in this paper are as follows. Section 2 describes the literature survey. Section 3 details the application of machine learning techniques in predicting the average marks.

S. Karthik Viswanath (✉) · P. B. Mohanram · K. Gowthaman · C. Arumugam
Department of Computer Science and Engineering, SSN College of Engineering, Chennai, Tamil Nadu, India
e-mail: karthikviswanath17072@cse.ssn.edu.in

P. B. Mohanram
e-mail: mohanram174310@cse.ssn.edu.in

K. Gowthaman
e-mail: krijeshan17078@cse.ssn.edu.in

C. Arumugam
e-mail: chamundeswaria@ssn.edu.in

S. Patnaik et al. (eds.), *Advances in Machine Learning and Computational Intelligence*, Algorithms for Intelligent Systems, https://doi.org/10.1007/978-981-15-5243-4_20

2 Literature Survey

Li et al. [1] proposed a novel online system for correcting and analyzing exam paper (OSCAEP) based on a combination of educational data mining (EDM) and network informatization. This provides means to analyze examination papers in depth and also provides valuable suggestions and advices to educators. He also adds that a large amount of data is needed for evaluating the learning effects and predicting accurate results from such a large amount of data is challenging.

Manjunath et al. [2] in his research work has made use of machine learning concepts like support vector machines, gradient boosting and classifiers like Naïve Bayes classifier and random forest classifier to predict and classify results of a class/group of students. The accuracy of the algorithms is determined by comparing the plots obtained using the algorithms with the plot of the actual results obtained by the students. With data and information increasing exponentially day by day management of information by traditional means is almost impossible and technologies like Big Data helps in managing such a massive amount of data.

Li and Yuan [3] made use of ML concepts like classification algorithms, neural network and feature extraction to predict academic records of students indulged in online environments. Initially, data preprocessing is done so as to avoid duplication of data, data losses and eradicate errors in the data followed by feature extraction so as to collect the necessary features for predicting the academic records of the students. Then, correlation analysis is done so as to find out the useful features as feature extraction outputs features that might not be of much use in predicting the academic records. To predict the academic records, he constructed a prediction model using neural network, K-nearest neighbor (KNN) and Gaussian regression. The accuracy varies according to the algorithm used in the model. He also claims that there are still problems in his study and the inclusion of more data attributes and necessary features could enhance the model.

Kabakchieva [4] has used data mining methods for predicting the student's performance, by using data mining algorithms, including two rule learners, a decision tree classifier, two popular Bayes classifiers and a nearest neighbor classifier. The WEKA software is used in the study. While analyzing 10,330 student's data from the Bulgarian University, the results indicated, J48 classifier classifies correctly about 2/3rd of the instances (65.94% for the tenfold cross-validation testing and 66.59% for the percentage split testing). Precision levels are very high for the Bad (83–84%), Good (73–74%) and Very good (69%) and are very low for the other two classes, Average (8–10%) and Excellent (2–3%).

In a case study done by Iqbal et al. [5] for the "Student Grade Prediction" in 2017, collaborative filtering (CF), matrix factorization (MF) and restricted Boltzmann machines (RBM) techniques have been used to systematically analyze a real-world data. The results indicate root mean squared error (RMSE) is comparatively low at 30% for the RBM technique. This shows an accuracy level of around 70% in RBM model.

3 Prediction Model

The prediction model to predict average marks is designed using python programming language and several of its machine learning packages like pandas, NumPy, etc. The dataset is of the form of a csv file that is read into a pandas data frame easily using the *readfromCSV* command. Then, the required matrices X and Y for input and output, respectively, are sliced from the data frame and stored separately.

The technique of supervised learning is used to predict the average marks for the given question paper. Supervised learning is the technique of learning the relationship between input and output based on the given dataset that has both input and output data. Since the relationship is learnt by the model based on the dataset user gives, it is called supervised learning. Several supervised learning algorithms like linear regression, logistical regression, K-nearest neighbors, etc., are being used to solve several ML problems. In this paper, the linear regression algorithm is used to predict the average marks of the student.

Here, normal equation and gradient descent methods are used to predict the average marks depending on the complexity level of the question paper. Prediction model tries to arrive at an equation of a polynomial that best fits the input dataset. Gradient descent can also be used to arrive at the value of weights that make up the equation of the polynomial.

The model is trained using the dataset generated in our university's internal exams. Each row in the dataset contains the marks allotted for each complexity level and the average marks obtained in our university for such a question paper. The features used by the model are K1, K2 and K3 and the output is average marks. K1 denotes easy questions, K2 indicates average questions and K3 denotes tough and complex questions.

Weightage is assigned based on the difficulty level. K1, K2 and K3 are assigned weights 2, 3 and 4, respectively. Average marks are arrived as a percentage of 100.

The dataset contains 1300 rows out of which 500 rows are taken for training the model, and the next 500 rows are taken for testing the model while the remaining data is used for cross-validation. The features list can be extended for any kind of examination that uses the Bloom's taxonomy [6].

3.1 Linear Regression

Linear regression is a supervised learning algorithm used for prediction. This algorithm is used to learn the relationship between the input and output variables by trying to fit a line that minimizes the error between the predicted output and the actual output. Equation (1) represents the equation of the line.

$$y = mx + c \tag{1}$$

In Eq. (1), y is the output, and x are the input with c being the intercept and m being the slope of the line. After the regression model is run, the slope value can be used to identify the equation of the line that best fits the training dataset. Initially, an arbitrary value of m is chosen and the output is predicted using a cost function and the error value is calculated. At each iteration, the value of m is altered by a small value called as the 'learning rate' and the cost function is calculated again. Eventually, the algorithm stops at the lowest value of error possible and that value is chosen as the slope of the equation that best fits the training data.

Normal Equation: Normal equation is an analytical approach to linear regression that involves the use of least-cost square function. The normal equation is an easy and simple method to minimize a function provided the number of features is minimum. The normal equation works using the vectorized input and output and involves inverting and transposing the input matrix. Equation (2) represents the normal equation formula.

$$\theta = \left(x^T x\right)^{-1}.\left(x^T y\right) \tag{2}$$

In Eq. (2), theta is the parameter matrix, x is the input matrix and y is the output matrix. The normal equation method is used to minimize the matrix theta in a single step without using gradient descent and the dilemma of choosing an optimum value of learning rate. But the main drawback of this algorithm is that the algorithm works best for matrices having up to 10,000 rows, but the cost of inverting the matrix becomes costlier as the number of rows in the matrix increases since the operation of inverting happens at $O(n^3)$. In this work, the matrix x contains the values for the different features of the model K1, K2, K3, etc. The matrix y contains the average marks for a particular combination of mark distribution. After the value of theta is found, the equation of the line is plotted against the dataset. The graph plotted using the normal equation is shown in Figs. 1 and 2.

Theta values obtained using normal equation are 0.4563, 0.4887 and 0.5187.

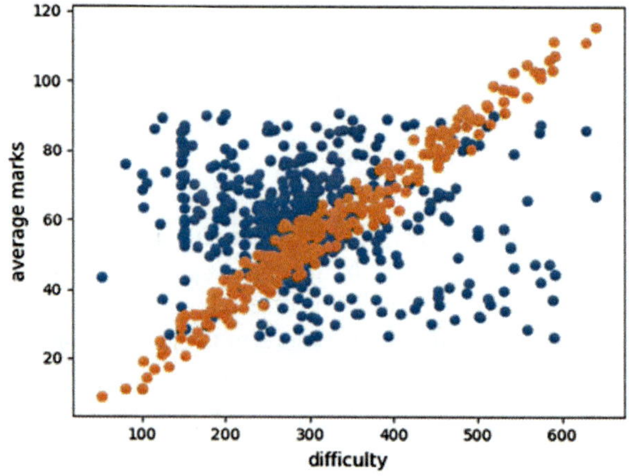

Fig. 1 Normal equation—graph plotted between actual (blue) and predicted marks (orange) for training data

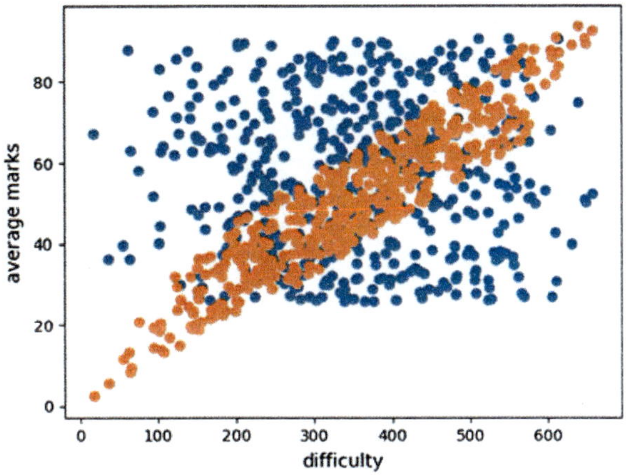

Fig. 2 **Normal equation**—graph plotted between actual and predicted marks for the test data

Gradient Descent: In gradient descent, the cost function will have its global minimum at some point. The purpose of gradient descent is to reach that point iteratively by taking small steps called the learning rate toward the global minimum. Equation (3) gives the formula for calculating the cost function.

$$J(\theta_i, \theta_j) = \frac{1}{2m} \sum_{i=1}^{m} (h(x) - y)^2 \tag{3}$$

In Eq. (3), θ_i and θ_j are the parameters for the line equation and m is the number of records in the dataset and $h(x)$ refers to the line equation itself. The matrix x contains the values for the different features of the model *K1*, *K2*, *K3*, etc. The matrix y contains the average mark for a particular combination of mark distribution. The slope of the cost equation is found and at each iteration the parameters are altered by a value equal to the learning rate alpha. Equation (4) represents the iterative step to obtain the minimum value of theta.

$$\theta_j = \theta_j - \alpha \frac{\partial}{\partial \theta_j} J(\theta_i, \theta_j) \tag{4}$$

The algorithm stops once the global minimum is reached. The learning rate α is chosen properly such that the algorithm does not overshoot the global minimum. Alternatively, the learning rate is varied proportionally such that once the error becomes low the learning rate is also reduced such that the global minimum can be reached without the danger of crossing it by using large learning steps.

The value of learning rate is chosen as 0.0001 in order to enable the cost function to converge to the global minimum. If the learning rate is raised to 0.001, then the cost function overshoots the global minimum and starts diverging. Thus, an optimal value of 0.0001 is chosen and the cost function converges to the global minimum.

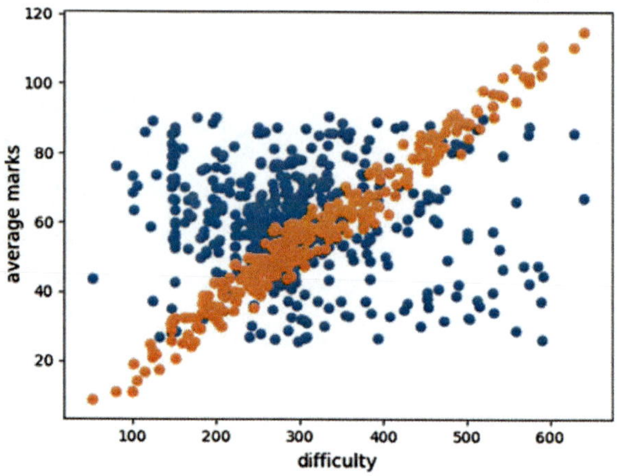

Fig. 3 Gradient descent—graph plotted between actual and predicted marks for training data

The number of iterations to run the gradient descent is fixed as 1000 since the cost function reaches the minimum by the 1000th iteration and there is no more change in the value of the cost function after the 1000th iteration. The model takes 1000 iterations to reach the minimum for the chosen learning rate of 0.0001.

At each iteration, the cost is calculated based on the current theta values, and the theta values are adjusted based on the first differential value of the cost function and the learning rate. Once the cost function reaches the global minimum, the slope of the cost function becomes close to zero and the theta values are not changed much. The value of the cost function at which the slope becomes zero is the global minimum and the value of theta is selected.

After the parameters for the equation are found using this algorithm, the equation of the line is finalized and the graph is plotted for input versus the predicted output. All the computations are done using the vectorized version of the aforementioned equations. The given dataset is split vertically into input and output matrices and the values of theta are initialized to some arbitrary values. After gradient descent, the values of theta are found out and stored in the matrix theta. Equation (5) is used to predict the output for the given input.

$$y = \theta * x + c \tag{5}$$

Theta values obtained using gradient descent are 0.4509, 0.48312 and 0.51327. The graph plotted using gradient descent is shown in Fig. 3 and 4.

4 Result

Table 1 gives the actual average marks and the marks predicted using the gradient descent and normal equation method. The methods are tested for their accuracy using

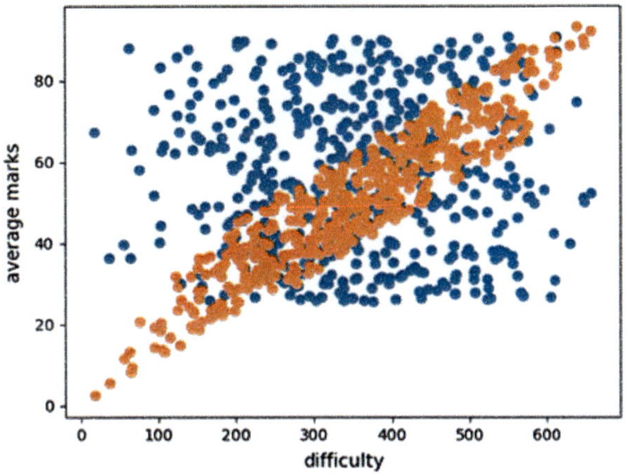

Fig. 4 Gradient descent—graph plotted between actual and predicted marks for test data

Table 1 Predicted versus actual results

S. No.	Gradient descent		Normal equation	
	Predicted results	Actual results	Predicted results	Actual results
1.	24.05	77.13	24.37	77.13
2.	35.69	62.35	36.16	62.35
3.	44.25	61.8	44.83	61.8
4.	41.81	56.23	42.35	56.23
5.	24.57	62.62	24.89	62.62
…	…	…	…	…
1303.	58.59	66.43	59.38	66.43
1304.	55.62	48.04	56.35	48.04
1305.	94.05	75.94	95.28	75.94

the performance metric root mean squared error value (RMSE). Low error indicates better performance with higher accuracy of the prediction model. From the RMSE values calculated for the training, test and cross-validation sets tabulated in Table 2, it is found that there is not much difference in the performance of gradient descent and

Table 2 RMSE/accuracy of predicted results

Dataset	Gradient descent		Normal equation	
	RMSE %	Accuracy %	RMSE %	Accuracy %
Training data	24.8016	75.1984	24.7994	75.2006
Test data	26.3138	73.6862	26.3120	73.6880
Cross-validation data	27.0278	72.9722	27.0271	72.9729

Table 3 RMSE value for data grouped based on marks

Marks	Gradient	Normal
0–40	29.46	30.14
41–60	18.80	18.12
61–80	27.69	27.38
81–100	37.72	37.28
Overall	27.27	27.26

normal equation. Both the methods have arrived at a high accuracy level of around 70%.

When the data is grouped based on the average marks, it is found that the model shows the best performance within the range 40–60 with an RMSE of around 18, while the performance in the range of 0–40 and 60–80 are in alignment with the population average. Only in the range of 80–100, the RMSE value is higher than the average around 37 as shown in Table 3. This is due to the fact that there is high variance in the first and the last mark sectors.

5 Conclusion and Future Work

We used an ML approach to find out the average marks based on the difficulty level of questions included in the question paper. Most of the earlier research in this area is based on the correlated data with the entrance test marks, marks in school final, marks in first year data, etc. Our approach is a new one adopted by us in predicting the marks with difficulty levels in the question papers. Our results yield the accuracy level of around 72–75%, which is in agreement with the earlier researchers. More data and directed research in this direction may open new and perfect results.

References

1. Y. Li.: An application of EDM: design of a new online system for correcting exam paper, in *The 13th International Conference on Computer Science & Education* (Colombo, Srilanka, 2018), pp. 335–340
2. S.K. Pushpa, T.N. Manjunath, T.V. Mrunal, S. Amartya, C. Suhas, Class result prediction using machine learning, in *International Conference on Smart Technologies for Smart Nation* (2017), pp. 1208–1212
3. L. Wang, A. Yuan, A prediction strategy for academic records based on classification algorithm in online learning environment, in *IEEE 19th International Conference on Advanced Learning Technologies* (2019), pp. 1–5
4. D. Kabakchieva, Predicting student performance by using data mining methods for classification cybernetics and information technologies. Cybern. Inf. Technol. **13**(1)

5. Z. Iqbal, J. Qadir, A.N. Mian, F. Kamiran, Machine learning based student grade prediction: a case study
6. Bloom's Taxonomy. https://en.wikipedia.org/wiki/Bloom%27s_taxonomy

Surveillance System to Provide Secured Gait Signatures in Multi-view Variations Using Deep Learning

Anubha Parashar, Apoorva Parashar, Vidyadhar Aski, and Rajveer Singh Shekhawat

1 Introduction

Biometrics is an important technique for uniquely identify the human. There are many techniques like face recognition, iris, fingerprint, etc. But all these either need high-resolution images or need human intervention. The valuable information collected by biometric systems wide range of applications in tracking, visual surveillance and studying different fields of life like disaster management, military applications and medical. The human recognition methods being used as of now like fingerprint recognition, facial recognition and iris recognition demand subjects' cooperation, physical contact at times and a lot of images from different views. The present algorithms cannot be applied in dynamic environment or to a subject not willing to cooperate as the system will not be able to make sense out of the given information. It is crucial to develop intelligent and new recognition systems that could provide security. There is a need for developing such robust systems that could help in the identification of people on the basis of their biometric behavior which includes their bodily moments and appearance [1]. These systems should be able to automatically collect and analyze the data and give out fair warnings before any incident takes place. Gait is the most efficient and effective biometric when it comes to monitoring individual subjects who are not very cooperative [2]. It is also regarded as the most

A. Parashar (✉) · V. Aski · R. S. Shekhawat
Manipal University Jaipur, Jaipur, India
e-mail: anubhaparashar1025@gmail.com

V. Aski
e-mail: vidyadharstjit@gmail.com

R. S. Shekhawat
e-mail: rajveersingh.shekhawat@jaipur.manipal.edu

A. Parashar
Maharshi Dayanand University, Rohtak, India
e-mail: apoorvaparashar0000@gmail.com

S. Patnaik et al. (eds.), *Advances in Machine Learning and Computational Intelligence*, Algorithms for Intelligent Systems, https://doi.org/10.1007/978-981-15-5243-4_21

satisfactory method for making surveillance system. This technique does not require the subject to touch and apparatus or go through heavy checking instead it captures their biometric while maintaining the privacy of the subject. Gait biometric collects the spatio-temporal information of the individual from a video stream and gathers information about the individual. Imitation or cheating can be done with different biometric methods but gait is operated from a distance which shows that gait is more reliable. Even though gait is the most efficient biometric method, it has some limitations [3]. The gait of a person is different in various conditions, like human locomotion (gait) can vary if a person is carrying bags or wearing a very heavy overcoat will differ from when that person is walking normally without these objects. There are some other external factors as well that can vary the gait [4] of an individual such as pace of walking, any injury to the individual, orientation of the camera and change in nature of ground surface.

2 Literature Review

Recognition of gait can be categorized into two groups; that is, appearance-based method and model-based methods.

In the appearance-based method, probe the sequences of gait is done without creating models of entire body of human outline structure and gives most of the productive results when the view is fixed.

On the other hand, the model-based methods tend to create replica of a human joint in order to estimate the gait parameters such as limb length, angular speed and trajectories.

Recognizing gait with changing view is tricky. Research regarding view change can be grouped into three categories. The first one [5] follows an approach of extracting a gait feature which is indifferent to change of view. An important procedure was put forth regarding this method in [6].

A markerless motion was used to depict the estimation of poses of lower limbs. The next step was to rebuild them in sagittal plane by revising the viewpoint. Using these rectified limbs, poses spatial displacements and angular measurements were obtained. If the view difference is large, this method and efficiently execute cross-view gait recognition but this method has certain limitations:

1. We cannot apply this method for frontal view as the poses of limbs cannot be traced easily;
2. For a markerless motion, the estimation of limb poses is not up to mark.

The second category [7, 8] is built upon projecting or mapping the connections of gaits across view. Before measuring the similarity, the relationship acquired by training normalizes gait features obtained from various views into an affiliated subspace. A simple non-cooperative camera system is used in this type of gait recognition system.

Correlation between gaits of various views across camera is decided by the learning process. If we are to compare this category with the first one, we can say that the second category is more effective and secure when we provide ample training samples to the learning process. The technique proposed in this research paper falls under this category [9–15].

3 Proposed Methodology

From Fig. 1, the proposed methodology can be seen, it is divided into five steps. First, the videos are converted into frames and then in second step the background is subtracted. In third step, the gait energy image is created by take the average of frames by given formula:

$$G(a, b) = 1/N \sum_{c=1}^{N} B(a, b, c)$$

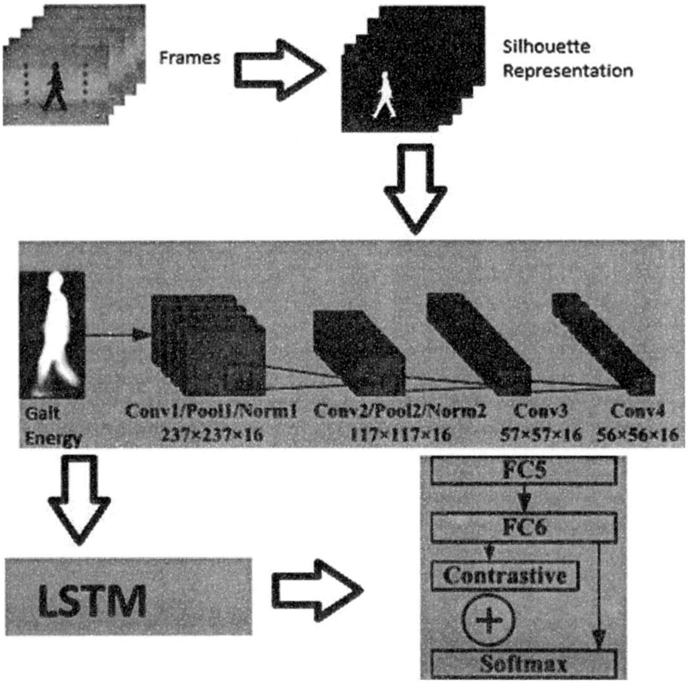

Fig. 1 Proposed framework

Fig. 2 Multi-view dataset

Here, *a*, *b* are *x* and *y* coordinates of an image, *N* are the total number of frames in one gait cycle, *c* represents the time frame with respect to frame.

In the fourth step, 3D CNN model is given to extract the spatial and temporal features. In the fifth step, these features are sent to LSTM layer in order to remember the features in all iterations. Finally, a fully connected layer is formed to and softmax activation function is used to get the final prediction.

Three-dimensional convolution neural network layer size is $237 \times 237 \times 16$. In the proposed network, gait energy images are compared with local regions, after this linear-projection technique is applied in order to calculate the variance between similar pairs of GEIs.

3.1 Dataset Used

CASIA-B consists of multi-view gait database. In this dataset, there exist 124 subjects and has 11 views. In order to provide variations in covariate condition, dataset consists of three variants: clothing, view angle and carrying bag is shown Fig. 2.
OU-ISIR consists of gallery for training and probe for testing sequences per subject. Each subject is divided into 5 angles, 55°, 65°, 75°, 85° and inclusion of all four angles as shown in Fig. 3 [5].

4 Performance and Discussion

In order to check the performance of the system different angles are taken in order to see the accuracy in CASIA-B dataset to see in the presence of various covariates. Figure 4 shows the performance of CASIA-B dataset in normal state. Figure 5 shows

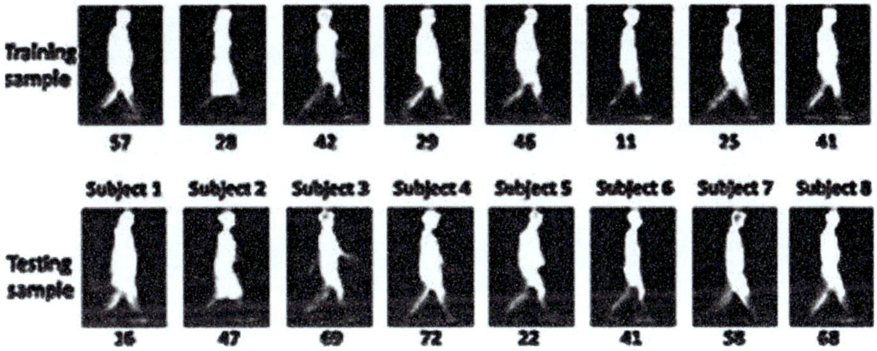

Fig. 3 OU-ISIR dataset

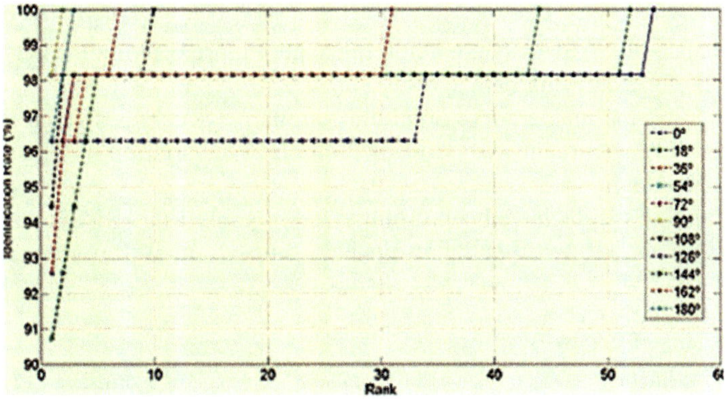

Fig. 4 Performance of CASIA-B normal state

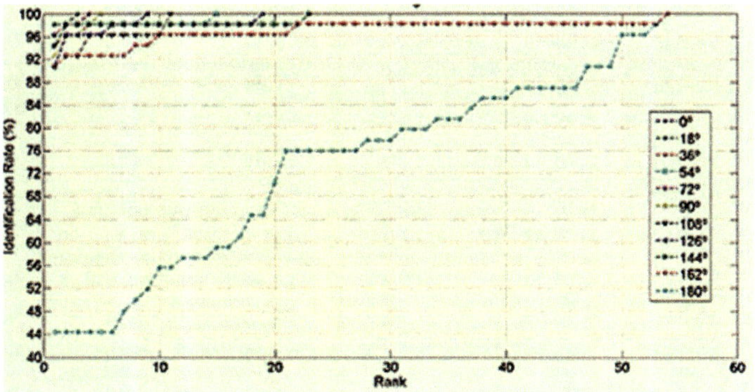

Fig. 5 Performance of CASIA-B clothing state

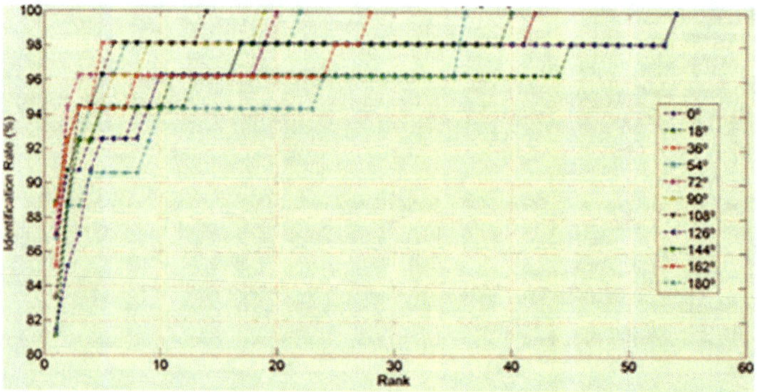

Fig. 6 Performance of CASIA-B carrying bag state

the performance of CASIA-B under clothing state. Figure 6 shows the performance of CASIA-B while carrying baggage.

Table 1 shows the match curve of all the cumulative accuracies given in Figs. 4, 5 and 6. Second experiment was performed on observe variation in view angles of different covariates.

Corresponding results have been summarized in Table 1 (CASIA-B dataset) and 2 (OU-ISIR dataset) shows that the results have outperformed from the state of art.

Table 1 Accuracy of CASIA-B

Angle in degrees	With bag		Overcoat		Normal	
	Deng. et al.	Our results	Deng. et al.	Our results	Deng. et al.	Our results
0	95.09	**96.64**	92.49	**94.35**	93.59	**94.14**
18	94.68	**94.49**	91.74	**90.30**	95.44	**96.14**
36	93.45	**94.49**	91.75	**94.44**	94.29	**95.14**
54	95.34	**95.69**	91.50	**98.35**	94.14	**96.04**
72	96.82	**96.44**	94.40	**95.09**	95.14	**96.05**
90	95.79	**97.59**	94.44	**94.08**	96.14	**99.05**
108	94.91	**95.63**	92.30	**95.29**	96.44	**97.14**
124	93.42	**94.58**	92.39	**95.30**	97.29	**98.29**
144	93.35	**94.63**	91.34	**94.15**	97.74	**96.14**
162	92.16	**93.78**	91.34	**94.59**	97.59	**98.14**
180	92.52	**93.49**	90.37	**91.44**	95.29	**96.04**

Table 2 Accuracy of OU-ISIR

Approaches	55°	65°	75°	85°	Overall
Shiraga et al. [6]	94.60	95.20	96.20	94.40	94.62
Our approach	**98.80**	**99.60**	**98.70**	**99.70**	**99.30**

5 Results

Recognition of human gait while carrying a bag falls under 82–95%. In overcoat, clothing recognition rate is 90–94%. For normal gait is in 94–99.70%.

6 Conclusion

Experimental result shows the proposed approach is more robust and efficient than the previous state of art. Further, it has produced better results than the previous model-free approaches. The major contributions of this research can be summarized as below:

1. Provides an efficient multi-variant model.
2. Proposed methodology gives a better result than the existing methods (that is with bag, overcoat and normal).
3. As the results are good in all the angles this system can actually be used for real surveillance for providing security.

References

1. M. Alotaibi, A. Mahmood, Improved gait recognition based on spe- cialized deep convolutional neural network. Comput. Vis. Image Underst. **164**, 103–110 (2017)
2. K. Bashir, T. Xiang, S. Gong, Gait recognition without subject cooper-ation. Pattern Recogn. Lett. **31**, 2052–2060 (2010)
3. M. Deng, C. Wang, T. Zheng, Individual identification using a gait dynamics graph. Pattern Recog. (2018)
4. M. Hofmann, G. Rigoll, Exploiting gradient histograms for gait-based person identification, in *2013 20th IEEE International Conference on Image Processing (ICIP)* (IEEE, 2013), pp. 4171–4175
5. H. Iwama, M. Okumura, Y. Makihara, Y. Yagi, The ou-isir gait database comprising the large population dataset and performance evaluation of gait recognition. IEEE Trans. Inf. Forensics Secur. **7**, 1511–1521 (2012)
6. K. Shiraga, Y. Makihara, D. Muramatsu, T. Echigo, Y. Yagi, Geinet: view-invariant gait recognition using a convolutional neural network, in *2016 International Conference on Biometrics (ICB)* (IEEE, 2016), pp. 1–8
7. D. Thapar, G. Jaswal, A. Nigam, V. Kanhangad, Pvsnet: palm vein authentication siamese network trained using triplet loss and adaptive hard mining by learning enforced domain specific features. arXiv preprint arXiv:1812.06271 (2018)

8. W. Kusakunniran, Q. Wu, J. Zhang, H. Li, Gait recognition under various viewing angles based on correlated motion regression. IEEE Trans. Circ. Syst. Video Technol. **22**, 966–980 (2012)
9. M. Hu, Y. Wang, Z. Zhang, D. Zhang, J.J. Little, Incremental learning for video-based gait recognition with lbp flow. IEEE Trans. Cybern. **43**, 77–89 (2013)
10. J. Yang, J.Y. Yang, Why can LDA be performed in PCA transformed space? Patt. Recogn. **36**(2), 563–566 (2003)
11. P.N. Belhumeur, J.P. Hespanha, D.J. Kriegman, Eigenfaces vs. fisherfaces: recognition using class specific linear projection. IEEE Trans. Patt. Anal. Mach. Intell. **19**(7), 711–720 (1997)
12. C. Liu, H. Wechsler, Robust coding schemes for indexing and retrieval from large face database. IEEE Trans. Image Process. **9**(1), 132–137 (2000)
13. B.S. Venkatesh, S. Palanivel, B. Yegnanarayana, Face detection and recognition in an image sequence using eigenedginess, in *Proceedings of Indian Conference on Computer Vision, Graphics Image Process* (2002), pp. 97–101
14. S. Fidler, A. Leonardis, Robust LDA classification by subsampling, in *Proceedings of International Workshop Statistical Analysis Computer Vision* (2003), pp. 97–104
15. T. Kanungo, D.M. Mount, N.S. Netanyahu, C.D. Piatko, R. Silverman, A.Y. Wu, An efficient k-means clustering algorithm: analysis and implementation. IEEE Trans. Patt. Anal. Mach. Intell. **24**(7), 881–892 (2002)

Usage of Multilingual Indexing for Retrieving the Information in Multiple Language

A. R. Chayapathi, G. Sunil Kumar, J. Thriveni, and K. R. Venugopal

1 Introduction

Multilingual information retrieval (MLIR) system manages the utilization of inquiries in a single language and recovers the reports in different dialects, where the translation of query plays an important role in the research of MLIR. In this work, language-independent indexing technology is used to process the text collections of English, Kannada and Hindi languages. We have used a multilingual dictionary dependent word-by-word query translation. The present advances cannot fulfill the needs of multilingual search for a keyword simultaneously, for not just issues like the establishment of semantic importance of various starting points, yet in addition, savvy gathering quick access, de-duplication, rectification and joining are thought about. This motivated us to design a system that displays results of the searched keyword in multi-languages according to users' reading preferences. This system will be developed to retrieve the result in multiple languages simultaneously. If we search the information of a keyword in English, the result of which is displayed in multiple languages simultaneously. We will use languages like Hindi and Kannada

A. R. Chayapathi (✉)
Information Science Department, Acharya Institute of Technology, Visvesvaraya Technological University, Bengaluru, Karnataka, India
e-mail: archayapathi@gmail.com

G. S. Kumar
Computer Science Department, Vijaya Vittala Institute of Technology, Visvesvaraya Technological University, Bengaluru, Karnataka, India
e-mail: gsuneel.k@gmail.com

J. Thriveni
Computer Science Department, Bangalore University, UVCE, Bengaluru, Karnataka, India
e-mail: thrivenijgowda@yahoo.co.in

K. R. Venugopal
Computer Science Department, Banglore University, Bengaluru, Karnataka, India
e-mail: venugopalkr@gmail.com

S. Patnaik et al. (eds.), *Advances in Machine Learning and Computational Intelligence*, Algorithms for Intelligent Systems, https://doi.org/10.1007/978-981-15-5243-4_22

to display the result. These languages are chosen as basic languages because of our proficiency in these languages. Later, the work can be extended to all Indian written languages.

The reason for this framework is to give users an approach to look through documents written in different languages for a given query. A few contrasts among monolingual and multilingual emerge if the user is knowledgeable with more than one language. So as to reflect varying understanding levels of user language, the UI must give differential showcase capacities. Interpretation into a few dialects is required when more than one user gets the outcomes. Contingent upon the client's degree of multifaceted nature, interpretation of different components at various stages can be given to users to a scope of data access needs, including keyword interpretation, term interpretation, title interpretation, abstract interpretation, explicit paragraph interpretation, caption interpretation, full document interpretation and so forth. The objective is to build a single search engine showing results in multiple languages at a time. Given that almost every Indian knows more than one language on an average, it is essential to cater to the information need in more than one language, information may be present in multiple languages exclusively, and the aim of this work is to facilitate the user with all these information across languages. The objectives of the proposed system include:

- Indexing the document (multilingual); retrieving, filtering.
- Presentation and summarization of information.
- Multilingual metadata; cross-language information retrieval.
- Morphological analysis and semantic parsing.
- Identify techniques for disambiguation and document segmentation.

The work to be done here is to build an indexing framework that can be used to index multilingual documents and to develop algorithms to effectively store the posting list for multilingual documents that provides an improved method to efficiently reduce the index size based on the vocabulary match across languages. The search engines like Google will give the result on only the requested language, for example, if we search in English, we can retrieve the information only in English if we want the same information in multi-language like Kannada, Hindi and soon then we have to give the keyword in respective languages.

2 Literature Survey

The various related works have been referred to as get strong knowledge in the field of work. The paper [1] clarifies the multilingual data recovery framework (MLIR) recovers the applicable data from numerous dialects because of a user query in a solitary source language. Adequacy of multilingual data recovery is estimated utilizing mean average precision (MAP). The fundamental element of multilingual data recovery is the score rundown of one language cannot be contrasted and other language score lists. MAP does not think about this component. We propose another measurement normalized distance measure (NDM) for estimating the adequacy of MLIR frameworks. In the paper [2], they have proposed design by altering an open-source system called Solr to construct a multilingual file. Arrangement choices are taken and usage difficulties looked in this procedure are clarified in detail. by examining the huge updates on various data sources and language inceptions, we concoct a fundamental hypothesis model and its calculation on news, which is fit for wise gathering, brisk access, de-duplication, adjustment and incorporation with news experiences. The work [3] presents new difficulties to customary IR innovations. This portrays how difficulties are looked at in FameIR, a multilingual and media IR shell planned as a guide to improving communication. WG methods give a brought together access to the Web sources (counting those in the hidden Web). They likewise enable access to a better granularity of data than the entire Web report. The paper [4] portrays a novel methodology dependent on one of the signals handling apparatuses in soft processing applied to Web data recovery, to be specific wavelet transform. The impact of two parameters, wavelet capacities on feature extraction and data recovery capacity of the calibration model was examined.

In the work [5] illustrates how to successfully address the issues of users, multilingual data recovery is a significant issue in regards to the persistently expanding volume of data in different dialects. This additionally addresses by ordering and recovery in a trilingual corpus English, French, Arabic. The work [6] delineated about building up a language-free strategy for disclosure of implicit information about patents from multilingual patent information sources. Conventional methods of multi- and cross-language patent recovery are generally founded on the procedure of interpretation. The work [7] demonstrated that retrieval effectiveness is more influenced by the translation direction (e.g., Italian-to-English, English-to-Italian) than the translation. This result demonstrates that a crucial role is played by the translation process.

The paper [8] deals with customizing an open-source implementation of the index to have multilingual documents. The work in [9], cross-language information retrieval (CLIR) relevant information for a query expressed in as native language. The results to the users are non-trivial tasks English transliterations of the given Kannada/Telugu word. The resulting multiple conversation choices for each query.

3 Architectural Design and Implementation

The framework architecture is a conventional order to deal with objects called frameworks, such that supports thinking about the basic properties of these items. Figure 1 demonstrates the architecture of the search engine. On admin side, the file indexing module allows the admin to set an index for each and every file that is uploaded. On the user side, the user will enter the query to be searched then the language translator module is used to determine the language used to detect the language used by the user and to determine its equivalents in other languages. The string searching algorithm

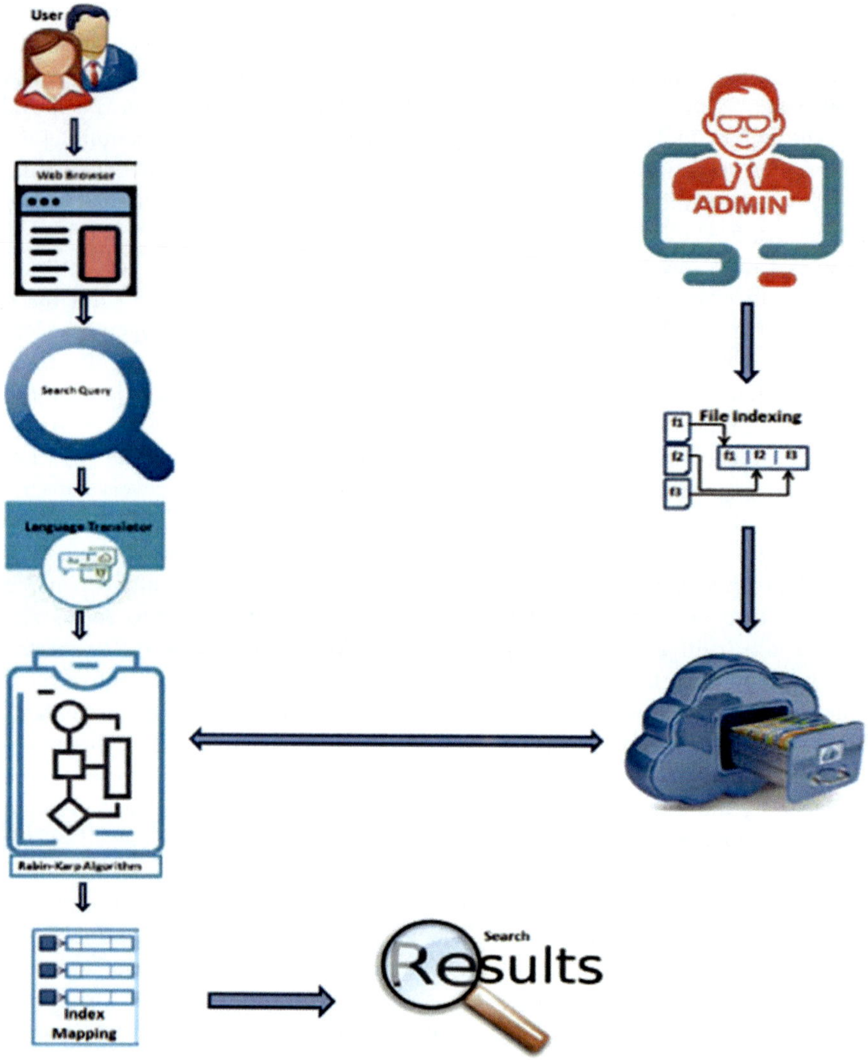

Fig. 1 Architecture of the search engine

Rabin–Karp algorithm that uses hashing to find anyone of a set of pattern strings in a searched text, which also seeks to speed up the testing of equality of the pattern to the substrings in the text by using a hash function is applied. The index mapping module is used to determine the index of the file according to the search query and finally displays the result in the target languages.

Figure 2 illustrates the exploded view of context-level design modules to be implemented for the MLIR framework. This shows a data system as a whole and emphasizes the way it interacts with external entities.

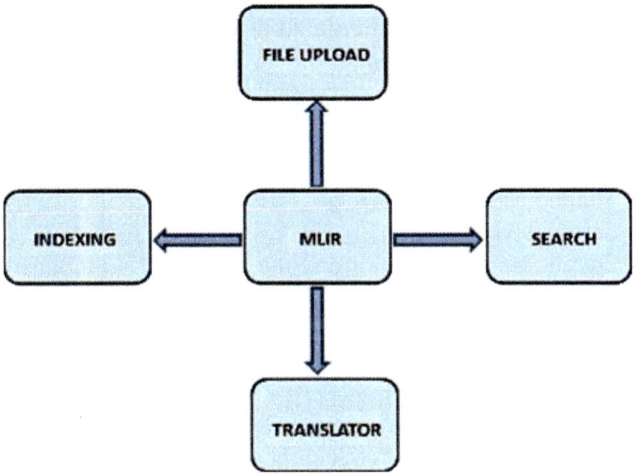

Fig. 2 Context-level design of MLIR framework

3.1 Rabin–Karp Algorithm for String Matching

It is concurred that the Knuth–Morris–Pratt (KMP) algorithm is quicker and better put compared with the Rabin–Karp algorithm for string matching. Rabin–Karp is simpler to implement on the off chance that we accept that a collision will never occur, however, if the issue you have is an average string looking KMP will be progressively steady regardless of what input you have. In any case, Rabin–Karp has numerous different applications, where KMP is not an alternative. With great hash age calculations that do exist today, it is conceivable that Rabin–Karp can yield extremely near 100% unwavering quality in finding a match. What's more, both have the complexity of $O(n)$. Both of the algorithms are linear, so just steady factors matter. In KMP, you may endure some cache misses and branch miss predicts. Remembering all these, for the most part, we have chosen Rabin–Karp's calculation for our usage. In the future work, we will attempt to build up the most extreme proficiency in design coordinating by utilizing and comparing both KMP and Rabin–Karp.

In the below Listing. 1 Rabin–Karp algorithm matches the hash value of the pattern with the hash value of the current substring of text, and if the hash values match, then only it starts matching individual characters. So Rabin–Karp algorithm needs to calculate hash values for (i) pattern itself and (ii) all the substrings of the text of length m. Since we need to efficiently calculate hash values for all the substrings of size m of text, we must have a hash function that has a particular property. Hash at the next shift must be efficiently computable from the current hash value and next character in text or we can say hash(txt[s + 1 … s + m]) must be efficiently computable from hash(txt[s … s + m − 1]) and txt[s + m], i.e., hash(txt[s + 1 … s + m]) = rehash(txt[s + m], hash(txt[s … s + m − 1]) and rehash must be $O(1)$ operation. The hash function used in Rabin and Karp calculates an integer value. The integer value for a string is the numeric value of a string. For example, if all possible characters are from 1 to 10, the numeric value of "122" will be 122. The number of possible characters is higher than 10 (256 in general) and pattern length can be large. So the numeric values cannot be practically stored as an integer. Therefore, the numeric value is calculated using modular arithmetic to make sure that the hash values can be stored in an integer variable. To do rehashing, we need to take off the most significant digit and add the new least significant digit for in hash value.

Listing 1. Usage of the Rabin–Karp Algorithm

```
package algorithm;
public class RabinKarp {
private RabinKarp(){ }
public static int contains(String source,String target)
{
    StringHashing.Hash sourceHash =new StringHashing.Hash(source);
    StringHashing.Hash targetHash =new StringHashing.Hash(target);
    int matches = 0;
    int tLength = target.length();
    int sLength =source.length();
    long tHash = targetHash.getStringHash();
    for (int i = 0; i + tLength &lt;= sLength; i++)
    {
        if (sourceHash.getHash(i, i + tLength - 1) == tHash)
        {
            if (equalsByChar(source, i, target))
            {           matches++;          }
        }}    return matches;
}
private static boolean equalsByChar(String source, int index, String target)
{
        if (source.length() - index &lt; target.length()) { return false; }
        for (int i = 0; i &lt; target.length(); i++)
        { if (source.charAt(i + index) != target.charAt(i))
            { return false;}}
}
return true;
public static class Alternative extends
StringHashing.AlternativeHash {
public static int contains(String source, String target)
{       int matches = 0;
        int tLength = target.length();
        int sLength = source.length();
        long tHash =getStringHash(target);
        long[] sPrefixHash = getPrefixHashes(source.toCharArray());
        for (int i = 0; i + tLength &lt;= sLength; i++)
{       if (getAltHash(sPrefixHash, i, i + tLength - 1) == tHash* pow[i])
        {           if (equalsByChar(source, i, target))
              { matches++; }      } }
return matches;
} } }}
```

4 Performance Evaluation Measures of IR Model

The evaluation performance of any framework is done by evaluating the records recovered in light of an inquiry as for their applicability score and figuring the suitable set-based measures. At the point, when this procedure is effective, it permits contrasts between the frameworks, hypothetical methodologies and foreseen functional utilization. In any case, each language innovation is not prepared to go under automatic measurement techniques.

The assessment of MLIR frameworks is cumbersome work. Client satisfaction is a significant and pivotal property of these frameworks. Evidently, a great MLIR framework ought to fulfill the data needs of the client. This property can be isolated into numerous components of performance. The nature of the recovered outcomes is simply founded on the three measurements: query translated, UI and system effectiveness. Client-oriented assessment is a tedious task and required a few resources. Along these lines, how far the inquiry resultant list causes the client to fulfill the necessary data need is another measurement and it is hard to assess. In this view, unreasonable efforts would be fundamental with respect to the client's individual features of searches and the inclination natures of various client's pertinence judgment of these searches. Many of the accessible areas are not relating to text retrieval, yet they are multimedia and multilingual recovery. All the evaluation activities normally utilized a measure named MAP for looking at the nature of the customary IR frameworks utilizing the standardized assortments and data needs. The pool is developed from the resultant list of a few systems and it confines the number of applicable archives which can be met.

5 Experimental and Result Analysis

The running environment is set up by installing the required IDE for developing working using java technologies. The algorithm mentioned in the design and implementation is developed to meet the objectives. As the work is still under process, we have attached it in Figs. 3, 4 and 5, few screenshots leading toward the required outcome. The system developed should be able to upload and process the input from the various language we set. The system must be able to translate the search queries

Fig. 3 Screenshot for login and searching the content as per the keywords

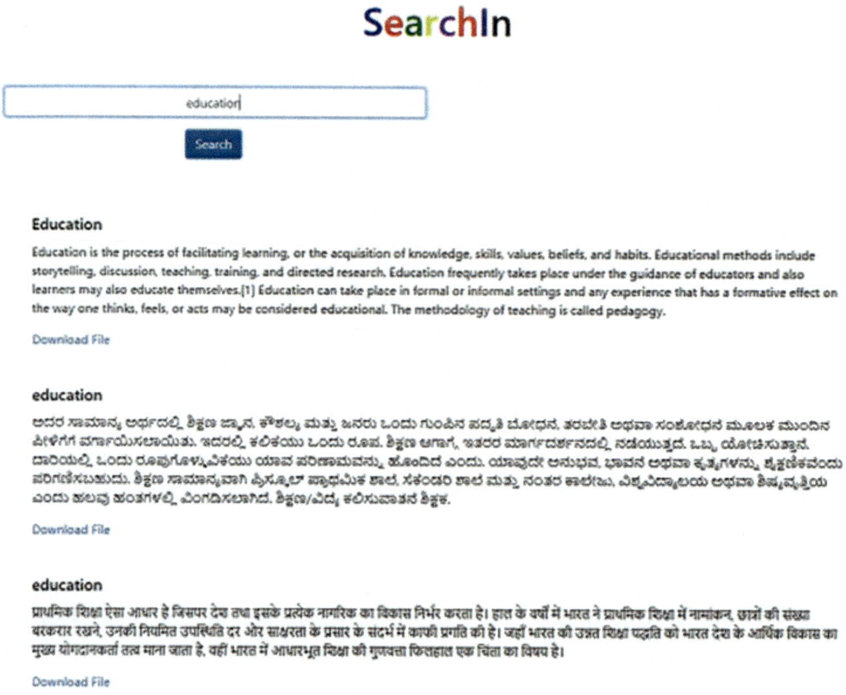

Fig. 4 Results for the searched keywords

into various languages we set. The system must be able to translate the search queries into the various language of the user's preference. In the initial work, our MLIR aims at three main languages such as English, Hindi and Kannada.

6 Conclusions

The work is implemented using the semantic-based methodology for multilingual information retrieval that has been executed for English, Kannada and Hindi. Multilingual information retrieval system manages the utilization of inquiries in a single language and retrieves the records in different dialects. The system utilizes multilingual lexicon-based word-by-word translation of queries. We utilized dialects like Hindi and Kannada to show the outcome. The algorithms that enable us to scan for keywords in documents of different dialects are introduced. The work will be reached out in a greater manner for taking care of documents in local Indian dialects and bigger word references.

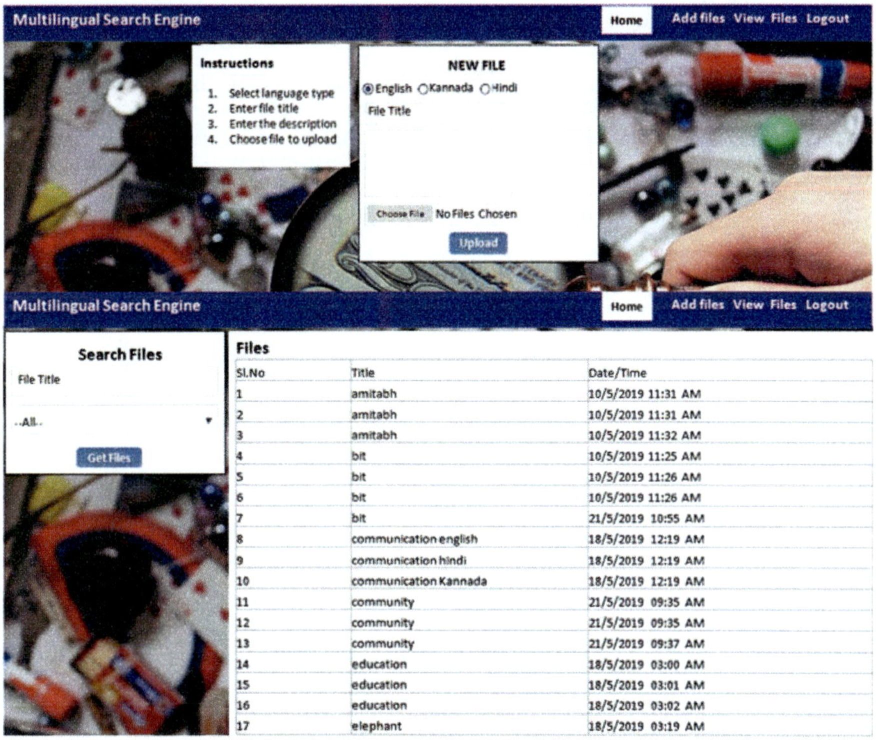

Fig. 5 Screenshot for uploading and searching the files

References

1. R. Korra, P. Sujatha, S. Chetana, M.N. Kumar, Performance evaluation of multilingual information retrieval (MLIR) system over information retrieval (IR) system, in *2011 International Conference on Recent Trends in Information Technology (ICRTIT)* (IEEE, 2011), pp. 722–727
2. A. Atreya, S. Chaudhari, P. Bhattacharyya, G. Ramakrishnan, Building multilingual search index using an open-source framework, in *Proceedings of the 3rd Workshop on South and Southeast Asian NLP* (2012), pp. 201–210
3. M. Gatius, M. Bertran, H. Rodríguez, Multilingual and multimedia information retrieval from web documents, in *Proceedings 15th International Workshop on Database and Expert Systems Applications* (IEEE, 2004), pp. 20–24
4. S.A. Al-Dubaee, N. Ahmad, New direction of applied wavelet transform in multilingual web information retrieval, in *2008 Fifth International Conference on Fuzzy Systems and Knowledge Discovery*, vol. 4 (IEEE, 2008), pp. 198–202
5. H. Aliane, An ontology based approach to multilingual information retrieval, in *2006 2nd International Conference on Information and Communication Technologies*, vol. 1 (IEEE, 2006), pp. 1732–1737
6. C.H. Lee, H.C. Yang, C.H. Wu, Y.J. Li, A multilingual patent text-mining approach for computing relatedness evaluation of patent documents, in *2009 Fifth International Conference on Intelligent Information Hiding and Multimedia Signal Processing* (IEEE, 2009), pp. 612–615
7. J.S. McCarley, Should we translate the documents or the queries in cross-language information retrieval? in *Proceedings of the 37th Annual Meeting of the Association for Computational*

Linguistics on Computational Linguistics, ACL '99 (Association for Computational Linguistics, Stroudsburg, PA, USA), pp. 208–214
8. R. Rahimi, A. Shakery, I. King, Multilingual information retrieval in the language modeling framework. Inf. Retriev. J. **18**(3), 246–281 (2015)
9. V.R. Mallamma, M. Hanumanthappa, Kannada and Telugu native languages to English cross-language information retrieval. Int. J. Comput. Sci. Inf. Technol. **2**(5), 1876–1880 (2011)

Using Bootstrap Aggregating Equipped with SVM and T-SNE Classifiers for Cryptography

Neeraja Narayanswamy and Siddhaling Urolagin

1 Introduction

A side channel attack [1] happens based on the information gained from exploiting the physical implementation of the system. This attack targets the recovery of the secret key from the electromagnetic signatures left after the implementation of the AES. There are two types of side channel attacks: Profiling and non-profiling. We focus on the mitigation of the profiling side channel attack is the most powerful kind of attack that comprises of two steps. Firstly, a duplicate of the target device is made and then exploits all the possible weaknesses of the copy. Then, with the information that is recovered, a key recovery attack is performed.

The other attempt was to record data from traditional key loggers [2] and extract commonly used passwords, so as to create a dictionary from which can then be used to attack web sites that require authentication in order to use some of their features. Key loggers come in different forms, but they are mainly divided into software and hardware key loggers. Software key loggers can be API based, Kernel based, or JavaScript based, or can use direct attacks like form grabbing and memory injection. Hardware-based key loggers can be keyboard overlays, firmware based or can analyze electromagnetic emissions and keyboard acoustics. There also have been attempts to use support vector machines (SVM) to use as intrusion detection systems (IDS) as they prove very useful in creating a profile of the target system [3]. Multiple ways of creating an SVM IDS have been explored, for example, creating a profile from WLANS [3], analyzing previous attacks, and using it as background software.

N. Narayanswamy · S. Urolagin (✉)
Department of Computer Science, BITS Pilani, Dubai Campus, Dubai, United Arab Emirates
e-mail: siddhaling@dubai.bits-pilani.ac.in

N. Narayanswamy
e-mail: f20160225@dubai.bits-pilani.ac.in

S. Patnaik et al. (eds.), *Advances in Machine Learning and Computational Intelligence*, Algorithms for Intelligent Systems, https://doi.org/10.1007/978-981-15-5243-4_23

This paper aims to combine the above two methods with an ensemble learner in order to provide an accurate copy of the system for white box penetration testing. The paper first introduces the principle concepts that the proposed algorithm uses. Primarily, bootstrap aggregating, an ensemble learner that can run two or more classifiers parallely which is ideal for this setup as it prevents over fitting, is used.

2 Bootstrap Aggregating

Bagging also called bootstrap aggregating is a type of ensemble learner that is used for classification and regression. Bagging can also be used in areas other than classification, for example, in this project, it is used to create an intrusion detection system with an automated key logger and packet analysis. The principle of bagging is to split the data and run it through different classifiers parallely. This model helps in accurate classification and is not susceptible to over fitting.

In each of this subset, the data is sampled uniformly with replacement. Let the size of a particular sample be k. This scenario may lead to repetitions of certain observations in some samples. When the sample size $k = n$, then, we can determine that a set has approximately 63.2% of unique or bootstrap samples [4]. We obtain m bootstrap samples and fit it into the m models; these models are then combined by average in the case of regression or voting in the case of classification.

In this project, the data is split as those collected by key loggers and those collected by packet sniffing. The data collected by the key logger is decipherd using t-SNE classification and the data collected by packet sniffing is sorted through using a SVM. We calculate the final output using by bagging as shown in Fig. 1.

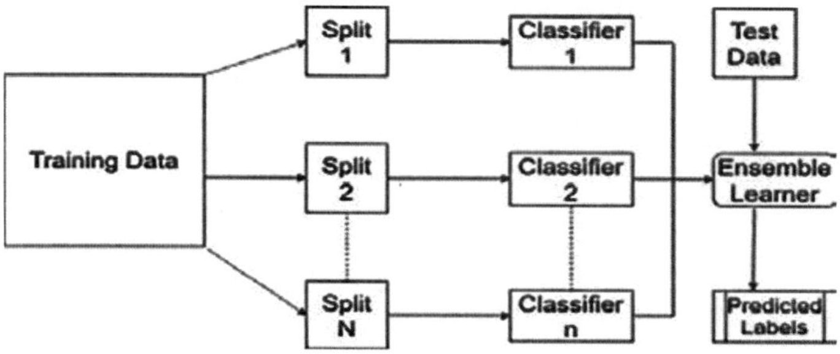

Fig. 1 Layout of the bagging algorithm

3 T-SNE Clusters

The T-distributed stochastic neighbor embedding is a visualization algorithm developed by Laurens van der Maaten and Geoffrey Hinton [5]. It is used to reduce the dimensionality of non-linear data. It is especially used to reduce the higher-dimensional data to either two-dimensional or three-dimensional data. The algorithm has two steps:

We initially construct a probability distribution in such a way that objects with a higher similarity have a higher probability to be grouped together than objects with lower probability. This is done over pairs of higher-dimensional objects.

We then construct a similar probability distribution over the lower-dimensional map so that the Kullback–Leibler divergence between the two distributions, with respect to their location on the map, is minimized.

Usually, the algorithm uses Euclidean distance as the base metric but it can be changed to fit the use of the programmer.

These t-SNE clusters are dependent on chosen parameters and sometimes may show a cluster in non-clustered data [6]. However, t-SNE is able to recover well-separated clusters when the correct parameters are chosen.

Figure 2 demonstrates the data collected from our key logger [2] and separated letters that often occur together into three-dimensional clusters. This data will further be reduced to two dimensions before being fed into the ensemble classifier.

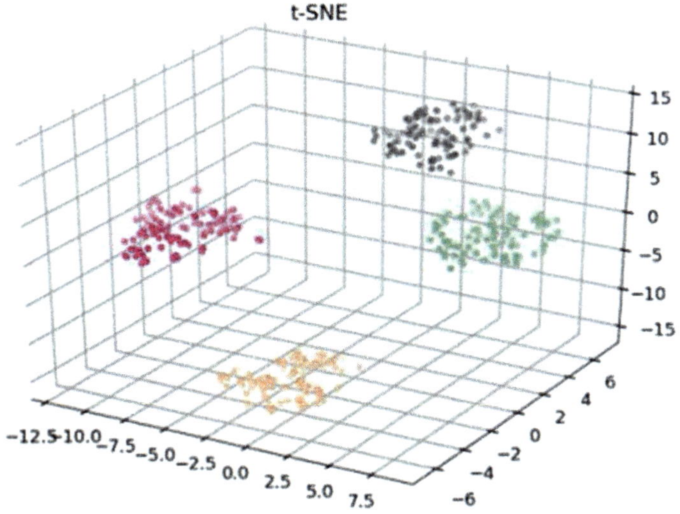

Fig. 2 Key logger data reclassified by t-SNE cluster

4 Support Vector Machines

A support vector machine [7] is a kind of linear classifier that uses a hyperplane to separate two classes of data with enough space to plot future points if there is any form of deviation from other points in a particular class.

The set can be of infinite dimension but the SVM plots these points in a finite-dimensional space. The hyperplane defines the center of the hard margin and the vector closest to it (support vector) defines the boundary of the hard margin. The other vectors should lie outside the hard margin.

If our test dataset has a vector that lies within the boundary of the hard margin and the hyperplane, then the SVM has to correct itself by making this vector the support vector and defining a new hard margin.

Our data is linearly separable as we are targeting requests from only one web site and separating it from the requests of the other web sites. The reason SVM was chosen was that not only was it an efficient classifier for our data but it also works well with other classifiers [8].

Figure 3 shows the plain data of unclassified packets. The green dots are those packets that are coming from the targeted web site; the Y-axis indicates IP range and the X-axis indicates URL address.

Figure 4 shows the classified data with only three support vectors. These packets are then used by the ensemble classifier to build a copy of the web site. The green dots represent those packets that are captured from the target web site and the rest represent those packets captured from other sites. We use the details of the target packet to build the web site piece by piece successfully.

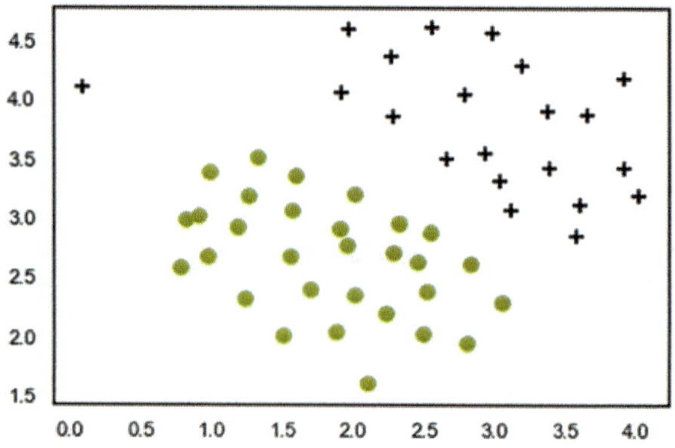

Fig. 3 Data captured from packets

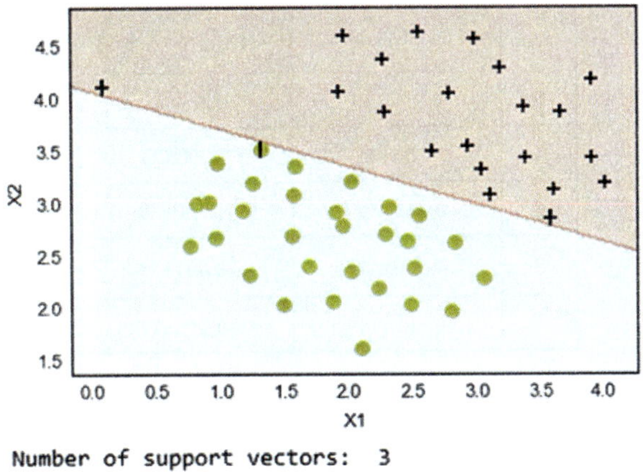

Fig. 4 Same data as Fig. 3 but now classified as SVM

5 Data Extraction

Web applications are used as they are most likely targets for cyber-attacks. This project uses key-loggers [9, 10] and packet sniffers [11]. API-based key loggers are used as they are freely available over the Internet to collect data that will then be fed into the t-SNE algorithm. These key loggers track API's that send get requests and record the data that the users send. These are then polled and sequences are found. This software runs as background applications instead of malware. This is normally done to obtain important pins and passwords from the user. A packet sniffer is used to record requests made by the web sites to the user and also the format of the protocols used. This software intercepts network traffic on the computer and makes a record of them. If a site does not use security settings like the https, then we need not decode the raw data else we need to use the proper RFC to make sense of the data captured. The relevant information is then extracted from the logs collected and stored them in a file that can be converted to workable data using the pandas library.

6 Methodology

With the analysis of the key logger and packet analysis, we find the occurrences of five letters and classify it as shown in Fig. 5. We then plot the unclassified data in a three D plot as shown in Fig. 6.

Using T-SNE clustering, we are able to reduce the dimensionality of our data into two dimensions as shown by Fig. 7. This classifies each key with the amount of occurrences and with the keys that they are used more often with. We also see the individual results of the SVM over the packets gathered, and from here, we can gather the exact frame work of the web site and where authentication is needed. These two results are put in the final ensemble classifier where a copy of the web site is made

	x	y	z	label
0	-0.296225	1.55966	-0.437544	0
1	1.209485	1.300883	-0.203428	0
2	0.415339	-1.750639	-0.864065	0
3	-0.573891	-0.041179	-1.297861	0
4	1.111044	-0.398614	0.831758	0

Fig. 5 Five random keywords taken from key logger data and tabulated

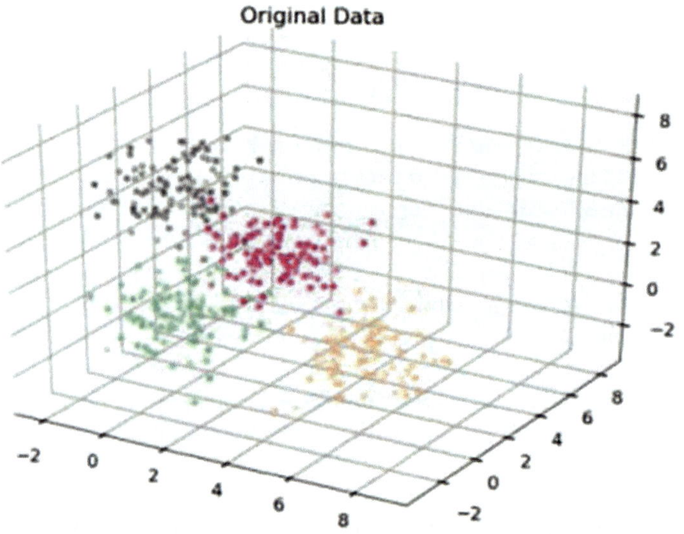

Fig. 6 Same data as figure but plotted on a 3D graph

along with a file of data needed for authentication like passwords. The time taken by this process is determined by the processing power of the system's GPU and the size of the web site. Figure 8 is a diagrammatic representation of the procedure that needs to be followed.

Figure 8 represents how we use our gathered data to train our SVM and t-SNE classifiers; the results of these classifiers are then input into the ensemble learner which then also tests the data to create a successful copy of the targeted web site. Before the data is put into the ensemble classifier, we do a split of 80–20 for test and training. This way, we not only get the Web profile but also access to various private areas of the web site, and thus, we can thoroughly analyze all the weaknesses that a particular Web application may have.

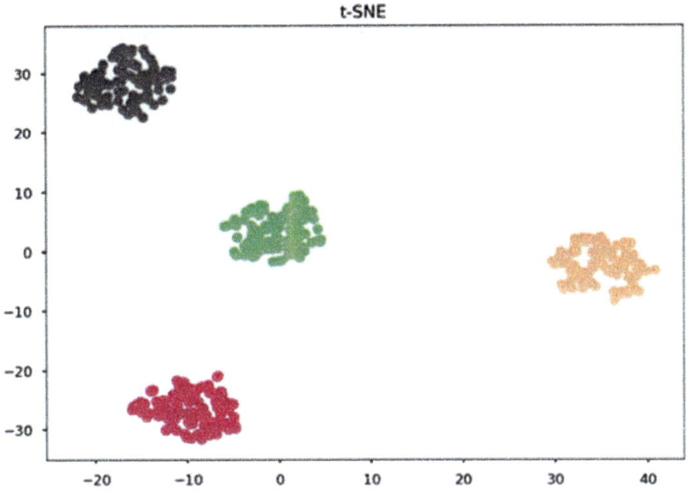

Fig. 7 Data in figure reduced to two dimensions via t-SNE clustering

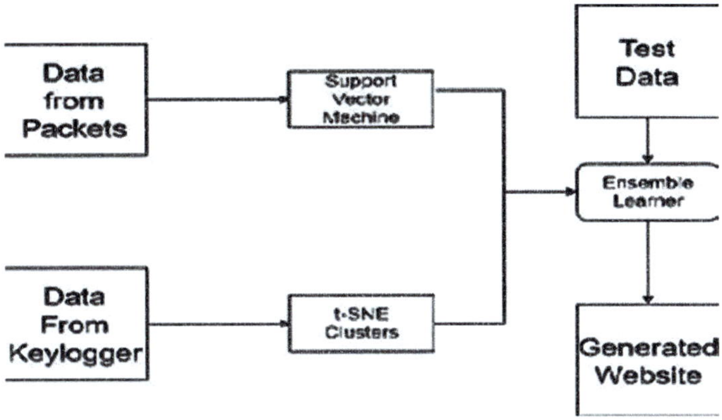

Fig. 8 Program block diagram

7 Experimental Results

Experiment results are conducted by collecting packet data from the application Wireshark. Wireshark captures all packets in the network. This data is then stored in the CSV file.Similarly, an online keylogger, that uses GET and POST Request API's, is used to collect authentication keys that are then stored in a CSV File. The CSV file is chosen as it is easier to interpret by the code using the Pandas Library.

First, we input the relevant data into the classifiers and send the result to the ensemble learner. The ensemble learner is chosen as the bagging classifier which uses the results to create a dummy web site with all its authentication keys.

Figure 9 plots the ensemble's learning curves for training and testing. The program projects an error rate of 24%. The smallest difference in the training and testing errors is detected when training is 20% complete. The larger the gap the less probability that the data is over fitted. However, we need to maintain a minimum gap of 10% to avoid under fitting.

Figure 10 demonstrates the difference in accuracy depending on the ensemble size. In our basic algorithm, we can obtain an accuracy of approximately 89.2% in the best case. In the worst-case scenario, we may get the accuracy only up till 64% and this phenomenon is seen with web sites that are too small.

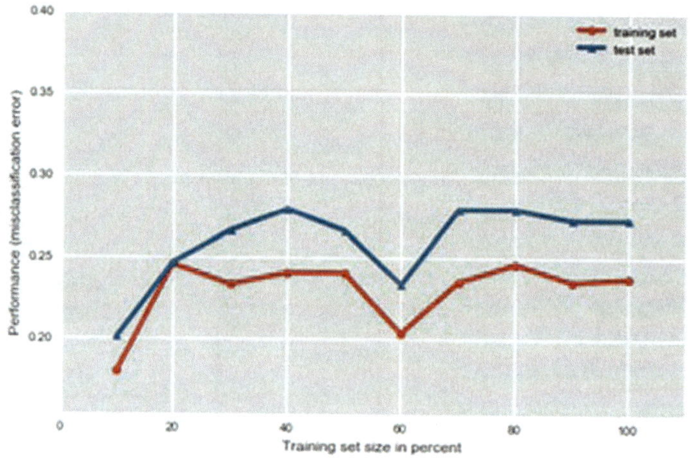

Fig. 9 Training and testing error

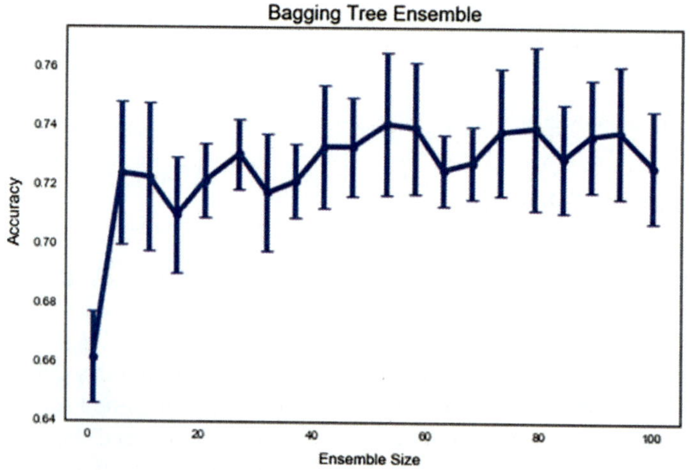

Fig. 10 Ensemble accuracy according to ensemble size

8 Conclusions

Thus, we are able to create a profile of a web site and obtain access to the database using a bootstrap aggregating ensemble tree with the relevant classifiers. This profile then determines the strength of the system and the keys used for authentication. We see that in the worst-case scenario up to 64% of the web site can be accurately remade. Black hat hackers need only 50% of the web site to find vulnerabilities; therefore, we need to patch any vulnerability found after creating this profile. This test should be conducted regularly so as to protect the program from new or improved threats. Although this algorithm is susceptible to the size and quality of data as well as the processing power of the system that it is running on, it can be used to create multiple copies that can help to determine the perfect ratio between program intelligence and security.

References

1. H. Maghrebi, T. Portigliatti, E. Prouff, *Breaking Cryptographic Implementations Using Deep Learning Techniques* (2016), pp. 3–26. https://doi.org/10.1007/978-3-319-49445-6_1
2. Z. Li, Z. Feng, J.D. Tygar, Keyboard acoustic emanations revisited, in *12th ACM Conference on Computer and Communications Security* (2005), pp. 373–382
3. N.M. Muamer, N. Sulaiman, Intrusion detection system based on SVM for WLAN. Procedia Technol. **1**, 313–317 (2012)
4. J.A. Aslam, R.A. Popa, R.L. Rivest, On estimating the size and confidence of a statistical audit, in *Proceedings of the Electronic Voting Technology Workshop (EVT '07)*, Boston, MA, 6 Aug 2007
5. L.J.P. van der Maaten, G.E. Hinton, Visualizing data using t-SNE. J. Mach. Learn. Res. **9**, 2579–2605 (2008)
6. A. Sanjeev, H. Wei, K.K. Pravesh, An analysis of the t-SNE algorithm for data visualization, in *31st Annual Conference on Learning Theory, Proceedings of Machine Learning Research*, vol. 75 (2018), pp. 1–8
7. S. Hou, Y. Zhou, H. Liu, N. Zhu, Exploiting support vector machine algorithm to break the secretkey. Radio Eng. **27**(1) (2018)
8. N. Chand, P. Mishra, R. Challa, E. Pilli, M. Govil, *A Comparative Analysis of SVM and its Stacking with Other Classification Algorithm for Intrusion Detection* (2016), pp. 1–6. https://doi.org/10.1109/icacca.2016.7578859
9. C. John, The evolution of malicious IRC bots. Symantec 23–24 (2005)
10. S. Tom, Web-based Keylogger used to steal credit card data from popular sites. Threatpost | The First Stop for Security News. 2016-10-06. Retrieved 2019-10-24
11. J.C. Kevin, *Law of Internet Security and Privacy*. (Aspen Publishers, 2003), p. 131. ISBN 978-0-7355-4273-0

Application of Apriori Algorithm on Examination Scores

M. A. Anwar, Sayed Sayeed Ahmed, and Mohammad A. U. Khan

1 Introduction

One of the biggest challenges facing educational institutions today is the exponential growth of educational data and how to investigate this data to improve the quality of managerial decisions [1]. The education institutions would like to know, for instance, which students will enroll in a particular course, and who will need assistance for graduation. The data mining enables organizations to uncover and understand hidden patterns in vast databases by using their current reporting capabilities. And these patterns are then built into data mining models and applied to predict individual behavior and performance with high accuracy. The analysis of the data collected by the institutions helped the management in reaching out effectively to those students or learner. In this way, resources and staff can be allocated by institutions more effectively. Data mining may also, for example, efficiently allocate resources with an accurate estimate of how many students will take action before he or she drops out.

Educational data mining (EDM) is an emerging discipline including, but not limited to information retrieval, recommender systems, visual data analytics, social network analysis (SNA), cognitive psychology, psychometrics, and so on. Its methods are often different from those methods from the broader data mining literature. What's more, EDM draws from several reference disciplines including data mining, learning theory, data visualization, machine learning, and psychometrics [2]. And this emerging field of EDM examines the unique ways of applying data mining techniques to solve educationally related issues.

This study demonstrates the ability of data mining algorithms to extract potentially useful information from a large amount of data consisting of marks awarded to students for each question in a final examination. We preferred *Apriori* algorithm

M. A. Anwar (✉) · S. S. Ahmed · M. A. U. Khan
Al Ghurair University, Dubai, UAE
e-mail: anwarma@yahoo.com

S. Patnaik et al. (eds.), *Advances in Machine Learning and Computational Intelligence*, Algorithms for Intelligent Systems, https://doi.org/10.1007/978-981-15-5243-4_24

for rule mining because of its effective and efficient association rule mining strength [3]. The algorithm has been applied here on a dataset of final exam marks to find association among scores of the questions in the same examination. It was verified that a student who correctly answered a given question is highly likely to score high on a related question down the question paper. This study can be extended in a straightforward manner to mine associations among the courses of a program that will support the decision maker to revise the existing strategies and help improve the learner's experience.

Section 2 presents literature review. In Sect. 3, we present the association rule mining concept and in Sect. 4, the steps involved in data mining are addressed. The discussion of results is presented in Sect. 5 and in the last section, conclusion and future work are briefed.

2 Literature Review

Data mining is an interdisciplinary subfield of computer science [4–6]. Data mining is the analysis step of the knowledge discovery in database process, or KDD [7]. Data mining techniques have their roots in machine learning, artificial intelligence, computer science, statistics, etc. [8]. And data mining is an exploratory process, but it can be used for confirmatory investigations [9]. It is different from other searching and analysis techniques because data mining is highly exploratory, where other analyses are typically problem-driven and confirmatory. Through the combination of an explicit knowledge base, sophisticated analytical skills, and domain knowledge, hidden trends and patterns are able to be uncovered. These trends and patterns form the predictive models that enable to assist organizations with uncovering useful information and then guide decision-making [10].

Educational data mining as a field for solving educationally related problems, at a high level, seeks solutions to improve methods for exploring the data, which usually has meaningful hierarchy at multiple levels, in order to discover new insights of how people learn in the context of these settings [11]. For instance, a student's college transcript may contain a temporally ordered list of courses taken by him or her, the grade that the student earned in each course, and information about when the student selected or changed his or her academic major or minor. We might also understand how different individuals engage with EDM system. Taken together, these learning analytics provide much useful information for the design of learning environments.

3 Association Rule Mining

Since its introduction in 1993, [12] the task of association rule mining has received a great deal of attention. Today, the mining of such rules is still one of the most popular pattern discovery methods in KDD. In brief, an association rule is an expression

$X \subseteq Y$, where X and Y are sets of items. The meaning of such rules is quite intuitive: Given a dataset D of transactions, where each transaction $T \subseteq D$ is a set of items, $X \subseteq Y$ expresses that whenever a transaction T contains X then T probably contains Y too. The probability or rule confidence is defined as the percentage of transactions containing Y in addition to X with regard to the overall number of transactions containing X.

That is, the rule confidence can be understood as the conditional probability $P(Y \subseteq X/X \subseteq T)$. The idea of mining association rules originates from the analysis of market-basket data where rules like "A customer who buys products $x_1$and x_2 will also buy product y with probability c%" are found where c is confident of the rule.

This direct applicability to business problems together with their inherent understandability (even for non-data mining experts) made association rules a popular mining method. Moreover, it became clear that association rules are not restricted to dependency analysis in the context of retail applications, but are successfully applicable to a wide range of business problems.

For dataset under investigation, we have used *Apriori* algorithm, which is an algorithm for frequent item set mining and association rule learning over transactional databases. It proceeds by identifying the frequent individual items in the database and extending them to larger and larger item sets as long as those item sets appear sufficiently often in the dataset. The frequent item sets determined by Apriori can be used to determine association rules which highlight general trends in the dataset.

4 Data Preprocessing

4.1 Dataset Information

The following dataset has been retrieved from University of California, Irvine (UCI) Machine Learning Repository [13], it was uploaded at September 24, 2015, by University of Genoa.

From the uploaded file, we have selected the dataset of the final marks of first year undergraduate students. The examination was held in two times (in two sheets) and some students took the exam two times. In both times, the exams addressed the same concepts but with different details. Some students who attended the course did not take the final exam, therefore, some Ids were missing in final grades and have been taken care of in data preprocessing. The questions of the final exam addressed the concepts of sessions of the course. So, the grades per question available were based on their reference to the session topics in addition to the total final grade. The field names indicate ES # of session. # of exercise (the total points dedicated to exercise). For example, "ES_1.1_Two" represents the Examination Session 1, Exercise 1, and score 2.

4.2 Data Cleaning

Some registered students for the course did not appear in the final examination, therefore their marks have been removed from the dataset. Question marks have been rounded to one decimal place because there was no marks that contained more than two decimal places. Marks has been converted from numeric to nominal to make them eligible for applying mining association rule since association rule mining support nominal data only in Weka.

5 Results and Discussion

Apriori algorithm was applied on the dataset using a publically available free data mining tool referred to as Weka [14]. The Weka available tool allows user to apply most of the data mining algorithms on a given dataset. The Weka provided a set of association rules with the minimum confidence of 90% and minimum support of 60%. For the analysis purpose, we chose to work with the top ten association rules as shown in Table 1.

For an instance, the first two rules generated in Table 1 are interpreted as follows. As per rule 1, a student who scores full marks in question 1.1 (ES_1.1), then there is a 100% likelihood to score the same in question 1.2 (ES_1.2). Similarly, according to rule 2, if a student scores full marks in question 3.2 (ES_3.2), then there is a 98% chance to score full marks in question 3.1 (ES_3.1).

The association determined by rule 1 becomes evident from by examining the question ES_1.1 and question ES_1.2 as shown in Appendix 1. The questions under

Table 1 Association rules, support 60% and confidence 90%

Rule No.	Antecedent	Consequent	Confidence (%)
1	ES_1.1_Two	ES_1.2_Three	100
2	ES_3.2_Two	ES_3.1_One	98
3	ES_3.1_One	ES_3.2_Two	98
4	ES_1.2_Three, ES_3.2_Two	ES_3.1_One	98
5	ES_1.2_Three, ES_3.1_One	ES_3.2_Two	98
6	ES_3.1_One	ES_1.2_Three	96
7	ES_1.2_Three	ES_3.1_One	96
8	ES_3.2_Two	ES_1.2_Three	96
9	ES_1.2_Three	ES_3.2_Two	96
10	ES_3.1_One, ES_3.2_Two	ES_1.2_Three	96

discussion are part of the examination question paper for the course "Digital Logic Design" of an undergraduate engineering degree program. According to the instructors involved in teaching that said course for several years have an agreed opinion that if a student, who is able do well on question ES_1.1 is expected to do the same for question ES_1.2.

The instructor can take advantage of such type of association rules to enhance their teaching methodology by spending more time on certain topics in the class while reducing time for the associated topics.

6 Conclusion

Apriori algorithm proved to be successful in finding the hidden association among questions of the same examination paper. The association rules generated in our student were letter verified by the expert instructors involved in teaching the related course. The study can further be extended for finding association rules among the courses in given baccalaureate program in any discipline.

Appendix A

[Sample questions extracted from "Digital Logic Design" Final Examination]

ES_1. Analysis of combinationa networks (Total 5 points).

Consider the following combinational network (Fig. 1):

ES_1.1. Fill out the following truth table (2 points) (Fig. 2).

ES_1.2. Fill out the following timing diagram (3 points) (Fig. 3).

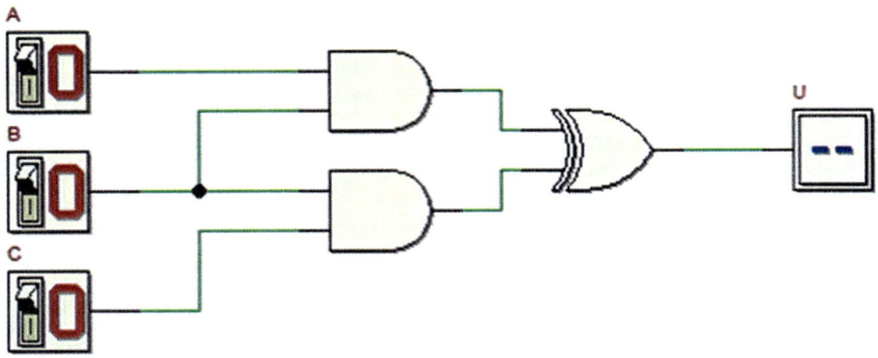

Fig. 1 Combinational network

Fig. 2 Truth table

A	B	C	U
0	0	0	
0	0	1	
0	1	0	
0	1	1	
1	0	0	
1	0	1	
1	1	0	
1	1	1	

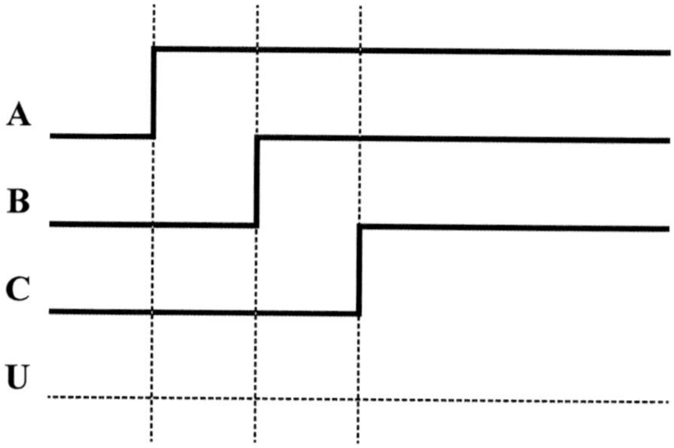

Fig. 3 Timing diagram

References

1. K. Koedinger, K. Cunningham, A. Skogsholm, B. Leber, An open repository andanalysis tools for fine-grained, longitudinallearnerdata, in *First International Conference on Educational Data Mining. Montreal*, Canada (2008), pp. 157–166
2. R. Baker, K. Yacef, The state of educational data mining in 2009: a review future visions. J. Educ. Data Min. **1**(1) (2009)
3. R. Agrawal, T. Imielinski, A.N. Swami, Mining association rules between sets of items in large databases, in *Proceedings of the 1993 ACM SIGMOD International Conference on Management of Data* (1993), pp. 207–216
4. Data Mining Curriculum. ACM SIGKDD. 2006–04-30. Retrieved 2014-01-27
5. C. Clifton, *Encyclopedia Britannica: Definition of Data Mining* (2010)
6. T. Hastie, R. Tibshirani, J. Friedman, *The Elements of Statistical Learning: Data Mining, Inference, and Prediction* (2009)
7. U. Fayyad, G. Piatetsky-Shapiro, P. Smyth, *From Data Mining to Knowledge Discovery in Databases* (1996). Retrieved 17 Dec 2008
8. M. Dunham, *Data Mining: Introductory and Advanced Topics* (Pearson Education, Upper Saddle River, NJ, 2003)
9. A. Berson, S. Smith, K. Thearling, *An Overview of Data Mining Techniques* (2011). Retrieved 28 Nov 2011, from http://www.thearling.com/text/dmtechniques/dmtechniques.htm

10. D. Kiron, R. Shockley, N. Kruschwitz, G. Finch, M. Haydock, Analytics: the widening divide. MIT Sloan Manag. Rev. **53**(2), 1–22 (2012)
11. EducationalDataMining.org. (2013). Retrieved 2013-07-15
12. R. Agrawal, T. Imielinski, A. Swami, Mining association rules between sets of items in large databases, in *Proceedings of the ACM SIGMOD International Conference on Management of Data (ACM SIGMOD'93)*, Washington, USA, May 1993
13. https://archive.ics.uci.edu/ml/index.php, last accessed on 30 Sept 2019
14. http://www.cs.waikato.ac.nz/ml/weka/, last accessed on 20 Sept 2019

Small Embed Cross-validated JPEG Steganalysis in Spatial and Transform Domain Using SVM

Deepa D. Shankar and Adresya Suresh Azhakath

1 Introduction

The data in today's world is diverse which comprise of banking transactions, military, hospital records, employee details, and so on. Hence, protecting this data is of prime importance since a major data leak can cause a lot of implications. Steganography is considered to be a superior technique because it helps message to remain undetectable. In steganography, the hidden message cannot be identified by the user [1]. Steganography means "secret writing." It uses certain algorithms or procedures to hide information into another medium in order to deny access to unauthorized users [2]. Steganography can work in two scenarios—to protect against detecting data and protect against removing data. The embedding can be done in text, audio, and video. The ability of identification is known as steganalysis. There are two types of steganographic detectors—Targeted and Blind. The former has an idea of the steganographic scheme used whereas the latter do not. Steganalysis procedures generally calculate the features of the images, which are not "normal" in assumed candidate images. If the method of steganalysis is interested in finding the existence of a secret message with a greater probability of success compared to that of randomized guessing, then the associated process is said to have failed [3]. Low embedding of 10% is used in this paper for JPEG images. The revolutionary idea under the work is to consider six separate kernels and four modes of sampling for feature-based steganalysis. The comparative study of classification is done using LSB matching which is a targeted steganographic scheme, and F5 which is a blind steganographic scheme. The classification is done using SVM. The support vector machine is a guided approach of

D. D. Shankar (✉)
College of Arts and Sciences, Abu Dhabi University, Abu Dhabi, UAE
e-mail: Deepadayashankar@gmail.com

A. S. Azhakath
Department of Computer Science, Bits Pilani, Dubai Campus, Dubai, UAE
e-mail: Sudee99@gmail.com

S. Patnaik et al. (eds.), *Advances in Machine Learning and Computational Intelligence*, Algorithms for Intelligent Systems, https://doi.org/10.1007/978-981-15-5243-4_25

machine learning statistics. Referring to a structural minimization, SVM aims on finding an optimal hyperplane in a kernel space, where training instances are linearly segregated [4]. The main task of the SVM method is in the selection of the kernel. It is well established. Once the kernel has been set, SVM classifiers have a user-defined parameter, but the kernel is a very large group that needs several parameters. Many works have been done on kernel limitation using previous knowledge, but an open testing topic remains the best option of a kernel for a particular problem [5].

2 Related Work

The effectiveness of steganalysis lies in the identification of stego and the exposure of objects. The appropriate features were extracted at the early phase of classification. With the guidance of a statistical model with only a best possible number of discrete cosine transform (DCT) coefficients, the ingrained data in JPEG images can be modeled with high precision. The methodology of using transform domain can be incorporated to obtain better result with nominal changes in the cover image [6]. Steganalysis is implemented in spatial domain as well, where the embedding is done directly into the pixel intensity of the image [7]. Among the different steganographic schemes, LSB matching and F5 are considered in this paper. The operation ± 1 is performed to change a bit in case of LSB matching. The addition and subtraction do not influence the hidden message. Xia et al. [8] proposed learning-dependent LSB matching steganalysis with SVM classifiers. Three essential histogram features of the images are being used to prepare the classifier. F5 algorithm makes use of matrix decoding to avoid the statistical attacks. Malathi and Gireeshkumar [9] conducted research to integrate a broad payload of various strategies related to LSB and F5. When statistical steganalysis is being considered, the analysis of features is very important. The analysis has two parts—feature extraction and classification. Liu et al. [10] proposed a steganalysis method based on fusing SVM classifiers. For the training of SVM classifier, various function subsets are used. Multi-classifier detection results are used to train a fusion classifier, and the fusion classifier may learn the relation and variety of sub-classifier detection results. Babu et al. [11] had done a detailed research on various methods of steganalysis, different methods of preprocessing based on the filtration, methods of feature extraction, and classification based on the machine learning to properly identify the image of steganography.

3 Methodology

Previous literature has opened a venue for exploration of machine learning in statistical steganalysis. Hence, the proposed work in the paper would help to bridge the gap found in the previous literature. The subsequent sections of this paper will throw light to the methodology and experimental results. As mentioned before, the

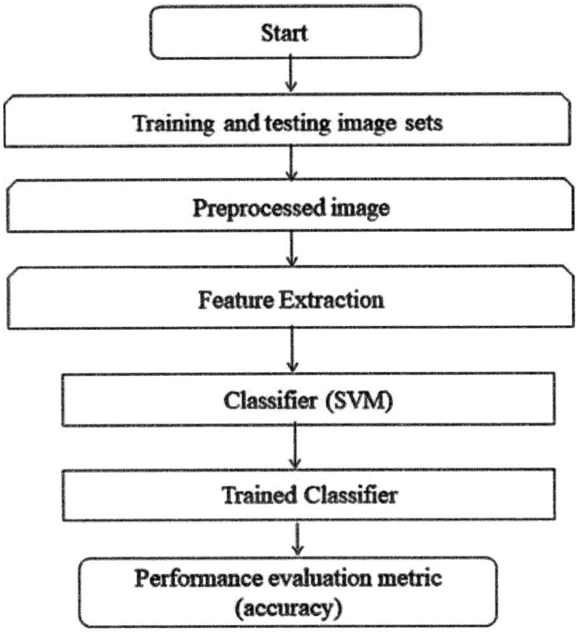

Fig. 1 System architecture of blind steganalysis with SVM

work is on JPEG images. This particular format is easy to store and distribute [12]. A small-scale embedding of 10% is considered for analysis. The raw images are transformed, and relevant features are mined. These outputs thus received are then served into the SVM. The image segmentation remains through 8 × 8 blocks, which is followed by feature extraction. The image values are normalized to advance the efficacy of the algorithm. The block diagram of the flow of data is given in Fig. 1.

4 Extraction of Features

4.1 Dataset

In this paper, two different databases—INRIA holiday dataset and UCID image dataset are considered. 1500 images from INRIA dataset are used as training images in this paper. The next set of data used for analysis is uncompressed color image database (UCID). The dataset used in the research contains 800 images which are being converted to JPEG without any compression. After conversion, each image is labeled from 1. The conversion helps to generate a different secret message, by using their label to be the key of a pseudo-random number generator. The image size is restricted to 256 × 256. INRIA holiday's image dataset are used as training dataset and UCID image dataset is used as test dataset.

Table 1 Extracted features

Order of features	Feature type	Number of features extracted
First order	Individual histogram	55
	Global histogram	11
	Dual histogram	99
Second order	Variance	01
	blockiness	02
	Co-occurrence	25
Markovian	–	81
Total features extracted		274

4.2 Extraction of Features

The features to be extracted should be selected so that it should not delete any important details pertaining to the image. Hence, a set of features are selected which would add up to 274. The individual histogram is created from DCT modes. The second-order features are inter-block dependent ones [13]. Co-occurrence, blockiness, and variance are the second-order features [14]. These calculations are caused by the dissimilarities in the position of coefficients—both row and column. The co-occurrence is calculated through the likelihood distribution of nearby JPEG constants.

Total number of distinct features to be extracted is shown in the Table 1.

4.3 Classification

If a machine learning model has to be designed with a set of data, it needs to be split into training dataset and test dataset. The model is trained through the training set. This would help to authenticate the test data [15]. 80% of data is generally taken as training set and the other 20% is used as test data. Classification is the process after the features are extracted. According to the features, the images are classified into different classes.

4.4 SVM Classification

SVM establishes a decision hyperplane in an N-dimensional space [16]. The neighboring data points are considered as support vectors. The space/distance between certain decision surface and also the adjacent data point shall be pointed to as margin. The SVM classifier operates with precision with regards to the separation margin,

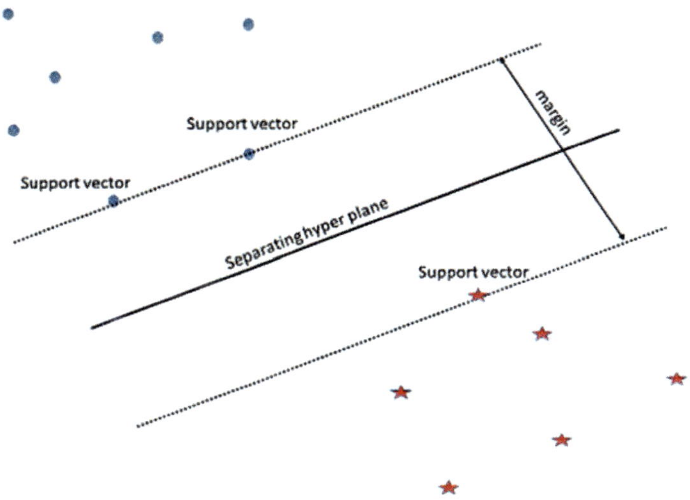

Fig. 2 SVM classification

well equipped to high-dimensional spaces, worthy of memory, and could be extended to both binary and multi-classifiers. With any of these classifiers, several kernels can indeed be selected. Therefore, in most literature, the SVM classifier is recommended for use as one of finest classifiers in the steganalysis realm [17, 18] (Fig. 2).

4.5 Cross-Validation

A database for evaluation is generally divided into training set and test set. The division is based on random assignment to avoid any bias. During experimentation, the training set will be larger than the test database. But in real-life scenario, the test data is used from the Internet which is very large than the training set. This difference in size can cause a variation in the performance. Therefore, learning and evaluation are done multiple times. This phenomenon is known as cross-validation. The amount of times is termed as the fold. By doing so, the resilience of the system could be evaluated by examining the statistics on detection performance. The cross-validation used in this paper has a value of $k = 10$.

4.6 Kernels

Kernels are used to calculate feature mapping of large dimensions. The paper uses nine types of kernels—linear, polynomial, multiquadratic, Epanechnikov, radial, and ANOVA. The Radial Basis Function kernel is represented as

$$k(x,\ y) = \exp\left(-g||x\ -\ y||^2\right). \tag{1}$$

If g has greater value, it introduces large variance, whilst the lower value makes smoother boundary with minimum variance.

The polynomial kernel function is symbolized mathematically by,

$$k\,(x,\ y)\ =\ (x * y + 1)^p \tag{2}$$

The polynomial degree is denoted as p.

The dot function is termed as

$$k(x,\ y) = x * y \tag{3}$$

The above kernel is the multiplication of x and y.

The ANOVA kernel, the output of which is popular in multidimensional situations, is characterized as

$$k(x,\, y) = \sum_{k=1}^{n} \exp\left(-\sigma\left(x^k - y^k\right)^2\right) \tag{4}$$

where σ can be derived from gamma, g, $g = \frac{1}{2\sigma^2}$.

5 Experimental Results and Discussion

LSB matching in spatial domain and F5 in transform domain is used for a comparison of classification results.

The details in Table 2 are extracted with six kernels and four sampling. Typically, the dot kernel has shown positive results for the LSB matching algorithm. The ROC curve for different kernels of the dataset is as shown in Fig. 3.

The details in Table 3 are extracted with the same kernels and samples as Table 2. The best results are obtained for stratified and automatic sampling for ANOVA (Fig. 4).

Table 2 SVM classification for LSB matching algorithm

	Linear	Stratified	Shuffle
Dot	10.60	54.94	54.65
Radial	5.95	11.02	11.32
Polynomial	11.16	53.93	53.34
ANOVA	8.52	48.48	48.63

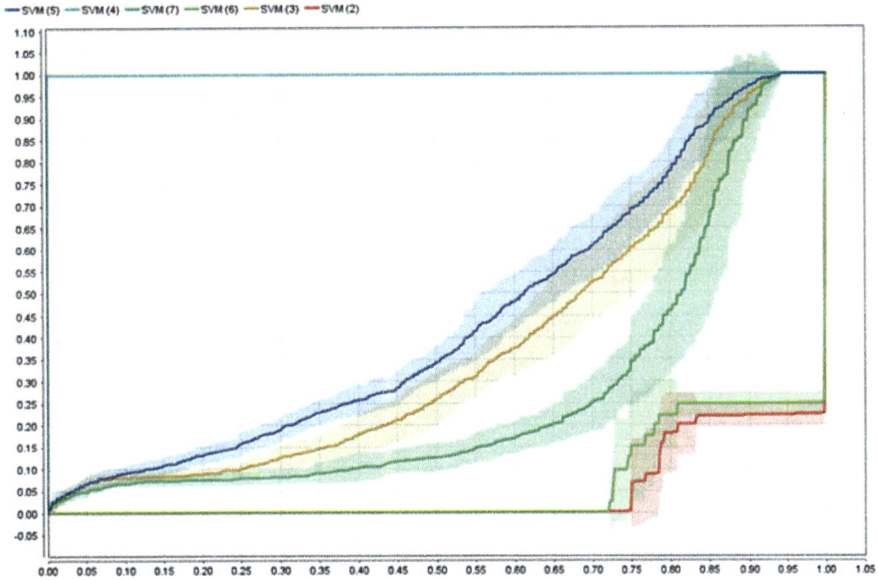

Fig. 3 ROC curve for SVM with LSB matching and different kernels

Table 3 SVM classification for F5 algorithm

	Linear	Stratified	Shuffle
Dot	87.53	93	92.94
Radial	10	50.3	48.57
Polynomial	50.04	81.27	81.81
ANOVA	90.54	92.32	92.35

6 Conclusion

Statistical steganalysis in JPEG images is employed in this paper. Different features which are fragile to embedding are used for the analysis. Firstorder, extended DCT, second order and Markov features are employed for analysis. The spatial domain with LSB matching and transform domain with F5 are the steganographic schemes used. SVM is employed to classify the image to be a cover or a stego. The training images used are 1500 images from INRIA holiday dataset and testing images of 800 are taken from UCID image dataset. Cross-validation is exercised in the paper for enhanced performance. Six various kernels and four kinds of samplings are employed. The results project the statement that the transform domain gives a better classification rate than a spatial domain. The dot kernel provides a decent performance with shuffle sampling for LSB matching as far as the kernels are concerned. The dot kernel adds a better rate with stratified and automatic sampling outcomes with F5. With distinct

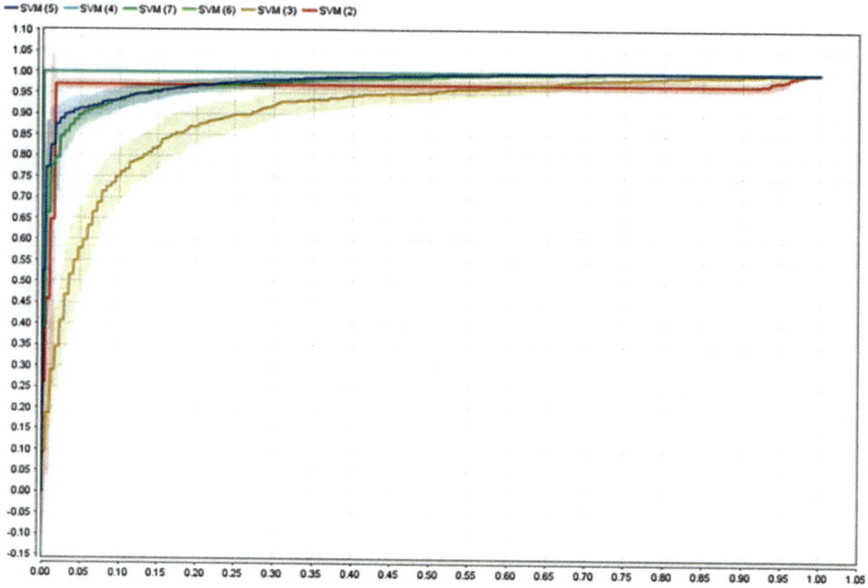

Fig. 4 ROC curve for SVM with F5 algorithm and different kernels

spatial and transform domain algorithms, further research can be performed and a comparative study can be launched.

References

1. B.R. Kumar, P.R. Murti, Data security and authentication using steganography. Int. J. Comput. Sci. Inf. Technol. **2**(4), 1453–1456 (2011)
2. V. Nagaraj, G. Zayaraz, V. Vijayalakshmi, Modulo based image steganography technique against statistical and histogram analysis. Netw. Secur. Cryptogr. 34–39 (2011)
3. J.A. Christaline, R. Ramesh, D. Vaishali, *Steganalysis with Classifier Combinations* (2014)
4. C. Cortes, V. Vapnik, *Support Vector Networks* (1995)
5. B. Scholkopf, A.J. Smola, *New Support Vector Algorithms* (1999)
6. A.A. Attaby, A.K. Alsammakand, M.F.M. Mursi Ahmed, Data hiding inside JPEG images with high resistance to steganalysis using a novel technique: DCT-M3. Ain Shams Eng. J. **9**(4) (2018)
7. M. Kalita, T. Tuithung, A comparative study of steganography algorithms of spatial and transform domain. IJCA Proc. Natl. Conf. Recent Trends Inf. Technol 9–14 (2015)
8. Z. Xia, L. Yang, S. Xingming, D. Sun, Z. Ruanand W. Liang, A learning-based steganalytic method against LSB matching. Steganography **20** (2011)
9. P. Malathi, T. Gireeshkumar, Relating the embedding efficiency of LSB steganography techniques in spatial and transform domains. Procedia Comput. Sci. **93**, 878–885 (2016)
10. P. Liu, C. Yang, F. Liu, X. Song, Improving steganalysis by fusing SVM classifiers for JPEG images, in *International Conference on Computer Science and Mechanical Automation (CSMA)* (2015), pp. 185–190

11. J. Babu, S. Rangu, A survey on different feature extraction and classification techniques used in image steganalysis. J. Inf. Secur. **08**(03) (2017)
12. V. Bhasin, P. Bedi, Steganalysis for JPEG images using extreme learning machine, in *Proceedings, SMC* (2013), pp. 1361–1366
13. A. Ashu, R. Chhikara, Performance evaluation of first and second order features for steganalysis. Int. J. Comput. Appl. **92**, 17–22 (2014)
14. L. Wang, Y. Xu, B. Du, Y. Ren, A posterior evaluation algorithm of steganalysis accuracy inspired by residual co-occurrence probability. Pattern Recognit. 106–117 (2019)
15. X. Hou, T. Zhang, Y. Wu, L. Ji, Combating highly imbalanced steganalysis with small training samples using feature selection 243–256 (2017)
16. B.D. Barkana, B. Yildirim, I. Saricicek, Performance analysis of descriptive statistical features in retinal vessel segmentation via fuzzy logic, ANN, SVM, and classifier fusion. Knowl. Based Syst. **118**, 165–176 (2017)
17. M. Castelli, L. Vanneschi, Á.R. Largo, *Supervised Learning: Classification, Encyclopedia of Bioinformatics and Computational Biology* (Elsevier, 2019), pp. 342–349
18. D. Shankar, V. Shukla, *Effect of Principal Component Analysis in Feature based Uncalibrated Steganalysis using Block Dependency*, SSRN online (2019)

Performance Analysis of Fruits Classification System Using Deep Learning Techniques

L. Rajasekar, D. Sharmila, Madhumithaa Rameshkumar, and Balram Singh Yuwaraj

1 Introduction

Fruit identification is a regular expression aimed at vision-based strategies designed for the recognition of fruits from videos or images. Deep learning approaches are commonly used in the identification of fruits in real time. Recent identification and recognition of fruit is mainly based on the use of convolutionary neural networks (CNN). Develop complete end-to-end fruit identification. In the field fruit detecting and recognizing, SSD, faster region CNN (F-RCNN), and you only look once (YOLO) are the three main methodologies [1]. The system uses DNN and Caffe deep learning library of OpenCV in this paper to detect and recognize fruits. The design uses the MobileNet software SSD algorithm to detect fruit. In our case, Movidius NCS, pretrained prototype (SSD-MobileNet) is trained and tested on the Raspberry Pi 3 with real-time video and by using neural compute pin [2]. The Movidius stick is an intelligent company's integrated machine intelligence platform. It has been designed to achieve high frame rates for low-power devices. Applications with low power that require real-time systems use Movidius, which enables inventor to deploy a neural network [3, 4]. Download the neural stick SDK and develop the graph using this tool. In fact, it is quite easy to generate graph files [5]. The SSD algorithm for fruit detection comprises of two segments: fruits feature collection and application of lesser convolution riddles for fruit detection. The riddles used to predict group notches then boxes for a secure array of evasion strongboxes on function maps. Later, through the use of non-maximum suppression algorithm, the final detection

L. Rajasekar
Bannari Amman Institute of Technology, Erode, India

D. Sharmila
Jayshree Ram Engineering College, Tiruppur, India

M. Rameshkumar (✉) · B. S. Yuwaraj
University of Leicester, Leicester, England, UK
e-mail: mithujaishu@gmail.com

S. Patnaik et al. (eds.), *Advances in Machine Learning and Computational Intelligence*, Algorithms for Intelligent Systems, https://doi.org/10.1007/978-981-15-5243-4_26

occurs [6]. So, in order to extract function maps, the SSD algorithm uses MobileNet architecture. MobileNet designed for resource-constrained gadgets such as smart phone and applications for embedded vision. MobileNet is related to efficient way of building lightweight deep neural networks using depth-separable convolutions [7]. By integrating the SSD model and MobileNet, achieving a firm besides proficient framework for detecting fruit based on deep learning is possible [8, 9]. Numerous articles are been identified that are attached to fruit detection systems. Various techniques are used for each detection application. Techniques based on F-RCNN are been tested to achieve an excellent KITTI accuracy. YOLO stays the fastest way toward getting simultaneous fruit identification through 45 fps on general purpose unit and an accuracy of 63.4% in VOC2007 dataset [10], but still it has weakness in recognizing small-sized fruits. Instead, by merging F-RCNN anchor box and using multi-scale functionality, this weakness is strengthened by SSD method [11, 12].

2 Literature Survey

Syal et al. [13] introduce minimum Euclidean distance-based segmentation technique for segmenting the fruit region from the input image. Ragit et al. [14] describe processing-based yield counting system, and health monitoring of citrus fruit is being processed. The system is being designed to automatically and accurately calculate the yield of citrus group tree along with moisture and temperature of the tree. Mustaffa et al. [15] focus on the identification of maturity of mango fruit. Raspberry Pi is a small computer, which is powerful enough to run an image processing algorithm is chosen for this system. Arivazhagan et al. [16] describe computer vision strategies used to recognize a fruit which rely on four basic features which characterize the object: intensity, color, shape, and texture. Experimental results on a database of about 2635 fruits from 15 different classes confirm the effectiveness of the proposed approach. Choudhary et al. [17] describe that fruits should be quickly and correctly differentiated from their surroundings for the fruit harvesting robot. Edge-based and color-based detection methods are generally used to segment images of fruits obtained under natural lighting conditions. The comparison results are shown in the segmented image results. Accordingly, a new mango detection method is proposed to position the centroid of mangoes. Lu et al. [18] present a detailed overview of various methods, i.e., preprocessing, segmentation, feature extraction, classification which addressed fruits and vegetables quality based on color, texture, size, shape, and defects. Kaur et al. [19] shed light on the advancements made in the automated agricultural industry. Digital image processing techniques are now widely used for maturity estimation of fruits and vegetables. It was observed that for achieving high accuracy, a compromise is to be made with high computational complexity.

3 Deep Learning Model

Deep learning is a ground-breaking division of machine learning model influenced by neural networks toward resolving many issues related to natural language processing and machine vision besides many other areas. This enables for automated detection that are very costly. An example says, deep learning can be used to recognize the different parameters in a fruits dataset, classify text, and convert audio to text. Recently, deep convolutionary neural networks are superior to people in fruit detection and recognition. Models of deep learning are designed in multiple layers. This model will get more information after adding multiple layers and need to prevent overfitting. Several techniques are used to counter overfitting, the common of which is data increase, regularization, and dropout [20]. Deep learning is based on the methods to learn, how to represent the information. CNN, deep belief networks, and deep auto encoder networks are the most common ones. The CNN model in the proposed system to build a system of fruit detecting and recognition as it is better to exert through manipulation of images. CNN's building blocks are pooling, convolution, and connected layers [20]. To obtain function maps like MobileNet [8], VGG [21], Inception [22], and ResNet [23], there are various suggested CNN architectures. We use MobileNet architecture in this paper for extracting function maps from the input frames. The methods Theano, Torch, and Caffe are available at many libraries. Caffe library is suggested for using in fruits recognition system. Caffe library consists of many benefits such as extraordinary performance, stress-free training, implementation, connections and easy to read source code [24].

4 Fruits Classification System

Detection of fruits is done using two processes, initially training a network required before recognition of fruit images. The suggested model detects the live streaming fruits image and achieves 80.23% mean average accuracy (MAP). The design uses the MobileNet software SSD algorithm for detecting fruits. CNN is equipped by loss function and bounding box [25, 26]. In order to classify non-fruits and fruits, selecting the loss function is highly challenging task. Corresponding technique of Intersection over Union (IoU) is used during the training period to relate the forecasts during ground truth trainings. Furthermore, non-maximum suppression (NMS) algorithm is used for removing multiple boundary boxes around a fruit during prediction time.

4.1 Training Stage

Image matching techniques. Matching strategy is answerable for deciding what boxes of default suit a predicted shape (ground truth). We choose from default

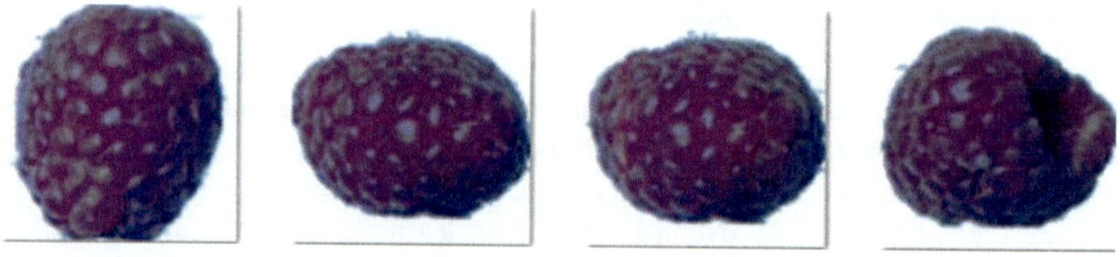

Fig. 1 Generating new samples with help of data augmentation

boundary boxes for each ground truth box, which vary by size, aspect ratio, and position. This begins with IoU and boxes by matching all ground truth. Consequently, our predictions in practice are more consistent and more variable.

Loss Function. It is a method for calculating the current model's poor performance that is useful for determining present returns besides expected productivity since this one be contingent on the bundle of learning. Equation (1) defines the loss function for the model.

$$L(x, c, l, g) = \frac{1}{N}(L_{\text{conf}}(x, c) + aL_{\text{loc}}(x, l, g)) \tag{1}$$

From (1), N denotes amount of positively matched evasion boxes and α denotes load loss of position. It creates an imbalance between the boundary box and the reality box on the surface.

Fruits Featured Maps and Default Boxes Aspect Ratio. For various fruit measurements, SSD approach recommends storing the fruit dataset to different sizes and then comparing the results. Nonetheless, simulation of same images along with the contribution of constraints diagonally to each fruit scale by means of the attributes maps with different layers in a solo network for prediction may be done.

Data Augmentation. Habitually speaking, CNNs need a large amount of training samples, as providing a dataset with enough samples is exactly difficult, it is better to use multiple types of data increase. For example, images are cut in various locations, also images are flipped steeply and straight where information misrepresentation is applied then clatters is added as shown in Fig. 1.

4.2 Fruits Identification

Single Short Detector. SSD method labels output space for prediction of bounding boxes to a list of default bounding boxes for each function map position across dissimilar aspect ratio and measures. At the time of training phase, this technique equals the fruit with default strong bounding boxes with different ratios. The feature map elements are associated with an amount of default boxes. A counterpart is considered for each and every strong box with an IoU more than 0.5. The basic method of SSD is to prediction of box counterbalances and category scores for a consistent array of default boundary boxes with small, convolutionary riddles that are used to attribute maps. To extract image attributes, the SSD algorithm uses MobileNet design. When a picture is made to pass on the architecture of MobileNet, SSD uses six additional layers of more added convolution as shown in Fig. 2.

For three of them, instead of four, we produce six predictions for every cell. With a total of 8732 predictions, using six layers are generated by the SSD method. Method SSD uses diverse attribute maps toward individually identify fruits. Additional layers may produce featured charts of various dimensions for the detection of fruits in different dimensions as the fruits may be of any dimension. An initial additional layer detects the smallest fruits, while a sixth additional layer detects the largest fruits. Any goal requires 20 scores and a strongbox for some category in Fig. 3.

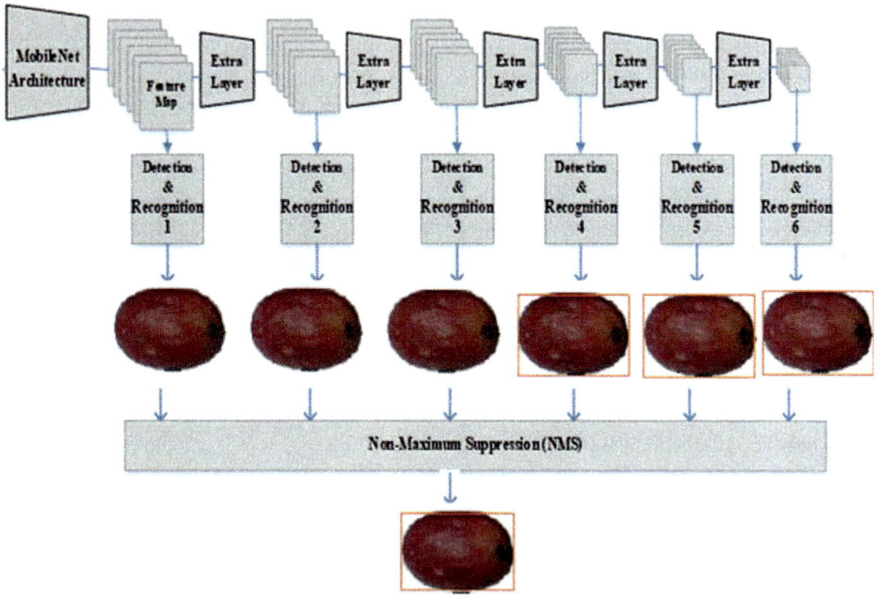

Fig. 2 Architecture of SSD

Fig. 3 Prediction of class score of grapes

MobileNet Architecture. The proposed system's basic architecture is the MobileNet network. It has been developed for machine visualization and also for transportable uses in general. This design carries the benefit of maximum reduced dimension of design and cost of computation. MobileNet technology can be exhibited on a smart device for various fruit and grain identification [8]. MobileNet is a proficient development which uses according to depth distinct convolutions to create feathery profound neural systems [27]. Starting here, MobileNet engineering is not quite the same as old style CNNs. In accordance with depth, detachable networks comprise of two layers: a depth layer (3 × 3) and point layer (1 × 1). A depth layer is for applying a channel for every information channel. At that point, the point layer uses 1 × 1 networks for the creation of a direct blend of the depth layer. Moreover, this network engineering use clumps standardization (BN) and rectified-linear unit (ReLU) for both depth-wise and point-wise convolutions. As of now Mobile net system has low computational cost but normally a few applications require the algorithm to be fast for embedded vision and smart phone applications.

Non-Maximum Suppression. Throughout the expectation period, a huge amount of boxes are created with the natural products everywhere. This is very important to shift through a large portion of a bouncing box through relating a technique called non-maximum suppression (NMS). The NMS method is mainly used to channel copy boxes everywhere the natural products. SSD calculation requests the expectation boxes through the certainty scores. Start after a top certainty scores forecast, SSD calculation keeps just the main five predictable boxes having an IoU more noteworthy than 0.5 through current expectation on behalf of a similar class which is dismissed.

5 Experiment Result

The proposed frameworks are intended for natural products discovery and acknowledgment progressively recordings by using profound learning strategies. Python language is used for executing the proposed framework as in the programming language. Additionally, a microcontroller used as inserted visualization stage and neural computing stick associated with SDK and introduced to manufacture an effective use. The introduced framework includes open-source microcontroller, neural computing stick, camera, and LCD. The products vital has introduced as shown in Fig. 4.

Table 1 shows the quality period for the identification and recognition of proposed fruits. As shown in Table 1, using the SSD-MobileNet system on open-source CPU, get one frames per second (FPS). This is not enough to introduce the identification and recognition of fruits in real time based on deep learning techniques. We use neural compute stick Movidius to solve this problem.

Most of the fruit detection systems are needs to be implemented in real time. On the other hand, techniques like SSD and YOLO work quickly, and the main issue is that they decrease expectation accuracy while techniques such as faster R-CNN achieve high accuracy but are a bit slower. YOLO and faster R-CNN are balanced

Fig. 4 Fruits and vegetables recognition system

Table 1 Table captions should be placed above the tables

Hardware	FPS
Open-source hardware (Rpi)	One
Open-source hardware with neural computing stick	Three

by SSD process. Using default boxes and multi-scale features, SSD speeds up the procedure by removing the need for the region proposal network and increasing precision. This can achieve average accuracy of 73% by using the SSD-MobileNet technique.

6 Conclusion

For embedded based vision implementation with high accuracy, an embedded fruit detecting and recognition system is proposed in this paper. Deep learning methods used to identify and recognize fruits in the proposed system. In a real-time video, the system will classify as well as recognize fruits. The fruits were discovered and categorized using the deep learning techniques and the DNN library of OpenCV based on the SSD-MobileNet algorithm. Raspberry Pi 3 used the devices in this study because it is an integrated network for inexpensive equipment and it has accurate specifications. By using Movidius neural compute ring, we improved fruit identification and recognition. This stick benefits from powerful device design for embedded based vision applications. This classification provided the whole thing exactly and quickly, it can process four frames per second using neural computing stick, and at the same time, the system gets only one frame per second without using neural computing stick. Such that, it is highly effective and efficient process by using Movidius NCS. In the future, it may work with a large amount of data of the maximum number of fruits in the context of different other model and transfer learning.

References

1. J. Huang, V. Rathod, C. Sun, M. Zhu, A. Korattikara, A. Fathi, I. Fischer, Z. Wojna, Y. Song, S. Guadarrama, Speed/accuracy trade-offs for modern convolutional object detectors. arXiv preprint arXiv:1611.10012 (2018)
2. OpenCV (2018), http://opencv.org/. Accessed 22 March 2018
3. Raspberry Pi (2015), https://www.raspberrypi.org. Accessed 4 May 2015
4. N.A. Othman, I. Aydin, A new IoT combined body detection of people by using computer vision for security application, in *2017 9th International Conference on Computational Intelligence and Communication Networks (CICN)*, Girne, 2017, pp. 108–112
5. Intel Movidius Neural Compute Stick (2017). https://developer.movidius.com/. Accessed 28 Sept 2017
6. MvNCCompile (2017). https://movidius.github.io/. Accessed 2017
7. W. Liu, D. Anguelov, D. Erhan, C. Szegedy, S. Reed, SSD: single shot multi-box detector. arXiv preprint arXiv:1512.02325 (2015)
8. A.G. Howard, M. Zhu, B. Chen, D. Kalenichenko, W. Wang, T. Weyand, M. Andreetto, H. Adam, MobileNets: efficient convolutional neural networks for mobile vision applications (2017)
9. P.L. Geiger, C. Stiller, R. Urtasun, Vision meets robotics: the kitti dataset. Int. J. Robot. Res. (IJRR) (2013)
10. M. Everingham, L. Van Gool, C.K.I. Williams, J. Winn, A. Zisserman, The PASCAL Visual Object Classes Challenge 2007 (VOC2007) Results. http://www.pascalnetwork.org/challenges/VOC/voc2007/workshop/index.html
11. W. Liu, D. Anguelov, D. Erhan, C. Szegedy, S. Reed, C.-Y. Fu, A.C. Berg, SSD: single shot multi-box detector, in *European Conference on Computer Vision* (Springer, Berlin, 2016), pages 21–37
12. S. Ren, K. He, R. Girshick, J. Sun, Faster R-CNN: towards real-time object detection with region proposal networks, in *NIPS* (2015)
13. A. Syal, D. Garg, S. Sharma, Apple fruit detection and counting using computer vision techniques, in *International Conference on Computational Intelligence and Computing Research* (IEEE, 2014)
14. B.P. Ragit, A.V. Bhingare, R.T. Bankar, An intelligent fruit counting system. Int. J. Innov. Sci. Modern Eng. (IJISME) (2015)
15. I.B. Mustaffa, S.F.B.M. Khairul, Identification of fruit size and maturity through fruit images using OpenCVPython and Rasberry Pi, in *International Conference on Robotics, Automation and Sciences (ICORAS)* (IEEE, 2017)

16. S. Arivazhagan, R.N. Shebiah, S.S. Nidhyanandhan, L. Ganesan, Fruit recognition using color and texture features. J. Emerg. Trends Comput. Inf. Sci. **1**(2) (2010)
17. P. Choudhary, R. Khandekar, A. Borkar, P. Chotaliya, Image processing algorithm for fruit identification. Int. Res. J. Eng. Technol. (IRJET) **4**(3) (2017)
18. S. Lu, Z. Lu, P. Phillips, S. Wang, J. Wu, Y. Zhang, Fruit classification by HPA-SLFN, in *8th International Conference on Wireless Communications & Signal Processing (WCSP)* (IEEE, 2016)
19. H. Kaur, B.K. Sawhney, A brief review on maturity level estimation of fruits and vegetables using image processing techniques. Int. J. Sci. Environ. Technol. **6**(6), 3407–3413 (2017)
20. M.A. Ponti, L.S.F. Ribeiro, T.S. Nazare, T. Bui, J. Collomosse, Everything you wanted to know about deep learning for computer vision but were afraid to ask, pp. 17–41. https://doi.org/10.1109/sibgrapi-t.2017.12
21. K. Simonyan, A. Zisserman, Very deep convolutional networks for large-scale image recognition. CoRR, vol. abs/1409.1556 (2014)
22. C. Szegedy, V. Vanhoucke, S. Ioffe, J. Shlens, Z. Wojna, Rethinking the inception architecture for computer vision
23. K. He, X. Zhang, S. Ren, J. Sun, Deep residual learning for image recognition. CoRR, vol. abs/1512.03385 (2015)
24. Y. Jia, E. Shelhamer, J. Donahue, S. Karayev, J. Long, R. Girshick, S. Guadarrama, T. Darrell, Caffe: convolutional architecture for fast feature embedding, in *MM 2014—Proceedings of the 2014 ACM Conference on Multimedia* (2014). https://doi.org/10.1145/2647868.2654889
25. T.-Y. Lin, M. Maire, S. Belongie, J. Hays, P. Perona, D. Ramanan, P. Dollár, C.L. Zitnick, Microsoft COCO: common objects in context (2014)
26. C. Szegedy, S. Reed, D. Erhan, D. Anguelov, Scalable, high-quality object detection. arXiv preprint arXiv:1412.1441 v3 (2015)
27. F. Chollet, Xception: deep learning with depthwise separable convolutions. arXiv preprint arXiv:1610.02357v2 (2016)

Early Detection of Locust Swarms Using Deep Learning

Karthika Suresh Kumar and Aamer Abdul Rahman

1 Introduction

Locust infestations have been feared and revered throughout history and are often viewed as natural disasters. Locust plagues have the potential to cause large-scale damage and pose a major threat to crop production. Even today, locust swarms can be spotted in many parts of the globe and continue to wreak havoc. This cataclysm has affected the livelihoods of individuals involved in the process of crop production and of those who are dependent on it for nourishment and subsistence.

In 2019 alone, a significant number of locust infestations have been recorded globally and to name a few include locust invasions in Mecca (Saudi Arabia) [1], Abu Dhabi (UAE) [2] and Sardinia (Italy) [3]. Sardinian Islands in Italy faced major crop and pasture losses after being hit by the worst locust swarm it had witnessed in 70 years. Many regions of Iran and Yemen face an alarming threat by locust swarms from time to time. Further, locust infestations need not be confined to a region due to the high mobility of locusts and can spread to other places at an alarmingly high erratic phase [4, 5]. Desert locusts are the deadliest of all the other locust species due to their ability to cover large distances in short durations of time. They are capable of flying up to 150 km a day with the wind [5]. The situation is exacerbated by their highly unpredictable flight paths. The locust infestation in Iran was not only damaging for the region but also posing an unprecedented threat to farmers of the nearby countries as well, especially the cotton farmers in Balochistan [6]. The desert locust outbreak that took place in West Africa in 2004 cost an estimate of $122 million to remedy the situation and ended up causing harvest losses of nearly $2.5 billion [7].

K. S. Kumar (✉) · A. Abdul Rahman
Birla Institute of Technology and Science Pilani, Dubai, UAE
e-mail: karthikaskumar1999@gmail.com

A. Abdul Rahman
e-mail: ar.aamer@gmail.com

S. Patnaik et al. (eds.), *Advances in Machine Learning and Computational Intelligence*, Algorithms for Intelligent Systems, https://doi.org/10.1007/978-981-15-5243-4_27

Locusts are short horned grasshoppers belonging to the family Acrididae [8] and are completely harmless under most conditions but become deleterious under certain weather patterns. Heavy and sporadic rain patterns trigger an increased secretion of serotonin in grasshoppers which leads to the eventual metamorphosis of grasshoppers to locusts. Locusts pose a danger when they transit from a solitary to a gregarious nature [9]. The solitary insects on forming small bands or groups incite behavioural and morphological changes, induced by vegetation, and exhibit gregarious characteristics and form swarms that cause havoc on crops.

Hence, identifying the locusts in the initial stage of band formation and keeping a track of the insects can help inhibit or avert potential swarm formations. Desert locusts by nature are found in some of the harshest environments in the world. Existing locust detection techniques involve manual inspections of prone areas and inputting the gathered information into a digital interface. Weather and soil data collected from satellites are also used as a part of locust monitoring [10]. These methods have led to significant success in early detection of locust activity. However, manual inspections can be a laborious and time-consuming task, and despite these measures, locust swarms continue to swarm in many parts of the world. Hence, the paper aims to contribute to early detection of locust swarm formation by automating the process of detection using deep learning techniques. This is thanks to the drastic improvements that have taken place over the last few years with regards to deep learning [11].

In reference to previous works, one of notable importance is the work published by Xia et al. [12] on insect classification. The paper focuses on insect image recognition using an improved VGG19 neural network architecture, and the proposed model had a successful precision rate of 0.899. Further, another paper by Nguyuyen et al. [13] is focused on pest detection, and it employs neural network to perform pest image recognition. This paper utilises adaptive neural network combined with convolutional neural network VGG16 and obtained an accuracy of 86% on all classes of the test set. Counting of Aphids by Chen et al. [14] used pre-trained VGG13 weights with different sized CNNs for the purpose of counting but employed an alternate method involving annotation and segmentation. This achieved a precision rate between 0.79 and 0.82 for different sized CNNs. Another work to note is the work of Xiong et al. [15], where they implemented colour image segmentation by pulse code neutral network for locust detection.

This paper draws inspirations from the above-mentioned previous research to implement locust detection utilising convolutional neural networks. As traditional machine learning algorithms show limitations when it comes to processing image data, deep learning is preferred in object detection scenarios. The proposed model implements a two-stage object detection algorithm, i.e., the faster RCNN model [16].

2 Background

2.1 *Deep Learning*

Deep learning models are composed of multiple processing layers that learn representations of raw data through multiple levels of abstraction. State-of-the-art deep learning models have shown exceptional accuracy in object detection and speech recognition, and have proved to be beneficial in other fields such as drug design and genomics. The back-propagation algorithm is used to draw structure from the dataset representations of the previous layers [17].

2.2 *Convolutional Neural Networks*

Convolutional neural network (CNN) holds a significant position in deep learning and is used for processing and analysing images or visual media. Convolutional neural network is multi-layered network and consists of blocks of convolutional and pooling layers, fully connected layers and one output layer. The convolutional layer extracts the features from an image using filters. The pooling layer further sub-samples the outputs from the convolutional layer and reduces the number of parameters [18].

The fully connected layer performs operation in the present layer neurons on inputs from previous layers and generates output. Over the years, many CNN models have been developed, and every newly developed model has a better and larger architecture with improved efficiency. AlexNet [19] is a popular neural network architecture which employs ReLU (rectified linear unit) for nonlinearity instead of sigmoid activation function. Hence, it trains faster and has a reduced overfitting characteristic. Another relevant architecture is the VGGNet [20]. It uses small size filters and hence reduces the number of parameters. It requires less training time and has six different configurations. Significant architectures include ResNet, GoogLeNet and MobileNet.

2.3 *Rcnn*

The fundamental step in RCNN [21] is the sliding window approach which constitutes going through the image with distinct-sized rectangles. Every image has a sliding window to look up and search at every position of the image. First, many small-segmented areas are obtained by non-object based segmentation. The bottom up method is then used, small-segmented areas are merged to form large-segmented areas, and region proposals are generated. Region proposals are just smaller parts of the original which we surmise to contain the objects of interest. There are 2 k region proposals generated by developed algorithms.

This is followed by the creation of feature vector for each region proposal and will represent the original image in a smaller dimension. Thus, the features are extracted by pre-trained neural network. An SVM classifier is used to classify the feature vectors and, in the process, outputs a confidence score.

2.4 *Fast RCNN and Faster RCNN*

Fast RCNN [22], as the name suggests, is a derivative of RCNN but builds up on the shortcomings of RCNN and is more efficient. The major difference though is feeding the input image to the CNN to obtain a convolutional feature map instead of feeding region proposals to the CNN. Further, the region of proposals is obtained from the feature map and warped into squares and is reshaped to a fixed size. The reshaping is done to feed it to a fully connected layer by an RoI pooling layer. A softmax layer is used to predict the offset values for the bounding box and class of the proposed region. Fast RCNN is comparatively better because the feature map is generated only after performing convolution operation only once per image.

Faster RCNN [16] consists of two networks, namely region proposal network which generates region proposals and another network that utilises the region proposals to detect the object. The main difference between faster RCNN and fast RCNN is that the former generates region proposals by selective search.

3 Methodology

3.1 *Data Collection*

The image datasets were manually collected and obtained from online sources and pre-processed to make training and testing datasets. The total images collected were 1522 and were subdivided into 1300 for the training set and 222 for testing dataset. Bounding boxes were manually inserted into each image. Shows a plot of the box area versus aspect ratio. Figure 1 shows a representation of the type of images that were used as training and testing datasets.

3.2 *Training and Evaluation of Dataset*

Double-shot object detection algorithms were found to work better than single-shot detection at the time of experiment. Faster RCNN object detectors are trained wherein the region proposal networks were combined with various state-of-the-art

Fig. 1 Image dataset

feature extraction architectures such as ResNet, VGG16, VGG19 and AlexNet. The maximum number of epochs was kept to five with a learning rate of 0.001.

The trained detector is then used on the test dataset to identify locusts and later evaluates to measure the performance. The detector is run on every image in the test set, and the results are collected. Finally, the detector is evaluated using average precision metric (1), and object detector evaluation functions are utilised to obtain average precision. The obtained precision is compared to assess the success rate of the different architecture on the object detector.

$$\text{Precision} = \text{TP}/(\text{TP} + \text{FP}) \tag{1}$$

$$\text{Recall} = \text{TP}/(\text{TP} + \text{FN}) \tag{2}$$

where

TP = True Positive
FN = False Negative
FP = False Positive.

4 Result and Inference

The different feature extraction architectures all gave similar levels of accuracy. All the detectors were trained at a learning rate of 0.001 and set to five epochs. The VGG16 gave the best accuracy at 83% which was 4% higher than the subsequent performing architecture GoogLeNet. The detectors detected the presence of locusts

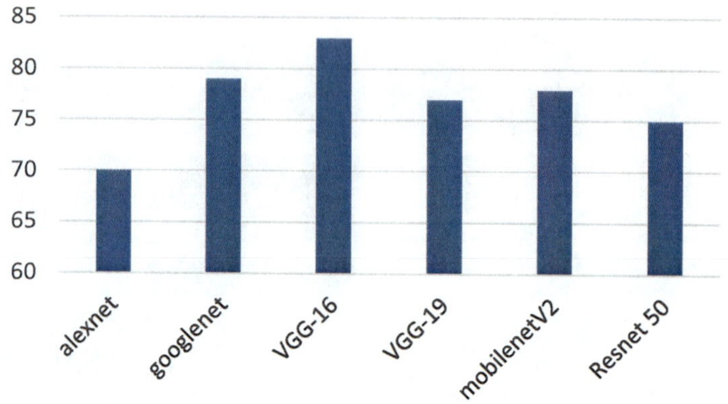

Fig. 2 Accuracy of different feature extraction architecture

in different environments with different lighting conditions giving the count of the number of the insect present in each individual frame as can be seen in figure, thus achieving and fulfilling the object of the model. The detectors were individually tested on the images, and 0.99% confidence levels were achieved on the images. The detector could successfully detect locust in the presence of other insects as well (Fig. 2).

Figure 3 gives a plot of the precision that the VGG16 RCNN detector had achieved when tested with 220 test image datasets. It was shown to have an average precision

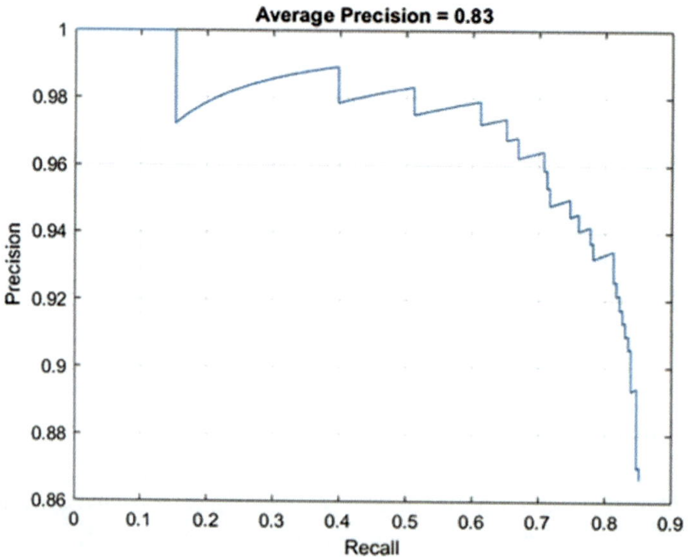

Fig. 3 VGG16 precision score

Fig. 4 Bounding boxes drawn around the locusts via the VGG16 detector

of 83%. The detectors were individually tested on the images, and 0.99% confidence levels were achieved on the images as can be observed in Fig. 4. The detector could successfully detect locust in the presence of other insects as well. Further, the trained model was able to identify locusts in a group and predict the count thus achieving and fulfilling the objective of the model, i.e., to detect the gathering of locusts in the early stages of swarm formation so that early action may be taken to minimise damage caused.

5 Conclusion

In conclusion, we can infer that the proposed model can aid in the detection of a band of locusts which is a prime factor contributing to formation of locust swarms. Thus, the model achieves the desired result by using deep learning techniques and achieves the highest precision under VGG16 architecture. For improved accuracy, image augmentation can be performed on the obtained database. Future works can employ the same model for other species of locusts and not only be limited to desert locusts. The proposed model can be applied to video streams or real-time image sequences obtained from unmanned aerial vehicles (UAVs) or other sources to regularly monitor locust populations and can make the existing manual work easier.

References

1. T. Stickings, Huge swarm of locusts descends on Mecca leaving worshippers covered in insects as cleaners battle to control the bugs at Islam's holiest site. Daily Mail, January 10, 2019. www.dailymail.co.uk/news/article-6577871/Huge-swarm-locusts-descends-Mecca.html. Last accessed 26 Oct 2019
2. A. Ahmad, Abu Dhabi's Al Dhafra area swarmed by locusts. Gulf News, Janurary 17, 2019. gulfnews.com/uae/health/abu-dhabis-al-dhafra-area-swarmed-by-locusts-1.61499251. Last accessed 26 Oct 2019
3. A. Amante, P. Graff, Sardinia hit by worst locust invasion for 70. Reuters, para. 1, June 11, 2019. www.reuters.com/article/us-italy-locusts/sardinia-hit-by-worst-locust-invasion-for-70-years-idUSKCN1TC1BY. Last accessed 27 Oct 2019
4. K. Cressman, Desert locust. in *Biological and Environmental Hazards, Risks and Disasters* (2016), pp 90
5. Food and Agriculture Organization of the United Nations: Locust Watch. Food and Agriculture Organization of the United Nations, 2019. www.fao.org/ag/locusts/en/info/info/index.html. Last accessed 26 Oct 2019
6. F. Ilyas, Locusts descend on parts of Sindh after attacking Balochistan. Dawn, para. 1, June 13, 2019, https://www.dawn.com/news/1487889/locusts-descend-on-parts-of-sindh-after-attacking-balochistan. Last accessed 26 Oct 2019
7. L. Brader et al., Towards a more effective response to desert locusts and their impacts on food security, livelihoods and poverty, in *Multilateral Evaluation of the 2003–05 Desert Locust Campaign* (2006), pp. 49
8. J. Simpson, A. Sword, Locusts. Curr. Biol. **18**, 364–366 (2008). https://doi.org/10.1016/j.cub.2008.02.029
9. M. Anstey, S. Rogers, S. Ott, M. Burrows, S. Simpson, Serotonin mediates behavioral gregarization underlying swarm formation in desert locusts. Science **323**, 627–630 (2009)
10. K. Cressman, The use of new technologies in desert locust early warning, in *Outlooks on Pest Management* (2008), pp. 55–59
11. I. Goodfellow, Y. Bengio, A. Courville, Deep learning (MIT Press, Cambridge, MA, 2016)
12. Xia, D., Chen, P., Wang, B., Zhang, J., Xie, C.: Insect Detection and Classification Based on an Improved Convolutional Neural Network (2018)
13. T. Nguyen, P. Hung, Pest detection on Traps using deep convolutional neural networks (2018)
14. J. Chen, Y. Fan, T. Wang, C. Zhang, Z. Qiu, L. He, Automatic segmentation and counting of aphid nymphs on leaves using convolutional neural networks (2018)
15. X. Xiong, Y. Wang, X. Zhang, Color image segmentation using pulse-coupled neural network for locusts detection (2006)
16. S. Ren, K. He, R. Girshick, J. Sun, Faster R-CNN: towards real-time object detection with region proposal networks, in *IEEE Transactions on Pattern Analysis and Machine Intelligence* (2015)
17. Y. LeCun, Y. Bengio, G. Hinton, Deep learning. Nature **521**, 436–444 (2015)
18. S. Albawi, T. Abed Mohammed, S. ALZAWI, Understanding of a convolutional neural network (2017)
19. A. Krizhevsky, I. Sutskever, G. Hinton, ImageNet classification with deep convolutional neural networks, in *Neural Information Processing Systems* (2012)
20. K. Simonyan, A. Zisserman, Very deep convolutional networks for large-scale image recognition (2014)
21. R. Girshick, J. Donahue, T. Darrell, J. Malik, Rich feature hierarchies for accurate object detection and semantic segmentation, in *Proceedings of the IEEE Computer Society Conference on Computer Vision and Pattern Recognition* (2013)
22. R. Girshick, Fast R-CNN (2015)

Prediction of lncRNA-Cancer Association Using Topic Model on Graphs

Madhavan Manu, Stephen Reshma, and Gopakumar G

1 Introduction

Recent advancements in genome sequence analysis identified that human genome mostly consists of non-coding RNAs (ncRNAs) that are not translated into proteins. The long non-coding RNAs (lncRNAs) are ncRNAs with length greater than 200 nucleotides [1]. Though the entire aspects of lncRNA functionalities are mostly unknown, studies [2] revealed their significant regulatory roles in many biological processes. The interactions of lncRNA with other molecules like DNA, protein, RNA, and with other ncRNAs (like microRNA) have a significant role in causing different types of cancers [3–5]. These interactions are often represented as networks and analyzed by tools derived from graph theory. Then, specialized graph mining algorithms can be applied to these graphs for various tasks like node classification, link prediction, and association mining [6].

At the same time, the topic model [7], a popular tool among the text mining community, is getting widespread attention in analyzing biological data [8]. The topic distributions of documents produced by algorithms such as LDA are more semantically interpretive. In recent years, many variants of LDA topic models are being used in bioinformatics for classification, clustering, and feature extraction on biological data [8]. The researchers also experimented the possibility of extending topic models for analyzing graph-structured data [6]. The independent assumption in

M. Madhavan (✉) · R. Stephen · G. Gopakumar
Department of Computer Science and Engineering, National Institute of Technology Calicut, Calicut, Kerala 673601, India
e-mail: manu_p150091cs@nitc.ac.in

R. Stephen
e-mail: reshmastep@gmail.com

G. Gopakumar
e-mail: gopakumarg@nitc.ac.in

S. Patnaik et al. (eds.), *Advances in Machine Learning and Computational Intelligence*, Algorithms for Intelligent Systems, https://doi.org/10.1007/978-981-15-5243-4_28

bag-of-word representation does not withhold in case of graphs and hence challenges the direct use of LDA in graphs.

In this work, we address the above issue by presenting a graph mining using LDA with extended graph-of-word (GoW) representation. Here, the interactions between lncRNAs, proteins, miRNAs, and known cancers form the graph. A node-weight matrix created from this interaction network based on the degree and level of nodes is used as the input to the topic model. We use our proposed method to predict cancer associations of 300 lncRNAs with 35 types of cancer.

The remainder of the paper is organized as follows. Section 2 gives key related works in this area. The methodology of the proposed work is discussed in Sect. 3. Section 4 presents the experiments conducted and their results. Section 5 concludes the paper with references to future works.

2 Related Works

The computational analysis of lncRNA includes identification from unknown transcriptome, functional annotation, and prediction of interaction with other molecules like proteins and miRNAs. The works [9–11] propose machine learning-based methods for classifying lncRNAs and protein-coding RNAs. Various network-based approaches for lncRNA analysis are surveyed in [5].

Topic models have been used with biological data in many previous works [8]. In [10, 12], a topic model is used for the classification of genomic sequences and gene expression data. The paper [13] predicts how specific biological pathways respond to a new drug using topic models.

There are LDA topic models used with data modeled as graphs (other than biological data), especially in social network analysis. The author-topic model [14] and its extensions graphs constructed from twitter hash tags for tweet mining [15] are few among such works. The graph topic model proposed in [6] by extending LDA for graph mining accepts data graphs as input. All the above methods require extra parameters to contain the additional information coming along with the graph structure of data.

In [16], to consider the dependence between terms in text documents, they modeled each document as a directed graph where nodes denote terms, edges term co-occurrences, and edge direction denotes term order. Instead of using term frequency, they used in-degree of each node in the graph. A similar representation of graphs for LDA is tried in our model, by redefining node weight for biological interaction graphs.

3 Method

The proposed system takes interaction networks prepared for each lncRNAs as input. Known association with cancer disease is also given as additional information to the system. Figure 1 depicts the overall working of the system. There are mainly three steps involved—constructing an interaction graph for each lncRNA from the input information, representing these constructed graphs as a node-weight matrix for LDA, and performing topic modeling and prediction using the LDA-derived topic distribution.

3.1 Constructing LncRNA Interaction Graph

An undirected graph is constructed for each lncRNA by considering its interactions with proteins, miRNAs and the known associations with the cancer types. In each graph, there are four kinds of nodes: (a) one lncRNA, (b) one or more protein, (c) one or more miRNA, and (d) one or more cancer nodes. If there exist a known interaction between two nodes, an edge is added.

Graph of word method proposed in [16] considers the degree as node weight. So, the node with more interaction got highest weight in the node-weight matrix. In

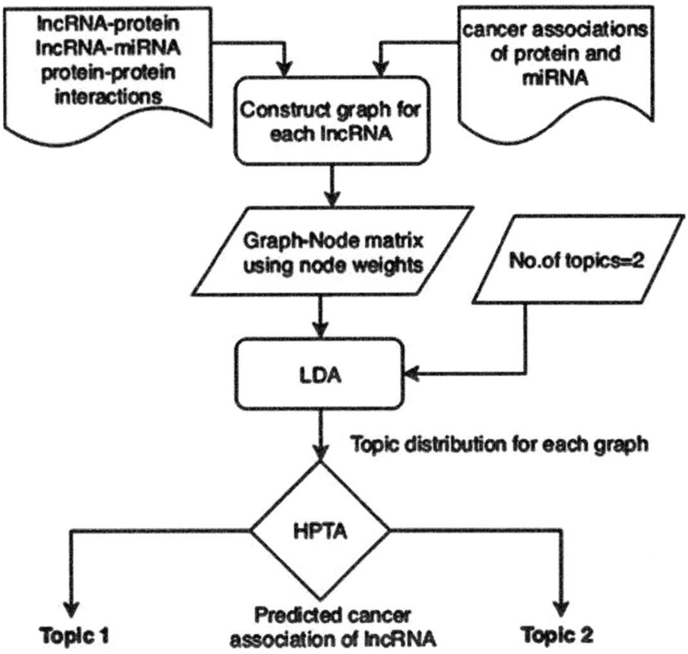

Fig. 1 Workflow of the proposed methodology

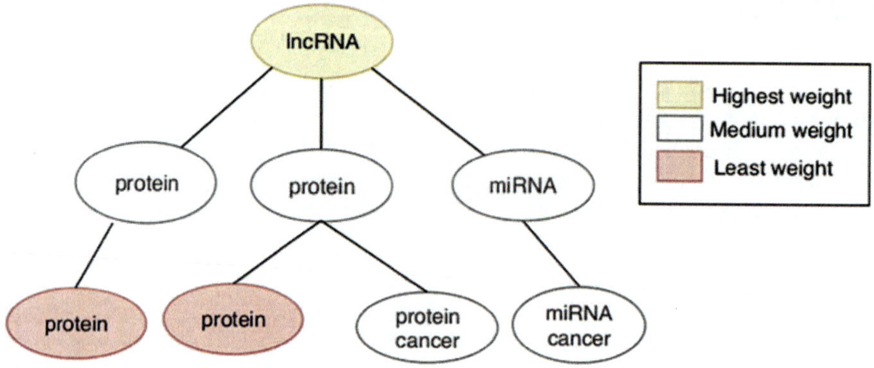

Fig. 2 Criteria for the node-weight assignment. LncRNAs have highest priority. Nodes having direct link with lncRNAs, and cancers come under second category. All other nodes come under third category of node weights

contrast to this method, the proposed work uses an extended node-weight method, which assigns a fixed weight to nodes based on their type and position in the interaction hierarchy. Since the network is prepared based on the lncRNA interaction, lncRNA node will be in the top and assign highest weight in the hierarchy. All nodes having direct interaction with lncRNA (lncRNA-protein and lncRNA-mRNA interacting nodes) are coming under the second-highest priority. All other nodes (protein-protein interactions, protein-miRNA interaction) have been assigned lowest weight in this representation. Figure 2 demonstrates this concept of node-weight assignment.

An interaction graph is constructed for each lncRNA in this way, and a set of 300 interaction graphs are thus generated. Graph constructed for HOTAIR lncRNA, which is known to be highly expressive in cancer patient samples, is given in Fig. 3.

3.2 *Document-Topic-Word Paradigm for LncRNA Graph*

The important step in adopting a topic model algorithm from text mining domain is identifying a document-topic-word analogy for the biological data. For text documents containing words, LDA uses bag-of-words (BoW) representation as input. Consider a set of N documents, constructed out of a vocabulary of size V, and let number of topics to be discovered is K. The input corpus is represented by BoW which is an $N \times V$ document-term matrix with value x_{dw} in the matrix representing the weight of the word w in document d. Analogous to this, we map lncRNA graphs as documents, graph nodes as words, and cancer association as topic. The node-weight matrix discussed in Sect. 3.1 serve as document-term matrix.

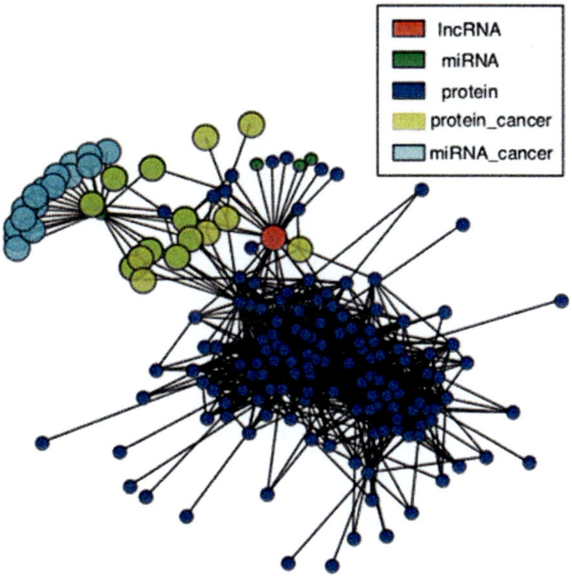

Fig. 3 An interaction graph constructed for HOTAIR lncRNA

3.3 *Prediction Using HPTA*

The final step in the proposed work is identification of lncRNAs as cancer-related or non-related, based on the topic distribution. We use number of topics K as 2, corresponding to the two classes. A document is assigned to a topic based on highest probable topic assignment (HPTA) method [1]. That is, each lncRNA graph is assigned a topic consonant with the highest probability in the document topic distribution. A majority voting scheme is applied to assign class labels for topic distribution. The topic z_j is assigned a class label c_i, if majority of the lncRNAs from the corpus S known to be in class c_i is assigned to topic z_j during HPTA. Equation 1 summarizes this concept.

$$C(z_i) = \arg\max_{c_j} \left(\sum_{k=0}^{|S|} f\left(d_k, c_j\right) \right) \quad (1)$$

where,

$$f\left(d_k, c_j\right) = \begin{cases} 1, \text{ if } d_k \in c_j \text{HPTA}(d_k) = z_i \\ 0, \text{ otherwise} \end{cases}$$

4 Experiments and Results

4.1 Dataset

The test set contains 150 known cancer-related lncRNAs and 150 non-cancer-related lncRNAs. The benchmark dataset is manually prepared by referring lncRNA2 cancer database [17]. Interaction data for lncRNA-protein interactions, lncRNA-miRNA interactions, and protein-protein interactions were collected from NPInter3.0 [18] and STRING databases [19]. Protein-cancer associations and miRNA-cancer associations were collected from human protein atlas [20], and dbDEMC2.0 [21], respectively. To avoid any possible bias in grouping, the size of both class of lncRNAs is kept equal.

4.2 Experimental Setup

We have experimented different combinations of node weights based on the degree, centrality, and hierarchical position of the nodes, and we empirically found that smaller weights gave better results. We assigned the highest weight (3) to the lncRNA node. Protein and miRNA nodes directly connected with lncRNA and both types of cancer nodes were given medium weight (2). All other nodes were given the least weight 1. We used Gensim implementation of LDA for topic modeling with number of topics 2. Also, the number of passes keeps as ten, and all other LDA parameters set to default values. The performance was compared with two different clustering algorithms: K-means and spectral clustering [22].

4.3 Results

The set of 300 lncRNA interaction graphs represented as a graph-node matrix was inputted to proposed LDA-HPTA method, K-means clustering and spectral clustering algorithms. The prediction results of three methods are given in Table 1.

Table 1 Prediction comparison of proposed method, K-means and spectral clustering

Method	Predicted cancerous (positive)	Predicted non-cancerous (negative)	Correctly predicted (out of 300)
Proposed method	174	126	218
K-means	297	3	149
Spectral clustering	261	39	131

Table 2 Performance comparison of three methods

Method	Precision	Recall	*F*-score
Proposed method	0.695	0.806	**0.747**
K-means	0.498	0.987	0.662
Spectral clustering	0.463	0.806	0.588

The predicted cancer associations of lncRNAs are compared with Lnc2Cancer 2.0 [17] database. From the results given in Table 2, the proposed model exhibits considerable improvement over K-means and spectral clustering in terms of precision and *F*-score.

Figure 4 shows receiver operating characteristic (ROC) curve and corresponding area under curve (AUC) values for the three methods, where our proposed model gives better AUC value.

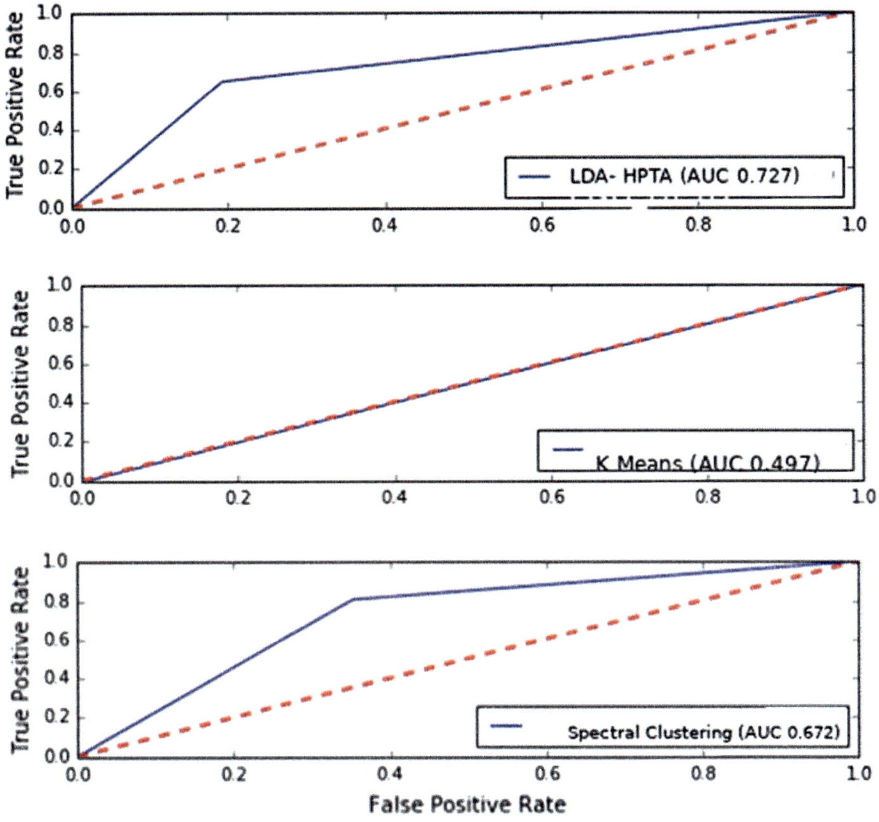

Fig. 4 ROC plot with AUC values for the three methods

5 Conclusion

This paper presented an LDA topic model on a set of lncRNA interaction graphs to predict cancer associations of lncRNAs. The two significant contributions of this work include a document-topic-word analogy to graph-structured data and a scheme for creation of node-weight matrix. The node-weight matrix computation of the graph considers the node type and its interaction level with lncRNAs.

With LDA-based HPTA, clustering the lncRNAs into cancerous or non-cancerous class was done. The results show the ability of this representation in capturing dependencies between nodes encoded as interactions. Since this is an initial work on using LDA topic model on biological datagraphs, we kept the notion of node weight simple by giving each type of node a fixed weight. More complex node-weight measures based on node degree and random walk weights could provide better results.

References

1. A.E. Teschendorff et al., HOTAIR and its surrogate DNA methylation signature indicate carboplatin resistance in ovarian cancer. Genome Med. **7.1** (2015)
2. H. Ma, Y. Hao, X. Dong, Q. Gong, J. Chen, J. Zhang, W. Tian, Molecular mechanisms and function prediction of long non-coding RNA. Sci. World J. (2012)
3. M.-C. Jiang et al., Emerging roles of lncRNA in cancer and therapeutic opportunities. Am. J. Cancer Res. **9**(7), 1354–1366 (2019)
4. A.M. Schmitt, H.Y. Chang, Long non-coding RNAs in cancer pathways. Cancer Cell **29**(4), 452–463 (2016)
5. H. Zhang, Y. Liang Y, S. Han, C. Peng, Y. Li, Long non-coding RNA and protein interactions: from experimental results to computational models based on network methods. Int. J. Mol. Sci. **20**(6), 1284 (2019)
6. J. Xuan, J. Lu, G. Zhang, X. Luo, Topic model for graph mining. IEEE Trans. Cybern. **45**(12) (2015)
7. D.M. Blei, A.Y. Ng, M.I. Jordan, Latent dirichlet allocation. J. Mach. Learn. Res. 3, 993–1022 (2003)
8. L. Liu et al., An overview of topic modeling and its current applications in bioinformatics, SpringerPlus 5.1 (2016)
9. J. Baek, B. Lee, S. Kwon, S. Yoon, LncRNAnet: long non-coding RNA identification using deep learning, Bioinformatics **34**(22), 15, 3889–3897 (2018)
10. M. La Rosa et al., Probabilistic topic modeling for the analysis and classification of genomic sequences. BMC Bioinformat. **16.6** (2015)
11. Q. Jiang, R. Ma, J. Wang, X. Wu, S. Jin, J. Peng, R. Tan, T. Zhang, Y. Li, Y. Wang, LncRNA2Function: a comprehensive resource for functional investigation of human IncRNAs based on RNA-seq data, in *BMC Genomics* (2015)
12. S.J. Kho, H.B. Yalamanchili, M.L. Raymer, A.P. Sheth, A novel approach for classifying gene expression data using topic modeling, in *Proceedings of the 8th ACM International Conference on Bioinformatics, Computational Biology, and Health Informatics* (ACM, 2017), pp. 388–393
13. N. Pratanwanich, P. Lio, Exploring the complexity of pathway–drug relationships using latent Dirichlet allocation. Computation. Biol. Chem. **53**, 144–152 (2014)
14. M. Rosen-Zvi, T. Griffths, M. Steyvers, P. Smyth, The author-topic model for authors and documents, in *Proceedings of the 20th conference on Uncertainty in artificial intelligence* (AUAI Press, 2004), pp. 487–494

15. Y. Wang, J. Liu, Y. Huang, X. Feng, Using hashtag graph-based topic model to connect semantically-related words without co-occurrence in microblogs. IEEE Trans. Knowl. Data Eng. **28**(7) (2016)
16. F. Rousseau, M. Vazirgiannis, Graph-of-word and TW-IDF: new approach to ad hoc IR. in *Proceedings of the 22nd ACM international conference on Information & Knowledge Management* (ACM, 2013)
17. Y. Gao, P. Wang, Y. Wang, X. Ma, H. Zhi, D. Zhou, X. Li et al., Lnc2Cancer v2. 0: updated database of experimentally supported long non-coding RNAs in human cancers. Nucleic Acids Research 47, D1028–D1033 (2018)
18. Y. Hao, W. Wu, H. Li, J. Yuan, J. Luo, Y. Zhao, R. Chen, NPInter v3. 0: an upgraded database of non-coding RNA-associated interactions. Database (2016)
19. C. von Mering, M. Huynen, D. Jaeggi, S. Schmidt, P. Bork, B. Snel, STRING: a database of predicted functional associations between proteins. Nucleic Acids Research **31**(1) (2003)
20. U. Mathias, C. Zhang, S. Lee, E. Sjöstedt, L. Fagerberg, G. Bidkhori, R. Benfeitas et al., A pathology atlas of the human cancer transcriptome. Science **357**(6352) (2017)
21. Z. Yang, L. Wu, A. Wang, W. Tang, Y. Zhao, H. Zhao, A.E. Teschendorff, dbDEMC 2.0: updated database of differentially expressed miRNAs in human cancers. Nucleic Acids Research **45**(D1) (2016)
22. A.Y. Ng, M.I. Jordan, Y. Weiss: on spectral clustering: analysis and an algorithm, in *Advances in Neural Information Processing Systems* (2002)

Sales Prediction Using Linear and KNN Regression

Shreya Kohli, Gracia Tabitha Godwin, and Siddhaling Urolagin

1 Introduction

Sales prediction plays a key role in building up a business. It is one of the most important parts of business intelligence. Sales prediction and forecasting give an insight into how a company should manage its workforce—labor, cash flow, and its resources. It is an evaluation tool that uses past and current sales data to predict future performance. Estimating future sales is an important part of the financial planning of any business. It enables companies to predict short-term and long-term performance. Accurate sales forecasts assist the companies in making informed choices which result in better supply chain management, an increase in profits, and better customer experiences. It is a crucial part of starting a new business as it helps in managing the available resources efficiently. Furthermore, with the help of data and gained insights, it becomes easier to understand consumer behavior. This gives way to the use of effective marketing techniques and can be used to selectively target the market. Some benefits of sales prediction for a business/company are that it can be used as a benchmark. It can also be used to help with planning. Demand and supply actions can be planned by looking at the forecasts.

Data analytics and predictive modeling are used for sales forecasting models. Predictive modeling is a process that uses information from data to determine the outcomes with data models. Many types of classifiers can be used to predict sales such as regression, K-nearest neighbor, decision trees, random forest, demand forecasting, classification methods, cluster analysis, and Bayesian classification. The

S. Kohli · G. T. Godwin (✉) · S. Urolagin
BITS Pilani, DIAC, Dubai, United Arab Emirates
e-mail: tabitha.godwin.tg@gmail.com

S. Kohli
e-mail: kohli.shreya27@gmail.com

S. Urolagin
e-mail: siddhaling@dubai.bits-pilani.ac.in

S. Patnaik et al. (eds.), *Advances in Machine Learning and Computational Intelligence*,
Algorithms for Intelligent Systems, https://doi.org/10.1007/978-981-15-5243-4_29

main classifier used in this paper is regression. The regression model equation might be as simple as $Y = a + bX$ in which case Y is your sales, the 'a' is the intercept, and the 'b' is the slope. With this model, we aim to correlate a variable that could be causing your sales to get better or worse. The two types of regression classifiers that are used in this paper are linear regression and KNN regression. The regression model has a few advantages. Firstly, the linear regression model is easy to interpret and understand due to its linearity. It can help businesses understand the relationship between various variables or factors affecting their profit. Furthermore, new data can be added easily which will not impact the accuracy of the KNN algorithm. It provides a powerful statistical method that can be used for data analysis.

This paper consists of a detailed literature review (Sect. 2), information regarding the dataset (Sect. 3). Then, there is a brief methodology (Sect. 4) where it tells how the data has been preprocessed and gives an insight into how feature selection has been done. Finally, we use linear regression and KNN regression models to train the dataset and extract the results. These results (Sect. 5) are then compared to conclude which regression analysis method is better for sales prediction on the Rossman dataset.

2 Literature Review

A large amount of data available in information databases becomes a waste until the useful information is extracted. Predictive analytics is known as the roof of advanced analytics—that is to predict future events. Predictive analytics is comprised of data collection and statistics, and deployment [1]. The rapid growth and advances of information technology enable data to be accumulated faster and in much larger quantities [2]. Predictive modeling is a combination of mathematical techniques that have in common the goal of finding a mathematical relationship between target with the purpose of estimating future values of those predictors and including them into the mathematical relationship to foretell future values of the target variable [3]. In every organization, the sales forecast is of utmost importance to help make decisions on every little and big detail, from budgets to spend to the labor required to profit, etc. Forecasting is a necessary task that helps ensure that an organization develops and plans successfully, but it is very much detested due to the amount of effort put into the process and how time-consuming the task can be. By feeding this data into such predictive analytic models, sales teams are now enabled with analytics-based insights and recommendations [4]. ML procedures have been gaining influence over time as interest in artificial intelligence has been growing [5]. Data mining means extracting information from the data, it means preparing data to gain the implied, prior unknown, potential and useful information, which can be represented as patterns [6].

Regression is an important analytical method used in teaching, economics, financing, etc., that computes the relationships between one dependent and one or more independent variable (s). The two primary kinds of regression are simple linear regression and multiple linear regression. Simple linear regression applies one independent variable to predict the result of the dependent variable, and multiple linear regression relates two or more independent variables to foretell the result of the dependent variable [7]. Sales forecast is rather a regression problem than a time series problem. Study shows that the utilization of regression approaches can often give us more reliable outcomes opposed to time series techniques [8]. Sales forecasting is an important aspect of almost all businesses these days. Companies, nowadays, are attempting to expand such abilities of forecasting and prediction to get the upper edge on their competitors. For example, a good forecasting model gives information to manufacturer about the right amount of inventory, workforce, or labor required to satisfy the demand for the product [9]. The main goal of sales prediction is to analyze how internal and external factors can affect weekly sales in the future for Rossmann stores. Sales prediction is carried out using various machine learning and data science algorithms [10]. From prior literature, it can be noted that there has already been intensive research on three major uses of sales prediction. First is the Microsoft Time Series algorithm. It provides us with optimized regression algorithms for forecasting continuous real-time values [11]. Second is spatial data mining for retail sales forecasting [12]. Support vector regression (SVR), a technique is used in designing regression models to predict the expected turnover. Built from prior expert knowledge along with analytic knowledge discovered during data mining processes, this model provides us with accurate results. Finally, a novel trigger model for sales prediction with data mining techniques [13] that focuses on how to forecast sales with more accuracy and precision. It is now said that companies that can accurately forecast sales can successfully adjust future production levels, resource allocation, and marketing strategies to match the level of anticipated sales. A regression model is used to forecast or predict the value of the dependent variable—sales, based on various independent variables [14].

3 Sales Dataset

The dataset used in this paper is the Rossmann Store Sales available on Kaggle [15]. Rossmann is a German drugstore chain with 3466 stores under them (Table 1).

It can be observed from Fig. 1a, that the sales production decreased from 39.21% in 2013 to 23.66% in 2015. About 20% of sales were observed on the first day whereas only 0.51% was on the last day of the week. It can be concluded that most of the products were sold on a Monday and Friday. The maximum number of customers was observed to be 256 M in 2013 and decreased by 108–148 M in the year 2015. Figure 1b depicts the sales with respect to the number of customers visiting in the years 2013, 2014, and 2015.

It is observed from Fig. 2 that most of the sales occur when the store is running a promotion. Similarly, the number of customers visiting the store is relatively high during the promotion time. The fourth quarter showed an increase in average number of buyers.

Table 1 Attributes of Rossmann store sales

Attribute name	Description
Id	It shows the ID of the form (Store, Date)
Store	It is a distinctive number for the stores
Sales	Indicates revenue for a particular day
Customers	Gives the number of buyers during a particular day
Open	Shows if the store was open (=1) or closed (=0)
State holiday	a = public holiday, b = Easter holiday, c = Christmas, and 0 = None
School holiday	This shows if the (Store, Date) was affected by the closure of public school
Store type	Indicates the 4 stores that are a, b, c, and d
Assortment	Gives an assortment level: a = basic, b = extra, and c = extended
Competition distance	It is the distance to the closest competitor store in meters
Competition open since [Month/Year]	Shows the month as well as the year the closest competitor store was started
Promo	This shows if a store runs a promotion on a particular day
Promo2	Indicates if a promotion is continuous for some stores, 0—the store is not involved and 1—the store is participating
Promo2 Since [Year/Week]	This indicates the year and week during which the given store took part in Promo2
Promo interval	It gives the consecutive intervals Promo2 began

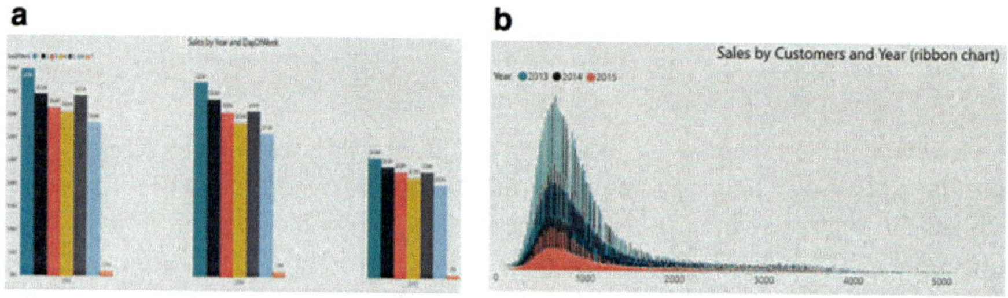

Fig. 1 Sales by (**a**) Year and days of the week. (**b**) Customers and year

Figure 3 depicts the distribution of sales and the distribution of customers. It can be observed that the standard deviation is smaller for the distribution of customers as the graph in Fig. 3a is narrower than the normal graph in Fig. 3b.

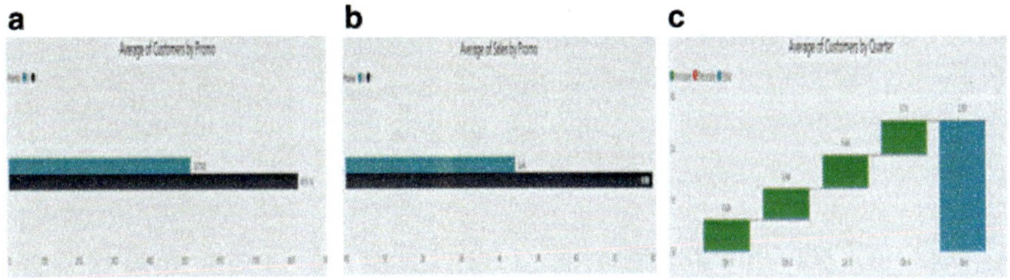

Fig. 2 (**a**) Average sales-promo, (**b**) average customers-promo, and (**c**) average customers-quarter

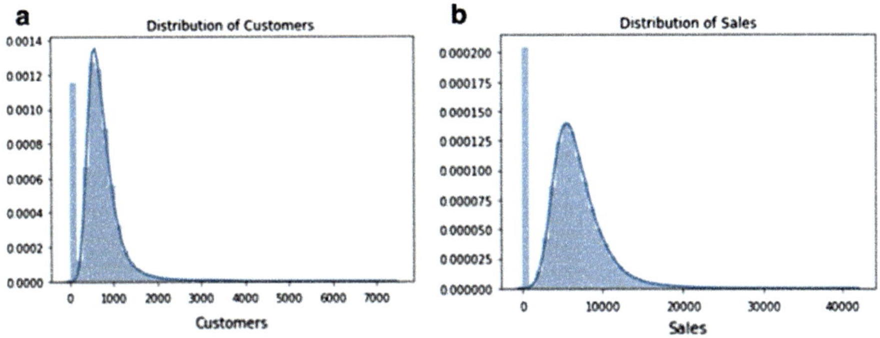

Fig. 3 (**a**) Distribution of customers and (**b**) distribution of sales

4 Methodology

The initial stage involves data collection and comprehension. This paper makes use of the Rossmann dataset from Kaggle. This data is then transformed into understandable form and the necessary features are selected. This is followed by predictive analysis using classifiers. Finally, the model is evaluated by applying various statistical methods (Fig. 4).

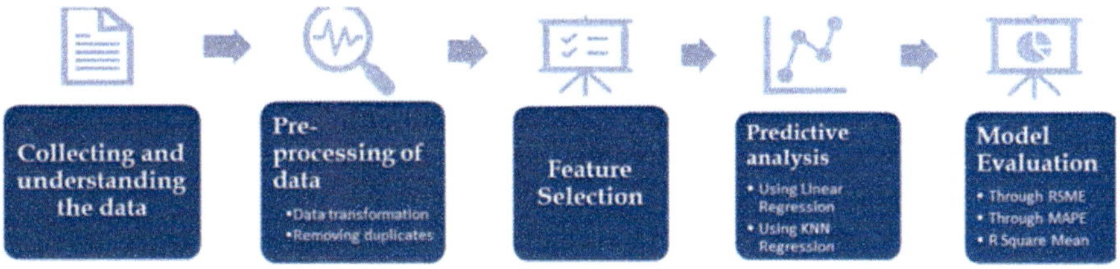

Fig. 4 Overall framework of sales prediction model used

4.1 Collecting and Understanding the Data

This is the process of gathering data and examining the dataset being used.

4.2 Data Preprocessing

Data preprocessing is a data mining method that includes converting raw data into an understandable form. This involves understanding the data, handling missing values, and removing duplicates.

4.3 Feature Selection

Feature selection is the process of finding the attributes or terms that are the most meaningful and extracting useful information from it.

4.4 Predictive Analysis

Linear Regression. Linear regression is a linear modeling approach to find the relationship between 1 or more independent variables (predictors) denoted as X and dependent variable (target) denoted at Y. Linear regression is all about finding the best fit line for the training as well as test data. The best fit line can be found by minimizing the distance between all data points and its distance to the regression line (by calculating the error (sum of squares error), we can find minimize distance), i.e., the distance between the points and the line should be minimum. This is done in a recursive method. The x value here varies between promo/customer/holiday, etc., and the y value here is sales.

K-Nearest Neighbor Regression. K-nearest neighbors is an algorithm that stores all available (previous) cases and uses that to predict the values based on a similarity measure. It uses 'feature similarity' to predict the values for test data/new data points. The value of the new point is assigned based on how closely it resembles other training data examples. KNN regression has two approaches. First is by calculating the average of the target of the K-nearest neighbors. Second is by computing an inverse distance weighted average of the K-nearest neighbors. KNN regression uses the same distance functions as KNN classification—Euclidean, Manhattan, and Minkowski. Initially, we try and eliminate all null values, replace missing values, so we can then use the data for the classifiers.

5 Experimental Setup

The dataset used here is Rossmann dataset, and scikit-learn library in Python was used for model selection, preprocessing, linear regression, and KNN. The matplotlib Python library was implemented for plotting graphs and data visuals. Furthermore, the sklearn.metrics module was applied for model evaluation using RMSE and MAPE.

To obtain the graph, k-fold cross-validation had been used, where k = number of folds. The dataset is divided into $k = 10$ subsets, and the holdout method is repeated k times thereby improving it. By doing so, a clear and accurate prediction of sales value was obtained, then termed as the predicted value from the above graph (Fig. 5).

The histograms on the diagonals in Fig. 6 illustrate the distribution of a single variable, whereas the rest of the graphs depict the relationship between the two features. Figure 6a shows the relationship between sales and promo while Fig. 6b shows the relationship between sales and number of customers.

Even though the RMSE and MAPE with training data for both models do not show much difference, we observe a larger value for test data (Table 2). It can be observed that KNN regression is an overfitting model. Regression model score is a metric that depicts the accuracy of each of the above classifiers. It can be noticed that linear regression has a slightly better model score when compared to KNN regression. Therefore, it can be concluded that linear regression is a better model to predict sales from the given dataset.

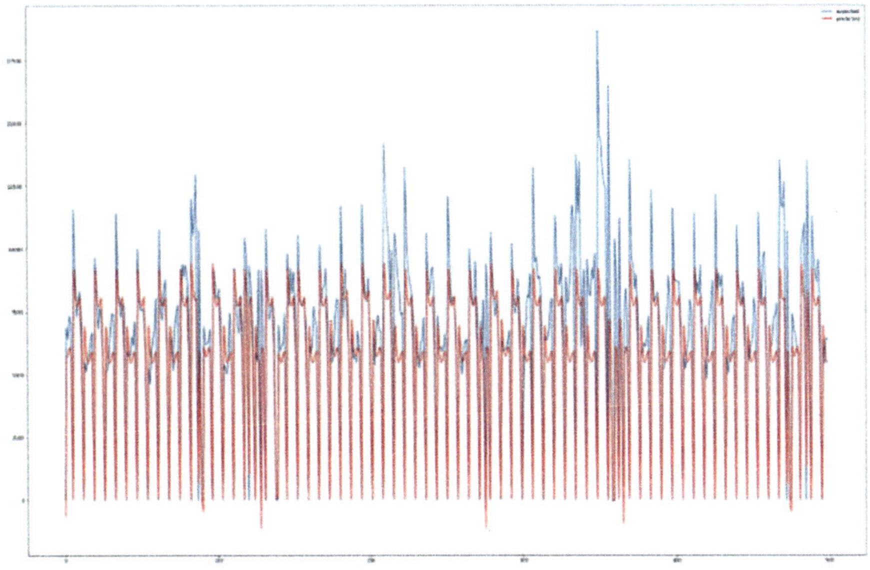

Fig. 5 Linear regression model

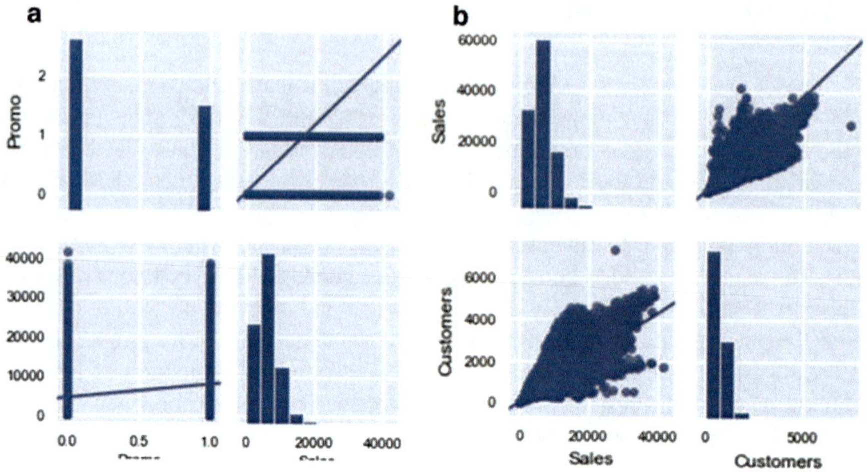

Fig. 6 (**a**) Regression line for sales and promo. (**b**) Regression line for sales and customers

Table 2 Results obtained

Model used	Model score (%)	RMSE	MAPE
Linear regression	72.19	Training data: 1742.51	Training data: 20.97
		Testing data: 1898.91	Testing data: 22.065
KNN regression	71.28	Training data: 1770.61	Training data: 21.83
		Testing data: 2546.79	Testing data: 31.40

6 Conclusion

KNN techniques are nonparametric hence mainly used in forestry problems like remote sensing. But parametric regression analysis, i.e., linear regression, has the advantage that it is easy to fit, we need to estimate only a small number of coefficients and is easy to interpret, whereas the statistical properties of KNN regression are explored lesser. In KNN regression, we observe that the difference between the training and testing error is higher than when compared to the linear regression model—which means that the KNN regression model works better for training data than testing data, thereby implying that it is an overfitting model. From this study, we can say that for the above-chosen dataset linear regression is a better model as testing and training errors for RMSE and MAPE are lesser, i.e., the difference between training and testing error is lesser for linear regression when compared to KNN regression. Therefore, it can be concluded that using linear regression we can accurately predict sales in the future. This will help companies/organizations plan their resources better and also helps in cost optimization—maximizing profit with minimum resources.

References

1. V. Kavya, S. Arumugam, A review on predictive analytics in data mining. Int. J. Chaos Control Modell. Simul. (IJCCMS) **5**(1/2/3) (2016)
2. R.R. Shelke, R.V. Dharaskar, V.M. Thakare, Data mining for supermarket sale analysis using association rule. Int. J. Trend Sci. Res. Dev. **1**(4). ISSN: 2456-6470
3. T. Wilson, S. Asthana, Predictive Modelling for Assessing the Sales Potential of the Customer. https://www.academia.edu/28362014/Predictive_Modelling_for_Assessing_the_Sales_Potential_of_the_Customer (2016)
4. J. Gonzalez, Sales Forecasting and the Role of Predictive Analytics, (July 18, 2017), [Online]. Avaliable: https://vortini.com/blog/forecasting-predictive-analytics
5. S. Makridakis, E. Spiliotis, V. Assimakopoulos, The accuracy of machine learning (ML) forecasting methods versus statistical ones: extending the results of the M3-competition (2017)
6. M. Xue, C. Zhu, Applied research on data mining algorithm in network intrusion detection 275–277. https://doi.org/10.1109/jcai.2009.25 (2009)
7. Y.M. Khaing, M.M. Yee, E. Ei, Forecasting stock market using multiple linear regression Aung. Int. J. Trend Sci. Res. Dev. (IJTSRD) **3**(5) (2019)
8. B.M. Pavlyshenko, Machine-learning models for sales time series forecasting, Lviv, Ukraine 21–25 August 2018, pp 3–11
9. G. Nguyen, Kedia, Jai, Snyder, Ryan, Pasteur, R., Wooster, R. Sales Forecasting Using Regression and Artificial Neural Networks. (2013)
10. A. Aima, WALMART sales data analysis & sales prediction using multiple linear regression in R programming Language, [Online], Available: https://medium.com/@arneeshaima/walmart-sales-data-analysis-sales-prediction-using-multiple-linear-regression-in-r-programming-adb14afd56fb (March 19)
11. P. Mekala, B. Srinivasan, Time series data prediction on shopping mall. Int. J. Res. Comput. Appl. Robot. **2**(8), 92–97 (2014). ISSN 2320-7345
12. M. Krause-Traudes, S. Scheider, S. Rüping, Spatial data mining for retail sales forecasting, in *11th AGILE International Conference on Geographic Information Science* (2008)
13. W. Huang, Q. Zhang, W. Xu, H. Fu, M. Wang, X. Liang, A novel trigger model for sales prediction with data mining techniques. Data Sci. J. **14**, 15 (2015). https://doi.org/10.5334/dsj-2015-015
14. E. Bank, How to develop & use a regression model for sales forecasting, Updated September 26, 2017. https://bizfluent.com/how-7298496-develop-regression-model-sales-forecasting.html. Accessed 4 Oct 2019
15. Rossmann Store Sales|Kaggle, Kaggle.com, 2019. [Online]. Available: https://www.kaggle.com/c/rossmann-store-sales/data. Accessed 07 Sept 2019

Image Captioning Using Gated Recurrent Units

Jagadish Nayak, Yatharth Kher, and Sarthak Sethi

1 Introduction

With the rise in social networking platforms, especially Web sites that dwell on image sharing, the process of automated image captioning is gaining traction for many reasons. Filtering obscene images and aiding the blind are some of the many reasons, and image captioning is pursued in today's world. This is a challenging task, as it is necessary to identify the relation between two different models—visual cues and natural language processing. The traditional way used sentence templates to match with pictures trained. These types of models do not work with new inputs. With the recent increase in the development of deep neural networks [1], this was made possible with the usage of convolutional neural networks to train images and recurrent neural networks to form grammatically correct words. There are various methods of image captioning [2], such as:

- Templates: This method detects objects and their features, later parsing sentences to learn the relation using models like conditional random fields that take into consideration nearby values of pictures and predict the output caption. This model fails to produce variable length sentences.
- Retrieval: These relate the test images with trained images and find similarities, allowing them to combine and produce captions accordingly. These highly depend on the training data. i.e., if we input any photograph that is mostly different than the trained data, this model fails.
- Neural networks: These produce the best results when compared to the various methods listed above by using a combination of convolution neural network (CNN) and LSTM type recurrent neural network (RNN) to learn the different relations of the models used in training.

J. Nayak (✉) · Y. Kher · S. Sethi
Birla Institute of Technology and Science Dubai Campus, Pilani, UAE
e-mail: jagadishnayak@dubai.bits-pilani.ac.in

S. Patnaik et al. (eds.), *Advances in Machine Learning and Computational Intelligence*, Algorithms for Intelligent Systems, https://doi.org/10.1007/978-981-15-5243-4_30

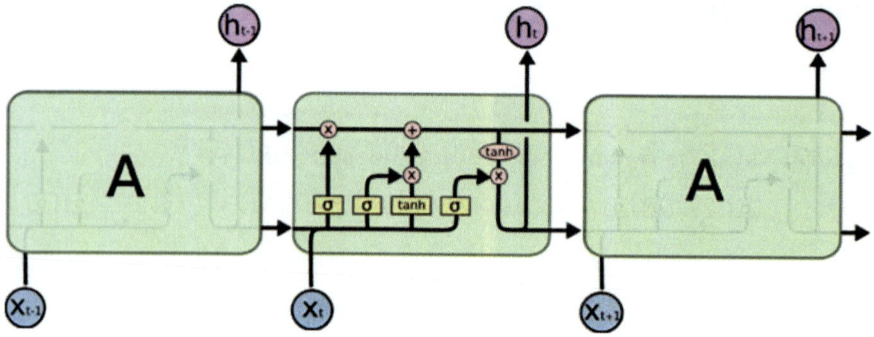

Fig. 1 LSTM RNN

Recurrent neural networks function by incorporating memory. In traditional neural networks, inputs were taken to be independent of each other. This fails when trying to predict words in a sentence. It is important to know the placement of words to make a meaningful sentence. The two important variants of RNNs are LSTMs and GRUs. Image captioning using deep networks have mostly been performed using long short-term memory (LSTM) and recurrent neural network (RNN) in conjunction with a convolutional neural network (CNN).

1.1 Long Short-Term Memory (LSTM)

The problem of vanishing gradients encountered when using RNNs can be solved with the help of LSTMs [3]. They are also used when more memory is required. We assume the cell states/neurons to be connected using a line. LSTM can add and remove information to the cell state using sigmoid layers where

0 = let nothing through
1 = let everything through. The LSTM is depicted in Fig. 1.

1.2 Gated Recurrent Units (GRU)

The GRU differs from the LSTM RNN by two means [4]:

- The forget gate is combined with the input gate to form an update gate.
- The hidden state/memory is combined with the cell state and is presented as the output.

The GRU is depicted in Fig. 2. There are two gates, mainly the reset and update gate. The reset gate determines how to accommodate new inputs with the past and the update gate must decide how much of the memory should be still stored. The

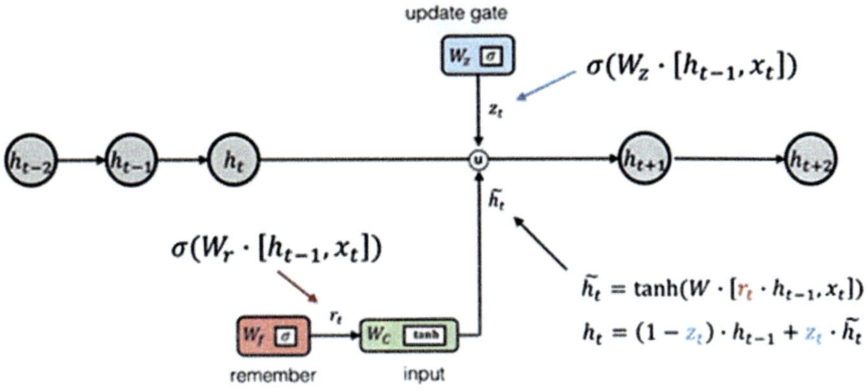

Fig. 2 GRU RNN

hidden state is calculated as shown in Eq. 1 where z refers to the update gate, r refers to the reset gate, h_t and h_{t-1} refer to the hidden states.

$$h_t = (1 - z_t) * h_{t-1} + z_t * h_t \tag{1}$$

2 Related Work

2.1 Show and Tell: A Neural Image Caption Generator

This model was the first model that used neural networks to caption images [5]. Being the first, it was also the most basic model and was fully end-to-end trainable. This meant that all the parameters in the entire model were trained simultaneously. Contemporary machine translation produced word-to-word outputs and reordered them to produce the caption, but with developments in machine learning, this model used RNNs, producing state-of-the-art performances. An encoder RNN is used to produce fixed length encodings which are fed into decoder RNNs as the first hidden state. This model replaced the encoder RNN with CNN features of the trained images. This method both increased the quality and decreased the output time of the results. They also used the BLEU evaluation metric to compare their results.

2.2 Deep Visual Semantic Alignments for Generating Visual Descriptions [6]

This technique wanted to learn the relations between parts of the sentences and the specific objects being described by them. They embedded both the image and sentence into a common space and the approach was tested. Using a multimodal recurrent neural network, they train the images and dataset. After this, they test this on their custom region-based annotation dataset. They introduced a ranking mechanism that relates a part of the sentence to the specific scene it describes. Higher the rank, the better the sentence describes the specific part of the image. They now map the sentence parts and the image parts on a neural network to train them and lay them on the same space in such a way that similar concepts occupy nearby spaces. Now, a score of similarity is used to define the relation between the sentence and the image. The dot products of all the regions defined in the image lead to the entire caption.

2.3 DenseCap: Fully Convolutional Localization Networks for Dense Captioning [7]

This method produces a dense captioning task in which regions of images are detected and a set of descriptions for each image are generated. Object detection, therefore, relates to descriptions whereas image captioning related to descriptions when the image region is the entire image. These techniques helped to introduce a dense localization layer that extracts activations inside the region of interest in the image. The developers used a VGG-16 model for this research and removed the last pooling layer. This localization layer takes the tensor and extracts the ROI from the identified regions. Now, these features for regions are passed through a fully connected layer with ReLu units and regularized dropouts. This converts each region into a 4096-dimensional vector. A final adjustment is done to this box and score and offsets calculated are again calculated. They used the METEOR metrics to compare their results.

2.4 Image Captioning with Deep Bidirectional LSTMS [8]

This method describes two new model variants of the deep bidirectional networks which perform as good as the previous models described without the usage of additional mechanisms like object detections. They used a CNN to learn the image features and a bidirectional LSTM to learn the sentence features. This technique is inspired by the human brain to use deeper Bi-LSTM layers and advances in CNN. End-to-end training is also carried out to reduce loss. LSTMs are already deep, but that depth is horizontal, in which weights are reused which limits the learning of more representative features. Vertical depth can be achieved by stacking multiple LSTMs or use a multilayer perceptron in between which helps in making the net deeper with increasing too much parameters. To check the outputs, they take the word which obtains the highest probability after passing it in both directions. They use one-hot encoding for word vectors. This method describes a variant of the bidirectional LSTM layer that includes two LSTM networks. These learn the encoding of words in both directions later correlating them with the CNN features to train the network. The final SoftMax layer produces a probability distribution over the words producing the caption. Using a multilayer perceptron makes it easier to train as it reduces the number of parameters.

2.5 Show, Attend and Tell: Neural Image Caption Generation with Visual Attention

Attention comes into play because representing an entire image using a vector does not help understand scenes, although this is the common approach [9]. Until now features from the last layer of the ConvNet were used to describe the image but this fails when more descriptive captions are required. More low-level features are desired, but this causes a heavy load in the amount of data that needs to be processed. This method produces a solution to this problem. Two types of attention are introduced, soft and hard. Instead of using the final fully connected layer of CNN, the lower convolutional layer vectors are used called them annotation vectors. Each annotation vector is supposed to belong to some specific part of the image (which makes sense as this is the basic nature of the CNN). For the decoder part, a slight variation of the widely used normal LSTM is proposed, in which a context vector is passed, which is to capture the visual information associated with a particular input location and is based on the previous generated word and the previous hidden state. The calculation of this context vector is what is different in hard and soft visual attention and is the dynamic representation of the relevant part of input image at that time step.

3 Methodology

3.1 Dataset Used

The FLICKR 8 k is used to train the network to caption images. This dataset contains 8000 images, each described by five captions. These captions are ranked according to how well they describe the pictures. The training dataset contains 6000 images, validation contains 1000, and testing also contains 1000 images.

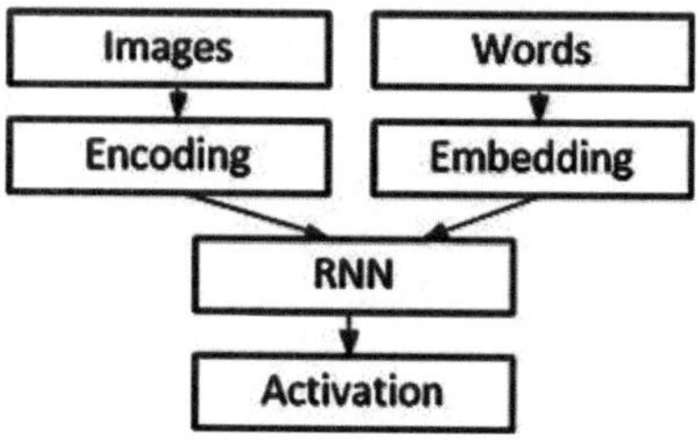

Fig. 3 Inject model

3.2 Model Architecture

To caption an image, a CNN is required to extract the features from the image (encoding), after which can be combined with word embeddings and fed to the RNN or combined with word encodings from the RNN and be used to generate captions. These lead to two types of models.

- Inject Model [10]—This model as shown in Fig. 3 prescribes the role of text generation to the RNN by providing it with both the encoded form of the image and word embeddings in order to predict sentences.
- Merge Model—The merge model, which is shown in Fig. 4, adds both the image encodings and the encoded form of the text descriptions and fed to a decoder model to generate the next word in the caption generated. The RNN is only used as an encoder in this model rather than text generation as described in the inject model.

The merge model was chosen over the inject model as it has been found to be more effective as compared to the other emphasizing that the RNN performs the best when used only for encoding the input. It would be very expensive with regard to time to train a CNN from scratch, therefore a model trained on ImageNet was used. VGG-16 was first used as the CNN model but proved to be slow to encode the images. Later when the InceptionV3 model was used, encoding time was reduced by 80.

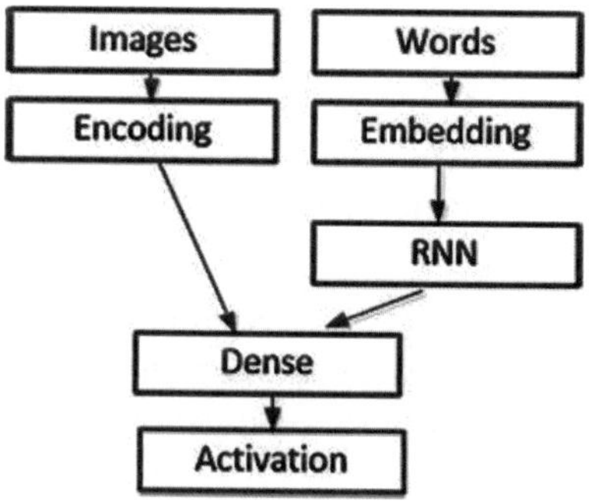

Fig. 4 Merge model

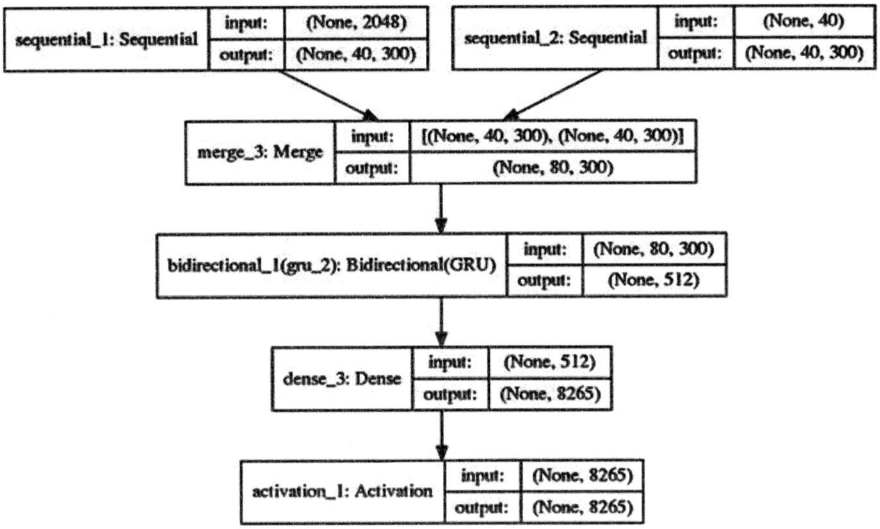

Fig. 5 GRU model

Table 1 Image captioning architecture parameters

Name	Value
Epochs	100
Optimizer	RMSPROP
Batch size	256
Training images	6120
Validation images	1000
Test images	1000

The proposed model is developed as shown in Fig. 5. The embedding layer was chosen to have a size of 300 after experimentation. This size produced word embeddings that were able to produce results with a low loss and good accuracy. The training was done on the created model shown in Fig. 5 with the parameters shown in Table 1.

4 Results and Conclusion

After training for 100 epochs, the loss was reduced to 1.98. The algorithm was tested on the testing images and the results are shown in Figs. 6 and 7. The 1000 testing images were run through the BLUE-4 metric. The proposed method produced a BLEU-4 score of 48.9 which was better as compared to previous methods on the same dataset [11]. The results are displayed as a graph in Fig. 8 and shown in Table 2.

Fig. 6 Two dogs are playing together in the grass and dog is running across the grass

Fig. 7 A dog playing with ball in a grassy field and two men are in a field

It is observed that GRU provides more than comparable results when used in an image captioning system. All the scores for the BLEU metrics fall within the higher range, making it an optimal choice when developing an image captioning model. When tested with the BLEU-4 metric and compared with the previous methods used, it is observed that the GRU method surpasses all of them. Due to a limitation in hardware, batch size was not able to be increased but doing so will result in better results as was seen when changing the batch size from 128 to 256. Also, feature localization will greatly improve the depicted results as proven in Karpathy et al. [12].

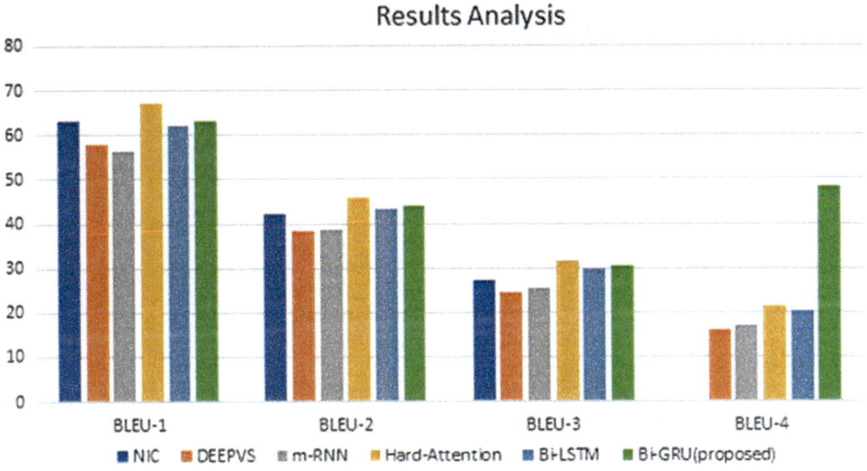

Fig. 8 Result analysis of LSTM RNN

Table 2 Score evaluation with BLEU-4

Methods/metrics	BLEU-4
RNN	4.86
RNN with image features	12.04
RNN with bidirectional mapping	14.10
GRU (proposed method)	48.9

References

1. S.J. Kim, H.R. Kim, Y.S. Kim, I.K. Lee, Building emotional machines: recognizing image emotions through deep neural networks. IEEE Trans. Multimed. **20**(11), 2980–2992 (2018)
2. V. Bakrola, P. Shah, S. Pati, Image captioning using deep neural architectures, in *2017 International Conference on Innovations in Information, Embedded and Communication Systems (ICIIECS)*, Coimbatore. IEEE, 17–18 March 2017, pp. 1–4
3. Y. Wu, L. Wu, C. Wan, J. Liu, Generative caption for diabetic retinopathy images, in *2017 International Conference on Security, Pattern Analysis, and Cybernetics (SPAC)*, Shenzhen, Embedded and Communication Systems (ICIIECS), Coimbatore, 2017, pp. 515–519
4. D. Zhang, M.R. Kabuka, Combining weather condition data to predict traffic flow: a GRU based deep learning approach, in *2017 IEEE 15th International Conference on Dependable, Autonomic and Secure Computing, 15th Intl Conf on Pervasive Intelligence and Computing, 3rd Intl Conf on Big Data Intelligence and Computing and Cyber Science and Technology Congress(DASC/PiCom/DataCom/CyberSciTech)*, Orlando, 2017, pp. 1216–1219
5. S. Bengio, O. Vinyals, A. Toshev, D. Erhan, Show and tell: a neural image caption generator, in *Proceedings of the IEEE Conference on Computer Vision and Pattern Recognition* (IEEE, 2015), pp. 3156–3164
6. A. Karpathy, L. Fei-Fei, Deep visual-semantic alignments for generating image descriptions, in *Proceedings of the IEEE Conference on Computer Vision and Pattern Recognition* (IEEE, 2015), pp. 3128–3137
7. A. Karpathy, J. Johnson, L. Fei-Fei, Densecap: fully convolutional localization networks for dense captioning. arXiv preprint, 2015, arXiv:1511.07571

8. C. Bartz, C. Wang, H. Yang, C. Meinel, Image captioning with deep bidirectional LSTMS. arXiv preprint, 2016 arXiv:1604.00790
9. R. Kiros, A. Courville, R. Salakhutdinov, R. Zemel, K. Xu, J. Ba, Y. Bengiol, Show, attend and tell: Neural image caption generation with visual attention. arXiv preprint, 2015, arXiv:1502.03044
10. K.P. Camilleri, M. Tanti, A. Gatt, Where to put the image in an image caption generator. arXiv preprint, 2018, arXiv:1703.09137
11. C.L. Zitnick, X. Chen, Learning a recurrent visual representation for image caption generation. arXiv preprint, 2014, arXiv:1411.5654
12. A. Joulin, A. Karpathy, L. Fei-Fei, Deep fragment embeddings for bidirectional image sentence mapping. arXiv preprint, 2014, arXiv:1406.5679

A BERT-Based Question Representation for Improved Question Retrieval in Community Question Answering Systems

C. M. Suneera and Jay Prakash

1 Introduction

Community question answering (CQA) systems, such as Quora and Yahoo! Answers, help people to get answers from experts on a variety of topics and to share their knowledge. One can post questions in natural language text, and subsequently, he gets answers, comments, or some additional questions to avoid ambiguity. The best answer selection is made by asker or by the system by analyzing the quality of answers using natural language processing (NLP) techniques and considering the up-votes made by other participants in the thread. A resolved question is adding to the archive as a pair of question and the best answer for future use.

Due to the proliferation of users and hence questions in the forums, the questioner may not get answers immediately. To tackle this, it is better to use the question–answer pair in the archive similar to the new question. The task of retrieving semantically relevant historical questions from the archive is called question retrieval. While computing the similarity between two questions, we need to resolve some challenges. Some of them are listed below.

1. A question can be asked in many ways, even without a single lexical match.
2. Two questions may have similar format, but the key concept on which the question is posing may differ.
3. Some questions contain the same named entities, but their relationship may be different.
4. Keeping named entities and relationships the same, the intent of the question may vary.

C. M. Suneera (✉) · J. Prakash
Department of Computer Science and Engineering, National Institute of Technology Calicut, Calicut, Kerala 673601, India
e-mail: suneera_p180047cs@nitc.ac.in

J. Prakash
e-mail: jayprakash@nitc.ac.in

S. Patnaik et al. (eds.), *Advances in Machine Learning and Computational Intelligence*, Algorithms for Intelligent Systems, https://doi.org/10.1007/978-981-15-5243-4_31

Apart from the above-mentioned challenges, the question written in the natural language needs to be represented to get a semantic representation before computing the similarity. In this paper, we use a sentence transformer fine-tuned on the BERT model to represent questions. We perform the question retrieval task by finding cosine similarity between new question and the question in the archive, then retrieved the top k most similar questions. We modified the BERT embedding by appending the vector with a topic value. To obtain topic value for questions, we extracted keywords using the rapid automatic keyword extraction (RAKE) [15] algorithm and then applied latent Dirichlet allocation (LDA) [1] for topic modeling. Evaluation of results is done on the Quora question pair dataset and compared the results with three question retrieval systems that use different question representations.

In Sect. 2, we look at some of the recent approaches that have been proposed to address the question retrieval in CQA and some state-of-the-art word embedding models. In Sect. 3, we describe the framework of our proposed approach. We described about the dataset and experiments that we conducted to evaluate the system in Sect. 4, and we conclude the paper in Sect. 5.

2 Related Works

Many works have been done so far on community question answering systems to tackle various challenges like question representation, best answer selection, evaluating the quality of questions and answers, historical query retrieval, and expert finding. Here, we are focusing on question retrieval based on semantic similarity.

2.1 Question Retrieval

Existing methods for query retrieval can primarily be classified as language model-based approaches, translation model-based approaches, topic modeling-based approaches, and neural network-based approaches [11].

Language model-based approaches are used by most of the earlier works in question retrieval. The approach makes use of large datasets to estimate the likelihood of each word given previous words. Some works use the metadata like category information to build the model for query retrieval [4]. The second approach uses a translation model trained on a large parallel corpus of texts. Here, translation probability of one question to another is calculated to measure question similarity [8, 9, 16].

As translation models consider question–answer pairs as parallel instead of heterogeneous, the model may retrieve low-quality answers. To overcome this issue, topic modeling-based approaches use the latent similarity of questions to perform question retrieval by mapping questions to topics regardless of the form of texts [6]. Ji et al. [7] proposed a question answer topic model (QATM) to learn the latent topics aligned across the question–answer pairs with the assumption that they share the same topic distribution.

Neural network-based approaches use artificial neural networks (ANNs) to learn semantic representation for question–answer pairs with the help of huge text corpus. Question retrieval using neural network models has shown good results compared with other approaches [17].

2.2 *Question Representation*

The most commonly used traditional vector representation is the bag of words (BOW) and its variations, which learn representation for a text, based on the frequency of words in the corpus. These representations can capture the relevance of a word in the text but fail to capture semantics [11]. Neural embedding makes use of the large dataset to get semantic word embedding. Word2Vec [10] is the well-known predictive embedding model that groups vectors of similar words in the vector space to represent the meaning mathematically.

Another word embedding model is proposed by Jeffrey et al. called Global Vectors for word representation (GloVe) [12], which is an unsupervised learning algorithm for obtaining vector representation. Facebook introduced a word embedding model called FastText [2], which build on top of Word2Vec. It represents a word as an n-gram of characters, which helps to embed out of vocabulary words and rare words.

Transfer learning has been successful in many NLP tasks in making use of language models that have trained on large datasets. The most significant breakthrough in this regard is embeddings from language models (Elmo) [13], which gives deep contextualized word representations for words in a text. It uses a bidirectional language model to embed the syntax and context-dependent semantics.

Bidirectional encoder representations from transformers (BERT) [5] is a language model released by Google, which is trained on a large unlabeled dataset from the entire Wikipedia and Book Corpus. The pre-trained model can fine-tune for different tasks to overcome the scarcity of task-specific labeled datasets and shows the state-of-the art results for many NLP tasks.

3 Proposed Approach

The core concept of our proposed approach is to transform questions to a semantic representation using the state-of-the-art language model BERT [5] before performing question retrieval. The framework of the approach is shown in Fig. 1. A query entered in natural language text and questions from the QA archive is passed through keyword extraction, topic modeling, and then produces a query representation by appending topic value with BERT embedding. The final step is to retrieve and rank relevant questions from archive based on the similarity with the new question.

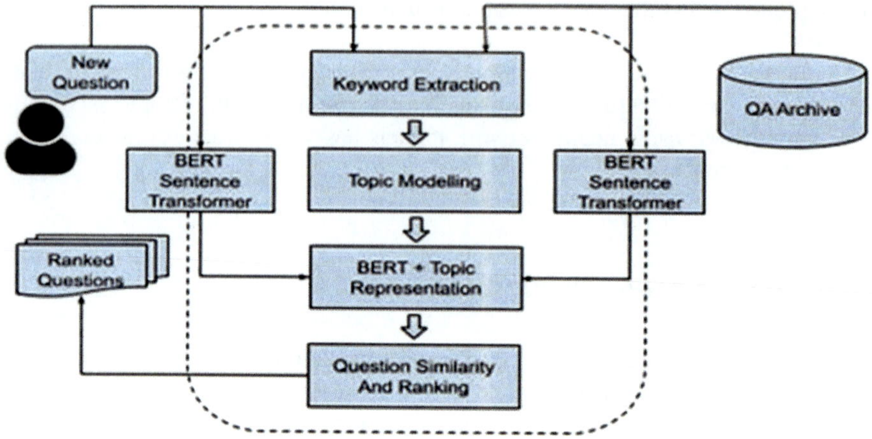

Fig. 1 Framework of the proposed approach

3.1 Keyword Extraction

We used the rapid automatic keyword extraction (RAKE) [15] algorithm to extract keywords from the question. The first step is to create a list of candidate keywords by splitting the whole text using stop words as word delimiter. The next step is to create a word co-occurrence graph from which a score is computed for each candidate keyword based on the frequency and degree of word vertices in the graph. Based on the frequency of co-occurrence, an adjoin task is performed to combine the keywords including the interior stop words. The score of this phrase is equal to the sum of score of member keywords. Finally, all keywords and phrases are ranked based on scores and topmost keywords will be extracted. A question q now can be represented as $q = \{k_1, k_2, \ldots, K_n\}$, where k is a keyword extracted by RAKE.

3.2 Topic Modeling

Topic modeling is used to classify a question to an abstract topic by statistically analyzing recurring patterns of words in the corpus. Here, we used the widely used topic modeling technique latent Dirichlet allocation (LDA) [1]. Each question is modeled as a Dirichlet distribution of T topics, and each topic is modeled as a Dirichlet distribution of words in the vocabulary. The number of topics is an input parameter to the model that should be chosen considering the characteristics of the corpus. Every question in the archive is assigned a topic value by the model learned on the question pair dataset.

3.3 Question Embedding Generation

We used a sentence transformer fine-tuned on the state-of-the-art language model BERT [5] to get semantically meaningful question embeddings. The model makes use of an attention mechanism called transformer, which contains stacked layers of encoders and decoders for task-specific translation. Each word in the input question can have a fixed-length representation containing word embedding, position embedding, and segment embedding. We used a BERT-based model fine-tuned on natural language inference (NLI) corpus [3], which gives a sentence embedding of size 768 obtained by performing mean pooling on word embeddings [14]. We appended the topic value obtained in the last step with the BERT embedding to get an improved representation of questions.

3.4 Question Similarity Computation and Ranking

We used the cosine similarity measure to compute the relatedness of two question vectors obtained through the previous steps. Cosine similarity score between two question vectors q_i and q_j can be defined as:

$$\text{Score}\left(q_i, q_j\right) = \frac{q_i . q_j}{\|q_i\| \left\|q_j\right\|} \tag{1}$$

Let q is the embedding of the new question, and q_i is that of a question from the archive. For each question q_i, we computed the *score* (q, q_i) and ranked them based on the highest score obtained.

4 Experimental Results and Discussion

This section presents description of datasets, results, and comprehensive discussion. The experiments have been performed in Python 3.7.

4.1 Dataset

To evaluate our system, we used a subset of Quora question pair dataset which is provided as part of the Kaggle competition. The original dataset for training contains 404,290 question pairs with a binary label (1 for similar and 0 for dissimilar). A repository of questions is created by taking 1500 questions from the dataset and we

Table 1 Samples from Quora question pair dataset

Id	qid1	qid2	question1	question2	is_duplicate
19	39	40	What is best way to make money online?	What is best way to ask for money online?	0
258399	508,578	508,579	What can we do for hair loss?	What are the best options for hair loss?	1

tested the retrieval system for 253 new questions taken from the dataset. Two samples of the dataset are shown in Table 1.

4.2 Evaluation Metrics

We used the mean average precision (MAP) and Precision@1 (P@1) measures for evaluating our approach. These two are the extensively used metric for evaluating question retrieval in community question answering systems.

4.3 Results and Discussions

We evaluated the performance of the proposed method and compared the results with three question retrieval systems. They used the following state-of-the-art embedding models to represent questions.

1. Word2Vec [10]: Continuous bag-of-words (CBOW) Word2Vec model is used to get the embedding for each word in the question and averaged them to get the question embedding.
2. FastText [2]: An n-gram-based model is build on top of Word2Vec, which helps to get embedding for unknown words in the questions.
3. BERT [5]: A sentence transformer fine-tuned on the general-purpose language model, bidirectional encoder representations from transformers (BERT), which is trained on a large corpus contains context of each word in the sentence in both direction.

We used the same question repository, test set, and similarity function to evaluate question retrieval systems using different question embeddings. The results are summarized in Table 2.

Results show that the proposed method (BERT + Topic) gives good result compared with other three approaches. The number of topics is a parameter that we need to decide before modeling the topic using LDA. We have tested the performance of the system by varying number of topics and we got good result when the number of

Table 2 Comparison of question retrieval systems using different question representations

	Word2Vec	FastText	BERT	BERT + Topic
Precision@1	0.3834	0.3438	0.7194	0.7312
MAP	0.7294	0.6316	0.7158	0.7326

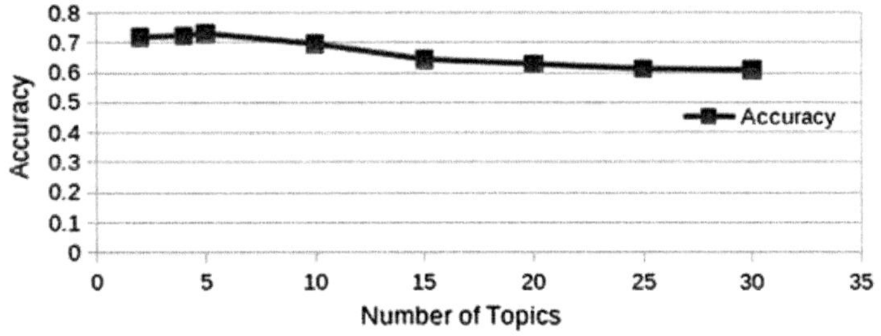

Fig. 2 Variation in the accuracy of the proposed approach for different number of topics

topics is 5. The performance variation of the proposed method for different number of topics is plotted in Fig. 2.

Here, we used a small repository of questions for search, and the accuracy of the result may deteriorate if we apply this approach on a large repository.

5 Conclusion

In this paper, we addressed the question retrieval problem in community question answering (CQA) systems. We have analyzed the impact of question representation on question retrieval tasks by using different embedding models, including the state-of-the-art language model BERT. Among these models, we observed that the result with BERT is considerably higher than the others. We also proposed a new framework for question retrieval. The model contains keyword extraction and topic modeling to get a topic value for the question posted by the user. This value is appended with the query vector obtained by a BERT sentence transformer to form a semantic embedding for the query. Every question in the archive is modeled by the same approach and calculated the cosine similarity with the new question. The top-ranked questions will be retrieved to the user. The experiments are conducted on a subset of the Quora question pair dataset, and the results show that the proposed question representation can improve the question retrieval task compared with other representations.

References

1. D.M. Blei, A.Y. Ng, M.I. Jordan, Latent dirichlet allocation. J. Mach. Learn. Res. **3**, 993–1022 (2003)
2. P. Bojanowski, E. Grave, A. Joulin, T. Mikolov, Enriching word vectors with subword information. Trans. Assoc. Comput. Linguistics **5**, 135–46 (2017)
3. S.R. Bowman, G. Angeli, C. Potts, C.D. Manning, A Large Annotated Corpus for Learning Natural Language Inference. arXiv preprint arXiv:1508.05326 (2015)
4. X. Cao, G. Cong, B. Cui, C.S. Jensen, *A Generalized Framework of Exploring Category Information for Question Retrieval in Community Question Answer Archives*, in Proceedings of the 19th international conference on World Wide Web. ACM (2010), pp. 201–210
5. J. Devlin, M.W. Chang, K. Lee, T.K. Bert, *Pre-training of Deep Bidirectional Transformers for Language Understanding*. arXiv preprint arXiv:1810.04805 (2018)
6. H. Duan, Y. Cao, C.Y. Lin, Y. Yu, *Searching Questions by Identifying Question Topic and Question Focus*, in Proceedings of ACL-08, HLT (2008), pp. 156–164
7. Z. Ji, F. Xu, B. Wang, B. He, *Question-Answer Topic Model for Question Retrieval in Community Question Answering*, in Proceedings of the 21st ACM International Conference on Information and Knowledge Management. ACM (2012), pp. 2471–2474
8. J. Jeon, W.B. Croft, J.H. Lee, *Finding Similar Questions in Large Question and Answer Archives*, in Proceedings of the 14th ACM International Conference on Information and Knowledge Management. ACM (2005), pp. 84–90
9. J.T. Lee, S.B. Kim, Y.I. Song, H.C. Rim, *Bridging Lexical Gaps Between Queries and Questions on Large Online Q&A Collections with Compact Translation Models*, in Proceedings of the Conference on Empirical Methods in Natural Language Processing. Association for Computational Linguistics (2008), pp. 410–418
10. T. Mikolov, K. Chen, G. Corrado, J. Dean, *Efficient estimation of Word Representations in Vector Space*. arXiv preprint arXiv:1301.3781 (2013)
11. N. Othman, R. Faiz, K. Smaïli, *Enhancing Question Retrieval in Community Question Answering Using Word Embeddings*, in Proceedings of the 23rd International Conference on Knowledge-Based and Intelligent Information & Engineering Systems (2019)
12. J. Pennington, R. Socher, C.D. Manning, in *Glove: Global Vectors for Word Representation* (2014). URL: https://nlpstanford.edu/projects/glove/. Accessed 11 Jan 2018
13. M.E. Peters, M. Neumann, M. Iyyer, M. Gardner, C. Clark, K. Lee, L. Zettlemoyer, *Deep Contextualized Word Representations*. arXiv preprint arXiv:1802.05365 (2018)
14. N. Reimers, I. Gurevych, *Sentence-BERT: Sentence Embeddings Using Siamese BERT-Networks*. arXiv preprint arXiv:1908.10084 (2019)
15. S. Rose, D. Engel, N. Cramer, W. Cowley, Automatic keyword extraction from individual documents, in *Text Mining: Applications and Theory* (2010)
16. X. Xue, J. Jeon, W.B. Croft, *Retrieval Models for Question and Answer Archives*, in Proceedings of the 31st Annual International ACM SIGIR Conference on Research and Development in Information Retrieval. ACM (2008), pp. 475–482
17. G. Zhou, Y. Zhou, T. He, W. Wu, Learning semantic representation with neural networks for community question answering retrieval. Knowl.-Based Syst. **93**, 75–83 (2016)

Kinect-Based Outdoor Navigation for the Visually Challenged Using Deep Learning

Anand Subramanian, N. Venkateswaran, and W. Jino Hans

1 Background

Visual impairments and other vision-related disorders afflict a significant number of people in today's world. There is a need for devices and systems that assist people with visual impairments in navigation. Outdoor mobility in particular presents a challenge, in terms of the multitude of obstacles faced by people when traversing from one place to another. The difficulty in navigation is often due to lack of information with regards to the environment and thus may prove to be dangerous for visually challenged people when they need to move through such environments. One step toward rectifying this problem lies in building assistive systems that can recognize commonly found objects in outdoor environments and communicate the necessary information to the person using it, in order to make their movements easier in outdoor surroundings.

2 Related Work

There has been significant research conducted into the design and construction of systems intended as navigational aids for people with visual impairments for indoor and outdoor purposes. The use of ultrasound sensors for aiding navigation was explored in [1]. Nassih et al. [2] implemented a setup whereby a RFID tag reader was attached

A. Subramanian (✉) · N. Venkateswaran · W. Jino Hans
SSN College of Engineering, Kalavakkam, Chennai, Tamil Nadu, India
e-mail: anand.subu10@gmail.com

N. Venkateswaran
e-mail: venkateswarann@ssn.edu.in

W. Jino Hans
e-mail: jinohansw@ssn.edu.in

S. Patnaik et al. (eds.), *Advances in Machine Learning and Computational Intelligence*, Algorithms for Intelligent Systems, https://doi.org/10.1007/978-981-15-5243-4_32

to a white cane along with a Braille interface. RFID tags with requisite information about location, postal codes, etc., were set up at various places, which were read by the tag reader to provide information. The Microsoft Kinect (version 1) [3] is a motion sensing device equipped with an RGB camera, an infrared camera, depth sensor and microphone and has been used in building assistive systems. Filipe et al. [4] introduced a system using the Kinect for indoor navigation where neural networks trained on line profiles were used to detect four classes—obstacle, no obstacle, upstairs and downstairs. Takizawa et al. [5] proposed a system using a Kinect connected to a white cane capable of detecting walking floors, chairs, upward stairs and downward stairs utilizing the depth maps from the Kinect, along with vibro-tactile feedback to indicate proximity. A navigation system for blind people using a recursive windowing-based mean method was proposed by Ali et al. [6] to detect obstacles.

3 Proposed Work

Most of the approaches mentioned above have focused on indoor navigation which is a significantly controlled environment. The previous works have taken advantage of this aspect by adding features to indoor environments that work in a predictable manner with their systems to help in mobility. On the contrary, external environments are far more dynamic and unpredictable, making the problem of tackling external mobility using assisted systems much more challenging. Some of the practical considerations are as follows:

Varying Nature Of External Environments. From a pedestrian's perspective, the nature of obstacles hindering their mobility in an external environment should be known to provide them with adequate insight. While it is quite impractical to attempt to glean information about every possible obstacle in this environment, it is possible to bring in some standardization to the obstacles based on their nature, as we have attempted here.

Processing Constraints. For a system to run intelligent decision-making algorithms based on sensory inputs, sufficient compute power must be necessary to process the inputs and convey the information to the pedestrian in an embedded environment while also being easily portable for usage. An object detection model is trained to recognize commonly found obstacles in an outdoor environment. To bring in an element of standardization, we restrict our detections to cars, pedestrians, motorcycles and bicycles. We believe that is adequately representative of most obstacles that a pedestrian would encounter. The system utilizes the Kinect to collect frames from an outdoor environment, which are provided to the object detection model. Based on the localization of the object in the image, the depth map provided by the Kinect is utilized to calculate distance between the person and the object, and pyTTS, a python wrapper for the Microsoft Speech API [7] is used to alert them through Bluetooth-paired earphones. The Kinect is also repurposed for portable use by utilizing a battery power bank for providing the requisite input of 12 V for its functioning. The Raspberry Pi (Model 3-B) serves as our processing unit, for reading frames from the

Kinect, as well as for onboard model inference and other processing. In order to tackle the processing constraints, we carry out extensive training, experimentation and benchmarking of three single shot detection (SSD) [8] models with the feature extractors being MobileNet V1 [9] with a Pooling Pyramid Network [10], MobileNet V2 [11] and an Inception network [12], in the embedded runtime environment, along with a RetinaNet model [13]. We also apply floating-point quantization to reduce the size of the final trained model and improve inference time for obtaining real-time performance.

4 Components

4.1 Microsoft Kinect

In order to set up the Kinect for use with the Raspberry Pi, we use the libfreenect library [14] provided by OpenKinect, with Python bindings to read the inputs provided by the Kinect, and integrate it into our pipeline for further processing. The libfreenect library allows us to read in input frames, both RGB and depth, using the Raspberry Pi. The input frames are of dimensions 640 × 480, and the equation specified by the OpenKinect documentation is used for the calculation of depth from the Kinect frames using the raw disparity values provided by the system:

$$d\ (\text{cm}) = 100/(-0.00307 * \text{Disparity} + 3.33) \tag{1}$$

Here, '*d*' refers to the distance and the 'Disparity' refers to the raw pixel values obtained from the depth maps from the Kinect. Noise pixels from the depth map were ignored, and the distance was calculated after obtaining the mean disparity from the pixels, using coordinates of the detected objects. From the mean disparity, the equation is used to infer the distance from the user with the conventional range of the sensor estimated to be 0.5–4 m.

4.2 Object Detection

The advent of convolutional neural networks has brought in several breakthroughs in the field of computer vision and deep learning, significantly impacting object detection and localization. Fundamentally, the SSD [8] is an object detector designed, keeping in mind real-time performance. The model operates by first extracting feature maps from a CNN and then applying 3 × 3 convolutional filters across each part of the image, across multiple scale feature maps, to detect objects. The original MobileNet [9] architecture utilizes depth separable convolutions to reduce computational time for performing convolution operations. The MobileNet V2 [11] added bottleneck residual layers with the intention of improving gradient propagation across the bottleneck layers with improved memory efficiency. The Pooling Pyramid Network [10] introduced a shared box predictor across different scales of feature maps, thereby attempting to improve the capacity of the model to incorporate information about the entire training data to stabilize the training process. In the InceptionNet [12], 1 × 1 convolution blocks are placed before the other convolution operators (3 × 3 and 5 × 5) to reduce the number of parameters of the model, while also replacing fully connected layers with a global average pooling layer. The RetinaNet [13] model comprises a standard backbone network with subnetworks for classification and predicting bounding box coordinates with the introduction of focal loss to address class imbalance in training data.

5 Workflow

First, the RGB and depth frames are read using the Kinect connected to the Raspberry Pi, which are passed on to the trained SSD model. Using the detections from the SSD, the regions are extracted and the depth for the detected object is calculated. The information is conveyed to the user's earpiece through the TTS system used (Fig. 1).

The Kinect was engineered for use with a portable power supply by modifying the power cord for connection with a portable power bank powered by AA batteries which provide a 12 V supply to the Kinect.

5.1 Kinect Depth Map

The Kinect depth map is represented as 11-bit integers, in 640 × 480 resolution. Using Eq. 1, we are able to calculate the distance between the Kinect and the obstacles based on the pixel values. To identify the regions of the Kinect depth map with objects of interest, we obtain the detections from the object detection model and

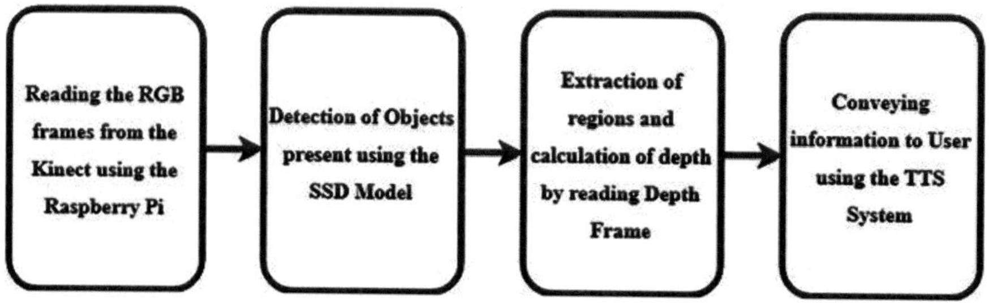

Fig. 1 Workflow of the system

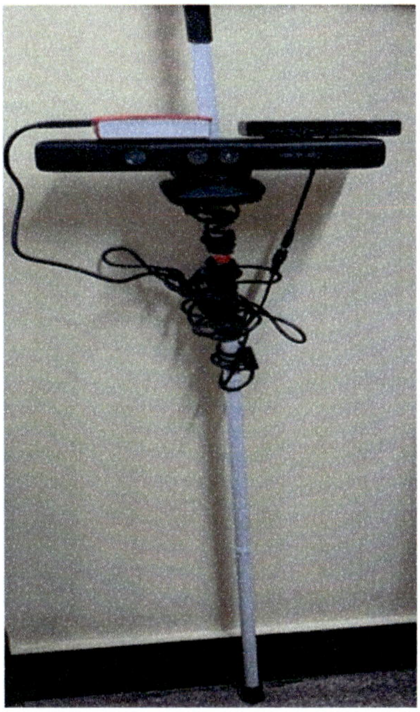

Fig. 2 Assembled system

localize the objects using the depth map. Once the distance is calculated, we automatically filter out pixels that are further than the Kinect's range since the Kinect reserves one bit for representing pixels away from the Kinect's range and provide the detected information to the user. The system is assembled on a white cane for usage, independent of any other specialized processing entity and fully functional on the Raspberry Pi. The TTS is integrated into the code as well, along with Bluetooth earphones to provide the details to the user (Fig. 2).

6 Experimental Results

6.1 Training the Model

We use the TensorFlow Object Detection API [15] to train all the models, by utilizing pre-trained weights for the purpose of transfer learning. The BDD100K dataset [16] was used for further training and fine-tuning. The BDD100K dataset consists of 100,000 frames taken from road perspectives comprising annotations of entities commonly found on roads, such as cars, pedestrians, motorcycles and bicycles. Our work is not intended as an entry in this competition. From the dataset, we utilize 20,000 images for training and utilize 4000 images from the validation set for testing our model. The classes taken into consideration are cars, bicycles, motorcycles and

Table 1 Frames per second and mean average precision of models

Models	Frames per second	Mean average precision
SSD-M (N)	0.8815	0.1251
SSD-M (Q)	1.4787	0.1033
SSD-I (N)	–	0.1149
SSD-I (Q)	0.3535	0.0992
SSD-P (N)	0.5376	0.0720
RetinaNet (N)	–	0.0866

pedestrians. All models, except the SSD-Pooling Pyramid Network, are trained for a total of 120 epochs on a laptop, with a Nvidia GeForce 1050 Ti GPU with 4 GB RAM. The SSD-Pooling Pyramid Network is trained for a total of 150 epochs to observe if performance increments could be gathered due to its initial poor performance. Post training, we calculate the mean average precision (mAP) with an intersection-over-union overlap of 0.5 for the testing set of 4000 images and benchmark the inference performance of all models, both quantized and non-quantized, in terms of frames per second (fps).The fps is measured by running the model on the Raspberry Pi.

Floating-point quantization is carried out using the TFLite toolkit provided by TensorFlow. By default, model weights are stored in the format of 32-bit floating-point values, and the resultant floating-point arithmetic involved in a forward pass can make computations expensive in an embedded runtime. By reducing the numeric precision of weight representation from 32-bit to 16-bit floating-point quantization, we achieve the twin purposes of faster runtime at inference as well as reduced model memory size at minimal costs to performance. For the purpose of representing the tabulation, SSD MobileNet V2 is referred to henceforth as **SSD-M,** SSD MobileNet V1 with Pooling Pyramid Network is referred to as **SSD-P** and SSD with Inception network is referred to as **SSD-I.** To represent quantized model, we add a **(Q)** to the above-mentioned abbreviation and a **(N)** to represent the non-quantized model. The RetinaNet model is included as an additional baseline to compare against the SSD models and is only benchmarked in its non-quantized form. For the purpose of measuring mAP, we use the implementation provided by Cartucho [17]. The SSD Pyramidal Pooling Network is not quantized, owing to its already optimized nature compared to MobileNet (Tables 1 and 2).

Table 2 Average precision of classes for each model

Classes	Model AP					
	SSD-M (N)	SSD-M (Q)	SSD-I (N)	SSD-I (Q)	SSD-P (N)	RetinaNet(N)
Cars	0.2686	0.2465	0.2654	0.2455	0.2516	0.2872
Pedestrians	0.0740	0.0573	0.0535	0.0497	0.0222	0.0326
Motorcycles	0.0712	0.0473	0.0320	0.0201	0.0061	0.0042
Bicycles	0.0865	0.0621	0.1085	0.0814	0.0082	0.0226

The SSD Inception network in its non-quantized form was unable to run on the Raspberry Pi, crashing during inference, but the quantized version was able to perform adequately with only minimal drop in mAP. The SSD Pooling Pyramidal Network processed one frame for every two seconds, much faster than the SSD Inception network, but at the cost of poor mAP. The RetinaNet model fared better compared to the other models in terms of detections on cars but had no tangible improvements in the other classes. It was unable to run on the Raspberry Pi, crashing during runtime. The SSD MobileNet V2 models displayed the best combination of accuracy and performance in real time, with the quantized version being the fastest of all models, while having a good mAP. Thus, the quantized and non-quantized SSD MobileNet v2 model was selected as candidates for deployment.

6.2 Outdoor Visualizations

In order to gauge the effectiveness of the combined setup, we also evaluate the final model and capture footage in a bustling road using a phone camera to validate the performance and visualize the detections of the model in a outdoor environment (Fig. 3).

We observe that the model is able to perform adequately well in such an environment and detect objects in the vicinity of the user, proving its ability to alert the user through the setup.

7 Conclusion

Thus, an outdoor navigation system utilizing the Microsoft Kinect, intended for people with visual impairments, is proposed. The system utilizes an object detector, an SSD model to read frames from the Kinect and predict the locations of cars, pedestrians, bicycles or motorcycles present in the vicinity of the user. From these detections, the distance from the obstacle to the user is calculated using the Kinect's depth maps, and the information is communicated to the user, through a TTS system and a pair of Bluetooth earphones, with the entire processing happening on a Raspberry Pi. Future work could involve training the model on a larger number of obstacles found in an outdoor setup and making the Kinect more modular through removing the outer casings to provide for a more durable system.

Acknowledgements The authors would like to thank the SSN Trust, SSN College of Engineering, Chennai, for its financial assistance in the development of this project.

Fig. 3 Visualizations from the model on footage captured with a phone camera in an outdoor setting

References

1. K. Kumar, B. Champaty, K, Uvanesh, R. Chachan, K. Pal, A. Anis, *Development of an Ultrasonic Cane as a Navigation Aid for the Blind People*, in 2014 International Conference on Control, Instrumentation, Communication and Computational Technologies (ICCICCT) (2014), pp. 475–479
2. M. Nassih, I., Cherradi, Y. Maghous, B. Ouriaghli, Y. Salih-Alj, *Obstacles Recognition System for the Blind People Using RFID*, IN 2012 Sixth International Conference on Next Generation Mobile Applications, Services and Technologies (2012), pp. 60–63
3. Kinect—Windows app development, https://developer.microsoft.com/en-us/windows/kinect. Last accessed 27 Mar 2019
4. V. Filipe, F. Fernandes, H. Fernandes, A. Sousa, H. Paredes, J. Barroso, Blind navigation support system based on microsoft kinect. Procedia Comput. Sci. **14**, 94–101 (2012)
5. H. Takizawa, S. Yamaguchi, M. Aoyagi, N. Ezaki, S. Mizuno, *Kinect Cane: Object Recognition Aids for the Visually Impaired*, in 2013 6th International Conference on Human System Interactions (HSI) (2013), pp. 473–478

6. A. Ali, M.A. Ali, *Blind Navigation System for Visually Impaired Using Windowing-Based Mean on Microsoft Kinect Camera*, in 2017 Fourth International Conference on Advances in Biomedical Engineering (ICABME) (2017), pp. 1–4
7. T. M. (aquent Llc), Microsoft Speech API (SAPI) 5.3. https://docs.microsoft.com/en-us/previous-versions/windows/desktop/ms723627(v%3dvs.85. Last accessed 2 April 2019
8. W. Liu, D. Anguelov, D. Erhan, C. Szegedy, S. Reed, C.-Y. Fu, A.C. Berg, SSD: *Single Shot MultiBox detector*, in Computer Vision (ECCV 2016). Springer International Publishing, Berlin (2016), pp. 21–37
9. A.G. Howard, M. Zhu, B. Chen, D. Kalenichenko, W. Wang, T. Weyand, M. Andreetto, H. Adam, *MobileNets: Efficient Convolutional Neural Networks for Mobile Vision Applications* (2017). arXiv preprint: arXiv:1704.04861
10. P. Jin, V. Rathod, X. Zhu, *Pooling Pyramid Network for Object Detection* (2018). arXiv preprint: arXiv: 1807.03284
11. M. Sandler, A. Howard, M. Zhu, A. Zhmoginov, L.C. Chen, *Mobilenetv2: Inverted Residuals and Linear Bottlenecks*. In: Proceedings of the IEEE Conference on Computer Vision and Pattern Recognition (2018), pp. 4510–4520
12. C. Szegedy, W. Liu, Y. Jia, P. Sermanet, S. Reed, D. Anguelov, D. Erhan, V. Vanhoucke, A. Rabinovich, *Going Deeper with Convolutions*, in Proceedings of the IEEE Conference on Computer Vision and Pattern Recognition (2015), pp. 1–9
13. L. Tsung-Yi, P. Goyal, R. Girshick, K. He, P. Dollár, *Focal Loss for Dense Object Detection*, in IEEE International Conference on Computer Vision (ICCV 2017) (2017), pp. 2999–3007
14. OpenKinect, OpenKinect/libfreenect. https://github.com/OpenKinect/libfreenect. Last accessed 2 April 2019
15. J. Huang, V. Rathod, C. Sun, M. Zhu, A. Korattikara, A. Fathi, I. Fischer, Z. Wojna, Y. Song, S. Guadarrama, *Others: Speed/Accuracy Trade-Offs for Modern Convolutional Object Detectors*, in Proceedings of the IEEE Conference on Computer Vision and Pattern Recognition (2017), pp. 7310–7311
16. F. Yu, W. Xian, Y. Chen, F. Liu, M. Liao, V. Madhavan, T. Darrell, *BDD100K: A Diverse Driving Video Database with Scalable Annotation Tooling* (2018). arXiv preprint: arXiv:1805.04687
17. Cartucho: Cartucho/mAP, https://github.com/Cartucho/mAP. Last accessed 4 May 2019

Prediction of Stock Market Prices of Using Recurrent Neural Network—Long Short-Term Memory

Haritha Harikrishnan and Siddhaling Urolagin

1 Introduction

The stock market is where the investors and traders who buy and sell shares of publicly held organizations. The prices of stocks revolve around the principles of demand and supply, and the fundamental goals of investing in the stock market are investment gain and ownership. Over a period of time, when the stock market value rises, the investors are able to get some profit from the organization they had invested in. Investing in the stock of an organization also implies taking an ownership stake in the organization that one would invest in. It is almost impossible to make prediction about the stock prices, owing to the volatility of various factors (interest rates, demand and supply, political scenario, etc.) that play a huge role in the movement of stock prices. Nevertheless, it is in fact, possible to devise an estimate of these stock prices. Stock prices never vary in isolation, i.e., the movement of one stock tends to have an avalanche effect on several other stocks [1]. This feature of stock price movement can be used as a significant tool to predict the prices of several stocks at the same time.

In this research work, the stock prices of five companies leading the technology industry (Apple. Inc, Cisco Systems Inc, International Business Machines Corporation (IBM), Intel Corporation and Microsoft Corporation.) are analyzed using LSTM to identify how the closing price could be predicted on analyzing the open, high and low price of a given stock. Section 2 of this paper gives the literature survey. In Sect. 3 of this paper, the implementation of the program is discussed. In Sect. 4, the results are discussed, and Sect. 5 gives the conclusion and the scope of future work.

H. Harikrishnan (✉) · S. Urolagin
Department of Computer Science, Birla Institute of Technology & Science, Pilani, Dubai Campus, Dubai International Academic City, Dubai, UAE
e-mail: f20160030@dubai.bits-pilani.ac.in

S. Patnaik et al. (eds.), *Advances in Machine Learning and Computational Intelligence*, Algorithms for Intelligent Systems, https://doi.org/10.1007/978-981-15-5243-4_33

2 Literature Survey

Autoregressive integrated moving average more commonly known as ARIMA [2] still continues to be one of the most applied methods in the time series analysis. At present, deep learning [3] is used frequently in financial time series forecasting. In [4], Qunzhuge et al. proposed a stock price prediction model implementing Bayesian classifier for LSTM network and emotional analysis. In [5], Zhu used fourteen indicators as input including the open, high, low and close price, related to technical analysis indicators, so as to learn the historical data of the S&P index and forecast the stock price. He found out that profit was higher than the ordinary trading strategy.

Apple. Inc, Cisco Systems Inc, International Business Machines Corporation (IBM), Intel Corporation and Microsoft Corporation are five of the top organizations in the tech industry. In [6], the authors state that Apple has a revenue of USD 265.595 billion and net income of USD 59.531 billion. In [7], the author states that Cisco has a revenue of USD 49.33 billion and net income of USD 110 million. The authors of [8] have mentioned that IBM has a revenue of USD 79.59 billion and net income of USD 8.72 million. In [9], it is stated that Intel has a revenue of USD 70.8 billion and net income of USD 21 million. It was stated in [10] that Microsoft has a revenue of USD 125.8 billion and net income of USD 39.2 million.

3 Prediction Using LSTM

The stock prediction system based on LSTM has three stages (as shown in Fig. 1):

1. Obtaining dataset and preprocessing
2. Construction of the model
3. Prediction, error calculation and accuracy evaluation.

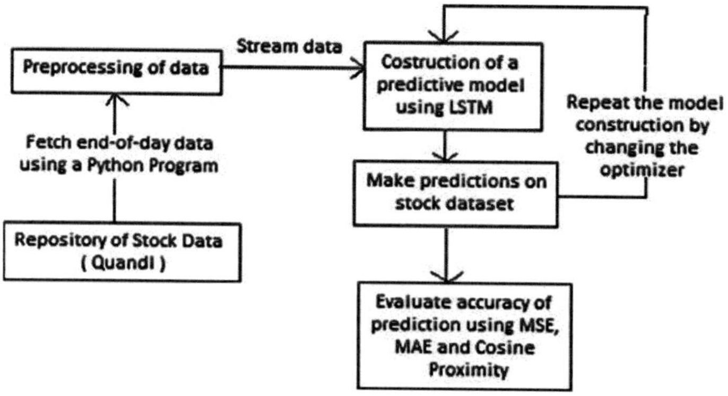

Fig. 1 LSTM-based stock price prediction model

3.1 Obtaining the Dataset and Preprocessing

The stock market data (end-of-the-day prices for five companies) was obtained from Quandl, which is a premier source for financial, economic and alternative datasets serving professionals interested in investments. The obtained dataset from Quandl contained twelve features; out of which the following five features had been selected: date of the observation, the opening price, the highest intra-day price reached, the lowest intra-day price reached and the closing price of the stock. These features were transformed in such a way so that it can be used for prediction by the model. First, the time series data is transformed into input–output elements for supervised learning. Then, the data is scaled into $[-1, +1]$ range.

3.2 Construction of the Model

The input data was split into training and testing sets by holdout method. By *k*-fold cross-validation, the given dataset was split into k sections where each section was used as a testing set at some point. But this method is not effective for a time series dataset as different sections of the dataset are used for forecasting; for instance, the first section (past dataset) may be kept as the test section and the rest as training set (future dataset). This is practically impossible. Hence, holdout method was used. Here, 70% of dataset was used as training set and 30% was used as the testing set. The LSTM model was fit on the training set, and its accuracy was evaluated on both training and testing set to understand how well the model has performed. The LSTM network was created with one input layer of thirty two neurons, ten hidden layers (one LSTM layer and nine dense layers) and one output layer (one neuron). The model was run using four different optimizers: Adam, Adagrad, RMSProp and SGD.

3.3 Prediction, Error Calculation and Accuracy Evaluation

After the LSTM model was fit to the training dataset, it was then used to predict the end-of-day (closing) stock price of any stock. The accuracy of this model was estimated using the following metrics: cosine proximity, mean squared error (MSE), mean absolute error (MAE) and R^2 error.

4 Experimental Results

The end-of-day stock prices of AAPL, CSCO, INTC, IBM and MSFT were downloaded from quandl.com. This dataset is updated every day with the latest value. Table 1 shows the stock prices for the years 2016 and 2017.

The graphs (as shown in Fig. 2) have been plotted to analyze the trends of the open, high, low and close price of each of the stocks. On plotting the graph Fig. 2a, it can be noticed that the open, high, low and close price of $AAPL increases slowly over time and then decreases. From the graphs in Fig. 2b, c, it can be noticed that the open, high, low and close price of $CSCO and $IBM increases slowly over time. On plotting the graph Fig. 2d, it can be noticed that the open, high, low and close price of $INTC increases to a certain period of time (for the first and second quarter of 2016) and then gradually decreases. From graph Fig. 2e, it can be noticed that the open, high, low and close price of MSFT gradually decreases showing a huge drop in stock prices over the two years.

In Fig. 3, it can be observed that the graph plotted while using Adam as optimizer shows the best prediction, while the graph plotted using SGD as optimizer turns out

Table 1 Sample of the stock price dataset obtained from Quandl

Company name	Stock	Date	Open	High	Low	Close
Apple Inc.	$AAPL	31-12-18	158.53	159.36	156.48	157.74
Cisco Systems Inc.	$CSCO	31-12-18	43.19	43.55	42.89	43.33
IBM	$IBM	31-12-18	113.33	114.35	112.4201	113.67
Intel Corporation	$INTC	31-12-18	47.09	47.48	46.55	46.93
Microsoft Corporation	$MSFT	31-12-18	101.29	102.4	100.44	101.57

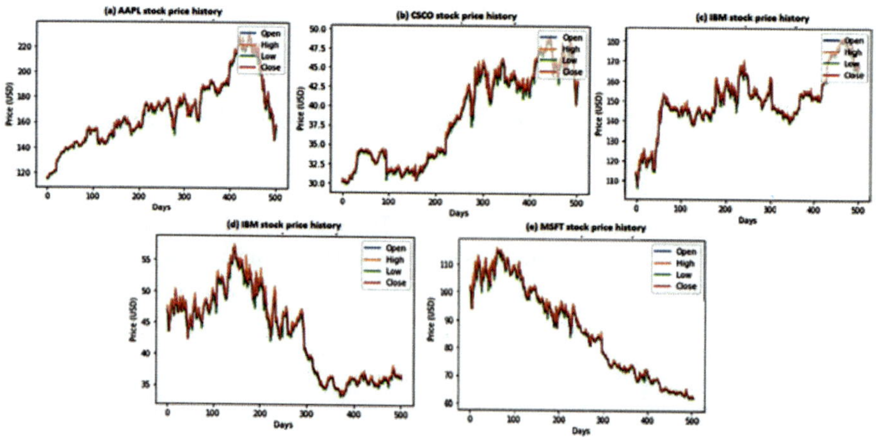

Fig. 2 Open, high, low and close price of different stocks

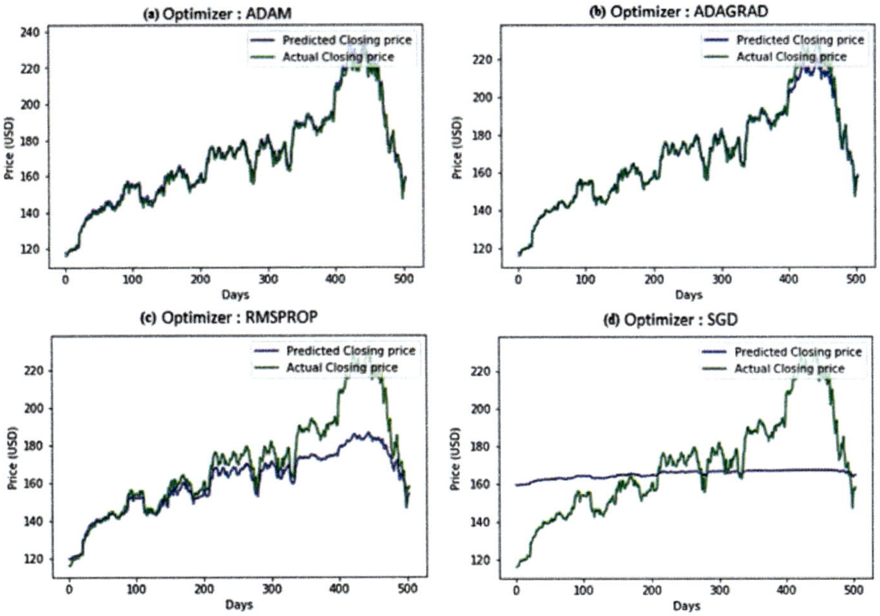

Fig. 3 Prediction graphs of $AAPL with different optimizers

to be not very good. Table 2 shows the accuracy rate of this model. The R^2 error is calculated for the most accurate optimizer, in the case Adam. It showed an error of 0.8498 with the train set and 0.4877 with the test set.

In Fig. 4, it can be observed that the graph plotted while using Adagrad as optimizer shows the best prediction, while the graph plotted using SGD as optimizer turns out to be not very good. Table 3 shows the accuracy rate of this model. The R^2 error is calculated for the most accurate optimizer, in the case Adagrad. It showed an error of 0.8023 with the train set and 0.7393 with the test set.

In Fig. 5, it can be observed that the graph plotted while using Adagrad as optimizer shows the best prediction, while the graph plotted using SGD as optimizer turns out to be poor. Table 4 shows the accuracy rate of this model. The R^2 error is calculated

Table 2 $AAPL—accuracy rates of the model

Optimizer	Train scores				Test scores			
	Adam	Adagrad	RMS Prop	SGD	Adam	Adagrad	RMS Prop	SGD
MSE	0.0005	0.0003	0.0054	0.0497	0.0024	0.0053	0.1286	0.261
MAE	0.0177	0.0114	0.0568	0.1793	0.0379	0.0566	0.3157	0.4488
Cosine proximity	−0.997	−0.997	−0.997	−0.997	−1	−1	−1	−1

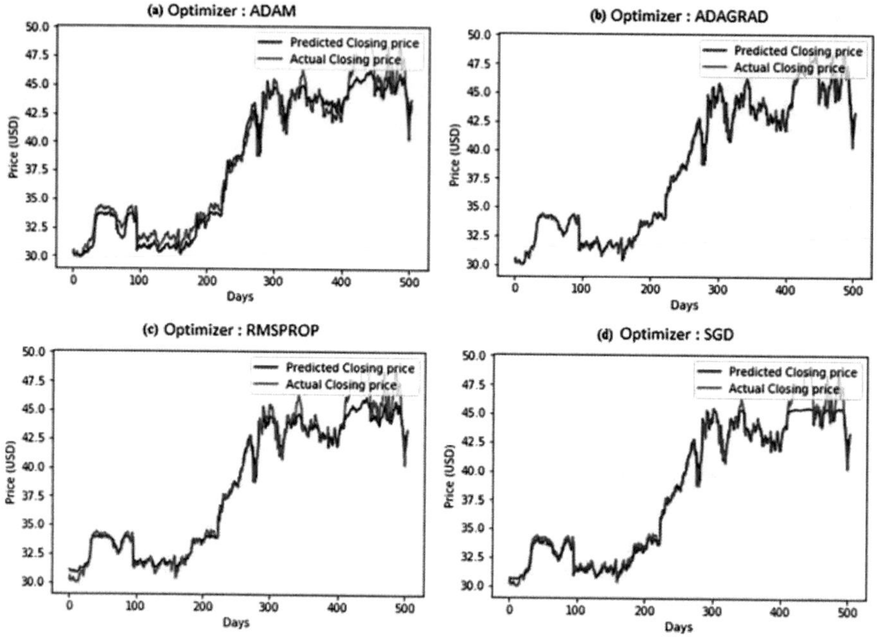

Fig. 4 Prediction graphs of $CSCO with different optimizers

Table 3 $CSCO—accuracy rates of the model

Optimizer	Train scores				Test scores			
	Adam	Adagrad	RMS Prop	SGD	Adam	Adagrad	RMS Prop	SGD
MSE	0.0021	0.0004	0.0008	0.0006	0.0065	0.0011	0.0076	0.0066
MAE	0.039	0.0131	0.0205	0.0188	0.0653	0.0254	0.0665	0.0568
Cosine proximity	−0.9801	−0.997	−0.997	−0.997	−1	−1	−1	−1

for the most accurate optimizer, in the case Adagrad. It showed an error of 0.9108 with the train set and −0.5156 with the test set.

In Fig. 6, it can be observed that the graph plotted while using Adam as optimizer shows the best prediction, while the graph plotted using SGD as optimizer turns out to be not very good. Table 5 shows the accuracy rate of this model. The R^2 error is calculated for the most accurate optimizer, in the case Adam. It showed an error of 0.8365 with the train set and 0.2801 with the test set.

In Fig. 7, it can be observed that the graph plotted while using Adam as optimizer shows the best prediction, while the graph plotted using SGD as optimizer turns out to be poor. Table 6 shows the accuracy rate of this model. The R^2 error is calculated for the most accurate optimizer, in the case Adam. It showed an error of 0.5629 with the train set and −41.97 with the test set.

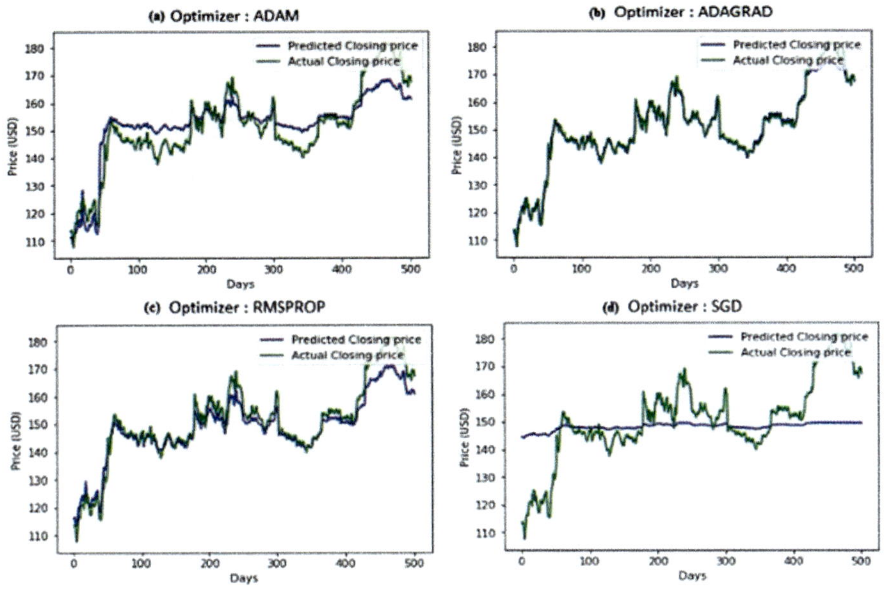

Fig. 5 Prediction graphs of $IBM with different optimizers

Table 4 $IBM—accuracy rates of the model

Optimizer	Train scores				Test scores			
	Adam	Adagrad	RMS Prop	SGD	Adam	Adagrad	RMS Prop	SGD
MSE	0.0084	0.0003	0.002	0.0328	0.0024	0.0025	0.0101	0.0883
MAE	0.0777	0.0128	0.0341	0.1251	0.1018	0.0358	0.0827	0.2405
Cosine proximity	−0.997	−0.997	−0.997	−0.997	−1	−1	−1	−1

5 Conclusion and Future Work

The LSTM network helped in predicting the end-of-the day stock price pretty accurately. Out of the four optimizers which were analyzed, Adam was the one which showed better results and outperformed the other optimizers. Future scope of this model would include analyzing the effectiveness of the model by modifying the training and testing ratios.

At its core, the stock market is a reflection of human emotions. Another potential extension of this stock prediction system would be to enhance it with a news feed analysis from any social media such as Twitter and Facebook where sentiments or emotions are measured and analyzed from the posts. These posts are a form of social sentiment indicator. As the popularity of social media increases, several people look at social media as a platform to voice their opinions and views on different topics.

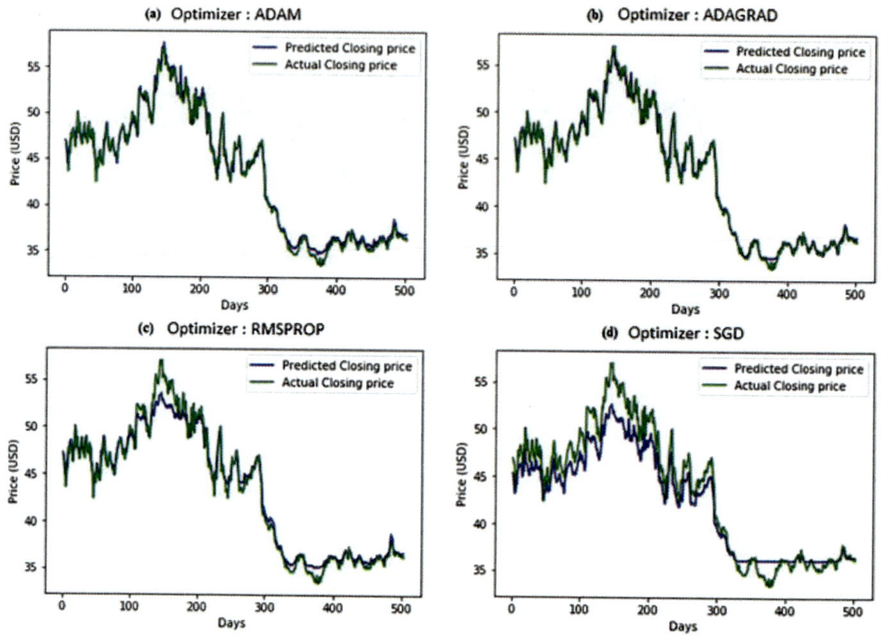

Fig. 6 Prediction graphs of $INTC with different optimizers

Table 5 $INTC—accuracy rates of the model

Optimizer	Train scores				Test scores			
	Adam	Adagrad	RMS Prop	SGD	Adam	Adagrad	RMS Prop	SGD
MSE	0.0007	0.0006	0.0021	0.0087	0.0006	0.0002	0.0006	0.0018
MAE	0.0208	0.0191	0.0033	0.0819	0.0198	0.0104	0.0193	0.0284
Cosine proximity	−0.994	−0.994	−0.994	−0.994	−0.7351	−0.7748	−0.735	−0.735

Individuals frequently post about organization's product, their business, services, etc. Such opinions provided by billions of users on a daily basis can be used as an indicator of consumer's outlook on various brands. Therefore, sentiment of the tweets can be analyzed to get a clearer and more accurate prediction of the stock prices.

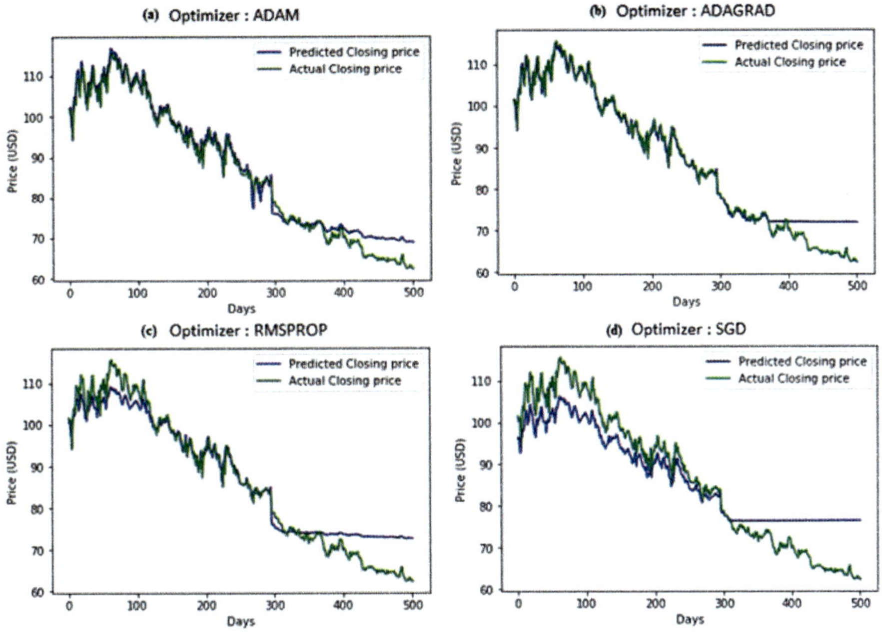

Fig. 7 Prediction graphs of $MSFT with different optimizers

Table 6 $MSFT—accuracy rates of the model

Optimizer	Train scores				Test scores			
	Adam	Adagrad	RMS Prop	SGD	Adam	Adagrad	RMS Prop	SGD
MSE	0.0011	0.0005	0.003	0.012	0.0083	0.0162	0.022	0.047
MAE	0.0255	0.0163	0.038	0.095	0.0801	0.1067	0.133	0.203
Cosine proximity	−0.997	−0.997	−0.997	−0.997	−0.47	0.6689	0.669	0.669

References

1. E. Soulas, D. Shasha, Online machine learning algorithms for currency exchange prediction. Tech. Rep. **31**, 1–13 (2013)
2. G.E.P. Box et al., *Time series analysis forecasting and control* (Wiley, New York, 2008)
3. Y. LeCun et al., Deep learning. Nature **521**(7553), 436 (2015)
4. Q. Zhuge, et al., LSTM neural network with emotional analysis for prediction of stock price. Eng. Lett. **25**(2), 167–175 (2017)
5. C. Zhu, J. Yin, Q. Li, A stock decision support system based on DBNs. J. Comput. Inform. Syst. **10**(2), 883–893 (2014)
6. Apple Consolidated Financial Statements. www.apple.com/newsroom/pdfs/Q4-FY18-Consolidated-Financial-Statements.pdf. Last accessed 20 Oct 2019
7. Cisco 2018 Annual Report. www.cisco.com/c/dam/en_us/about/annual-report/2018-annual-report-full.pdf. 20 Oct 2019

8. IBM Annual Report 2018. www.ibm.com/annualreport/assets/downloads/IBM_Annual_Report_2018.pdf. Last accessed 20 Oct 2019
9. Intel Annual Report 2018. https://s21.q4cdn.com/600692695/files/doc_financials/2018/Annual/Intel-2018-Annual-Report_INTC.pdf. Last accessed 20 Oct 2019
10. Microsoft Annual Report 2018. https://www.microsoft.com/en-us/Investor/earnings/FY-2019-Q4/press-release-webcast. Last accessed 20 Oct 2019

Identifying Exoplanets Using Deep Learning and Predicting Their Likelihood of Habitability

Somil Mathur, Sujith Sizon, and Nilesh Goel

1 Introduction

Kepler Space Mission, originally planned for a 3.5 years mission duration, went onto provide meaningful data and fueled stellar discoveries for just under ten years. Named after astronomer Johannes Kepler, Kepler Space Telescope was launched into an Earth-trailing heliocentric orbit and was designed to survey and discover a region of our galaxy for signs of habitable exoplanets and provide estimates about the amount of such planets in our galaxy. Kepler Space Telescope observed over 155,000 stars photometrically with incredible precision and is responsible for discovering nearly 2300 planets along with a couple of thousands planet candidates as well (Fig. 1) [1].

Most of the first list of planets were discovered using study of threshold crossing events (TCEs) and were compiled heterogeneously. This is also known as the transit method—analyzing transit signals around a star to detect periodic variations of flux or brightness. Next step was to classify whether the variation in these signals was due to a transiting exoplanet, Astrophysical False Positives (AFPs) or instrument malfunctions.

In the next section, we will outline the different ML approaches previously used for vetting of exoplanets and to predict the habitability of potential candidates.

S. Mathur (✉) · S. Sizon · N. Goel
Birla Institute of Technology & Science, Pilani, Dubai Campus, Dubai International Academic City, Dubai 345055, UAE
e-mail: somilmathur@outlook.com; f20160271@dubai.bits-pilani.ac.in

S. Patnaik et al. (eds.), *Advances in Machine Learning and Computational Intelligence*, Algorithms for Intelligent Systems, https://doi.org/10.1007/978-981-15-5243-4_34

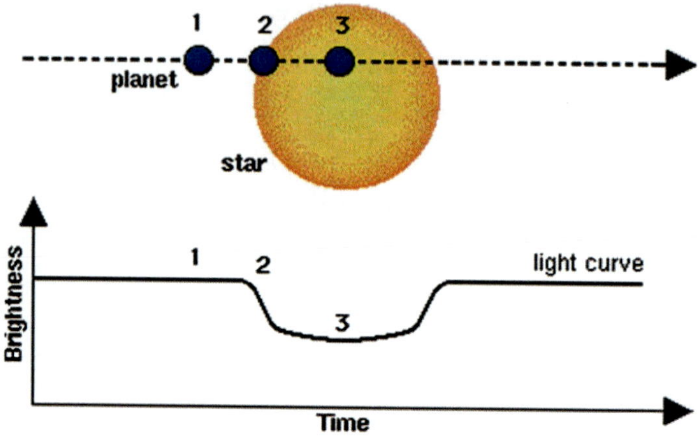

Fig. 1 Transit method

2 Similar Studies in the Past

2.1 Machine Learning Approaches

Classification proved to be a particularly challenging task as there were a lot of stellar objects that have comparable radii to that of planets [2]. And so, detecting false positives had to be done manually. However, as the mission matured and list of candidate planets increased, there was a shifting focus toward large-scale studies [3]. A project known as Robovetter was proposed that used a decision tree design to mimic the manual procedure of humans to remove false positive signals. Robovetter was used considerably and was able to produce entire catalogs of exoplanets without any human intervention.

Soon, many others began to experiment with machine learning to train a model to correctly identify exoplanets with high accuracy. Autovetter [4] was one such project that used a random forest model to consign exoplanets. Neural networks have also been used for studies related to planetary science and atmospheric classification.

2.2 Habitability of Exoplanets

Amongst astronomers and biologists alike, it is widely accepted that one of the key features for the evolution of life is liquid water [5]. This means that understanding the thermal characteristics of a planet becomes crucial, and for any exoplanet to harbor life, it should be located at an optimal distance from its parent star. This range of optimal distance is colloquially termed as The Goldilocks' Zone. Goldilocks' Zone for any star depends on its size, age, radiation and density. However, an estimated range for an average star is between 0.95 and 1.96 AU [6]. AU stands for astronomical units and is the distance between Earth and Sun. In the late twentieth century, Kasting [7] proposed a model to calculate the habitable zone for the first time.

There has been little work carried out in terms of using machine learning to classify and predict habitability of exoplanets. Therefore, using above literature, it is possible to build a meaningful classification model to predict the habitability of planets. Through this paper, we aim to perform this classification and categorize them into mesoplanets, psychroplanets and non-habitable planets.

3 Methodology

The implementation of our model follows a knowledge discovery in databases (KDD) approach. This approach follows a typical outline namely data cleaning, data selection, data transformation, pattern evaluation and knowledge representation. KDD is essentially an iterative process, here evaluation measures can be enhanced, mining can be refined, and new data can be integrated and transformed in order to get different and more appropriate results.

3.1 Method for Classification of Exoplanets Using Transit Signals

The first part of our proposed study is the implementation of a deep learning model to identify and classify exoplanets using properties of their transit signals. This is done using a convolutional neural network.

Before feeding our model with data, it is imperative we conduct a preliminary step of data preprocessing. This is because the quality of data and meaningful information that is derived from this step has a direct impact on the capability of our model to learn. We carry out this step in the following stages:

1. *SMOTE*—Synthetic minority Oversampling technique is required to correct the imbalance in data. This is done to reduce the overreliance on majority class values
2. *Normalization*—In order to compare related values, it is important to have a uniform scale of values. There are many features in the given input set with different ranges and normalization helps to bring them to a common scale.
3. *Smoothing*—For this stage, we make use of Savitsky Golay filter and Gaussian filter. Both filters provide robustness even in a noisy environment. By averaging out adjacent datapoints, we increase the precision of the data without distorting the signal tendency.
4. *Standardization*—Standardizing the dataset makes the training process more well behaved as the numerical conditions of the optimization problems are improved (Fig. 2).

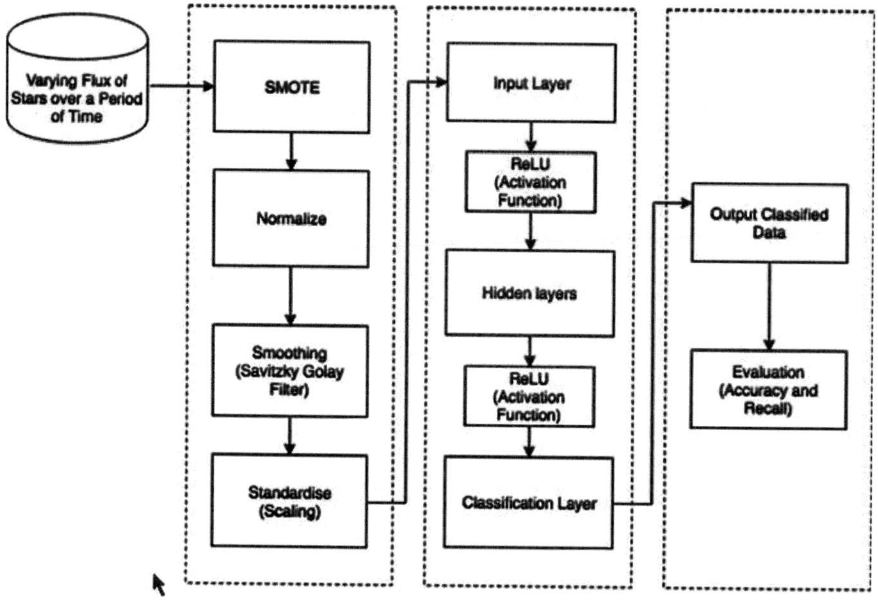

Fig. 2 Flow diagram for exoplanet classification process

3.2 Method for Predicting Habitability of Exoplanets

The dataset for this task has been obtained from the Planetary Habitability Laboratory's (PHL) exoplanet catalog [8]. To conduct this task, little preprocessing is required. The dataset is already cleaned but requires categorization of serial data.

This model is developed using KNN, random forest and SVM. The model classifies the exoplanet into the following:

1. *Psychroplanet*—'Psychro' means cold and psychroplanets are planets that have relatively low surface temperatures (−50 to 0 °C). While it is considered a habitable planet, it is unsuitable for most terrestrial life. Most famous example of a psychroplanet is Mars [9].
2. *Mesoplanets*—Mesoplanets have a size between Mercury and Ceres [10]. It is also considered a habitable planet is usually suitable for terrestrial life. They have thermal temperatures ranging from 0 to 50 °C.
3. *Non-habitable planet*—These are all the planets that do not fall into the above two categories. They are considered to be completely inhabitable. Gas giants fall into this category be default.

4 Implementation

4.1 Deep Learning Model for Classification of Exoplanets

Moving forward following the preprocessing steps detailed in Sect. 3.1, we are ready to feed our data into our CNN. We acquired our training set of labeled TCEs form the Autovetter Planet Candidate Catalog which is available on the NASA Exoplanet Archive.

We have used a one-dimensional convolutional neural network along with a max pooling layer. All the hidden layers in the architecture use ReLU. ReLU is a linear rectifier activation function that helps to eliminate negative values and provides a linear graph. For our output layer, we use a sigmoid function. Sigmoid function is the most common activation function in CNNs. The sigmoid function gives an 'S' shaped curve and has a finite limit. We use this function because we do not want our predictor to simply classify the exoplanets as habitable or not. We would also like to predict the probability of each outcome.

Hyperparameters. The hyperparameters used for our model are chosen in order to optimize the values of the performance metrics. The learning rate is taken as 0.0015, epoch value is 150, and the batch size is 64 (due to the high number of input dataset and low standard deviation). There is no drop-out rate.

Performance Metrics. Since we cannot accurately judge the requirement of the number of hidden layers to be used, we assign this number based on trial and error. This also helps us to maximize the efficiency of our performance metrics. We use the following metrics for this basis:

1. *Accuracy*—Ratio of right classifications
2. *Recall*—Ratio of true planets with correct classification (completeness)
3. *Precision*—Ratio of TCEs classified as exoplanets that actually are planets (reliability)
4. *Area Under Receiver (AUC)*—This metric provides us with random value predictions. It is the probability that an arbitrarily chosen planet has a higher predicted score than an arbitrarily chosen AFP.

Optimizer. In order to minimize the cost function of an algorithm, we run multiple iterations with different weights and biases associated with the neurons. This is called gradient descent optimization. There are multiple available gradient descent algorithms; for our model, we make use of Adam Optimizer (Adaptive Moment Estimation). Adam is a stochastic gradient descent algorithm that uses the squared gradients to scale the learning rate and takes advantage of momentum by using moving average of the gradient instead of gradient itself. This helps in minimizing the cross-entropy cost function over the training data.

Along with the above optimization, we have also improved our dataset by applying random horizontal reflections to the TCEs during training.

4.2 *Prediction of Exoplanet Habitability Probability*

As mentioned in Sect. 3.2, this task required little preprocessing of data. One important requirement was to categorize all the serial data using data exploration techniques. For the probability estimation regarding the habitability of exoplanets, we use Naïve Bayes algorithm. In order to obtain meaningful estimation from this algorithm, it is important to provide categorized data with defined splits between different types of fields. Given dataset contains fields such as radius of planet, mass, density and thermal temperature. This serial data is converted into categorized data by examining the distribution of data and then passed to Naïve Bayes algorithm.

Probability estimation using Naïve Bayes Algorithm. Above-obtained dataset is now randomly split into training data and test data. We have split it in 80:20 ratio, respectively. The model is then trained using the 'P_Habitable' predictor with 1 relating to habitable and 0 relating to non-habitable. We then test the model and evaluate the model's performance using testing data.

The evaluation of the model and the subsequent insights are discussed in further sections.

5 Evaluation

5.1 *Evaluation of Convolutional Neural Network*

The classification of exoplanets based on the TCEs and light curves using the deep learning model is evaluated in three different scenarios—without preprocessing, with Savitsky Golay filter and with Gaussian filter. Following we will discuss all the three cases (Table 1; Figs. 3, 4 and 5).

Table 1 Evaluation of exoplanet classification with and without preprocessing

	Without preprocessing	Savitsky Golay filter	Gaussian filter
Accuracy (training)	0.9746	0.9134	0.7452
Accuracy (testing)	0.8912	0.8899	0.7183
Set Error (training)	0.0012	0.0012	0.0919
Set Error (testing)	0.0054	0.0054	0.124
Precision (training)	0.0	0.0	0.017
Precision (testing)	0.0	0.0	0.029
Recall (training)	0.0	0.0	1
Recall (testing)	0.0	0.0	1

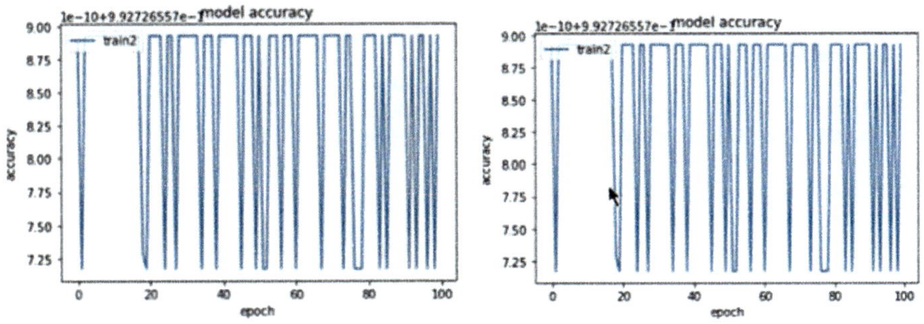

Fig. 3 Model accuracy and model loss without preprocessing

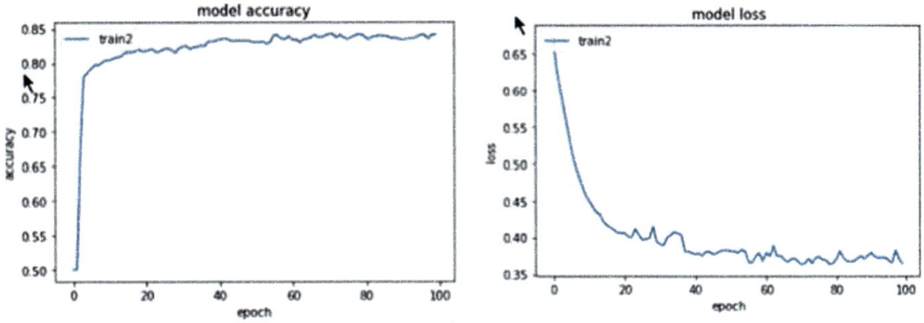

Fig. 4 Model accuracy and model loss with Savitsky Golay filter

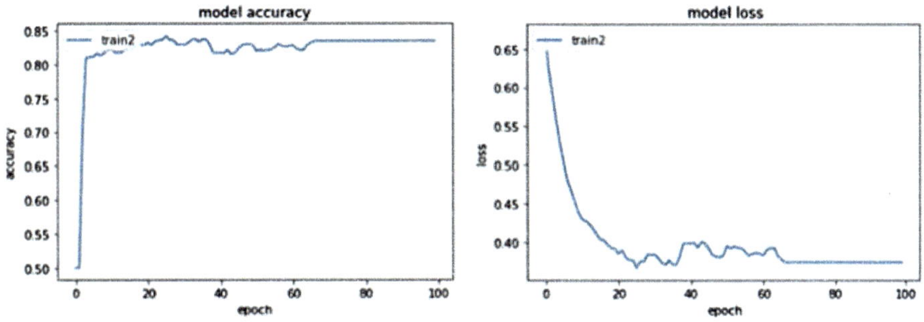

Fig. 5 Model accuracy and model loss with Gaussian filter

5.2 Evaluation of Probability of Exoplanet Habitability

As previously outlined, we have used Naïve Bayes algorithm for this task. The results are described below based on the various characteristics of exoplanets and the parent star (Fig. 6).

Based on the above findings, the following information can be inferred:

```
$P_ZoneClass
   P_ZoneClass
Y            Cold            Hot           Warm
  0 0.005102040816 0.962018140590 0.032879818594
  1 0.000000000000 0.000000000000 1.000000000000

$P_MassClass
   P_MassClass
Y          Jovian      Neptunian      Subterran    Superterran         Terran
  0 0.218253968254 0.260770975057 0.022108843537 0.295351473923 0.203514739229
  1 0.000000000000 0.000000000000 0.009793253536 0.290533188248 0.699673558215

$P_CompositionClass
   P_CompositionClass
Y             gas           iron     rocky iron    rocky water      water gas
  0 0.483560090703 0.003968253968 0.488662131519 0.015306122449 0.008503401361
  1 0.000000000000 0.000000000000 0.989118607182 0.010881392818 0.000000000000

$P_AtmosphereClass
   P_AtmosphereClass
Y   hydrogen rich   metals rich no atmosphere
  0 0.14795918367 0.75907029478 0.09297052154
  1 0.15669205658 0.84330794342 0.00000000000

$P_Mass.EU.
   P_Mass.EU.
Y          0 to 1        1 to 5       5 to 10      10 to 20      Above 20
  0 0.05045351474 0.30272108844 0.17290249433 0.16893424036 0.30498866213
  1 0.08269858542 0.63329706202 0.17736670294 0.09140369967 0.01523394995

$P_Radius.EU.
   P_Radius.EU.
Y          0 to 1      1 to 2.5     Above 2.5
  0 0.04421768707 0.46088435374 0.49489795918
  1 0.06964091404 0.93035908596 0.00000000000
```

Fig. 6 Habitability probabilities I

- *P_ZoneClass* relates to the area around the parent star. We can clearly see that the likelihood of an exoplanet being habitable is the highest in the WARM region, whereas it is significantly lower in the COLD and HOT regions.
- *P_MassClass* relates to the approximate mass of the exoplanet in question. We see here that the Terran class has the highest probability for habitability with nearly 70%. Second highest class is Superterran with 29% probability of habitability.
- *P_CompositionClass* relates to the composition of the planets. We can infer here that the requirement for habitability strikingly correlates with rocky iron type composition of the planet's interiors.
- *P_AtmosphereClass* relates to the atmospheric composition of the exoplanet. We can see that planets having metals rich atmosphere are most likely to be habitable.

- *P_Mass* provides information about the mass of the planet in units of Earth's mass. (EU). The highest probability for planet to be habitable is when the planetary mass is between 1 to 5 EU. In other cases, the probability is low.
- *P_Radius* relates to the radius of the exoplanet. Radius between 1 and 2.5 times Earth radii has the highest probability for habitability.

5.3 Evaluation of Classification Model for Exoplanet Habitability

In this section, we will discuss the results from the three classification models for classifying type of exoplanet namely KNN, SVM and random forest (Tables 2 and 3).

KNN. This model yields accuracy of nearly 88%. Again, since *p*-value in this model is less than 0.05, the model is statistically significant. Kappa value is given to be almost 76%, and the no information rate is lesser than the accuracy.

Table 2 Results of KNN, SVM, random forest model classification

	Prediction	Reference		
		Mesoplanet	Non-habitable	Psychroplanet
KNN	Mesoplanet	19	0	13
	Non-habitable	0	583	1
	Psychroplanet	84	8	185
SVM	Mesoplanet	1	0	0
	Non-habitable	0	585	1
	Psychroplanet	102	6	198
Random forest	Mesoplanet	47	0	43
	Non-habitable	0	710	1
	Psychroplanet	69	0	202

Table 3 Overall statistics for KNN, SVM, random forest models

	KNN	SVM	Random forest
Accuracy	0.881299	0.8779395	0.8945896
95% CI	(0.8582585, 0.9017868)	(0.8546457, 0.8986952)	(0.8746451, 0.9123356)
No Information rate	0.6618141	0.6618141	0.6623134
P-value [Acc > NIR]	<0.0000000000002204	<0.0000000000002204	<0.0000000000002204

SVM. This model yields accuracy rate of around 88%. The Kappa value is similar to that of the model presented by KNN at approximately 75%. As expected, the no information rate is lesser than the accuracy rate.

Random Forest. The accuracy obtained from this model is nearly 90%, and it is also interesting to note that the p-value is less than 0.05 which indicates that the model is statistically significant. Other metrics can also be observed from the data given above.

6 Discussion

Through this paper, we have implemented various machine learning models to achieve more accurate results in classifying exoplanets and predicting the probabilities of their habitability. The deep learning model presented in this paper was tested with three different approaches namely without preprocessing, with Savitsky Golay filer and with Gaussian filter. Our results show that by using these filters and performing normalization and standardization, we could improve the performance of the model. The recall on using Gaussian filter was 1, while with Savitsky Golay filter, it was 0.3.

We then performed prediction analysis using Naïve Bayes algorithm to determine the probabilities of habitability of exoplanets. Our results in this analysis accord well with our expectations and were supported by the given literature.

Lastly, we built three classification models to categorize the type of exoplanets into mesoplanets, psychroplanets and non-habitable planets. This area of research has not been previously explored in depth and could yield useful insights to future researchers studying the type of planets found in interstellar star systems.

References

1. M. Johnson, in *How Many Exoplanets Has Kepler Discovered?* (2015). URL: https://www.nasa.gov/kepler/discoveries
2. K. Rice, The detection and characterization of extrasolar planets. Challenges **5**(2), 296–323 (2014)
3. F. Fressin, G. Torres, D. Charbonneau, et al. ApJ. **766**, 81 (2013). D. Foreman-Mackey, T.D. Morton, D.W. Hogg, E. Agol, B. Schölkopf, AJ. **152**, 206 (2016)
4. S.D. McCauliff, J.M. Jenkins, J. Catanzarite et al., ApJ **806**, 6 (2015)
5. S. Seager, Exoplanet habitability. Science **340**(6132), 577–581 (2013). URL: http://science.sciencemag.org/content/340/6132/577
6. R.K. Kopparapu, R. Ramirez, J.F. Kasting, V. Eymet, T.D. Robinson, S. Mahadevan, R.C. Terrien, S. Domagal-Goldman, V. Meadows, R. Deshpande, Habitable zones around main-sequence stars: new estimates. Astrophys. J. **765**(2), 131 (2013)
7. J.F. Kasting, D.P. Whitmire, R.T. Reynolds, Habitable zones around main sequence stars. Icarus **101**(1), 108–128 (1993)
8. Planetary Habitability Laboratory, in *HEC: Description of Methods Used in the Catalog*. URL: http://phl.upr.edu/projects/habitable-exoplanets-catalog/methods

9. P.B. Price, A habitat for psychrophiles in deep antarctic ice. Proc. Nat. Acad. Sci. **97**(3), 1247–1251 (2000). URL: http://www.pnas.org/content/97/3/1247
10. *A Thermal Planetary Habitability Classification for Exoplanets* (n.d.). URL: http://phl.upr.edu/library/notes/athermalplanetaryhabitabilityclassificationforexoplanets

Use of Artificial Intelligence for Health Insurance Claims Automation

Jaskanwar Singh and Siddhaling Urolagin

1 Introduction

The amount of significant information in this world is way more than a human mind can process, yet we need human intelligence to logic and reason every workflow. Machine learning is designed to solve that crisis, and it takes the speed and computation power of a machine and intelligence of the human mind. Combined these two are unbeatable. The healthcare industry is a major, and one of the most important parts of the world is consistently evolving as artificial intelligence and becomes significant in today's digital era. Recently, AI advancements have introduced numerous developments in the field of health care which have not only helped reduce the amount of money spent in the industry but are also time-efficient, ultimately leading to saving lives or serving better.

Every year, an estimate of $30 billion is lost to fraudulent health insurance claims [1]; as health insurance is viewed as a major component of the healthcare industry, it becomes an easy target for criminals. Healthcare industry is an integral component of every human's life and must be made affordable. Automating the health insurance platforms, in particular, will significantly contribute to the field of AI technology. Further, the automation will help the industry save millions by proactive management, improved case management, targeted investigation, and fast settlement [1]. The aim of this research is to comprehensively look for an efficient method to automate the health insurance claim approvals than the ones in existence to improve operational efficiency. The study will pay focused attention to the ongoing insurance mechanisms and components of customer requirements, the process of judging a

J. Singh · S. Urolagin (✉)
Birla Institute of Technology & Science, Pilani, Dubai Campus, Dubai International Academic City, Dubai 345055, UAE
e-mail: siddhaling@dubai.bits-pilani.ac.in

J. Singh
e-mail: f20160075@dubai.bits-pilani.ac.in

S. Patnaik et al. (eds.), *Advances in Machine Learning and Computational Intelligence*, Algorithms for Intelligent Systems, https://doi.org/10.1007/978-981-15-5243-4_35

claim, and finally analyzing the factors and patterns leading to an outcome to use them further in constituting an automated system that will help humans do better and faster while reducing the settlement time as machine learning algorithms are superior to manual traditional predictive models (they recognize patterns in unstructured and semi-structured data as well). This system will provide timely medical assistance. In sect. 2 related work is discussed, the preprocessing, feature selection, classification methods and metric are elaborated in sect. 3, Experimental results are covered in sect 4 followed by conclusion given in sect. 5.

2 Related Work

Ever since the beginning of artificial intelligence, the goal has always been to make the human life easier and with that mindset, and AI has grown for decades in every industry and direction to finally reach a point where it turns everything it touches to gold. Now as we discuss the past and growth of AI, here's some similar work is done in the past to advance and achieve new levels in the field of health care as people in the past era faced major challenges in meeting the right perspective of their patients in a timely manner and in the provision of accurate services that can best fit patients' needs and requirements. In 2019, Johnson and Khoshgoftaar studied the impact of different sampling techniques in machine learning and used deep neural networks as the baseline algorithm in the case of an imbalanced class [2]. Duman and Sağıroğlu, in 2017, in their paper explained about the types of health insurance fraud and compare and analyze the methods previously used to detect that misuse of the insurance [3]. Bauder and Khoshgoftaar, in 2017, conducted a comparison study among supervised, unsupervised and hybrid ML approaches to figure out the best one for fraud detection [4]. Bauder et al. in 2016, also addressed the misuse of medical insurance and constructed a system to avoid it. They used naive Bayes algorithm and gaining an *F*-score over 0.9 [5]. In 2017, they tested their model on real-world fraudulent cases by using three strategies: feature selection, grouping, and removing overlapping specialties [6]. In 2012, Kirlidog and Asuk gathered healthcare claims data from a Turkish insurance company and used machine learning techniques to detect patterns and extract subsequent knowledge to bring out the anomalous cases for further investigation [7]. In 2012, Shin et al. used a scoring system to draw out the abusive utilization of medical insurance based on extracted information from electronic medical insurance claims. They even quantified the degree of abuse along with the categorization of problematic providers [8]. M. Kumar et.al [9] have developed a system to predict health insurance claims using machine learning. The application of machine learning in medicine is elaborated in [10]. A system is developed for detecting insurance fraud in [11]. The application of data mining techniques are used to detect fraud claims [12] and data mining has been applied for healthcare [13]. A database derived from multiple insurers is used in [14] to find fraud claims.

Table 1 Distribution of records in dataset

Total number of claim records	Number of accepted claim records	Number of rejected claim records
8808	7481	1327

3 Methodology

3.1 Dataset

The dataset used for this study is closer to real-life health insurance pseudo claims data generated in scientific way. It has a detailed description of the patients along with the specifications of the claim represented in 60 columns and over 8000 rows. The target column is STATUS which has three possible outcomes: paid, processed, or pending. Here we consider paid and processed as a claim accepted and pending status as a claim rejected. Other relevant information about the dataset is as follows. Table 1 clearly shows the imbalanced class situation of the dataset.

3.2 Preprocessing Data

This involved feature reduction where any and all features that have no correlation with the target label will be removed under bases, such as *empty or constant value columns* (discarded because they are irrelevant and only take up computation power without having any real effect toward the outcome*), duplicate rows* (considered noise in machine learning), *null values* (dropped because they hinder efficiency).

Once all the useless columns and rows are removed, we encode the object columns using a label encoder.

The remaining rows are split for training and testing purposes. The ratio for the split is 75:25; i.e., 75% of the considered dataset is used for training and the remaining 25% is used for testing.

We use synthetic minority over-sampling technique (SMOTE) which creates synthetic (not duplex) samples of the class in minority in order to get the number of records of minor class equal to the number of records of the major class.

3.3 Feature Selection

When we are done applying feature reduction techniques on the dataset, all the irrelevant features and columns are removed, and a very small ratio of relevant features that have any correlation (be it negative or positive) are taken further in the process. Out of 60 columns in the given dataset, only 8 are taken further to train

Table 2 Description of selected features

Master contract	Name of the client (employer)
Branch	The branch office/location of the client
Gender	The gender of the patient (employee) making the claim
ICD chapter	ICD stands for international classification of diseases. It is a system provided by the World Health Organization used by health professionals to classify and record the diagnosis
Claim type	Type of claim made
Total amount claimed	The total amount claimed by the person making the claim
Total payable	The total amount payable by the insurance
Status	The target label that classifies the approval or rejection of the claim. The possible outcomes of this label are paid, processed, and pending. Paid and processed are considered to be approved, and pending is taken as rejected

the model, the rest were either constant throughout the dataset and had too many unique values to be useful or had too many null values (empty) to be considered. The columns that did take part in the model training are listening and described in Table 2.

3.4 Classification Methods

K-nearest neighbors: KNN, a simple and efficient machine learning algorithm, works on the assumption that similar objects lie close, i.e., in terms of machine learning, similar objects are neighbors. The K in the name indicates the number of closest neighbors to be considered while deciding the class/category of the object under examination and is provided as a parameter in the implementation phase. Once K has a value, we find out those K-nearest neighbors and take a vote. The class with major votes wins and is considered as the final outcome.

When applied on our dataset, the model boosted the performance and provided impressive results. Some of the important formulae used in the computation of the outcome are listed below:

1. Distance metrics:

$$D(x,p) = \sqrt{\sum_{i}^{n}(x_i - p_i)^2} \tag{1}$$

   where x is the query data record and p is the current record from dataset.
2. Distance weighting:

$$W(x, p_i) = \frac{\exp(-D(x, p_i))}{\sum_{i=1}^{k} \exp(-D(x, p_i))} \tag{2}$$

where $D(x, p_i)$ is the distance, x is the query record and p_i is the ith record.

3. Error:

$$\text{Error} = \pm \sqrt{\frac{1}{k-1} \sum_{i=1}^{k} (y - y_i)^2} \tag{3}$$

Artificial Neural Network: The network of this deep learning algorithm is built on neurons forming multiple layers of mathematical computations each for a different task. The most basic neural network consists of three layers, namely input, hidden, and output layer. The input layer takes in information processes it in a way that becomes handy for the output layer, and then output layer responds to the information fed and processed. Application of neural network on the insurance claims dataset provided us with good results but not as good as compared to KNN.

Mathematical equations used in the algorithm are as follows:

Sigmoid Activation Formula:

$$f(x) = \frac{1}{\left(1 + \mathrm{e}^{(-1*z)}\right)} \tag{4}$$

Random Forest: Random forest, an improvised version of decision trees, like the name suggests a collection of tree models or decision tree models to be exact. The entire model of the random forest operates on the working of individual decision trees working as an ensemble where each decision tree comes up with their own set of rules to provide an outcome (given a training dataset and a target). Once the building blocks of the forest are implemented, the rest is motivated by wisdom of crowd; i.e., the outcomes of each classifier are taken into consideration and the majority label wins. When applied on the given dataset with these parameters, the model gave us average results as compared to KNN, ANN, and SVM. The following are the formulas used for computation of node impurity:

$$\sum_{i=1}^{c} fi(1 - fi) \quad \sum_{i=1}^{c} -fi \log(fi) \tag{5}$$

Support Vector Machine: Support vector machine algorithm, a discriminative, hyperplane separating classifier, can be described as a linear classification, model that caters both linear and non-linear data problems. It operates on the idea of a line or a hyperplane that distinguishes between the labeled class. At first, it attempts to plot every record reserved for training on a n-dimensional space (n being the number of unique classes in target label) and then draws a line to divide them in sort of clusters. The line or hyperplane can also be considered as a boundary for each class; the incoming record is classified where it fits best on the line.

Application of SVM on Dataset: The results turned out to be better than random forest but not as good as KNN and ANN. Some important mathematical formulas used in the computations are as follows:

$$\text{Length of Vector} = \sqrt{x_1^2 + x_2^2 + x_3^2} \tag{6}$$

$$\text{Direction of vector} = \left\{ \frac{x_1}{|X|}, \frac{x_2}{|X|}, \frac{x_3}{|X|} \right\} \tag{7}$$

$$\text{Equation of straight line} = y = ax + b \tag{8}$$

3.5 Accuracy/Performance Metrics

The performance metrics used in this research confusion matrix, precision, recall, $F1$ score, and ROC score. But since the dataset is imbalanced, we will majorly focus on precision, recall, and $F1$ score rather than ROC curve as sometimes it can be misleading

Precision: The formula to compute precision is as follows:

$$\text{Precision} = \frac{\text{tp}}{\text{tp} + \text{fp}} \tag{9}$$

where tp = true positive and fp = false positive.

Recall: The formula to calculate recall is:

$$\text{Recall} = \frac{\text{tp}}{\text{tp} + \text{fn}} \tag{10}$$

where tp = true positive and fn = false negative.

F1 Score: the formula for F1 Score is:

$$F1 = 2 \times \frac{\text{Recall} \times \text{Precision}}{\text{Recall} \pm \text{Precision}} \tag{11}$$

4 Experimental Results

The experiment of this study holds two main objectives,

Table 3 Comparison of the 04 implemented algorithms

Classifier	$F1$-score	ROC	Accuracy
K-nearest neighbors	0.926	0.925	0.970
Artificial neural networks	0.96	0.925	0.9298
Random forest	0.95	0.897	0.9220
Support vector machine	0.96	0.91	0.9293

1. To automate the process of healthcare insurance claims settlement.
2. To find the best/optimal solution for objective no 1.

Therefore, we performed an experiment with four different implementations. In order to compare the results and find the best one, the data preprocessing and preparation were similar in each case, and the training models or classifiers were different. Once the experiments were performed, we tabulated the important values in Table 3.

The information presented in the table above provides us with a few things to observe, (i) The nature of dataset was quite distinctive and that is why out of all four classifiers, rule-based classification beat the other three in terms of accuracy, (ii) The size of the dataset did not allow ANN, a self-improving model (works best on bigger datasets) to perform that well even though its ROC score tried to match the one of KNN, (iii) Limited number of relevant features gave random forest a very deficient opportunity to expand in terms of quantity of trees constructed and that resulted in average performance, (iv) Like every other model, SVM requires training phases as well which requires a bigger dataset, hence, not so best results.

Figures 1, 2, 3 and 4 illustrate the graphic representation of the confusion matrix of each classifiers, i.e., the count for each of the following: the number of records that

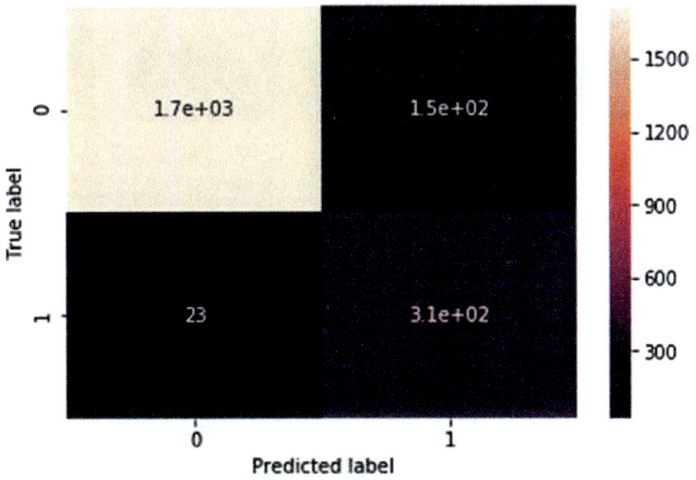

Fig. 1 Confusion matrix for KNN

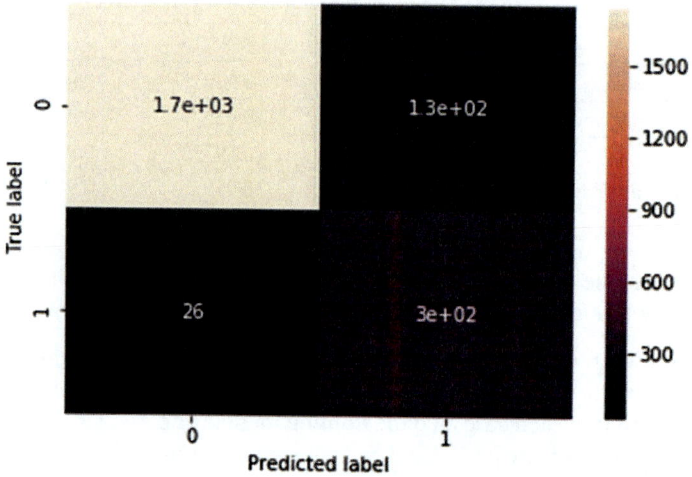

Fig. 2 Confusion matrix for ANN

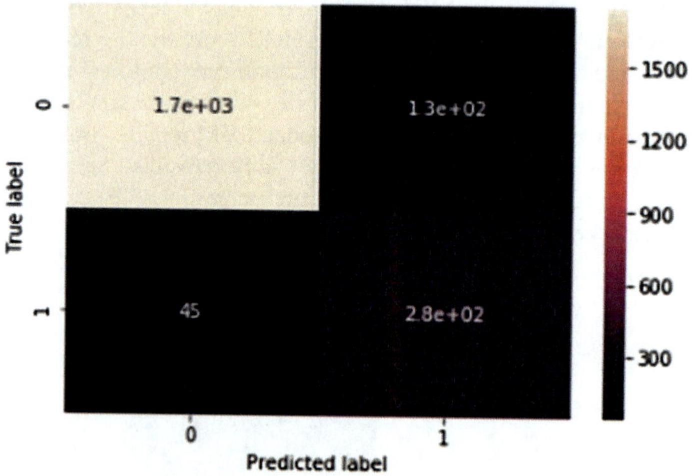

Fig. 3 Confusion matrix for random forest classifier

were originally true and were predicted false by the model; the number of records that were originally true and were predicted true by the classifier; the number of test records that were false originally and were predicted the opposite; and the number of records that were false in nature and were predicted correctly. These counts help us calculate precision, recall, $F1$-scores, and finally accuracy.

Figures 5, 6, 7 and 8 give us the graphical depiction of ROC curves built for each of the four implementations and their performance, since the score differs minor, the

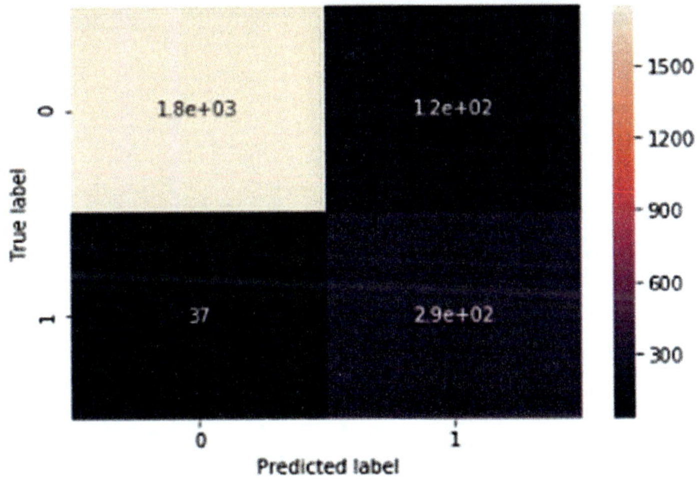

Fig. 4 Confusion matrix for SVM

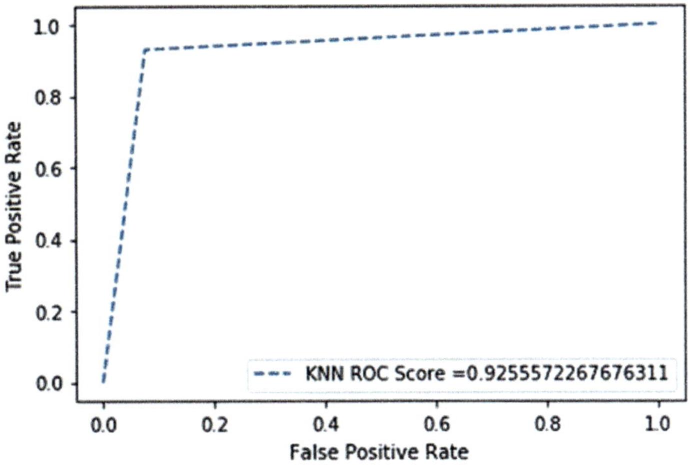

Fig. 5 ROC curve for KNN classifier

curves in the figures seem to be alike; but in truth, two of them (Figs. 5 and 6) show better results than the others.

5 Conclusion

In this study, we gathered that the performance of a model depends upon the nature of dataset be it the number of records, number of features or relevance to the targeted

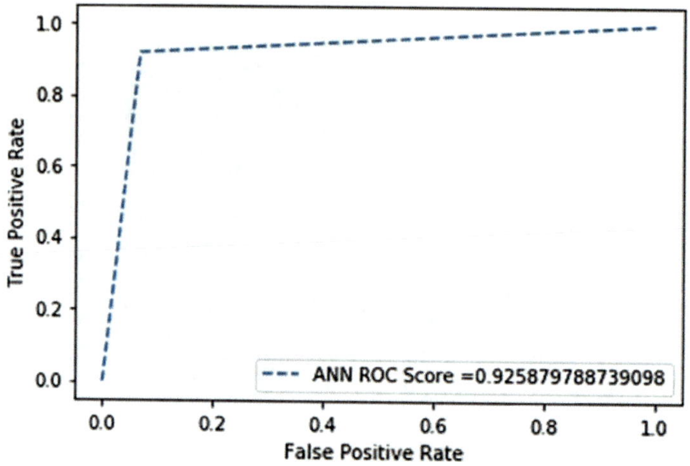

Fig. 6 ROC curve for ANN classifier

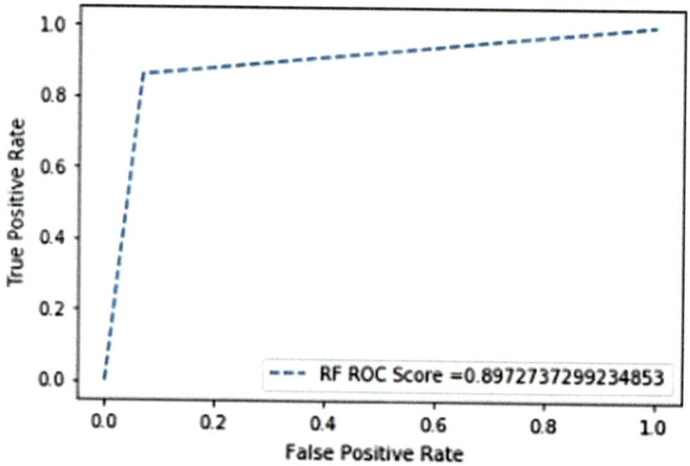

Fig. 7 ROC curve for RF classifier

goal. We constructed and executed four different implementations on the dataset in hope that it will automate the outcomes of the claims when provided with a record of features that are included in this study. During the experiment, we observed the results on predictive models such as KNN, ANN, Random forest and SVM, when compared among each other, KNN proved itself to be the superlative for this dataset. The development of all four of them establishes that machine learning (ML) and artificial intelligence (AI) are beginning to link critical technological tools to real-world problems of day to day that can effectively enable industry providers to become a better player and a performer in gaining customer satisfaction and operational

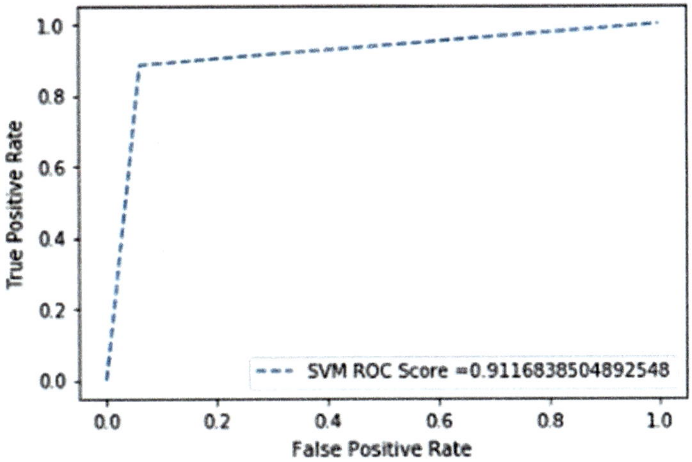

Fig. 8 ROC curve for SVM classifier

management along with sparing of time, labor, human error prevention and eventually saving lives and tons of money. Another pathway to consider in the conclusion of the study is toward the healthcare community, to develop a trustworthy relationship in the use of artificial intelligence and machine learning tools that will improve and increase as the AI progresses.

References

1. R. Malhotra, S. Sharma, *Machine Learning in Insurance* (2018)
2. J.M. Johnson, T.M. Khoshgoftaar, *Deep Learning and Data Sampling with Imbalanced Big Data* (2019)
3. E.A. Duman, S. Sağıroğlu, *Health Care Fraud Detection Methods and New Approaches* (2017)
4. R.A. Bauder, T.M. Khoshgoftaar, *Medicare Fraud Detection Using Machine Learning Methods* (2017)
5. R.A. Bauder, T.M. Khoshgoftaar, A. Richter, M. Herland, *Predicting Medical Provider Specialties to Detect Anomalous Insurance Claims* (2016)
6. R.A. Bauder, T.M. Khoshgoftaar, M. Herland, *Medical Provider Specialty Predictions for the Detection of Anomalous Medicare Insurance Claims* (2017)
7. M. Kirlidog, C. Asuk, *A Fraud Detection Approach with Data Mining in Health Insurance* (2012)
8. H. Shin, H. Park, J. Lee, W.C. Jhee, *A Scoring Model to Detect Abusive Billing Patterns in Health Insurance Claims*
9. M. Kumar, R. Ghani, Z.-S. Mei, *Data Mining to Predict and Prevent Errors in Health Insurance Claims Processing*, in Proceedings of the 16th ACM SIGKDD International Conference on Knowledge Discovery and Data Mining (2010)
10. Ultra-modern medicine: examples of machine learning in healthcare (2019)
11. System and method for detecting healthcare insurance fraud (2014)
12. V. Rawte, G. Anuradha, *Fraud Detection in Health Insurance Using Data Mining Techniques* (2015)

13. H.C. Koh, G. Tan, *Data Mining Applications in Healthcare* (2005)
14. M. Tyler, N. Basant, P. Robin, S. Rahman, *Healthcare Insurance Claim Fraud Detection Using Datasets Derived from Multiple Insurers* (2010)

Sentiment Analysis and Prediction of Point of Interest-Based Visitors' Review

Jeel Patel and Siddhaling Urolagin

1 Introduction

Tourism industry is connected with people travel to different parts of the world for enjoyment, social connection, and business purposes. Tourism is closely related to transportation, hotels, and hospitality. Tourists across the world express their opinion about their visits to various social media. Understanding the emotions and opinions expressed by tourist is highly required to improve the services of tourism-related industries. The natural language processing and text mining can be used to identify the keywords which are most significant to realize emotions expressed by visitors. Various researchers have highlighted their observation of statistical results from Internet-based information taking advantage of different suppositions for investigation techniques.

Kumari et al. [1] tested a model to discover the extremity of suppositions utilizing SVM. The exhibition of SVM has determined based on exactness, review f-measure, and the precision of 90.99% is accomplished. Wang et al. [2] proposed the origination of opinion processing and their essential issues. Further, the single-model calculation is utilized to arrange the substance as emotional or goal. The issue related to multimodal examination strategies is talked about.

Anjaria et al. [3] present the novel methodology of abusing the client impact factor to anticipate the result of a political decision result. Authors likewise propose a crossover approach of separating feeling utilizing immediate and circuitous highlights of Twitter information dependent on Support Vector Machines (SVM), Naive

J. Patel (✉) · S. Urolagin
Department of Computer Science, Birla Institute of Technology & Science, Pilani, Dubai Campus, Dubai, UAE
e-mail: jeeltpatel@gmail.com

S. Urolagin
e-mail: siddhaling@dubai.bits-pilani.ac.in

S. Patnaik et al. (eds.), *Advances in Machine Learning and Computational Intelligence*, Algorithms for Intelligent Systems, https://doi.org/10.1007/978-981-15-5243-4_36

Bayes, Maximum Entropy, and Artificial Neural Network-based administered classifiers. Shahnawaz et al. presented the procedure of conclusions examination in the research which has been used for understanding the inclination and sentiments from the information. It is additionally used to discover the frame of mind of the author toward the specific point whether it is positive, negative, or neutral. It has been recognized in [4] that the sentiment analysis of opinions expressed by user is important. In their work, various approaches for sentiment analysis are discussed along with open problems and issues. Deficiency exactness and powerlessness to perform well in various space and execution are the primary issues in the present systems.

Bhadane [5] centers around the different techniques utilized for arranging a content given and put the feelings communicated in it into the negative or positive notion. They actualized a lot of strategies for perspective order and extremity distinguishing proof of item audit utilizing AI (SVM) joined with area explicit vocabularies. Their outcomes show that the proposed systems have accomplished about 78%. Fernandes et al. [6] proposed a sentiment method, which gave the data of level of every extremity. The proposed work result can be used by the purchasers and merchants to build their efficiency. Arora et al. [7] proposed a technique to know the ubiquity of advanced cell brands and their working frameworks based on client surveys gathered from twitter.

Bali et al. [8] exhibited a way to deal with mine client's assessments. What's more, the outcome got from the examination is used by vender to create retailing systems. Bhargava [9] investigated the utilization of sentiment analysis has been expanded pointedly. They bolstered the English vernacular which assumes a noteworthy job for the investigation in this field. In this paper, they proposed a technique by which different dialects can be separated by any for the investigating assumptions and can likewise perform notion examination.

Patil [10] et al. discussed that preprocessing strategy ought to be viewed as exceptionally useful in the content mining technique. They concentrated on stop word evacuation and slang words to expel unique characters. A content with elements is finished by utilizing Stanford NER and CRF. NER model isolates the information in classification name, location, organization. In order to arrange the feelings of the tweets into the Positive, Negative and Neutral, Tyagi [11] proposed an upgraded information lexicon that will perform information pre-handling tasks. The information gathered is of three prevalent portable brands from the period of March to June that are rough 60,000 tweets and those tweets are put away into the content record.

Jandailet al. [12] proposed one approach for dissecting the survey for the client supposition and moreover giving insight concerning the various levels, issues, applications, and AI strategies of conclusion mining. In our research work, the reviews from visitors have been collected from Internet and the sentiment analysis of reviews is carried out. The sentiment classification based on TF-IDF features techniques along with classifiers such as SVM and Naïve Bayes is developed.

Rest of the paper is sorted out as pursues: In Sect. 2, the model for sentiment classification is described; In Sect. 3, experimental results are given; the conclusion is covered in Sect. 4.

2 Sentiment Classification of Tourism Data

The sentiment analysis and prediction of point of interest-based reviews have been carried out as shown in Fig. 1. The tourism data is collected from Internet sources such as tour-pedia.org. The tour-pedia.org contains mainly two types of information about tourism (i) places and (ii) reviews about the places. This information collected from various sources of the Internet such as Facebook, web site about booking, Foursquare, and other sites. The dataset consists of information about various locations such as Dubai, Berlin, Rome, etc. The review data consists of attributes as given in Table 1.

The review data is subjected to preprocessing and filtering. We prepare the data more suitable for further steps by removing noise, links, emoji's, etc. The feature extraction from the review text is the next step. We utilized term frequency-inverse document frequency (TF-IDF) [13]—based method to collect the features. This measure reflects the significance of a word to documents in a given dataset. This feature is computed as follows:

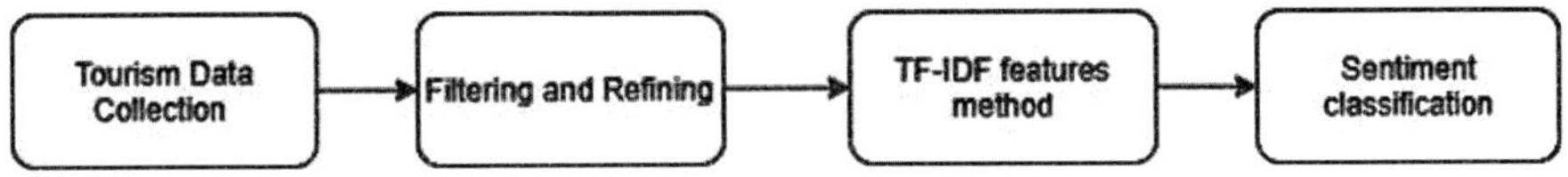

Fig. 1 Architecture of sentiment prediction

Table 1 Attributes of review dataset

Field	Description
Id	The unique identifier of the review
Text	The text of the review
language	The language of the review
source	One among GooglePlaces, Foursquare, Facebook
rating	Rating expressed by the user. Range is between 1 and 5
Time	Date of the review
wordsCount	Number of words of the text
analysis.kaf	The result of the OpeNER pipeline in KAF
analysis.json	The result of the OpeNER pipeline in KAF-JSON
polarity	The polarity of the review
place.id	id of the place associated to the review
place.name	Name of the place associated to the review
place. location	Location of the place associated to the review
place.category	Category of the place associated to the review
authorName	The name of the review author

$$\mathrm{TF}(t, d) = \frac{\mathrm{Count}(t)}{\mathrm{Terms}(d)} \tag{1}$$

Here, TF provides the occurrences of a term t in a given document, which is computed using (1). This indicates the frequency to total terms in a document.

$$\mathrm{IDF}(t) = \frac{\mathrm{Count}(d)}{|D|} \tag{2}$$

The IDF of a term t measures how many times t is occurring in different documents d of a given dataset D. This is computed using (2)

$$\mathrm{TF} - \mathrm{IDF}(t, d) = \mathrm{TF}(t, d) \times \mathrm{IDF}(t) \tag{3}$$

TF-IDF is the multiplication of TF and IDF of a term t and it measures the importance of a term and this weighing technique is used in this research work.

The sentiment of a review indicates the emotional content expressed by visitors on a given topic. The sentiment analysis lies at the core part of any business enhancement, improving service, feedback analysis, improving the visitors' turnaround. Moreover, in the tourism industry, the understanding sentiment of visitors and incorporating them is very important. The dataset collected from tour-pedia.org on reviews has computed positive and negative sentiment. We utilized the review text-based TF-IDF features to develop classifiers. The prediction results of the classifiers are either positive or negative based on the sentiment of visitors.

3 Experimental Results

We gather the dataset for various areas survey under the class of point of interest (POI) from tour-pedia.org. The dataset consists of various attributes which are shown in Table 1. One of the attributes of these contents is a positive and negative estimation. A further advance is content handling. The preprocessing of review text is carried out by filtering links, emojis, etc. Following further, we do highlight extraction; we do term weighting. The term weighting technique utilized is the TF-IDF strategy. The significance expands relative to the occurrences of a word showing up in the dataset; however, it is balanced by the recurrence of the word in writings. Varieties of the TF-IDF weighting strategy are frequently utilized. Term weighting is doling out an incentive to each term got from the dataset. The extremity plans to discover the significance of a term in speaking to a sentence. We utilize Naive Bayes and Support Vector Machine (SVM) for the classification of review sentiments. In this examination, we partition the information into two classes state, positive class and negative class. Each datum incorporates into the positive class has named "1" while the negative class has the marked "−1".

Figure depicts (see Fig. 2a) available number of samples of positive and negative classes. In a bar graph, the *x*-axis shows sentiments (positive and negative) for the text; the *y*-axis shows the total number of positive and negative texts for the entire dataset. For positive texts, more than 2000 are positive texts are present in dataset, while for negative texts, around 700 texts are in the dataset. Figure 2b shows most sentences fall between approximately 3–22 words but it is fair to say that the majority of text falls between 1 and 25 words. This is no wonder accounting that data has a limit on how many characters one can use in a message. Overall, it looks like 1–20 words cover more than 80% of all sentences which makes this dataset set a good training candidate. Progressively, the positive texts are not same as there are negative ones which do not seem like a large enough distinguish to cause any concern now. Figure 2b presents the word count distribution for collected texts.

The following (see Fig. 3) shows a word cloud depicts the most common words in the entire tour-pedia.org dataset after preprocessing data. Keywords "Dubai" and "Best" are very leading but some knowledge would help make sense of why some of the other words are there. Some words in the dataset do use very strong language. Also, some words in the dataset do use non-English language. As we know that the customers responded with the text. A word cloud will help us, for better understanding the themes and topics addressed from user's review. The words like "Best," "Dubai," "check," "visit," etc. are commonly found in the survey responses. This suggests that the positive sentiments expressed within the responses are assignable to attributes. The word cloud provides us with a deeper understanding of the responses and the sentiments expressed with them. Table 2 shows the number of texts with sentiment labeled using the two approaches. We are taking into consideration neutral those texts do not express any opinion.

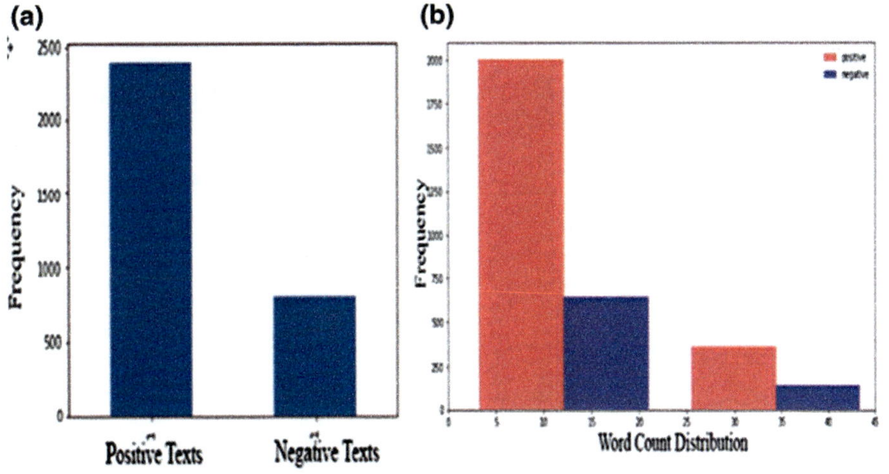

Fig. 2 **a** Number of positive texts (1) & negative texts (-1) and **b** Word count distribution of reviews positive (1) & negative (−1) reviews

Fig. 3 Word cloud formation

Table 2 Number of reviews with sentiments class

Positive	Negative	Total
2368	788	3156

Above graphs (see Fig. 4a) show the output of the frequency of the first 25 words. The top-most common words count available for texts for both positive and negative sentiments. On the *x*-axis, samples of the common word occurring in the dataset, *y*-axis counts the occurring of the samples. The highest common word occurring has the count more 2000 in the entire dataset. Next, (see Fig. 4b) shows the log–log plot for the frequency of the word which is like the previous frequency graph but includes all words and is plotted on a base 10 logarithmic scale which helps us visualize the rapidly diminishing frequency of words as their rank drops. The word distribution present in this data dictionary is a very common phenomenon in large

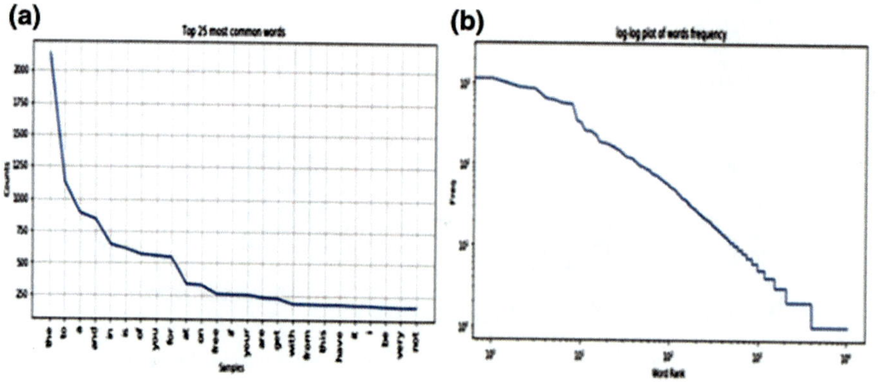

Fig. 4 **a** Top-most common words frequency **b** Log–log plot of word frequency

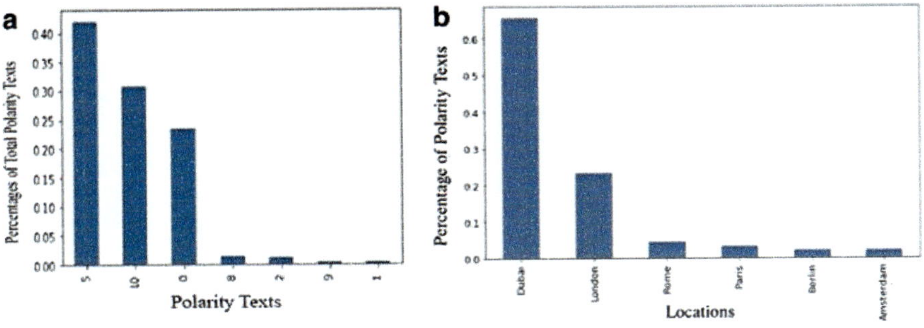

Fig. 5 **a** Percentage of polarities. **b** Percentage of origin texts

samples of words as shown in the figure, where the most frequent word will occur about twice as often as the second most frequent word, three times as often as the third most frequent word, etc. Words such as 'i', 'and' & 'is' are highly used in human utterances. These kinds of words mainly appear equally in both negative and positive conform expressions and such that they bring small instances that can be incorporated in the model so we can get to rid them off the path.

Above graphs (see Fig. 5a) shows the bar graph plot for the polarity for texts. The graph shows the x-axis shows the different types of polarities for reviews; the y-axis shows the percentage of occurrences polarities for the texts. Polarity with "5" with more than 40% of texts, while at the same time polarity "1" having the least percentage of texts with 0.02. Next, Fig. 5b shows the percentage of origin of texts for different locations. Dubai is having the highest percentage of texts with more than 60% of texts and Amsterdam with the least percentage of texts. From this plot, we can see several tourists from different parts of the world like visit Dubai and share their opinions for the different tourist spots in Dubai. Also, London is having second stand having 25% of texts sharing opinions, followed with Rome and Paris sharing the same number of percentage of texts with 5%, Berlin with 2% of texts, Amsterdam with 1% of texts; from all the above plot, we can say Dubai has the number of opinions from different parts of the countries.

Matrix shows (see Fig. 6) the distribution of polarities visited by travelers from represented locations. The bar graphs represents, various locations on the x-axis and different polarities shown in the y-axis. Focused on the 6 locations from the above analysis as well as polarity and locations mostly since based on the previous analysis. Concerning locations, one can see that tourists travel to Dubai the most.

The results accomplish for sentiment identification with the established classifiers. Hence, we can conclude the classifier that implements the SVM method overcomes the Naive Bayes one presenting the results. Table 3 shows the results obtained for sentiment detection by classifiers. The results obtained from the classifiers were summarized through the cross-validation process using the entire annotated tweets dataset. The accuracy for the SVM classifier was 0.80 while the Naive Bayes classifier was 0.74, which makes the SVM classifier better. The F-Measure, also better using

Fig. 6 Distribution of polarities visited by travelers from represented Location

	Dubai	London	Rome	Paris	Berlin	Amsterdam
polarity_0	478	155	24	34	39	12
polarity_1	4	6	1	12	11	9
polarity_2	25	13	1	26	17	11
polarity_3	0	12	0	18	6	10
polarity_4	0	19	0	16	6	2
polarity_5	837	353	56	49	15	37
polarity_6	10	20	0	14	4	3
polarity_7	30	10	0	24	4	6
polarity_8	27	7	8	13	4	11
polarity_9	7	4	0	12	3	6
polarity_10	696	191	41	31	15	17

Table 3 Comparison of developed classifiers

Classifier	Accuracy	Class	Precision	*F*-score
SVM	0.80	Positive	0.81	0.84
		Negative	0.70	0.71
Naïve Bayes	0.74	Positive	0.89	0.84
		Negative	0.61	0.46

the SVM classifier. Thus, we can finish up the classifier which actualizes the SVM method outflanked the Naive Bayes one introducing the best outcomes.

4 Conclusion

We have performed and looked at two methodologies for supposition examination for a situation learn about surveys composed by the vacationers. The principal approach utilizes Naive Bayes classifiers and the subsequent one uses SVM classifiers. The outcomes got by the SVM feelings classifier show a precision of 80% for the recognition of the assessment extremity. The Naive Bayes assessment classifier shows a precision of 74% of estimation extremity. The outcomes demonstrated an F1-score for a negative notion of 45%, for F1-score for a positive estimation of 86%. The outcomes that appeared by both grouping approaches are viewed as acceptable for area surveys or messages. We utilized TF-IDF as highlight extraction strategies and utilizing Naive Bayes, SVM calculations for characterization techniques. By utilization of the SVM classifier builds the precision by 5% which demonstrates great outcomes.

One of the fundamental shortcomings recognized is worried about the distinguishing proof of the element alluded to by the conclusion identified in the writings. Even though the gathered writings are identified with encountering the travel industry, the feelings communicated in the messages may allude to different substances. Likewise, it will enthusiasm to research connections crosswise over surveys substances that target improving precision for identification of the feeling extremity. Therefore, it will be conceivable to distinguish which angles concerning the audits or messages were viewed as positive or negative. Later, we might want to improve the precision of our investigation by performing bigger scale explores by having a bigger dataset and gather surveys from more sites.

References

1. U. Kumari, A. K. Sharma, D. Soni, Sentiment analysis of smartphone product review using SVM classification technique, in *International Conference on Energy Communication Data Analytics and Soft Computing (ICECDS)* (2017), pp. 1469–1474
2. Y. Wang, Y. Rao, L. Wu, A review of sentiment semantic analysis technology and progress, *13th International Conference on Computational Intelligence and Security (CIS)* (2017), pp. 452–455
3. M. Anjaria, R.M.R. Guddeti, Influence factor-based opinion mining of twitter data using supervised learning, in *Sixth International Conference on Communication Systems and Networks (COMSNETS)*, (2014) February 2014, p. 10
4. S.P. Astya, Sentiment analysis: approaches and open issues. Int. Conf. Comput. Commun. Autom. **9**, 1–5 (2017)
5. A. Balahur, *Sentiment Analysis in Social Media Texts* (June 2013), pp. 120–128
6. R. Fernandes, R.D'Souza, Analysis of product Twitter data though opinion mining, *India Conference (INDICON)* (2016), pp. 1–5
7. D. Arora, K.F. Li, S.W. Neville, Consumers' sentiment analysis of popular phone brands and operating system preference using Twitter data: a feasibility study, in *29th International Conference on Advanced Information Networking and Applications* (2015), pp. 680–686
8. A. Bali, P. Agarwal, G. Poddar, D. Harsole, N.M. Zaman, Consumer's sentiment analysis of popular phone brands and operating system preference. Int. J. Comput. Appl. **155**(4), 4 (2016)
9. Y.S. RupalBhargavaand, MSATS: multilingual sentiment analysis via text summarization. IEEE **9**(8), 97–110 (2017)
10. S. Patil, P.V. Wangikar, P.K. Jayamalini, Tweet Data Preprocess. Segment. NER **8**(1), 2075–2079 (2017)
11. N. Tyagi, S. Ahmad, A. Khan, M.M. Afzal, Sentiment analysis evaluating the brand popularity of mobile phone by using revised data dictionary **7**(3), 53–61 (2018)
12. R.R.S. Jandail, A proposed novel approach for sentiment analysis and opinion mining. Int. J. UbiComp, **5** (2014)
13. A. Alessa, M. Faezipour, Preliminary flu outbreak prediction using twitter posts classification and linear regression with historical centers for disease control and prevention reports: prediction framework study. JMIR Public Health Surveill **5**(2), e12383 (2019)

Soil Analysis Using Clustering Algorithm in Davao Region

Oneil B. Victoriano

1 Introduction

Soil test or soil analysis is conducted for one of several possible reasons. These reasons are to estimate the plant nutrients [1], soil quality [2], and the other may be done for engineering (geotechnical) investigations [3]. Soil testing is an important [4] mechanism to determine the nutrient [5] needed of plants and for environmental evaluations. Soil testing is the basis for fertilization requirement [6], because some soils are naturally lacking in plant nutrients. Most literatures in soil testing chemical characterization include level of acidity (pH), phosphorus (P), potassium (K) and nitrogen (N) [7, 8].

There are a number of data mining [9] and machine learning applied to agricultural soil data [10], but most of the literature available and surveyed are on classification and prediction. Clustering in soil test analysis is studied based on the dataset for a specific context; clustering in soil test analysis dataset can find significant knowledge discovery.

The research used *X*-means and *K*-means clustering algorithm applied to soil test analysis dataset. The dataset was preprocessed and applied clustering algorithm to chemical attributes pH, N, P and K and location. The study finalized its finding by describing and comparing centroid produced by the algorithm, and checked on the performance of the clustering algorithm used.

The significance of this study is to add knowledge discovery in the area of soil analysis using *X*-means and *K*-means algorithm. Additionally, DARFOXI [11] benefited the outcome of the study for soil analysis, because the study created a localized study of soil analysis in Davao Region and other neighboring areas and evaluated current soil analysis reports by equating to the discovered clusters.

O. B. Victoriano (✉)
University of Mindanao, Davao City, Philippines
e-mail: ovictoriano@umindanao.edu.ph

S. Patnaik et al. (eds.), *Advances in Machine Learning and Computational Intelligence*, Algorithms for Intelligent Systems, https://doi.org/10.1007/978-981-15-5243-4_37

2 Review of Related Literature

2.1 *Soils Analysis Dataset*

Dataset used in the related studies was from soil extracted data like moisture content, clay content, liquid limit, plastic limit, plastic index, and consistency index [12]. Some related studies' datasets are extracted using visible and neuro-infrared spectroscopy tool with soil organic matter(SOM) contents and reflectance spectra measured in the laboratory [13, 14]. Spatial patterns datasets [15–17], soil visual assessment dataset [18], satellite imagery dataset, and spatial data [19–22] are also from related studies. Fuzzy *k*-means clustering is potentially better suited for spectral data and multivariate character [23].

2.2 *Clustering in Soils Data*

Most of the clustering algorithms used in the studies were *k*-means [24–26], fuzzy taxonomy and fuzzy clustering [27], fuzzy k-means clustering, principal components fuzzy c-means clustering [28], and principal component analysis (PCA) [26]. To name a few researches in clustering, there is research in digital mapping of taxonomic soil units [27, 28] and evaluation and management the soil fertility [29, 30].

2.3 K-*Means and* X-*Means Algorithm*

K-means clustering is a clustering algorithm that assigned each object exclusively to one of a set of clusters. Objects are group in one cluster because of the measure of distance between each objects. *K*-means first define centroid in random by Euclidean distance or similar measure. Heuristic is used in *X*-means to determine the number of centroids. The algorithm starts with a minimum number of centroids, then repeatedly use random centroids that makes sense according to the data [31].

3 Methodology

3.1 *Data Gathering*

Dataset is the transactional soils test reports filed by the Department of Agriculture Region Field Office XI (DARFOXI)—Regional Soils Laboratory's encoder. The study used spreadsheet soil test report, an aggregated report from soil nutrient, results of analysis, nutrient requirements and fertilizer recommendation reports.

3.2 Data Preprocessing

Data files are not formatted in a single template; automation to extract is not available due to individual client transactions are in different form and style. Data was reformatted and re-encoded for unique keywords. The dataset's aggregated columns were separated to new columns and missing values were filled out. From the original transaction of 3978 rows, the dataset was expanded to 5870 rows.

3.3 Clustering

The study used Rapid Miner Studio to experiment on *X*-means and *K*-means clustering algorithm. Replace missing values and normalization was performed as preprocessing techniques.

The X-means operator was used with the following parameters: *k* min—2 while *k* max—60; measure type is numerical measures; numerical measure is Euclidean distance; and maximum run is 10. The study also included checking the cluster distance performance operator for performance evaluation of centroid based clustering methods. Two performance measures are supported: Average within cluster distance and Davies–Bouldin index.

The clustering (*K*-Means) operator was used with the following parameters: set add cluster attribute, add as label, remove unlabeled, *k* is 4, max runs is 10, measure types are Bregman divergences, divergence is square Euclidean distance, max optimization steps is 100. The same parameters with the other operators (besides K-means) listed above were used.

4 Results and Discussions

4.1 Cluster Centroid Analysis

Using X-means algorithm, Cluster_0 is the cluster with the lowest values of level of acidity (pH), phosphorus (P) and potassium (K), while Cluster_1 is the cluster with the lowest values of nitrogen (N). The highest values of pH, N and P are in Cluster_3, while Cluster_2 locations are with the highest values of *K* (see Table 1).

In using K-means algorithms, lowest values of level of acidity (pH), phosphorus (P) and potassium (K) are in Cluster_2, while lowest values of nitrogen (N) is in

Table 1 *X*-means algorithms' centroid table

Attribute	Cluster_0	Cluster_1	Cluster_2	Cluster_3
pH	*−0.85141*	0.73324	0.74138	**0.91620**
N	0.05720	*−0.07498*	−0.05649	**1.73956**
P	*−0.16280*	−0.12534	0.57618	**9.71778**
K	*−0.33926*	−0.22782	**2.19177**	0.25791

Table 2 *K*-means algorithms' centroid table

Attribute	Cluster_0	Cluster_1	Cluster_2	Cluster_3
pH	−0.63999	**1.03840**	*−0.63610*	0.90155
N	8.21051	**0.94898**	−0.07327	*−0.08159*
P	0.42495	**10.03965**	*−0.16710*	0.08403
K	**0. 17180**	0.51695	*−0.34219*	0.47478

Cluster_3. The highest values of pH, N and P are in Cluster_1, while Cluster_0 locations are with the highest values of K (see Table 2).

Results in both algorithms have similarities, *X*-means' Cluster_0 and *K*-means' Cluster_2 both have the lowest values of pH, P, and K; while *X*-means' Cluster_1 and *K*-means' Cluster_3 both have the lowest value of *N*. Similar centroid results in Cluster_2 of *X*-means and Cluster_0 of *K*-means, both have the highest value of *K*. While, *X*-means' Cluster_3 and *K*-means' Cluster_1 are equated due to have the same number of centroids with highest chemical element. The similarity of clusters centroids will be further discussed in the number of sample occurrences.

4.2 Average Within Centroid Distance of Clusters

In *X*-means, the average distance within the centroid is 2.052. The average distance of each samples within each other from its centroid is shown in Table 3. Cluster_1 has the best value from its centroid or best performing. Both Cluster_1 and Cluster_2 are relatively acceptable distance. While Cluster_3 is the least performing.

In K-means, the average distance within the centroid is 2.066. Cluster_2 has the best value or best performing. Cluster_3 is relatively acceptable distance. Cluster_0 and Cluster_1 points are with far from each other.

Table 3 Average distance of each samples from its centroid

Algorithm	Cluster_0	Cluster_1	Cluster_2	Cluster_3
X-means	**2.012**	**1.103**	**4.156**	35.741
K-means	23.763	34.188	**0.946**	**2.740**

Comparing *X*-means and *K*-means results, *X*-mean algorithm has good clustering mechanism than of the *K*-means results. The overall average of centroid distance of points within centroid distance, *X*-means has the value less than with that of *K*-means. Also, there are three (3) clusters in *X*-means with least values than that of *K*-means with only two (2) clusters.

4.3 Statistically Difference

The study used Anova to check the results of the clustering algorithm is statically difference. In the *X*-means algorithms, three (3) attributes the pH, P, and K have zero (0) *P*-values; while nitrogen (N) attribute has 7.66952E-20 P-value. All P-values are acceptable due to all values are less than 0.05. In the *K*-means algorithm, three (3) attributes the pH, N, and P have zero (0) *P*-values; while potassium (K) attribute has 5.120E-142 *P*-value. All *P*-values are acceptable due to all values are less than 0.05.

4.4 Frequency Distribution

In cluster centroid analysis using X-means, majority of locations have findings with lower levels of level of acidity (pH), phosphorus (P) and potassium (K). There are no locations having high levels of chemicals; values on X-means' Cluster_3 and Cluster_2 with high levels of pH, N, P, and K do not have significant percentage count on the sample data (see Table 4).

In *K*-means results, Cluster_2 is with samples greater than fifty percentage count on locations having lower level of pH, P, and K; while Cluster_3 results are locations have lower level of N. There are no high levels of chemical on any locations since Cluster_0 and Cluster_1 do not have significant percentage count (see Table 5).

Comparing similar locations on both algorithms results, *K*-means clustered seven (7) locations with that of *X*-means clustered only five (5) locations with low levels of pH, P, and K; *K*-means clustered four (4) locations with that of *X*-means clustered only three (3) locations with lower levels of *N*. Combining these results pointed out that Caraga Region, Comval, Davao City, Luzon and Region 9 samples have lower levels of pH, P, and K; while Davao del Norte and Davao Oriental samples have lower levels of *N*. Both results do not have significant samples with high values of pH, N, P and K.

Table 4 X-means frequency distribution table of clustered location attributes

Location	Cluster_0		Cluster_1		Cluster_2		Cluster_3		Total
	Count	%	Count	%	Count	%	Count	%	Count
?	26	87	4	13					30
Caraga Region	130	70	52	28	4	2			186
Comval	592	64	307	33	22	2			921
Davao City	847	59	460	32	110	8	7	0	1424
Davao del Norte	339	28	702	59	158	13	1	0	1200
Davao del Sur	267	35	310	41	188	25			765
Davao Occidental	21	37	10	18	24	42	2	4	57
Davao Oriental	138	25	397	71	26	5			561
Luzon	52	80	9	14	4	6			65
Region 10	78	48	46	29	9	6	28	17	161
Region 12	249	46	239	44	54	10			542
Region 9	22	76	6	21	1	3			29
Visayas	14	38	23	62					37
Grand Total	2775	46	2565	43	600	10	38	1	5978

Table 5 *K*-means frequency distribution table of clustered location attributes

Location	Cluster_0		Cluster_1		Cluster_2		Cluster_3		Total
	Count	%	Count	%	Count	%	Count	%	Count
?	24	80	4	13	2	7			30
Caraga Region	140	75	44	24	2	1			186
Comval	692	75	226	25	3	0			921
Davao City	1017	71	397	28	6	0	4	0	1424
Davao del Norte	472	39	714	60	14	1			1200
Davao del Sur	372	49	392	51			1	0	765
Davao Occidental	23	40	32	56			2	4	57
Davao Oriental	245	44	315	56	1	0			561
Luzon	57	88	8	12					65
Region 10	84	52	43	27	6	4	28	17	161
Region 12	291	54	234	43	17	3			542
Region 9	22	76	7	24					29
Visayas	18	49	19	51					37
Grand Total	3457	58	2435	41	51	1	35	1	5978

5 Conclusion and Recommendation

Soil analysis with the use of data mining and machine learning is common in modern agriculture. Soil test is far more advance from obtaining samples, mapping, and analysis.

The study proves that knowledge discovery in the area of soil testing analysis can be created by clustering algorithms. We have proved that *X*-means and *K*-means results have good division of clusters in the context area. The cluster's centroid average distance within and among its observations is far more acceptable. The performance of each algorithm is checked and all *p*-values are acceptable due to all values are less than 0.05. The good result of this study is the frequency distribution table to cluster different location with knowledge discovery of particular soil chemical characterization.

Due to time constraints, the proponent will experiment more on clustering the chemical characteristics on each type of crops, soil type, and soil test request date submission. There will be a series of researches on the dataset, and one of it is looking into possibility of soil fertilizer recommendation prediction study in the near future.

References

1. V. Suneetha, Nutrient analysis of soil samples from various places. J. Chem. Pharmaceut. Res. **7**(3), 291–293 (2015)
2. E. Bunemann, G. Bongiorno, Z. Bai, R. Creamer, G. De Deyn, R. de Goede, L. Brussaard, Soil quality—a critical review. Soil Biol. Biochem. **120**(January), 105–125 (2018)
3. E. Rabot, M. Wiesmeier, S. Schlüter, H. Vogel, Soil structure as an indicator of soil functions: a review. Geoderma **314**, 122–137 (2018)
4. The Importance of Testing Soil Quality. https://www.agrocares.com/en/importanceofsoiltesting/. Last accessed 2018/10/12
5. E. Manjula, S. Djodiltachoumy, Data mining technique to analyze soil nutrients based on hybrid classification. Int. J. Adv. Res. Comput. Sci. **8**(0976), 505–510 (2017)
6. B. Rama Krishna, B. Satyanarayana, Agriculture soil test report data mining for cultivation advisory. Int. J. Comput. Appl. **6**(2), 11–16 (2016)
7. M. Chiranjeevi, R. Nadagoudar, Analysis of soil nutrients using data mining techniques. Int. J. Recent Trends Eng. Res. **4**(7), 103–107 (2018)
8. F. Nájera, Y. Tapia, C. Baginsky, V. Figueroa, R. Cabeza, O. Salazar, Evaluation of soil fertility and fertilisation practices for irrigated maize (Zea mays L.) under Mediterranean conditions in Central Chile. J. Soil Sci. Plant Nutr. **15**(1), 84–97 (2015)
9. V. Rajeswari, K. Arunesh, Analysing soil data using data mining classification techniques. Indian J. Sci. Tech. 9(19), (2016)
10. M. Devi, Enhanced crop yield prediction and soil data analysis using data mining. Int. J. Modern Comput. Sci. (IJMCS) ISSN: 2320–7868 (Online), **4**, 6, December, (2016)
11. Regional Soil Laboratory Services, http://car.da.gov.ph/index.php/2-uncategorised/45-soil-laboratory-services. Last accessed 2018/10/12
12. B.T. Pham, L.H. Son, T.A. Hoang, D.M. Nguyen, D. Tien Bui, Prediction of shear strength of soft soil using machine learning methods. CATENA: An Interdiscip. J. Soil Sci. Hydrol. Geomorphol. Focus. Geoecol. Landsc. Evol. **166**(April), 181–191 (2018)

13. Y. Hong, S. Chen, Y. Zhang, Y. Chen, L. Yu, Y. Liu, Y. Liu, H. Cheng, Y. Liu, Rapid identification of soil organic matter level via visible and near-infrared spectroscopy: Effects of two-dimensional correlation coefficient and extreme learning machine. Sci. Total Environ. Int. J. Sci. Res. Environ. Relat. Humankind **644**, 1232–1243 (2018)
14. F. Terra, J. Demattêb, R.A. Viscarra Rossel, Proximal spectral sensing in pedological assessments: vis–NIR spectra for soil classification based on weathering and pedogenesis. Geoderma: Global J. Soil Sci. **318**(February 2017), 123–136 (2017)
15. V. Khosravia, F.D. Ardejanib, S. Yousefic, A. Aryafarc, Monitoring soil lead and zinc contents via combination of spectroscopy with extreme learning machine and other data mining methods. Geoderma: Global J. Soil Sci. **318**(January), 29–41 (2018)
16. C. Brungard, J. Boettinger, M. Duniway, S. Wills, T. Edwards, Machine learning for predicting soil classes in three semi-arid landscapes. Geoderma: Global. J. Soil Sci. **239**, 68–83 (2015)
17. S. Khanala, J. Fultonb, A. Klopfensteinb, N. Douridasc, S. Shearerb, Integration of high resolution remotely sensed data and machine learning techniques for spatial prediction of soil properties and corn yield. Comput. Electron. Agric. **153**(April), 213–225 (2018)
18. G. Bondia, R. Creamerb, A. Ferraric, O. Fentona, D. Wall, Using machine learning to predict soil bulk density on the basis of visual parameters: Tools for in-field and post-field evaluation. Geoderma: Global J. Soil Sci. **318**(January), 137–147 (2018)
19. D. Vermeulen, A.V. Niekerk, Machine learning performance for predicting soil salinity using different combinations of geomorphometric covariates. Geoderma: Global J. Soil Sci. **299**, 1–12 (2017)
20. J. Heila, V. Häringb, B. Marschnerb, B. Stumpe, Advantages of fuzzy *k*-means over *k*-means clustering in the classification of diffuse reflectance soil spectra: a case study with West African soils. Geoderma: Global J. Soil Sci. **337**(August 2018), 11–21 (2019)
21. J. Pohjankukka, H. Riihimaki, P. Nevalainen, T. Pahikkala, J. Ala-Ilomaki, E. Hyvonen, J. Varjo, J. Heikkonen, Predictability of boreal forest soil bearing capacity by machine learning. J. Terrramech. **68**, 1–8 (2016)
22. N.R. Regmia, C. Rasmussen, Predictive mapping of soil-landscape relationships in the arid Southwest United States. CATENA: An Interdiscipl. J. Soil Sci. Hydrol. Geomorphol. Focus. Geoecol. Landscape Evol. **7**, 917–918 (2018)
23. B.A. Stevenson, S. McNeill, A.E. Hewitt, Characterising soil quality clusters in relation to land use and soil order in New Zealand: An application of the phenoform concept. Geoderma: Global J. Soil Sci. **239**, 135–142 (2015)
24. R.B. Palepu, R.R. Muley, An analysis of agricultural soils by using data mining techniques. Int. J. Eng. Sci. Comput. **7**(10) (2017)
25. E. Hot, V. Popovic-Bugarin, Soil data clustering by using *K*-means and fuzzy *K*-means algorithm. 23rd Telecommunications Forum Telfor (TELFOR), **8**(1), 890–893 (2015)
26. A. Kumar, N. Kannathasan, A novel soil profile feature reduction model using principal component analysis. Indian J. Sci. Tech. **8**(29) (2015)
27. M. Horáček, P. Samec, J. Minár, The mapping of soil taxonomic units via fuzzy clustering—a case study from the Outer Carpathians, Czechia. Geoderma: Global J. Soil Sci. **326** (October 2017), 111–122 (2017)
28. L. Yang, A. Zhu, Y. Zhao, D. Li, G. Zhang, S. Zhang, L. Band, Regional soil mapping using multi-grade representative sampling and a fuzzy membership-based mapping approach. Pedosphere **27**(2), 344–357 (2017)
29. M. Kommineni, S. Perla, D. Yedla, Survey Using Data Min. Tech. Soil Fert. **7**, 917–918 (2018)
30. N. Insozhan, V. Parthasarathy, Evaluation and management of soil fertility. Int. J. Pure Appl. Mathem. **117**(8 Special Issue), 11–14 (2017)
31. Rapid Miner Studio 9.0.1 Documentation, (2016)

Gold Tree Sorting and Classification Using Support Vector Machine Classifier

R. S. Sabeenian, M. E. Paramasivam, Pon Selvan, Eldho Paul, P. M. Dinesh, T. Shanthi, K. Manju, and R. Anand

1 Introduction

Image classification is a process in computer vision that can classify an image according to its statistical properties and features. Image classification analyzes the numerical and statistical properties [1] of various image features and organizes data into labels or class. Machine learning classification algorithms typically employ two processes, i.e., *training* and *testing of samples* [2]. Firstly, statistical properties image features are taken and a unique label is assigned to each class, i.e., *training class*. In testing stage, these feature space or trained model are compared with testing features and used to classify image. Image classification involves the process of combining images of similar class using machine intelligence. SVM is one of the most common supervised machine learning algorithms used for image classification. SVM involves complex data transformations and finds out a hyper-plane separating two classes.

A Support Vector Machine (SVM) [3, 4] is a machine learning classifier used for bifurcating two different classes to items. SVM can be multiclass or binary classifier. The data points are given labeled training data (supervised learning); the gist of SVM involves in finding an optimal hyper-plane separating the class boundaries. So SVM can be visualized as an optimization problem to find the class boundaries with trade of error [5]. Consider a binary classifier; the hyper-plane is a dividing line of two classes. Any point in left of this plane falls into class I and on right falls into class II. It finds out a line/hyper-plane in multidimensional space that separate outs classes.

R. S. Sabeenian · M. E. Paramasivam · E. Paul (✉) · P. M. Dinesh · T. Shanthi · K. Manju · R. Anand
Sona SIPRO, Sona Signal and Image PROcessing Laboratory, Department of Electronics and Communication Engineering, Sona College of Technology, Salem 636005, Tamil Nadu, India
e-mail: eldhopaul@sonatech.ac.in

P. Selvan
Curtin University, Dubai International Academic Block, Dubai, United Arab Emirates

S. Patnaik et al. (eds.), *Advances in Machine Learning and Computational Intelligence*, Algorithms for Intelligent Systems, https://doi.org/10.1007/978-981-15-5243-4_38

In real-time applications, there is always a trade of in finding an optimal hyper-plane which finding perfect class for thousands of training data set. So there is a need of data preprocessing which is called regularization parameter. There are two terms involved in regularization, i.e., parameter regularization and gamma. These are tuning parameters in SVM classifier. Varying these parameters, we can achieve flexibility in nonlinear classification along with more accuracy and computational time. Tuning parameters are of SVM are

- Kernel
- Regularization
- Gamma

1.1 Kernel

Finding of hyper-plane can be visualized as transformation problem using linear algebra. For linear kernel, the equation for prediction for a new input using the inner dot product between the training vector (x) and its support vector (x_i) which is calculated as

$$G(x) = A_0 + \sum b_i \times (x, x_i$$

The above equation involves finding the inner products of new input vector (x) along with all its support vectors in training data. The coefficients A_0 and b_i are obtained from the training data by using the learning algorithm. Higher-order functions like polynomial and exponential kernels [6] are used for calculating the plane of separation in multi dimension environments. Solving the higher-order kernels in higher dimension is a tedious task.

1.2 Regularization

The regularization parameter helps in optimization parameters and helps to avoid the misclassification of training samples. For large values of regularization parameter K, optimization will choose a smaller margin in the hyper-plane. If the hyper-plane chosen satisfy conditions of classification, i.e., data points are correctly classified. Conversely, for a narrow value of K, the optimizer will took larger margin separating hyper-plane. Optimal hyperplane is chosen based on the distance between two class boundaries.

Fig. 1 Child images

1.3 Gamma

The parameter gamma defines how far the data samples from the hyper-plane. Low gamma implies the data points far away from hyper-separation line. Whereas higher gamma implies that data points are close to hyper-plane. The closer data points in the training data should be taken precisely because these data points can cause more misclassification of samples. The lower gamma points are more away from the separation line and will not have an impact in misclassification. Figure 2 shows a gold tree along with several child items. A gold tree consists of different child items connected to the main cast (Fig. 1).

2 Gold Tree

Gold tree consists of cast of various child ornaments connected together to form a tree structure. There are different gold tree for various ornaments which helps in batch production. Each child item has unique designs and fine details embedded on it (Fig. 2).

2.1 Pendant

The pendant is a piece of jewelry like chain which hangs around the neck. There are various pedants available from short pedants to long pedants with unique designs.

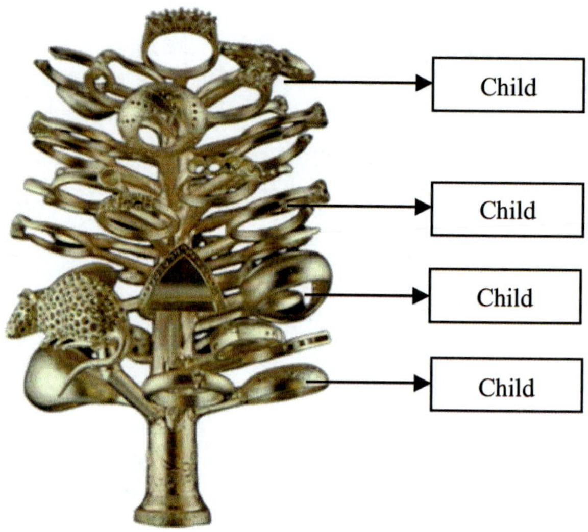

Fig. 2 Gold tree with child items

2.2 *Necklace Pendant*

A necklace pendant is a fancy ornament which is attached to the necklace which is kept around the neck.

2.3 *Letter Pendant*

Letters are also used in the ornaments to make it as a perfect work. The beauty of the ornament can be described by these letters. The letters will intake the beautiful memories and make it more precious.

2.4 *Fancy Pendant*

These types of pendant are available with different shapes, color and also with variety of stones. Each fine art makes the ornaments more attractive.

2.5 *Earring Pendant*

The beauty can be expressed through earrings. Diamond, rubies, white gold, etc. make it as more attractive.

3 Proposed Work

The proposed method involved in finding the class of child items using machine intelligence. The proposed algorithm can be divided into different steps

- Child image acquisition
- Data cleaning/preprocessing
- Splitting of data in test and train data
- GLCM feature extraction
- Training of data
- Validating the results

Firstly, the sample of child images is taken using a smart camera with high-resolution images. The acquired child images may be with different size, lacks of contrast, noise, etc. All these artefacts are removed in the preprocessing stage and made to unique size. Images are split into test and train images in 25:75 ratios from the data base.

A Gray-level Co-occurrence Matrix (GLCM) method is used to extract the features from the test and train images. The GLCM features are explained below. The extracted features are then fed to the SVM classifier. Support Vector Machine is used to differentiate the two classes (hyper-plane) in effective manner. An SVM involves in mapping of points in multiple classes given according to the data labels.

The hinge function loss associated to the SVM is given by

$$C(x, y, f(x)) = \begin{cases} 0, & \text{if } y * f(x) \geq 1 \\ 1 - y * f(x), & \text{else} \end{cases}$$

x and y are data points and $f(x)$ is the hinge function. The minimal loss function associated to hinge function is given by

$$\min_{W} \lambda||w||^2 + \sum_{i=1}^{499}(1 - y_i x_i, w)$$

W is regularization parameter used for tuning the SVM model. The gradient of the above equation gives the direction of error rate which can be used for estimating the step size in the direction of error rate.

$$\text{Gradient, } G = \frac{\partial}{\partial w_k}\lambda||w||^2 = 2\lambda w_k$$

$$\frac{\partial}{\partial w_k}(1 - y_i x_i, w) = \begin{cases} 0, & \text{if } y_i x_i, w \geq 1 \\ -y_i x_{ik}, & \text{else} \end{cases} \tag{4}$$

where data points are x and y and W is a regularization parameter of the SVM model.

3.1 Training of Samples

Training of child items consists of three main steps.

- Image acquisition
- Selection of the region of interest (ROI).
- Classification of images.

3.2 Feature Extraction

GLCM [3, 7–9] is Gray-level Co-occurrence Matrix used for extracting features from the child images. GLCM provides various features which are described below. It provides the measure of repetition local patterns in image intensity distribution. It is commonly used statistical measurement for texture analysis and classification. GLCM [7] is based on the repetition of same gray-level values throughout the image. The image is divided into sub-blocks with respect to textures features, i.e., coarse and fine textures of image. Intensity of image plays a vital role in estimating the GLCM features.

3.3 GLCM Features

Contrast. Contrast [10] is a measure the differences in pixel intensities presents in an image. Image lacks contrast means there will not be visible variations in the edges. If the pixel values are distributed uniformly across all intensities range, the contrast of image will be good.

Energy. Energy [10] describes the measure of uniformity in the image.

Entropy. Entropy [10] is the measure of average information content in the image.

Mean. Mean [11] is the average value of the total number of data.

Variance. Variance [11] is the difference of pixel values from the mean value.

Standard deviation. The standard deviation [11] is the square root of variance of data samples.

Maximum and minimum pixel value. Maximum pixel value is the largest number of pixel values present in image, whereas minimum pixel value is the lowest number of pixel values present in an image.

$$\text{Maximum pixel value} = \text{Max}[X(i, j)] \tag{5}$$

$$\text{Minimum pixel value} = \text{Min}[X(i, j)] \tag{6}$$

Homogeneity. Homogeneity [10] is the measure of uniformity in the pixel distribution of an image. If the composition of pixel distribution is same, the image is said to be homogeneous

Correlation. The relation of variables at different time intervals is called correlation. It can be auto-correlation or cross-correlation [10].

$$G[m, n] = p[m, n] * r[-m, -n] \tag{7}$$

Cross-correlation is a measure of different variables of a function at different time intervals.

$$f(x, y) = \int_{-\alpha}^{\alpha} f^{a}(\tau) g(t + \tau) d\tau \tag{8}$$

4 Confusion Matrix

The results of the training data can be validated by metric parameters like confusion matrix. A confusion matrix is a two way table that consists of correctly classified and misclassified data points. Accuracy is the ratio of sum of true positive + true negative with sum of true positive, true negative, false negative, false positive (Tables 1 and 2).

Table 1 GLCM feature values of different child classes

GLCM features								
S. No.	Mean	VAR	SD	Contrast	Entropy	Max	Min	Class name
1	118.24	54.62	7.34	200	0.005	255	25	CH1
2	124.9	55.73	7.41	194	−0.011	255	64	CH2
3	124.34	55.72	7.41	210	0.0051	254	49	CH3
4	125.64	51.8	7.14	200	−0.004	255	26	CH4
5	118.61	58.74	7.62	200	−0.002	255	24	CH5
6	115.07	48.13	6.91	199	0.026	254	55	CH6
7	113.74	44.5	6.65	202	−0.027	254	57	CH7
8	115.58	41.51	6.45	199	−0.033	254	55	CH8
9	117.55	38.28	6.16	200	−0.025	255	57	CH9
10	85.018	38.63	6.16	213	0.002	247	34	CH10
11	115.88	34.15	5.89	190	−0.04	252	56	CH11
12	116.61	40.44	6.63	192	0.011	255	57	CH12
13	117.71	45.83	6.35	195	0.033	254	57	CH13
14	88.914	43.28	6.25	209	0.003	247	38	CH14
15	116.3	46.34	6.78	190	0.002	244	56	CH15

5 Conclusion

The proposed algorithm involves sorting of the child items from a gold tree. The proposed algorithm has good classification accuracy and adaptability. Sorting of the child items of the gold tree was done with the help of an SVM classifier. This classification method using SVM provides good accuracy and adaptability. The test result shows the high precision of the proposed system and further modified using deep learning.

Table 2 Confusion matrix of different child classes

Class Name	CH1	CH2	CH3	CH4	CH5	CH6	CH7	CH8	CH9	CH10	CH11	CH12	CH13	CH14	CH15
CH1	**29**	2	1	0	1	0	0	0	0	0	0	1	0	0	1
CH2	4	**48**	1	0	0	1	4	2	0	0	1	0	0	0	1
CH3	2	1	**50**	1	2	1	1	0	0	0	0	0	0	0	0
CH4	0	1	0	**30**	2	1	1	1	1	0	0	0	0	0	0
CH5	0	1	0	2	**50**	1	2	2	2	1	0	0	0	0	0
CH6	0	0	0	0	1	**52**	2	1	1	1	0	0	1	1	0
CH7	0	2	1	1	1	0	**55**	0	1	0	1	1	0	0	0
CH8	2	2	0	0	0	0	0	**57**	1	0	0	1	0	1	0
CH9	0	0	1	0	1	0	1	0	**62**	0	0	0	0	0	0
CH10	1	1	0	1	0	0	0	0	0	**38**	1	1	0	0	0
CH11	1	4	1	1	2	0	0	0	0	0	**29**	0	0	0	0
CH12	0	0	1	0	0	0	0	0	0	0	0	**29**	1	1	1
CH13	0	1	0	1	0	1	0	1	0	0	0	0	**40**	1	0
CH14	0	0	0	0	1	0	1	0	1	0	0	0	0	**35**	1
CH15	1	0	0	0	0	0	0	0	0	1	0	1	1	1	**40**

Accuracy = 86.78%

References

1. V. Vapnik, *The Nature of Statistical Learning Theory* (Springer-Verlag, NewYork, 1995)
2. B.E. Boser, I.M. Guyon, V.N. Vapnik, A training algorithm for optimal margin classifiers, in *Proceedings of Fifth Ann. Workshop Computational Learning Theory* (ACM Press, New York, 1992), pp. 144152
3. Yiming Yang, An evaluation of statistical approaches to text categorization. Inf. Retr. **1**(1–2), 69–90 (1999)
4. T. Joachims, *Making large-scale SVM learning practical* (No. 1998, 28). Technical report, SFB 475, (1998)
5. M. Park, K. Yoon, Learning and selecting confidence measures for robust stereo matching, in *IEEE Transactions on Pattern Analysis and Machine Intelligence*, vol. 41, no. 6, pp. 1397–1411, 1 June 2
6. B. Scholkopf, A. Smola, K.-R. Muller, Nonlinear component analysis as a Kernel eigenvalue problem. Neural Comput. **10**, 1299–1319 (1998)
7. B. Fan et al., A performance evaluation of local features for image-based 3D reconstruction. IEEE Trans. Image Process. **28**(10), 4774–4789 (2019). https://doi.org/10.1109/TIP.2019.2909640
8. Gilles Burel, Dominique Carel, Detection and localization of faces on digital images. Pattern Recogn. Lett. **15**(10), 963–967 (1994)
9. R.C. Gonzalez, R.E. Woods Pearson, A textbook of "Digital Image Processing", Education, Inc. Third Edition (2004)
10. A.K. Jain, Reference book of Fundamentals of Digital Image Processing, Pearson Education, Inc. (2002)
11. J.C. Russ, *The Image Processing Handbook*, CRC Press (2007)

Computational Intelligence

Dynamic Economic Dispatch Using Harmony Search Algorithm

Arun Kumar Sahoo, Tapas Kumar Panigrahi, Jagannath Paramguru, and Ambika Prasad Hota

1 Introduction

In the ongoing era, the load demand of the power system is profoundly stochastic in nature and increasing regularly for which cost of generation is to be controlled around all the hours in a day. The cost-effective operation of system depends on the total demand being applicably pooled amid generating units with concern to limit the gross generation cost. The financial plan of power provider, the prime practical setup and planning of electrical power generation system are vital to the power production. With gigantic interconnected power system, the mash of vitality and persistent increment in costs, it is required to minimize operational charges of power provided. Reducing the fuel cost for connected units can decrease operational expenses. Economic aspects of load to be dispatched by the connected units are scheduled to accomplish ideal operational cost [1]. The ideal working expense is to be acquired by considering the limitation on system operation to affirm the system constraints, accordingly maintaining a strategic reserve from the breakdown of system subjected to unexpected problems. Dynamic economic dispatch (DED) works by online planning of generators by finding the prime era, thereby accomplishing requested supply adjust over a given interim of time with ideal working expense in various systems considering different constraints of operation. Ramp rate limitations may influence the operational choice on hourly basis [2]. The issue addressed in the paper is the test of the right allotment of load to the accessible units to achieve the load demand. The cost function without valve point effect is treated as

A. K. Sahoo (✉) · A. P. Hota
IIIT, Bhubaneswar, India
e-mail: c116003@iiit-bh.ac.in

T. K. Panigrahi
PMEC, Berhampur, India

J. Paramguru
KIIT, Berhampur, India

S. Patnaik et al. (eds.), *Advances in Machine Learning and Computational Intelligence*, Algorithms for Intelligent Systems, https://doi.org/10.1007/978-981-15-5243-4_39

convex function with quadratic cost function. Considering the practical constraints such as valve point loading effect and ramp rate limit, the convexity behaves as non-convexity. Heuristic technique approaches [3] are simple to apply with faster computational performance at optimum price [4]. Different upgrading strategies are utilized to make a solution of the DED issue, such as particle swarm optimization (PSO) [5], gravitational search algorithm (GSA) [6], genetic algorithms (GA) [7] and simulated annealing (SA) [8]. However, the performance of these techniques is significantly influenced by the parameters. This paper shows an optimum solution by applying the improved version of HS technique than some other techniques.

2 Problem Formulations

The DED problem is to schedule the power among the committed generator to satisfy the load demand for respective time period. The DED is formulated as a nonlinear and complex optimization problem considering several constraints. The cost function is given in Eqs. 2 and 3 for the DED problem, whereas Eq. 1 shows the cost function for static economic dispatch.

Let Ci be the charge of generating energy by a unit. Therefore, cost for the i units be

$$C = \sum_{i=1}^{N} C(i)\text{Rs/h} \tag{1}$$

$$\text{Min}C\left(P_g\right) = \sum_{t=1}^{\text{NT}} \sum_{i=1}^{\text{NG}} C(P_{it}) \tag{2}$$

where,

$$(P_{it}) = a_i P_{it}^2 + b_i P_{it} + c_{it} \tag{3}$$

NT total time period, ai, bi are ci are the coefficients
NG no. of generated units

Equation 3 shows the cost function of DED not considering valve point loading and Eq. 4 shows the cost function with valve point loading.

$$C(P_{it}) = a_i P_{it}^2 + b_i P_{it} + c_{it} + \left|d_i \sin\left(e_i\left(P_{i\text{min}} - P_{it}\right)\right)\right| \tag{4}$$

2.1 Demand Load Balancing Constraint

$$P_D = \sum_{i=1}^{N} P_{it} + P_L \tag{5}$$

P_D = load demand, P_{gi} = total generated load at different time, P_L = transmission loss

$$P_{Lt} = \sum_{j=1}^{n}\sum_{i=1}^{n} P_{it} B_{ij} P_{jt} + \sum_{i=1}^{n} B_{i0} P_{it} + B_{00} \tag{6}$$

2.2 Generator Constraint

The output power of the generator is maintained within the upper bound and lower bound [3].

$$P_{i\min} < P_{it} < P_{i\max} \tag{7}$$

2.3 Ramp rate Constraint

The online action for generating units is constrained by ramp rate bounds. These bounds have an impact on the operational decisions. The current scheduling may disturb the future scheduling as generation increases due to ramp rate bounds.

$$\begin{cases} P_{i,t} - P_{i,t-1} \le UR_i \\ P_{i,t-1} - P_{i,t} \le DR_i \end{cases}$$
$$i = 1, 2, 3 \ldots, N.t = 2, 3 \ldots T \tag{8}$$

3 Harmony Search Algorithm

Currently, Geem et al. proposed a music-inspired HS meta-heuristic algorithm for searching actual process of harmony. Harmony in music is analogous to the optimization process and the process of improving the harmony. HS algorithm improvises the process to optimize the global and local systems. A lot of meta-heuristic algorithms are proposed based on population such as evolutionary algorithms which contain genetic algorithm, evolutionary strategies, DE algorithm, HS algorithm, etc. And the algorithms based on swarm contain particle swarm optimization, bees algorithms, ant colony optimization, etc., over the last period. The opening values for the decision variables are not required in this algorithm. This algorithmic process uses a stochastic-based search process on the memory of the harmony considering rate and adjusting rate of the pitch to overlook derived information [9, 10]. Processes and performances on music require an ideal state of harmony and strong minded by artistic estimation and optimization procedures produce the finest state, determined by objective function value. Harmony search makes the process as the following [11]:

1. Each decision variable is referred from each musician.
2. Decision variable's value range is referred from musical instrument's pitch.
3. The solution vector at certain iteration is referred from harmony of music at a certain time.
4. The objective function is referred from audience's aesthetics.

3.1 Harmony Search Technique Steps

a. **Initialization of Harmony Memory (HM)**

HM is created [11] by considering solution matrix with dimensions as that in HMS. All the elements in harmony memory matrix signify one solution. Here, the solutions are stochastically created and again arranged by ordering in a reverse way to HM, constructed on their values depending upon the objective function like,

$$f(a_1) \le f(a_2) \le f(a_3) \ldots \le f(a_{\text{HMS}})$$

$$\text{HM} = \begin{bmatrix} a_1^1 & a_2^1 & \cdots & a_N^1 \\ a_1^2 & a_2^2 & \ddots & a_N^2 \\ a_1^{\text{HMS}} & a_2^{\text{HMS}} & & a_N^{\text{HMS}} \end{bmatrix}$$

b. **Improvise New Harmony**

New harmony vector is improvised by HS process,

$$a_i^{'} = a_1^{'}, a_2^{'}, a_3^{'}, \ldots, a_N^{'}$$

Initialization of HMCR, PAR_{max} and PAR_{min} is processed and determined whether in the limit. The new value is updated,

$$a_i^{'} \in \begin{cases} a_i^{'} \in \{a_i^1, a_i^2, a_i^3, \ldots, a_i^{\text{HMS}}\} \text{w.pHMCR} \\ a_i^{'} \in Ai \quad \text{w.p}(1 - \text{HMCR}) \end{cases}$$

By tuning the entire decision variable, a search process is added to the new harmony vector.

$$a_i^{'} = a_1^{'}, a_2^{'}, a_3^{'}, \ldots, a_N^{'}$$

From HM taking PAR operator,

$$a_i^{'} \leftarrow \begin{cases} \text{Pitch adjusted w.p PAR} \\ \text{No change w.p } (1 - \text{PAR}) \end{cases}$$

Any produced arbitrary number rd $\varepsilon[0, 1]$ is inside possibility limit of PAR, and a new assessment variable ($a_i^{'}$) is attuned on the given equation:

$$a_i^{'} = \left(a_i^{'} + rd()\right) \times bw$$

Here, *bw* is a random bandwidth of the distance.

c. **Updating the harmony memory**

The HM was updated by the new vector created $a_i^{'} = a_1^{'}, a_2^{'}, a_3^{'}, \ldots, a_N^{'}$, and each objective function is considered for new harmony vector f (a′). If the objective function value of new vector is good than the previous harmony vector, then it is stored in HM; otherwise, this new vector is overlooked.

d. **Checking the stopping condition**

The process of iteration in the above steps is finished when the extreme number of iteration is touched. Lastly, the harmony memory vector with the best value is selected and is considered as the finest solution to the problem.

Figure 1 shows the flow chart of the harmony search algorithm.

4 Result and Analysis

The dynamic load dispatch with load variation in a dynamic sequence is analysed by some optimization technique for the IEEE 10 unit system. The output power for the hourly dispatched corresponding to the demand for the harmony search technique and invasive weed optimization technique was shown in Tables 1 and 2, respectively. The harmony search technique is applied for the system, and the total cost based on hourly load dispatch is obtained and is presented in Table III. After the dynamic load dispatch was obtained with an implementation of HS technique, the economics of the solution dispatch obtained is economic than the other techniques applied to

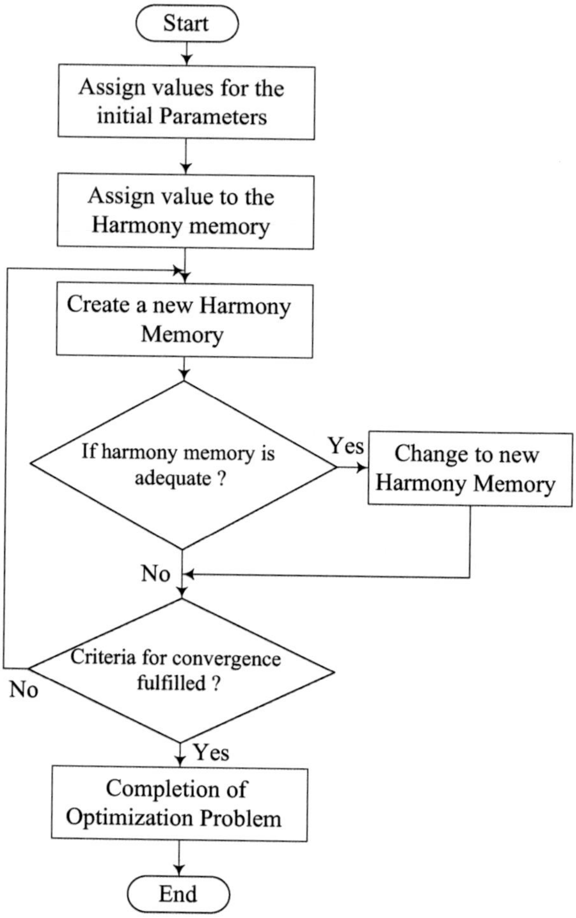

Fig. 1 Flow chart for harmony search

the system [12, 13]. Figure 2 shows the convergence curve by applying the harmony search technique and invasive weed optimization technique with the iteration taken. It was identified that harmony search technique is giving optimum value with a faster converging time.

Tables 1 and 2 present the generation of power at different hours of a day to match the demand considering the various constraints by harmony search technique and invasive weed optimization technique, respectively. Comparing the cost of dynamically economic dispatch with same numbers of iterations, harmony search technique is giving the optimum cost for the generation considering the constraint as shown in Table 3.

Table 1 Generator schedule for 24 h using HS technique (hourly load dispatch)

Hour	P1, MW	P2, MW	P3, MW	P4, MW	P5, MW	P6, MW	P7, MW	P8, MW	P9, MW	P10, MW
1	227.444	226.031	86.345	63.195	126.547	125.824	56.535	47.478	21.596	55
2	225.933	222.307	81.775	60.011	122.429	123.839	91.646	82.802	44.253	55
3	224.225	221.919	186.515	120.077	125.776	125.218	93.285	84.543	21.437	55
4	303.279	223.313	188.512	125.101	126.721	124.767	127.190	85.607	46.503	55
5	300.986	310.446	194.098	117.248	122.721	120.394	126.929	84.385	47.788	55
6	304.733	309.812	306.866	123.575	171.021	124.087	125.255	75.815	31.831	55
7	380.970	310.699	292.707	120.731	173.736	124.389	128.660	85.174	29.929	55
8	381.429	399.268	293.122	122.102	172.895	121.820	118.877	85.250	26.231	55
9	456.292	397.083	300.316	180.742	171.907	125.194	127.247	85.043	25.173	55
10	456.263	396.408	296.789	244.523	221.341	132.542	128.673	116.336	24.121	55
11	460.113	457.939	315.184	241.171	225.722	128.276	127.948	86.739	47.902	55
12	457.327	458.344	322.677	295.167	220.151	150.428	128.747	86.850	45.304	55
13	457.327	458.344	322.677	295.167	220.151	150.428	128.747	86.850	45.304	55
14	381.476	395.872	292.044	239.156	174.386	122.460	128.16	84.158	51.281	55
15	377.881	314.454	293.415	240.036	147.092	121.778	92.983	85.586	47.769	55
16	302.172	308.528	182.893	182.058	173.728	120.070	93.541	84.508	51.497	55
17	301.913	308.456	182.201	179.520	121.145	104.660	95.683	85.055	46.362	55
18	305.061	308.827	290.814	177.358	172.472	122.029	125.901	48.740	21.794	55
19	379.640	309.029	307.666	240.756	127.491	124.644	118.192	84.796	28.782	55
20	457.245	459.938	338.439	169.950	221.770	130.392	129.257	86.585	23.421	55
21	454.199	396.598	296.088	121.651	215.812	122.257	128.324	84.518	49.549	55
22	379.515	396.723	194.246	118.165	171.137	110.843	92.429	84.574	25.363	55
23	303.133	311.017	191.319	119.540	124.157	58.024	92.048	47.786	29.971	55
24	227.495	309.460	87.074	61.621	119.431	114.732	91.575	83.735	33.874	55

Table 2 Generator schedule for 24 h using IWE technique (hourly load dispatch)

Hour	P1, MW	P2, MW	P3, MW	P4, MW	P5, MW	P6, MW	P7, MW	P8, MW	P9, MW	P10, MW
1	226.63	135.86	186.94	72.91	73.12	123.37	56.86	85.31	20.00	55
2	226.62	135.00	167.54	118.74	122.82	122.44	56.53	85.31	20.00	55
3	226.62	215.00	152.27	118.39	172.68	156.20	56.53	85.31	20.00	55
4	303.25	222.27	202.44	163.90	172.89	124.40	56.53	85.31	20.00	55
5	303.25	285.15	177.30	178.55	222.60	96.31	56.53	85.31	20.00	55
6	379.87	313.19	192.64	176.10	172.71	122.45	56.53	114.92	44.60	55
7	303.25	393.19	224.83	226.10	174.27	131.61	58.30	115.45	20.00	55
8	303.25	396.80	261.62	241.17	172.73	122.38	88.30	85.45	49.31	55
9	379.87	396.80	304.68	241.25	222.60	122.45	93.06	88.27	20.01	55
10	379.87	460.00	303.86	256.50	224.97	123.64	99.87	118.27	50.01	55
11	456.50	396.80	306.98	300.00	228.90	132.20	129.61	120.00	20.03	55
12	456.50	459.81	308.26	299.62	222.60	122.45	129.59	116.14	50.03	55
13	379.87	460.00	313.03	296.56	172.73	127.01	129.59	86.14	52.06	55
14	379.87	396.80	321.70	246.56	128.55	128.46	129.69	115.31	22.06	55
15	303.25	340.52	312.59	230.76	172.75	126.23	129.59	85.31	20.00	55
16	303.17	309.41	258.22	180.76	122.81	122.28	127.03	55.31	20.00	55
17	303.25	309.69	187.29	180.83	126.29	123.61	97.03	47.00	50.00	55
18	379.87	316.80	189.49	185.98	172.73	136.72	93.06	47.00	51.33	55

(continued)

Table 2 (continued)

Hour	P1, MW	P2, MW	P3, MW	P4, MW	P5, MW	P6, MW	P7, MW	P8, MW	P9, MW	P10, MW
19	379.87	396.80	243.97	235.98	172.77	123.06	100.21	47.00	21.33	55
20	456.79	399.39	301.27	254.01	222.63	125.99	129.70	77.00	50.22	55
21	456.50	388.42	253.54	229.14	222.57	122.02	129.59	47.00	20.22	55
22	379.87	308.43	185.62	180.26	172.73	122.45	129.51	47.00	47.13	55
23	303.25	228.43	142.35	169.86	122.87	111.68	99.51	47.00	52.05	55
24	303.25	171.23	191.31	120.71	88.31	61.68	93.45	47.00	52.06	55

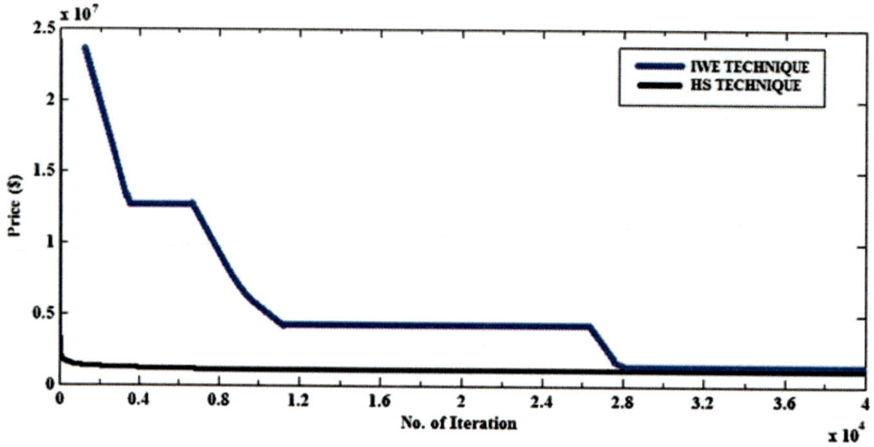

Fig. 2 Convergence curve for the HS technique

Table 3 Cost comparison

S. No.	Methods	Cost ($)
1.	Harmony search technique	1020273.682345
2.	Modified invasive weed optimization [12]	1035630.223367
3.	Invasive weed optimization	1039743.254276
4.	Deterministically guided PSO [13]	1049167.000000

5 Conclusion

Using harmony search technique in dynamic economic dispatch, the result is optimized and the result is also satisfying all the constraints. The main focus of this paper is to survey and summarize the applications of HS for solving the DED problems. It was found to converge to the optimum result at a faster rate. The method requires primitive mathematical operators, so is computationally inexpensive in terms of both memory requirements and speed. In the harmony search technique, the optimal solution is obtained by successive iteration.

References

1. J. Wood, B.F. Wollenberg, *Power Generation, Operation, and Control* (Wiley, New York, NY, USA, 2012)
2. D.P. Kothari, J.S. Dhillon, *Power System Optimization*, 2nd edn. (PHI Learning Private Limited, India, 2011)
3. M.F. Zaman, S.M. Elsayed, T. Ray, R.A. Sarker, Evolutionary algorithms for dynamic economic dispatch problems. IEEE Trans. Power Syst. **31**(2), 1486 (2016)

4. D.C. Walters, G.B. Sheble, Genetic algorithm solution of economic dispatch with valve point loadings. IEEE Trans. Power Syst. **8**(3), 1325–1331 (1993)
5. I. Selvakumar, K. Thanushkodi, A new particle swarm optimization solution to nonconvex economic dispatch problems. IEEE Trans. Power Syst. **22**(1), 42–51 (2007)
6. R.K Swain., K.C. Meher, U.C. Mishra, Dynamic economic dispatch using hybrid gravitational search algorithm. *Power, Control and Embedded Systems (ICPCES)*, 2012 2nd International Conference on. IEEE (2012)
7. C. Chao-Lung, Improved genetic algorithm for power economic dispatch of units with valve-point effects and multiple fuels. IEEE Trans. Power Syst. **20**(4), 1690–1699 (2005)
8. D.N. Simopoulos, S.D. Kavatza, C.D. Vournas, Unit commitment by an enhanced simulated annealing algorithm. IEEE Trans. Power Syst. **21**(1), 68–76 (2006)
9. X.S. Yang, Harmony search as a metaheuristic algorithm, in Music-inspired harmony search algorithm: theory and applications, ed. by Z. W. Geem, *Studies in Computational Intelligence,* vol. 191 (*Springer*, Berlin, 2009), pp. 1–14
10. M.A. Osama, R. Mandava, The variants of the harmony search algorithm: an Overview. Artif. Intell. Rev. **36**, 49–68 (2011)
11. K.S. Lee, Z.W. Geem, A new structural optimization method based on the harmony search algorithm. Comput. Struct. **82**(9–10), 781–798 (2004)
12. R. Sharma, N. Nayak, K. R. Krishnananda, P.K. Rout, Modified invasive weed optimisation with dual mutation technique for dynamic economic dispatch. Published *2011 IEEE*
13. T. Aruldos, A. Victoirea, A. Ebenezer Jeyakumar, Deterministically guided PSO for Dynamic Dispatch Considering valve-point effect, in *IEEE transaction on power system* 25 November 2004

Packing Density of a Tori-Connected Flattened Butterfly Network

Md. Abdur Rahim, M. M. Hafizur Rahman, M. A. H Akhand, and Dhiren K. Behera

1 Introduction

At present, the researchers are attempting several ways to achieve more and more computational power. To adjust the intensive demand of computational power, different methodologies are considered such as the clock pulse, multi-core and many-core processor as well as multi-node massively parallel computer (MPC) system [1]. All above techniques improved the computational power, but MPC systems exceed the limit of others. It consists of millions of nodes [2, 3], and the node of the MPC itself is a multi-core processor. These huge numbers of node cannot be connected by conventional connection topologies, for that apposite method is needed. In MPC, interconnection network plays a fateful role to connect those huge numbers of nodes and it is also a reliable alternative process which can be interconnected together with different types of networks. The use of interconnection network with MPC systems has become tremendous and increases performance radically.

The overall performance of MPC such as fault tolerant [4], inter-node communication, and network on chip depends on its interconnection network. However,

Md. Abdur Rahim · M. A. H Akhand
Department of Computer Science and Engineering, Khulna University of Engineering & Technology, Khulna, Bangladesh
e-mail: mailtoabdurrahim@gmail.com

M. A. H Akhand
e-mail: akhand@cse.kuet.ac.bd

M. M. Hafizur Rahman
Dept. of Computer Networks & Communications, College of Computer Science & Information Technology, King Faisal University, Al Ahsa 31982, Saudi Arabia
e-mail: mhrahman@kfu.edu.sa

D. K. Behera (✉)
Indira Gandhi Institute of Technology, Sarang, Odisha, India
e-mail: dkb_igit@rediffmail.com

S. Patnaik et al. (eds.), *Advances in Machine Learning and Computational Intelligence*, Algorithms for Intelligent Systems, https://doi.org/10.1007/978-981-15-5243-4_40

due to the lack of enough connectivity, the MPC industry community does not get attracted to the potential attention of the hierarchical interconnection network. As mentioned earlier, MPC systems are consisted by huge number of nodes. Therefore, optimal diameter degree (ODD) network is needed for obtaining the credibility of a hierarchical interconnection networks. The midimew-connected mesh network (MMN) [5] is the better option of ODD; it supports small distance parameters, less number of wire required connecting nodes, low cost, and constant node degree [6]. The midimew network is enwrapped by the 2D mesh network where one dimension enwraps like Tori-connected and the other dimension enwraps diagonally. However, in MMN, long diagonal links bring some limitation by which routing algorithm becomes complex and long length wires in VLSI realization [7]. To avoid those limitations, reinvestigate the MMN network.

Considering investigation, remove the 2D mesh BM from the MMN network and integrate flattened butterfly network in it. 2D torus network is considered for the higher-level TFBN to have the less number of long links because long links are not desirable for VLSI implementation. It also speeds up the inter-node communication. Then, we can achieve a new hierarchical interconnection network that is Tori-connected flattened butterfly network (TFBN) [8]. The hierarchical interconnection network TFBN is a combination of flattened butterfly network denoted as a basic module (BM) and the torus network denoted as a higher-level interconnection network which are connected hierarchically. Despite all of those research works, the thing is essential and important to study how case to be implemented of the mentioned TFBN network in VLSI realization.

In this consideration, the packing density is one of the essential parameters that determine the implementability of mentioned interconnection network. The key objective of this paper is to study packing density of the mentioned TFBN over other networks. The remainder section of the paper is arranged as follows: In Section II, the basic architecture of the TFBN is described. The main contribution of this paper, packing density, and the conclusion of this study are discussed in Section III and Section IV, respectively.

2 Tori-Connected Flattened Butterfly Network

Tori-connected flattened butterfly network (TFBN) is a hierarchal interconnection network formed with basic module (BM) and higher-level networks in a hierarchal fashion. In this paper, 2D TFBN is considered and the static network architecture of it is discussed. The BM can be defined as Level_1, and higher-level networks can be defined as Level_2 network, Level_3 network, and so on.

2.1 Basic Module

The basic module of a TFBN is a 2D flattened butterfly network consists of ($2^n \times 2^n$) grid where total numbers of row and column are 2^n and nodes 2^{2n}, n is the value of positive integer. There is no obligatory to select the value of n, here taking $n = 2$ for excellent granularity. In Fig. 1, the basic modules (BM) are imprinted which are (4×4) and contain 2^{n+2} free ports. Those free ports and links are used to connect higher-level interconnection of a TFBN. Figure 1 shows free ports of BMs that are denoted as Level_1.

Considering the value of $n = 2$, a (4×4) BM will have $2^{2+2} = 16$ free ports. For $r = 0$, four (i.e., $4(2^r) = 4(2^0) = 4$) free ports will be available for inter-level connectivity: 2 for vertical and 2 for horizontal. For higher-level network, two adjacent BMs are connected with bidirectional link which is fixed with ongoing and outbound links. Vertical links consist of vertical in and vertical out. On the other hand, horizontal links consist of horizontal in and horizontal out.

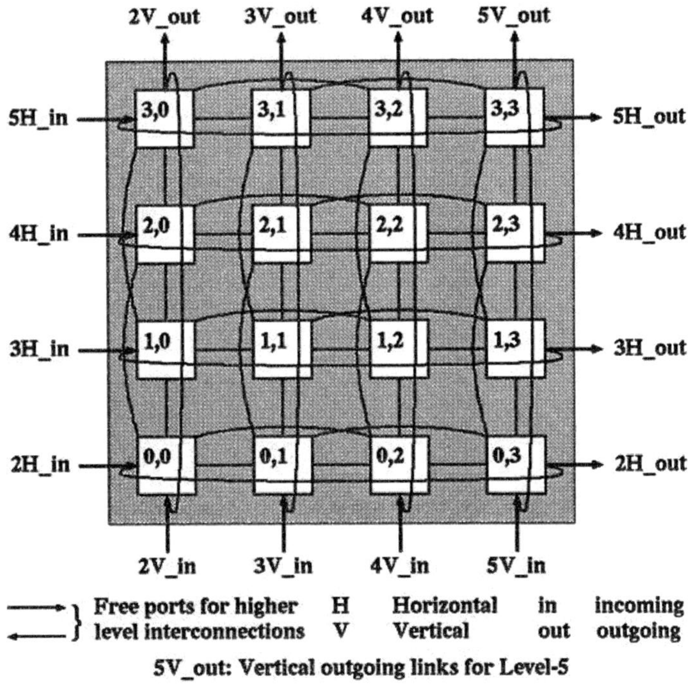

Fig. 1 Basic module of a TFBN

2.2 Higher-Level Network

The number of lower-level network is 2^{2n} which are inserted in one higher-level networks; this connection process remains for connecting every node of first level to higher level. In Level_2, ($2^n \times 2^n$) torus networks are connected with BM where every torus network is connected with $2^{2n} = 16$ BM. In the same way, the single network of Level_3 is connected with 16 networks of Level_2; thus, TFBN is constructed which is shown in Fig. 2. Removing the complexity of Fig. 2 is not imprinted of BMs links. Level_1 contains $4 \times (2^r) = 2^{r\pm2}$ free link where $2(2^r)$ are vertical and $2(2^r)$ are horizontal. Here, r belongs to $\{0, 1\dots n\}$, and r is the symbol of inter-level connectivity whose minimal and maximum values are $r = 0$ and $r = n$.

TFBN is described as TFBN (n, L, r) whose BM is ($2^n \times 2^n$) and where levels hierarchy is denoted as L and inter-level connectivity r. In this paper, $n = 2$ for excellent granularity. Observing TFBN (2, L, r) can be calculated $L_{\max}$ from ($2^n \times 2^n$) basic modules, applying the equation $L_{\max} = 2^{n-r} + 1$. For example, taking $r = 0$ and $n = 2$, $L_{\max} = 2^{2-0} + 1 = 5$. Thus, the highest value is 5 for (4×4) BM, free ports and links. In Level_L TFBN, the total number of nodes is $N = 2^{2nL}$ and those can be interconnected with highest number of nodes by a TFBN (n, L, r). $N= 2^{2n(2\hat{}(n-r)+1)}$ when maximum hierarchy $L_{\max} = 2^{m-r} + 1$. It means at least one million nodes can be interconnected when $m = 2$ and $r = 0$ which can be constructed by massively parallel computer system.

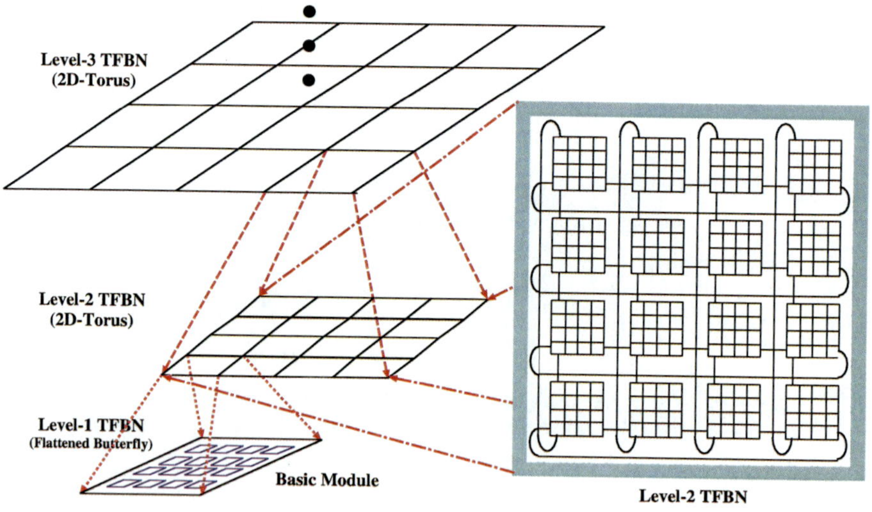

Fig. 2 Higher-level interconnection of a TFBN

Table 1 Topological analysis of different interconnection networks

Network	Degree	Diameter	Packing density
2D_Mesh	4	30	2.13
2D_Torus	4	16	4
TESH (2,2,0)	4	21	3.04
TTN (2,2,0)	6	15	2.84
MMN (2,2,0)	4	17	3.76
TFBN (2,2,0)	8	10	3.2

3 Topological Analysis

Basically, multi-node massively parallel computer (MPC) system performance depends on several fundamental technology and their implementation outcomes. To study the perfectness of any interconnection network, static appraisal is broadly used in particular for a MPC system. Some static performance metrics are focused in this section to classify the performance and effectiveness of the TFBN considering packing density.

3.1 Cost Parameters

An interconnection network's exact cost depends on its total number of interconnection links and the cost of particular node. The mutual cost of the processor and memory calculates the cost of particular node and also the cost of I/O interfaces of the specific node equivalent to its node degree. The node degree determines the router cost which is one of the significant cost parameters in a node. Table 1 illustrates the node degree where node degree of TFBN is 8, node degree of TTN is 6, and the remaining networks node degree is 4. Therefore, the router cost is not the same for all considered networks.

3.2 Distance Parameters

The MPC system design influences the distance between two nodes. To illustrate the distance parameters, there are some predefined scales such as meter denotes the system-level hierarchy distance, cm denotes the board level, and mm denotes the chip level. However, hop distance is used for measuring two nodes distance which are static. We have obtained the distance parameter that is hop distance using shortest path algorithm via computer simulation. The highest value of hop distance in a network is called diameter. All considered networks diameter is shown in Table 1, where we have noticed that mesh network holds the highest diameter value and the TFBN holds lowest.

3.3 Packing Density

The product of degree and diameter is called the static cost of any interconnection network. The ratio between the total number of nodes and this static cost is known as packing density [9]. For very-large-scale integration (VLSI) layout, the lower the chip area and the higher the packing density are essential. Actual packing density will depend upon the actual VLSI realization of any interconnection networks, and actual VLSI realization is cost handy. However, before the real implementation of any interconnection network in the VLSI chip, static analysis of packing density gives a significant result whether the interconnection network is suitable or not. This is why, the static analysis of packing density for any interconnection network or hierarchical interconnection network is crucial for the next step of performance analysis. The packing density of any interconnection is defined by the following equation.

$$\text{Packing Density} = \frac{\text{Total Number of Nodes}}{\text{Diameter} \times \text{Degree}} \tag{1}$$

3.4 Some Generalization

TFBN stands at lowest diameter and highest degree, and that high node degree results in slightly low packing density than that of MMN and torus network. However, the diameter is extremely lower than that of MMN. The maximum hop distance in a shortest path among all distinct pairs of nodes in an interconnection network is known as diameter. And the diameter yields an idea about the maximum latency in that network. This is why, the diameter is broadly used for the comparison of static network performance of different interconnection and hierarchical interconnection networks.

The distance parameter, diameter, of mesh and torus networks is assessed by applying their specific formula. Also, we have evaluated the diameter of the proposed TFBN along with its rivals MMN [5], TTN [9], and TESH [10] network by applying computer simulation. Due to high diameter of mesh network, the packing density of mesh network is low. The packing density of TTN and TESH network is also a bit low because of slightly high diameter of TESH network and high degree of TTN. The proposed TFBN results in high packing density because of low diameter even though the degree is high.

The different types of network's packing density are determined considering 256 nodes which are illustrated in Table 1. It is depicted that packing density of a TFBN is higher than that of TTN, TESH, and mesh networks and also lower to some extent to that of MMN.

4 Conclusion

In this paper, we have viewed the packing density of the TFBN in detail. For determining the effectiveness, we also consider this parameter of mesh, torus, TESH, TTN as well as MMN and compared with TFBN's one, and excellence of TFBN is determined by the comparison. The TFBN's interconnection links reduce the hop distance and remove long diagonal links that improve the distance parameters compared with others, and it also brings the high packing density. Due to high node degree of TFBN, the packing density of TFBN is slightly lower than that of other networks considered in this paper. In the next steps of our study, we are intending to evaluate the cost-effectiveness and time cost-effectiveness of TFBN.

The Tori-connected flattened butterfly networks (TFBN) architecture and its packing density are illustrated in this paper. We have the plan to study and work in the future with: (1) Evaluation of TFBN for higher-dimensional static network and (2) evaluation of MMN where 2D mesh is replaced by torus network and higher-level network as a flattened butterfly network.

References

1. B.S.P. Mishra, S. Dehuri, Parallel computing environments: a review. IETE Technical Review **28**(3), 240–247 (2011)
2. V. Puente, J.A. Gregorio, R. Beivide, F. Vallejo, A low cost fault tolerant packet routing for parallel computers, in *Proceedings of the 17th IEEE/ACM international Parallel and Distributed Processing Symposium (IPDPS)*, April, 2000
3. P. Beckman, Looking toward exascale computing, keynote speech, in *International Conference on Parallel and Distributed Computing, Applications and Technologies (PDCAT08)*, Dunedin, New Zealand, December 2008
4. F.J. Andujar, J.A. Villar, J.L. Sanchez, F.J. Alfaro, Jos Duato, N-dimensional twin torus topology. IEEE Trans. Comput. **64**(10), 2847–2861 (2015)
5. Md. Rabiul Awal, M.M. Hafizur Rahman, M.A.H. Akhand, A new hierarchical interconnection network for future generation parallel computer, in *Proceedings of the 16th International Conf. on Information and Communication Technology (ICCIT)*, pp. 314–319, 2014
6. M.R. Awal, M.M. Hafizur Rahman, M.A.H. Akhand, A new hierarchical interconnection network for future generation parallel computer, in *Proceedings of 16th Int'l. Conference on Computers and Information Technology* (Khulna, Bangladesh, 2013), pp. 314–319
7. M.R. Awal, M.H. Rahman, R. Mohd Nor, T.M. Bin Tengku Sembok, M.A. Akhand, Architecture and network-on-chip implementation of a new hierarchical interconnection network. J. Circ. Syst. Comput. **24**(02), 154 (2015)
8. M.H. Sohaini, M.M. Hafizur Rahman, R.M. Nor, T.M. Tengku Sembok, M.A.H. Akhand, Yasushi Inoguchi, A low hop distance hierarchical interconnection, in *Network*, *Proceedings of the 2nd International Conf. on Electrical Information and Communication Technologies (EICT)* (2015), pp. 43–47

9. M.M. Hafizur Rahman, Y. Inoguchi, Y. Sato, S. Horiguchi, TTN-a high performance hierarchical interconnection network for massively parallel computers. IEICE Trans. Inf. Syst. **E92-D**(5), 1062–1078 (2009)
10. V.K. Jain, T. Ghirmai, S. Horiguchi, TESH: a new hierarchical interconnection network for massively parallel computing. IEICE Trans. Inf. Syst. **E80-D**(9), 837–846 (1997)

A Study on Digital Fundus Images of Retina for Analysis of Diabetic Retinopathy

Cheena Mohanty, Sakuntala Mahapatra, and Madhusmita Mohanty

1 Introduction

Diabetes is one of the most chronic diseases in the world. India is among the top 3 countries in terms of diabetic population. One of the major effects of this disease is on human eye leading to diabetic retinopathy. As blood sugar level increases, the blood vessels of the retina get damaged and this causes diabetic retinopathy. It is one of the major causes of vision loss worldwide and also the reason of impaired vision in diabetic patients between 25 and 74 years of age [1]. By 2030 [2], the number of diabetic patients will be at least 366 million. Clinical studies have shown that treatment which includes surgery can reduce the risk of blindness by more than 90% if diabetic retinopathy is detected and treated early.

Figure 1a represents a normal and healthy retinal image with normal macula, retinal blood vessels, and optic nerve. Figure 1b indicates a diabetic retinopathy retinal image showing the presence of hemorrhages, exudates, and cotton wool spot.

During the initial stages of diabetic retinopathy, there are no symptoms. The advanced stages of diabetic retinopathy cause blurred vision, distortion, and vision loss. Therefore, diabetic retinopathy must be detected in early stages. Hence, machine learning [3] plays a very important role for the detection of diabetic retinopathy in early stages [4].

C. Mohanty
Biju Patnaik University of Technology, Rourkela, Odisha, India
e-mail: cheena.mohanty@gmail.com

S. Mahapatra (✉) · M. Mohanty
Trident Academy of Technology, Biju Patnaik University of Technology, Bhubaneswar, Odisha, India
e-mail: mahapatra.sakuntala@gmail.com

M. Mohanty
e-mail: madhusmita.mohanty@tat.ac.in

S. Patnaik et al. (eds.), *Advances in Machine Learning and Computational Intelligence*, Algorithms for Intelligent Systems, https://doi.org/10.1007/978-981-15-5243-4_41

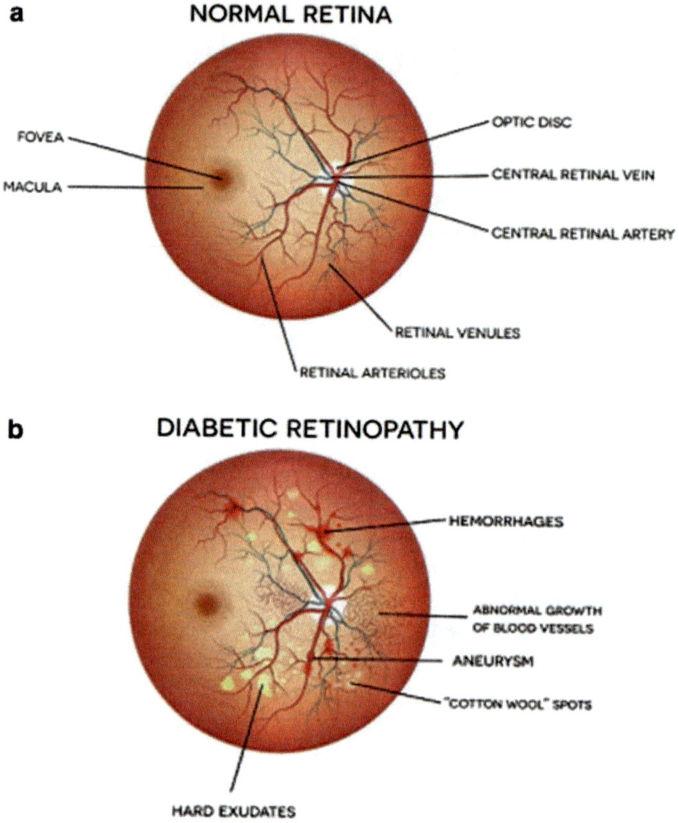

Fig. 1 **a** Normal retina, **b** Diabetic retinopathy

Generally, the color fundus images are used for the detection of diabetic retinopathy. The various important feature components of diabetic retinopathy are microaneurysms, hemorrhages, exudates, and blood vessels. Diabetic retinopathy is categorized into four stages—mild, moderate, severe proliferative, and non-proliferative. Edge detection is one of the foremost steps in image analysis, image processing, and image pattern recognition. Various applications of edge detection are computer vision, image segmentation, medical diagnosis, etc.

Skewness is a measure of the asymmetry of the probability distribution of a real-valued random variable. The skewness value can be positive or negative or even undefined.

The remaining sections have been organized as follows: Sect. 2 presents reviews on recent literatures. In Sect. 3, the methodology has been proposed. Section 4 contains results and discussion, and Sect. 5 presents the conclusion.

2 Literature Review

Tsai et al. [5] have proposed a study which will help children to understand the concept of healthy eating behavior by using digital image interface design. This could improve the cognitive performance of children for healthy diet. The main aim of this study is to conduct multi-level surveys for identifying children's cognitive gap in image of food. Shailesh Kumar and Basant Kumar [6] in their paper presented an improved diabetic retinopathy detection scheme by extracting accurate area and accurate number of microaneurysms from color fundus images. Dutta et al. [7] proposed a model for detection of diabetic retinopathy. They proposed an automated model to identify the components of DR. They have presented a deep learning model which can classify the features as blood vessels, exudates, and hemorrhages. Weighted fuzzy C-means algorithm has been used to identify the target class. Raman et al. [8] proposed a methodology which focuses on enhancing images, filtering noise, and detection of blood vessels, extracting exudates and microaneurysms and classifying various stages of diabetic retinopathy as mild, moderate, severe NPDR, and PDR by using machine learning. Cheloni et al. [9] presented a systematic review based on PRISMA guidelines. Anacan et al. [10] have used screening methods to distinguish between a healthy retina and a diseased one. By using local binary patterns and statistical texture analysis classification between diabetic retinopathy, glaucoma and healthy fundus images have been done. They also studied the efficiency of the algorithm. Prasad et al. [11] proposed a model for classification of diabetic retinopathy using fuzzy neural network classifier. They have compared the results against various performance metrics. Evangelia et al. [12] presented a model for the identification of people with diabetes who are at the risk of developing diabetic retinopathy using patient characteristics and clinical measurements. Syeda et al. [13] have presented a work that investigates techniques for the detection of diabetic retinopathy through exudates from color fundus images that include elimination of optic disk and detection of exudates using image processing methods.

3 Methodology

The proposed method starts with data acquisition [14], proceeding with image preprocessing, followed by segmentation and edge detection, and finally, feature extraction and classification.

Step 1 The image is taken from database and is loaded for image acquisition and preprocessing using MATLAB.

Step 2 The RGB image is then converted into a grayscale image for further processing.

Step 3 For proper statistical feature extraction of the image, the grayscale image is then converted into binary image.

Step 4 For removing the artifacts from the image, median filtering technique is applied.
Step 5 Edge detection of the image is done by using various operators like Sobel operator, Prewitt operator, and Laplacian of Gaussian operator.
Step 6 Finally, the feature extraction of the image is done.
Step 7 The above steps are repeated for identifying the statistical parameters for the rest of the images.

3.1 Preprocessing

One of the first steps in the methodology is preprocessing. The purpose of preprocessing is to extract all possible regions from input retinal image without any background errors. The steps involved are RGB to grayscale, grayscale to binary image, and noise removal [15].

3.1.1 RGB to Grayscale

The original input image is converted to grayscale image providing intensity information.

3.1.2 Grayscale to Binary Image

Binary image is used as input for feature extraction process that helps in generating unique feature to distinguish several components of diabetic retinopathy.

3.1.3 Noise Removal

Raw data or images cannot be directly used for any kind of research purpose or machine learning. As camera images generally contain noise in the form of distortion, blur, etc., therefore, median filter is used as one of the methods to remove noise. Compared with other filtering methods like Gaussian filtering, median filter is more suitable for precise preservation of images.

3.2 Edge Detection

Various operators have been used for edge detection [16, 17].

3.2.1 Canny Edge Detector

It is a technique to extract useful structural information from different vision objects and reduce the amount of data to be processed. It has been widely used in computer vision systems.

3.2.2 Sobel Operator

The operator is based on convolving the image with a small, separable, and integer-valued filter in both horizontal and vertical directions. It is a discrete differentiation operator, computing an approximation of the gradient of the image intensity function. At each point in the image, the result of the operator is either the corresponding gradient vector or the norm of the vector. The Sobel operator is used in image processing and computer vision where it creates an image emphasizing edges.

3.2.3 Prewitt Operator

It is a discrete differentiation operator computing an approximation of the gradient of the image intensity function. The operator calculates the gradient of the image intensity at each point giving the direction of the largest possible increase from light to dark and the rate of change in that direction.

3.2.4 Laplacian Edge Detector

The edge can be sharpened or enhanced by taking second-order derivative of image intensity. Edge detection by second-order derivative operator corresponds to the detection of zero-crossing. It is widely used in second-order derivative operator. Smoothing of the image intensity is done by convolving it with a digital mask corresponding to Gaussian function. The Laplacian mask is applied on the smooth image intensity profile and after the zero-crossings in the image are calculated subjected to Laplacian second-order derivative.

In a similar way, Roberts and zero-cross operators have been used.

3.3 Feature Extraction

Feature extraction starts from an initial set of measured data and builds derived values intended to be informative and non-redundant [18].

One of the statistical features extracted is skewness and corner feature. Skewness is used to calculate the lack of symmetry for the probability distribution of random variables. It is measured in terms of positive, negative, and normal.

4 Results and Discussion

This paper proposed and compared different edge detection techniques such as Canny edge detector, Sobel operator, Prewitt operator, and Laplacian of Gaussian edge detector for analysis of diabetic retinopathy using images of retina. The whole experimental setups are completed using MATLAB @ 8.4 software. The statistical parameters or the performance of matrices is very much significant for the calculation of pixel-pixel based on four values such as true positive (TP), true negative (TN), false positive (FP), and false negative (FN). The most significant statistical measures are sensitivity (Se) and specificity (Sp), which are responsible for improving the accuracy (Ac) to evaluate the overall system performance.

$$\text{Sensitivity}\,(\%\,) = \frac{\text{TP}}{\text{TP} + \text{FN}} \times 100 \tag{1}$$

$$\text{Specificity}\,(\%\,) = \frac{\text{TN}}{\text{TN} + \text{FP}} \times 100 \tag{2}$$

$$\text{Accuracy}\,(\%\,) = \frac{\text{TN} + \text{TP}}{\text{TN} + \text{TP} + \text{FP} + \text{FN}} \times 100 \tag{3}$$

The validation of experimental works is completed by statistical measurements like accuracy (Ac %), sensitivity (Se %), and specificity (Sp %). Figure 2 represents the outputs of different stages of preprocessing. First of all the original image of Fig. 1, is converted to RGB image which is shown in Fig. 2a. Then the RGB image is converted to grayscale image as shown in Fig. 2b. After that the binary image is generated from grayscale image as shown in Fig. 2c and before the edge detection the noise is removed from the image by using median filtering method which is shown in Fig. 2d.

Figure 3a–f illustrates the images obtained after edge detection using zero-cross detector, Laplacian of Gaussian detector, Roberts operator, Prewitt and Sobel operators, respectively. It can be seen that Sobel operators provide more accurate edge detection like indicating the blood vessels, hemorrhages, and exudates [19, 20] of a digital retina image as compared to other operators, as Sobel operators have slightly superior noise suppression characteristics. Second to Sobel operator is the Prewitt edge detector. Table 1 shows the different measured values using various edge detection techniques. From this analysis, Canny edge detector obtained accuracy, sensitivity, and specificity of 86%, 88%, and 88%, respectively. The Prewitt operator obtained accuracy, sensitivity, and specificity of 91%, 89% and 92%, respectively. The Laplacian edge detector obtained accuracy of 89%, sensitivity of 88%, and specificity of 90%.

Figure 4 indicates the points marked on the grayscale image for corner feature extraction. Finally, Table 1 shows that the Sobel operator outperforms in all respect compared to all other edge detection techniques used for study of retina image for analysis of diabetic retinopathy. Figure 5 illustrates the image output of various

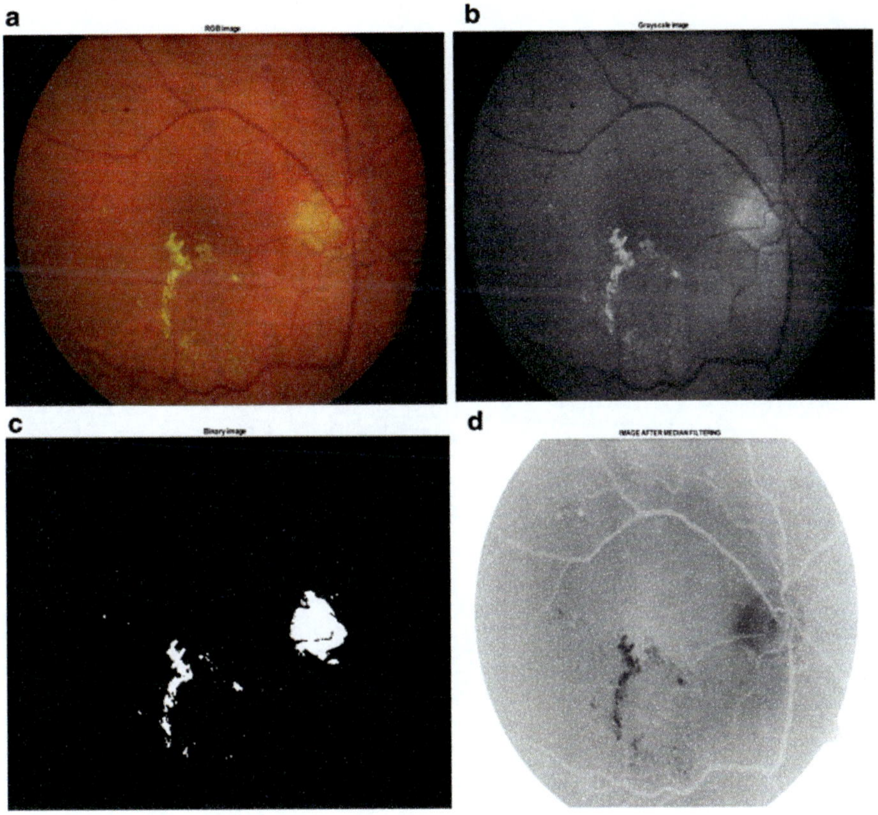

Fig. 2 **a** RGB image, **b** Grayscale image, **c** Binary image, **d** Image noise removals

operators, and it can be observed that Canny edge detector and Sobel operator outputs are clear and distinct. Figure 6 presents the skewness corner plot of Prewitt, Roberts, Laplacian of Gaussian, and zero-cross edge detectors.

5 Conclusion

We conducted a simple study of retina images of patients suffering from diabetic retinopathy [21]. The most common cause for retinal disease is diabetes. Detection and classification of deformation in diabetic retinopathy is highly demanding piece of work as in most of the cases, it is asymptomatic [22]. In this paper, we have discussed how fundus images can be used for analysis using various image edge detection techniques. The various edge detection operators are helpful in achieving the desired classification of features of diabetic retinopathy like augmented blood vessels, exudates, and hemorrhages. In this proposed work, there are four edge detection techniques

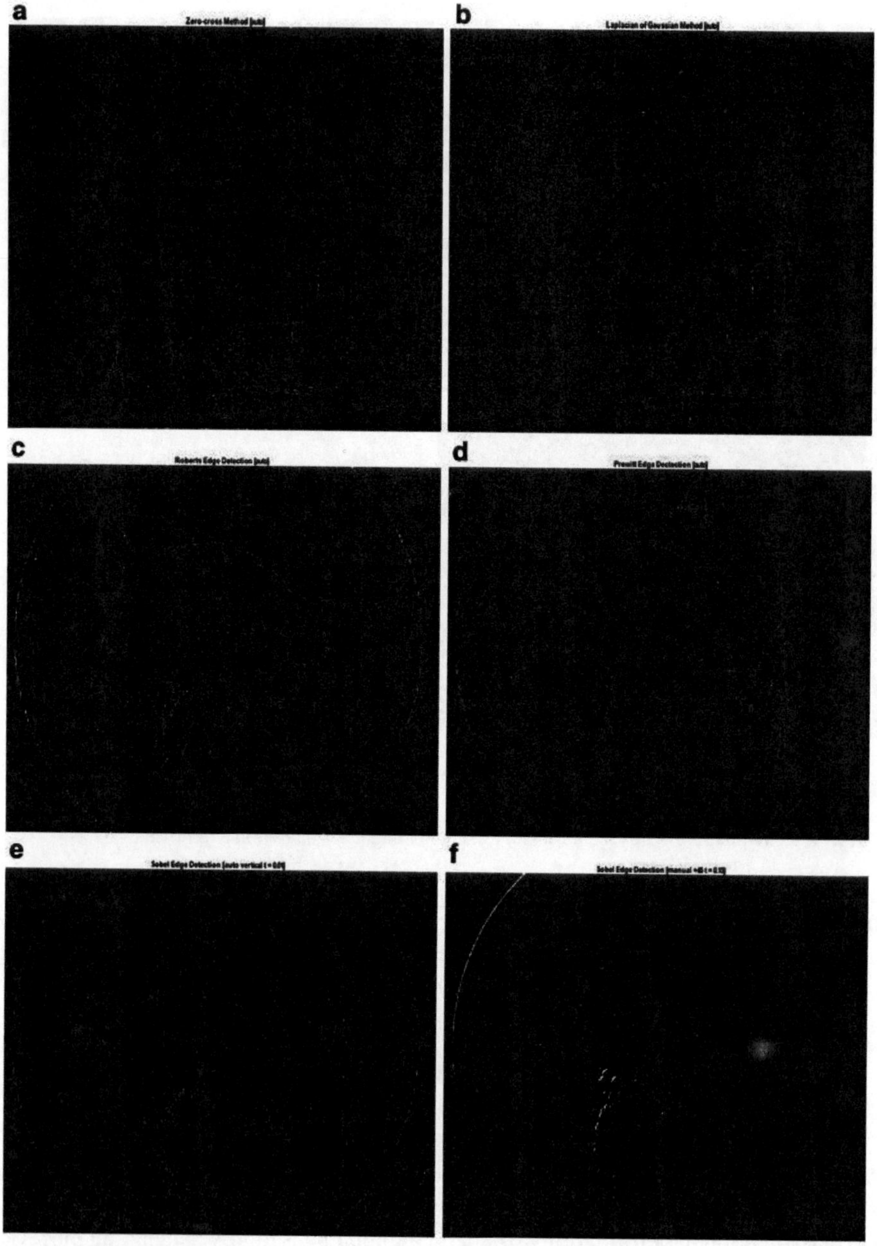

Fig. 3 **a** Edge detection using zero-cross, **b** Laplacian of Gaussian method, **c** Roberts edge detection, **d** Prewitt edge detection, **e** Sobel edge detection (auto), **f** Sobel edge detection

Table 1 Observation of statistical parameters in the proposed method

Edge detection techniques	Statistical measures		
	Ac (%)	Se (%)	Sp (%)
Canny edge detector	86	88	88
Sobel operator	95	94	96
Prewitt operator	91	89	92
Laplacian edge detector	89	88	90

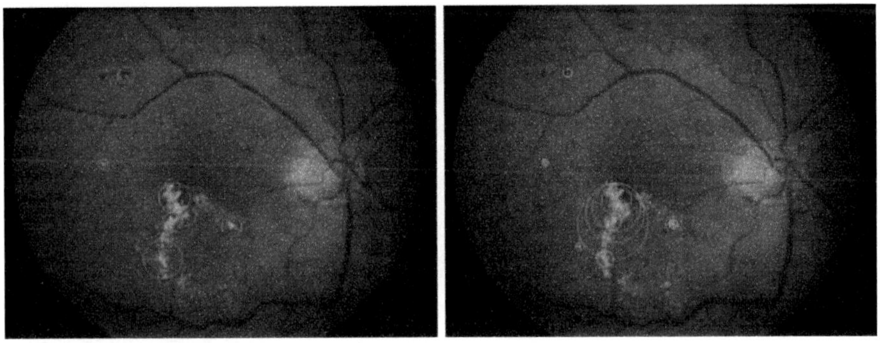

Fig. 4 Corner feature extraction

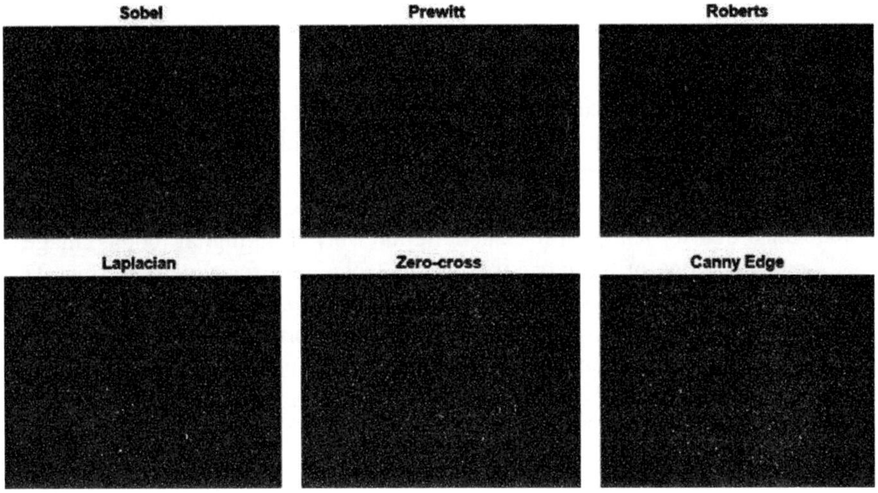

Fig. 5 Image output using various operators

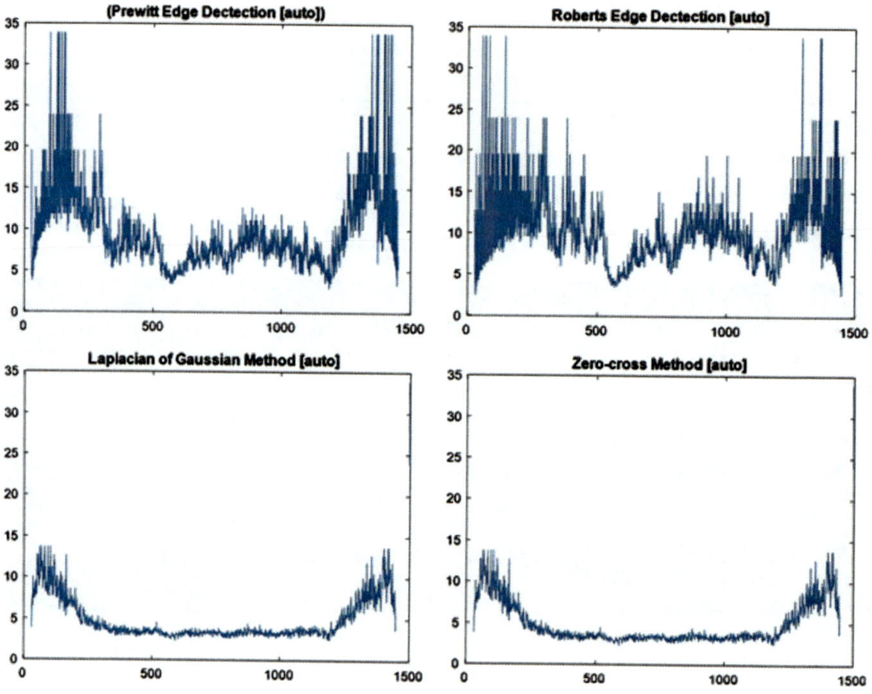

Fig. 6 Skewness corner plots of various operators

imposed to study digital fundus image for analysis of diabetic retinopathy. The statistical measures accuracy, sensitivity, and specificity are obtained. Finally, the Sobel operator outperforms in terms of accuracy, sensitivity, and specificity of 95%, 94%, and 96%, respectively, as compared to other existing techniques. Therefore, it can be concluded that for analysis of diabetic retinopathy from digital fundus images, Sobel operator is one of the most appropriate techniques for edge detection. The work elucidated in this paper reveals just a small fraction of the tremendous potential that image processing with machine learning can be provided in the classification and analysis of diabetic retinopathy. So, this method can be implemented for clinical diagnosis of diabetic patients in various medical applications in the future. Although a lot of work has been done in this field, detection of diabetic retinopathy in initial stages and providing a patient-friendly model is still a challenge.

References

1. Diabetic retinopathy Classification and Clinical features, Available at https://www.uptodate.com/contents/diabetic-retinopathy-classification-and-clinical–features
2. M.S. Ahmed, B. Indira, A survey on automatic detection of diabetic retinopathy. Int. J. Comput. Eng. Tech. **6**, 36–45 (2015)

3. H. Priya et al., Computer aided diagnosis methods of diabetic retinopathy using fundus images, in *International Conference on Circuits and Systems in Digital Enterprise Technology. IEEE, Kottayam* (2018). https://doi.org/10.1109/iccsdet.2018.8821200
4. R. Ghosh, K. Ghosh, S. Maitra, Automatic detection and classification of diabetic retinopathy stages using CNN, in *4th International Conference on Signal Processing and Integrated Networks*, IEEE, Noida, (2017), pp. 550–554. https://doi.org/10.1109/spin.2017.8050011
5. C.W. Tsai et al., A research of digital image based cognitive learning systems in applications of preventive medicine-An example of Redding elementary school, San Francisco, in *4th International Conference on Signal and Image Processing*. IEEE, Wuxi (2019), pp. 846–849. https://doi.org/10.1109/siprocess.2019.886
6. S. Kumar, B. Kumar, in Diabetic retinopathy detection by extracting area and number of micro aneurysm from colour fundus images, in *5th International Conference on Signal Processing and Integrated Networks*. IEEE, Noida (2018), pp 359–364. https://doi.org/10.1109/spin.2018.8474264
7. S. Dutta, C. Manideep, S.M. Basha, R.D. Caytiles, S. Iyengar, Classification of diabetic retinopathy images by using deep learning models. Int. J. Grid Distrib. Comput. **11**, 89–106 (2018). https://doi.org/10.14257/ijgdc.2018.11.1.09
8. V. Raman, P. Then, P. Sumari, Proposed retinal abnormality detection and classification approach: computer aided detection for diabetic retinopathy by machine learning approaches, in *8th International Conference on Communication, Software and Networks*. IEEE, Beijing (2016), pp. 636–64. https://doi.org/10.1109/iccsn.2016.7586601
9. R. Cheloni, A.S. Gandolfi, C. Signorelli, A. Odone A, Global prevalence of diabetic retinopathy: protocol for a systematic review and meta-analysis. BMJ Open 9 (2019)
10. R. Anacan, et al., Retinal disease screening through statistical texture analysis and local binary patterns using machine vision, in *10th International Conference on Humanoid, Nanotechnology, Information Technology, Communication and Control, Environment and Management*. IEEE, Baguio City (2018), pp. 1–6. doi: 10.110p/HNICEM.2018.8666278
11. D. Prasad, L. Vibha, K.R. Venugopal, Multistage classification of diabetic retinopathy using fuzzy neural network classifier. ICTACT J. Image Video Process. IJIVP **8**, 1739–1746 (2018)
12. E. Kotsiliti et al., A classification model for predicting diabetic retinopathy based on patient characteristics and clinical measurements. J. Model. Ophthalmol. **4**, 69–85 (2017)
13. E. Syeda, et al. Detection of exudates in diabetic retinopathy: a review, in *International Conference on Electrical, Electronics and Optimization Techniques*. IEEE, Chennai (2016), pp. 2063–2068. https://doi.org/10.1109/iceeot.2016.7755052
14. DIARETDBI database
15. www2.it.lit/fi/project/imageret/diaretb1
16. C. Bhardwaj, S. Jain, M. Sood, Appraisal of pre-processing techniques for automated detection of diabetic retinopathy, in *5th International Conference on Parallel, Distributed and Grid Computing*. IEEE, Solan (2018), pp. 734–739. https://doi.org/10.1109/pdgc.2018.8745964
17. S. Ahmed, Comprative study among sobel, prewitt and canny edge detection operators used in image processing. J. Theor. Appl. Inf. Tech. **96**, 6517–6525 (2018)
18. A. Sharma, M. Ansari, R. Kumar, A comparative study of edge detectors in digital image processing, in *4th International Conference on Signal Processing, Computing and Control*. IEEE, Solan (2017), pp. 246–250. https://doi.org/10.1109/ispcc.2017.8269683
19. R. Sahoo, Extraction of different features to detect diabetic retinopathy from retinal fundus image: a review. J. Manag. IT Eng. **8**, 292–305 (2018)
20. K. Palavalasa, B. Sambaturu, Automatic diabetic retinopathy detection using digital image processing, in *International Conference on Communication and Signal Processing*. IEEE, Chennai (2018), pp. 72–75.https://doi.org/10.1109/iccsp.2018.8524234
21. P. Kokare, Wavelet based automatic exudates detection in diabetic retinopathy, in *International Conference on Wireless Communication, Signal Processing and Networking*. IEEE, Chennai (2017), pp. 1022–1025. https://doi.org/10.1109/wispnet.2017.8299917
22. R. Shalini, S. Sasikala, A survey on detection of diabetic retinopathy, in *2nd International Conference on I-SMAC*. IEEE, India (2018), pp. 622–630. https://doi.org/10.1109/i-smac.2108.8653694

Design of Mathematical Model for Analysis of Smart City and GIS-Based Crime Mapping

Sunil Kumar Panigrahi, Rabindra Barik, and Priyabrata Sahu

1 Introduction

> In matters of justice and the rule of law, an ounce of prevention is worth significantly more than a pound of cure . . . Prevention is the first imperative of justice.

In current years, the cities are progressively facing more complex challenges, due to rapid growth in urban populations as well as the variety of technical and infrastructure-oriented problems. These problems are more challenges to make functionality in what a city was livable, the problems like crime control, parking, waste management, traffic congestion, and deteriorating infrastructure.

In smart city, crime analysis has turned into a big challenge, which needs a strong determination of research on crime and its investigation, finding evidence, and crime preventions.

Crime mapping and analysis play an important role in a smart city and effectively work for crime representation and visualization and take action adequately to the problem of criminality. The GIS plays smart solutions for smart city and an effective function in mapping of crime. In this paper, we formulate a mathematical model for smart city and GIS to recognize the hot spots in addition to support the expansion of investigation. The major challenges are integrations of resources and the present procedural approach toward investigation for crime mapping and the advancement of safe and secure city strategies. Chapter 2 contains literature review on GIS,

S. K. Panigrahi (✉) · R. Barik
School of Computer Science and Engineering, KIIT University, Bhubaneswar, India
e-mail: ctcsunil@gmail.com

R. Barik
e-mail: rabindra.mnnit@gmail.com

P. Sahu
Department of Computer Science Engineering and Application, IGIT, Sarang, India
e-mail: priyabsahu@gmail.com

S. Patnaik et al. (eds.), *Advances in Machine Learning and Computational Intelligence*, Algorithms for Intelligent Systems, https://doi.org/10.1007/978-981-15-5243-4_42

Chap. 3 proposes mathematical model for smart city, Chap. 4 describes GIS-based crime analysis, and Chap. 5 presents conclusion.

2 Previous Work Study

2.1 Literature Review on GIS

In 1986, Burrough and McDonnell [1] defined GIS as "a powerful tool capable of storing, retrieving, transforming, and displaying spatial data for a particular purpose." In 1992, Goodchild [2, 3] defines GIS and his scholars in 1999 Forer, Unwin, and in 2005 Longley [4] elaborates the meaning of GIS as a computer system which is capturing, managing, integrating, manipulating, analyzing, and displaying geographically referenced data. In 1997, Wilson [5] describes the GIS as a model which is extremely effective for mass torts and the class actions.

In 2005 the Dischinger [6] and in 2001 Wallace Chamberlayne considered GIS as a tool to the mainstream for investigation officer as a crime analyst. In 2007, Reis [7] defines GIS as an evidence medium which helps the judges, juries, and litigators in understanding the real fact and environmental relationships presented during trial. And in 2002, Markowitz [8] explained, visualizing and resolving complex environmental and legal problems we can use the GIS as a powerful tool. In the same year, Suggs drew attention in environmental crime. In 2006, Lewis [9] recommended identifying and analytically mapping in the minority population distributions where most of the crime and criminal activity are suspected to exist. In 2003, Wilson [5] defines that GIS can be used as a tool for investigative analysis in crime scenes. Manheim [10] in 2006 examined GIS as the utility for forensic taphonomy tool. And Burdette [11] 2007 applied GIS in a marine for mapping. Aschenback [12] 1991 described GIS as enhancing litigation through the availability of new information capable of providing answers to previously unanswerable questions. Wilson in 2003 demonstrated the usefulness of GIS as an investigative analysis tool, and Smith [13, 14] in 2001 described the application of GIS in Kosovo.

3 Proposal of Mathematical Model for Smart City

The city defined by most of the dictionary was based on the populations, and smart is a concept that describes how efficiently the task is completed with respect to time, money, and accuracy.

3.1 Properties of the Smart City

Currently, the cities are classified into four categories depending upon two important parameters, developed or emerged economy and legacy or new city [15–25]. Most cities cannot pay directly for "smartness," and more often, the cities cannot even evaluate the basic infrastructure, and the innovation in many situations has to be led also by private capital with a focus on interventions that pay for themselves most of the authors focus on.

The basic information is derived from different sources and formulates to the following equations. As the cities are basically focusing on better service providers to its citizen, but due to rapid urbanizations, they fail to provide effective service to its citizen, so the cities utilize the technology to make the city as a smart. We are trying to here identify the parameters that influence the smartness of the city, and Figure 1 contains the sets of all parameters of different properties of smart city.

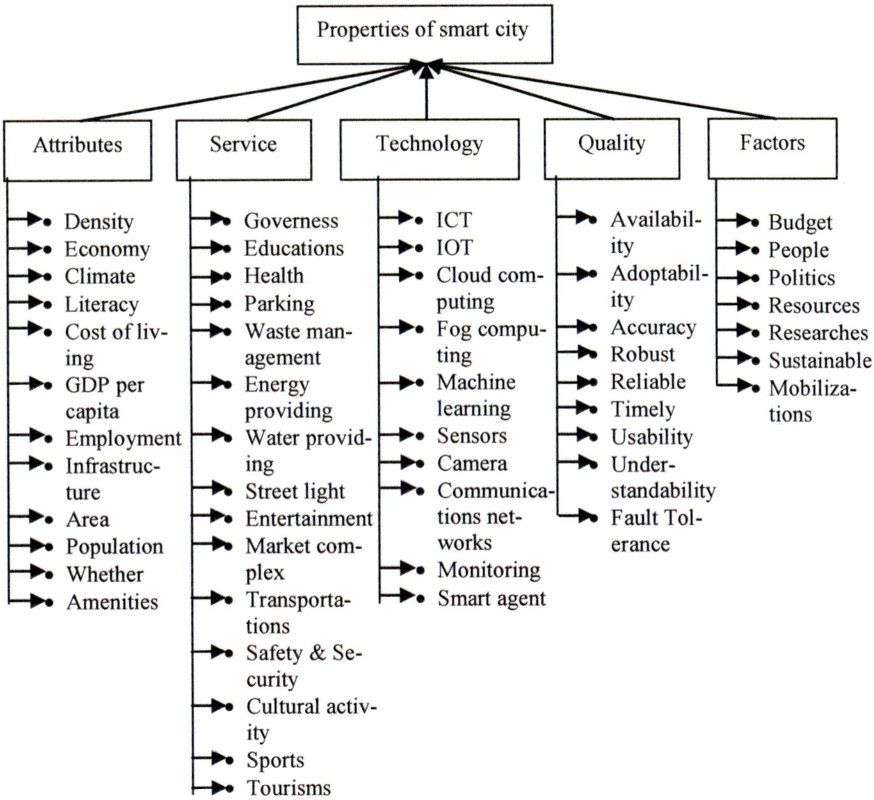

Fig. 1 All the properties and parameters of a city in modeular form

- **Attribute of a city**: An attribute is an element that has a value and is associated with a city; in other words, an attribute is a characteristic given to a city.
- **Service of a city**: A service is a valuable action, deed, or effort performed to gratify a need or to fulfill a demand that stays in the city. Services are one of the two key components of economics, the other being goods.
- **Technology applied for smartness to a city**: Technology is the purposeful application of information in the design, actions oriented, and utilization of goods and services, in city to do the activities faster and accurately.
- **Quality of a smart city**: Quality is an indicator of how accuracy is maintained in service in city to make it smart. In an information technology service, quality is sometimes taking major role to make a city smart. The standard of something as measured against other things of a similar kind; the degree of excellence of something, a distinctive.
- **Factors that affect for implementing of smartness in a city**: Factors are influencing the city as smart; without the help of factors, it is difficult to make a city smart.

3.2 Algorithm Design for Analysis of Smart City

From the above conclusion, we define the city in the form of set theory.

Let C be the set of parameters for smart city:

$$C = \{A, S, T, Q, F\}$$

Where

A set of attributes of the city
S set of services provided by the city
T set of technology applied to a city
Q set of quality factors for smartness of city
F set of factors that affect the smartness

For Smartness (Sm) of a city, we can evaluate

$$\mathrm{Sm}(C, S) = \sum (A, T, Q, F) \tag{1}$$

Let C be the city and R is the set of requirements for smartness (it is the combinations of service and technology). Consider the city attribute is constant.

DO (city to smartness)
THEN it needs

Assessment (attributes and factors)
Planning (attributes and service)
Consultant (attributes, service, technology and factors)

Smart Solutions (service, technology and quality)
Cost (technology and factors)
Implementations (technology, quality, factors)

ENDO (till it satisfies all the properties of the city).
The attribute and service are limited. When we talk about the technology, the technologies have limitations and bounded performance indicator.

4 GIS Based Crime Analysis

A geographical information system (GIS) is a tool for identifying, locating capturing, storing, analyzing, and managing the raw data and its related attributes, which are spatially referenced to the Earth. Crime analysis mapping (CAM) is the combination of crime analysis techniques with the geographic information system in order to focus on the spatial perspective of criminal activities and other law enforcement operations [26].

Crime analysis mapping can be used for problem-solving tool, and it can also be used as investigation purpose. Three factors are considered to produce accurate and proper crime analysis maps: (i) the map and its purpose, (ii) the map audience, and (iii) the data types that are used in the map as shown in the case study (Velasco and Boba 2000). These factors usually determine the nature of the map that will be used and the presentation method.

4.1 Analysis of Crime in the Smart City

The crime analysis divided [27] and investigation into five workflow stages as shown in Figure 2: (1) All data related to crime are stored; (2) proper design of database for processing and storing crime data; (3) the operations like searching, retrieving, and collecting additional information for crime analysis; (4) clues identified after analyzing database; and (5) necessary action carried out using information to prosecute (or defend) individuals.

For smartness in safety and security of citizen in that city

Smartness (city, safety, and security) = {Location as attribute, crime control is service, using GIS as a technology, quality in crime investigation, limited resource and manpower is factor}.

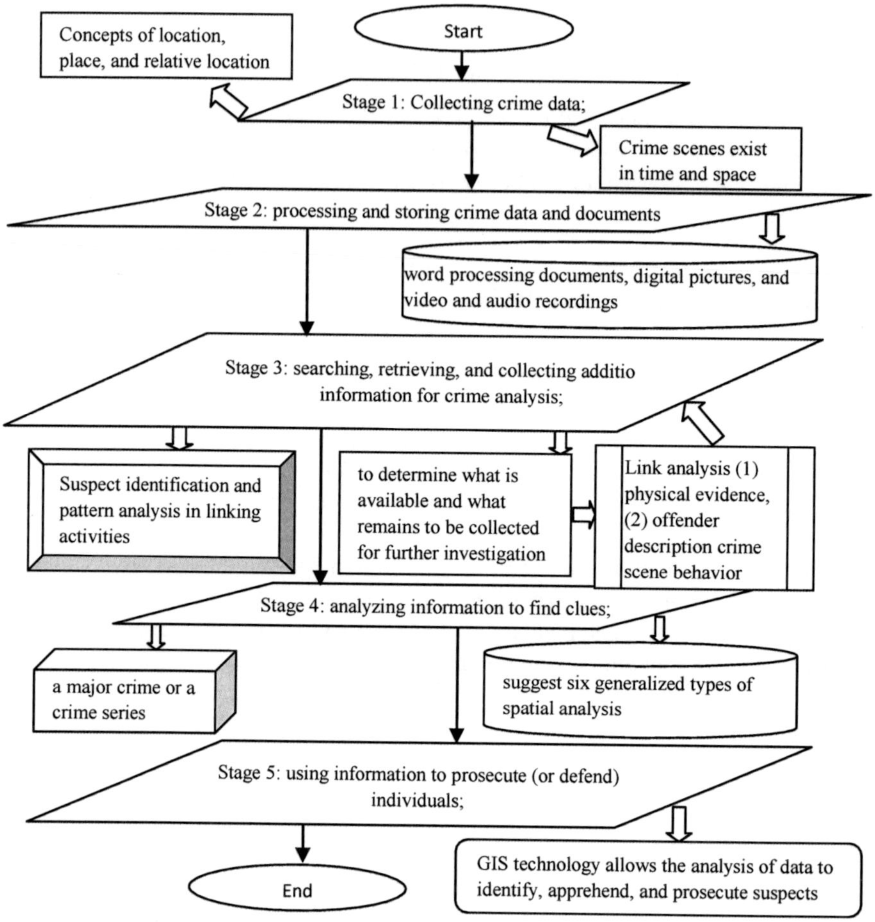

Fig. 2 Flowchart of investigations process and data storing

4.2 Case Study for Bhubaneswar City

For example, considering the Bhubaneswar city for theft of vehicles may have significantly increased. To determining the area where the crime activity has increased the most, crime analysis mapping has been used. The theft of vehicle incidents showed high rates during the last years. As per the police record, which part of the city is mainly affected by thefts from vehicle incidents is shown in Fig. 3b, c. The patrolling officers can use the information to enhance surveillance in that region as shown in Figure 3d. Furthermore, as per the locations shown in Figure 3e, the community policing officers can aware the citizen, how to stop auto theft-related crimes in their neighborhoods. The purpose of the analyses is to identify the more problematic areas of Bhubaneswar for theft of vehicle for the purpose of initiating an education

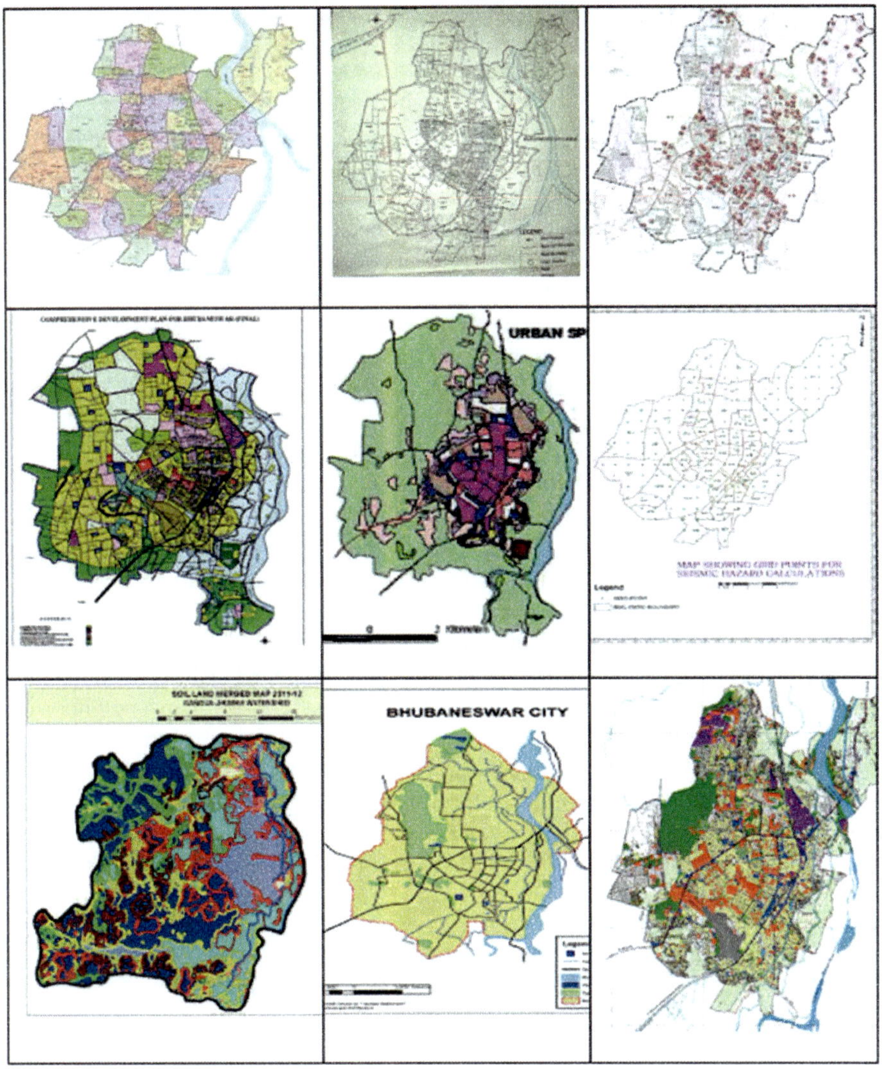

Fig. 3 Crime mapping and analysis. **a** Bhubaneswar urban area division, **b** most of the market area, **c** theft area of vehicle, **d** need the patrolling, **e** awareness required for vehicle theft, **f** cluster division of theft, **g** red zone for theft, **h** Patrolling path, **i** expansions of Bhubaneswar area for crime mapping

campaign. As shown in Fig. 3f, g, population density, city area, and the tabular crime data are used for crime data analysis. Because the purpose of the map is to decide which part of the city has the most thefts from vehicle, a map of the whole city, portraying streets, has been used in Fig. 3h, i. The study region is large, so it is not required to use detailed maps, like land-use parcels or streets, since point symbols will cover a lot of the land-use parcels and streets.

The map is messy, and it is not easy to identify accurately the high affected areas, for which the purpose of making the map. Moreover, point data, which are geocoded to the crime same location of Bhubaneswar, will be stacked on top of each other and there would be no way to decide if more than one theft from vehicle took place at a specific place. Thus, this map does not precisely describe theft from vehicle incidents, and it does not meet the requirements of the audience because it is almost impossible to discriminate the incident from another incident.

Basically, the theft of vehicle occurred in locations of marketplaces, educational institute, office building, etc. Moreover, when a problem is recognized and analyzed, the presentation of a map becomes important, and it is important to set up maps that are suitable for the purpose [26, 28]. For instance, a comprehensive map of every hot spot location may be required for police patrol officers and investigation officers who need specific information that is the focus.

Using crime analysis mapping gives one the ability to help in all steps of the problem and solutions sequence. The purpose of a crime analysis of map is to assist in the recognition of a specific problem; crime analysts may map chosen data to discover patterns of activity that have, until that time, gone unnoticed. For instance, to decide if there is a pattern to commercial burglary incidents, analysts may map a past month's data for every commercial burglary that happened in the city. Analyzing for clusters of activities may identify a group of incidents occurring in a neighboring area. Mapping for successive months may show pattern changes. In this case, the crime analysis mapping purpose is to discover a problem, i.e., in a specific area, "Is there a sudden increase in commercial burglary incidents?" Studying a particular problem in depth is another purpose. Crime analysis mapping can be used to carry out a complete spatial analysis of the issue to examine it more closely from different viewpoints by using a mixture of data.

5 Conclusion

This mathematical model is used to evaluate the smartness of the city. The service and technology evaluate the smartness of city. The police department used the mathematical model to identify the current crime rate in Bhubaneswar city. The algorithm generates the hot spots for patrolling. The model generates the frequent crime patterns, using the mathematical algorithm. The result is presented using the GIS to find the root for patrolling, controlling, and avoidance of crime hot spots. The researchers are able to develop a better algorithm which can generate crime hot spots and the density of crime in different areas using different mapping techniques. More sophisticated algorithm can be designed for crime analysis.

References

1. P.A. Burrough, R.A. McDonnell, *Principles of geographical information systems* (Oxford University Press, Oxford, 1986)
2. M.F. Goodchild, Geographic information science. Int J Geogr Inform Sci **6**(1), 31–45 (1992)
3. M.F. Goodchild, Geospatial technologies and homeland security research, in D.Z. Sui (ed) (2008)
4. P. Longley, M.F. Goodchild, D.J. Maguire, D.W. Rhind, *Geographical information systems and science*, 2nd edn. (Wiley, New York, 2005)
5. A.C. Wilson, S.G. Jones, M.A. Smith, R. Liles, Tracking spills and releases: high-tech in the courtroom. Tulane Environ Law J **10**, 371–394 (1997)
6. S.S. Dischinger, L.A. Wallace, Geographic information systems: coming to a courtroom near you. Colo Lawyer **34**(4), 11–23 (2005)
7. D.G. Ries, Computer presentations by lawyers in the conference room, classroom, and courts. Pa Bar Assoc **78**, 56–70 (2007)
8. K.J. Markowitz, Legal challenges and market rewards to the use and acceptance of remote sensing and digital information as evidence. Duke Environ. Law Policy Forum **12**(2):219–264 (2002). Marshall P (2013) USDA's high-res view of fraud. http://gcn.com/articles/2013/10/09/gcn-award-usda-fraud-detection.aspx. Accessed 14 Oct 2013
9. B.C. Lewis, Changing the bathwater and keeping the baby: exploring new ways of evaluating intent in environmental discrimination cases. Saint Louis Univ Law J **50**, 459–516 (2006)
10. M.H. Manhein, G.A. Listi, M. Leitner, The application of geographic information systems and spatial analysis to assess dumped and subsequently scattered human remains. J. Forensic Sci. **51**(3), 469–474 (2006). https://doi.org/10.1111/j.1556-4029.2006.00108.x
11. L.G. Burdett, J.D. Adams, W.E. McFee, The use of geographic information systems as a forensic tool to investigate sources of marine mammal entanglement in fisheries. J. Forensic Sci. **52**(4), 904–908 (2007). https://doi.org/10.1111/j.1556-4029.2007.00466.x
12. R.J. Aschenbach, Geographic information systems as a decision making tool. Ohio State Law J **52**, 351–368 (1991)
13. D.G. Smith, Kosovo: applying GIS in an International Humanitarian Crisis. ArcUser, July–September 2001 (2001)
14. D. Smith, The neighborhood context of police behavior, in Reiss A, Tonry M (eds) Communities and crime (University of Chicago Press, Chicago (1986), pp. 313–341
15. https://e.huawei.com/in/solutions/industries/smart-city
16. https://www.accenture.com/in-en/insight-smart-cities
17. https://www.xenius.in/smart-city/
18. https://www.silverspringnet.com/solutions/smart-cities/
19. https://in.nec.com/en_IN/products/public-safety-security/smart-city
20. http://www.verizonenterprise.com/view/factsheets/11416/smart-technologies-and-smart-city-solutions-factsheet
21. https://smartcitysolutions.eu/en/smart-city-solutions/
22. http://www.mindteck.com/smartcity-solutions
23. http://www.rolta.com/solutions/smart-city-solutions/
24. https://www.efftronics.com/smart-city-solutions
25. http://www.mavensystems.com/smart-city.html
26. R. Boba, *Introductory guide to crime analysis and mapping* (U.S Department of Justice, Washington, D.C., 2001)
27. J.L. Zhao, H.H. Bi, H.C. Chen (2003) Collaborative workflow management for interagency crime analysis, in: Intelligence and security informatics, proceedings. Lecture notes in computer science, vol. 2665 (Springer, Berlin, 2003) pp. 266–280
28. D.E. Brown, The regional crime analysis program (ReCAP): a framework for mining data to catch criminals, in *IEEE International Conference on 11–14 Oct, 1998*, vol. 3 (1998), pp. 2848–2853

Pattern Storage and Recalling Analysis of Hopfield Network for Handwritten Odia Characters Using HOG

Ramesh Chandra Sahoo and Sateesh Kumar Pradhan

1 Introduction and Related Work

Content-addressable memories (CAMs) are associative memories that can retrieve patterns even if input patterns are partially and/or noisy as like human brain. The basic principle of CAMs is the equilibrium points of nonlinear dynamical systems which are also known as attractors. Auto-associative CAMs have been proven well-known massively parallel operations which is robust to noisy input. Hopfield network [1, 2] is a best-known auto-associative recurrent network with binary threshold nodes that guaranteed to converge a local minima. In Hopfield model, patterns are stored locally as connection strength between processing units until a stable state is reached which represents memorized patterns. Several modifications [3–10] are done so far for improving storage capacity as well as retrieving efficiency. The network is initialized with an input pattern and then is updated asynchronously and in parallel which leads to converge to most likely stored pattern due to the dynamical behaviour of the neurons. In this study, we used HOG feature extraction method to extract the reduced feature set of image that is to be stored in and recall from Hopfield network. HOG feature extraction method [11–15] is a very efficient and popular feature descriptor for object detection; however, many researchers have been using it for pattern recognition and classification problems. In HOG feature descriptor gradient orientations of an image are counted by dividing into small connected regions. Odia language is the official language, spoken by more than 50 million people in the Indian state of Odisha. In modern Odia alphabets, there are 12 vowels, 35 simple consonants, 10 digits and

R. C. Sahoo (✉) · S. K. Pradhan
Department of Computer Science & Applications, Utkal University, Bhubaneswar, India
e-mail: rsahoo22@gmail.com

S. K. Pradhan
e-mail: sateesh1960@gmail.com

S. Patnaik et al. (eds.), *Advances in Machine Learning and Computational Intelligence*, Algorithms for Intelligent Systems, https://doi.org/10.1007/978-981-15-5243-4_43

nearly about 200 composite characters. In this proposed work, we focused only on vowels and consonants to measure the recall efficiency of Hopfield model through extracted HOG features.

2 Features of Odia Script and Challenges

Odia script is one of the oldest and popular scripts among various Indian regional languages. Odia script is originally developed from Kalinga script as descendants from Brahmi scripts. Odia script is written left to right as like other regional scripts used in India, and these characters are called Aksara. In modern Odia script, there are 12 vowels as shown in Fig. 1, 35 simple consonants as shown in Fig. 2, 10 numerals as shown in Fig. 3 and nearly 200 composite characters called Juktakshyara. Most of these characters are rounded and have similar shape in their upper part, and there

Fig. 1 Odia vowels with English transliteration

Fig. 2 Odia consonants with English transliteration

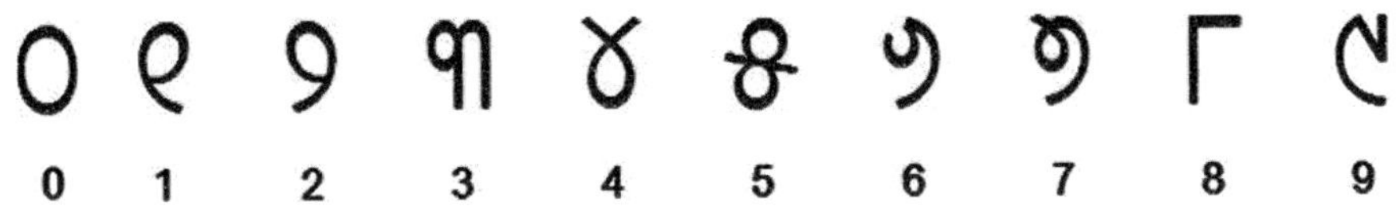

Fig. 3 Odia numerals with their English numerals

is no reference line as in other Indian regional languages like Hindi and Bangla. Due to this similar shape of these characters, recognition is a more challenging task to perform.

3 Proposed Method

3.1 Dataset

For this proposed work, we considered NIT Rourkela handwritten database as very few standard databases are available for Odia script. There are 47 folders, out of which 12 are for vowels and 35 for consonants. Each folder consists of 320 handwritten Odia character images written by 160 individuals and each of size 81×81 pixels. We consider 100 images from each class for our work to describe the recall efficiency.

3.2 Experimental Details

The proposed method demonstrated storage and recalling efficiency of Odia handwritten characters in Hopfield neural network using HOG features. Steps followed for this work are as follows: (1) preprocessing, (2) feature extraction using HOG descriptor and (3) measuring the storage and recall efficiency by Hopfield network.

3.3 Preprocessing

Some basic preprocessing steps have been followed to improve the performance of the network. Firstly, images are resized to dimensions of 32×32 pixels to get reduced feature vector for lesser training time in the network. Secondly, Gaussian filter has been applied to remove noise from the reduced images. And lastly, binarization and normalization processes are applied to filtered images and the passed to feature extraction step.

3.4 Feature Extraction Through Histogram of Oriented Gradients (HOG)

Basically, HOG is a feature descriptor method mostly used for object detection in computer vision. It focuses on shape or structure of an object and is different from edge feature detection methods as it provides edge features with its direction whereas edge features only provide edge features. HOG divides the image into smaller regions and calculates gradients and orientation of each region by producing a vector of features. Steps for obtaining HOG feature are described in the following section.

- Calculate gradients in x and y directions of each region (say R).
  Gradients are the small changes in x and y directions and are calculated as follows:

$$\text{Change in}x \text{ direction, } R_x(r, c) = R(r, c+1) - R(r, c-1) \tag{1}$$

$$\text{Change in}y \text{ direction, } R_y(r, c) = R(r-1, c) - R(r+1, c) \tag{2}$$

  where r and c are row and column of the sub-region of image.
- Calculate the magnitude and orientation for each pixel value by applying Pythagoras theorem as follows:

$$\text{Total gradient magnitude, } G_M = \sqrt{R_x^2 + R_y^2} \tag{3}$$

  Now, calculate the orientation (direction) for the same pixel as follows:
  $\tan(\theta) = \frac{R_y}{R_x}$ and the value of angle will be

$$\theta = \tan^{-1}\left(\frac{R_y}{R_x}\right) \tag{4}$$

- Calculate histogram of gradients for each sub-blocks of 8 × 8
  We divide the image 8 × 8 non-overlapping cells and generate the histogram for each cell. Each cell produces a 9 × 1 vector of histogram gradients features.
- Normalize gradients in 16 × 16 cell
  After getting the features of each 8 × 8 cell, we combine four 8 × 8 cells to get a 16 × 16 cell for normalization. As each 8 × 8 cell has a 9 × 1 vector, we get four such cells to get a 36 × 1 vector and to normalize; we divide each of these values by the square root of the sum of squares of the values as follows:
  Obtain the root of sum of squares as $K = \sqrt{v_1^2 + v_2^2 + \cdots + v_{36}^2}$, where $v_1, v_2, \ldots, v_{36}$ are the values of vector obtained from a 16 × 16 block after merging four 8 × 8 cells.
  Now to get the normalized vector, divide all the values of 36 × 1 vectors by *K as*

$$\text{Normalized vector } V_{\text{norm}} = \left(\frac{v_1}{K}, \frac{v_2}{K}, \ldots, \frac{v_{36}}{K}\right) \tag{5}$$

- Obtain the features for the complete image
  After getting normalized features of all 16 × 16 blocks, we combined all these to get the features of whole image. In our case, we have four 16 × 16 blocks, and hence, we get 144 features for one complete image.

These features are now trained (learn by Hebbian rule) in Hopfield network, and then, one can recall them with or without noise.

3.5 *Hopfield Neural Network*

Hopfield neural network is a recurrent network which can be used as a content-addressable memory to memorize patterns by its internal states. In this model of n binary neurons, each neuron can have $z_i \in \{0, 1\}$or $\{+1, -1\}$ where neurons i and j are coupled symmetrically weights $w_{ij} = w_{ji}$ with no self-connection, i.e. $w_{ii} = 0$. In the training phase, it stores the pattern as energy attractor by which when we try to recall some pattern, the stored patterns attract the presented pattern towards them. Two basic phases of Hopfield network are storage and retrieve phases. Several learning rules have been proposed like Hebbian learning, pseudo-inverse learning, delta learning and Storkey learning rule for Hopfield network for storing and recalling of patterns. Here, we present the Hebbian learning rule to present the recall efficiency of Odia handwritten characters. Storage and retrieval phases are explained in the following section. Storage and retrieval in Hopfield network are presented in the following algorithm.

Suppose Hopfield network consists of N fully connected neurons having no self-connection with each neuron i has state:

$$x_i, i \in \{1, 2, \ldots N\} \tag{6}$$

To store P patterns of $X = (X^1, X^2, \ldots, X^P) \in \{-1, 1\}^N$, then patterns are stored into the Hopfield network by Hebbian rule is as follows:

$$w_{ij} = \frac{1}{N}\sum_{p=1}^{P} x_i^p x_j^p \text{for} i \neq j \text{ and } w_{ii} = 0 \tag{7}$$

where x_i^p is the ith unit of pattern p, $i = \{1,2,\ldots, N\}$ *and* $p = \{1,2,\ldots, P\}$.

For retrieval of an input pattern (say test_i) with or without noise is presented to the network at time $t = 0$.

Present the test pattern (*say* test_i) to the network at *time* $t = 0$ as

$$y_i(t = 0) = \text{test}_i, \forall i = \{1, 2, \ldots, N\} \tag{8}$$

Now, state of neuron *i*, (*chosen randomly*) is updated using

$$y_i(t+1) = f\left(\sum_{i=1}^{N} w_{ij} y_i(t)\right), \forall j = \{1, 2, \ldots, N\} \tag{9}$$

where f is hard-limit nonlinearity function given as

$$f(\theta) = \begin{cases} 1, & \text{if} \quad \theta \geq 0 \\ -1, & \text{if} \quad \theta < 0 \end{cases}, \text{ where } \theta \text{ is the threshold value.} \tag{10}$$

Repeat the above step in Eq. (9) until the output from the node remains unchanged or reached to a stable state.

Hopfield network guarantees to retrieval of a pattern because of its energy function associated with each state of the network, which either decreases or remains the same upon repeated updating of neurons that leads to local minimum. This local minimum in the energy function is the stable state which eventually converges to the desired retrieval of pattern. The energy function $E(x)$ of a state x is given as follows:

$$E(x) = -\frac{1}{2} x W x^{\mathrm{T}} + \phi x^{\mathrm{T}} \tag{11}$$

And, it can also be written as

$$E(x) = -\frac{1}{2} \sum_{i=1}^{N} \sum_{j=1}^{N} w_{ij} x_i x_j + \sum_{j=1}^{N} \phi x_i \tag{12}$$

4 Discussion and Conclusion

The proposed model demonstrated the recalling efficiency of handwritten Odia characters (vowels, consonants and numbers) in Hopfield model with HOG feature extraction technique. The simulation work has been implemented to obtain the recalling efficiency with original images as well as with 10, 20, 30 and 40% noisy images. New images that were not used to train the network have also been tested for recalling. Results we obtained with different noise percentages are presented in Table 1. It describes that without noise, the recalling rate is 100% and the regression plot diagram is shown in Fig. 4. To recall noisy images of stored patterns from the network, 10–40% external noises were introduced into the original images. Recalling efficiency decreases as noise percentages are increased. It has been observed that even with 40% noise, nearly 60% of the patterns are recalled correctly. Regression plot graphs for

Table 1 Pattern recall efficiency (in %) of Hopfield model through HOG features

No. of patterns trained	Noise in percentages				
	No noise	10% noise	20% noise	30% noise	40% noise
50% of NIT database (50 from each class)	100	93.43	87.05	72.17	60.5
Full database (100 from each class)	100	93.02	86	71.33	58.87

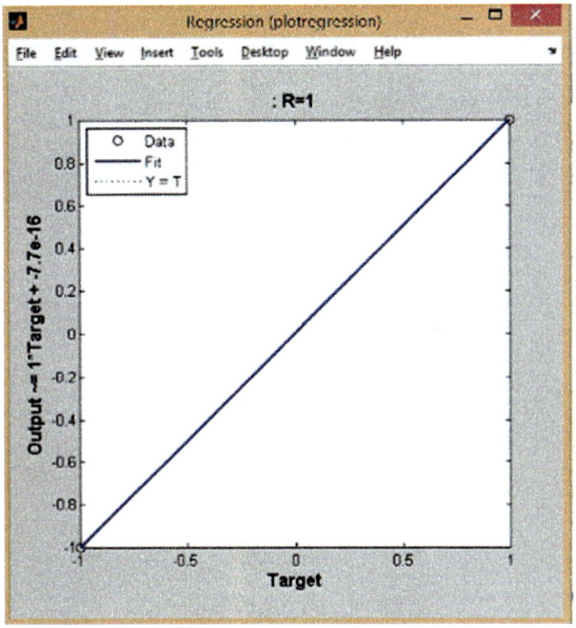

Fig. 4 Regression plot graph for recalling patterns without noise (originally stored patterns)

10, 20, 30 and 40% of noise are shown in Fig. 5, Fig. 6, Fig. 7 and Fig. 8 respectively. Then, we tested for some new patterns from each class which are not used for training and the result is shown in Table 2. For this, we divide the whole dataset into training and testing sets. Here, we used 90% for training and 10% for testing. Result shown in Table 2 demonstrated that only 25 patterns out of 470 patterns were not correctly recalled with nearly 95% accuracy without any noise, and with 40% noise, nearly 59% patterns are correctly recalled.

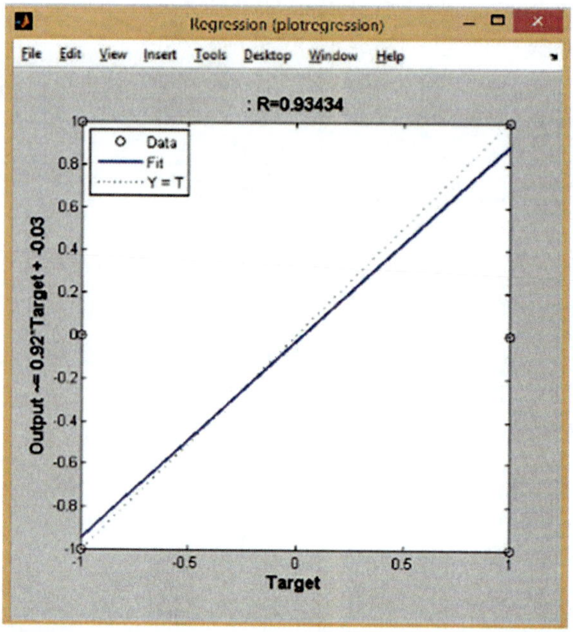

Fig. 5 Regression plot graph for recalling patterns with 10% noise

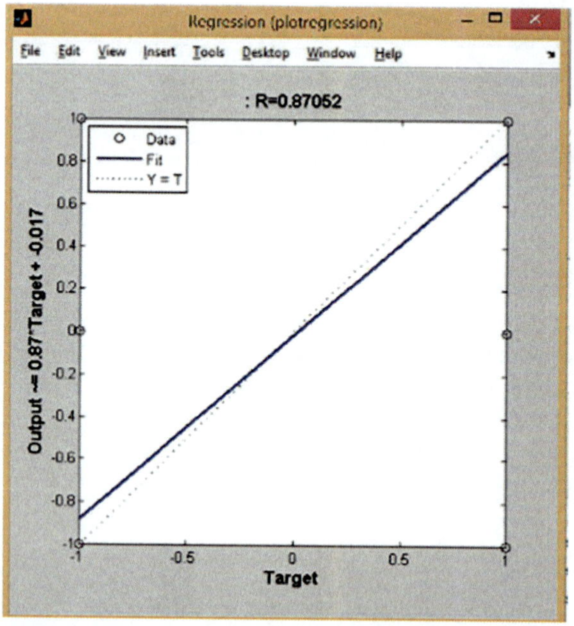

Fig. 6 Regression plot graph for recalling patterns with 20% noise

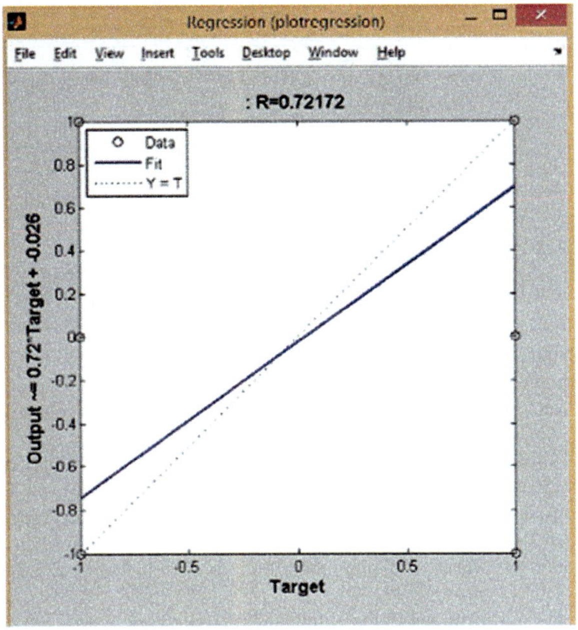

Fig. 7 Regression plot graph for recalling patterns with 30% noise

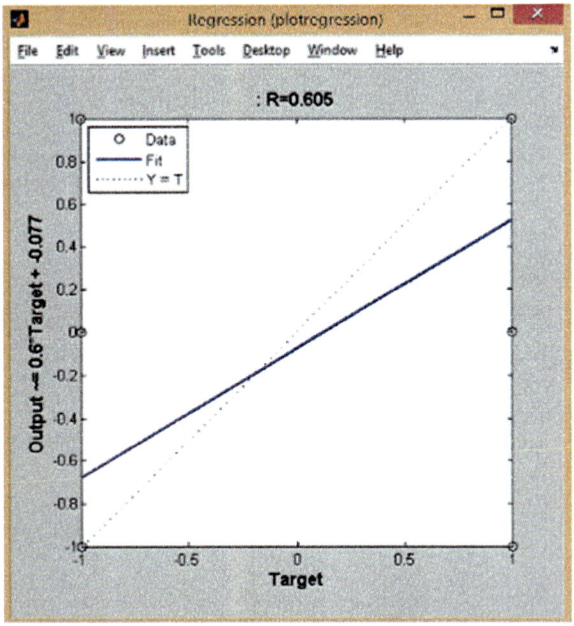

Fig. 8 Regression plot graph for recalling patterns with 40% noise

Table 2 Pattern recall efficiency (in %) of Hopfield model through HOG features with partitioning data

Database	Train–test partition	Noise in percentages				
		No noise	10% noise	20% noise	30% noise	40% noise
NIT handwritten Odia database	90–10	94.66	90	80.66	70.87	58.66

References

1. J.J. Hopfield, Neural Networks and Physical Systems with Emergent Collective Computational Abilities. Proceedings of the National Academy Sciences, USA **79**, 2554–2558 (1982)
2. J.J. Hopfield, Neural networks and physical systems with emergent collective computational abilities. Proc. Nat. Acad. Sci. USA **81**, 3088–3092 (1984)
3. D.J. Amit, H. Gutfreund, H. Somopolinsky, Storing Infinite number of patterns in a spin-glass model of neural networks. Phys. Rev. Lett. **55**(14), 461–482 (1985)
4. D.J. Amit, *Modeling Brain Function: The World of Attractor Neural Networks* (Cambridge University Press, New York, NY, USA, 1989)
5. S. Haykin, Neural Networks: A Comprehensive Foundation, Upper Saddle River: Prentice Hall, Chap 14(1998), p. 64
6. Z. Zhou, H. Zhao, Improvement of the Hopfield neural network by MC-adaptation rule. Chin. Phys. Letters **23**(6), 1402–1405 (2006)
7. H. Zhao, Designing asymmetric neural networks with associative memory. Phys. Rev. **70**(6), 066137 (2004)
8. M. Kawamura, M. Okada, Transient dynamics for sequence processing neural networks. J. Phys. A: Math. Gen. **35**(2), 253 (2002)
9. D.J. Amit, Mean-field using model and low rates in neural network, in *Proceedings of the International Conference on Statistical Physics,* 5–7 June (Seoul Korea, 1997), pp. 1–10
10. A. Imada, K. Araki, Genetic algorithm enlarges the capacity of associative memory, in Proceedings of the Sixth International Conference on Genetic Algorithms (1995), pp. 413–420
11. N. Dalal, B. Triggs, Histograms of oriented gradients for human detection. IEEE Comput. Soc. Conf. Comput. Vision Patt. Recogn. **1**, 886–893 (2005)
12. P. Dollar, S. Belongie, P. Perona. The fastest pedestrian detector in the West, in *Proceedings of the British Machine Vision Conference* (BMVA Press, 20100, pp. 68.1–68.11
13. P.F. Felzenszwalb, R.B. Girshick, D. McAllester, D. Ramanan, Object detection with discriminatively trained part-based models. IEEE Trans. Pattern Anal. Mach. Intell. **32**(9), 1627–1645 (2010)
14. F. Porikli, Integral histogram: a fast way to extract histograms in Cartesian spaces. IEEE Conf. Compute. Vision Patt. Recogn. **1**, 829–836 (2005)
15. C. Vondrick, A. Khosla, T. Malisiewicz, A. Torralba, HOGgles: visualizing object detection features, in International Conference on Computer Vision (2013) pp. 1–

A Distributed Solution to the Multi-robot Task Allocation Problem Using Ant Colony Optimization and Bat Algorithm

Farouq Zitouni, Saad Harous, and Ramdane Maamri

1 Introduction

The multi-robot task allocation (MRTA) problem is a very important aspect of multi-robot systems, in which a set of elementary tasks must be assigned optimally to some robots. The MRTA problem is frequently present in several applications of multi-robot systems, such as reconnaissance [1, 2], unmanned search and rescue operations [3, 4], transport and delivery of goods [5, 6] and autonomous exploration [7, 8]. The MRTA problem can informally be defined as follows: "First, we suppose two distinct groups of robots and tasks, respectively. Then, we should assign tasks to robots while optimizing a certain objective function, e.g. minimize the makespan". It is important to mention that the MRTA problem is NP-hard [9] and finding its optimal solution in linear time is not often feasible. Over the last decade, several solutions have been proposed to solve different MRTA problems. Usually, they are categorized into two families: centralized and decentralized [10]. Generally, approaches of the first family provide optimal solutions; however, they are not robust. Conversely, approaches of the second family are usually robust; however, no guarantee is provided on the quality of their solutions. Currently, many researchers prefer decentralized approaches for two

F. Zitouni (✉)
Department of Computer Science, Kasdi Merbah University, Ouargla, Algeria
e-mail: farouq.zitouni@univ-constantine2.dz

S. Harous
Department of Computer Science and Software Engineering, UAE University, Abu Dhabi, United Arab Emirates
e-mail: harous@uaeu.ac.ae

R. Maamri
Department of Computer Science, Abdelhamid Mehri University, Constantine, Algeria
e-mail: ramdane.maamri@univ-constantine2.dz

F. Zitouni · R. Maamri
LIRE Laboratory, Abdelhamid Mehri University, Constantine, Algeria

S. Patnaik et al. (eds.), *Advances in Machine Learning and Computational Intelligence*, Algorithms for Intelligent Systems, https://doi.org/10.1007/978-981-15-5243-4_44

main reasons: (i) decentralized approaches are more adequate for MRTA problems, due to their distributed nature; and (ii) since most MRTA problems are NP-hard, then finding the optimal solution is not a good alternative. Distributed approaches divide the environment into logical regions [1, 11] and use market-based techniques [12] to allocate tasks to robots. In real-life scenarios, tasks might require the cooperation of many robots to accomplish them. Therefore, it is important to provide techniques that allow robots to select tasks, while considering both their preferences and task requirements.

We propose a distributed solution to the MRTA problem in the domain of search and rescue missions. It is based on a well-known distributed architecture [13], which is an established benchmark for distributed MRTA problems [14]. The solution has two main phases. During the first phase, we use ant colony optimization algorithm [15] to allow each UAV to select some tasks. During the second phase, we use bat algorithm [16] to obtain conflicts-free assignments between UAVs. We compared the solution to an exhaustive method [17]. Obtained results show the effectiveness of our solution, in terms of execution time. Finally, we found that quality of our solutions is very close to the optimal one (the difference is only 3.17%).

The remainder of paper is structured as follows: Sect. 2 presents some related work. Section 3 explains the proposed solution. Section 4 provides obtained results. Section 5 gives a conclusion and highlights some future perspectives.

2 Related Work

During the last decade, the MRTA problem became a very hot topic in robotics, and several good solutions have been proposed to solve it. They are based on several techniques, such as constrained/unconstrained optimization, metaheuristics and market-based algorithms [18]. Usually, they can be divided into two categories: centralized and decentralized/distributed. Centralized approaches use a single entity that has a global view on the system [19]. Generally, the temporal complexity is reduced, because communications and coordination between robots are minimal. However, scalability and robustness are very often affected, due to two main reasons: (i) computing load on the central entity increases according to the number of robots and tasks; and (ii) if the central entity fails, then the whole system will fail too. Decentralized approaches overcome these limitations.

Gerkey and Matarić [9] proposed a formal taxonomy for distributed approaches. It uses three indicators: task, robot and allocation. The first one indicates whether a task can be single-robot (SR) or multi-robot (MR): i.e. it requires one or several robot(s) for its accomplishment. The second one indicates whether a robot can be single-task (ST) or multi-task (MT): i.e. it can perform one or several task(s) simultaneously. The third one indicates whether assignments are instantaneous (IA) or time-extended (TA): i.e. allocations of tasks to robots are static or dynamic. This taxonomy was implicitly adopted by same authors [5] to minimize the use of resources, makespan and communications. Their MRTA algorithm uses an auction-based technique and commitment to tasks. In [20, 21], the taxonomy of Gerkey and Matarić [9] has been extended to include (i) dependencies and constraints on tasks' scheduling; and (ii) precedence, synchronization and time windows constraints on tasks. In [22], authors proposed a market-based algorithm to solve a MRTA problem with constraints on tasks. Authors of [7] focused on formation of ad hoc teams of heterogeneous robots in MRTA problems. They used several techniques: auctions, learning, clustering and genetic algorithms [3, 4]. Other auction-based MRTA algorithms have also been described in [23, 24]. Authors of [11] proposed an algorithm that allocates a huge number of tasks. It combines centralized and decentralized techniques, to increase efficiency and reduce communications. It separates robots and tasks into strongly connected partitions and operates on them simultaneously. The same authors [25] improved the Hungarian method, by considering uncertainties' effects on task allocation process. The algorithm allocates robots to tasks, while calculating an interval that represents allocation tolerances, with respect to some external forces. Forces are modelled by uniform distribution probabilities and can disrupt or invalidate allocation values. Authors of [26, 27] proposed an incremental allocation system based on Hungarian algorithm that produces an optimal solution.

Simultaneous tasks' and motions' planning is considered in [28]. Authors argue their choice by the fact that these two aspects cannot be dissociated: e.g. a failure of a physical motion makes the task planning step unnecessary. They use a multi-graph to encode robots' hardware capabilities. Then, a Markov decision process is implemented to guide robots, by integrating feasibility probabilities in the decision-making process. MRTA problems have also been studied in multi-vehicle routing domain. Authors of [29] studied some routing policies on vehicles performing tasks with priorities. Authors of [30] projected the MRTA problem on Singapore's taxi distribution system. They used the available centralized infrastructure and proposed a distributed model, where computers embedded on taxis act on behalf of drivers and the environment is divided into logical regions. Taxis in the same regions negotiate to find a solution that minimizes the average waiting time of customers and the number of empty taxis. Similarly, authors of [6] compared their stochastic trajectory planning solution to the Singapore road network, by integrating historical traffic and travel data. Authors of [8] described a cooperative scenario of foraging tasks, using pheromone to indicate their execution rates. Hence, four heuristics-based approaches have been proposed to increase the efficiency. Authors of [1] integrated the MRTA problem into a mobile data collection network, using several collectors with/without interferences between allocated sectors and frequency bands. Authors of [31] proposed a stochastic clustering auction search, which uses simulated annealing, to explore an allocation space. Authors of [32] generalize the competitive analysis of the online weighted bipartite matching problem for tasks' groups. Authors of [17, 33–35] proposed several solutions to the MRTA problem, using different techniques and metaheuristics, such as Q-learning, Powerset algorithm, ant colony optimization, genetic algorithms, firefly algorithm, quantum genetic algorithms and artificial bee colony.

3 Proposed Solution

3.1 Mathematical Model and Architecture of the Problem

We propose a solution to the multi-robot task allocation problem. We target search and rescue missions performed by some unmanned aerial vehicles (UAVs) [36–38]. Some survivors are stuck into an environment and require supports, such as drugs (D), food (F) and evacuations (E). The UAV can offer some supports: drugs, food and/or evacuations. Each survivor is supposed to be assigned to the most suitable UAVs. We define a set $S = \{s_1, \ldots, s_n\}$ of n survivors and a set $U = \{u_1, \ldots, u_m\}$ of m UAVs. The aim is to rescue as many survivors as possible. The following assumptions summarize the configuration of our multi-robot task allocation problem:

1. A task (i.e. survivor) corresponds to some UAVs visiting its location and offering required supports to the concerned survivor. For an UAV, visiting the location of a survivor and offering needed supports is seen as the accomplishment of the considered task. Thus, we target scenarios where (i) each UAV can rescue only one survivor at a time and (ii) some survivors may require multiple UAVs.
2. The rescue order of survivors (priorities) is crucial: i.e. usually, we start with survivors having critical conditions.
3. UAVs perform their tasks asynchronously.
4. A task is performed when all required UAVs have already visited its location.
5. Locations of UAVs and survivors are shared and supposed to be a common knowledge between UAVs. Thus, we target scenarios where assignments are instantaneous.
6. Finally, we omit search and rescue time and we just consider navigation time between survivors.

The objective is to maximize the number of allocated tasks (rescued survivors) and minimize the navigation time between survivors. The previous problem can be formulated as follows. Given an undirected graph $G = (V, E)$, where V is the set of nodes and E is the set of edges. The structure of G is illustrated by Fig. 1 for a set $S = \{s_1, s_2, s_3, s_4\}$ of four survivors.

Each node $i \in V \backslash \{b, h\}$ has an associated profit p_i and represents a survivor, with the exception: nodes b and h represent base and hospital, respectively. The time that takes an UAV to travel from node i to node j is defined by d_{ij}. Initially, all UAVs are located at node b. As we said, the time to serve a node is negligible. If an UAV moves from node i and reaches node j at time t, then a revenue of $\left(p_j - d_{ij}\right)$ is obtained. Thus, the goal is to select an ordered subset of nodes (i.e. tour), such that visiting them one by one and maximizing the sum of all revenues. Finally, each node must be visited only once by an UAV, and at the end of this visit, it either goes to the hospital or returns to the base. For each UAV, the number of visited nodes must be computed. We denote by k the number of nodes whose revenues are collected by a certain UAV. For the ease of notation, we write $K = \{1, \ldots, k\}$. Finally, for each $i \in V$, $j \in V' = V \backslash \{b, h\}$ and $l \in K$, we define the variable $y_{\langle i,j,l \rangle}$ as follows.

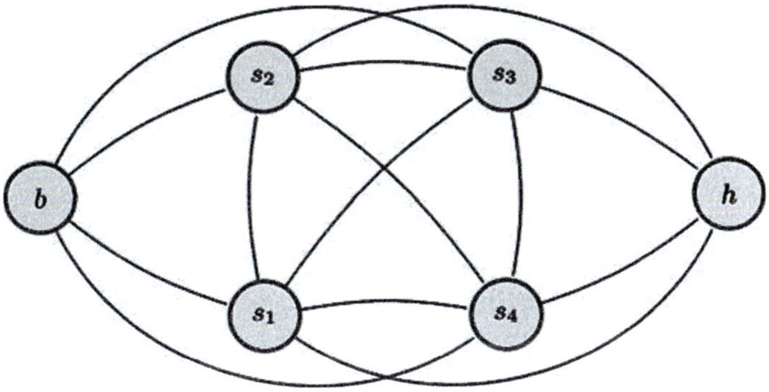

Fig. 1 Structure of G for a set $S = \{s_1, s_2, s_3, s_4\}$ of four survivors

$$y_{\langle i,j,l\rangle} = \begin{cases} 1, \text{ if segment}(i, j)\text{is used as} l\text{th edge} \\ 0, \text{ otherwise} \end{cases} \quad (1)$$

Equation 1 means that if $y_{\langle i,j,l\rangle} = 1$, then (i, j) is the lth edge of the path. Finally, the mathematical model is given by Eqs. 2, 3, 4, 5, 6 and 7.

$$\max \sum_{i \in V} \sum_{j \in V'} \sum_{l \in K} (p_j - d_{ij}) y_{\langle i,j,l\rangle} \quad (2)$$

Subject to

$$\forall j \in V' : \sum_{i \in V} \sum_{l \in K} y_{\langle i,j,l\rangle} \leq 1 \quad (3)$$

$$\forall l \in K : \sum_{i \in V} \sum_{j \in V'} y_{\langle i,j,l\rangle} = 1 \quad (4)$$

$$\forall j \in V', \forall l \in K \backslash \{k\} : \sum_{i \in V} y_{\langle i,j,l\rangle} - \sum_{j \in V'} y_{i,j,l+1} = 0 \quad (5)$$

$$\sum_{j \in V'} y_{0,j,1} = 1 \quad (6)$$

$$y_{\langle i,j,l\rangle} \in \{0, 1\}, \forall i \in V, \forall j \in V', \forall l \in K \quad (7)$$

where

- Equation 2: we try to maximize the sum of all revenues.
- Equation 3: assures that each node can only be visited once.
- Equation 4: dictates that k nodes different from the depot must be visited.
- Equation 5: ensures the connectivity between nodes.
- Equation 6: assumes that the departure is from the depot.

- Equation 7: all $y_{\langle i,j,l\rangle}$ must be binary.

The consensus-based bundle algorithm (CBBA) [13] is a fully distributed multi-agent task allocation algorithm, and it uses a two-phase architecture. During phase one, called inclusion phase, each UAV uses a greedy-based strategy to form a bundle of tasks. During the second phase, called consensus phase, UAVs try to resolve different conflicts between them. It is worth pointing out that CBBA is an established benchmark for comparing performances of distributed task allocation problems [14]. The two-phase architecture of CBBA is illustrated by Algorithm 1 (it is run independently on each UAV). In this work, we adopt this architecture for the allocation of tasks. Next sections describe inclusion and consensus phases.

Algorithm 1: The fully distributed MRTA algorithm run on each UAV.

```
1 t ← 1;
2 while not converged do
3 |   Inclusion phase: construct a list of tasks;
4 |   Consensus phase: communicate and resolve conflicts;
5 |   t ← t + 1;
6 end
```

3.2 Inclusion Phase

During this phase, each UAV employs a greedy-based strategy to select some survivors from the set S: hence, a path form node b to node h or b is traced in G. The goal is to build a bundle of survivors for each UAV. For example, if we consider the graph G of Fig. 1, a certain UAV might have as path $\{b, s_1, s_3, h\}$, which means that its bundle task contains survivors s_1 and s_3. This phase has a low computational complexity, because of the use of a greedy-based strategy [14]. However, as each UAV selects survivors without any communication with other UAVs, then there are likely many conflicts between them, which reduces the efficiency. We propose a method to construct survivor bundles for each UAV, using fuzzy logic (FL) [38] properties and ant colony optimization (ACO) algorithm [15]. First, each UAV constructs its own graph G and computes variables p_j and d_{ij} of Eq. 2. Then, the ACO algorithm [15] is applied, on the graph G, to determine a path from node b to node h or b (i.e. bundle of survivors). Equations 8 and 9 show how to calculate these variables.

$$p_j = \alpha e^{-\Phi_j(t)} \tag{8}$$

The coefficient α limits the maximum value of p_j. The term $\Phi_j(t)$ returns the time elapsed between discovering a survivor j and his/her rescue: a smaller value of $\Phi_j(t)$ means a larger value of p_j.

$$d_{ij} = \beta\left(1 - e^{-\frac{H}{\gamma}}\right) \tag{9}$$

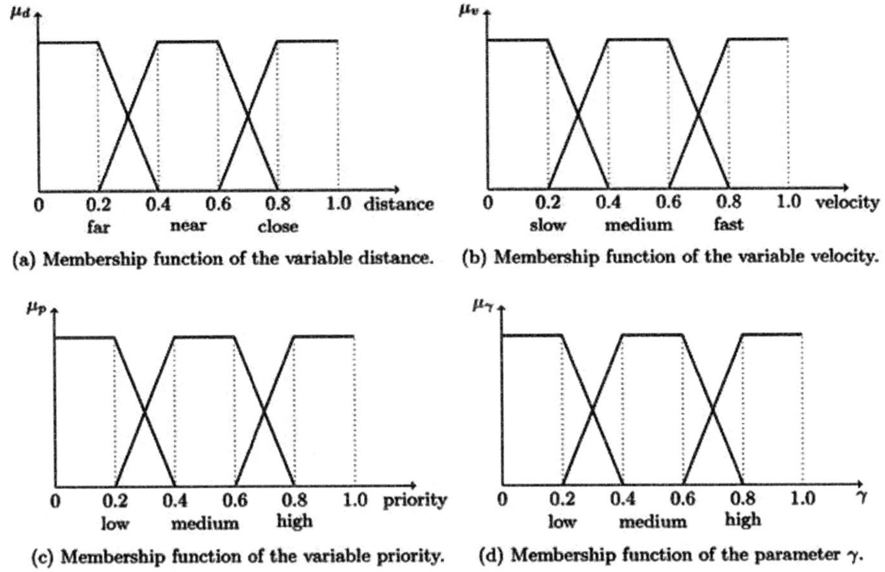

(a) Membership function of the variable distance. (b) Membership function of the variable velocity.

(c) Membership function of the variable priority. (d) Membership function of the parameter γ.

Fig. 2 Different membership functions used to compute the parameter γ

The coefficient β limits the maximum value of d_{ij}. The parameter H represents Hamming distance between supports needed by survivor j and offered by considered UAV. The parameter γ defines growth rate of term $e^{-\frac{H}{\gamma}}$, and it is computed using FL properties as follows.

- All distance, velocity and priority values are normalized into intervals [0, 1].
- A fuzzification is applied on distance, velocity and priority values, respectively, in order to transform their numerical values to linguistic variables. To do this, membership functions defined in Fig. 2a–c are used.
- Linguistic variables are passed to the fuzzy inference system, and all relevant fuzzy rules are extracted. A fuzzy rule has the following format: IF (distance IS far/near/close) AND (velocity IS slow/medium/fast) AND (priority IS low/medium/high) THEN (γ IS low/medium/high).
- A defuzzification is performed to merge all relevant rules and obtain a numerical value for γ. We use "OR" FL operator to merge common linguistic variables and gravity centre method to perform defuzzification.

When the graph G is built for every UAV, each one runs an ACO algorithm on its own graph G to determine a path from node b to node h or b that maximizes the sum of revenues $\left(p_j - d_{ij}\right)$. This path represents its bundle of tasks.

3.3 Consensus Phase

During this phase, UAVs communicate and resolve any conflict among their survivor bundles. We propose a method to resolve conflicts between bundles of UAVs, using bat algorithm (BA) [16]. At the end of this phase, we obtain conflicts-free bundles and each UAV is assigned to survivors belonging to its list. Algorithm 2 describes different steps of consensus phase.

Algorithm 2: Steps of consensus phase run on each UAV.

foreach $u \in U$ **do**
  send my survivor bundle to u;
end
wait until receiving all survivor bundles;
construct the matrix of survivor bundles as in Example 1;
initialize the bat population $M_d = [m^d_{pq}]$ and $V_d = [v^d_{pq}]$ $(d = 1, \dots, D, p = 1, \dots, (|U| \times 3), q = 1, \dots, (|S| \times 3))$;
initialize frequencies $F_d = [f^d_{pq}]$, pulse rates r_d and the loudness A_d;
while *not stopping criteria* **do**
  generate new solutions by adjusting frequency F_d using Equation 12;
  update velocities V_d and solutions M_d using Equations 13 and 14, respectively;
  if *$(rand > r_d)$* **then**
    select a solution among the best solutions;
    generate a local solution around the selected best solution using Equation 15;
  end
  generate a new solution by flying randomly;
  discretize the search space using Equations 16 and 17, respectively;
  if *$(rand < A_d$ and $f(M_d) < f(M^*))$* **then**
    accept the new solutions;
    increase r_d and reduce A_d using Equations 18 and 19, respectively;
  end
  rank the solutions and find the current best solution M^*;
end
foreach $u \in U$ **do**
  send my best solution M^* to u;
end
wait until receiving all best solutions;
find the global best solution and consider it as solution to the MRTA problem;

Example 1. Let $S = \{s_1, s_2, s_3, s_4\}$ be a set of four survivors and $U = \{u_1, u_2, u_3\}$ be a set of three UAVs. We suppose that u_1, u_2 and u_3 have [1, 0, 1], [0, 1, 1] and [1, 1, 1] as support vectors, respectively. Also, we assume that u_1, u_2 and u_3 have $\{s_2, s_3, s_4\}$, $\{s_1, s_4, s_3\}$ and $\{s_3, s_1, s_4, s_2\}$ as survivor bundles, respectively. The matrix of survivor bundles is shown in Table 1. Each cell, in the last row of the matrix, contains UAVs that might offer the corresponding support. This set will be denoted as Θ. If Θ contains several UAVs, there is a conflict. Therefore, an accepted solution should have only sets Θ containing one UAV. This simple example has $(1 \times 2 \times 2) \times (2 \times 1 \times 2) \times (2 \times 2 \times 3) \times (2 \times 2 \times 3) = 2304$ different solutions.

Equation 10 defines a variable $z_{i,j,l,u}$ as follows: if $z_{i,j,l,u} = 1$, then (i, j) is the lth edge of the path done by UAV u. Hence, Eq. 11 expresses the objective function used to evaluate the quality of solutions.

$$z_{i,j,l,u} = \begin{cases} 1, & \text{if segment}(i, j) \text{is used as } l\text{th edge by UAV } u \\ 0, & \text{otherwise} \end{cases} \tag{10}$$

Table 1 Matrix of survivor bundles

		s_1			s_2			s_3			s_4		
		D	F	E	D	F	E	D	F	E	D	F	E
u_1	D	0	0	0	1	0	0	1	0	0	1	0	0
	F	0	0	0	0	0	0	0	0	0	0	0	0
	E	0	0	0	0	0	1	0	0	1	0	0	1
u_2	D	0	0	0	0	0	0	0	0	0	0	0	0
	F	0	1	0	0	0	0	0	1	0	0	1	0
	E	0	0	1	0	0	0	0	0	1	0	0	1
u_3	D	1	0	0	1	0	0	1	0	0	1	0	0
	F	0	1	0	0	1	0	0	1	0	0	1	0
	E	0	0	1	0	0	1	0	0	1	0	0	1
		$\{u_3\}$	$\{u_2, u_3\}$	$\{u_2, u_3\}$	$\{u_1, u_3\}$	$\{u_3\}$	$\{u_1, u_3\}$	$\{u_1, u_3\}$	$\{u_2, u_3\}$	$\{u_1, u_2, u_3\}$	$\{u_1, u_3\}$	$\{u_2, u_3\}$	$\{u_1, u_2, u_3\}$

$$\max \sum_{i \in V} \sum_{j \in V'} \sum_{l \in K} \sum_{u \in U} \left(p_j - d_{ij}\right) z_{i,j,l,u} \tag{11}$$

$$F_d = F_d^{\min} + \left(F_d^{\max} - F_d^{\min}\right) \beta_d \tag{12}$$

$$V_d^{t+1} = V_d^t + \left(M_d^t - M^*\right) F_d \tag{13}$$

$$M_d^{t+1} = M_d^t + V_d^{t+1} \tag{14}$$

$$M_{\text{new}} = M_{\text{old}} + \sigma \in A_d^t \tag{15}$$

$$S\left(v_{pq}^d\right) = \frac{1}{1 + e^{-v_{pq}^d}} \tag{16}$$

$$m_{pq}^d = \begin{cases} 1, \ u_p \in \Theta_q \wedge S\left(v_{pq}^d\right) \in \left[a_p, b_p\right] \\ 0, \ u_p \notin \Theta_q \vee S\left(v_{pq}^d\right) \notin \left[a_p, b_p\right] \end{cases} \tag{17}$$

$$r_d^{t+1} = r_d^0\left(1 - e^{-\gamma t}\right) \tag{18}$$

$$A_d^{t+1} = \alpha A_d^t \tag{19}$$

where

- $\beta_d \in [0, 1]$: random matrix drawn from a uniform distribution.
- $F_d^{\min}$: frequency minimum.
- $F_d^{\max}$: frequency maximum.
- σ: random number drawn from a Gaussian normal distribution $N(0, 1)$.
- ε: scaling factor.
- α_p and β_p: lower and upper bounds used to choose an UAV from the set Θ_q
- α and γ: constants.

4 Simulation and Result Discussion

We developed a software to generate randomly 10 datasets: $\{D1, D2, D3, D4, D5, D6, D7, D8, D9, D10\}$. Each one contains two tables representing attributes of UAVs and survivors, respectively. Tables 2 and 3 show the structure of UAVs' and survivors' attributes. We used 10 datasets to (i) target the main complexity source of MRTA problems: i.e. the size of problem; and (ii) observe, qualitatively and quantitatively, the efficiency and scalability of proposed solution.

Table 2 Structure of UAVs' attributes

ID	Location	Supports	Velocity
...	...	...	...

Table 3 Structure of survivors' attributes

ID	Location	Supports	Priority
...	...	...	...

Columns of Table 2 are explained as follows: (i) UAVs' identifier, (ii) spatial location, (iii) supports and (iv) velocity. Columns of Table 3 are explained as follows: (i) survivor' identifier, (ii) spatial location, (iii) supports and (iv) priority. Table 4 summarizes configurations of used datasets.

Each configuration was run 10 times, to minimize the effect of noise. In each one, our solution was compared to an exhaustive method [17], in terms of makespans and objective function values. It is worth pointing out that the principle of exhaustive methods is very simple: they consider all possible feasible solutions and then select the optimal one. All configurations have two inputs: UAVs' and survivors' tables. The expected outputs are the best allocations that maximize the sum of all revenues defined by Eq. 11. We implemented a simulator using Java programming language and JADE platform. All simulations were run on a DELL laptop Intel(R) Core (TM) i7-2640 M CPU@ 2.80 GHz 2.80 GHz, RAM 8.00 Go. Figure 3 summarizes obtained makespan and objective function values.

Figure 3a shows a comparison between makespan values, according to allocation methods and datasets. We observe that the exhaustive method is effective for datasets$\{D1, D2, D3, D4, D5\}$, but makespan values increase significantly for the other datasets. The main reason for such rapid increase is the size of datasets. On the other hand, we notice that the proposed solution has near-constant makespan values, even for large datasets: i.e. scalable. Figure 3b shows a comparison between objective function values, according to allocation methods and datasets. We clearly observe that objective function values generated by the proposed solution are very close to the optimal ones. Hence, we can say that our solution is efficient. In summary, datasets' sizes and allocation method have a direct impact on allocation qualities. In most real-life situations, such as search and rescue missions, it is often unavoidable to consider a compromise between these factors.

Table 4 Configurations of used datasets

	Datasets									
	$D1$	$D2$	$D3$	$D4$	$D5$	$D6$	$D7$	$D8$	$D9$	$D10$
$\|U\|$	5	10	15	20	25	30	35	40	45	50
$\|S\|$	10	20	30	40	50	60	70	80	90	100

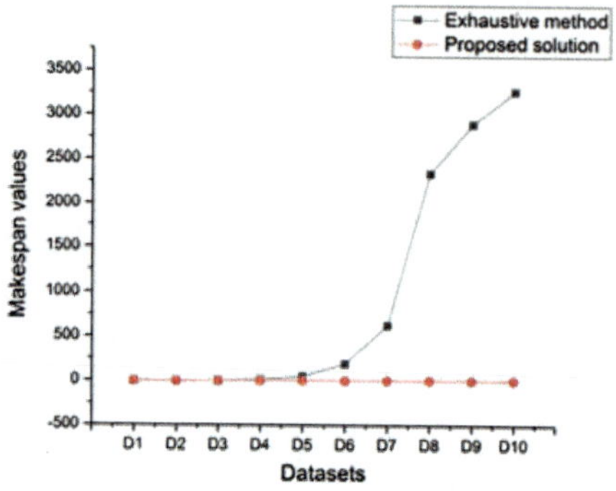

(a) Makespan values.

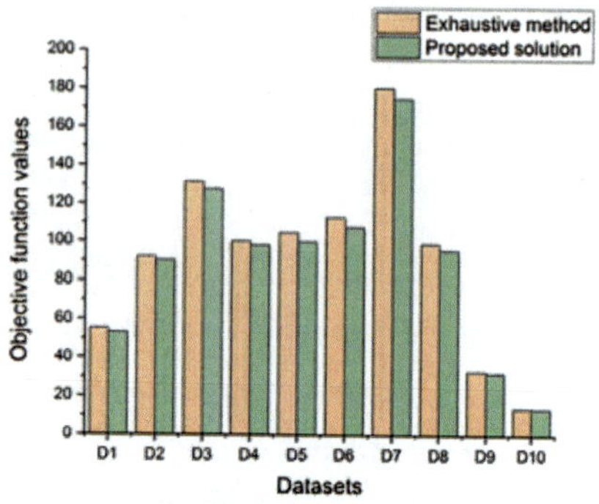

(b) Objective function values.

Fig. 3 Comparative study, in terms of makespans and objective function values

5 Conclusion

We proposed a distributed solution to the multi-robot task allocation problem, in the domain of search and rescue of survivors. The solution is based on a fully distributed architecture [13], which is an established benchmark for distributed multi-robot task allocation problems [14]. It has two main phases: construction of a bundle of tasks for each UAV and assignment tasks to UAVs without conflicts. The comparative study shows that the proposed solution is effective and scalable, in terms of makespan values. Finally, we found that quality of our solutions is very close to the optimal ones (the difference is only 3.17%), in terms of objective function values. In future, we plan to introduce temporal and spatial constraints on the problem, since they are two crucial features in the field of search and rescue of survivors. Also, the number of membership functions can be increased to strengthen the rule base.

References

1. D.C. Güner, E. Modiano, Dynamic vehicle routing for data gathering in wireless networks, in *49th IEEE Conference on Decision and Control (CDC)*. IEEE (2010), pp. 2372–2377
2. William Lenagh, Prithviraj Dasgupta, and Angelica Munoz-Melendez. A spatial queuing-based algorithm for multi-robot task allocation. *Robotics*, 4(3):316-340, 2015
3. E. Gil Jones, M. Bernardine Dias, A. Stentz, Learning-enhanced market-based task allocation for oversubscribed domains, *2007 IEEE/RSJ International Conference on Intelligent Robots and Systems*. IEEE (2007), pp. 2308–2313
4. E. Gil Jones, M. Bernardine Dias, A. Stentz, Time-extended multi-robot coordination for domains with intra-path constraints. Autonomous Robots **30**(1), 41–56 (2011)
5. B.P. Gerkey, M.J. Mataric, Sold!: Auction methods for multirobot coordination. IEEE Trans. Robot. Autom. **18**(5), 758–768 (2002)
6. S. Lim, D. Rus, Stochastic motion planning with path constraints and application to optimal agent, resource, and route planning, in *2012 IEEE International Conference on Robotics and Automation*. IEEE (2012), pp. 4814–4821

7. E.G. Jones, B. Browning, M.B. Dias, B. Argall, M. Veloso, A. Stentz, Dynamically formed heterogeneous robot teams performing tightly-coordinated tasks, in *Proceedings 2006 IEEE International Conference on Robotics and Automation, 2006. ICRA 2006.* IEEE (2006), pp. 570–575
8. P. Dasgupta, Multi-robot task allocation for performing cooperative foraging tasks in an initially unknown environment, in *Innovations in Defence Support Systems-2*. Springer (2011), pp. 5–20
9. B.P. Gerkey, M.J. Mataríc, A formal analysis and taxonomy of task allocation in multi-robot systems. The International Journal of Robotics Research **23**(9), 939–954 (2004)
10. A. Khamis, A. Hussein, A. Elmogy, Multi-robot task allocation: a review of the state-of the art, in *Cooperative Robots and Sensor Networks 2015* (Springer, 2015), pp. 31–51
11. L. Liu, D.A. Shell, Multi-level partitioning and distribution of the assignment problem for large scale multi-robot task allocation. Robotics: Science and Systems VII (MIT Press: Cambridge, MA, USA, 2011), pp. 26–33
12. M.B. Dias, R. Zlot, N. Kalra, A. Stentz, Market-based multirobot coordination: a survey and analysis. Proc. IEEE **94**(7), 1257–1270 (2006)
13. H.L. Choi, L. Brunet, J.P. How, Consensus-based decentralized auctions for robust task allocation. IEEE Trans. Robot. **25**(4), 912–926 (2009)
14. X. Chen, P. Zhang, G. Du, F. Li, A distributed method for dynamic multi-robot task allocation problems with critical time constraints. Robot. Auton. Syst. **118**, 31–46 (2019)
15. M. Dorigo, M. Birattari, *Ant colony optimization* (Springer, 2010)
16. Yang, X-S, A new metaheuristic bat-inspired algorithm. In *Nature inspired cooperative strategies for optimization (NICSO 2010)* (Springer, 2010), pp. 65–74
17. F. Zitouni, R. Maamri, Fa-setpower-mrta: a solution for solving the multi-robot task allocation problem, in *IFIP International Conference on Computational Intelligence and Its Applications* (Springer, 2018), pp. 317–328
18. S.H. Liu, Y. Zhang, H.Y. Wu, J. Liu, Multi-robot task allocation based on swarm intelligence. J. Jilin Univ. (Engineering and Technology Edition) **40**(1), 123–129 (2010)
19. S. Ahmed, T. Pongthawornkamol, K. Nahrstedt, M. Caesar, G. Wang, Topology-aware optimal task allocation for publish/subscribe-based mission critical environment. In *MILCOM 2009–2009 IEEE Military Communications Conference* (IEEE, 2009), pp. 1–7
20. G. Ayorkor Korsah, A. Stentz, M. Bernardine Dias, A comprehensive taxonomy for multi-robot task allocation. Int. J. Robot. Res. **32**(12), 1495–1512 (2013)
21. E. Nunes, M. Manner, H. Mitiche, M. Gini, A taxonomy for task allocation problems with temporal and ordering constraints. Robot. Auton. Syste. **90**, 55–70 (2017)
22. R. Zlot, A. Stentz, Market-based multirobot coordination for complex tasks. Int. J. Robot. Res. **25**(1), 73–101 (2006)
23. X. Li, D. Sun, J. Yang, Networked architecture for multi-robot task reallocation in dynamic environment, in *2009 IEEE International Conference on Robotics and Biomimetics (ROBIO)* (IEEE, 2009), pp. 33–38
24. Maitreyi Nanjanath and Maria Gini. Repeated auctions for robust task execution by a robot team. *Robotics and Autonomous Systems*, 58(7):900–909, 2010
25. L. Liu, D.A. Shell, Assessing optimal assignment under uncertainty: an interval-based algorithm. Int. J. Robot. Res. **30**(7), 936–953 (2011)
26. L. Liu, D.A. Shell, Tunable routing solutions for multi-robot navigation via the assignment problem: a 3d representation of the matching graph, in *2012 IEEE International Conference on Robotics and Automation* (IEEE, 2012), pp. 4800–4805
27. L. Liu, D.A. Shell, A distributable and computation-flexible assignment algorithm: from local task swapping to global optimality, in *Proceedings of Robotics Science Systems* (2013), pp. 257–264
28. I.A. Sucan, L.E. Kavraki, Accounting for uncertainty in simultaneous task and motion planning using task motion multigraphs, in *IEEE International Conference on Robotics and Automation* (IEEE, 2012), pp. 4822–4828
29. M. Pavone, S.L. Smith, F. Bullo, E. Frazzoli, Dynamic multi-vehicle routing with multiple classes of demands, in *2009 American Control Conference* (IEEE, 2009)

30. K.T. Seow, N.H. Dang, D.-H. Lee, A collaborative multiagent taxi-dispatch system. IEEE Trans. Autom. Sci. Eng. **7**(3), 607–616 (2009)
31. K. Zhang, E.G. Collins, A. Barbu, An efficient stochastic clustering auction for heterogenous robot teams, in 2012 IEEE International Conference on Robotics and Automation (IEEE, 2012), pp. 4806–4813
32. L. Luo, N. Chakraborty, K. Sycara, Competitive analysis of repeated greedy auction algorithm for online multi-robot task assignment, in 2012 IEEE International Conference on Robotics and Automation (IEEE, 2012), pp. 4792–4799
33. F. Zitouni, R. Maamri, An adaptive protocol for dynamic allocation of tasks in a multi-robot system, in 2016 International Conference on Advanced Aspects of Software Engineering (ICAASE) (IEEE, 2016), pp. 128–133
34. F. Zitouni, R. Maamri, Cooperative learning-agents for task allocation problem, in *Interactive Mobile Communication, Technologies and Learning* (Springer, 2017), pp. 952–968
35. F. Zitouni, R. Maamri, S. Harou, Fa-qabc-mrta: a solution for solving the multi-robot task allocation problem. Intell. Serv. Robot. **1–12**, 407 (2019)
36. J. Turner, Q. Meng, G. Schaefer, A. Whitbrook, A. Soltoggio, Distributed task rescheduling with time constraints for the optimization of total task allocations in a multirobot system. IEEE Trans. Cybern. **48**(9), 2583–2597 (2017)
37. A. Whitbrook, Q. Meng, P.W.H. Chung, Reliable, distributed scheduling and rescheduling for time-critical, multiagent systems. IEEE Trans. Autom. Sci. Eng. **15**(2), 732–747 (2017)
38. L.A. Zadeh, Fuzzy logic. Computer **21**(4), 83–93 (1988)

Collective Intelligence of Gravitational Search Algorithm, Big Bang–Big Crunch and Flower Pollination Algorithm for Face Recognition

Arshveer Kaur and Lavika Goel

1 Introduction

Optimization is the procedure of finding or searching for the best solution possible. Before the introduction of heuristic algorithms, only mathematical optimization techniques existed which get stuck in the local minima and require the search space derivation making them inefficient to solve the real-world problems [1]. In contrast, heuristic algorithms focus on finding the near optimal solution to the problem in a satisfiable amount of time instead of the best solution. But as per 'No-Free Lunch (NFL)' theorem, there does not exist any single algorithm which is capable of solving all the existing optimization problems [1]. So, we propose a new algorithm where the variants of BB-BC, GSA and characteristics of FPA are used in their better versions.

Face recognition problem has become popular these days in research field as it has been used in various social networking Web sites for tagging various people in photos, in smart locks of the mobiles, security access authentication, crowd surveillance, etc. Feature selection is a major part of the face recognition application which can highly impact the efficiency of the system [2]. It can be achieved by extracting the grayscale values of the pixels of the picture, constructing a matrix and extracting eigenvectors or we can say eigen faces. The major part is to find eigenvectors such that they represent different dimensions and contribute enough in representing the training face vectors and have been extracted using the BB-BC algorithm.

A. Kaur (✉) · L. Goel
Department of CSIS, Birla Institute of Technology & Science, Pilani, Pilani 333031, India
e-mail: p20170432@pilani.bits-pilani.ac.in

L. Goel
e-mail: lavika.goel@pilani.bits-pilani.ac.in

S. Patnaik et al. (eds.), *Advances in Machine Learning and Computational Intelligence*, Algorithms for Intelligent Systems, https://doi.org/10.1007/978-981-15-5243-4_45

2 Related Work

A lot of research has been carried in face recognition to find out various methods to achieve image representation and recognition. Face recognition using eigen faces has been proposed in [2] in 1991. The faces are represented as a combination of eigenvectors in an eigen space. Other approaches for face recognition include face representation using binary patterns [3] and labeled graphs [4]. Particle swarm optimization (PSO) method was proposed in [5] which searches the best known solution in candidate solution space and is dependent on best position of the particle as well as whole swarm's known till then [6]. Other variants of the algorithm include chaotic version of PSO [7] for data clustering and improved PSO [8] in which the swarm is divided into sub-swarms and PSO is applied on each one of them.

GSA assumes that each candidate solution has a certain mass and communicates with different candidates because of Newtonian law of gravity and the concept of mass communication [9]. A new operator 'Disruption' is suggested for enhancing the exploitation and exploration proficiency of the traditional GSA [10]. BB-BC optimization algorithm is influenced from the expansion and transformation of universe [11]. UBB-CBC [12], a variant of BB-BC, uniformly initializes the population in the first phase in order to increase the diversity of search space and adds chaos to the second phase for avoiding untimely convergence toward the representation point.

3 Existing Methodologies

3.1 Big Bang–Big Crunch Algorithm

BB-BC (based on a theory of evolution) works in two stages, the big bang stage and the big crunch stage [11]. The first stage is used for the purpose of initialization of the population. The second stage works as the convergence operator having multiple inputs and single output called as center of mass. It narrows the candidates to form a point that can be considered as a representative of all the points initialized in the first step. The center of mass is computed using Eq. 1.

$$p^c = \sum_{i=1}^{M} \frac{1}{f^i} p^i \Big/ \sum_{i=1}^{M} \frac{1}{f^i} \tag{1}$$

where p^i is a candidate member in the search space, f^i is the value of fitness function.

After the big crunch phase, the algorithm will repeat the big bang step for next iteration in which new candidates are generated by adding or subtracting a normal random number followed by the big crunch step.

3.2 Uniform Big Bang–Big Crunch

This approach is an improved variant of big bang–big crunch optimization. Chaos is used to crunch the points to a single point that can represent all of them on the basis of theory of centre of mass or minimum cost strategy in the second phase of big crunch. The chaos added in the big crunch phase leads to better convergence and helps in avoiding premature convergence [12]. A base set or candidates are created in such a way that every possible dimension is covered.

The new population of next iteration is created in the big bang phase of iteration using some chaotic map. Various chaotic maps that can be used in the chaotic big crunch phase include Henon map, logistic map, tent map, sinus, gauss, sinusoidal iterator and circle map. These maps produce the chaos between 0 and 1.

3.3 Chaotic Local Search

The search is conducted in all the dimensions taking current best probable solution as the base point and radius of the neighborhood as 'r'. A candidate is selected along each dimension $Y_g^m(t)$ where m represents the dimension and t represents the iteration. After each iteration, the best solution is represented by minimum or maximum fitness value of these candidates along each dimension [14]. The candidates are updated after every iteration using the Eq. 2.

$$\begin{cases} y_g^m(t) = y_g^d(t-1) & \text{if } d \neq m \\ y_g^m(t) = y_g^d(t-1) + r * \left(2 * \text{chaos}^d(t) - 1\right) & \text{if } d = m \end{cases} \tag{2}$$

where $\text{chaos}^d(t)$ is the chaos variable for dth dimension in tth iteration. The radius 'r' is updated after every iteration using $r = p * r$ where p is the shrinking factor.

3.4 Gravitational Search Algorithm

Newton's gravitational law specifies that the force between two objects is given by the product of the mass of objects divided by the distance square between them. Acceleration of agent i at time t in dimension d is calculated using Eq. 3:

$$a_i^d(t) = F_i^d(t)/M_{ii}(t) \tag{3}$$

where M_{ii} is the inertial mass of agent i. The formulae to calculate velocity and position of the agents are given in Eq. 4.

$$v_i^d(t+1) = rand_i * v_i^d(t) + a_i^d(t) \text{ and } x_i^d(t+1) = x_i^d(t) + v_i^d(t+1) \tag{4}$$

3.5 Chaotic Gravitational Search Algorithm

In this technique, chaos is added in order to improve the performance of classic GSA exhibiting stochasticity and ergodicity. Stochasticity refers to the exhibition of randomness in a system. Chaos exhibits a 'seemingly-random' behavior, i.e., it appears to be random but is actually deterministic. Ergodicity represents the property of a system to go through all the points in a given space within sufficient time. The force between the objects is calculated using Eq. 5 (modified version of GSA).

$$F_i^d(t) = \sum_{j=1, j\neq i}^{N} \text{chaos}_j F_{ij}^d(t) \tag{5}$$

3.6 Flower Pollination Algorithm

Flower pollination algorithm (FPA) [13] works on the basis of survival of the fittest. It is an iterative process which achieves optimal solution by use of abiotic (local) and biotic (global) pollination (modification). The insects use levy flight method to take large distance jumps randomly and drop the pollens at a distance far away from the parent flower. Levy flight helps in improving the diversity of the population and increases the chances of obtaining the global optima. The levy function mainly takes three parameters and is represented as $x = \text{levy}(n, m, \text{beta})$ where n depicts the number of steps to be taken, m represents the dimensions and beta is the strength factor having value between 1 and 2.

4 Proposed Methodology

4.1 Hybrid GSA-BB-BC with Chaotic Local Search

The proposed approach consists of various phases namely: initialization, traversal, big crunch and chaotic local search each taken from different nature-inspired computational intelligence techniques. Initialization is done using the UBB-BC approach, traversal proceeds like the CGSA, next is the big crunch phase taken from BB-BC algorithm for convergence toward heavier mass and the last phase of chaotic local search is used for the exploitation purpose. The phases are explained below in more detail.

Initialization. For the purpose of initialization, a modified version of Eq. 3 of UBB-BC approach is used. Initially, two candidate objects, O1 and O2, are set with n as the length of the candidate and initial dividing factor as 1. A chaotic variable c

is used in place of random variable r where c and $c \neq \{0, 0.25, 0.5, 0.75, 1\}$. If the candidate is out of the defined search space, only the random weighted value is taken from its original value to make it within the search space, i.e., $c_i(\text{new}) = c_i(\text{old}) \times w_i$. $c_i(\text{old})$ represents the candidate out of the search space, and w_i is weight between 0 and 1. The value of d is incremented after every 2^d candidate is generated. The process continues until the desired population size is reached. All the particles or candidates generated are assigned random mass values between 0 and 1.

Traversal. In this phase, the force acting on the particles of the population by the rest of the population is determined. Acceleration of all the particles is calculated (Eq. 3), and accordingly, their velocities are updated using the chaotic version of GSA, large jumps are added in the acceleration with the help of levy flight concept used in FPA. The equation for velocity change is modified as shown in Eq. 6. The position vector of each particle is updated using this velocity and the previous position vector (Eq. 4).

$$v_i(t+1) = \text{chaos}_i v_i^d(t) + a_i^d(t) \times \text{levy(iteration no, dim, beta)} \tag{6}$$

Big Crunch Phase. In this phase, we find the representative point in the search space using Eq. 1. The center point represents the best solution found in the current iteration. If the difference between the center of mass in few consecutive iterations is small, it indicates that the algorithm is converging toward the global best solution. The process is stopped if the difference between the distances of the centers in consecutive iterations is below the initially determined threshold.

Chaotic Local Search. It is the process of exploiting the current best solution in order to find a better solution. The process takes place in multiple iterations, and if a solution is found that is better than the current best, the iterations are halted, and the obtained solution would be considered as the global best. The search radius is decremented after every iteration for converging the algorithm onto a solution. This goes on till the search radius gets smaller than a pre-defined threshold.

Pseudo Code.

(1) Initialize population using uniform BB-BC
(2) While termination condition is not met. Do

   i. For all candidate
      a. Calculate force acting upon each candidate using Eq. 5
      b. Calculate acceleration of each candidate using Eq. 3
      c. Update velocity of each candidate using Eq. 6
      d. Update positions of each candidate using Eq. 4
      e. If the candidates go out of the search space, modify the candidates as per equation $c_i(\text{new}) = c_i(\text{old} \times \text{w}_\text{i}$.
      End-for
   ii. Find out the current global best agent using Eq. 1.
   iii. For all dimensions, do
      a. Initialize chaotic variable
      End-for

iv. while termination criteria are not satisfied do
   a. for every dimension d do
      compute candidate fitness of $X_g^d(k)$end-for
   c. pick up the local optima $X_g(k)$
   c. compare with current global best
   d. if better solution found, update global best
   e. update the search radius
   f. update candidates using the Eq. 2.
   end-while
v. Update masses of candidates.

4.2 Application to Face Recognition

Principal component analysis (PCA) is used to extract the eigen faces. Initially, we divide the available dataset into training and testing datasets. PCA is used on training dataset to extract the eigenvectors. The number of eigenvectors would be equal to the number of samples in the training data out of which top 'k' eigenvectors are considered. The dataset is read in the form of grayscale values for each picture. The average face vector is created by calculating the average value of each pixel position in the picture. Then, each picture is represented as its difference from the average picture. The average face vector is subtracted from each face vector in the training dataset.

Now, each face vector in our dataset should be represented as a weight vector of the eigenvectors. For this, the eigenvectors matrix in the higher dimension should be multiplied with the difference matrix calculated earlier (After subtracting the average face). The result would be a $k \times 1$ vector for each dataset sample representing its components along each eigenvector. The k eigenvectors in the eigenvector space would be considered as the initial points for the next part of our algorithm. The fitness function of these eigenvectors is the sum of its corresponding weight across all the training set vectors divided by the total number of vectors. Each eigenvector is considered as an object with mass, and its velocity and acceleration are calculated. In each iteration, the centre of mass is calculated. A chaotic local search is performed with the centre of mass as the reference to find out the vector that gives the best fitness function value in that local search space. The obtained fitness value is compared with the least one among the existing eigenvectors. If the new solution is better, it will replace the least eigenvector. Positions, masses, accelerations, fitness values are updated after each iteration. This goes on until a termination condition is met.

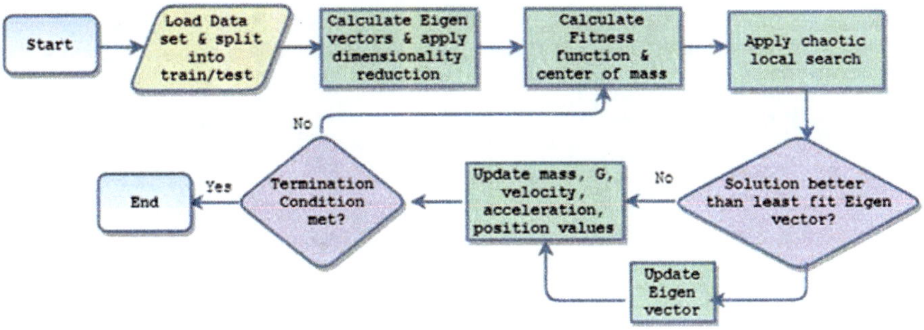

Fig. 1 Diagrammatic representation of the proposed approach

5 Experimental Results

The proposed methodology was applied for a face recognition scenario (flowchart given in Fig. 1). The dataset used for this test was the ORL dataset. There are ten distinct images for each class (person), and the total number of classes is 40. Thus, it comprises 400 images in total. Each image is of size 92 × 112 pixels. Each pixel can have 256 possible gray level values.

A dataset is divided in 90:10 ratios for train and test part. The maximum number of iterations was set to one third of the number of eigen faces. The number of components was given as 75 making number of iterations as 25. Principal component analysis was performed on the dataset, and eigen faces were extracted. The top 75 eigen faces were used for classification purposes. Iterations in chaotic local search are 25. The sample of the eigen faces extracted is given in Fig. 2.

The performance of the output is calculated by generating an ROC curve that plots false positives against true positives (shown in Fig. 3). The precision and recall values of different class labels have been noted. The varying precision values of 0.96–1.0 were observed and shown in Fig. 4. The picture depicts each class (numbered from 1 to 40), their support in the test set and the results obtained. Comparison of the algorithm with existing algorithms is given in Table 1.

6 Conclusion

The proposed hybrid algorithm combines the best exploration and exploitation features of the individual algorithms (GSA, BB-BC and FPA). It is tested on the ORL benchmark dataset for the face recognition application and performs comparable to existing algorithms. The achieved precision for the application is 96–100 (%). In future, the algorithm can be tested on other more challenging datasets like Extended Yale B, Grimace, MUCT, Faces 9 and Faces 96 for analyzing its efficiency on more complex datasets.

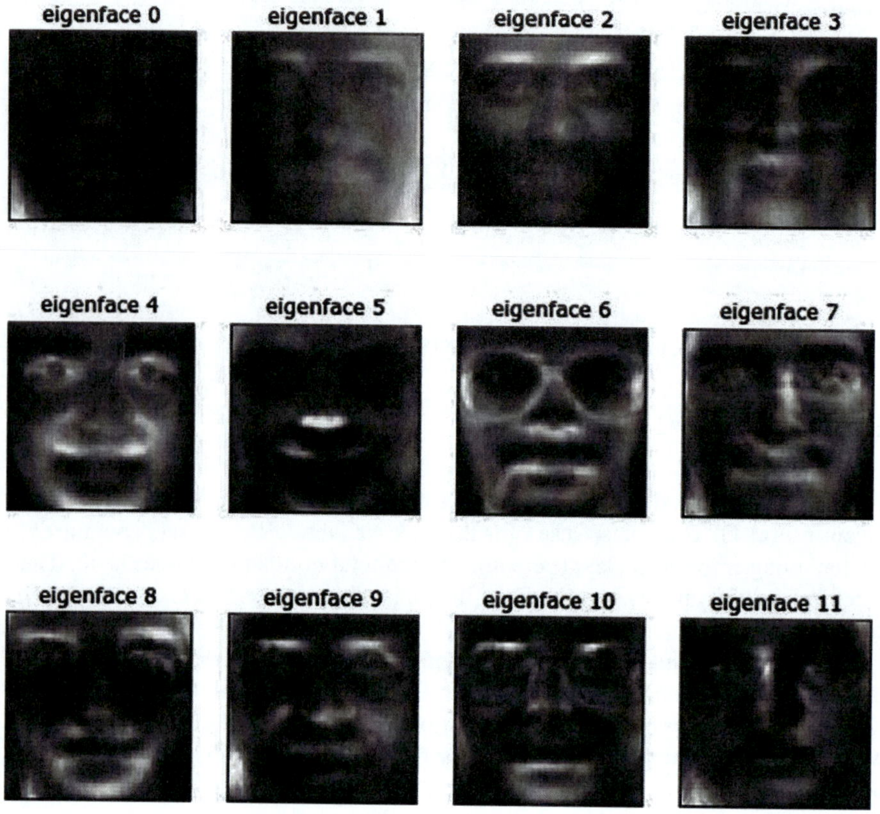

Fig. 2 Grayscale images of the eigen faces

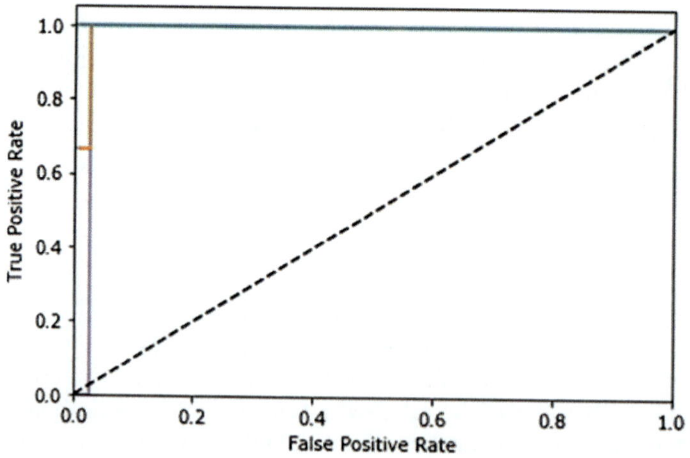

Fig. 3 ROC curve for multiple classes

	precision	recall	f1-score	support
0	0.67	1.00	0.80	2
1	1.00	1.00	1.00	1
2	1.00	1.00	1.00	1
3	1.00	1.00	1.00	1
4	1.00	1.00	1.00	2
5	1.00	1.00	1.00	2
6	1.00	1.00	1.00	1
7	1.00	1.00	1.00	2
8	1.00	1.00	1.00	1
9	1.00	1.00	1.00	3
10	1.00	1.00	1.00	1
11	1.00	1.00	1.00	1
12	1.00	1.00	1.00	1
13	0.00	0.00	0.00	1
14	1.00	1.00	1.00	1
15	1.00	1.00	1.00	1
16	1.00	1.00	1.00	1
17	1.00	1.00	1.00	1
18	1.00	1.00	1.00	2
19	1.00	1.00	1.00	3
20	1.00	1.00	1.00	2
21	1.00	1.00	1.00	1
22	1.00	1.00	1.00	1
23	1.00	1.00	1.00	1
24	1.00	1.00	1.00	1
25	1.00	1.00	1.00	1
26	1.00	1.00	1.00	1
27	1.00	1.00	1.00	3
avg / total	0.96	0.97	0.97	40

	precision	recall	f1-score	support
0	1.00	1.00	1.00	2
1	1.00	1.00	1.00	1
2	1.00	1.00	1.00	1
3	1.00	1.00	1.00	1
4	1.00	1.00	1.00	2
5	1.00	1.00	1.00	2
6	1.00	1.00	1.00	1
7	1.00	1.00	1.00	2
8	1.00	1.00	1.00	1
9	1.00	0.67	0.80	3
10	1.00	1.00	1.00	1
11	1.00	1.00	1.00	1
12	1.00	1.00	1.00	1
13	1.00	1.00	1.00	1
14	1.00	1.00	1.00	1
15	1.00	1.00	1.00	1
16	1.00	1.00	1.00	1
17	1.00	1.00	1.00	1
18	1.00	1.00	1.00	2
19	0.00	0.00	0.00	0
20	1.00	1.00	1.00	3
21	1.00	1.00	1.00	2
22	1.00	1.00	1.00	1
23	1.00	1.00	1.00	1
24	1.00	1.00	1.00	1
25	1.00	1.00	1.00	1
26	1.00	1.00	1.00	1
27	1.00	1.00	1.00	1
28	1.00	1.00	1.00	3
avg / total	1.00	0.97	0.98	40

Fig. 4 Performance measures with precision 0.96 (left) and 1.0 (right)

Table 1 Comparison with other algorithms

Technique	GSA	GA	ACO	Our proposed algorithm
Accuracy (%)	75–77	97.5	99	96–100

References

1. Seyedali Mirjalili, Moth-flame optimization algorithm: a novel nature-inspired heuristic paradigm. Knowl.-Based Syst. **89**, 228–249 (2015)
2. T.A. Matthew A. P. Pentland, Face recognition using eigenfaces, in *Proceedings CVPR'91., IEEE Computer Society Conference on Computer Vision and Pattern Recognition* IEEE (1991)
3. A. Timo, A. Hadid, M. Pietikainen, Face description with local binary patterns: Application to face recognition. IEEE Trans. Patt. Anal. Mach. Intell. **28**(12), 2037–2041 (2006)
4. W. Laurenz et al., Face recognition by elastic bunch graph matching. IEEE Trans. Patt. Anal. Mach. Intell. **19**(7), 775–779 (1997)
5. E. Russell, J. Kennedy, A new optimizer using particle swarm theory. *MHS'95., Proceedings of the Sixth International Symposium on Micro Machine and Human Science, 1995* (IEEE, 1995)
6. Erik Cuevas, Humberto Sossa, A comparison of nature inspired algorithms for multi-threshold image segmentation. Expert Syst. Appl. **40**(4), 1213–1219 (2013)
7. L.Y. Chuang, C.J. Hsiao, C.H. Yang, Chaotic particle swarm optimization for data clustering. Expert systems with Applications **38**(12), 14555–14563 (2011)
8. J. Yan et al., An improved particle swarm optimization algorithm. Appl. Math. Comput. **193**(1), 231–239 (2007)
9. R. Esmat, H. Nezamabadi-Pour, S. Saryazdi, GSA: a gravitational search algorithm. Inf. Sci. **179**(13), 2232–2248 (2009)
10. S. Sarafrazi, H. Nezamabadi-Pour, S. Saryazdi, Disruption: a new operator in gravitational search algorithm. ScientiaIranica **18**(3), 539–548 (2011)

11. O.K. Erol, I. Eksin, A new optimization method: big bang–big crunch. Adv. Eng. Softw. **37**(2), 106–111 (2006)
12. B. Alatas, Uniform big bang–chaotic big crunch optimization. Commun. Nonlinear Sci. Numer. Simul. **16**(9), 3696–3703 (2011)
13. X.-S. Yang, Flower pollination algorithm for global optimization, in *International conference on unconventional computing and natural computation* (Springer, Berlin, Heidelberg, 2012)
14. C. Choi, J.-J. Lee, Chaotic local search algorithm. Artif. Life Robot. **2**(1), 41–47 (1998)

Fuzzy Logic Based MPPT Controller for PV Panel

Mahesh Kumar, Krishna Kumar Pandey, Amita Kumari, and Jagdish Kumar

1 Introduction

As non-renewable energy sources are limited in quantity and depleting fast, demand for renewable energy sources is increasing. Renewable energy plays a significant role in supplying the increasing demand for power. Due to an increase in electronic gadgets and appliances, the quality of power must also be maintained. Over the past few decades, PV-based power generation has emerged as a promising source of energy. The photovoltaic system directly transforms solar energy into electricity [1].

There are different methods of MPPT such as reference cell method, sampling method, P & O method and incremental conductance method. Depending on their sophistication, rate of conversion and performance, each process has its own merits and demerits. Due to its low complexity, fast conversion speed and high efficiency, the FLC is used to track the MPP [2–4].

The MPPT technique is implemented by controlling the duty cycle of MOSFET-based DC-DC Boost converter, which is being connected to solar PV system. The FLC generates the desired duty cycle to track MPP and is given to the MOSFET/IGBT in the form of pulses to obtain the desired output voltage from the boost converter. For improved efficiency and better utilization of the PV system, it is generally integrated with grid and energy storage system. The proposed model can be integrated with DC/AC micro-grid for various applications [5, 6].

M. Kumar · K. K. Pandey · A. Kumari (✉) · J. Kumar
Punjab Engineering College, Sector 12, Chandigarh 160012, India
e-mail: amitakumari@pec.ac.in

M. Kumar
e-mail: mik.kumar5@gmail.com

K. K. Pandey
e-mail: kpkrishna224@gmail.com

S. Patnaik et al. (eds.), *Advances in Machine Learning and Computational Intelligence*, Algorithms for Intelligent Systems, https://doi.org/10.1007/978-981-15-5243-4_46

2 Proposed Model

The system includes the boost converter connected to PV panel implementing MPPT to get the desired output voltage. FLC is proposed and implemented to track the MPP from the PV panel. The output voltage is maintained constant irrespective of change in temperature and irradiance. Block diagram is drawn in Fig. 1.

2.1 PV System

PV array system is a solid-state device containing PV cells comprising semiconductor material such as silicon or germanium. As light falls on them, the PV cell transforms the solar energy directly into electrical energy. Its characteristic is dependent on number of series and parallel solar cells. The output current and voltage of the PV array depend on the series and shunt resistance [7, 8].

Figure 2 shows characteristics curve of PV module where,

V_{mp} Peak voltage point

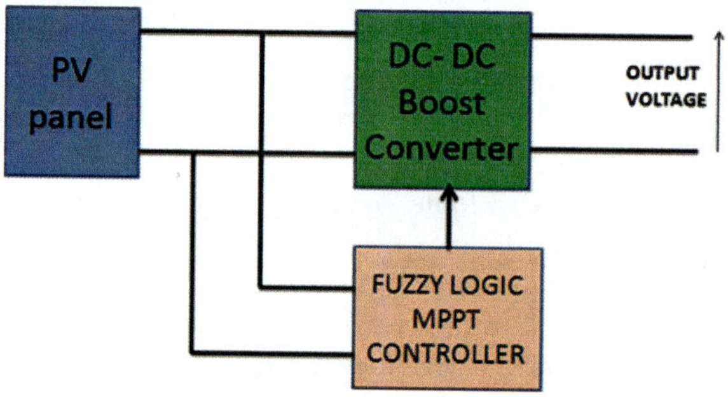

Fig. 1 Proposed model

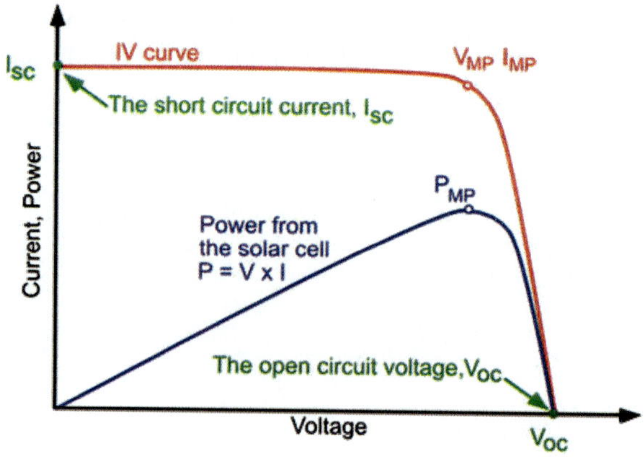

Fig. 2 Characteristic curves of PV module

Table 1 Specification of single PV module

PV module specifications	
Parameters	Values
V_{oc} (open circuit voltage)	36.3 V
I_{sc} (short circuit current)	7.84 A
V_{mpp} (MPP voltage)	29 V
I_{mp} (MPP current)	7.35 A
N (cells per module)	60

I_{mp} Peak current point
P_{mp} Peak power point.

Tables 1 and 2 show the specifications of PV module and array used. The specified power, voltage and current change with irradiance level.

Figure 3 shows V-I curve of the panel at irradiance level of 1000, 800 and 1200 W/m^2 at 25 °C. It behaves as a constant current source initially and as a constant voltage source ahead of MPP.

The P-V curve of the panel at different irradiance levels is shown in Fig. 4. The maximum power occurs at a particular point pertaining to specified irradiance level.

When dP/dV = 0, the peak power point is reached.

If dP/dV > 0, the duty cycle increases the output voltage of the PV array until it reaches the peak power point.

Table 2 Specification of PV array

PV array specifications	
Parameters	Values
No. of parallel strings in module	8
No. of series strings in module	8
Panel rating	13.64 kW

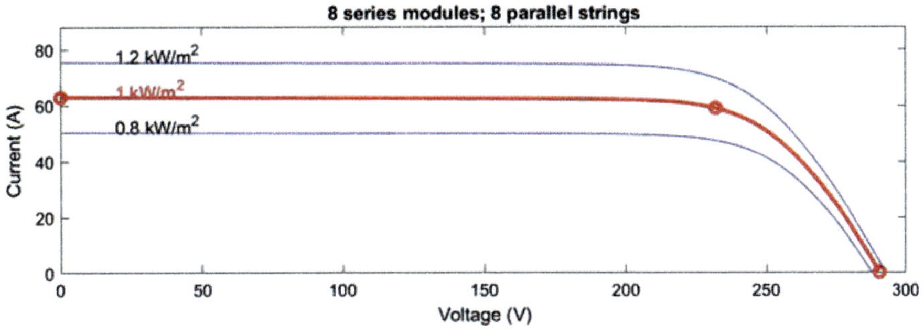

Fig. 3 V-I curve at different irradiance

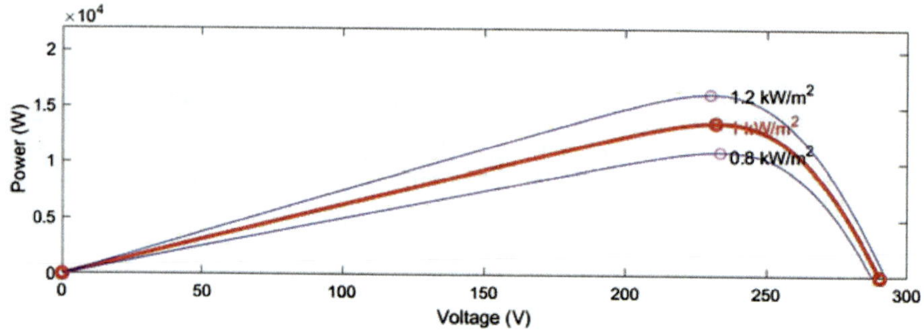

Fig. 4 P-V curve at different irradiance

If dP/dV < 0, duty cycle decreases the voltage of PV output until it reaches the peak power point.

2.2 *Boost Converter*

DC-DC boost converter increases the voltage applied to its input terminal to specified value depending upon its duty cycle D. This consists of a DC voltage, an inductor, a condenser, a diode and a semiconductor switch. The duty ratio of the switch is varied accordingly to regulate the output voltage. It operates in two modes when switch is closed; inductor stores energy, diode is reversed biased, and capacitor discharges through load, [8] and when the switch is open, the inductor transfers its energy to the capacitor and the capacitor charges. [9]. Figure 5 shows boost converter model.

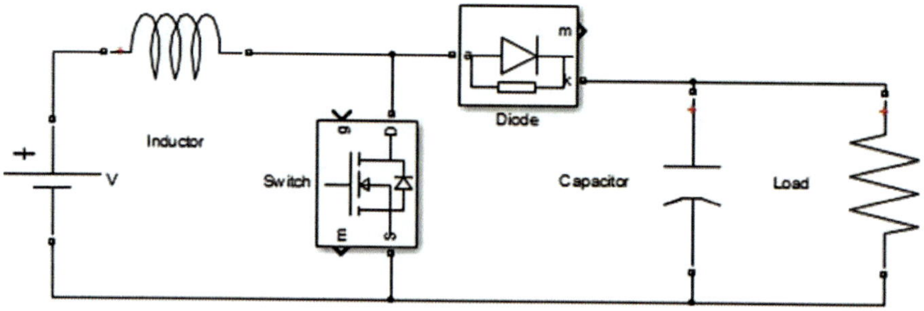

Fig. 5 Boost converter model

3 Fuzzy Logic Controller

Figure 6 shows block diagram of proposed model, which includes PV system connected to the DC-DC boost converter using FLC for MPPT.

3.1 *Fuzzification*

The error E (input 1) and change in error (input 2) are calculated through (1) and (2):

$$E(k) = \frac{P(k) - P(k-1)}{I(k) - I(k-1)} \tag{1}$$

$$CE(k) = \Delta E \tag{2}$$

The output of FLC (Duty Ratio) is given by (3):

$$D(k) = D(k-1) + \Delta D \tag{3}$$

Equations (1), (2), (3) depict how to find error, change in error and desired duty cycle.

Seven membership functions are made according to the desired duty ratio. The fuzzification block works on AND principle, and desired output is maintained using fuzzy rules. The values of both error and change in error lie between [-1, 1]. Figure 7 depicts output duty ratio membership functions. It lies in the range [0, 1].

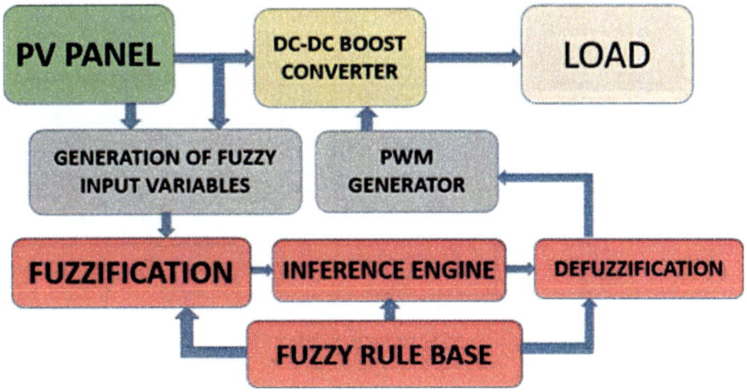

Fig. 6 Block diagram of proposed model

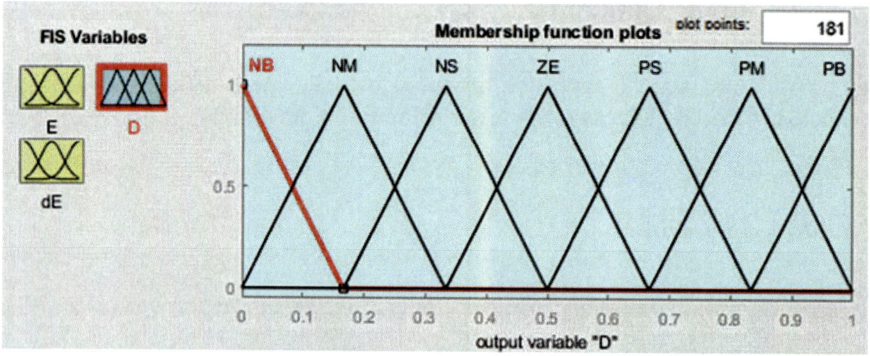

Fig. 7 Membership function of output (duty ratio)

3.2 Fuzzy Inference Engine

After observing the behavior of the PV system, the vague rules are made and picked. Vague decisions are made for variance in irradiance and temperature on the basis of fuzzy rules to achieve MPP of the PV system. Syntax IF/THEN is used for law. There are 49 MPPT fuzzy logic rules described in Fig. 8.

dE \ E	NB	NM	NS	ZE	PS	PM	PB
NB	NB	NB	NB	NB	NB	NB	NB
NM	NM	NM	NM	NM	NM	NM	NM
NS	NS	NS	NS	NS	NS	NS	NS
ZE	ZE	ZE	ZE	ZE	ZE	ZE	ZE
PS	PS	PS	PS	PS	PS	PS	PS
PM	PM	PM	PM	PM	PM	PM	PM
PB	PB	PB	PB	PB	PB	PB	PB

Fig. 8 Fuzzy rule base

3.3 *Defuzzification*

Centroid method is used in defuzzification to achieve the optimal duty ratio value for the boost converter. The optimum value of duty cycle is given as a control signal to the switch of the boost converter.

4 Simulation Results

The modeling and simulation of the proposed model are successfully tested in MATLAB/Simulink. Figure 9 shows the proposed system model (Table 3).

Figure 10 depicts variation in irradiation levels, while Fig. 11 shows the power of input side of boost converter for variation in irradiance at constant temperature of 25 °C. The proposed Simulink model contains PV module, boost converter and FLC. Solar irradiance and constant temperature act as inputs to the PV module. Figure 12 shows input and output voltages with irradiance change. The converter is used to step up the input voltage to desired value using MPPT controller, i.e., fuzzy logic controller. The input and output voltages almost remain constant with change in irradiance.

Table 4 depicts maximum power, voltage and current values to be tracked, while Table 5 shows the actual tracked values. The results show very small difference between these values, and hence, satisfactory results are obtained.

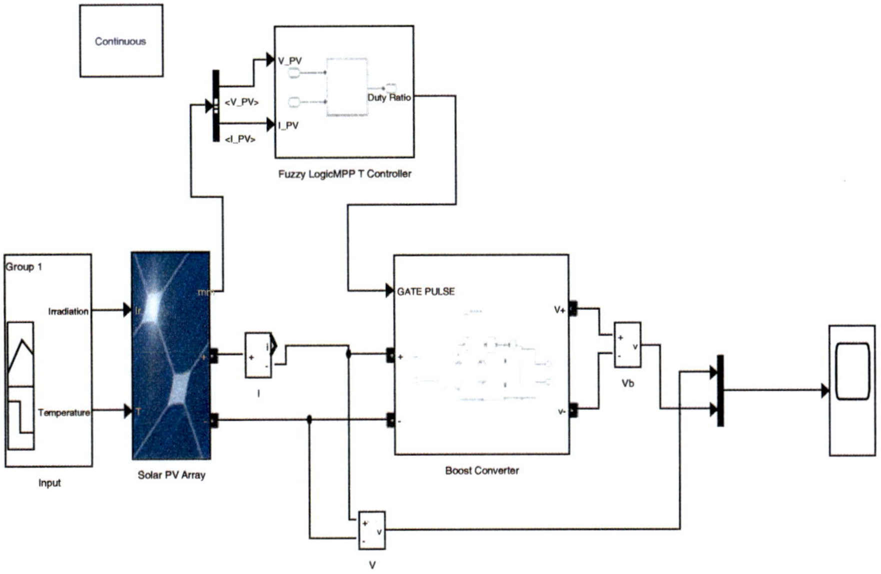

Fig. 9 MATLAB model of proposed system

Table 3 Specification of boost converter

Boost converter specifications	
Parameters	Values
Input voltage	232 V
Output voltage	510 V
Current ripple factor	5%
Voltage ripple factor	10%
Switching frequency	10 kHz
Inductor	11 mH
Capacitor	1000 μF
Load resistor	20 Ω

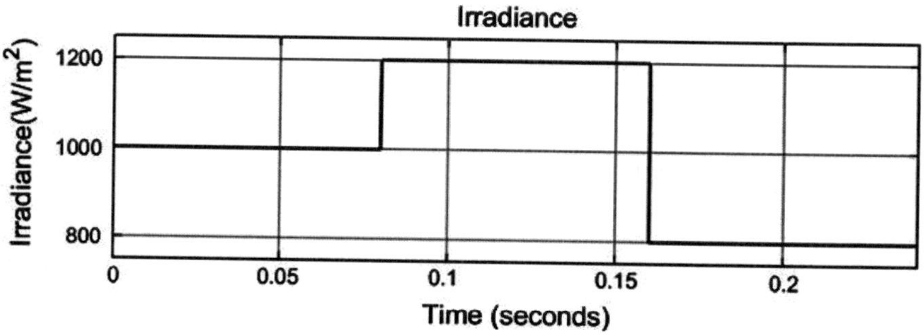

Fig. 10 Irradiance versus time

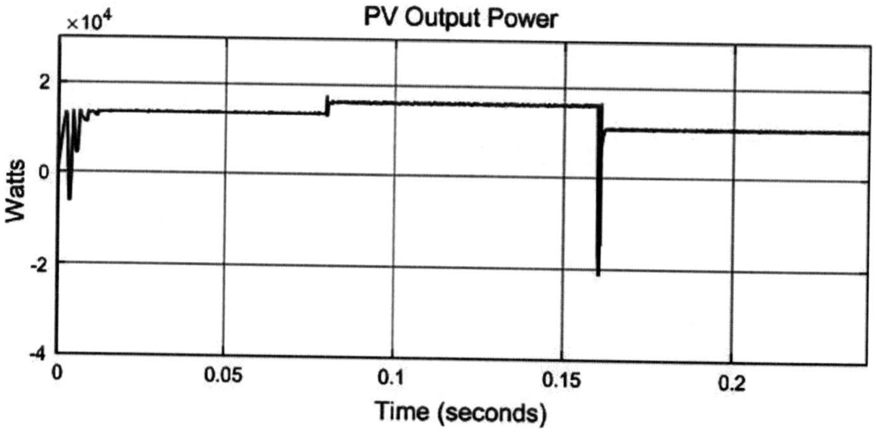

Fig. 11 PV array output power versus time

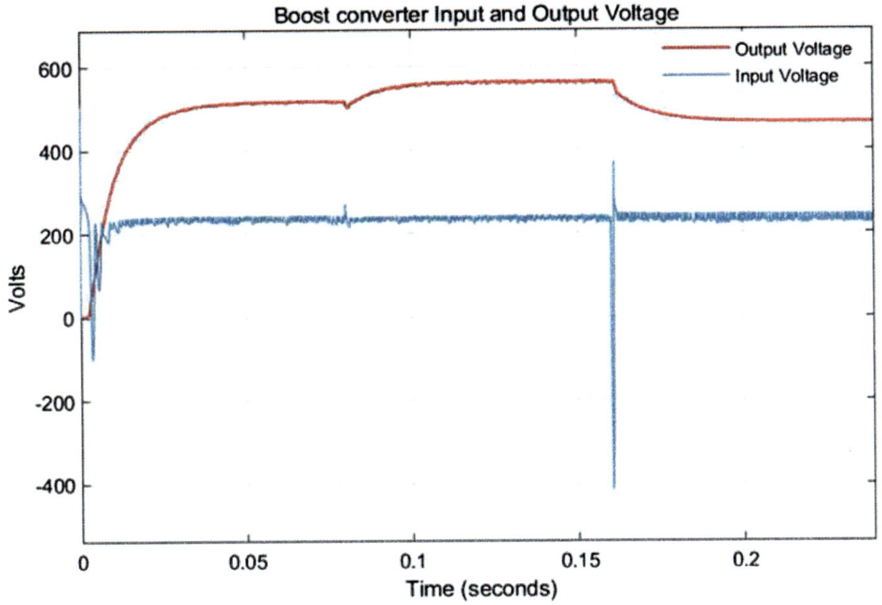

Fig. 12 Input (I/P) and output (O/P) voltage of boost converter

Table 4 Values of power, voltage and current to be tracked

Irradiance (W/m^2)	Maximum power (P_m, to be tracked) (kW)	Voltage (V_{mpp}, to be tracked) (V)	Current (I_{mpp}, to be tracked) (A)
1000	13.64	232	58.8
1200	16.18	230.1	70.33
800	10.99	233.5	47.08

Table 5 Tracked values of power, voltage and current

Irradiance (W/m^2)	Maximum power (P_m) (kW)	Voltage (V_{mpp}) (V)	Current (I_{mpp}) (A)
1000	13.60	228.3	59.55
1200	16.12	229.95	70.10
800	10.92	229.2	47.64

5 Conclusions

In MATLAB/Simulink, the model was successfully modeled and simulated. With the switch in irradiance levels and fixed temperature, the fuzzy logic controller successfully extracts the maximum energy from the PV module. The boost converter stage increases the PV voltage from the panel to the desired level, and the voltage output is within defined limits. The boost converter's input and output power are nearly tracked at maximum value and demonstrate limited variance due to some losses in switching.

References

1. K. Ding, X. Bian, H. Liu, T. Peng, A MATLAB Simulink based PV module model and its application under conditions of non-uniform irradiance. IEEE Trans. Energy Convers. **27**(4) (2010)
2. D.S. Morales, in *Maximum Power Point Tracking Algorithms for Photovoltaic Applications*. School of Science and Technology, Aalto University (2010)
3. W. NianCHun, S. Zao, K. Yukita, Y. Goto, K. Ichiyanagi, Research of PV model and MPPT methods in MATLAB, in *Power and Energy Engineering Conference* (2010)
4. T. Eshram, L. Patrick, P.L. Chapman, Comparison of photovoltaic array maximum power point techniques. IEEE Trans Energy Convers **22**(2), 439–449 (2007)
5. K. Ravichandrudu, S.K. Fathima, P.Y. Babu, G.V.P Anjaneyulu, Design and performance of a bidirectional isolated DC-DC converter for renewable power system. Int. J. Electr. Electron. Eng. **7**(2), 81–87. e-ISSN: 2278-1676, p-ISSN: 2320-3331 (2013)
6. F. Iov, M. Ciobotaru, D. Sera, R. Teodorescu, F. Blaabjerg, Power electronics and control of renewable energy systems. IEEE Trans. Ind. Electron. **55**(7), 1–27 (2007)
7. I.H. Altas, A.M. Sharaf, A Photovoltaic array simulation model for Matlab-Simulink GUI environment, in *International Conference on Clean Power*, pp. 341–345 (2007)
8. T. Salmi, M. Bouzguenda, A. Gastli, A. Masmoudi, MATLAB/Simulink based modelling of solar photovoltaic cell. Int. J. Renew. Energy Res. **2**(2) (2012)
9. Zhang, J, Dissertation, in *Bidirectional DC-DC Power Converter, Design Optimization, Modeling and Control* (Blacksburg, Virginia, 2008)

IOT and Cloud Computing

A Survey on Cloud Computing Security Issues, Attacks and Countermeasures

Deepak Ranjan Panda, Susanta Kumar Behera, and Debasish Jena

1 Introduction

The National Institute of Standards and Technology (NIST) defines cloud as a model that enables convenient on-demand network access to a shared pool of configurable computing resource, e.g., network, storage, hardware, applications, etc., that can be rapidly allocated, scaled as well as released with minimum management effort or service provider intervention [1].

Different cloud computing research have emerged in recent years, including significant advances. Cloud computing has become widely adopted in both the private and public sectors due to the practicality of its services, which can potentially add convenience at different levels. Cloud computing enables resources to be shared in a pool that can be rapidly provisioned and can be offered to the user with minimal interaction of the service provider. Providing security to the cloud environment, convenient data storage and providing efficient computational service are the primary concern for a cloud computing environment [2].

Cloud computing provides computing services which includes servers, storage, databases, networking, software, analytics and intelligence over the Internet which reduces physical installations and maintenance of the system. This results the overall cost reduction and enhancement of system efficiency. The consumers whose data and services are being processed in the cloud have to completely rely on the cloud service provider (CSP) for the privacy and security of their information. The mutual trust is

D. R. Panda · S. K. Behera (✉)
Department of MCA, College of IT & Management Education, Bhubaneswar, Odisha, India
e-mail: citesusanta@gmail.com

D. R. Panda
e-mail: panda.dpak@gmail.com

D. Jena
Information Security Laboratory, IIIT Bhubaneswar, Bhubaneswar, Odisha 751003, India
e-mail: debasish@iiit-bh.ac.in

S. Patnaik et al. (eds.), *Advances in Machine Learning and Computational Intelligence*, Algorithms for Intelligent Systems, https://doi.org/10.1007/978-981-15-5243-4_47

achieved to some extent through service-level-agreement, but a considerable number of cloud-related security issues becomes unavoidable and should be handled by either the CSP or the users themselves.

Data plays an important role while any security is concerned. Cloud computing has to enforce added data security as data in this environment is distributed in nature and of multi-tenant architecture [3]. Cloud services entirely depend on Internet which are vulnerable to different attacks and security threats. Potentially, severe attacks such as SQL injection [4] to get unauthorized access to other customer's data, insecure Web application program, data breaches, denial-of-service attacks and data losses are the growing security concerns over the past years [5].

Cloud delivers on-demand IT services through virtualization. The concept of hardware virtualization in IaaS and programming level virtualization for PaaS allows appropriate degree of customization, security, isolation and manageability for delivering IT services on demand. Server consolidation with virtualization also allows sharing of resources of a single physical server by number of applications simultaneously. So, virtual machines construct can be considered as the entire back end for cloud-related services, and at the same time, it induces threats like unexpected phishing, side-channel attack, VM rollback attack, VM escape attack, etc. Therefore, cloud security is a combination of data security and corresponding virtual machine security [3].

This paper presents a survey of the security of cloud computing focusing on different security issues, attacks and existing solutions for this emerging technology. The rest of the paper is structured as follows. Section 2 presents fundamentals of cloud computing including cloud service model and deployment model. Section 3 presents security requirements and issues in cloud; Sect. 4 includes different attacks and recommends solutions. Section 5 suggests future works for the intelligent and multicloud environment, and finally, Sect. 6 concludes the paper.

2 Fundamentals of Cloud Computing

A cloud computing model is composed of five essential characteristics, three service models and four deployment models [1]. This section provides a brief overview of cloud computing service models, deployment models and their respective advantages and disadvantages.

2.1 *Cloud Service Model*

The cloud architecture is generally classified into three cloud service models: infrastructure-as-a-service (IaaS), the lowest layer, which provides fundamental infrastructure for the other layers; platform-as-a-service (PaaS), the middle layer, which provides an environment for developing and hosting users' applications; and

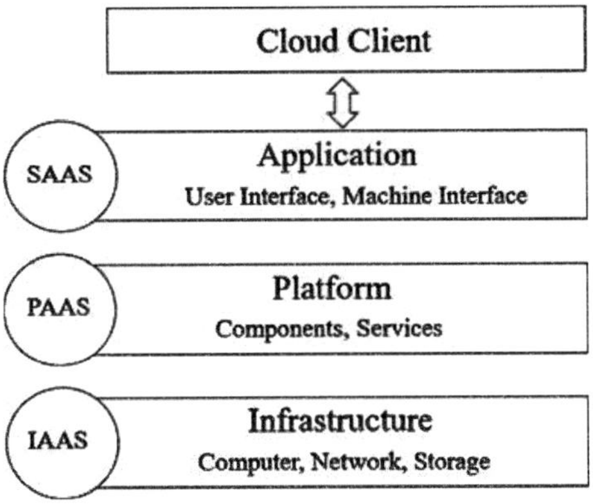

Fig. 1 Service model in cloud architecture

software-as-a-service (SaaS), the upper layer, which provides an application layer that works as a service on demand [6]. This architecture follows a bottom-up approach [7, 8] as shown in Fig 1.

2.1.1 Software as a Service (SaaS)

SaaS is the topmost layer in the cloud stack which includes the software services [9]. It allows users to access softwares/applications that are running on a cloud, and it delivers the service over the Internet. The advantages of SaaS in computational requirements include no software installation and maintenance on user's personal computers. The users depend on the service providers for security as the infrastructure, and the execution platform lies completely within the range of the service provider [3].

2.1.2 Platform as a Service (PaaS)

PaaS is a delivery of computing platform over the Internet. It is a layer beneath the SaaS. It is an integrated cloud-based environment that allows developers to efficiently code required applications, run and manage them on the PaaS layer [10]. The service providers grant infrastructure such as network, servers, operating system or storage to the developers. Although the cloud service provider controls the computing infrastructure in PaaS, the users have certain control over the deployed application. In comparison to Saas, PaaS model provides greater security to the users [3, 11].

2.1.3 Infrastructure as a Service (IaaS)

IaaS is the lowest layer that provides virtual delivery of computing infrastructure which includes hardware, networking, operating system and storage services. It allows users to utilize complete resources without purchasing physical equipments. The infrastructure is managed wholly by the cloud service provider. In this model, users have some control over operating system, deployed services and some selected part of the network. Thus, user side security control is more in IaaS than the previous models [3, 10].

2.2 *Cloud Deployment Model*

There are four main deployment models for cloud computing proposed by NIST [1]—public clouds, private clouds, hybrids clouds and community clouds.

2.2.1 Public Clouds

Public clouds are owned and operated by a third-party cloud service providers, where hardware and software resources are publicly shared among various users over the Internet. As a third-party public cloud service provider manages and monitors this environment, there are some security issues and such clouds are not ideal for sensitive data [4, 12].

2.2.2 Private Cloud

A private cloud [13, 14] is the one in which the infrastructure of hardware and software is maintained on a private network. Cloud services are provided exclusively for a single organization, and the cloud is owned by either the organization or by a third-party, located on or off premises. It is not shared with another organization. Private cloud solves the security issues of public cloud, but at same time, it is very expensive. This is usually not suitable for small-to-medium size businesses and is mostly used by large enterprises.

2.2.3 Community Cloud

Cloud services are provided exclusively for a community of organizations that have common cloud requirements. The cloud is managed by either the organizations or by a third-party, located on or off premises. The drawback of this model lies in the fact that there are still a number of unanswered questions regarding service outages, contractual and security implications, i.e., issues regarding data being spread across multiple organizations and multiple domains [14].

2.2.4 Hybrid Cloud

The cloud infrastructure is a composition of two or more distinct cloud infrastructures (private, community or public) that remain unique entities. The resources are shared among the clouds in hybrid approach. By using a hybrid approach, companies can maintain control of an internally managed private cloud while relying on the public cloud as and when needed [12]. Hybrid cloud offers the benefit of cost and scaling like public clouds, while the security and control of private clouds are also taken care of. The issues which pose threat on the hybrid cloud include the data privacy and integrity concerns, while data flows from public to private environment or vice versa since privacy controls in the public cloud environment vary significantly from the private cloud [14].

3 Security Requirements and Issues in Cloud Computing

3.1 Security Requirements

Confidentiality, integrity and availability are major goals of computer security. Confidentiality means secrecy of the data items. The user's data contains many sensitive information. It is vital to protect users' data stored at cloud servers. Similarly, when we send a piece of information to cloud or retrieve a piece of information from the cloud, we need to conceal it during transmission. The threat of data compromise increases in the cloud, due to the increased number of parties, devices and applications involved that leads to an increase in the number of points of access [15] information needs to be changed constantly. However, integrity requirement demands that it must be changed by legitimate users through authorized mechanisms only. In otherwords, integrity requires that data is modified by authorized parties within the access rights constraints. Finally, availability means that service as well as resources of cloud is provided to its legitimate users in most uninterrupted manner.

3.2 Security Issues in Cloud Computing

Security is a concern at all levels of cloud hardware, software and communication. Security issue arises due to poor authentication policy as well as data defects in cryptographic methods. Major security issues as identified by [6] are embedded security, application, client management, cloud data storage, cluster computing and operating system.

3.2.1 Embedded Security

Embedded security attack is centered around problem that can arise during deployment virtual machine monitoring as the host machine has the power of updating and changing any resources in the virtual machine [16].

3.2.2 Application

The number of users in the cloud and software that is developed in the cloud or provided as a service is ever increasing. The abundance of code is attributed as primary concern of security in cloud [17].

3.2.3 Client Management

Confidentiality of client data is one of the major requirements of cloud security. The basic mechanism followed is data encryption. However, it is possible to predict the key of encryption with either brute force or cryptanalysis attacks. Hence, algorithms that are resistant to such attacks are to be employed. Another type of attack can be authentication attack, where an attacker is able to capture authentication sequence and logins to the cloud as if he is the legal user and via at liberty to conduct active attack by viewing or modifying the user's data. Strong authentication plays a central role to stop illegal access by intruders [18, 19].

3.2.4 Cloud Data Storage

Data is stored in data centers owned by cloud service providers. The security of data at the cloud centers lies with the service providers, rather than the user. Management of information and records of the users are controlled by legal contracts and enforced by laws; many of which are ill-equipped to cope with the affordances of the new technologies [20]. Moreover, users has no role in creation of contract, whereas the service provider exercises monopoly in formulating terms and conditions.

3.2.5 Cluster Computing

Clustering helps to process data in parallel fashion resulting higher throughput, but security challenges are increasing too [21].

3.2.6 Operating System

Security issues are dominant in mobile operating systems where attacker is trying to exploit vulnerabilities in the OS to launch attacks that can steal useful user information. The security issues arise out of usage of many VMs, servers across networks or multiple operating systems.

4 Security Attacks and Countermeasures in Cloud Computing

A cloud computing scenario can be modeled using three different classes of participants: clouders (users), services and the cloud provider (Fig. 2). Every interaction in a cloud computing scenario can be addressed to two entities of these participant classes. In the same way, every attack attempt in the cloud computing scenario can be detailed into a set of interactions within this three-class model. For instance, between a user and a service instance, one has the very same set of attack vectors that exist outside the cloud computing scenario. Hence, talking about cloud computing security means talking about attacks with the cloud provider among the list of participants [17]. This does not require the cloud provider to be malicious himself; it may also just play an intermediate role in an ongoing combined attack [31]. Figure 2 shows all possible six attack surfaces. (a) Service-to-user (b) user-to-service (c) cloud-to-service (d) service-to-cloud (e) cloud-to-user (f) user-to-cloud.

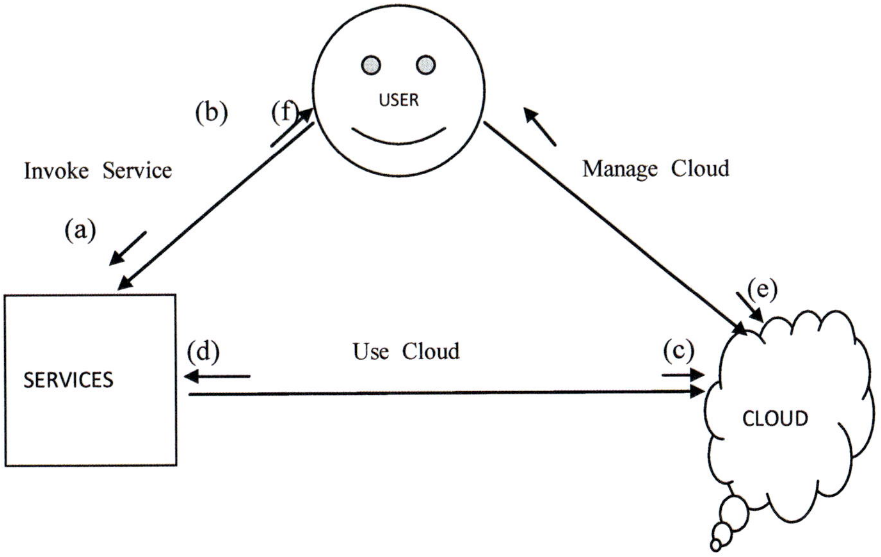

Fig. 2 Attack surfaces in the cloud

Further, attacks can be of two types—passive attack and active attack. In passive attacks, the attacker is observing the data content, and analysing traffic, whereas in active attack, the attacker is modifying the data content or obstructing the normal usage of resources for the legitimate users. Countermeasures refer to any action either remedial or preventive to deal with the attacks. The attacks in cloud can be categorized at the service models as follows: SaaS layer attacks, PaaS layer attacks and IaaS layer attacks.

4.1 SaaS Layer Attacks

The major issues identified by [22] at SaaS layer are data ownership, data access, data locality, identity management and authentication. The popular attacks are DoS, authentication attack and SQL injection attack. DoS attack is known as denial-of-service attack which is consider as negation of availability. DoS is an active attack in which communication can fail, and even the user may not be able to connect to service provider as the attacker exploits vulnerability in the protocol to flood the servers with undue service requests which ultimately consume the bandwidth and deprive the legitimate user of cloud services. DDoS is known as distributed denial-of-service attack where attacker takes control of many nodes in cloud to plant a malicious code and sends signal to all the zombies to lunch DDoS attack upon a targeted victim node. The effect of DDoS is devastating [23, 24]. Authentication attack endangers privacy and integrity of data item stored in the cloud. Attackers may try brute-force attack (trying all possible key) or a dictionary attack by trying all keys offered by a dictionary of all possible attack [22].

4.2 PaaS Layer Attacks

The major attacks outlined by [22] at PaaS layers are phising, port scanning and man-in-the-middle attack and metadata spoofing attack. Phising is an attack to lure the customer with lucrative offer to click on a link to divert the controller to attacker site. It is also defined in a different context through social engineering attack where the customer is contacted over phone in impressive manner to supply his login or other credentials for some legitimate transaction from the service provider [25]. Port scanning attack is the study of application environment through open ports and further launch of attacks based on the knowledge gained by exploiting vulnerabities [26, 27]. Man-in-the-middle attack is an attack where the insecure channel between parties is exploited to make them belive that they are computing with each other, but actually, they are communicating with attacker. The insecure channel is used to lunch this attack [28]. Metadata spoofing is to discover metadata stored in WSDL file. Meta data is data about data and can be useful for attacker to gain knowledge about sensitive data of the user.

4.3 IaaS Layer Attacks

The major attacks identified are side-channel attack, VM rollback attack and VM escape attack. Side-channel attack is centered on study of parameters like time, cache, heat, etc., consumed for the cryptographic algorithm execution. The information gathered by measuring these parameters is retrieved from the cryptographic software that is neither the plaintext nor the ciphertext resulting from the encryption process [29]. VM rollback attack is to gain control over another user's VM by using an old snapshot of victim user without the knowledge of the latter. Moreover, the attacker can change permission granted to the user using rollback, a permission control module [30]. In VM escape attack, the attackers attempt to break down guest operating systems or gain access to the memory so that they can access the hypervisor [25]. The summary of attacks and possible countermeasures for each service model is listed in Table 1.

5 Future Work

We recommend the following future works in the area of cloud computing security. (i) According to Gartner, a multicloud strategy will be the common approach taken by 70% of enterprises by the end of the year 2019. In this new environment, attack surface is expanded; hence, future research is to find intensive and specified technology, processes and cultural innovation to achieve security and flexibility of business computations, (ii) data at rest in the cloud centers is vulnerable to data breach. As per a recent survey conducted by Sky High, only 9.4% of cloud provides are encrypting data at rest. The users have no knowledge that who is accessing their data. It is necessary to find appropriate mechanisms so that data at rest is also safe. Future research can address ethical and legal issues to deal with security of data at rest, (iii) there is an increase of investment in artificial intelligence and machine learning by the enterprises, to formulate business strategy, improve operational efficiency and accelerate decision making. Handling sensitive data in machine learning is a critical task as the concept of ownership breaks down with machine learning data sets comprising of millions of rows of users data. Research is required to apply machine learning for detecting attacks in the multicloud environment and obtaining useful knowledge while preserving privacy of data items.

6 Conclusion

Cloud computing can bring various benefits for organizations. At the same time, it brings many security issues and challenges. In this paper, we made a survey on various attacks at various layers of cloud. The surface of attacks is changing with

Table 1 Types of attacks and counter measures

Attacks	Affected cloud services	Countermeasures
DoS	SaaS, PaaS and IaaS	• Using multilevel authentication and authorization • Using signature-based approach • Using an intrusion detection or intrusion prevention system
Authentication	SaaS	• Using strong passwords and a better authentication mechanism • Applying service provisioning markup language, secure assertion markup language and extensible access control markup language standards to secure federated identities • Encrypting communication channels to secure authentication tokens
SQL injection	SaaS	• Avoiding use of dynamically generated SQL in the code • Using a proxy-based architecture to dynamically detect and extract user input
Phishing	SaaS, PaaS and IaaS	• Using secure web links • Identifying spam e-mails • Not clicking on unknown/doubtful URLs
Port scanning	SaaS, PaaS and IaaS	• Using a time-independent feature set • Using packet counts and neural networks • Using firewalls
Man in the middle	SaaS, PaaS and IaaS	• Requiring a proper secure socket layer architecture • Using robust encryption and decryption algorithm • Using an intrusion detection system
Metadata spoofing	SaaS and PaaS	• Encrypting information about service functionality and other details • Requiring strong authentication to access files
Cross-VM	IaaS	• Using a virtual firewall • Using encryption and decryption
VM rollback	IaaS	• Using suspend and resume
VM escape	IaaS	• Monitoring hypervisor activities • Requiring VM isolation • Using a secure hypervisor • Configuring the host/guest interactions

the introduction of multicloud computing. The artificial intelligence and machine learning integrated cloud are posing more serious security issues. It is vital to make preparations to address security issues in the intelligent multicloud environment.

References

1. P.M. Mell, T. Grance, The NIST definition of cloud computing, in Computer Security Publications from the National Institute of Standards and Technology (NIST) SP 800145 (National Institute of Standards & Technology, Gaithersburg, 2011)
2. Security Guidance for Critical Areas of Focus in Cloud Computing, Cloud Security Alliance [online]
3. S. Basu et al., Cloud computing security challenges and solutions-a survey, in *2018 EEE 8th Annual Computing and Communication Workshop and Conference (CCWC)* (Las Vegas, NV, 2018), pp. 347–356. https://doi.org/10.1109/CCWC.2018.8301700
4. T.S. Chou, Security threats on cloud computing vulnerabilities. Int. J. Comput. Sci. Inf. Technol. **5**(3), 79 (2013)
5. R. Chowdhury, Security in cloud computing. Int. J. Comput. Appl. **96**(15), 0975–8887 (2014)
6. L. Alhenaki, A. Alwatban, B. Alamri, N. Alarifi, A survey on the security of cloud computing, in *2019 2nd International Conference on Computer Applications and Information Security (ICCAIS)* (Riyadh, Saudi Arabia, 2019), pp. 1–7. https://doi.org/10.1109/CAIS.2019.8769497
7. D.Q.L. Shilpashree Srinivasamurthy, *Survey on Cloud Computing Security*. Indiana University, US
8. S. Kumar, R. Goudar, Cloud computing—research issues, challenges, architecture, platforms and applications: a survey. Int. J. Future Comput. Commun. 356–360 (2012)
9. Software as a Service (SaaS). Cloud Taxonomy. Open crowd
10. W. Huang, A. Ganjali, B. Kim, S. Oh, D. Lie, The state of public infrastructure-as-a-service cloud security. ACM Comput. Surv. **47**, **4**, **68**, 31 (2015)
11. M. Boniface et al., Platform-as-a-service architecture for real-time quality of service management in clouds, in *5th International Conference on Internet and Web Applications and Services* ICIW) (Barcelona, Spain: IEEE), pp. 155–160. https://doi.org/10.1109/iciw.2010.91
12. W.C.N. Nimit Kaura, C.A.L. Lal, Survey paper on cloud computing security, in *International Conference on Innovations in Information, Embedded and Communication Systems (ICIIECS)*
13. There's No Such Thing As A Private Cloud. Information Week. 30 June 2010
14. S. Goyal, Public versus private versus hybrid versus community—cloud computing: a critical review. Int. J. Comput. Netw. Inf. Secur. **6**, 20–29 (2014). https://doi.org/10.5815/ijcnis.ii2014.03.03
15. D. Zissis, D. Lekkas, Addressing cloud computing security issues. Future Gener. Comput. Syst. **28**(3), 583–592 (2012)
16. Y. Jiang, C. Perng, T. Li, R. Chang, Selfadaptive cloud capacity planning, in *Proceedings of the 2012 IEEE Ninth International Conference on Services Computing (SCC)* (IEEE, 2012), pp. 73–80
17. M. Ouedraogo, S. Mignon, H. Cholez, S. Furnell, E. Dubois, Security transparency: the next frontier for security research in the cloud. J. Cloud Computing **4**(1), 1–14 (2015)
18. B. Sumitra, C.R. Pethuru, M. Misbahuddin, A survey of cloud authentication attacks and solution approaches. Int. J. Innov. Res. Comput. Commun. Eng. 2(10) (2014)
19. N. Fotiou, A. Machas, G.C. Polyzos, G. Xylomenos, Access control as a service for the Cloud. J. Internet Serv. Appl. (2015). ISSN 1869–0238
20. C. Rogers, L. Duranti, Ethics in the cloud. J. Contemp. Archival Stud. **4**(2) (2017)
21. J.-M. Kim, J.-K. Moon, B.-H. Hong, An effective resource management using clustering schemes, 10–21. https://doi.org/10.14257/astl.2013.43.03

22. S. Subashini, V. Kavitha, A survey on security issues in service delivery models of cloud computing. J. Netw. Comput. Appl. **34**(1), 1–11 (2011)
23. G. Somani, M.S. Gaur, D. Sanghi, M. Conti, R. Buyya, DDoS attacks in cloud computing: issues, taxonomy, and future directions. Comput. Commun. **107**, 30–48 (2017)
24. T.K. Subramaniam, B Deepa, Security attack issues and mitigation techniques in cloud computing environments. Int. J. UbiComp **7**(1), 1–11
25. A. Singh, K. Chatterjee, Cloud security issues and challenges: a survey. J. Netw. Comput. Appl. **79**, 88–115 (2017)
26. P. Deshpande, S.C. Sharma, S.K. Peddoju, A. Abraham, Security and service assurance issues in cloud environment. Int. J. Syst. Assur. Eng. Manage. **9**(1), 194–207 (2018)
27. A. Akbarabadi, M. Zamani, S. Farahmandian, J.M. Zadeh, S.M. Mirhosseini, An overview on methods to detect port scanning attacks in cloud computing, p. 6
28. A. Singh, D.M. Shrivastava, Overview of attacks on cloud computing **1**(4), 3 (2012)
29. S. Anwar et al., Cross-VM cache-based side channel attacks and proposed prevention mechanisms: a survey. J. Netw. Comput. Appl. **93**, 259–279 (2017)
30. P. Mishra, E.S. Pilli, V. Varadharajan, U. Tupakula, Intrusion detection techniques in cloud environment: a survey. J. Netw. Comput.
31. N. Gruschka, M. Jensen, Attack surfaces: a taxonomy for attacks on cloud services, in *2010 IEEE 3rd International Conference on Cloud Computing* (Miami, FL, 2010), pp. 276–279. https://doi.org/10.1109/cloud.2010.23

Mitigating Cloud Computing Cybersecurity Risks Using Machine Learning Techniques

Bharati Mishra and Debasish Jena

1 Introduction

The use of cloud computing is on the rise due to its elastic, on-demand and pay-per-use model. Business start-ups no more need to be concerned about the investment in IT infrastructure and software licensing. Instead, they can focus on their core business goals. According to a report by Gartner [1], the cloud services industry shall grow exponentially by 2022. However, there is an alarming growth of cyberattack incidents on cloud customers [2]. They reported that most attacks on cloud include brute force password guessing attack or stolen credential attack, attacks against known vulnerability in software, web application attacks like cross-site scripting, SQL injection attack and IoT attack. They further confirmed the fact that attackers are searching for opportunities of vulnerability rather than targeting specific organizations. Start-ups in a quick mode to launch their product do not put forward an adequate budget for security controls, and as a result, their customers become easy prey of the cybercriminals. Skybox Security, a computer security company, reported [3] that the reason for the rise in cloud cyberattacks is due to the rapid growth of vulnerabilities in cloud containers. Cloud containers are lightweight virtual machines that can be deployed quickly and easily. But with the ease comes the security lapses. If the old container image has the vulnerability, it can be used to spread over several other cloud deployments within seconds. They also reported that a container vulnerability known as CVE-2019-5736 has been discovered earlier this year which was created by an actor who created a rogue image that can get the administrative privilege and in turn get control over a physical server. Most of the container run-time systems like Docker and Kubernetes got affected by the vulnerability. Also, cloud service providers like Amazon's cloud and Google cloud customers got affected. Thus, cyberattacks on cloud systems are increasing at an alarming rate. This motivates researchers to develop

B. Mishra (✉) · D. Jena
IIIT Bhubaneswar, Bhubaneswar, Odisha 751003, India
e-mail: bmbharati@gmail.com

S. Patnaik et al. (eds.), *Advances in Machine Learning and Computational Intelligence*, Algorithms for Intelligent Systems, https://doi.org/10.1007/978-981-15-5243-4_48

techniques to detect cyberattacks in the cloud and take preventive measures to protect the users. Machine learning techniques can be used to monitor the cloud system and detect cyberattacks and vulnerabilities before any damage happens.

The rest of the paper has been organized as follows. In Sect. 2, the reasons for cyberattack in the cloud have been discussed. Section 3 discusses research papers working on machine learning techniques for cyberattacks in the cloud. The concluding remark has been provided in Sect. 4.

2 Reasons for Cyberattack in Cloud

In a report provided by Cloud Security Alliance [4], the top reasons for cyberattack in the cloud have been identified which are shown in Fig. 1. In this section, the reasons are discussed in detail.

1. **Abuse and Nefarious Use of Cloud Computing**

Cloud provides storage as well as computes services to its clients which are cheap, scalable and pay per use. This feature can be exploited by malicious users who can use these platforms to perform computations which can be used to break security controls like password guessing, attack vector generation, flooding the network, credit card data stealing, etc.

2. **Insecure Application Programming Interfaces**

Cloud APIs are available for the public with a nominal fee. There is no auditing mechanism in place about the use of these APIs. If the API has some vulnerability,

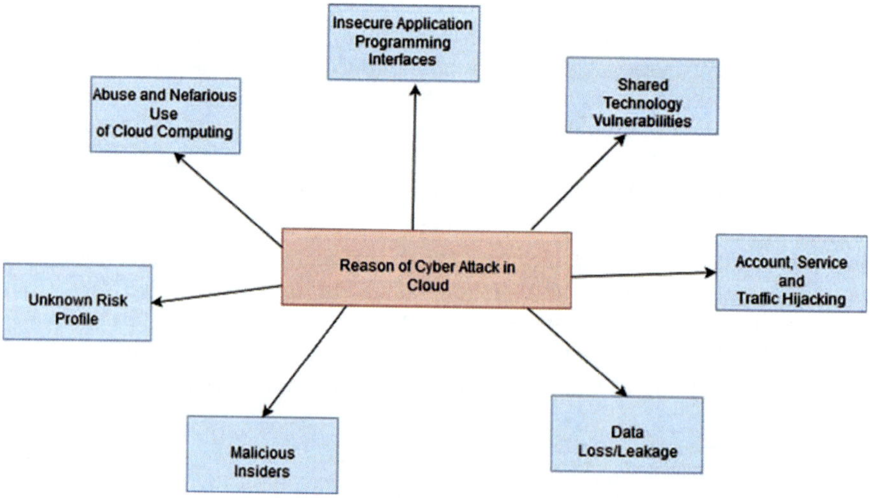

Fig. 1 Reasons for security breaches in cloud

it can be exploited by the attackers to carry out attacks on the cloud platform where genuine users' resources are residing.

3. **Malicious Insiders**

Cloud providers are not investing in the proper monitoring of their employee's activities. Moreover, companies do not expose the security breaches that happen inside the company premises as a company policy. Hence, cloud users do not have any idea about any data theft or loss due to malicious insiders.

4. **Shared Technology Vulnerabilities**

Cloud computing allows the sharing of computing and storage resources through virtualization technique. As a result, it may be possible that rival's data reside in the same servers. One VM may exploit the vulnerabilities due to software or misconfiguration to mount an attack on rivalry VM.

5. **Data Loss/Leakage**

Data is stored in dispersed locations in the cloud and replicated across multiple nodes. Due to misconfiguration, there can be leakage of data or data may be inadvertently deleted by server administrators during VM migration.

6. **Account, Service and Traffic Hijacking**

Cloud providers use user ID and password or OAuth mechanism to authenticate users. Weaker passwords or forgotten OAuth configurations may lead to weaker authentication and hence account, service or traffic hijacking.

7. **Unknown Security Profiles**

Due to privacy reasons, cloud providers do not allow auditing of server logs by the third party for security flaws. This keeps the cloud users in dark about any security events.

3 Use of Machine Learning Techniques to Mitigate Cyberattacks

In this section, research papers working on cybersecurity issues in the cloud have been analyzed. It is found that machine learning techniques have been used to detect DoS attacks, classify cyberattacks, detect security incidents from server logs, network intrusion detection, user behaviour anomaly detection, and predict cloud security incidents from historical data. In Table 1, the details of the research papers have been presented.

Table 1 Research papers on cloud cybersecurity

Paper	Year	Issue detected/addressed	Cloud platform	Preventive measure/result
Ristenpart et al. [5]	2009	Cross-VM side channel attack	Amazon EC2	– Data blinding techniques can be used
Rocha and Correia [6]	2011	– Virtual machines can be relocated by the attacker		The problems can be solved using the following techniques – Cryptography – Trusted computing – Distributed trust
Khorshed et al. [7]	2011	They have used machine learning techniques to classify the cyberattacks on the cloud	Experimental set-up of cloud environment using hypervisors, such as Xen, VMWare, Hyper-V	Concluded that SVM with polynomial kernel gave the best result for attack classification
Chonka et al. [8]	2011	They are able to detect two types of DoS attack, HTTP and XML		They have proposed a methodology to find the source of DoS attacks using a back propagation neural network
Modi et al. [9]	2012	Network intrusion detection in cloud		Integration of Bayesian classifier and snort-based network intrusion detection system
Li et al. [10]	2012	Distributed intrusion detection system for the cloud		Use of neural network to design the distributed IDS for cloud
Chiu et al. [11]	2012	User behaviours anomaly detection in cloud		– Anomaly detection techniques have been used.
Kumar et al. [12]	2016	IDS system for cloud to detect and classify DoS attacks in the cloud environment		– Successfully classifies the DoS attacks to various categories – Use of one-class support vector machines (SVM) algorithm

(continued)

Table 1 (continued)

Paper	Year	Issue detected/addressed	Cloud platform	Preventive measure/result
Zekri et al. [13]	2017	DDoS detection for cloud		C.4.5 algorithm is used along with signature detection techniques to detect DDoS flooding attacks
Masetic et al. [14]	2017	Threat classification for cloud		Studied the various machine learning algorithms like popular supervised learning algorithms such as support vector machine (SVM), random forest (RF), artificial neural networks (ANN), decision tree (DT), Na¨ıve Bayes (NB) and their relevance in attack classification
Masetic et al. [15]	2017	SYN flood attack detection		Use of SVM to classify the attacks with 100% accuracy
Roumani and Nwankpa [16]	2019	Cloud incident prediction		Use of neural network and time series method to build a hybrid model to perform incident prediction from historical data

4 Conclusion

Incidents of cyberattack on cloud is on the rise. Without proactive measures by the cloud service providers, they shall lose the user trust and cloud cannot be utilized for trusted computing. More work needs to be performed in finding suitable machine learning techniques to detect and classify cyberattacks in the cloud, and appropriate security controls need to deployed by cloud providers in order to gain the confidence of customers. To mitigate, configuration-related cyberattacks, configuration manager tools like VMware's vCenter Configuration Manager (VCM) [17] can be installed by cloud service provider. It shall provide a detailed report on the configuration changes on the VMs owned by the customer. This shall enable the customer to verify the vendor's conformance to policy or regulations such as Sarbanes–Oxley Act (SOX), Payment card Industry Data Security Standard (PCI DSS), Health Insurance Portability and Accountability Act (HIPAA) and Federal Information Security Management Act (FISMA). To perform anomaly detection, it is not possible for an administrator of cloud system to analyze the cloud data traffic and detect attacks.

This task can be automated using machine learning techniques. ML can be used to process all the events occurring in the cloud environment and find anomaly patterns to detect cyberattacks. Anomaly detection is a clustering and classification problem.

Hence, machine learning can be used to solve this. Leveraging this technique, Google has launched a cybersecurity company called "Chronicle" [18], which shall bring about products to mitigate cybersecurity attacks using machine learning.

References

1. P. Releases, Gartner forecasts worldwide public cloud revenue to grow 17.5 percent in 2019. https://www.gartner.com/en/newsroom/press-releases/, https://www.google.com/search?q=2019-04-02-gartner-forecasts-worldwide-public-cloud-revenue-to-g&ie=utf-8&oe=utf-8&client=firefox-b-ab
2. A. Blog, Threat intelligence report. https://www.armor.com/threat-intelligence/
3. J. Su, Why cloud computing cybersecurity risks are on the rise: Report. https://www.forbes.com/sites/jeanbaptiste/2019/07/25/, https://www.google.com/search?q=why-cloud-computing-cybersecurity-risks-are-on-the-rise-report%2F%239d6ef9562109&ie=utf-8&oe=utf-8&client=firefox-b-ab
4. J. Archer, A. Boehme, D. Cullinane, P. Kurtz, N. Puhlmann, J. Reavis, Top threats to cloud computing, version 1.0, Cloud Security Alliance (CSA) (2010)
5. T. Ristenpart, E. Tromer, H. Shacham, S. Savage, Hey, you, get off of my cloud: exploring information leakage in third-party compute clouds, in *Proceedings of the 16th ACM conference on Computer and communications security* (ACM, 2009), pp. 199–212
6. F. Rocha, M. Correia, Lucy in the sky without diamonds: stealing confidential data in the cloud, in *2011 IEEE/IFIP 41st International Conference on Dependable Systems and Networks Workshops (DSN-W)* (IEEE, 2011), pp. 129–134
7. M.T. Khorshed, A.S. Ali, S.A. Wasimi, Trust issues that create threats for cyberattacks in cloud computing, in *2011 IEEE 17th International Conference on Parallel and Distributed Systems* (IEEE, 2011), pp. 900–905
8. A. Chonka, Y. Xiang, W. Zhou, A. Bonti, Cloud security defence to protect cloud computing against http-dos and xml-dos attacks. J. Netw. Comput. Appl. **34**(4), 1097–1107 (2011)
9. C.N. Modi, D.R. Patel, A. Patel, R. Muttukrishnan, Bayesian classifier and snort based network intrusion detection system in cloud computing, in *2012 Third International Conference on Computing, Communication and Networking Technologies (ICCCNT'12)*, (IEEE, 2012), pp. 1–7
10. Z. Li, W. Sun, L. Wang, A neural network based distributed intrusion detection system on cloud platform, in *2012 IEEE 2nd international conference on Cloud Computing and Intelligence Systems*, vol. 1 (IEEE, 2012), pp. 75–79
11. C.-Y. Chiu, C.-T. Yeh, Y.-J. Lee, Frequent pattern based user behavior anomaly detection for cloud system, in *2013 Conference on Technologies and Applications of Artificial Intelligence* (IEEE, 2013), pp. 61–66
12. R. Kumar, S.P. Lal, A. Sharma, Detecting denial of service attacks in the cloud, in *2016 IEEE 14th International Conference on Dependable, Autonomic and Secure Computing, 14th International Conference on Pervasive Intelligence and Computing, 2nd International Conference on Big Data Intelligence and Computing and Cyber Science and Technology Congress (DASC/PiCom/DataCom/CyberSciTech)* (IEEE, 2016), pp. 309–316
13. M. Zekri, S. El Kafhali, N. Aboutabit, Y. Saadi, Ddos attack detection using machine learning techniques in cloud computing environments, in *2017 3rd International Conference of Cloud Computing Technologies and Applications (CloudTech)* (IEEE, 2017), pp. 1–7

14. Z. Masetic, K. Hajdarevic, N. Dogru, Cloud computing threats classification model based on the detection feasibility of machine learning algorithms, in *2017 40th International Convention on Information and Communication Technology, Electronics and Microelectronics (MIPRO)* (IEEE, 2017), pp. 1314–1318
15. Z. Mašetić, D. Kečo, N. Doğru, K. Hajdarevi´c, Syn flood attack detection in cloud computing using support vector machine, TEM J. **6**(4), 752 (2017)
16. Y. Roumani, J.K. Nwankpa, An empirical study on predicting cloud incidents. Int. J. Inf. Manage. **47**, 131–139 (2019)
17. VMware vCenter Configuration Manager (VCM). https://www.vmware.com/support/vcm/doc/help/vcm-57/Content/Core_CS/Introduction.htm
18. Chronicle, The New Cybersecurity Firm From Google. https://solutionsreview.com/security-information-event-management/5-things-to-know-about-chronicle-the-new-cybersecurity-firm-from-google/

A Modified Round Robin Method to Enhance the Performance in Cloud Computing

Amit Sharma and Amaresh Sahu

1 Introduction

In these days, cloud computing is one of the hot market and revenue generating technologies. Cloud computing is a distributed computer storage network where resources like applications, services and software can be shared through Internet [1]. These components are available to clients on demand basis. The service providers always try to attract more consumers with its services, and the consumers always try to consume the services with less cost and without interruption.

There are generally three major types of components available in cloud computing [2].

1. SaaS, 2. PaaS, 3. IaaS (Fig. 1).

Based on customer requirements, there are four types of cloud deployment models [2].

1. Public cloud, 2. Private cloud, 3. Hybrid cloud, 4. Community cloud (Fig. 2).

Optimization is the process where a system is modified to make some features of it work more, producing results or less number of resources use. A computer program is said to be optimized when it runs faster, runs in less memory requirements or having less energy consumption [3].

Few examples of optimization are minimization of cost in production in oil refinery where available resources are raw materials, labour, etc., and production target must be accomplished. Minimization of waiting time for patients in emergency room before they are addressed by the doctor, here resources are doctors, the rooms, the patients, the nurses and the equipment, etc.

A. Sharma (✉)
Department of Statistics, Utkal University, Bhubaneswar, Odisha, India
e-mail: sharma.sharma582@gmail.com

A. Sahu
ABIT College, BPUT, Cuttack, Odisha, India

S. Patnaik et al. (eds.), *Advances in Machine Learning and Computational Intelligence*,
Algorithms for Intelligent Systems, https://doi.org/10.1007/978-981-15-5243-4_49

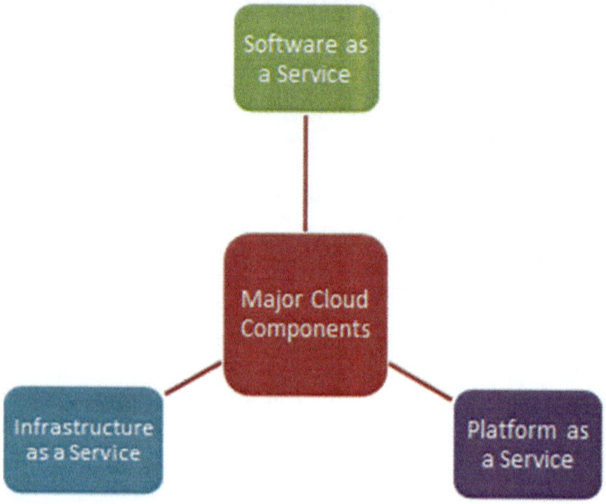

Fig. 1 Major cloud components

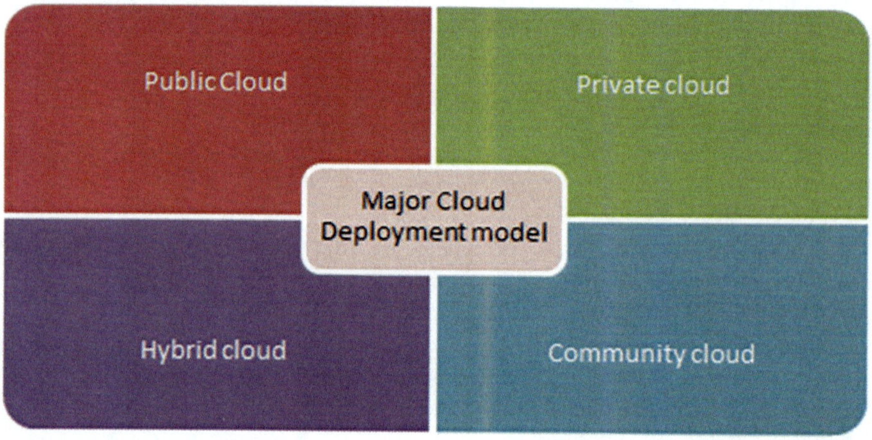

Fig. 2 Major cloud deployment model

In order to provide uninterrupted services to consumer, sometimes providers fail to provide the services due to unavoidable circumstances which lead to the service-level agreement (SLA) desecration among the service providers and consumers. The SLA break happens due to many factors like proper resource allocation, more power consumption, security breach and cost of each service [4].

The service providers are always in constant pressure for providing quality of services (QoS). QoS can be defined as availability, reliability, performance by IaaS, PaaS or SaaS. SLA depends on QoS [5]. The growth of cloud has gradually decreased due to reduction in performance of available resources in cloud. The aspects to minimize the execution time and customer's requirement cost allocation of resources have a great role. Depending on the task scheduling algorithms, throughput and quality of services of services provider improve. Load balancing between the available resources is an important factor to get maximum benefits. Many factors that affect the performance of task scheduling algorithm are CPU utilization, throughput, response time, turnaround time, waiting time, fairness, cost, etc. [6].

Scheduling is a method by which work is assigned to resources to finish execution. Scheduling makes a system multitasking with a single CPU and avoids starvation. The goal of scheduler is to maximize throughput, minimize waiting time and minimize response time. There are two types of scheduling in cloud computing: (1) job scheduling and (2) task scheduling. Job scheduling is specific to user and here system assigns specific jobs. It requires more computing resources and high-performance parallel processing procedures. Task scheduling is performed using heuristics model where task arrives at the same time or different time [2] (Fig. 3).

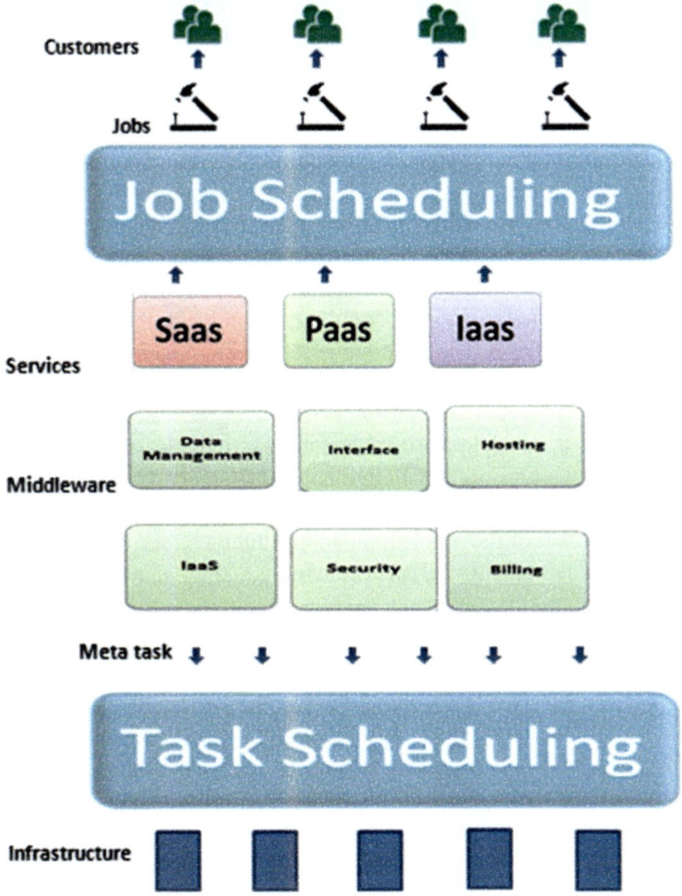

Fig. 3 Provisioning model of cloud

Round Robin algorithm works on task scheduling process where each job is allotted with time slot or quantum. If the job is not completed with the time slot, it interrupts the job and comes after the other job which is arrived in the quantum time. This principle of Round Robin makes fair for all tasks. If the time quantum is very short, it causes many context switches and decreases the CPU efficiency. If the time quantum is very large, the Round Robin acts as a First Come First Scheduling. So, the time quantum is a very import parameter in Round Robin algorithm [7].

In this proposed technique, we have proposed an algorithm to calculate a time quantum value by taking average of burst time of tasks which arrive at the same time that we have described in details in proposed model section. We have also compared the result with the original Round Robin in result and discussion section, and finally, our conclusions and future work are discussed in conclusion section.

2 Literature Survey

Day by day cloud market is growing very fast and taking major technology part of IT industry. Due to this, there are lots of researches have been done in past years on reducing execution time, make span, optimization on resource allocation, cost optimization, increase in throughput on cloud to increase the quality of services of service provider to customer. In virtual machines, the tasks are assigned by the agent brokers, requested by the client for system execution. So many algorithms are introduced for task scheduling in virtual machines. The popular algorithms are shortest job first, shortest job remaining first, priority scheduling, Round Robin algorithm, multilevel queue scheduling, multilevel feedback queue scheduling and many more. In [8], the authors have suggested priority impact scheduling algorithm (PISA) which is compared with the FIFO algorithm and has more 26.7% success response than FIFO. In PISA, the client has to set the priority. The authors Jindal in [6] have suggested an algorithm in which waiting time of individual task reduced significantly. Based on the waiting time, the tasks are prioritized as task scheduling is the backbone of cloud environment. The result of the algorithm is optimum on waiting time for individual task. The quality of services (QoS) increases by reducing waiting time of each task and bottleneck use of number of virtual machine concurrently. In [9], the authors have proposed Hungarian algorithm-based binding policy (HABBP) policy for resolving issue of resource allocation in virtual machines. HABBP is worked on load balancing techniques. It reduces the total execution time by 54.73% than the simplex algorithm on computational time. This policy can be used in open stack cloud management platform like smart parking, drought affected area of Africa, urban areas, etc. In [10–15], the authors have reviewed on different types of algorithm and benefits of usage in cloud. The authors have suggested on how to optimize the task scheduling, maximum resource utilization, minimization of cost, maximum throughput and many more. In quality of services, the client requests for resources and resources are allocated to the client. Effective methods are used to avoid SLA non-compliance between service provider and cloud user. The authors have suggested for improvement of Round

Robin scheduling algorithm in [16, 17]. A new version of RR is suggested to which distributed the load balancing in cloud. Both response time and waiting time are substantially shorter. In [18], the authors have suggested on uniform load distribution in cloud. This paper has proposed best load distribution on two approaches, first one hybrid predict earliest finish time heuristic using ant colony optimization and the second one is hybrid heterogeneous earliest finish time heuristic using ant colony optimization. In [19], authors have tried to reduce cost and processing scheduling time and proposed a particle swarm optimization algorithm which is based on small position particle principle. The experiment results also give optimal solution, less running time and assigning large tasks in less time. In [20], authors have suggested ant colony optimization (ACO) algorithm for task scheduling by comparing with other scheduling algorithms. The objective of this analysis is to minimize the make span of a given set of task and as it works on random optimization search will be used to assign task to virtual machines. In [21], authors have suggested particle swarm optimization (PSO) algorithm in cloud computing which aims on speed and accuracy. To improve execution time transferring time and execution cost of the task, multi-objective particle swarm optimization model for optimizing task scheduling was developed. In [22], authors have suggested stochastic integer programming to solve resource scheduling problem in cloud. The stochastic integer programming is solved based on Grobner bases theory and provided experimental result which says significant results of improvement of SLA between service providers and consumers. In [23] authors have suggested a cat swarm optimization (CSO) based on heuristic resources scheduling algorithm to schedule task to available resources. The experiment also shows a good result to minimize the cost in assigning task to resources and over the particle swarm optimization algorithms. In [24], authors have suggested resources allocation based on grasshopper optimization algorithm and compared with the evolutionary algorithm and got the best result by using GOA. It also compares with the exiting GA algorithm and SEIRA and gives the optimization results of task scheduling. In [25], authors have suggested on improving resource allocation by genetic algorithms which works on fitness function and genes mutation. Cloud entities are considered as genes and have scheduling schemes and provide fitness solution with time to time.

In [7], the authors have suggested a SRDQ algorithm which is a combining shortest job first (SJF) and Round Robin (RR) having a dynamic variable task time quantum. The SRDQ algorithm is based on two keys, one is for dynamic task quantum and other is to split the ready queue into sub queues. The first queue is for short job, and second queue is for long jobs. The authors have done experiment in CloudSim environment (3.0.3) with two different versions SJF&RR with dynamic quantum (SRDQ) and SJF&RR with static quantum (SRSQ) in three virtual machines. The experimental results also give that the proposed algorithm has good result in minimizing turnaround and waiting times.

In [26], the authors have reviewed many task scheduling algorithms like RR, MaxMin, MinMin, FCFS, MCT, PSO GA and modified Round Robin (MRR) algorithm. The authors have also verified MRR algorithm using CloudSim toolkit, and the results show that the average waiting time is less than the original Round Robin method.

3 Proposed Model

Till now, Round Robin scheduling algorithm is widely used in task scheduling in CPU and cloud computing as well. It is a pre-emptive scheduling, and it specially designs for sharing operating system. The CPU switches to another job between the processes when the time quantum expires.

The first process is chosen by the CPU scheduler from ready queue. A timer is laid to break after one time quantum and releases the process. Below are two things that occur in Round Robin.

1. If the process CPU burst time < one time quantum, then the CPU dispatches the process releases wilfully.
2. If the CPU burst time of the currently running process > time quantum, then the process will be put at tail of the ready queue.

By introducing **modified Round Robin** scheduling algorithm, we have tried to minimize the performance parameters like waiting time and turnaround time in cloud.

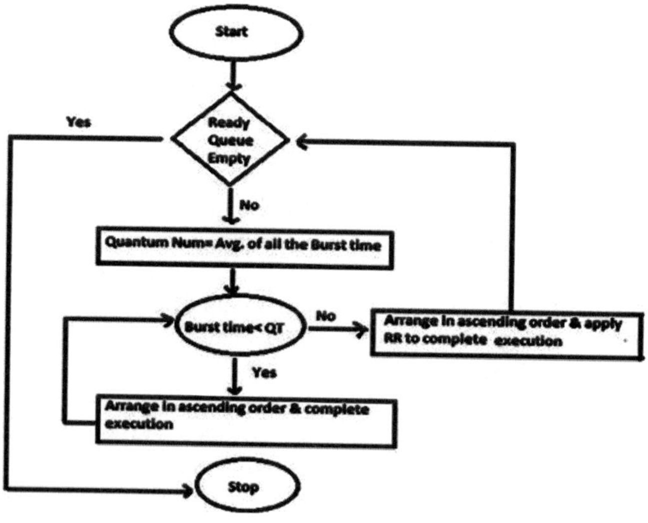

Fig. 4 Flow chart of modified round robin

4 Approach

Step 1 Arrange all the process in order to their burst time.
Step 2 Compute the average of all the burst time.
Step 3 Consider the quantum number is equal to the average of burst time.
Step 4 The tasks are less than the quantum numbers. Arrange these tasks in ascending order, and complete the execution.
Step 5 The tasks are greater or equal to the quantum number. Arrange in ascending order, and complete the execution.
Step 6 Execute all the process until the time quantum expires for each process.

5 Result and Discussion

In cloud computing, multiple requests are received by the users at the same time and different times. The Round Robin task scheduling algorithm works on different arrival time as well as same arrival time. Here, we have considered the case where all tasks are arrived at the same time.

Table 1

No. of tasks	Burst time
T1	14
T2	8
T3	6
T4	4

6 Using Original Round Robin Algorithm

Quantum Number: 5

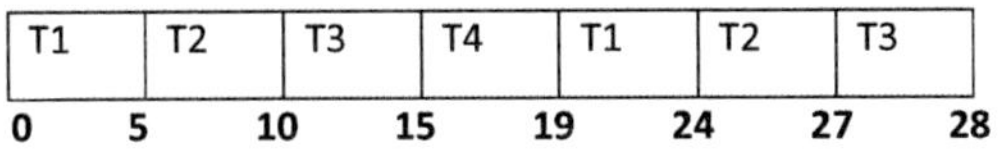

Table 2

No. of tasks	Waiting time	Turnaround time
T1	14	24
T2	19	27

(continued)

(continued)

No. of tasks	Waiting time	Turnaround time
T3	22	28
T4	15	19

Average waiting time in original Round Robin: 17.5
Average turnaround time in original Round Robin: 24.5.

7 Using Modified Round Robin Algorithm

Quantum Number: Average Burst Time (i.e. 8)

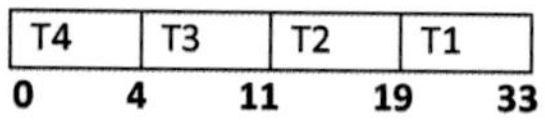

Table 3

No. of tasks	Waiting time	Turnaround time
T1	19	33
T2	11	19
T3	4	11
T4	0	4

Average waiting time in modified Round Robin: 8.5
Average turnaround time in modified Round Robin: 16.75.

From Figs. 5 and 6, it is cleared that the average waiting time and the turnaround are significantly lower than the original Round Robin. The average waiting time is reduced by 28.57%, and the average turnaround is reduced by 49.25% when compared with the original Round Robin.

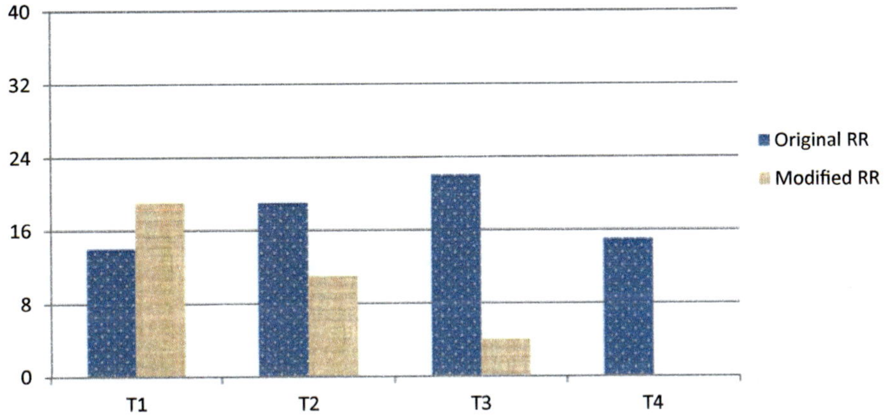

Fig. 5 Waiting times comparison

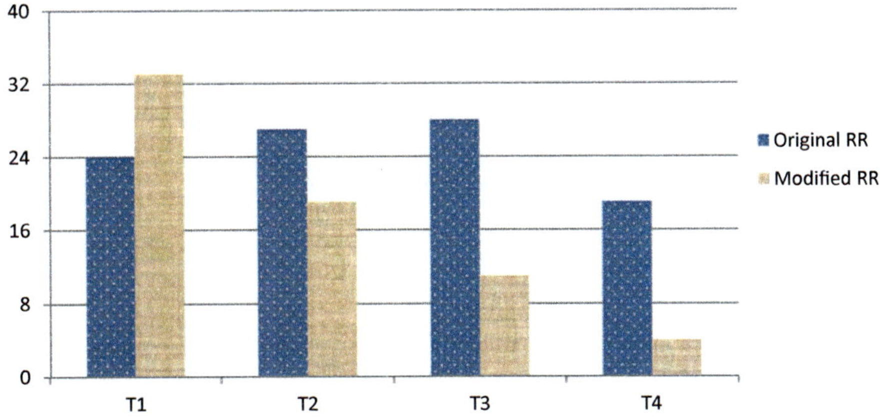

Fig. 6 Turnaround times comparison

8 Conclusion

In this paper, the scheduling of different tasks arriving at the same time is evaluated. Round Robin scheduling algorithm is well known for its fairness to the other processes. Based on the time quantum and arrival time, it will give chances to all process to avoid starvation. Factors like waiting time, response time and turnaround time must be taken into consideration in order to take the scheduling decision. In the above proposed model, we have calculated quantum number for tasks arriving at the same time and compared the results with the original Round Robin. The results shown in Figs. 4 and 5 have considerably reduced the waiting time and turnaround time. The waiting time and turnaround time are considerably reduced by 28.57% and

49.25%, respectively. The proposed algorithm may not efficiently work on the tasks are arriving at different time interval.

Hereafter, the researchers intend to precede their experiments in finding a better calculation methodology for quantum number for tasks arriving at different time.

References

1. S.K. Ghosh, IIT Kharagpur, Cloud computing, NPTEL
2. T. Erl, R. Puttini, Z. Mahmood, Cloud computing concepts, technology and architecture (2013)
3. Optimization (Computer Science), https://simple.m.wikipedia.org/wiki
4. M. Alhamad, T. Dillon, E. Chang, SLA-Based trust model for cloud computing, in *2010 13th International Conference on Network-Based Information Systems* (Takayama, 2010), pp. 321–324. https://doi.org/10.1109/nbis.2010.67
5. D. Ardagna, G. Casale, M. Ciavotta et al., Quality-of-service in cloud computing: modeling techniques and their applications. J. Internet Serv. Appl. **5**, 11 (2014). https://doi.org/10.1186/s13174-014-0011-3
6. A. Jindal, Optimization of task scheduling algorithm through QoS parameters for cloud computing. MATEC Web Conf. **57**, 02009 (2016). https://doi.org/10.1051/matecconf/20165702009
7. S. Elmougy, S. Sarhan, M. Joundy, A novel hybrid of Shortest job first and round Robin with dynamic variable quantum time task scheduling technique. J. Cloud Comput. **6**, 12 (2017). https://doi.org/10.1186/s13677-017-0085-0
8. H. Wu, Z. Tang, R. Li, A priority constrained scheduling strategy of multiple workflows for cloud computing, in *2012 14th International Conference on Advanced Communication Technology (ICACT)* (PyeongChang, 2012), pp. 1086–1089
9. S. Akintoye, A. Bagula, Optimization of virtual resources allocation in cloud computing environment (2015). https://doi.org/10.1109/afrcon.2017.8095597
10. N. Singla, S. Bawa, Review of efficient resource scheduling algorithms in cloud computing. IJARCSSE **3**(8) (2013), ISSN: 2277 128X
11. A.K. Bardsiri, S.M. Hashemi, A review of workflow scheduling in cloud computing environment. Int. J. Comput. Sci. Manage. Res. **1**, 348–351 (2012)
12. V. Kaur, G. Kaur, Cloud computing scheduling algorithms: a review. Int. J. Comput. Sci. Netw. **4**(2) (2015), ISSN: 2277-5420
13. S. Varshney, S. Singh, A survey on resource scheduling algorithms in cloud computing. Int. J. Appl. Eng. Res. IJAER **13**(9), 6839–6845 (2018), ISSN 0973-4562
14. P. Lijin, S. SivaSathya, K.S. Guruprakash, Survey on resource allocation techniques in cloud computing. Int. J. Eng. Technol. UAE. **7**, 823–828 (2018)
15. A. Singh, M. Malhotra, A comparative analysis of resource scheduling algorithms in cloud computing (2013)
16. P. Sangwan, M. Sharma, A. Kumar, Improved round robin scheduling in cloud computing. Adv. Comput. Sci. Technol. **10**(4), 639–644 (2017), ISSN 0973-6107
17. S. Mittal, S. Singh, R. Kaur, *Enhanced Round Robin Technique for Task Scheduling in Cloud Computing Environment*. Springer Singapore (2017)
18. A. Kaur, B. Kaur, Load balancing optimization based on hybrid Heuristic-Metaheuristic techniques in cloud environment. J. King Saud Univ. Comput. Inf. Sci. (2019), ISSN 1319-1578
19. L. Guo, S. Zhao, S. Shen, C. Jiang, Task scheduling optimization in cloud computing based on heuristic algorithm. J. Netw. **7** (2012). https://doi.org/10.4304/jnw.7.3.547-553
20. M. Tawfeek, A. El-Sisi, A. Keshk, F. Torkey, Cloud task scheduling based on ant colony optimization. Int. Arab J. Inf. Technol. **12**, 64–69 (2013). https://doi.org/10.1109/ICCES.2013.6707172

21. F. Ramezani, J. Lu, F. Hussain, Task scheduling optimization in cloud computing applying multi-objective particle swarm optimization, pp. 237–251 (2013). https://doi.org/10.1007/978-3-642-45005-1_17
22. Q. Li, Y. Guo, Optimization of resource scheduling in cloud computing, in *12th International Symposium on Symbolic and Numeric Algorithms for Scientific Computing*, pp. 315–320 (2010). https://doi.org/10.1109/synasc.2010.8
23. S. Bilgaiyan, S. Sagnika, M.N. Das, Workflow scheduling in cloud computing environment using cat swarm optimization, in *Souvenir of the 2014 IEEE International Advance Computing Conference, IACC* (2014). https://doi.org/10.1109/iadcc.2014.6779406
24. J. Vahidi, M. Rahmati, Optimization of resource allocation in cloud computing by grasshopper optimization algorithm, in *2019 5th Conference on Knowledge Based Engineering and Innovation (KBEI)* (Tehran, Iran, 2019), pp. 839–844. https://doi.org/10.1109/kbei.2019.8735098
25. S. Javed, W. Manzoor, N. Akhtar, K. Zafar, Optimization of resource allocation scheduling in cloud computing by genetic algorithm. Fast Res. J. FRJ (2015)
26. R.A. Khurma, H. Harahsheh, A. Sharieh, Task scheduling algorithm in cloud computing based on modified round robin algorithm. J. Theor. Appl. Inf. Technol. **96**, 5869–5888 (2018)

Channel Capacity Under CIFR Transmission Protocol for Asymmetric Relaying

Brijesh Kumar Singh and Mainak Mukhopadhyay

1 Introduction

The technological frontier of the next century is the dream of wireless communications that enable the exchange of information between people or phones. Keeping these goals into our mind, we will have to go for new technologies to increase the performance of various wireless communication systems.

Medium between transmitter and receiver is known as fading channel which is always a weak link for any wireless system. In general, fading channel is modelled using probabilistic functions like CDF, PDF and MGF. Due to probabilistic nature of fading channel, perfect estimation of the performance of wireless system at receiver is not possible. However, it is possible to present analytical expressions for probability of error (SER and BER) and capacity of channel by investigating the CDF, PDF and MGF of different fading channels. Further, these expressions compute the performance degradation due to the impairment of the wireless system.

Communication relays are considered very important towards developing next generation communication systems. Communication relays can significantly improve coverage and throughput performance. Even though, it consumes radio resources, yet it can significantly improve network efficiency and service availability

B. K. Singh (✉) · M. Mukhopadhyay
Birla Institute of Technology, Deoghar Campus, Mesra, India
e-mail: bksingh2003@gmail.com

M. Mukhopadhyay
e-mail: mainak@bitmesra.ac.in

S. Patnaik et al. (eds.), *Advances in Machine Learning and Computational Intelligence*, Algorithms for Intelligent Systems, https://doi.org/10.1007/978-981-15-5243-4_50

2 Literature Review

Radio-frequency spectrum is very limited, whereas the demand of various services is growing at a speedy pace [1], so because of this reason, higher value of spectral efficiency is needed. Better spectral efficiency of the wireless communication system can be achieved in many different ways such as by having lesser cell area, lesser reuse distance of cochannel and by using better schemes of channel allocation [2]. Depending on the knowledge of the channel side information, we have complex expressions of channel capacity under fading and these channel capacity expressions are in the form of either channel variation in frequency or in time [3–6]. Goldsmith and Varaiya analyzed the capacity for Rayleigh fading channels for different adaptive transmission strategies [7].

In [8], for Nakagami-m fading, channel capacity utilizing dual diversity was examined by Khatalin and Fonseka. For Rayleigh fading, the capacity in various conditions of diversity along with the rate of adaptation and transmission protocol was studied in [9]. Some different fading channels like Rician, Weibull and Hoyt fading channels have been explored in [10, 11]. Channel capacity for various diversity plans and distinct level of adaptation and transmitting power strategies for the correlated Rayleigh fading channel were obtained in [12]. MGF-based method was advocated in [13, 14] for the calculation of capacity with ORA strategy just through the use of the numerical methods. A unique approach based on MGF was found out in [15] for the assessment of channel capacity for different level adaptations and power transmission schemes.

For mixed Rice/Rayleigh fading environment, amplify-and-forward relaying system with variable gain amplification was used and the implementation of polarization diversity was done in [16]. In [17], relay was used in three different modes (decode-and-forward mode, amplify-and-forward mode, hybrid decode-amplify-forward mode) for analysis of outage probability. To improve both, range and performance gain, in [18], multi-hop relaying and cooperative relaying were combined. In this paper, for performance analysis, we have taken asymmetric relaying system having Rayleigh and $k - \mu$ fading channel.

Remaining paper is organized as given below. Section 3 describes the system model. For this system, method to derive closed-form solution of channel capacity under CIFR transmission scheme is discussed in Sect. 4. Numerical analysis is discussed in Sect. 5, and finally, Sect. 6 presents the conclusion and future scope.

3 System Model

Figure 1 shows a relay structure consisting of a relay R with a source S and a destination D. Initially, source transmits the information and this information is received by relay. After receiving the message, relay decodes the message and then removes the noise and then forwards it towards destination. In this scheme of relaying,

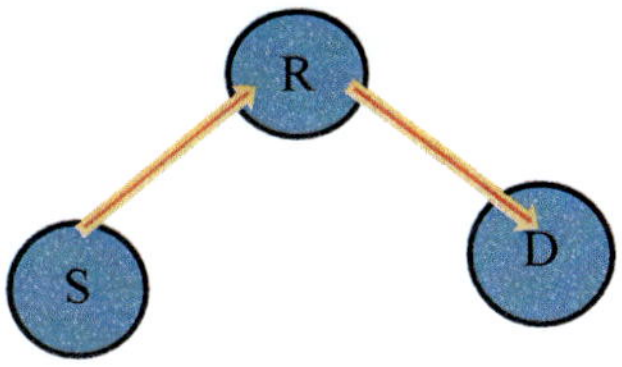

Fig. 1 Relaying system with dual-hop scheme

fading channel taken between source S to relay R and between relay R to destination D are Rayleigh fading channel and $k - \mu$ fading channel.

Fading channel having Rayleigh distribution is given by PDF

$$f(\gamma_1) = \frac{1}{\gamma_1} e^{\frac{-\gamma}{\bar{\gamma}_1}} \tag{1}$$

where γ_1 and $\bar{\gamma}_1$ are instantaneous and average SNR for S-R link. Fading channel with $k - \mu$ distribution is defined by the following PDF.

$$p(\gamma_2) = K_1 \gamma_2^{\frac{(\mu-1)}{2}} e^{-\mu \frac{(1+k)\gamma_2}{\overline{\gamma_2}}} I_{\mu-1}\left(2\mu \sqrt{\frac{k(1+k)\gamma_2}{\overline{\gamma_2}}}\right) \tag{2}$$

where $\overline{\gamma}_2$ and γ_2 are average and instantaneous SNR for $k - \mu$ fading channel. Here, the value of K_1 is given by

$$K_1 = \frac{\mu(1+k)^{\frac{(\mu+1)}{2}}}{k^{\frac{(\mu-1)}{2}} e^{\mu k} \overline{\gamma_2}^{\frac{(\mu-1)}{2}}} \tag{3}$$

4 Closed-Form Expression of Channel Capacity Under CIFR Transmission Scheme

In this section, for performance analysis, capacity under CIFR scheme has been derived. From [19], per unit bandwidth channel capacity in case of CIFR scheme is given as

$$C_{\mathrm{CIFR}} = \mathrm{Log}_2\left(1 + \frac{1}{\int\limits_0^\infty M(s)\mathrm{d}s}\right) \tag{4}$$

Here, $M(s)$ is MGF of overall SNR γ of the system.

Let

$$I_3 = \int_0^\infty M(s)\mathrm{d}s \tag{5}$$

So,

$$C_{\mathrm{CIFR}} = \mathrm{Log}_2\left(1 + \frac{1}{I_3}\right) \tag{6}$$

From the table of integral series [20], we know that

$$\int_0^\infty x^{\lambda-1} e^{-\beta.x} \gamma(v, \alpha x)\mathrm{d}x = \left(\frac{\alpha^v \Gamma(\lambda + v)}{v(\alpha + \beta)^{\lambda+v}}\right) {}_2F_1\left(1, \lambda + v; v + 1; \frac{\alpha}{\alpha + \beta}\right) \tag{7}$$

and

$${}_2F_1(a, b; c; x) = \sum_{n=0}^{\infty} \frac{(a)_n (b)_n}{(c)_n} \frac{x^n}{n!} \tag{8}$$

and

$$\int_0^\infty x^{\lambda-1}(1 + x)^{-\alpha+v}(x + \beta)^{-v}\mathrm{d}x = \begin{pmatrix} B(\alpha - \lambda, \lambda)_2 \\ \times F_1(v, \alpha - \lambda; \alpha; 1 - \beta) \end{pmatrix} \tag{9}$$

After putting value of $M(s)$ from [21] into Eq. (5) and then using Eqs. (7), (8) and (9) for further calculation, we get new expression of I_3 as given below.

$$I_3 = \left(\begin{array}{c} \left(\begin{array}{c} \frac{A\left(\frac{\mu(1+k)}{\bar{\gamma}_2}\right)^{m+\mu} \Gamma(m+\mu+1)}{m+\mu} \sum_{n=0}^{\infty} \left(\frac{\mu(1+k)}{\bar{\gamma}_2}\right)^n \\ (B((n+m+\mu-1), 2)) \left({}_2F_1 \left(\begin{array}{r} n+m+\mu+1, \\ n+m+\mu-1; \\ n+m+\mu+1; \\ 1 - \frac{\mu(1+k)}{\bar{\gamma}_2} \end{array} \right) \right) \end{array} \right) \\ - \left(\begin{array}{c} \frac{A\left(\frac{\mu(1+k)}{\bar{\gamma}_2}\right)^{m+\mu} \Gamma(m+\mu+1)}{m+\mu} \sum_{n=0}^{\infty} \left(\frac{\mu(1+k)}{\bar{\gamma}_2}\right)^n \\ (B((n+m+\mu-1), 2)) \left({}_2F_1 \left(\begin{array}{r} n+m+\mu+1, \\ n+m+\mu-1; \\ n+m+\mu+1; \\ 1 - \frac{\mu(1+k)}{\bar{\gamma}_2} - \frac{1}{\bar{\gamma}_1} \end{array} \right) \right) \end{array} \right) \end{array} \right) \tag{10}$$

Finally, using Eqs. (10) and (6), we get Eq. (11) as given below.

$$C_{\mathrm{CIFR}} = \mathrm{Log}_2\left(1 + \frac{1}{\left(\left(\begin{array}{l}\frac{A(K_2)^{K_3}\Gamma(K_5)}{K_3}\\ \sum\limits_{n=0}^{\infty}(K_2)^n B(K_7,2)\\ \times\left({}_2F_1(K_6,K_7;K_6;1-K_2)\right)\end{array}\right) - \left(\begin{array}{l}\frac{A(K_2)^{K_3}\Gamma(K_5)}{K_3}\\ \sum\limits_{n=0}^{\infty}(K_2)^n B(K_7,2)\\ \times\left({}_2F_1\left(\begin{array}{l}K_6,K_7;K_6;\\ 1-K_2-\frac{1}{\bar{\gamma}_1}\end{array}\right)\right)\end{array}\right)\right)}\right) \quad (11)$$

Thus, Eq. (11) represents channel capacity under CIFR transmission protocol for asymmetric relaying system having Rayleigh fading channel between source S to relay R and $k-\mu$ fading channel between relay R to destination D.

5 Numerical Analysis

Numerical analysis for the expression of C_{CIFR} is provided in this section. We have drawn plots of C_{CIFR} w.r.t. SNR for various values of k and μ. Here, Fig. 2 shows the plot of C_{CIFR} for value of k equal to 0.5. For getting different curves in Fig. 2, we have taken values of fading parameter μ equal to 1.0, 1.4, 1.8 and 2.2. From this

Fig. 2 C_{CIFR} with $k = 0.5$

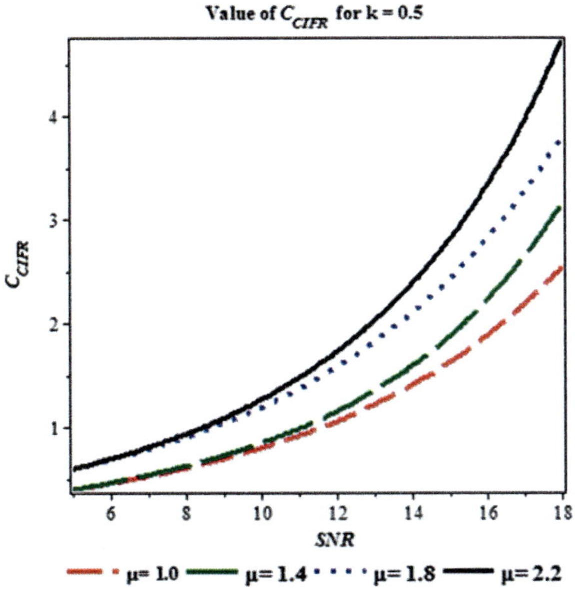

figure, it is clear that for a particular value of fading parameter k, when we increase the values of other fading parameter μ, there is increase in the value of C_{CIFR}.

In Figs. 3, 4, 5, we have plotted C_{CIFR} after taking different values of fading parameter k. Value of k for Figs. 3, 4, 5 are 0.8, 1.1 and 1.3 and for getting different

Fig. 3 C_{CIFR} with $k = 0.8$

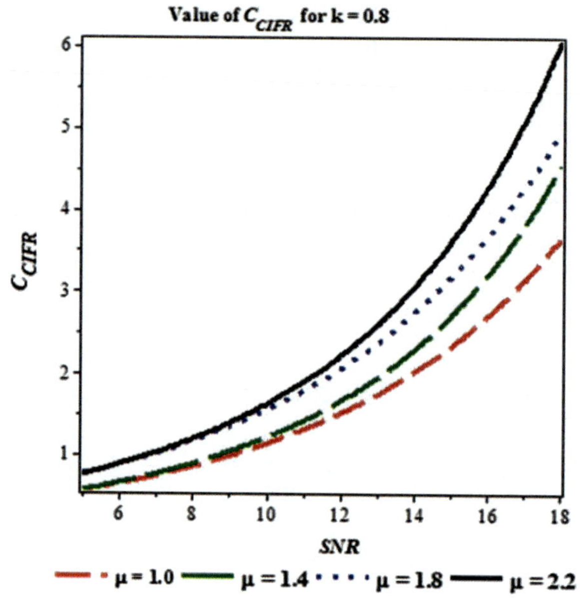

Fig. 4 C_{CIFR} with $k = 1.1$

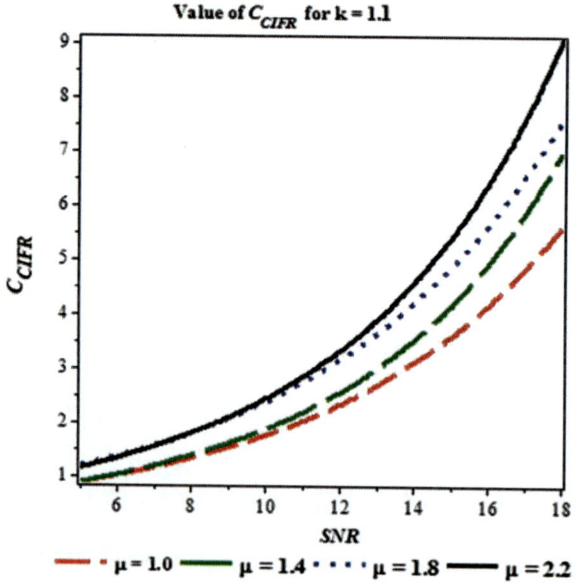

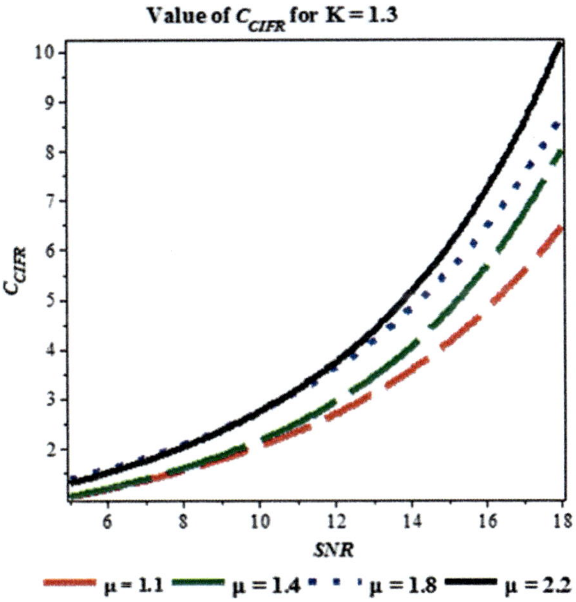

Fig. 5 C_{CIFR} with $k = 1.3$

curves in these three figures also, we have taken value of μ equal to 1.0, 1.4, 1.8 and 2.2. From Figs. 3, 4, 5, we can also observe that the value of C_{CIFR} increase when there is rise in the value of fading parameter μ. Using Figs. 2, 3, 4 and 5, we can conclude that as we are increasing the value of fading parameter k, and there is also increase in value of C_{CIFR}. We can also conclude from all of these four figures that with the increase in value of average SNR, there is increase in the value of C_{CIFR}.

6 Conclusion and Future Scope

In fading environment, relay is very essential for increasing the performance of wireless communication system. For asymmetric relaying system having Rayleigh fading channel between source to relay station and $k - \mu$ fading channel between relay station and destination, this paper analyses the effect of signal-to-noise ratio (SNR) on channel capacity under channel inversion with fixed rate (CIFR) transmission protocol. Closed-form expression of channel capacity under CIFR transmission protocol (C_{CIFR}) has been evaluated and then we have plotted various curves for C_{CIFR} w.r.t. SNR after taking different values of fading parameters k and μ into consideration. We can conclude from Figs. 2, 3, 4 and 5 that the value of C_{CIFR} enhances as we increase the value of fading parameter μ. We can also conclude that there is increase in the value of channel capacity under channel inversion with fixed rate if we are increasing the values of fading parameter k. Also, with the increase in the values of average SNR, there is increase in the values of C_{CIFR}. Thus, for drastically enhancing the coverage and throughput performance of wireless communication systems, the use of relays has been found very critical. In future, we can also evaluate various other parameters of performance measure. For improving coverage and throughput performance, this system can be used for implementing wireless communication system in fading environment.

References

1. K. Pahlavan, A.H. Levesque, Wireless data communications. Proc. IEEE **82**(9), 1398–1430 (1994)
2. G.L. Stüber, *Principles of Mobile Communications* (Kluwer, Norwell, MA, 1996)
3. M.K. Simon, M.S. Alouini, Digital communication over fading channels, vol. 95 (John Wiley & Sons, 2005)
4. John G. Proakis, Masoud Salehi, *Digital communications*, vol. 4 (McGraw-hill, New York, 2001)
5. P. Varzakas, G.S. Tombras, Spectral efficiency of a cellular MC/DS-CDMA system in Rayleigh fading. Int. J. Commun Syst **18**(8), 795–801 (2005)
6. P. Varzakas, G.S. Tombras, Spectral efficiency of a single-cell multi-carrier DS-CDMA system in Rayleigh fading. J. Franklin Inst. **343**(3), 295–300 (2006)
7. A.J. Goldsmith, P.P. Varaiya, Capacity of fading channels with channel side information. IEEE Trans. Inf. Theory **43**(6), 1986–1992 (1997)
8. S. Khatalin, J.P. Fonseka, Capacity of correlated Nakagami-m fading channels with diversity combining techniques. IEEE Trans. Veh. Technol. **55**(1), 142–150 (2006)
9. M.S. Alouini, A.J. Goldsmith, Capacity of Rayleigh fading channels under different adaptive transmission and diversity-combining techniques. IEEE Trans. Veh. Technol. **48**(4), 1165–1181 (1999)
10. N.C. Sagias, G.S. Tombras, G.K. Karagiannidis, New results for the Shannon channel capacity in generalized fading channels. IEEE Commun. Lett. **9**(2), 97–99 (2005)
11. S. Khatalin, J.P. Fonseka, On the channel capacity in Rician and Hoyt fading environments with MRC diversity. IEEE Trans. Veh. Technol. **55**(1), 137–141 (2006)
12. R.K. Mallik, M.Z. Win, J.W. Shao, M.S. Alouini, A.J. Goldsmith, Channel capacity of adaptive transmission with maximal ratio combining in correlated Rayleigh fading. IEEE Trans. Wireless Commun. **3**(4), 1124–1133 (2004)
13. K.A. Hamdi, Capacity of MRC on correlated Rician fading channels. IEEE Trans. Commun. **56**(5), 708–711 (2008)
14. R.C. Palat, A. Annamalai, J.H. Reed, An efficient method for evaluating information outage probability and ergodic capacity of OSTBC system. IEEE Commun. Lett. **12**(3), 191–193 (2008)
15. M. Di Renzo, F. Graziosi, F. Santucci, Channel capacity over generalized fading channels: a novel MGF-based approach for performance analysis and design of wireless communication systems. IEEE Trans. Veh. Technol. **59**(1), 127–149 (2009)
16. M. Delibasic, M. Pejanovic-Djurisic, Performance improvement of relay system in Rayleigh/rice fading using polarization diversity, in *40th International Conference on Telecommunications and Signal Processing (TSP)* (IEEE, 2017), pp. 233–236
17. Z. Liu, Y. Yuan, L. Fu, X. Guan, Outage performance improvement with cooperative relaying in cognitive radio networks. Peer-to-Peer Netw. Appl. **10**(1), 184–192 (2017)
18. H.K. Boddapati, M.R. Bhatnagar, S. Prakriya, Performance of cooperative multi-hop cognitive radio networks with selective decode-and- forward relays. IET Commun. **12**(20), 2538–2545 (2018)

19. V.K. Dwivedi, G. Singh, Marginal moment generating function based analysis of channel capacity over correlated Nakagami-m fading with maximal-ratio combining diversity. Prog. Electromagnet. Res. **41**, 333–356 (2012)
20. I.S. Gradshteyn, I.M. Ryzhik, Table of integrals, series, and products. Academic press (2014)
21. B.K. Singh, M. Mukhopadhyay, Performance analysis over asymmetric Rayleigh and $k - \mu$ fading channel for dual hop decode-forward relaying-(submitted for publication). IJE Lett.

Maximizing the Lifetime of Heterogeneous Sensor Network Using Different Variant of Greedy Simulated Annealing

Aswini Ghosh and Sriyankar Acharyya

1 Introduction

1.1 Motivation

In WSN, sensor nodes are powered by battery and battery cannot be replaced or recharged easily. So energy usage should be reduced to enhance lifetime of sensor. In WSN, coverage problem can be subdivided into two main sub-problems, one is area coverage problem and another one is target coverage. Sensor nodes locations are the basic input for the algorithms that check coverage of the network [1].

Sensor node deployments are mainly of two types, random deployment and deterministic deployment. If details of the region are not known or regions are inaccessible, then random deployment is suitable. For example, in battlefield surveillance, random deployment of sensor nodes would be used. The most common way of extending the lifetime in random deployment is scheduling the sensor nodes in such a way that only a subset of sensor nodes that are enough in number to satisfy coverage constraint need to be active at a time [2].

Dividing sensors into some distinct sets is a method to increase lifetime of sensor network. There should be guarantee that every set must cover all targets. These distinct sets are activated one by one in such a way that only one set is active at time. In addition to increasing lifetime increase and decreasing the number of active sensors, the following two must be satisfied. First one, selected sensors must guarantee target

A. Ghosh (✉)
MTNL, Mumbai, India
e-mail: aswinimtnl@gmail.com

S. Acharyya
Department of Computer Science and Engineering, MAKAUT, Kolkata, West Bengal 700064, India
e-mail: srikalp8@gmail.com

S. Patnaik et al. (eds.), *Advances in Machine Learning and Computational Intelligence*, Algorithms for Intelligent Systems, https://doi.org/10.1007/978-981-15-5243-4_51

detection of the given area and second one, sensor should also be able to connect to the center via super node.

In three ways, a sensor network can be deployed. (1) random [2, 3] (2) controlled placement [1, 4–8] (3) uniform distribution [9].

In random method, placing a seed node at the origin network generation is done. If the environment is unknown, then sensors are randomly distributed over a wide sensing area. When random placement is used, then sensors are randomly dropped from aircraft.

Second type of sensor distribution is called controlled placement; if in advance terrain properties are determined, then sensors can be carefully deployed on a sensor field. Controlled placement mainly used to meet the requirements of different levels of services. Third type of sensor distribution is called the uniform distribution. Here, using uniform distribution, sensor nodes are randomly distributed using their X- and Y-coordinates. Sometimes, it is called a rectangular distribution.

1.2 Contribution

Here, in this paper, greedy simulated annealing is used to maximize number of sensor groups by selecting sensor node in each group optimally. Each sensor group must cover all targets. So sensor node will get maximum rest. This greedy algorithm finds out minimum number of sensor per group, so it minimizes energy consumption. Minimizing energy consumption leads to maximizing network lifetime.

In the fifth section of this paper, proposed grouping algorithm is given. In fourth section, sensor network protocol design is discussed. The last section is about total results. Fifth section is also responsible for selecting monitoring sensor using different types of SA. It is confirmed that using super node with higher energy and communication range increases lifetime of network.

2 Literature Survey

In WSN, a number of approaches have been proposed to increase lifetime of sensor. All methods can be broadly classified into four: (1) Controlled mobility (2) topology control (3) data reduction (4) duty cycling. Sensor node turns off and on in duty cycling approach. When sensor nodes are turned off, then less energy dissipation required and when sensors are in active mode then more energy dissipated. Using turn on and, sensor save energy in duty cycling approach. Cluster-based topology is proposed in [10]. In cluster-based approach, sensor nodes change their role, sometimes, it acts as cluster head and sometimes, it acts as cluster member node. In controlled mobility method, mobile base station, data mules, and mobile relays approaches are used. In mobile base station approach, a highly powerful base station collects data from other node by moving around whole WSN [11–14]. In this method, differential power consumption of sensors is reduced. But drawback of this method is higher latency due to slow speed of base node. In data reduction approach grouping of sensor node are formed. In this method, grouping is formed in such a way that group can detect all targets. Here, for sending data to base station, all sensor of that group are not used, and only sensors which are in shortest path are used. But, in those methods, grouping was not formed in greedy way. Our approach shows that it outperforms all previous approach. It is better than duty cycling because our approach select the sensor in very greedy way, which sensor will be on and which sensor in off mode. Here, in this approach, there is no problem of higher latency like base station approach. Like cluster approach, there is no significant changes of network proposed in our case. Finally, it is better than grouping sensor in data reduction approach, because this approach, because it always use shortest path to send data.

3 Basic Preliminary

In point coverage approach, aim is to cover a set of randomly distributed points. In Fig. 1, in a square region, a set of sensor node and a set of target node are deployed uniformly. Timing protocol is a grouping protocol which is used in sensor networks. This protocol runs in two phase, first one is initiation phase and second one is execution phase. Here, group is formed from common sensor node and super node or relay node. Each group consists of some super node and some monitoring sensor node. Group formation in this protocol is done using fit function designed in protocol. In initiative phase, some of sensor node sends their information to their neighbor node and in executive phase, data is reached to destination via super node. Data reception and data transmission occur in executive phase.

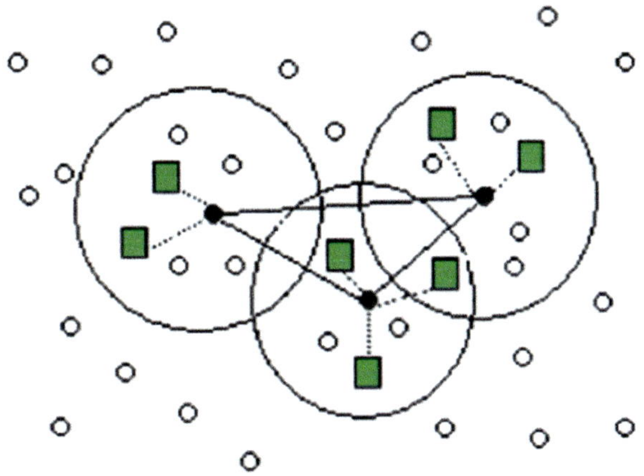

Fig. 1 Set of sensor node and target node

4 Proposed Method

Here, problem is designing a protocol in such a way so that energy consumption gets decreased and as a result, life of sensor gets increased. In this protocol, main focus is given to maximum uses of energy of common sensor.

In protocol design, location of sensor node and number of time using them will be considered. Distance between sensor node and super node (relay node) and how many times a sensor node is used have an important role in energy consumption of that group. Therefore, relation should be established between these two parameters and their energy consumption.

Considering network provisions first, we will discuss the problem. In our network, there are N sensor nodes, S1 to Sn. We have M super node Su1 to SuM ($M < N$). Selected one group stays active when other sensor nodes stay in sleep mode.

4.1 Network Provision for Our Problem Statement

K target is composed of super node and common sensor node placed randomly in network. Super node and sensor node are distributed uniformly. Following conditions should be hold by schedule activities of sensor node after running the algorithm.

- All targets must be covered.
- Sensor nodes S1 to Sn performing monitoring task must be uniformly deployed.
- All super node Super 1 to Super M to will be uniformly deployed.
- There will be chosen sets of node from C1 to CJ. Each set must be active set of sensor node which formed in each round by protocol.
- Each set CJ must cover all K target.

Here, our target is to divide sensors in active and inactive groups. Active group must guarantee connectivity and coverage. The main objective of this algorithm is to maximize number of groups so as a result, lifetime of sensor gets maximized. Common sensor has low energy E1 and limited processing power, But super node has higher energy, longer lifetime, and greater processing power.

- All super nodes are connected to each other so that at least one path exists between any two super node
- Each active sensor in group via relay node must be connected to super node to send its own data to super Node
- Initial energy of sensor node EI, communication range RC, and sensing range RS (RS $\leq$ RC).

4.2 Sensing Nodes Selection Algorithm

At the beginning of each performance, round grouping algorithm is executed. It consists of two sections. The first section is for selecting of active sensor node and second section is responsible for collecting data from neighborhood node and sending through relay node.

In first section, one of Cj group is formed in such a way that above-mentioned provisions are satisfied. When one group is active, then another sensor stay in sleep mode uses little energy. Sleeping node will be evaluated in next phase. This evaluation is done by considering a series of the physical factors of sensors during a round

5 Implementation

5.1 Simulated Annealing

Simulated annealing (SA) algorithm is a probabilistic optimization technique. It is used to find out the approximate global optimum of a function to be optimized. It is a metaheuristic technique for global optimization in a vast search space for an optimization problem (Fig. 2).

```
Create a initial random  solution γ
E_old=cost_Of(γ);
for (tmp=tmp_max; tmp>=tmp_min; tmp=next_tmp (tmp) ) {
for (j=0; j<j_max; j++) {
Successor_function(γ);
E_new = cost_Of(γ);
Delta = E_new-E_old;
if (Delta>0)
if (random () >= exp(-Delta/K*tmp);
undo_function (γ);
else E_old=E_new;
else E_old=E_new;}}
```

Fig. 2 Simulated annealing algorithm

```
Create random initial solution γ
calculate cluster center C1 of solution Y
calculate cluster center C2 of targets
calculate distance between C1 and C2.
if(distance<=sensor range)
E_old=cost_of (γ);
else go to step 1.
for(tmp=tmp_max;tmp>=tmp_min;tmp=next_tmp(tmp) ) {
for(j=0;j<j_max; j++ ) {
Successor_function(γ);
E_new=cost_of(γ);
Delta=E_new-E_old;
if(Delta>0)
if(random() >= exp(-Delta/K*tmp);
undo_function(γ);
else E_old=E_new;
else E_old=E_new;}}
```

Fig. 3 Algorithm for cluster-based greedy simulated annealing

5.2 Cluster Center Based Greedy Simulated Annealing

In this technique, above described simulated annealing algorithm is modified in such a way so that it becomes greedy. In this method, center point is calculated for randomly placed target, while random group selection of sensor is done then it is checked weather distance between center point of selected sensor group and center point of targets is less than sensing range or not. If distance is less than sensing range, then group has higher probability to be connected cover (Fig. 3).

5.3 Grid-Based Greedy Simulated Annealing

In this algorithm, area, where sensor node will be deployed, is divided into fixed number of grid. After selecting group of sensor randomly, it is checked whether this group is covering all grid cell or not. If it covers all grid cells, then selected group is chosen for next processing with higher probability (Fig. 4).

```
1. Create random initial solution γ
2. Divide the area into equal number of grid
3. Check each grid is covered by sensor or not
4. if Each grid is covered
5. Then Eold=cost_of(γ);
6. else follow step1
3. for(tmp=tmpmax;tmp>=tmpmin;tmp=next_tmp(tmp) ) {4. for(j=0;j<jmax; j++ ){
5. Successor_function(γ);
6. Enew=cost_of(γ);
7. Delta=Enew-Eold;
8. if(Delta>0)
9. if(random() >= exp(-Delta/K*tmp);
10.undo_function(γ);
11.else
12.Eold=Enew;
13.else
14.Eold=Enew; }}
```

Fig. 4 Algorithm for grid-based greedy simulated annealing

5.4 Voronoi Diagram-Based Greedy Simulated Annealing

In this algorithm, after selecting random sensor group, voronoi diagram is drawn. If half or more of voronoi cells are bounded, then this sensor group is selected with higher probability. In below, simple simulated annealing every new solution will be selected based on probability received from voronoi diagram (Fig. 5).

```
1. Create random initial solution γ
2. Eold=cost_of(γ);
3. for(tmp=tmpmax; tmp>=tmpmin;tmp=next_tmp(tmp) ) {
4. for(j=0;j<jmax; j++ ) {
5. Successor_function(γ);
6. Enew Voronoi=cost_of(γ);
7. Delta=Enew-Eold;
8. if(Delta>0)
9. if(random() >= exp(-Delta/K*tmp);
10.undo_function(γ);
11. else
12. Eold=Enew;
13. else
14. Eold=Enew;}}
```

Fig. 5 Algorithm for voronoi diagram-based greedy simulated annealing

6 Result

In below Fig. 6, common sensor node, super node, and target node are shown. Common sensor is displayed by circle node, Super node is shown by +sign, and target with star sign.

MATLab has been used to implement the proposed algorithm. Algorithms in [2, 12, 15–18] are compared with our proposed algorithm. Our algorithm was superior to them.

In simulation cluster based greedy, grid based greedy, and Voronoi diagram-based greedy SA algorithm is applied to select the monitoring sensors, super node to form a connected cover in a point coverage network.

Experiment was performed on a 200 * 200 environment. Common sensor node, super node, and target are distributed randomly. We have considered energy of super node three times greater than common sensor node. In every round, common sensor and super sensor dissipate some energy. We have tested with different seven data sets. In every data set, we have provided number of common sensor, number of super node, sensor range, and number of target. It is given in Table 1.

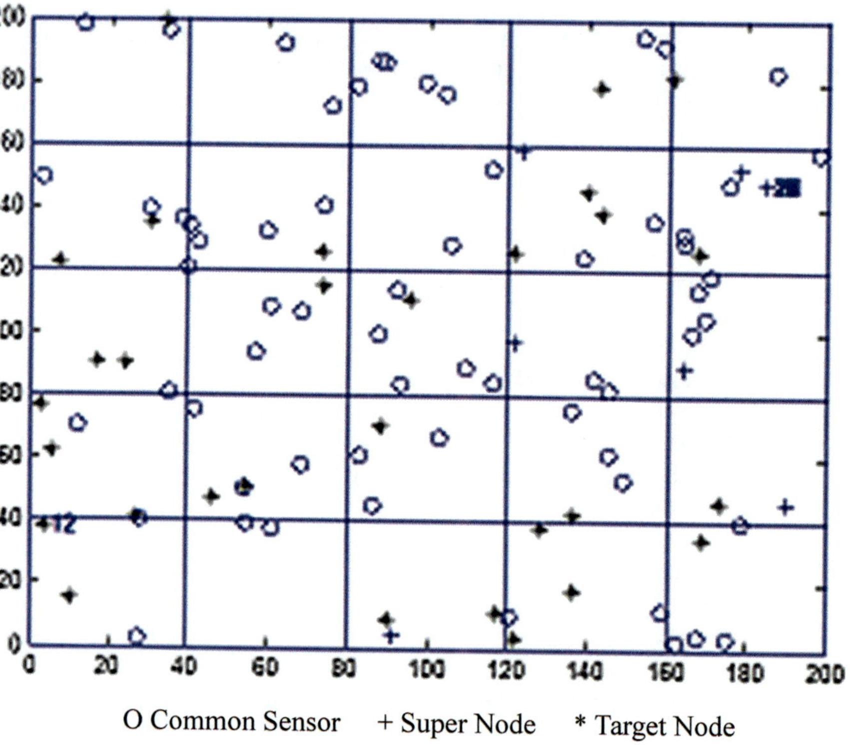

Fig. 6 Random deployment of common sensor, super node, and target

Table 1 Test case of input parameters

	Set1	Set2	Set3	Set4	Set5	Set6	Set7
Common sensor	80	70	60	80	70	70	80
Super node	20	30	10	30	10	40	30
Sensor range	60	40	70	90	80	90	70
Target point	20	40	10	30	20	10	10

Table 2 Comparative study of different algorithms

	Number of rounds	Average number of connected cover	Average time elapsed (in sec)
Simulated annealing	10	15	31.37
Grid-based greedy simulated annealing	10	21	135.82
Cluster center-based greedy simulated annealing	10	14	23.32
Voronoi diagram-based greedy simulated annealing	10	11	636.99

If the target was observed by several sensor, the sensor that sees more targets is selected for sensing.

To reduce activation interface node between target observer node and super node, Dijkstra algorithm is used.

We run the algorithm 10 times for all seven set of inputs; it is compared with simple SA. It is clearly showing that grid-based greedy algorithm gives better result comparative to simple SA. Here, set of connected cover is more comparative to other algorithms. Connected cover means a set of monitoring sensor which can cover the target and also satisfy that the number of sensors is minimum. In Table 2, data is given. As the number of connected cover increases, sensor gets more rest and lifetime gets extended.

7 Conclusion

In this paper, we have proposed a new greedy method to select monitoring sensors in each round in heterogeneous wireless sensor networks. In the proposed algorithm, a set sensor node is chosen which can cover a set of target node. In a very greedy way, set of monitoring sensor is chosen. According to simulation results, it is observed that this greedy-based greedy algorithm gives maximum number of such set which can cover the target. It can be used for optimization of a large-scale wireless sensor network and it is able to provide a good result. Prolonging network lifetime is the outcome of using these greedy simulated annealing.

8 Future Scope

Here, in this paper, we have considered only random deployment of sensor. Algorithm we have developed will work for control placement and uniform distribution. In greedy simulated annealing, sometimes, old move visited many times but do not gives better result, and if we construct queue to keep history of visited node, some improvement may take place in respect of time to execute program.

References

1. B. Carbunar, A. Grama, J. Vitek, *Distributed and Dynamic Voronoi Overlays for Coverage Detection and Distributed Hash Tables in Ad-hoe Networks (ICPADS 2004)* (Newport Beach, CA, 2004), pp. 549–559
2. M. Cardei, M.T. Thai, Y. Wu, W. Li, Energy-efficient target coverage in wireless sensor networks, in IEEE INFOCOM (2005)
3. B. Wang, K. Chua, V.C. Srinivasan, W. Wang, Information coverage in randomly deployed wireless sensor networks. IEEE Trans. Wireless Sens. Netw. **6**(8) (2007)
4. C. Chen, J. Yu, *Designing Energy-Efficient Wireless Sensor Networks with Mobile Sinks* (ACM, USA, 2006)
5. C.-F. Huang, Y.-C. Tseng, The coverage problem in a wireless sensor network, in *Proceedings of ACM International Workshop on (WSNA)*, pp. 115–121 (2003)
6. J. Lu, J. Wang, T. Suda, Scalable coverage maintenance for dense wireless sensor networks. EURASIP J. Wireless Commun. Netw. (2007)
7. S. Slijepcevic, M. Potkonjak, Power efficient organization of wireless sensor networks, in *Proceedings of IEEE International Conference on Communications*, pp. 472–476 (2001)
8. M.K. Watfa, S. Commuri, An energy efficient and self-healing 3-dimensional sensor cover. IJAHUC **3**(1) (2008)
9. Y. Xu, X. Yao, A GA approach to the optimal placement of sensors in wireless sensor networks with obstacles and preferences, in IEEE CCNC (2006)
10. M.R. Mundada, N. Thimmegowda, T. Bhuvaneswari, V. Cyrilraj, Clustering in wireless sensor networks: performance comparison of EAMMH and LEACH protocols using MATLAB. Adv. Mater. Res. **705**, 337–342 (2013)
11. C. Chen, J. Ma, Designing energy-efficient wireless sensor networks with. mobile sinks, in *ACM International Workshop on (WSNA)*, pp. 343–349 (2006)
12. A.S. Rostami, M.R. Tanhatalab, H.M. Bernety, S.E. Naghibi, Decreasing the Energy Consumption by a New Algorithm in Choosing the Best Sensor Node in Wireless Sensor Network with Point Coverage, (CICN) (2010)
13. A.S. Rostami, K. Nosrati, H.M. Bernety, Optimizing the parameters of active sensor selection and using GA to Decrease energy consumption in point coverage wireless sensor networks, (WCSN) (2010)
14. D. Tian, N.D. Georganas, A node scheduling scheme for energy conservation in large wireless sensor Networks. Wireless Comm. Mob. Comput. **3**, 271–290 (2003)
15. W. Awada, M. Cardei, *Energy Efficient Data Gartering in Heterogeneous Wireless Sensor Networks* (IEEE, WiMob, 2006)

16. Z. Liu, Maximizing Network Lifetime for Target Coverage Problem in Heterogeneous Wireless Sensor Networks, MSN 2007, LNCS 4864, pp. 457–468 (2007)
17. A.S. Rostami, H.M. Bernety, A.R. Hosseinabadi, A novel and optimized algorithm to select monitoring sensors by GSA. ICCIA (2011)
18. J. Liu, H. Shen, Characterizing data deliver-ability of greedy routing in wireless sensor networks. IEEE Trans. Mob. Comput. 99 (2017)

Fire Detection and Controlling Robot Using IoT

Mohit Sawant, Riddhi Pagar, Deepali Zutshi, and Sumitra Sadhukhan

1 Introduction

A robot is a machine that works automatically and has the ability to perform tasks similar to human beings. Multiple experiments have proven that robots can be beneficial in multiple fields such as medicine [1], rehabilitation [2, 3], rescue operation [4, 5], and industry [6]. Robots used in industries are mufti-function manipulators designed for more specialized materials, divisions, gadgets or devices through various programmed movements which help them perform various tasks [7]. The demand for a single system that can control, communicate, and integrate various robots regardless of their function and specifications in the Fourth Industrial Revolution (4IR) has increased many folds. Many robotics projects are now embedded with machine learning algorithms [8, 10] along with IoT in order to improve the performance of the robots.

The introduction of robotics to machine learning will not only help reduce injuries and deaths of firefighters but also increase safety, productivity, and the quality of the given task [11]. Robots may be classified depending upon their function or if they are user-controlled or autonomous.

The field of firefighting is extremely dangerous, and there have been countless devastating losses because of the lack in technological advancements in this field. Also, the current methods used for fighting fire are inefficient and inadequate as they

M. Sawant · R. Pagar · D. Zutshi · S. Sadhukhan (✉)
Rajiv Gandhi Institute of Technology, University of Mumbai, Maharashtra, India
e-mail: sumitrassadhukhan@gmail.com

M. Sawant
e-mail: mohitssawant@gmail.com

R. Pagar
e-mail: pagar.riddhi5@gmail.com

D. Zutshi
e-mail: deepalizutshi@gmail.com

S. Patnaik et al. (eds.), *Advances in Machine Learning and Computational Intelligence*, Algorithms for Intelligent Systems, https://doi.org/10.1007/978-981-15-5243-4_52

rely heavily humans who are prone to errors, no matter how extensively they have been trained.

While a variety of firefighting robots have been developed, most of them only help firefighters monitor fire from a distance and not contribute greatly to putting out the fire. The major work done on firefighting robots is based on early detection of fire. With this project, we aim at detecting as well as controlling small-scale fires. Using a variety of sensors and cameras, fire can be detected, and with the help of the IoT and Wi-Fi module, we can control our robot manually using mobile phones. A camera mounted on the robot helps the user monitor the situation and Raspberry Pi to know the status of the robot. After the fire has been detected, the owner is notified on their phone. Fire and smoke can be detected using a thermal sensor and an ionization smoke detector. A camera mounted on the robot gives a live feed to the viewer. Here, the robot overcomes the problem of hitting the obstacle by sensing the obstacle and moves in an obstacle-free direction to reach its destination.

Two currently available firefighter robots are Thermite and Fire Rob that have been deployed on multiple occasions in the industry. Thermite is a belly packed remote-controlled firefighting robot that can be controlled from as far as 400 m. It can deliver up to 1200 GPM of water or 150 psi of foam. The dimensions of the robot are 187.96 cm × 88.9 cm × 139.7 cm. It powers up to 25 bhp using a diesel engine. The main component in the Thermite robot is a mufti-directional nozzle that is backed by a pump that can deliver 600 GPM. It provides infrared videos in real-time. It is deployed in extremely dangerous situations such as fire on planes, in chemical planes, or even in nuclear reactors. Fire Rob is another firefighting robot controlled by a single operator via remote control. It eliminates danger to firefighters but extinguish fires from up to 55 m away. Two onboard tanks are capable of carrying 1800 L of water and 600 L of foam. Fire Rob has been built to withstand temperatures as high as 250 °C and thermal radiation of up to 23 kW/m for 30 min.

This project has major applications in the prevention of fire leading to loss of valuable data, in server rooms for immediate action in case of fire, in areas of a high probability of explosion (like kitchen, gas stores, etc.) or in an environment requiring permanent operator's attention.

2 Literature Review

There are a variety of ideas and implementations of mobile robots that operate in hazardous situations, which are introduced, enhanced, and upgraded from time to time. With new and emerging technologies, there are a lot of enhancements and upgrades being performed on the available systems. In the paper titled "Development of a Network-based Autonomous Firefighting Robot," movable robot consists of sensors like LM35 and Arduino with sonar, and range sensors are used to detect the fire and distances on its way toward fire. They have used the Wi-Fi module which in some cases depends on connection and if the connection is not working properly the whole system will suffer. In one of the systems, two wheels made of nylon and a caster ball are used. The water container has the capacity to contain at least 1L water. It is made of strong cardboard that has water-resistant property. In another system, the BOT is designed to detect fire, gas as well as an intruder. In one of the systems, BOT is made up of an acrylic sheet which is a bad conductor of heat and beneficial to protect the internal circuit of BOT. The use of more sensors decreases the speed of BOT also increases the cost for the same as compared to other implementations. Obstacle avoidance in real time is a mandatory feature for vehicles in an unknown environment [9]. In another model, the user can control the robot by using the Bluetooth module. The Bluetooth module communicates with the android application by using a driving motor, Arduino mega, voltage divider, tires, Bluetooth, and motor driver [12].

3 Methodology

The methodology has been divided into three main categories. The first category is the mechanical and structural schematics of the robot, followed by a detailed description of the hardware requirements and then finally the program design. The required parts were assembled and experimented to find the optimal distance required to extinguish the fire successfully [13].

3.1 Mechanical/Structural Schematics

A structure of the robot was created using Auto CAD for better illustration as shown in Fig. 1.

For the robot to have good movement, two wheels were installed on each of the rear sides and a caster wheel is installed at the front of the robot. The wheels stabilize the robot and help it take a 360° turn. The body of the robot and the circuits are protected

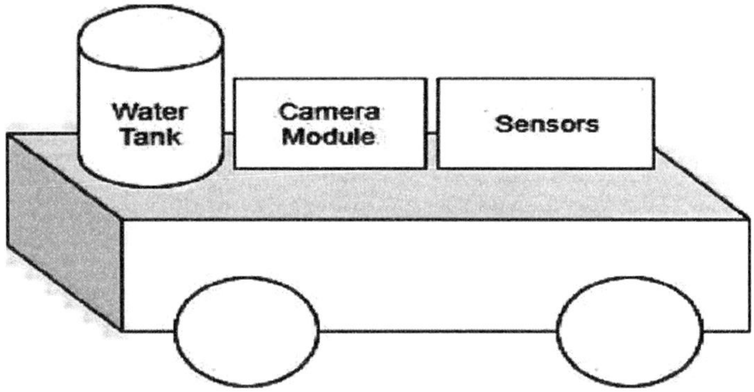

Fig. 1 Structure of the robot

from the fire with the help of acrylic sheets which can resist temperatures of up to 200 °C. The chassis contains slots and holes, making it easy to install and mount electrical components on to the body of the robot.

The ultrasonic infrared sensor installed in front of the robot help in avoiding obstacles and detecting fire, respectively. An HD camera and a thermal camera mounted the robot help determine a path, as well as changes in temperature and can be viewed by the user on their mobile device.

3.2 Hardware Requirements

The electronic components form the most vital part of this project. Several types of sensors, DC motors with wheels, cameras, and a water pump. Figure 4 illustrates the working of the robot, its operations as well as the various components of the robot. A Raspberry Pi 4 is used which is connected to all the other components. Infrared sensors and ultrasonic sensors are used for obstacle avoidance. A thermal camera and a normal HD camera are used for surveillance. Flame sensors are installed in rooms where camera vision cannot penetrate. GSM module is used to notify the user about the fire outbreak through SMS service. Depending upon the severity of the fire, the fire department is also notified.

Raspberry Pi 4: It is much faster than previous versions and has the ability to output 4 K video at 60 Hz or power dual monitors. It has an in-built Wi-Fi module and Bluetooth which does not require any extra hardware. It has C-type slots. The power requirement for pi 4 is 4 A, 5 V.

GSM Module: It is a chip used to establish communication between a mobile device and a GSM system. It has a power supply circuit and a communication interface (Figs. 2, 3 and 4).

Ultrasonic Sensor: A crucial aspect of this project involves the detection and avoidance of obstacles and objects that may hinder the path of the robot. The sensor

Fig. 2 Raspberry Pi

Fig. 3 GSM module

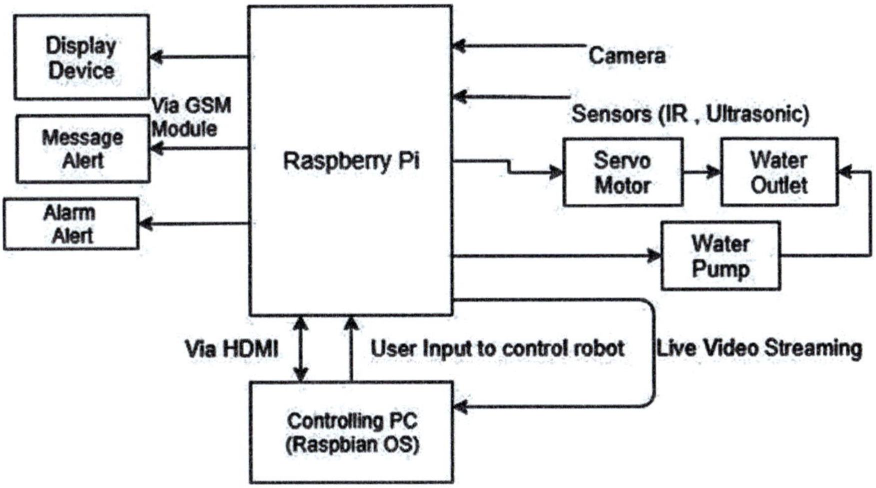

Fig. 4 Block diagram showing the working of the robot

should be inexpensive, compact, and functional to help the robot reach its destination successfully. The sensor must also provide the robot enough time and distance for it to react appropriately. These requirements are met by an ultrasonic sensor. The HC-SR04 ultrasonic sensor used in this project helps determine a range between 2 and 400 cm by transmitting ultrasonic waves to the environment and calculating the time taken by the waves to hit an object and reflect back to the sensor (Fig. 5).

Fig. 5 Ultrasonic sensor

Fig. 6 Infrared sensor

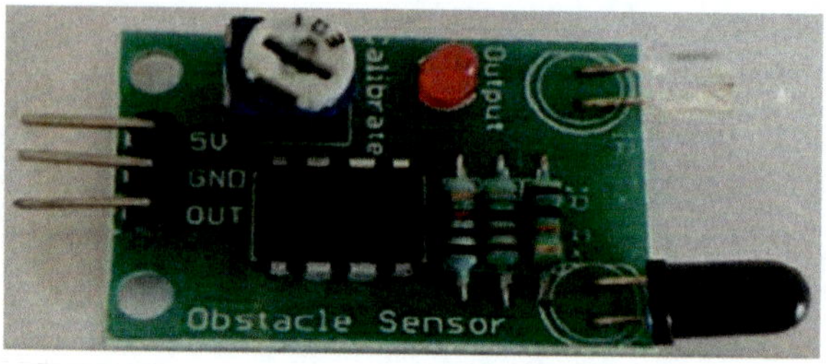

Infrared Sensors: Infrared (IR) sensors are used to detect the presence of objects in front of them. The presence of an object is indicated by an output of 3.3 V, and in case there are no objects present in front of it, it gives an output of 0 V. This can be done with the help of an IR pair (transmitter and receiver), and the transmitter emits IR rays which get reflected in case an object is present in front of it. Depending on the feedback, the output is made higher or lower (Fig. 6).

Flame Sensors: Flame sensors are designed to detect the presence of flames or fire and thus allow flame detection. In the case of detection of a flame, responses such as sounding of alarms, deactivation of fuel lines, or activation of a fire suppression system can be chosen. In industrial furnaces, they are used to confirm the presence of a flame to ensure proper working of the furnace. Sometimes, a flame detector is faster and more accurate than a smoke or heat detector in detecting the presence of open flames (Fig. 7).

DC Motor with Wheel: DC motors with wheels are used to enable the movement of the robot in this project. It works at around 5–10 V and the suitable current for this motor is 73.2 mA. It is installed at the rear wheels of the robot (Fig. 8).

Water Pump: The water pump plays a crucial role in the successful functioning of this robot as it helps to extinguish the fire by emitting a powerful spray of water at the flame/fire. For this project, a small, lightweight water pump has been used which creates low noise, is highly effective, and utilizes minimal power. The pump works at 6 V with a working current of 0.8 A (Fig. 9).

Thermal Camera and Camera: AMG883 Thermal camera has been mounted on the body of the robot and gives the user a live feed of the environment and the temperature changes. Along with this, the Raspberry Pi camera module is used which

Fig. 7 Flame sensor

Fig. 8 DC motor

Fig. 9 Water pump

can help as a surveillance camera. It provides a resolution of 1080p, 720p, and 640 × 480p 60/90. When not being used for the detection of fire and obstacle avoidance, this camera acts as a surveillance camera and can be used to detect the presence of intruders and act as a security mechanism.

4 Control Programming

Movement of the Robot

The moment of the robot is controlled using socket programming. The moment of the robot is controlled using an app. In socket programming, the robot and the controlling device are connected to the same IP address. In this, the robot acts as a client who listens to the server (Figs. 10, 11 and 12).

5 Results

The firefighting robot has been developed with the objective of locating and extinguishing fires. The robot has the ability to locate fires using flame sensors, ultrasonic sensors, and an infrared sensor. The flame sensor functions to detect the location of the fire, whereas the ultrasonic sensor and the infrared sensor function to detect the presence of obstacles in the path taken by the robot. The sensors are connected to Raspberry Pi, which controls the functioning of the DC motors.

Fig. 10 Program for the movement of the robot

```
set pins 33, 11, 13, 15, 29, 31
set UDP Port=5050 Fig. 11. Maze for Experiment
get UDP_IP andUDP_P ORT usingSocket
while (true)
set bu er data size 1024
raw=data
if raw== \forward"
pins 33,13 =True
pins 11,15 =False
print \Robot move forward"
else if raw== \stop"
pins 33,13,11,15 =False
print \Robot Stop"
else if raw== \backward"
pins 33,13 = False
pins 11,15 =True
print \Robot move Backward"
else if raw== \right"
pins 33,15= True
pins   11,13   =False
print \Robot move right"
else if raw== \left"
pins 33,15 = False
pins 11,13 =True
print \Robot move left"
else print \stop
```

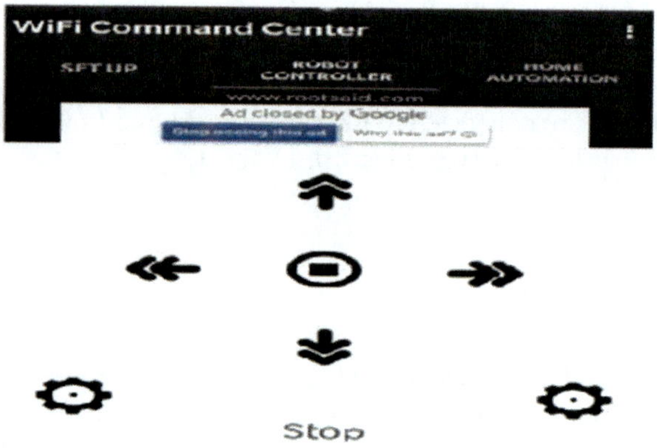

Fig. 11 Interface for controlling the robot

Fig. 12 Camera surveillance

A thermal camera is mounted on the robot which continuously detects the temperature change in the surroundings. If it senses temperature above the threshold value, then its relay gets ON, and hence, it goes to the fire-affected area and then extinguishes the fire by sprinkling water on it.

Also, a thermal camera can capture fire in its visible range. If there is some obstacle, then the camera fails. To overcome this, we use flame sensors. Flame sensors are installed in a different section where the camera cannot reach. When the flame sensor detects fire or smoke, it triggers the GSM module which is mounted on Raspberry Pi. At the same time, it also gives the coordinates of that location. So that, the robot can be moved to fire-infected area.

6 Conclusion

Various models of firefighting robots have been successfully developed and deployed that can be easily controlled remotely. Among other features, it includes the ability to detect fire at different locations automatically and reach these locations with ease because of a compact and lightweight structure. These robots also possess the ability to avoid objects and structures with the help of an ultrasonic sensor. The user can extinguish fires remotely using a mobile device connected to the robot as well as monitor the surrounding conditions of the robot during the process of firefighting with the help of a camera that is connected to the mobile device. From the experimental results, the robot is able to successfully sense smoke and fire. In conclusion, the project entitled "Fire Detection and Controlling Robot Using IoT" has achieved its aim and objective.

7 Future Scope

A variety of systems have been developed to help provide assistance in case of a fire. The current systems, however, have limitations and can be improved further by adding features or changing the design. The current models are not able to successfully function when it needs one floor to another, i.e., it cannot climb stairs. This poses a challenge if the building is multistory. Another limitation of the current system is that it majorly relies on the user's commands to perform various functions. The robot can be automated so as to improve its decision-making capabilities and decide on an appropriate course of action. The current systems also make use of water to put out fires. This becomes dangerous if the fire has been caused by electrical malfunctions. The design of the robot can be improved to mount a fire extinguisher which can be used to extinguish any type of fire. Also, developed models only aim at helping firefighters monitor fire or sometimes even put out the fire, but the robots can be modified to help people trapped inside burning structures. Fire and heat are not the only things that can harm people, toxic gases are also released when substances burn. A robot can carry oxygen masks or filters to help people breathe. It is also important to make sure that the body of the robot is well insulated to tolerate high temperatures. The current models have a very short working time which can definitely be improved. The sensors used in current models have a very short distance field, and the fire could be recognized at a distance not more than 1.5 m away. Also, the use of low-efficiency on-board computer only allows the robot to carry out main tasks without any extensions. The absence of optical means makes it difficult for the robot to perceive its environments. With technological advancements, these limitations can be overcome to create a robot that extensively helps in extinguishing fires and saving human lives.

References

1. S. Jeelani et al., Robotics and medicine: a scientific rainbow in hospital. J. Pharm. Bioallied Sci. **7**(Suppl 2), S381–S383 (2015)
2. M. Aliff, S. Dohta, T. Akagi, Simple trajectory control method of robot arm using flexible pneumatic cylinders. J. Robot. Mechatron. **27**(6), 698–705 (2015)
3. M. Aliff, S. Dohta, T. Akagi, Trajectory control of robot arm using flexible pneumatic cylinders and embedded controller, in *IEEE International Conference on Advanced Intelligent Mechatronics (AIM)*, 2015, pp. 1120–1125
4. C. Xin, D. Qiao, S. Hongjie, L. Chunhe, Z. Haikuan, Design and implementation of debris search and rescue robot system based on internet of things, in *International Conference on Smart Grid and Electrical Automation (ICSGEA)*, 2018, pp. 303–307
5. M. Yusof, T. Dodd, Pangolin: a variable geometry tracked vehicle with independent track control, in Field Robotics, pp. 917–924
6. C.-P. Day, Robotics in industry—their role in intelligent manufacturing. Engineering **4**(4), 440–445 (2018)
7. J. Lee, G. Park, J. Shin, J. Woo, Industrial robot calibration method using denavit—Hatenberg parameters, in *17th International Conference on Control, Automation and Systems (ICCAS)*, 2017, pp. 1834–1837

8. N.S. Sani, I.I.S. Shamsuddin, S. Sahran, A.H.A. Rahman, E.N. Muzaffar, Redefining selection of features and classification algorithms for room occupancy detection. Int. J. Adv. Sci. Eng. Inf. Technol. **8**(4–2), 1486–1493 (2018)
9. T.L. Chien, H. Guo, K.L. Su, S.V. Shiau, Develop a multiple interface based firefighting robot, in *2007 IEEE International Conference on Mechatronics, IEEE*, 2017, pp. 1–6
10. N.S. Sani, M.A. Rahman, A.A. Bakar, S. Sahran, H.M. Sarim, Machine learning approach for bottom 40 percent households (B40) poverty classification. Int. J. Adv. Sci. Eng. Inf. Technol. **8**(4–2), 1698–1705 (2018)
11. J.-H. Kim, S. Jo, B.Y. Lattimer, feature selection for intelligent firefighting robot classification of fire, smoke, and thermal reflections using thermal infrared images. J. Sens. **2016**, 13 (2016)
12. H.I. Darwish, Developing a Fire Fighting Robot, 2010
13. M. Aliff, N.S. Sani, A. Azinal, Development of Fire Fighting Robot, IJACSA, 2019

DDoS Prevention: Review and Issues

Shail Saharan and Vishal Gupta

1 Introduction

The Internet revolutionized the way of communication systems, information sharing, entertainment, inexhaustible learning, social interaction, E-commerce, etc. Any organization which is dependent on the Internet connectivity for its proper functioning can be harmed easily by its network infrastructure being targeted, and disruptive attackers do the same. They often use Internet protocols and infrastructure to launch attacks on connected victims. One such form of attack is distributed denial-of-service attack (DDoS attack). In this attack, an attacker, in a distributed fashion, disrupts the availability of network resources by overwhelming the target network with attack packets. DDoS attacks are third on the list of network security attacks, and they continue to grow every year [1, 2], therefore, it is still a major threat.

DDoS defense mechanisms are a set of techniques or tools used to defend networks attached to the Internet and consist of modules such as traffic monitoring and analysis, prevention, detection, and mitigation [3]. Different taxonomies are suggested based on the defense mechanisms of DDoS attacks. Kalkan et al. [4] categorized DDoS defense into detection and prevention, and prevention further into capability-based and filtering-based mechanisms. Filtering as a DDoS countermeasure means to allow or deny a packet entry. Based on this and a few other studies, DDoS defense strategies can be classified as one or more of the following:

- DDoS attack detection—Once the attack has taken place, strategies falling under this realm are capable of detecting the attack with probability P, where P depends upon the particular detection mechanism. Here, the network is allowed to be susceptible to DDoS attacks. Based on different heuristics, these attacks are detected

S. Saharan · V. Gupta (✉)
Birla Institute of Technology and Science (BITS), Pilani, India
e-mail: vishalgupta@pilani.bits-pilani.ac.in

S. Saharan
e-mail: p20170404@pilani.bits-pilani.ac.in

S. Patnaik et al. (eds.), *Advances in Machine Learning and Computational Intelligence*, Algorithms for Intelligent Systems, https://doi.org/10.1007/978-981-15-5243-4_53

once the network is already under attack, and then corrective measures are taken. These heuristics vary from network traffic analysis, request/response analysis, unusually slow network performance, unavailability of Web resources, etc.

- DDoS attack mitigation—Once the attack is detected, strategies falling under this realm are capable of mitigating it with probability *P*, and within time *T*. Again, the value of *P* and *T* depends upon the particular mitigation mechanism. Once the attack is detected, DDoS mitigation techniques or tools are used to mitigate the impact of such an attack. These techniques are primarily based on rate limiting, tracing the attacker and blocking it, etc.
- DDoS attack prevention—In general, prevention means not allowing something from happening in its first place. DDoS defense strategies falling under this realm have used the word "prevention" with different meanings and contexts. This paper further elaborates on it later.

DDoS detection and mitigation, to a greater extent, have an unambiguous meaning and definition. However, the word prevention in the context of reflection-based amplification DDoS attacks is used with different meanings in the literature. In this paper, we define two definitions of DDoS prevention, ideal prevention, and true prevention. We also present the review of different prevention techniques and argue about whether they are actually preventing the DDoS attacks or not. To the best of our knowledge, there is no comprehensive study dealing with this topic.

2 Ddos Attack Prevention: Definition and an Extensive Survey

DDoS prevention is used in the literature within three different contexts: (a) early detection techniques, (b) reactive techniques, and (c) proactive techniques.

(a) Early detection—As stated by Nguyen et al. [5], it is difficult to protect from DDoS sometime after the attack has started. So, early detection is defined as detecting the attack on an earlier stage, or pre-attack stage, before all of the network resources are consumed.

(b) Reactive techniques—These techniques first detect the attack, whether it is taking place or not, and if it is taking place, then mitigating it. Therefore, primarily, this includes detection and then mitigation.

(c) Proactive techniques—Proactive techniques are those techniques that are implemented before the attack takes place or before it harms the victim.

Table 1 shows the classification of various DDoS defense techniques available in the literature in the above three categories.

Table 1 Summary of DDoS defense techniques

Author	DDoS defense mechanisms	Classification
Nguyen et al. [5]	Early detection using KNN classifier	Early detection
Kaushal et al. [6]	Early detection in WSN	Early detection
Trung et al. [7]	Fuzzy inference system using SDN	Reactive
Yadav et al. [8]	For MANET using trust, reputation, and honesty	Reactive
Oo et al. [9]	Hidden semi-Markov model	Reactive
Navaz et al. [10]	Entropy	Reactive
Jema et al. [11]	Protection using machine learning and traffic authentication	Reactive
Zhang et al. [12]	IP-behavior analysis	Reactive
Kim et al. [13]	Request/response match	Reactive
Jin et al. [14]	Hop-count filtering	Reactive
Osanaiye et al. [15]	OS fingerprinting	Proactive
Luo et al. [16]	Identifier/locator separation	Proactive
Ferguson et al. [17]	Ingress filtering	Proactive
Park et al. [18]	Allowed IP packets on a link	Proactive
Li et al. [19]	Source periodically informing neighboring nodes	Proactive
Keromytis et al. [20]	Secure overlay points and secret servlets	Proactive
Kim et al. [21]	Score of packets	Proactive
Freiling et al. [22]	Honeypots deployed	Proactive
Duan et al. [23]	Route mutation to protect critical links	Proactive
Luo et al. [24]	Dynamic path identifiers	Proactive

2.1 What is Prevention?

Generally, all the DDoS attacks have the following characteristics: (a) IP spoofing, which refers to spoofing the destination IP address of the network traffic. (b) Destination IP address routing, which refers to the underlying network property using which routing always happens on the basis of destination IP address. (c) BOTs, which refers to the machines whose security can be compromised and can be controlled by an attacker. Based on these characteristics, any DDoS prevention technique should prevent the victim targeted by an attacker from being overwhelmed by the attack traffic. This can be achieved by preventing one or more conditions, as stated above.

Ideal Prevention. As defined by BB Gupta et al. [25], ideal prevention can be considered as "Attack prevention methods which try to stop all well known signature-based and broadcast-based DDoS attacks from being launched in the first place or edge routers, keeps all the machines over Internet up to date with patches and fix security holes." Later the authors argue that, with this definition of prevention, DDoS attacks are always vulnerable to attack types for which signatures and patches do not exist in the database.

True Prevention. Here, we define True prevention as a method of preventing the attack by making the network self-sufficient. For it, let:

- B be the network bandwidth of the network in which an attacker, or a bot in control of an attacker, resides.
- V be the victim of DDoS attack
- I_V be the IP address of victim V
- I_A be the IP address of an attacker

True prevention is defined as a set of techniques embedded into the network routers which prevent the attack traffic to reach V, even though the destination IP address of network packets belonging to DDoS attack is I_V, and automatically mitigates the attack within some constant time T where T is directly proportional to B.

2.2 Early Detection

This section provides a survey of techniques that falls under the category of early detection. Nguyen et al. [5] divided network status into three classes: Pre-attack phase, attack phase, and normal network. For early detection, classification is done using KNN classifier using nine features. The proposed technique by Kaushal et al. [6] claims to prevent flooding attacks. The methodology includes limiting the number of transmissions of each node based on the number of neighboring nodes. Here, normal packets can also suffer due to limitations imposed on the transmission rate.

2.3 Reactive Techniques

This section provides a survey of techniques that claims to be DDoS prevention techniques but actually falls under the category of Reactive ones. Trung et al. [7] proposed to use hard decision thresholds along with the fuzzy inference system to detect DDoS attacks in an SDN environment. It zeroes out three parameters that can be a criterion of DDoS attacks and defines some hard decision boundaries for classification. The problem with the proposed methodology is that it is not prevention since it is taking action after the attacker has initiated the attack, and the victim gets affected. Yadav et al. [8] proposed honesty and reputation scores for each node to detect a possible DDoS attack among the mobile ad hoc networks. This approach again is not a preventive measure as it is a reactive scheme. Moreover, it does not state how the adjoining node will know that its immediate neighbor has modified the IP address or has misrouted the traffic. Oo et al. [9] proposed a packet classification approach to DDoS prevention using hidden semi-Markov model. It involves collecting packets every second, extracting features, applying a classification algorithm, followed by applying the hidden semi-Markov model algorithm. The proposed methodology has a reactive approach and does not deal with prevention as such. Moreover, it is computationally expensive and memory-intensive since it requires keeping track of packets from each source IP to each destination IP. Navaz et al. [10] propose an entropy-based system and anomaly detection system for providing DDoS detection in a cloud-based environment. It uses a multithreaded intrusion detection system (IDS) to process large amounts of data, which then passes the monitored alerts to a third-party service, maintained by cloud service provider. IDS assigns a threshold value for the packets. The entropy is calculated using port, IP addresses, and flow size as inputs, and a normalized entropy value is compared with a threshold value for final classification. This approach is also reactive as it materializes only after the attack has begun. Jema et al. [11] proposed a machine learning-based classifier to classify the traffic. It pays major stress on how problematic the false positives generated by the IDS are and devises a method to reduce them. The idea is to use two additional Web servers (bait server and decoy server), along with the real Web server. All the normal traffic passes through bait server first, which after validation passes through to a real Web server. The invalidated traffic passes directly to the decoy server, which presents another layer of authentication to remove the number of false positives. Although it might reduce the false positives, it is still a reactive approach and takes place after only after some damage has already been done. Based on IP behavior analysis, Zhang et al. [12] proposed an approach by creating records for each user's sending and receiving traffic and judging if the behavior meets normal standards. To detect the abnormal behavior of each IP, a nonparameter CUSUM algorithm is applied. Classification of a user as an attacker, normal user, or victim is based on a decision algorithm. This paper also follows a reactive measure. To prevent DNS amplification attacks, Kim et al. [13] proposed a framework using software-defined network (SDN). The basis of this approach is matching request and response packets of a DNS query. At the client-side, a DNS response is only accepted when there is a request for it; otherwise, the packet is dropped. The problem with this approach is that it is not mitigated in the network itself, and the attack is allowed to reach the victim's machine. Jin et al. [14] proposed an approach using hop counts. These are calculated with the help of the TTL field present in the IP header. This technique can predict abnormal behavior by finding the difference between the number of hops taken and the approximate number of hops actually required between source and destination addresses. Packets with abnormal hop counts will be dropped. Again, this is not prevention as the attack traffic will reach the victim. In addition, the initial TTL may not be same for all the cases, and the attacker can manipulate the hop counts.

2.4 *Proactive Techniques*

Many authors have defined ways that proactively prevent DDoS attacks. Ferguson et al. [17] proposed ingress filtering, which prohibits an attacker within the originating network from launching an attack using a spoofed source IP address. This technique may not work for complex topologies and has limitations of ingress filtering in general. Osanaiye et al. [15] use the underlying operating system (OS) fingerprinting for prevention against IP spoofing. The detection scheme is based on matching the OS of the spoofed IP packet with OS of true source. Such a scheme is limited to the detection of spoofed packets and dropping them, only works in the case if the spoofed source and actual source have different OSs, and also has extra communication overhead. Route-based distributed packet filtering (RDPF) proposed by Park et al. [18] proposes prevention and traceback. Prevention is based on allowing a pre-defined combination of source and destination IP addresses on a particular link. This technique cannot prevent intelligently spoofed IP addresses, does not support dynamic changes in the topologies, and if BGP is used for configuration, then hijacking in a BGP session can mislead routers. Li et al. [19] proposed Source Address Validity Enforcement (SAVE) in which the source periodically informs about itself by sending messages to all directly connected nodes and solves RDPF dynamic routing problem. Through this approach, each router will know valid IP addresses that can arrive on it. Hence, it prevents attacks from invalid IP addresses. Still, this technique cannot stop attacks from valid IP addresses and causes an increase in memory and computational costs in routers. Keromytis et al. [20] proposed a method using Secure Overlay Services. Authentication of the packets is done at secure overlay points, which forward the packets through overlay nodes to beacon nodes. Beacon nodes forward these packets to a secret servlet (or secret node). Finally, these secret servlets send the packets to destination nodes. For this approach, scalability can become a problem as secret servlets, beacons, and other specially designated nodes need to maintain state for each target. Duan et al. [23] proposed a proactive routing mutation technique, focusing mainly on infrastructure DDoS attacks. The proposed approach is based on the assumption that the path leading to public servers has a few critical links from which the major traffic passes. Congesting these links will majorly affect the connectivity of these servers and degradation in service. So, the attacker also targets these critical links to attack the infrastructure. The proposed solution uses proactive route mutation techniques to reduce exposure and improve the availability and flexibility of public servers. This methodology is for preventing the attacks targeting the critical links and not directly a victim. So, even if the flow is switched to non-critical links, it will not matter as the victim will be affected. Kim et al. [21] proposed an approach in which a packet score is kept for each packet, and a packet is called a legitimate packet if its value is below a threshold. This approach cannot filter low volume traffics, and the attacker can mimic the characteristics of a legitimate packet. Luo et al. [24] proposed prevention against DDoS attacks with dynamic path identifiers (PID) that identify the path between network entities as inter-domain routing objects. In the case of dynamic PIDs, periodic update of PIDs takes place and new PIDs are installed into adjacent domains (into data plane) for packet forwarding. This would require changes in underlying Internet routing architecture. To launch DDoS attacks, the attacker also uses bots that are compromised hosts in the control of an attacker. Preventing all the hosts from becoming bots is one of the effective techniques. Luo et al. [16] proposed a technique by using identifier locator separation, which separates the identity of a node from its location.

This prevents creation of the botnets, thus preventing the attack. Freiling et al. [22] stated that for a coordinated automated activity by many hosts, which may result in DDoS attacks, command and control (C&C) is needed. So to prevent DDoS attacks, this remote control mechanism needs to be infiltrated and stopped. The technique requires deploying the honeypots to gather information about how C&C operates, how and what commands it sends, etc. The base of [16, 22] is to prevent botnets creation, but DDoS attacks can still happen using booter services [26].

3 Conclusion and Future Work

DDoS attacks disrupt the normal connectivity of an organization with the Internet world. A lot of work is done focusing on detection and mitigation, but very little with a focus on prevention, and that too with different meanings of the word "prevention." In this paper, we defined the characteristics of a typical DDoS attack, defined two definitions of "DDoS prevention," and argued that future work should focus on practically feasible prevention techniques, which conforms to these two definitions. Future work should focus upon introducing/modifying the present network functionality so that network itself can prevent DDoS attacks from happening. In case an attack takes place, the network should itself mitigate it within some constant time interval, as stated in the definition of "*true prevention.*" We expect this can be achieved by avoiding any of the three conditions, as listed in Sect. 2.1. In any case, the victim should not be affected.

References

1. B. Alex, B. Christiaan, P. Eric, M. Niamh, S. Raj, S. Craig, S. ReseAnne, S. Dan, S. Bing, McAfee labs threats report, 2018 March, Last accessed 2018 June 19
2. K. Oleg, B. Ekaterina, G. Alexamder, DDoS Attacks in Q3 2018, 2018 Oct, Last accessed 2018 Dec 22
3. S. Bhatia, S. Behal, I. Ahmed, Distributed denial of service attacks and defense mechanisms: current landscape and future directions, in *Versatile Cybersecurity* (Springer, Cham, 2018)
4. K. Kalkan, G. Gür, F. Alagöz, Filtering-based defense mechanisms against DDos attacks: a survey. IEEE Syst. J. **11**(4), 2761–2773 (2016)
5. H.V. Nguyen, Y. Choi, Proactive detection of DDoS attacks utilizing k-NN classifier in an anti-DDoS framework. Int. J. Electr. Comput. Syst. Eng. **4**(4), 247–252 (2010)
6. K. Kaushal, V. Sahni, Early detection of ddos attack in WSN. Int. J. Comput. Appl. **134**(13), 0975–8887 (2016)
7. P. Van Trung, T.T. Huong, D. Van Tuyen, D.M. Duc, N.H. Thanh, A. Marshall, A multi-criteria-based DDoS-attack prevention solution using software defined networking, in *2015 International Conference on Advanced Technologies for Communications (ATC)* (IEEE, Oct 2015), pp. 308–313
8. N. Yadav, V. Parashar, Trust or reputation base encryption decryption technique for preventing network from DOS attack in MANET, in *2016 International Conference on Inventive Computation Technologies (ICICT)*, vol. 1 (IEEE, Aug 2016), pp. 1–6

9. K.K. Oo, K.Z. Ye, H. Tun, K.Z. Lin, E.M. Portnov, Enhancement of preventing application layer based on DDoS attacks by using hidden semi-Markov model, in *Genetic and Evolutionary Computing* (Springer, Cham, 2016), pp. 125–135
10. A.S. Navaz, V. Sangeetha, C. Prabhadevi, Entropy based anomaly detection system to prevent DDoS attacks in cloud. arXiv preprint arXiv:1308.6745 (2013)
11. J.D. Ndibwile, A. Govardhan, K. Okada, Y. Kadobayashi, Web Server protection against application layer DDoS attacks using machine learning and traffic authentication, in *2015 IEEE 39th Annual Computer Software and Applications Conference*, vol. 3 (IEEE, July 2015), pp. 261–267)
12. Y. Zhang, Q. Liu, G. Zhao, A real-time DDoS attack detection and prevention system based on per-IP traffic behavioral analysis, in *2010 3rd International Conference on Computer Science and Information Technology*, vol. 2 (IEEE, July 2010), pp. 163–167
13. S. Kim, S. Lee, G. Cho, M.E. Ahmed, J.P. Jeong, H. Kim, Preventing DNS amplification attacks using the history of DNS queries with SDN, in *European Symposium on Research in Computer Security* (Springer, Cham, Sept 2017), pp. 135–152
14. C. Jin, H. Wang, K.G. Shin, Hop-count filtering: an effective defense against spoofed DDoS traffic, in *Proceedings of the 10th ACM conference on Computer and Communications Security* (ACM, Oct 2003), pp. 30–41
15. O.A. Osanaiye, Short paper: IP spoofing detection for preventing DDoS attack in cloud computing, in *2015 18th International Conference on Intelligence in Next Generation Networks* (IEEE, Feb 2015), pp. 139–141
16. H. Luo, Y. Lin, H. Zhang, M. Zukerman, Preventing DDoS attacks by identifier/locator separation. IEEE Netw. **27**(6), 60–65 (2013)
17. P. Ferguson, RFC-2267: Network ingress filtering: defeating denial of service attacks which employ IP source address spoofing, Jan 1998
18. K. Park, H. Lee, On the effectiveness of route-based packet filtering for distributed DoS attack prevention in power-law internets, in *Proceedings of the 2001 Conference on Applications, Technologies, Architectures, and Protocols for Computer Communications (SIGCOMM'01)*. ACM, New York, NY, USA, 15–26 2001. http://dx.doi.org/10.1145/383059.38306
19. J. Li, J. Mirkovic, M. Wang, P. Reiher, L. Zhang, SAVE: Source address validity enforcement protocol, in *Proceedings. Twenty-First Annual Joint Conference of the IEEE Computer and Communications Societies*, vol. 3 (IEEE, June 2002), pp. 1557–1566
20. A.D. Keromytis, V. Misra, D. Rubenstein, SOS: an architecture for mitigating DDoS attacks. IEEE J. Sel. Areas Commun. **22**(1), 176–188 (2004)
21. Y. Kim, W.C. Lau, M.C. Chuah, H.J. Chao, PacketScore: a statistics-based packet filtering scheme against distributed denial-of-service attacks. IEEE Trans. Dependable Secur. Comput. **3**(2), 141–155 (2006)
22. F.C. Freiling, T. Holz, G. Wicherski, Botnet tracking: exploring a root-cause methodology to prevent distributed denial-of-service attacks, in *European Symposium on Research in Computer Security* (Springer, Berlin, Heidelberg, Sept 2005), pp. 319–335
23. Q. Duan, E. Al-Shaer, S. Chatterjee, M. Halappanavar, C. Oehmen, Proactive routing mutation against stealthy Distributed Denial of Service attacks: metrics, modeling, and analysis. J. Defense Modell. Simul. **15**(2), 219–230 (2018)
24. H. Luo, Z. Chen, J. Li, A.V. Vasilakos, Preventing distributed denial-of-service flooding attacks with dynamic path identifiers. IEEE Trans. Inf. Forensics Secur. **12**(8), 1801–1815 (2017)
25. B.B. Gupta, R. Joshi, M. Misra, Distributed denial of service prevention techniques. arXiv preprint arXiv:1208.3557 (2012)
26. L. Krämer, J. Krupp, D. Makita, T. Nishizoe, T. Koide, K. Yoshioka, C. Rossow, Amppot: monitoring and defending against amplification DDos attacks, in *International Symposium on Recent Advances in Intrusion Detection* (Springer, Cham, Nov 2015), pp. 615–636

Early Detection of Foot Pressure Monitoring for Sports Person Using IoT

A. Meena Kabilan, K. Agathiyan, and Gandham Venkata Sai Lohit

1 Introduction

The Internet of things is a system which consists of the combination of the interaction between computing devices, mechanical and digital machines, objects, animals or humans with the unique ability of identifiers and a way to transfer data over a network without any means of humans. IoT have helped to overcome a huge leap in evolution of the human species. Some says that IoT have saved as several years to attain a modern or complete smart city. Today, over 5 billion people use smartphones and it is predicted that it will increase to 26 billion by the year 2025. IoT technology can be seen in several practical applications like healthcare, environment monitoring and many more. Improving the performance in the field of medical management and health maintenance holds a biggest demand in today's era. The need of providing endowment concern to the patient with less medical cost as well as dealing with the availability of nursing staff is a primary issue. Recent development, in the field of Internet of things (IoT) is enhancing and improving healthcare as well as biomedical field [1]. Movement examination and determination is the study of human locomotion. These are used to estimate the nature of walk, plan, and treat individuals with conditions affecting their ability to walk. It is mainly needed in monitoring the stepping patterns after any orthopaedics surgical operations during rehabilitation.

A sport is something with thrill and excitement. Currently, technologies have been made to increase the performance of the athletes and to make a better audience's experience. With the advancement of wireless and sensors technology, it has become

A. Meena Kabilan (✉)
Department of CSE, Sri Sairam Engineering College, Chennai, Tamilnadu, India
e-mail: meenakabilan.cse@sairam.edu.in

K. Agathiyan · G. Venkata Sai Lohit
Department of E&T, Amity School Engineering and Technology, Amity University, Noida, UP, India
e-mail: agathian3@hotmail.com

S. Patnaik et al. (eds.), *Advances in Machine Learning and Computational Intelligence*, Algorithms for Intelligent Systems, https://doi.org/10.1007/978-981-15-5243-4_54

an easy opportunity to analyze all the relevant information of the athletes and gain the valuable insights [2].

Currently, sportsmen are facing a lot of problems in overcoming an injury. Injury will almost happen unconsciously and it is very difficult to avoid them. Several injuries will take a very long time to recover. But we can do something to recover from the injury rapidly. So, device is required in order to analyze the regular activity of the athlete and to determine the potential risk of injury [3]. With increase in technology and advancement in the biomedical instrumentation, new devices have been designed, manufactured, and tested for the monitoring of rehabilitation activities after surgical operations that can be caused due to injury. Before the technological advancement, the proper way of an exercise depends only on visual analysis/manual methods. There are number of external measurement system for gait monitoring, which is an important part in rehabilitation procedure. Among them, walking aid is most common. If the walking aid pattern is consider, then it provides the quantitative data for gait analysis, and these data provides useful information to the orthopaedic so that they can properly instruct the patient's to use it correctly at early stage. The conventional rehabilitation method, i.e., proper way of exercise by the patient also depends on visual analysis/manual methods. This method has disadvantage of human errors such as the perception of the loads on the patient's lower limbs, the movement, the drag, the synchronism, etc. So, there is a need of making an in-sole shoe which not only gives quantitative data for gait analysis but also give feedback to the patient for correct use of it [4].

This paper focuses on the injuries caused due to the footwork of the athlete. Normally, when an athlete during their workout or while running put equal pressure on both foots. But when an injury occurs on the foot like ankle twist or sprain, then there will be a difference of pressure on their foot. The device will record the pressure from exact position on the foot. For example, if an athlete is facing an ankle twist, then there will be a lot of pressure on the side of the foot. These data will help us to identify the proper injury and will also give us the proper methods to recover from them. All these data will be stored in cloud, which can be helpful in the future. So in this, we propose IoT device which constantly monitors the athlete's footwork while they run, so that when an injury occurs, these can be analyzed immediately and proposes a counter measurement to recover from the injury. This technique is completely depends on the sensors and the analytical data we provide to for recovering from an injury. Later, it transmits all its data to the cloud for storing and it processes, and if necessary, send the feedback to the doctor or the coach through the mobile application to inform about the current situation of the athlete.

The rest of the paper is organized as follows: Sect. 2 contains the related work on foot pressure detection, while Sect. 3 gives about the existing technology on the IoT in sports. Section 4 provides the design of the device. Section 5 covers the working principles of the proposed system. Section 6 provides the result and discussion. Finally, conclusion of the paper gives the directions for future work.

2 Related Work

IoT in sports is considered as a hot research topic today. Several researchers have suggested various methodologies and used many different microcontrollers for the foot pressure monitoring system.

De Silva et al. [1] have used tactile sensors combining with inertial measurement units to identify human postures, localizing and detect falls in real time. They used a transmitter which is located in the waist belt to transmit the data from the shoe to the waist belt. Through waist belt, it is then transmitted to the mobile phones. Majumder et al. [2] have developed three biomechanical models for three different gait events and implemented in smartphones to analyze gait. They used a smart shoe worn sensors system to validate the model and the real-time detection of abnormality in users gait patterns. The result from three different data sets is also presented to show that their approach provides a high rate of classification correctness in distinguishing between normal and abnormal walking patterns. Their system may also find multiple applications in gait behaviour detection for people with various disabilities who are at a high risk of falls.

Similarly, Malvadeet et al. [3] have proposed an electronic in-sole system which was designed in a shoe using force sensing resistor (FSR) sensor that monitors pressure wirelessly, thus saving the information in an array of sensor network for transferring data into CPU. Four sensors are placed, one at the heel, one at the lateral part, one at the metatarsal head, and one at the anterior. The plantar pressure is used to analyze pressure distribution. The signal from the pressure sensor will be processed by microcontroller and sent to mobile via Bluetooth module. ElSaadany et al. [4] presented a multisensory system that investigates the walking patterns to predict a cautious gait in stroke patient. In their study, a smartphone built-in sensor and an IoT shoe with a Wi-Fi communication module was used to discreetly monitor in-sole pressure and accelerations of the patient's motion. Their proposed system can also warn the user about their abnormal gait and possibly save them from forthcoming injuries from fear of falling. The system may also find multiple applications in gait behaviour detection for people with various disabilities who are at a high risk of falls related injuries with location information.

These applications are not only used in the detection of foot injury and monitoring gait. They are also used in human detection for security. For example, Luke Russell et al. [5] proposed by mounting photoreceptors and magnetic sensors in strategic locations such as a doorframe. These may provide a possibility of knowing much new information about people passing the sensors, using sensors people already welcome, integrated in their homes. Even though with much application, we still required a low cost, high impact method of using Bluetooth-linked microcontrollers to connect information from distributed sensors. The requirement of much more advanced system is still increasing in demand.

3 Existing System

IoT have been merged with the sports for the past one decade. At first, IoT focused its technology towards the better experience of the audience. Fan Engagement has been the successful victory for the IoT. It encourages the fans to view their favourite teams and athletes like never before. Many companies have invested several billions just to attract the audience into watching the game. The technology gives the audience a wide range of knowledge about their favourite game. Later, its technology focuses on the player development. These technologies focus on researching an individual athlete performance. The analytical measures which are for used on the athlete are almost effective. This creates a greater impact to the athletes. It also focuses the player safety. Embedded devices such as smart insoles and built-in chips give teams an abundance of data that helps to keep players healthy and fit. These devices are mostly monitors the athletes cardio. Today, almost every device keeps the record of the results and is combined with the statistical team in order to keep track of the athletes physic. But these monitors are mostly on the upper part of the body. There are very less devices to monitor the foot work of an athlete.

Today, devices are made to count the steps and distance covered by a person with the help of GPS. Even there are devices to monitor the pressure of the foot. But the problems with these devices are that they are wired with the equipment. Few devices are not even capable of storing the result. It takes some delay on displaying the results. So to counterattack these entire problems and also to add some more features in order to benefit the athletes, this monitoring system is made.

4 System Design

The block diagram of the system is as shown in Fig. 1. It shows the basic building blocks of the system. Figure 1 shows the basic structure of IoT-Based Feet Pressure Monitoring System. Sensors will used to collect the data through controller and it will send to the memory. The cloud helps to analyze the data for further clinical studies using IoT (Fig. 2).

The components used in this device are:

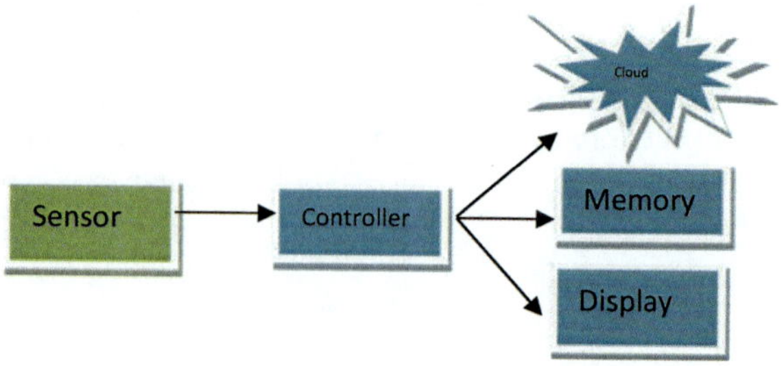

Fig. 1 Block diagram of system

Fig. 2 Prototype of Arduino board

(A) Sensors: The sensors used are the basic load sensors. We have used six load cells placed at different pressure point locations of the foot for acquisition of data. These six points are the most significant pressure points for all fragile athletes.

(B) Controller: The controller used is Arduino Uno. The Arduino software is easy-to-use yet flexible enough for advanced users. It runs on Mac, Windows, and Linux. Arduino boards are relatively inexpensive compared to other micro-controller platforms. Most microcontroller systems are limited to Windows. Arduino has a simple, clear programming environment. The Arduino software is published as open source tools, available for extension by experienced programmers. It has an extensible hardware.

(C) IoT and Cloud: The IoT and Cloud are the next interface used to transmit data to the doctor. The Internet of things (IoT) is the network of physical devices, vehicles, buildings, and other items like embedded with electronics, software, sensors, actuators, and network connectivity that enable these objects to collect and exchange data. The IoT allows objects to be sensed and/or controlled remotely across existing network infrastructure. Cloud communications are Internet-based voice and data communications where telecommunications applications, switching and storage are hosted by a third-party outside of the organization using them, and they are accessed over the public Internet. Cloud services are a broad term, referring primarily to data centre-hosted services that are run and accessed over an Internet infrastructure. We have used the ESP 8266 Model for Wi-Fi communication.

(D) Memory: The memory unit is interfaced onto the Arduino board for storing the data that has been recorded over a period of time. This data can be used for further references. The memory could also be stored on the application of the Android phone.

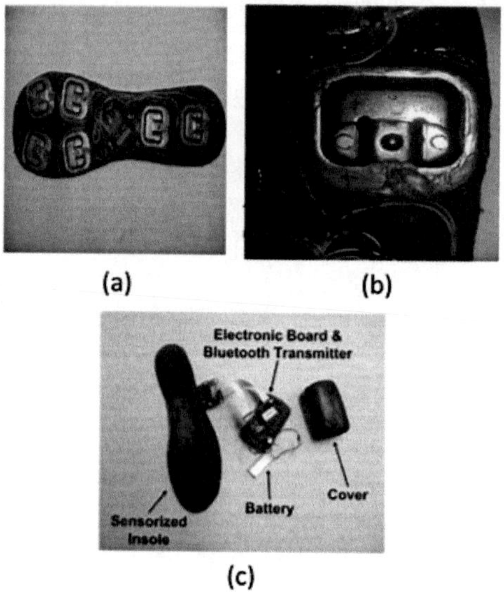

Fig. 3 **a** Six load cells on the sole of the shoe. **b** Placing load cells in the sole. **c** Load cells for the monitoring pressure

5 Working Principles

The IoT-Based feet Pressure Monitoring System is a wearable technology used to gather, aggregate, and transfer the data to the user which can then also be transferred to the doctor in charge of monitoring the feet pressure pattern of the patient. The foot pressure monitoring system can be used to analyze the pattern of pressure applied the machine learning algorithm of supervised learning. In this algorithm part, classification is used to predict the outcome of a given sample when the output variable is in the form of categories. This learning uses labelled training data to learn the mapping function that convert input variables into output variables. It also generate accurate outputs when the given inputs of the athlete while he/she is standing or walking or even running. There has always been a need for real-time acquisition system to measure the pressures exerted at various points of the foot to avoid further stresses inside the tissues or at the interface with the bones.

This analyzed data is made available to the doctor through the Cloud interface on the Arduino board. The doctor can then analyze this data and accordingly develop a comfortable sole or shoe for the patient which adjusts smoothly and gradually along the contours of their foot for an effortless and painless walk (Fig. 3).

6 Results and Discussion

Amount of pressure is applied at each point where the load cells placed has shown in Fig. 4. Using that we can easily identify the pressure variation during the static and dynamic conditions of sports person. The input data is captured through sensors

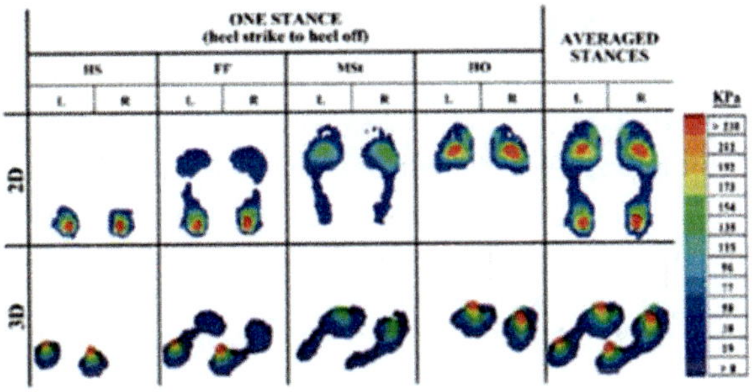

Fig. 4 Amount of pressure

from both legs. The collected data is transmitted over controller which will be stored in database. The stored data has analyzed by classification techniques for further clinical studies. Based on pattern identification, athletics can easily identify their own foot pressure style and correct themselves for their success.

Figure 5 explains the difference of two pressure points with equal time slots. The heel sensor data graph has shown in Fig. 6.

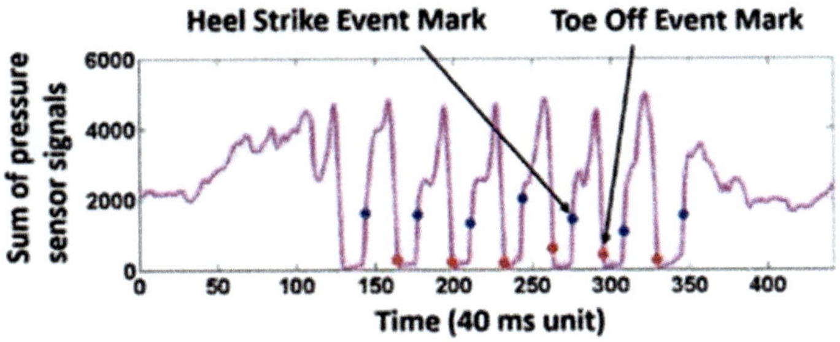

Fig. 5 Comparison of pressure points

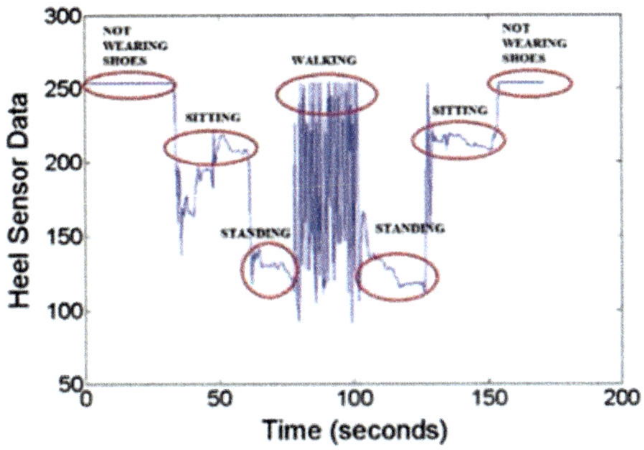

Fig. 6 Heel sensor data graph

7 Future Plan

The plans for future developments of this device include the size shrinkage and cost optimization. We also intend to extent the IoT connectivity by incorporating a Wi-Fi module right onto the Arduino board so as the seamlessly upload the walking pattern records to the cloud without the aid of the forth connected Android phone. Aim to make this device as a stand-alone package with its results, data, and records. Also, it always available from ant device connected over the internet, of both, the medical practitioner as well as the user.

References

1. A.H.T.E. De Silva et al., Development of a wearable tele-monitoring system with IoT for bio-medical applications, in *IEEE 5th Global Conference on Consumer Electronics* (2016)
2. A.J.A. Majumder et al., Your walk is my command: gait detection on unconstrained smart-phone using IoT system, in *IEEE 40th Annual Computer Software and Applications Conference (COMPSAC)* (2016)
3. P.S. Malvade et al., IoT based monitoring of foot pressure using FSR sensor, in *International Conference on Communication and Signal Processing (ICCSP)* (2017)
4. Y. ElSaadany et al., A wireless IoT system towards gait detection in stroke patients, in *IEEE International Conference on Pervasive Computing and Communications Workshops* (2017)
5. L. Russell et al., Personalization using sensors for preliminary human detection in an IoT environment, in *International Conference on Distributed Computing in Sensor Systems* (2015)
6. https://www.allerin.com/blog/how-iot-can-be-the-real-mvp-in-the-future-of-sports
7. https://www.researchgate.net/publication/304295757_Architecture_of_an_IoT-based_system_for_football_supervision_IoT_Football
8. http://www.ijarcsms.com/docs/paper/volume4/issue6/V4I6-0083.pdf

Development and Simulation of a Novel Approach for Dynamic Workload Allocation Between Fog and Cloud Servers

Animesh Kumar Tiwari, Anurag Dutta, Ashutosh Tewari, and Rajasekar Mohan

1 Introduction

Cloud computing has become part and parcel of the online industry for many years now. Giving users access to data centres over the Internet [1], cloud computing soon became the backbone of many sectors in the economy. But over the years, the constant rise in number of users resulted in increased load on the cloud servers [2] hindering with latency and in turn reducing the quality of service (QoS). This thus led to the introduction of servers smaller than and much closer to the user, by CISCO called FOG computing [3]. The fog architecture as shown in Fig. 1 involves adding a fog layer between the user and the cloud servers acting as a "descended cloud" [4].

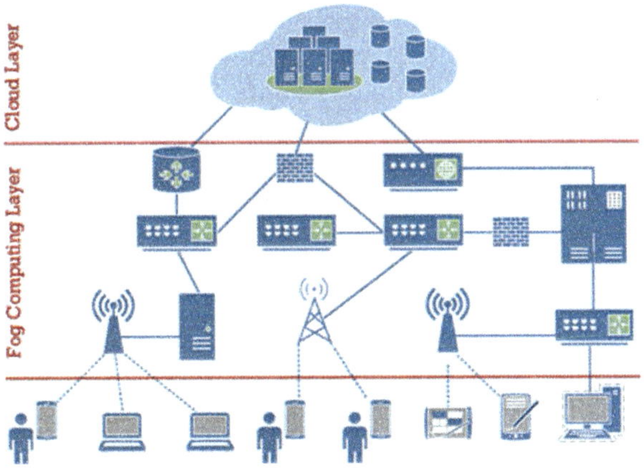

Fig. 1 Fog architecture

A. K. Tiwari · A. Dutta · A. Tewari · R. Mohan (✉)
Department of Electronics and Communication, PES University, Bengaluru, India
e-mail: rajasekarmohan@pes.edu

S. Patnaik et al. (eds.), *Advances in Machine Learning and Computational Intelligence*, Algorithms for Intelligent Systems, https://doi.org/10.1007/978-981-15-5243-4_55

Workload distribution between the fog and the cloud still remains an issue, and with this paper, we try to bring forward a solution to the same. We aim to lower the latency and increase system efficiency while maintaining a healthy task priority to success rate ratio. To achieve this, we went ahead with the introduction of a controller which will play a vital role in our system. It has two integral parts, the "sequencer" which sequences the incoming tasks efficiently and the "allocator" which by making use of information from the cloud and the fog layer schedules the tasks between them. The proposed algorithms for both the components of the controller have been discussed in further sections in detail along with the practical results obtained after simulating the model.

The remainder of the paper has been organized as follows: Sect. 2 provides a summary of the related works and literature. Section 3 describes our proposed methodology and algorithms. Section 4 looks at the results obtained during the experiments. Section 5 concludes the paper.

2 Related Works

Myriad amount of research has been done on fog-cloud computing. During the literature survey, we came across papers and journals that have done their best to make a paradigm shift in the field of work load allocation between fog and cloud servers. However, there are still some areas in the field that are left uncharted. In this paper, we have tried to incorporate those areas as the crux of our project. In [2], authors joined the dots between fog servers and power usage and latency. The usage of VMs in [1] was inspirational while developing our sequencer algorithm. Lee et al. [4] focus on topology of distribution of fog servers and use a fixed structure to carry out their experiments. The authors in [5] introduce a controller of their own into their network which follows a different strategy than ours. Khakimov et al. [6] make strategic assumptions regarding the functioning of their CPU which we integrated into our algorithms too. Al-Khafajiy et al. [7] focus on sharing workload between fog servers which served as the basis for our allocator algorithms.

3 System Model

The infrastructure of fog computing designed here aims to reduce overall delay in the network and optimize load on fog devices using heuristic methodologies. The framework sees the introduction of a controller which deploys the developed algorithms on the incoming tasks by sequencing them and then allocating the tasks amongst the fog devices and the cloud. In this section, we look deeply into the architecture proposed by us.

Let U be a task set, $U = \{1, 2, 3, \ldots, N\}$ and u_i, $i \in \{1, 2, \ldots, N\}$, represent the ith task. u_i has four attributes denoted by $u_i = \left\langle u_i^{\mathrm{cyc}}, u_i^{\mathrm{pri}}, u_i^{\mathrm{mem}}, u_i^{b.w}\right\rangle$, where

- u_i^{cyc} is the CPU cycles required by the task for its completion. Ranging from 1 to 1000.
- u_i^{pri} is the priority of the incoming task. This ranges from 1 to 100.
- u_i^{mem} is the memory needed by the task ranging from 1 to 1024.
- $u_i^{b.w.}$ is the required bandwidth for the task.

These numbers were arrived upon to help generalize the results and keeping in mind the granularity of the factors involved.

3.1 Proposed Model

Sequencer. The sequencer works on the batch-mode principle where the incoming tasks are operated upon in batches and then moved to the next station in line. Here tasks are arranged amongst each other, first with respect to their task priorities and then their requirement of CPU cycles. The tasks are then sent to the allocator one after another. The number of tasks in a batch is predefined.

Allocator. Allocator runs two sets of algorithms on the incoming tasks. Based on the parameters of the tasks, it decides whether to offload it to the cloud or the fog. In case of the latter, it also decides the most suitable fog server for a given task.

3.2 Compute Resource Block (CRB)

Before we get into details about the algorithms that run in the controller, we must first define a very important factor used here called "compute resource block" henceforth referred to as CRB. The CRB is calculated for every incoming task and even the available fog servers. The calculation of CRB is heuristic in nature meaning certain assumptions made here were obtained after multiple trials, and the more optimal values were chosen to be a part of the calculations.

CRB Of Incoming Tasks. For the calculation of the CRB for every task, two attributes, namely CPU cycle requirement (u_i^{cyc}) and the memory requirement (u_i^{mem}), of the task are taken into consideration. The task CRB ($\mathrm{CRB}_{\text{-t}}$) equation is as follows:

$$\mathrm{CRB_t} = 0.65 * u_i^{\mathrm{cyc}} + 0.35 * u_i^{\mathrm{mem}} \tag{1}$$

The choice of this proportion has been arrived at after extensive experimentation of various combinations of values.

CRB Of Fog Servers. For the fog servers, the average CRB is calculated. This is the average CRB that a fog server can handle. Certain assumptions made during the calculations are as follows:

a. Average u_i^{cyc} of the incoming task is 500
b. Average u_i^{mem} of the incoming task is 512
c. The upper limit of the main memory utilization in fog servers is 85%
d. $\mathrm{CRB_f}$ (threshold fog CRB) is 1.5 times the $\mathrm{CRB_i}$ in order to provide some leeway to the tasks.

$$\begin{aligned}\mathbf{CRB_i} &= 0.85 * 0.65 * u_i^{\mathrm{cyc}} + 0.35 * u_i^{\mathrm{mem}} \\ &= 0.85 * 0.065 * 500 + 0.35 * 512 = \mathbf{455.45} \\ \mathbf{CRB_f} &= \mathrm{CRB}_i + \mathrm{CBR}_i/2 = \mathbf{683.175}\end{aligned}$$

3.3 Proposed Algorithms

Sequencer. The sequencer obtains the priority (u_i^{pri}) and the CPU cycle requirement (u_i^{cyc}) of each task. It creates multiple priority lists and sends the tasks with same priorities to their respective lists. While it is doing that it also arranges the content of the lists with respect to their CPU cycle requirements from lowest to highest. Once these operations are run on all tasks in the batch, the tasks starting from the list with highest priority are sent to the "allocator" as in Fig. 2.

Allocator. The allocator runs two sets of algorithms on the incoming tasks, namely

a. Fog versus Cloud allocation algorithm
b. Inter-fog allocation algorithm.

Fog Versus Cloud Allocation Algorithm. This algorithm is responsible for deciding whether a task should go the fog servers or should it directly be sent to the cloud. This decision is made based on the calculated CRBs of the task and the available fog servers. This acts a preliminary screening of tasks.

Step 1: Begin
Step 2: Tasks arrive into the allocator
Step 3: u_i^{cyc} and u_i^{mem} of the task are obtained by the allocator
Step 4: CRB of task ($\mathrm{CRB_t}$)is calculated
Step 5: $\mathrm{CRB_t}$ is compared to the threshold CRB of fog servers ($\mathrm{CRB_t}$)
Step 6: if $\mathrm{CRB_t} > \mathrm{CRB_f}$
    task directly sent to cloud
    else
    task made ready to be sent to fog servers
Step 7: loop back to Step 2 for new task till all tasks in batch have arrived
Step 8: End

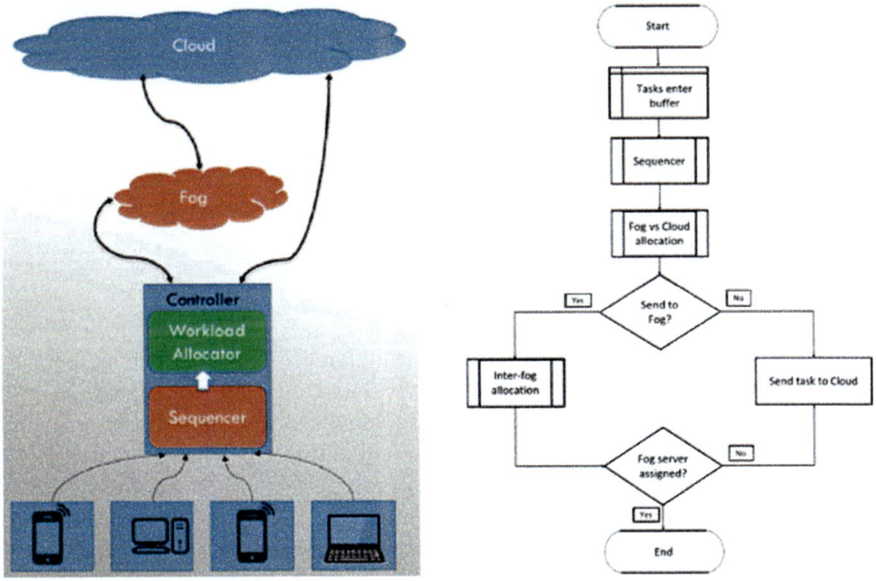

Fig. 2 System model and flowchart

Inter-Fog Allocation Algorithm. If the Fog vs Cloud algorithm decides that a particular task has to be sent to the fog servers, this algorithm helps in deciding which of the available fog servers is best capable of handling the task. There are measures taken to send the task to cloud if the criteria required by the task are not met by any of the servers, this is done to lower failure rate of the tasks. Algorithm is based on the principle of network bandwidth availability and hop count between two servers and the variation in the CRB.

Step 1: Begin
Step 2: Obtain $u_i^{b.w}$ and CRB_t of task
Step 3: Go to first fog server in network
Step 4: if $CRB_t < CRB_f$
        apply Dijkstra's algorithm to that node
    else
        go to next node
    else if all nodes tested then send task to cloud
Step 5: Compare $u_i^{b.w}$ and bandwidth available at branches in that path
Step 6: if $u_i^{b.w}$ < branch bandwidth
        execute task in that fog server
    else
        go to next node
    else if all nodes tested then send task to cloud
Step 7: End

4 Simulation and Results

The proposed algorithms were deployed on two networks with 3 and 5 fog servers, respectively, and the simulations were carried out with varying number of tasks in a batch.

4.1 *Average Workload Distribution*

Figure 3 represents the average workload distribution between the cloud and the fog servers based on the percentage of tasks assigned to them. Here A%, B%, C%, D%, E% refer to the % of tasks assigned to fog servers A, B, C, D and E, respectively. "Cloud-a %" refers to the % of tasks sent to cloud directly, whereas "Cloud-b%" refers to the % of tasks sent to the cloud after failing to find a suitable fog server.

We notice that the distribution of tasks among the fog servers is equal for both the models, and a total of 29% of tasks were sent to the cloud in comparison to 34% of 3-fog model, with the reduction as expected. Thus, we can say that by increasing the number of fog servers in the network, we have lowered the load on the cloud.

4.2 *Results of the Allocator Algorithms*

In Fig. 4, we see that as the number of tasks in the batch increases, the number of tasks that get sent to the cloud after the inter-fog allocation also increases rapidly. This happens due to the depletion of resources in the available fog servers. Thus, to maximize the fog utilization while maintaining a healthy response time, based on

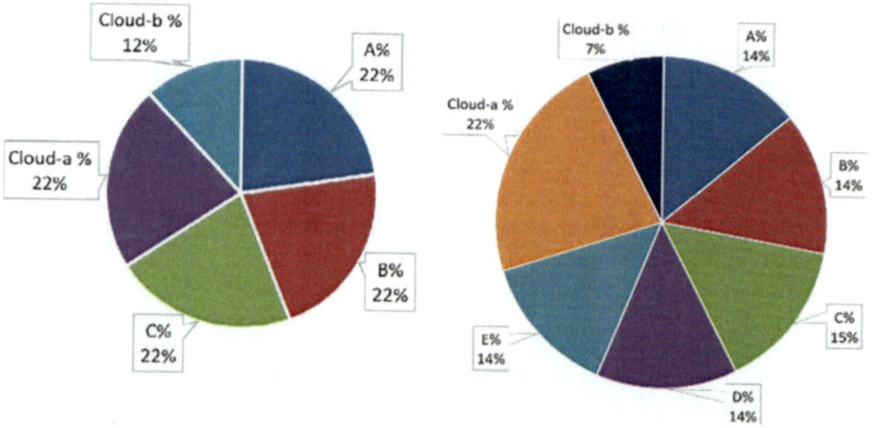

Fig. 3 Average workload distribution in 3 and 5 fog server models

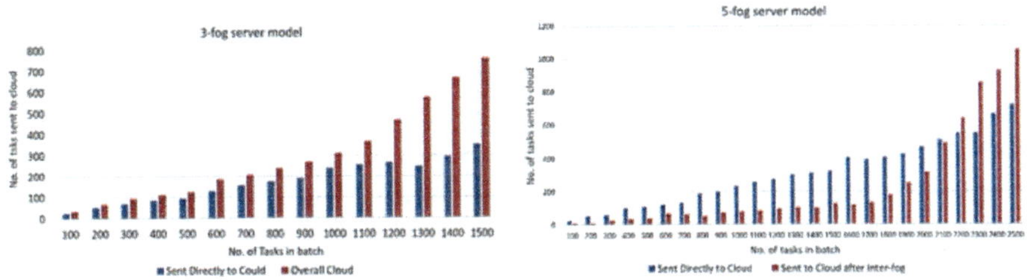

Fig. 4 Variation in no. of tasks being sent to cloud via the two allocator algorithms

the resources available, a viable batch count has to be chosen. For example, just by inspecting the graph below in Fig. 4a, one can say that the optimum batch count for the given 3-fog model is around 1000 (in the real world this number will be much larger due to larger resources available). In Fig. 4b for the 5-fog server model, the optimum batch count seems much higher at 2000. This is due to the added resources of the 5-fog network. This indicates that with the increase in number of fog servers in an infrastructure, the batch count can also be increased allowing more tasks to be processed together.

4.3 *Success Rate*

In order to check the efficiency of the proposed algorithms, deadlines for the tasks were introduced. For the 3-fog server model, in Fig. 5, it is observed here that after

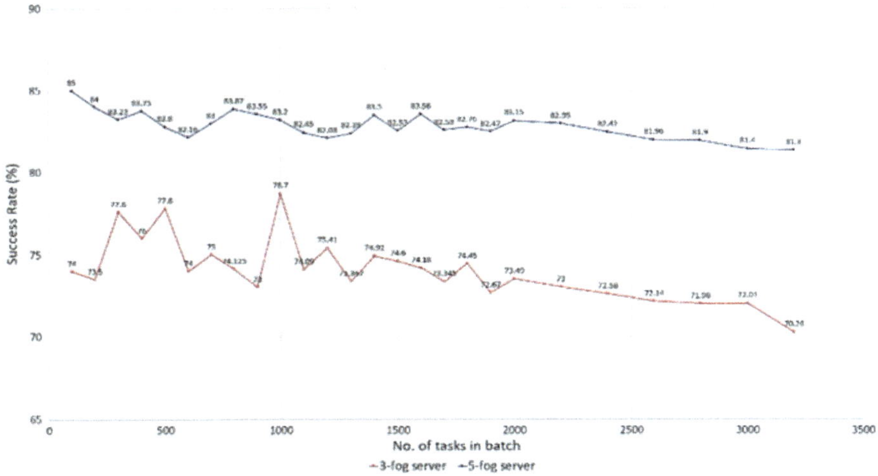

Fig. 5 Comparison of success rate (%) of tasks in 3-fog and 5-fog server models

about 1200 batch tasks, the success rate starts to decline; this is because a higher number of tasks are now being sent to the cloud as per Fig. 4a, which results in more time taken by them to be allocated, thus leading to them not meeting their assigned deadlines. And for the 5-fog server model, a similar decline is seen after the 2000 tasks mark. The success rate of the 5-fog server model is higher than the 3-fog server model, but this too experiences a drop in success rate; this is due to the fact that since there are more fog servers, and most of those are getting utilized by the tasks, the time taken by other tasks to scan all available fog servers increases leading to failures. This is a trade-off that occurs in such a scenario.

5 Conclusion

In this paper, a set of three new priority-driven algorithms were developed that aim to improve the distribution of workload between fog and cloud servers. Said algorithms were studied and analysed and were then tested on two separate fog networks. The results obtained from the experiments seem promising as they show adequate workload allocation occurring in the network.

References

1. P.Y. Zhang, MengChu Zhou, Dynamic cloud task scheduling based on a two-stage strategy. IEEE Trans. Autom. Sci. Eng. **15**(2), 772–783 (2018)
2. R. Deng, L. Rongxing, C. Lai, T.H. Luan, H. Liang, Optimal workload allocation in fog-cloud computing toward balanced delay and power consumption. IEEE Internet Things J. **3**(6), 1171–1181 (2016)
3. A. Jain, P. Singhal, Fog computing: driving force behind emergence of edge computing, in *Proceedings of SMART-2016, IEEE Conference ID: 39669, 5th International Conference on System Modeling & Advancement in Research Trends* (Moradabad, India, 25–27 Nov 2016)
4. J.-H. Lee, S.-H. Chung, W.-S. Kim, Fog server deployment considering network topology and flow state in local area networks, in *The 9th International Conference on Ubiquitous and Future Networks (ICUFN)* (Milan, Italy, 2017)
5. A. Banerjee, M.S. Chishti, A. Rahman, R. Chapagain, Centralized framework for workload distribution in fog computing, in *3rd International Conference for Convergence in Technology (I2CT)* (Pune, India, 2018)
6. A. Khakimov, A. Muthanna, M.S.A. Muthanna, Study of fog computing structure, in *2018 IEEE Conference of Russian Young Researchers in Electrical and Electronic Engineering (EIConRus)*
7. M. Al-Khafajiy, T. Baker, A. Waraich, D. Al-Jumeily, A. Hussain, IoT-fog optimal workload via fog offloading, in *IEEE/ACM International Conference on Utility and Cloud Companion (UCC Companion)*, 2018
8. P. Sangulagi, A.B. Sutagundar, Context aware information classification in fog computing, in *Second International Conference on Advances in Electronics, Computer and Communications* (Bangalore, India, Feb 2018)

A Novel Strategy for Energy Optimal Designs of IoT and WSNs

Rajveer Singh Shekhawat, Mohamed Amin Benatia, and David Baudry

1 Introduction

Computing and communication technologies have merged together in the form of IoT leading to development of smart and innovative solutions for a variety of applications. The two major aspects that have helped a universal deployment of IoT applications are low cost and small size packed with huge computing power and aided by high data rate wireless communication. A number of IoT applications demand devices operating on battery, thus restricting the useful life of the solution as most scenarios do not allow replacement of batteries. Most of the times, we desire a target period of survivability of IoT devices and thus are challenged to design and develop IoT devices meeting the expected lifetime. The optimization can also reduce the cost and size of IoT nodes with just a sufficient battery. Researchers kept their focus only on design/development [1, 2] phase, but the proposal encompasses the complete lifecycle of the IoT applications. In this approach, right from design/development stage to deployment and finally the operational level, focus on best use of available battery power is maintained. This derives from the fact that about 50–60% energy used by networked devices is spent by communication system, whereas about 30% is consumed by encryption function [3]. The optimal location of nodes impacts the smooth intercommunication as well as reduces excessive power transmission by transmitters on nodes, thus saving on energy. Once an acceptable deployment strategy is achieved, further optimization of three sub-systems can be attempted during operation time. This three-pronged approach can help us to address the smart system design and deployment for achieving the targeted lifetime for the given application.

R. S. Shekhawat (✉)
School of Computing & IT, Manipal University Jaipur, Jaipur 303007, India
e-mail: rajveersingh.shekhawat@jaipur.manipal.edu

M. A. Benatia · D. Baudry
LINEACT Laboratory - CESI, 80 rue Edmund Halley, Saint Etienne du Rouvray 76808, France

S. Patnaik et al. (eds.), *Advances in Machine Learning and Computational Intelligence*,
Algorithms for Intelligent Systems, https://doi.org/10.1007/978-981-15-5243-4_56

2 The Technologies' Choices for Battery-Powered IoT

IoT is a result of integration of data communication, embedded systems, cybersecurity, etc., for a variety of applications. In this section, we shall present the technologies and paradigms prevalent in data collection, data communication, and encryption so as to illustrate the choices available to developers.

Data acquisition and data pre-processing are highly dependent on the nature of sensed variables and the requisite sampling frequency. The sparsity of signals and redundancy across the sensors can be utilized by compressed sensing [4, 5], adaptive sensing [6] and soft sensing [7] apart from traditional methods of sub-sampling, encoding/compression, tweaking resolution, etc., for reduction of overall data to be handled.

Data communication and related infrastructure are next choice for deploying IoT applications. One can choose one of the protocols from IPv4 or IPv6, IEEE 802.11.5, 6LoWPAN, BLE, LoRa, MQTT, etc., for IP layers. The application layer choices are CoAP, AMQP, WebSocket, XMPP, etc. Table 1 shows the list of protocols available for use layer-wise.

Securing the data and access to the nodes in IoT is perhaps the most challenging task. A large number of attacks on major installations get through the large attack surface aided by relatively weak security in IoT networks. Lightweight cryptography to ensure data security is the most suited approach to reduce energy use by IoT devices. The trust management for human/robotic access of IoT network is another emerging dimension. Table 1 also lists, as part of session layer, protocols used for securing IoT applications. There are vast choices that a developer of IoT applications has at his disposal for networking protocols and must make an intelligent choice for a given application. One may refer to [8] which provides more details about the protocols and their parameters that can be suitably optimized.

The major emphasis of this paper is on the premise that a majority of IoT devices are likely to be powered by battery which must last long enough. This poses a challenge to the developers to ensure that the IoT nodes do not starve of power as

Table 1 List of popular protocols suitable for parameter optimization

Layer name	Protocols
Physical/data link (MAC + LLC)	802.15.4, MQTT, CoRE, DDS, AMOP, LoRaWAN, DigiMesh, ANT+, IEEE 802.15.6, DASH7, Bluetooth Low Energy, Z-Wave, DECT/ULE, NFC, WirelessHART, Wireless M-Bus, HomePlug GP, 3G/LTE, LTE-A
Network Layer (Encap + Routing)	6LowPAN, 6TiSCH, 6LoThread, AODV, AODv2, LOADng, LEACH, CARP, CORPL, ZigBee, ISA 100.11a, RPL, WiMAX
Session layer (+Security)	EEE 1888.3, TCG, Oath 2.0, SMACK, EDSA, Ace, ISASecure, DTLS, Dice, X.509, Block Chains, SASL, OTrP
Application layer	CoAP, XMPP, AMQP, WebSocket

the battery charge depletes. The choice of replacing the battery or energy harvesting is not considered viable. Many such applications using battery are smart agriculture, smart city, smart buildings, smart health, etc. For these, there is an alternative way to mitigate the battery life concern by estimating the battery life of an application upfront and deploying only suitable battery that would last through the lifecycle of the application. The challenge however is to design, develop and deploy systems with a given power budget. An approach aligned with this thinking is highlight of this paper and elaborated below.

Design Time There is good amount of work reported on the high-level system design employing hardware software co-design principles [9–11]. However, energy consumption has been just one of the design goals which acquired increased importance after emergence of WSNs and IoT. One can start with assignment of energy budgets for major functions of IoT node upfront to ensure requisite lifetime of the node. The top-level energy budget is then further hierarchically broken down as targets for subsystems, e.g. data acquisition/processing, data communication and encryption. The full life of system can be ascertained by assessing the average operational energy consumption.

Many optimization objectives are application specific like lifetime, network size (number of nodes), node types (edge, router, gateway, etc.), data rate, network span (spatial spread). There are plenty of choices of microcontrollers offering many low-power modes, and developing software to make use of these can be daunting task. Low-power AFE [5] blocks can be used for data acquisition and so also peripherals to match low-power protocols. The following paragraphs speak of these choices in a little more detail.

Data Acquisition. Sampling time can be decided as per application need and nature of signal rather than based on first principles. Similarly, the resolution of analog-to-digital converters can be dependent on desired accuracy for the application rather than selecting ideally. Signal-agnostic sensing like compressed sensing and matrix completion for sparse signals can help to a great extent to reduce the load on host processor and also lower payload for data exchange. One may also optimize pre-processing needs to suit the nature of signals and those demanded by application use. For example, the room temperature can be easily maintained comfortable by sensing temperature in integral degree centigrade rather than its fractions which shall also save on cost of temperature sensors which may not be very precise.

Data Communication. The first and foremost is the choice of data rate that is sufficient for achieving the target life of application. Here, the physical layer parameters like the spectrum band, the modulation and encoding (optimize optional parameters) schemes used can provide a great mileage for low power. Similarly, some of the protocol parameters of data link layer do offer sufficient flexibility to make a good choice of them to allow less time to wait, transmit, etc. The available routing protocols to suite application with optimal parameters and routing table sizes help in a significant way, and the literature is flushed with many such reports [12].

Data Encryption. The mechanism to select scheme of encryption along with suitable key types (symmetric, non-symmetric) and size (number of bits) offers a significant advantage to realize low power objective. The smaller key sizes can be used for a smaller time period which is just enough so as to not allow the hackers to get their hands on the keys and changing the keys frequently.

Deployment Time The configuration and location of devices that should be taken care to suit the deployment scenario can be the basis of a successful application of IoT. In want of improper location of nodes, many projects have failed especially in smart energy area of which one of the authors has witnessed. The site-specific node placement (line of sight or in-direct) can greatly help achieve meet the life expectancy of the IoT application. Salient aspects of the approach are briefly discussed below.

Data Acquisition. The sensors need to be placed at most desirable place so that these can sense the physical variables accurately. This, however, many a time may counter the node placement for ease of communication. A compromise has to always be evolved in such situations.

The task of deployment is daunting in residential premises, office spaces and hospitals where walls and other objects interfere with radio-frequency communications. The latter phenomena make smooth operation difficult without missing crucial data exchanges. Evolving trouble-free localization of sensor nodes, intermediate (cluster centres, routers) nodes and gateways poses a very complex optimization problem.

Data Communication. The energy consumption of nodes to transmit would depend significantly on the distance of other nodes from it, the receiver sensitivity of the nodes being fixed. A number of approaches [13–16] have been reported in the literature which are successful to some extent in achieving the objectives of the location problem. The placement gets further complicated due to presence of walls and objects around which attenuate and reflect the transmitted RF power. The best way to handle the configuration is to model the working environment, but only very crude approximation is possible and thus, fine-tuning of placements has to be done at deployment time only. The potential applications for this optimization are smart hospitals, smart metering [17, 18] and smart buildings [19, 20].

Data Encryption. There is not much scope for the deployment time improvements for the encryption methods. However, the overhead of key exchanges would be affected by the node placement and the communication sub-system energy consumption.

Operation Time There are many possible options that can be exercised to still optimize the power consumption of an IoT node during operation time. Good scope exists in all functional areas of the node that are key consumers of energy. We shall delineate below some specific aspects for each functional block that can easily be manipulated to arrive at sufficient gains in overall life of a node.

Data Acquisition. The actual sensor data reduction/compression can be based on many considerations. The sensor data value variations are way different from our perception at the time of design, and thus, methods adapting to real-time sensor variable value variation and sampling it dynamically make a lot of sense. Adaptive sensing can be implemented in many different ways [21].

Another redundancy in collected sensor data can be overlapping values from neighbouring sensors, and thus, taking care of these duplicate values from the data communication point of view is another important aspect which needs to be taken into account. Some of the approaches are discussed in [22] at length.

Data Communication. There exist many approaches to reduce the communication burden, e.g. reducing data being communicated involving compressed sensing and packing of sensed data in packets and minimizing the frequency of communication affecting certain parameters of protocols. The data communication needs and thus traffic patterns can be different than envisaged during requirement phase and implemented in the design, and thus, data aggregation [23–25] methods play an effective role to reduce data traffic. The variation of data communication protocol parameters thus needs to dynamically adapt in real time [4]. Yet another aspect that can reduce the data comms load is the frequency of data to be sent as dictated by the user of the data and set thru a policy during configuration/deployment time [18].

Data Encryption. Having decided on smaller size key, dynamic key generation (for every time T) can be one good approach. The time T can be decided to be less than the time taken for brute force attack to break the key. Thus, before a hacker succeeds, we would have changed the key(s). Key distribution overhead on communication can still be insignificant compared to overall improvement in time spent by nodes for encrypting messages, and thus, IoT system can succeed in securing data communication with a much lower power budget.

3 A Case Study—Smart Buildings

In the following paragraphs, how the proposed approach and methods can be used for a better implementation of smart buildings are highlighted.

Embedding sensors in buildings and in appliances and machines contained in them are to report different information related to lighting, air conditioning, energy use, comfort level and air quality. The energy-efficient operation and health monitoring of buildings using such sensors and systems is the key objective [13, 15] of many projects. The data acquisition, reporting and analysis systems must therefore also emphasize the low-energy operation [26]. The modern technological advancements have made this desire feasible. However, most of such systems have been applied to large buildings only where the relative costs of such systems are much smaller than the cost of buildings. It will, however, make more sense if such systems can be made viable for small- and medium-size buildings which exist in larger numbers. Emergence of Internet of things (IoT) has made this objective easier to achieve. One can develop systems which can consist of a mix of wireless sensor nodes and IoTs [14]. Second factor that can allow buildings to be made smart is automation of design, deployment and operation of sensor nodes and related sub-systems [20] which have been the theme of this paper.

The proposed scheme starts with design stage itself focusing on least energy usage by the sub-systems aided by optimal deployment of sensor nodes, intermediate nodes and gateways, and finally the energy-efficient operation. There is thus an assumed lifetime of system which is desired so that the system components can be re-energised before next cycle. So energy budget is one of the design goals. The top-level energy budget is then hierarchically divided as targets for sub-systems like data acquisition, data processing and data communication [25]. These energy targets can be further broken down to another level. This stops at hardware and software blocks of smallest functionality. The deployment strategies can be another major contributor to minimize use of energy by identifying optimal placement of nodes and other infrastructural entities [20]. The task of deployment is much simpler in open spaces like smart agriculture but equally daunting in residential premises and office spaces where walls and other objects interfere with radio-frequency communications resulting in absorption, reflection and refraction. The latter phenomena make it very difficult to predict smooth operation of the deployed nodes without missing crucial data exchanges. Evolving trouble-free localization of sensor nodes, intermediate nodes and gateways poses a very complex optimization problem [16]. A number of approaches have been reported in the literature which are successful to some extent in achieving the objectives of the location problem in addition to the ones that have been pursued by members of the project team [20]. The latter shall be further refined and applied to real-life problems like smart agriculture and smart buildings.

The last leg of optimization focuses on operational aspects of the smart systems which involve actual communication of data. Some of the aforesaid methods can reduce the data acquisition, communication and encryption burden on the node(s). Overall, modelling of operational energy utilization and its minimization can be stated as a complex nonlinear optimization problem. Use of prevalent soft computing techniques like genetic algorithms, ant colony optimization, simulated annealing, etc., can help arrive at usable solutions. Such optimization methods also help during design and deployment as well of the WSN and IoT-based applications.

4 Conclusion

The three-step approach to address the low-power development of IoT-based applications can be very effective to meet the desired lifetime of the system along with lower cost and maintenance. With the rise in number of IoT applications, this approach can also be optimally used to ensure just enough security of operations and nodes so as not to compromise the back-end systems. Lately, majority of attacks have been perpetrated through less secure IoT as security had been always looked at a costly proposition for simpler IoT devices. A methodology based on the proposed approach is under development and test by authors. The early signs of prototype so developed speak well of the soundness of the method, and benefits can accrue for achieving low-power IoT applications meeting target lifetime. The platform being developed is also proposed to include monitoring of battery health dynamically so that unexpected energy profile of battery can be accounted for estimating actual life of IoT node(s).

References

1. A.S. Shah, A review on energy consumption optimization in IoT. Information (MDPI) **10**, 108 (2019)
2. B. Martinez et al., The power of models: modeling power consumption for IoT devices. IEEE Sens. J. 5777–5789 (2015)
3. de Meulenaer, Giacomo et al.: On the energy cost of communication and cryptography in wireless sensor networks, in *IEEE Conference on WIMOB* (2008)
4. L.S. Powers et al., Low power real-time data acquisition using compressive sensing, in *Proceeding SPIE 10194, Micro- and Nanotechnology Sensors, Systems, and Applications IX* (2017)
5. Chen F. et al, A Signal Agnostic Compressed Sensing Acquisition System for Wireless and Implantable Sensors, CSCICC, 2010
6. T. George, A.K. Dutta, M. Saif, Islam, Micro and Nanotechnology Sensors, Systems, and Applications IX (Anaheim, CA, USA, April 09, 2017)
7. L. Fortuna, S. Graziani, A. Rizzo, M.G. Xibilia, Soft Sensors for Monitoring and Control of Industrial Processes (Springer, 2007)
8. D.E. Culler, Open Standards Reference Model (2011). http://people.eecs.berkeley.edu/~culler/
9. D. Gajski, F. Vahid, S. Narayan, J. Gong, Specification and Design of Embedded Systems (Prentice Hall, 1994)
10. T. Givargis, F. Vahid, Embedded System Design: A Unified Hardware/Software Introduction (Wiley, 2006)
11. A. Gerstlauer, D.D Gajski, S. Abdi, Embedded Systems—Modelling Synthesis and Verification (Springer, 2009)
12. N.A. Pantazis et al., Energy-efficient routing protocols in wireless sensor networks: a survey. IEEE Commun. Surv. Tutorials. **15** (2), 2Q (2013)
13. A. Guinard, A. McGibney, D. Pesch, A wireless sensor network design tool to support building energy management, in *Proceedings of the First ACM Workshop on Embedded Sensing Systems for Energy-Efficiency in Buildings* (ACM, 2009), pp. 25–30
14. N. Heo, P.K. Varshney, Energy-efficient deployment of intelligent mobile sensor networks. IEEE Trans. Syst. Man Cybernetics-Part A: Syst. Humans **35**(1), 78–92 (2005)
15. A. Nacci, *A, Methods and tools for effective smart buildings deployment, doctoral dissertation* (Milan Polytechnic, Italy, 2016)
16. Y. Song, C. Gui, X. Lu, H. Chen, B. Sun, A genetic algorithm for energy-efficient based multipath routing in wireless sensor networks. Wireless Pers. Commun. **85**(4), 2055–2066 (2015)
17. A. Keshtkar, S. Arzanpour, P. Ahmadi, Smart residential load reduction via Fuzzy logic, wireless sensors and smart grid incentives. Energy Build. **104**, 165–180 (2015)
18. M.W. Ahamad et al., Building energy metering and environmental monitoring—a state of art review and directions for future research. Energy Build. **120**, pp. 85–102 (2016)
19. C. Talon, IOT bridging the gap for intelligent small and medium sized buildings. Navigant Res. (2016)
20. M.A. Benatia, M.H. Sahnoun, D. Baudry, A. Louis, A. El-Hami, B. Mazari, Multi-objective WSN deployment using genetic algorithms under cost, coverage, and connectivity constraints. Wireless Pers. Commun. **94**(4), 2739–2768 (2017)
21. Makhoul et al., Data reduction in sensor networks performance evaluation in a real environment. IEEE Electron. Syst. Lett. (2017)

22. G.M. Dias, B. Bellalta, S. Oechsner, *Using data prediction techniques to reduce data transmissions in the IoT* (IEEE World Forum on IoT, Reston, VA, USA, 2016)
23. B. Pourghebleh, N.J. Navimipour, Data aggregation mechanisms in the internet of things: A systematic review of the literature and recommendations for future research. J. Netw. Comput. Appl. **97**, 23–34 (2017)
24. D. Feng et al., A survey of energy efficient wireless communications. IEEE Commun. Surv. Tutorials **15**(No1), 1Q (2013)
25. M.B. Krishna, M.N. Doja, Multi-objective meta-heuristic approach for energy-efficient secure data aggregation in wireless sensor networks. Wireless Pers. Commun. **81**(1), 1–16 (2015)
26. H. Wang, H.E. Roman, L. Yuan, Y. Huang, R. Wang, Connectivity, coverage and power consumption in large-scale wireless sensor networks. Comput. Netw. **75**, 212–225 (2014)

Multichannel Biosensor for Skin Type Analysis

V. L. Nandhini, K. Suresh Babu, Sandip Kumar Roy, and Preeta Sharan

1 Introduction

The skin is the largest organ in the body. Skin acts as a waterproof, insulating shield, guarding the body against extremes of temperature, damaging sunlight, and harmful chemicals. It also exudes antibacterial substances that prevent infection and manufactures vitamin D for converting calcium into healthy bones. Skin color is due to melanin, a pigment produced in the epidermis to protect us from the sun's potentially cancer-causing ultraviolet rays. Dark-skinned people produce more numerous and deeper-colored melanin particles [1]. Skin cancer is the most common of all human cancers. Skin cancers are of three major types: basal cell carcinoma, squamous cell carcinoma, and melanoma. The vast majority of skin cancers are BCCs or SCCs. While malignant, these are unlikely to spread to other parts of the body. They may be locally disfiguring if not treated early. Like many cancers, skin cancers start as precancerous lesions [2]. These precancerous lesions are changes (dysplasia) in the skin that are not cancer but could become cancer over time [3]. Dysplastic changes in skin:

Actinic keratosis—red or brown patch.
A nevus—abnormal moles.

V. L. Nandhini (✉)
Department of ECE, Govt. SKSJTI, Bangalore, India
e-mail: sunandi7276@gmail.com

K. Suresh Babu
Department of ECE, UVCE, Bangalore, India

S. K. Roy
Department of ECE, AMC Engineering College, Bangalore, India

P. Sharan
Department of ECE, The Oxford College of Engineering, Bangalore, India

S. Patnaik et al. (eds.), *Advances in Machine Learning and Computational Intelligence*, Algorithms for Intelligent Systems, https://doi.org/10.1007/978-981-15-5243-4_57

Due to their relative lack of skin pigmentation, Caucasian populations generally have a much higher risk of getting non-melanoma or melanoma skin cancers than dark-skinned populations. People with skin type Asian and Dark can usually safely tolerate relatively high levels of sun exposure without enhanced skin cancer risk as reported in WHO URL [4].

A skin cancer screening is a visual exam of the skin. The screening checks the skin for moles, birthmarks, or other marks that are unusual in color, size, shape, or texture. Certain unusual marks may be signs of skin cancer. The screening can cause misleading results if the effected skin area color matches that of skin. Another limitation of visual inspection is the human dependency of interpretation. To reaffirm the presence of skin cancer, a costly skin biopsy procedure is required. In the proposed work, we are presenting a novel approach to scientifically identify the skin type based on the refractive index (RI) of the skin. This will help to avoid an error that may crop up due to wrong interpretation during visual inspection. The research work started with a literature survey to understand the state-of-the-art technology available and potential research areas. We found there is a need for research for the early detection of cancer. Our next aim was to identify the right low-cost technology that can be easy to use and portable. We opted to go with the design of nanoscale device using integrated optics based on PhC.

2 Photonic Crystal

Photonic sensing is a new accurate measurement technology for biosensing applications [5]. Photonic devices are the components that are used for manipulation, detection, and creation of light, and a PhC is a periodic optical nanostructure, and it affects the motion of photons. PhC, in which the RI changes periodically, provides an exciting new tool for the manipulation of photons and has received keen interest from a variety of fields [6]. PhC can be fabricated for one, two, or three dimensions (see Fig. 1).

Good propagation of light always results in high selectivity. Two dimensional (2D) PhC are used as waveguide to modify the flow of light. PhC is the periodic structure that controls light in ways that are analogous to how electrons are controlled in semiconductors. They exhibit photonic band gap (PBG) that is used

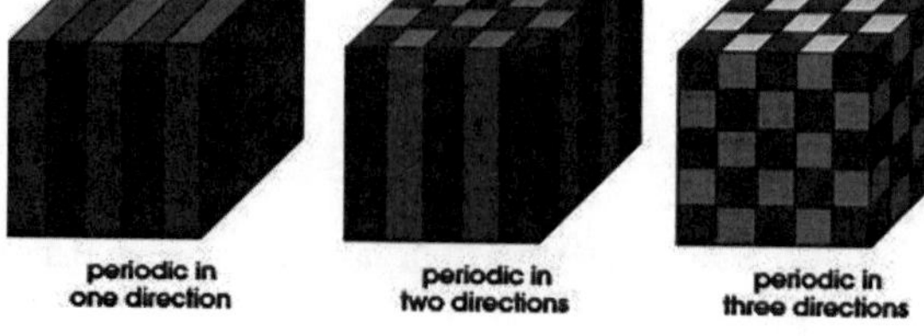

Fig. 1 Photonic crystal structures **a** 1D, **b** 2D and **c** 3D

to inhibit spontaneous emission, form all-dielectric mirrors, form waveguides, and also, they exhibit useful dispersion properties such as super-prisms, negative refraction, and dispersion compensation. The undergraduate-level textbook [7] on PhC and electromagnetism in periodic (or partially periodic) geometries on the scale of the wavelength of light provides great insight on the working principle of PhC. Optical microcavities that confine light in high-quality resonance promise all of the capabilities required for a successful next-generation microsystem bio-detection technology [8]. For PhC-based sensors, the analysis is mainly done by two methods: two-dimensional finite difference time domain (2D-FDTD) and plane wave expansion (PWE) approaches. The propagation of electromagnetism waves can be simulated by the 2D-FDTD method, and the PBG can be calculated with the PWE approach. Most of the PhC biosensors can detect chemical and biochemical molecules. Biosensors can be designed according to different PhC structures such as photonic crystal fibers, PhC-based waveguides, and PhC-based resonators [9–13]. In a PhC, the light propagation is controlled by introducing defect (point defect and line defect) in the dielectric structure to alter the PBG. Point defect traps the light, whereas a line defect creates the waveguide in the PhC.

Figure 2 shows the schematic diagram of PhC-based sensor circuit. The sensor is made to dip in the sample, and therefore, the air in rod in air or holes in slab is replaced by the sample. The light is passed through one end of the sensor and is detected at the other end by the detector. The input given to the sensor will be a Gaussian pulse and will interact with the sample. Gaussian profile of light source consists of intensity distribution (intensity as a function of radial distance from the beam center) which is given by the Gaussian function as given in Eq. 1.

$$I(r) = I_0 e^{\left(-\frac{2r^2}{\omega_0^2}\right)} \tag{1}$$

where I_0 is the intensity at the beam center and r is the radial distance from the beam center.

ω_0 is the radius at which the intensity drops to a value of $1/e^2$, or 13.5%, or its peak value. Since, in mathematical terms, the intensity of a Gaussian beam never drops completely to zero, it is standard to define the beam radius as the distance from the beam center to the $1/e^2$ point. The propagation of light varies depending upon the dielectric constant of the sample. PhC-based sensor utilizes this property of the PhC structure.

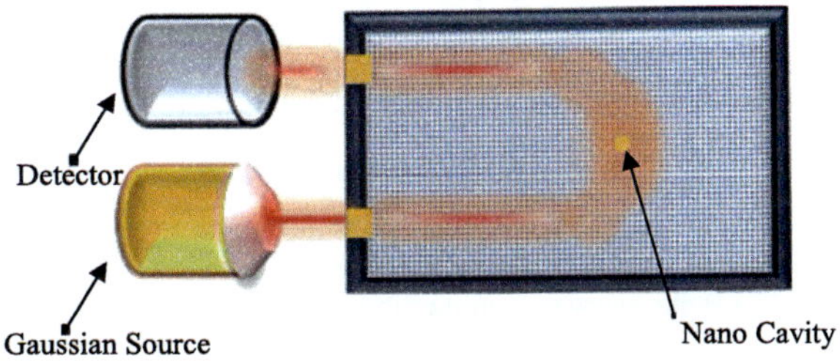

Fig. 2 Schematic image of photonic crystal-based sensor circuit

3 Multichannel Cavity-Based PhC Sensor Design

The structure of this biosensor consists of a hexagonal lattice of air holes in a silicon slab and its operation based on two waveguides that are incident waveguide and output monitor. Nanocavity that has been created in this lattice is for sensing purpose. Pulse is applied to the incident waveguide, and then, the resonant mode of a resonator is excited, and the output signal is recorded at the end of the output monitor. The nano-cavity is of 0.18 mm, and others around are 0.22 mm in the sensor.

Figure 3a shows structure of the single-cavity biosensor with RI changes. Figure 3b shows the 3D layout design of the sensor; the nanocavity has a dimension of 0.18 mm, and all other crystals are 0. 22 mm. However, for a multichannel biosensor, the layout design is reorganized to accommodate multiple cavities. Based on the result obtained, we found that nanocavity 0.22 mm shows distinct shifts to differentiate different skins. As shown in Fig. 4a, we designed the multichannel with nanocavity 0.22 mm. Figure 4b shows 3D view of the design.

We have created a biosensor with three cavities or point defects. The use of biosensor is that it can sense three RI simultaneously. The cavities can be shifted or

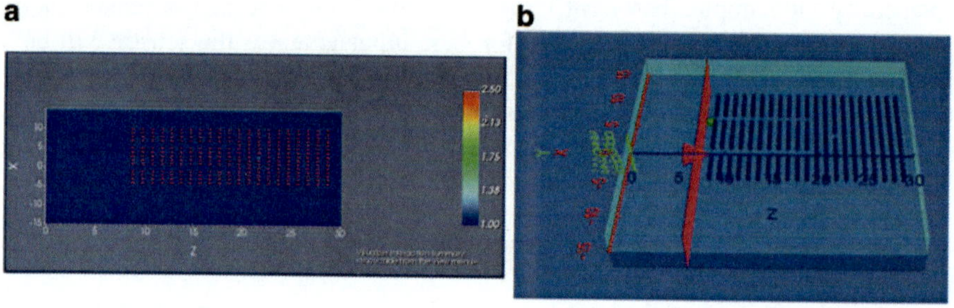

Fig. 3 **a** Structure of the single-cavity biosensor with RI changes, **b** 3D view of biosensor

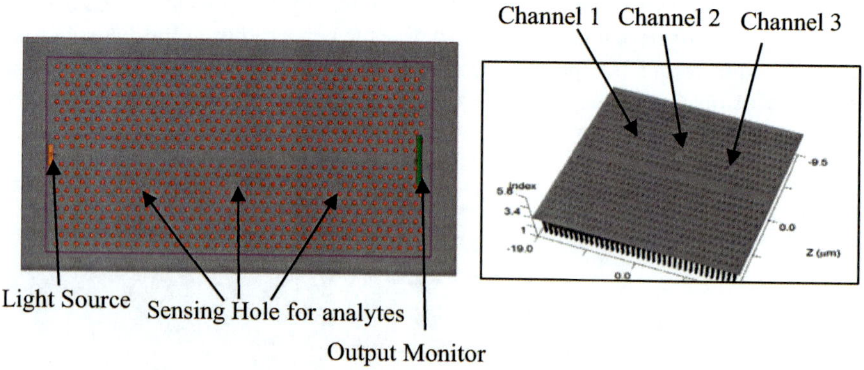

Fig. 4 **a** Preliminary structure of three sensing holes PhC biosensor, **b** 3D view

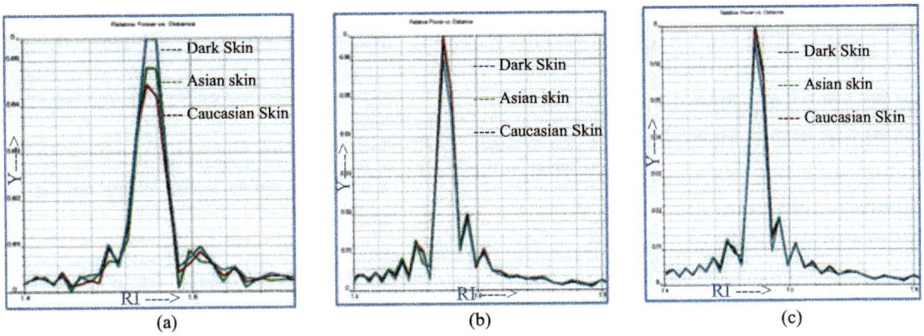

Fig. 5 Transmission spectrum for nanocavity **a** 0.22 mm, **b** 0.20 mm and **c** 0.18 mm. **b** shows the transmission spectrum of biosensor for nanocavity 0.20 mm and **c** shows the transmission spectrum of biosensor for nanocavity 0.18 mm

altered in Fig. 4 shown, by shifting or changing the properties of the sensing hole, the wavelength shifts, and new resonant wavelength can be obtained. First, the RI of the first channel is changed by giving three values (1.36, 1.44, 0.94) simultaneously; no changes are done in channel 2 and channel 3; the wavelength is observed, and then, for the second channel, the refractive index is changed with the same three values, making the first channel to be air as of the third channel; another wavelength is observed, and the third channel is also computed in the same manner, keeping the second and first channel RI to be air, that is 1. Then, all three wavelengths can be merged to observe the effect.

4 Result and Discussion

4.1 *Single Channel*

The single-channel structure of biosensor based on micro-resonator to detect small changes in the RI is shown in Fig. 4. This was used to first analyze the dimension of the nanocavity that will provide the best sensitivity. As can be seen in Fig. 5a that the structure with nanocavity 0.22 mm shows distinct spectra for three types of skins.

4.2 *Multichannel*

With the encouraging result for a single channel, further, we continued with simulation with three cavities as shown in Fig. 4.

Channel 1. Figures 6 and 7 show the result of three types of skins—Dark Skin, Asian Skin, and Caucasian skin on Channel 1. Table 1 shows corresponding amplitude vs wavelength data.

Channel 2. Figures 8 and 9 show the result of three types of skins—Dark skin, Asian skin, and Caucasian skin on Channel 2. Table 2 shows corresponding amplitude vs wavelength data.

Channel 3. Figures 10 and 11 show the result of three types of skins—Dark skin, Asian skin, and Caucasian skin on Channel 3. Table 3 shows corresponding amplitude vs wavelength data.

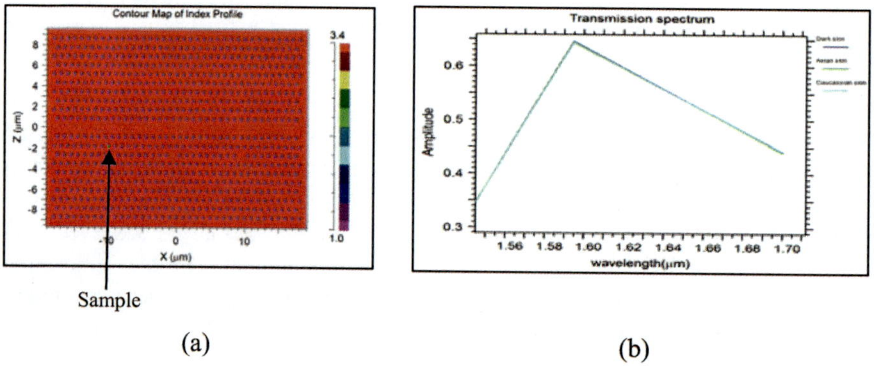

Fig. 6 Results for channel 1 in point defect PhC biosensor **a** RI distribution, **b** 3 RI values

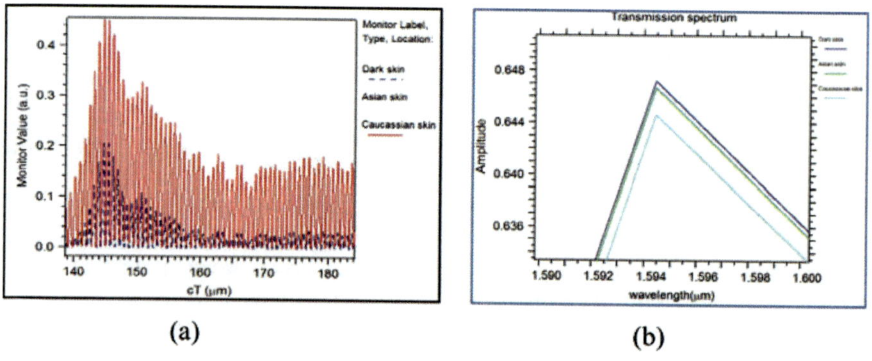

Fig. 7 **a** Monitor output of three different RI values merged, **b** amplitude versus wavelength

Table 1 Amplitude versus wavelength for channel 1

Types of skin	Amplitude	Wavelength (μm)
Dark skin	0.300	1.5922
Asian skin	0.296	1.5923
Caucasian skin	0.295	1.5926

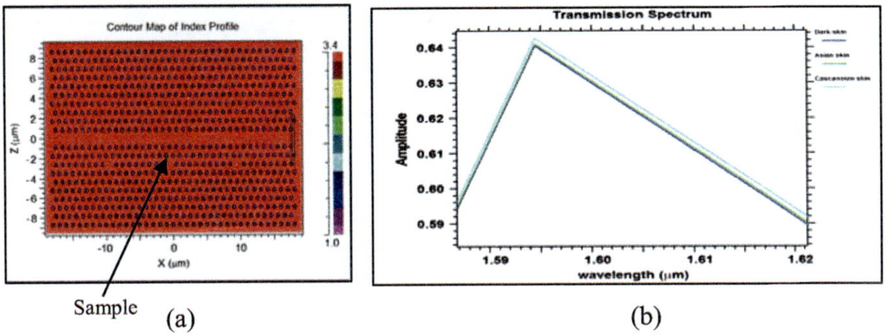

Fig. 8 Results for channel 1 in point defect PhC biosensor **a** RI distribution, **b** 3 RI values

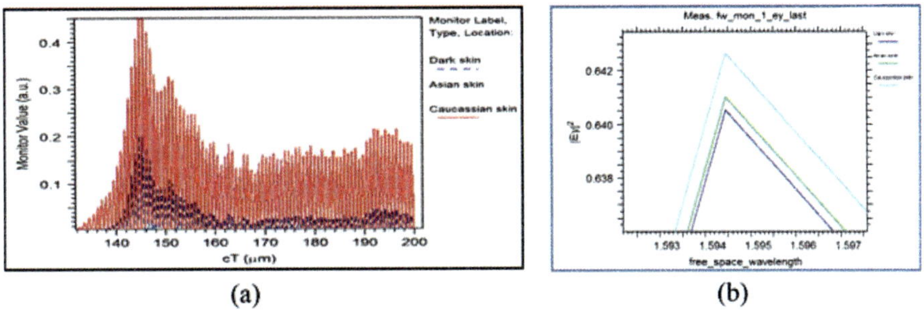

Fig. 9 **a** Monitor output of three different RI values merged, **b** amplitude versus wavelength

Table 2 Amplitude versus wavelength for channel 2

Types of skin	Amplitude	Wavelength (μm)
Dark skin	0.640	1.5943
Asian skin	0.641	1.5944
Caucasian skin	0.643	1.5945

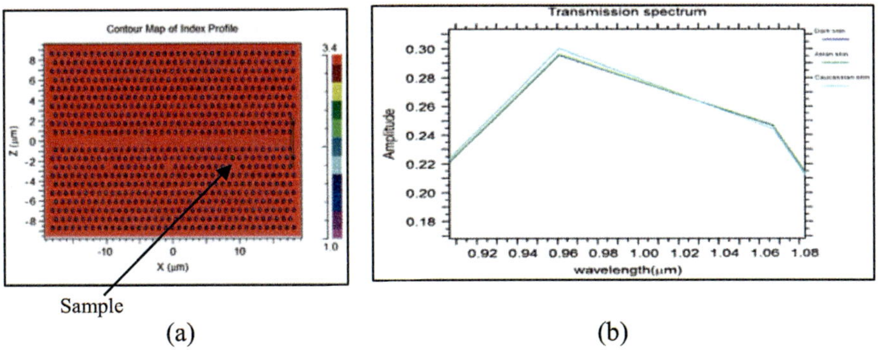

Fig. 10 Results for channel 1 in point defect PhC biosensor **a** RI distribution, **b** 3 RI values values merged **b** amplitude versus wavelength

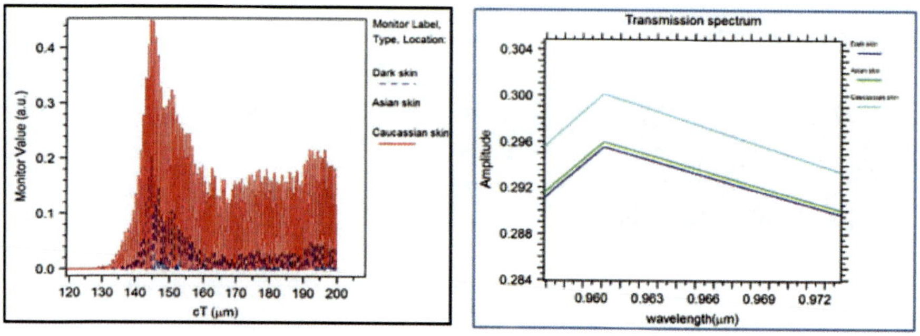

(a) (b)

Fig. 11 **a** Monitor output of three different RI values merged, **b** amplitude versus wavelength

Table 3 Amplitude versus wavelength for channel 3

Types of skin	Amplitude	Wavelength (μm)
Dark skin	0.300	0.960
Asian skin	0.296	0.960
Caucasian skin	0.295	0.960

5 Conclusion

We reported a single cavity and multi-cavity on a single PhC to realize a highly sensitive, compact biosensing system for skin analysis. When the size of the nano-sensitivity structure is 0.22 mm, then all the three skin types exhibit distinct graphs. The moderate shift in wavelength is obtained according to various amplitude levels. From the graph, it is clear that as and when the sensing hole diameter is further decreased to 0.02 nm, then there will be an overlapping of the curves for all the three different types of skin. When further hole size decreases, then we cannot distinguish between Dark and Asian skin as the content of the melanin pigment present is more. We can conclude that by using the nanocavity structure of differing hole sizes, differentiation of types of skin can be achieved. Our results reveal new possibilities in the field of skin type analysis. Multiplexing of the resonant wavelengths using a photonic heterostructure and increasing the detection sensitivity will further extend the sensing applications.

References

1. L.L. Chan, A label-free photonic crystal biosensor imaging method for detection of cancer cell cytotoxicity and proliferation. Apoptosis **12**(6), 1061–1068 (2007)
2. N. Susuma, Recent progresses and future prospects of two and three dimensional photonic crystals. J. Light Wave Technol **24**(1), 4554–4567 (2006)

3. D.J. John, G.J. Steven, N.W. Joshua, D.M. Robert, Photonic crystals molding the flow of light, 2nd edn. (Princeton University press, 2008)
4. F. Vollmer, L. Yang, Label-Free Detection with High Q Microcavities: A Review of Bio sensing Mechanisms For Integrated Devices and Nanophotonics, vol. 1(3–4), pp. 267–291 (2012)
5. Poonam Sharma, Preeta Sharan, Photonic crystal based ring resonator sensor for detection of glucose concentration for biomedical applications. Int. J. Emerg. Technol. Adv. Eng. (IJETAE) **4**(3), 2250–2459 (2014)
6. M. Bilkish, S. Preeta, H. Zakir, Modelling and simulation of pressure sensitivity of Bragg grating sensor for structural health monitoring application. Int. J. Adv. Comput. Res. **1**(2), 73–77 (2014)
7. V.L. Nandhini, K.R. Sandip, K. Suresh Babu, Detection of malignant tissue using metal dielectric interface based plasmonic biosensor. J. IJITEE **8** (6S4) (2019)
8. A. Mishra, A. Basu, V. Tyagi, Silicon Photonics and High Performance Computing, vol. 718 (2018)
9. V.L. Nandhini, K. Suresh Babu, K. Sandip Roy, Photonic crystal based micro interferometer biochip for early stage detection of melanoma. Pertanika J. Sci. Technol. **26** (3), 1505–1512 (2018)
10. S. Irani, Pre-cancerous lesions in the oral and maxillofacial region: a review with special focus on etiopathogenesis. Iran. J. Pathol. **11** (4), 303–322 (2016)
11. https://www.who.int/uv/faq/skincancer
12. S. Poonam, S. Preeta, Design of PC based biosensor for detection of glucose concentration in urine. IEEE Sens. J. **15**(2), 1035–1042 (2015)
13. https://www.cancerresearchuk.org/our-research-by-cancer-type, Cancer Research, UK

LoBAC: A Secure Location-Based Access Control Model for E-Healthcare System

Ashish Singh and Kakali Chatterjee

1 Introduction

The healthcare system is the dynamic interconnected components which lead to health services in different places. It includes a variety of health services and facilities to individuals, families, and communities with the purpose to promote, restore, and maintain their health status. The main objective of the E-healthcare system is to provide healthcare services in a broad range with the help of the Internet. The services are available in a secure shared manner at anytime, anywhere.

These healthcare services are vulnerable from different types of threats and attacks such as unauthorized access to the E-health system, misuse of access privileges, bypassing sensitive information, sensitive data leakage, and unauthorized modification in the patient records. These threats and attacks are possible due to remote locality of the healthcare services and accessed through the Internet without the knowledge of user location. Several security solutions have been proposed for the security of these healthcare services and data. One of the security mechanism presents in the healthcare system is access control [1–3] which controls the access of the healthcare data and resources with the help of security policies and rules. Discretionary, mandatory, role-based, and attribute-based access control models are some traditional access control models (ACM) used in the present healthcare system. But these models do not provide location assurance and dynamically control the access of the user [4]. Due to this, several security issues may be arising such as abuse access to patient files, insecure access devices, easily hacked online medical devices, unauthorized access

A. Singh (✉) · K. Chatterjee
Department of Computer Science & Engineering, National Institute of Technology Patna, Patna (Bihar), India
e-mail: ashish.cse15@nitp.ac.in

K. Chatterjee
e-mail: Kakali@nitp.ac.in

S. Patnaik et al. (eds.), *Advances in Machine Learning and Computational Intelligence*, Algorithms for Intelligent Systems, https://doi.org/10.1007/978-981-15-5243-4_58

to healthcare computers, malware and phishing attacks, and improper disposal of old hardware.

To address the above-mentioned security issues, we have proposed a new secure ACM in which the access is not only based on the identity of the user but also check the location of the requester. While providing the access, integration of the location of the user in the ACM will increase the usability and availability of the system. The achieved strong location assurance may solve the security issues that arise due to abuse access to patient files, insecure access devices, and easily hacked online medical devices.

The work's major findings are the following:

- We have proposed an ACM which uses the location of the user for controlling the access in the healthcare system.
- It also provides secure location assurance.
- The implemented LoBAC solution restricts the access of the user if the system detects unauthorized user location.

The remainder of the paper is arranged as follows: In Sect. 2, we have discussed several exiting works present in this field. In Sect. 3, the proposed LoBAC solution is described. The implementation of the proposed model is discussed in Sect. 4. In the last Sect. 5, the paper is closed with the conclusion.

2 Related Works

In this part of the paper, we have discussed lots of work present in the literature for access security in the healthcare system [5–11]. The literature and analysis of the work give us a direction for the development of our proposed model.

Ardagna et al. [12] proposed access control policies which support the location for controlling the access of the user. They also highlight interoperability and uncertainty issues in the LABC model. They define location-based predicates in terms of position-based, movement-based, and interaction-based.

Xue et al. [13] proposed an ACM in which location is integrated with the access control policy for the security of the cloud storage system. The attributes of that user are used for determining the access privileges. This technique provides more fine-grained and flexible ACM in terms of location awareness and attributes of that user.

Ulltveit-Moe and Oleshchuk et al. [14] proposed an ACM that uses the role and location of the user for mobile security. They implement their model with the help of GeoXACML language. The proposed location-aware RBAC provides location-dependent access control as well as enhance the security of the system.

Baracaldo et al. [15] proposed an access control framework for the security of real-world applications. The proposed Geo-social-RBAC uses the concept of location in the RBAC model while providing access to the users. The model integrates the geographic and social dimensions of the user in the access location for granting access.

Rao et al. [16] describe secure privacy-preserving policies for securing of location-based services. The proposed FakeIt system provides a mechanism that satisfies user security and privacy requirements as well as provides a decision-making context before sharing their data.

Yisa et al. [17] suggested a conceptual ACM for business application systems that secure the data and resources against unauthorized access and data infringement. In this approach, each user is registered with the username and password at a specific location.

Kim et al. [18] proposed an access control technique in heterogeneous wireless networks for mobile services. They used the concept of Naive Bayesian classifier for the estimation of network status. According to network status, access will be controlled.

To maximize the distance of co-channel transmitters in vehicular ad hoc networks, Martin-Vega et al. [19] proposed a distributed location-based access model. The geolocation of the vehicles is used for resource allotment in VANETs.

Sukmana et al. [20] introduced an extended version of location-based access control which includes Internet-based location. For authorization and access control decisions, the Internet-based location converted physical location using the IP geolocation technique. The proof of concept is implemented on a business cloud storage solution.

In 2012, Zhang et al. [21] suggested an authentication and authorization model for mobile transactions. The location of the user through the smartphone is fetched, and the fetched location will be used for authentication and authorization purpose.

3 Proposed LoBAC Model

In this part of the paper, the proposed LoBAC model is presented for the security of the healthcare system. The proposed ACM is based on the location of the user. In this solution, access to the requested data will be changed if the location of the user does not meet with the registered location. *LocationManger* is responsible for fetching the location of the user and verify the process. Figure 1 shows the architecture of the proposed LoBAC model. In this proposed model, healthcare users, authentication server, location verifier, access control module, and healthcare storage server are the basic components.

Our proposed LoBAC model takes the following steps when a user wants to access the data from the healthcare application:

1. First, it is mandatory for each accessing user, registered with the Web-based login interface (healthcare application) in the form of (e-mail, password) for accessing healthcare services and data.

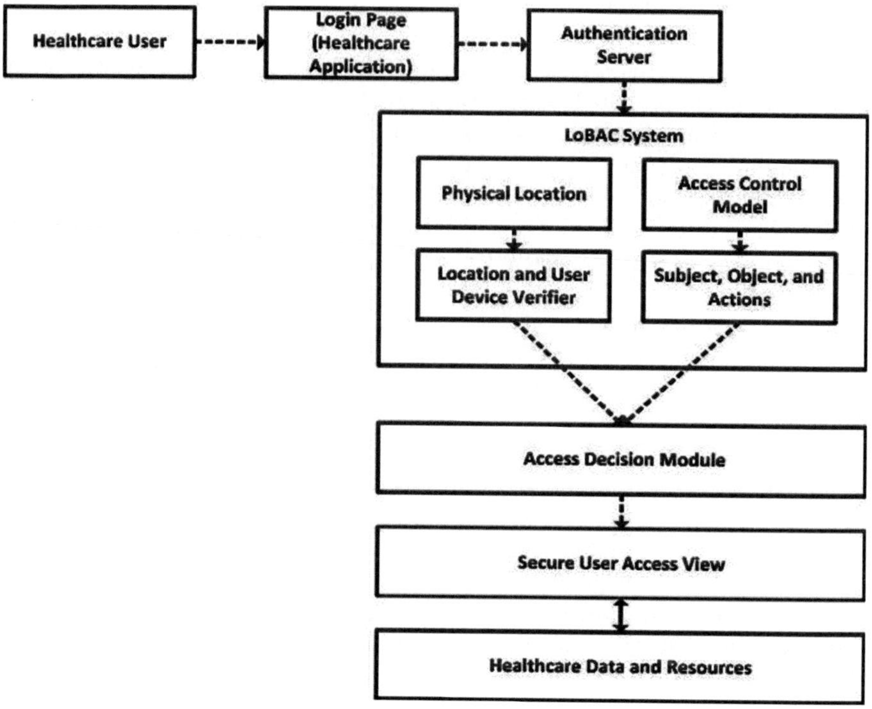

Fig. 1 Proposed LoBAC model for the healthcare system

2. The user enters her/his login credentials in the healthcare application login page for the authentication purpose.
3. If the user credentials are valid and authentication is successful, then the control goes to the LoBAC system for knowing the location of the user.
4. Through GPS, her/his current physical location is fetched.
5. The location and user device verifier, verify the location of the user and ACM subject, object, and action of that user.
6. After the successful completion of the process, the control goes to the access decision module.
7. The access decision module takes access grant decisions and provides healthcare data and services through a secure access interface.

Definition: ACCESS CONTROL RULE In the LoBAC system, an access control rule is expressed by using the following tuples. All these tuples are used for making an access decision.

<HU, HO, HA, HL>

- HU refers to a healthcare user that wants to access the data from the healthcare application.
- HO is healthcare objects (medical data).

- HA is the action taken by a user performed on the object.
- HL is the current location of the user.

Definition: ACCESS CONTROL PREDICATES The access control predicates for taking a decision about an access request is expressed as:

- <user = registered user (doctor/patient) $\wedge$ Object = medical data $\wedge$ action = read/write/append $\wedge$ location = registered location (true) > $\rightarrow$ access = granted
- <user = registered user (doctor/patient) $\wedge$ Object = medical data $\wedge$ action = read/write/append $\wedge$ location = unregistered location (false) > $\rightarrow$ access = denied
- <user = unregistered user (doctor/patient) $\wedge$ Object = medical data $\wedge$ action = read/write/append $\wedge$ location = registered location (true) > $\rightarrow$ access = denied
- <user = unregistered user (doctor/patient) $\wedge$ Object = medical data $\wedge$ action = read/write/append $\wedge$ location = unregistered location (false) > $\rightarrow$ access = denied.

4 Implementation Results

In this part of the paper, we have discussed the implementation and results of the proposed LoBAC model.

For this, first, we have built a prototype based on our proposed model to identify the location of the user location. We have used longitude and latitude from the GPS for the identification of the user's location. The implementation has been done using Android Studio Platform with Java, Android, and XML programming languages. Cloud firebase server is used for storing the user credentials and other information such as prescription, problems, hospital name, doctor name, and patient name. The user log database contains the other user information such as username, user ID, e-mail ID, longitude, latitude, and specialized. The cloud firebase is shown in Fig. 2.

For the implementation purpose, we have registered five doctors and five patients. All the analysis and enforcement of rules are applied to these users. To access the system, it must be required that the user is registered within the system. For securing the system, we have provided different views to each registered user under a specific category. The doctor can see the patient list and appointment including own information from own registered location. The patient can also see diet, health tips, doctor's prescription, doctor's suggestion, and online order from own registered location.

In the implemented system, *LocationManager* will be responsible for providing the user location. Figure 3 shows the code for fetching the current user location in the form of (latitude, longitude). After fetching the location, the location range is verified with the registered location. If both are matched, then the user will be able to access the data.

While the user is registered within the system, the current latitude and longitude are also registered within the system. This location is further used for granting access. The access view of the doctor shows the current patient list, check out the patient list, appointment, etc. The location verifier page of the registered user shows that

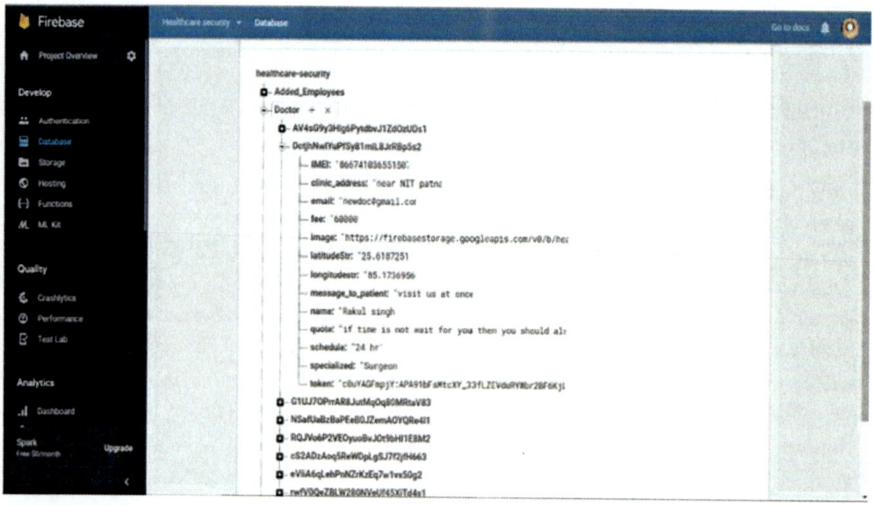

Fig. 2 Cloud firebase

Fig. 3 Location manager and location verifier

the user credentials are verified and the user is valid. But for accessing the data, it is mandatory for user to verify own location. In this page, if the user location is found within range, then the user will be able to access the data; otherwise, the access view is denied.

5 Conclusion and Future Works

In this paper, we have discussed the present E-healthcare system, its advantages, and present security issues. After that, the present access control solutions are also discussed. To overcome the present security issues, a new access control solution named LoBAC has been proposed. The model uses the location of the user for controlling the access of the user. The implementation of the model demonstrated that the proposed model will restrict access if the found location is unregistered. This model will not be feasible in the case of when a legitimate user is outside from the leaving/registered location and he wants to access the data for a different purpose. In the future, we will try to add more constraints while the user accesses the healthcare data. We will also investigate more complex cases and security issues for the E-healthcare system.

References

1. P. Samarati, S.C. de Vimercati, Access control: policies, models, and mechanisms, in *International School on Foundations of Security Analysis and Design* (Springer, Berlin, Heidelberg, 2000), pp. 137–196
2. Ravi S. Sandhu, Pierangela Samarati, Access control: principle and practice. IEEE Commun. Mag. **32**(9), 40–48 (1994)
3. X. Jin, R. Krishnan, R. Sandhu, A unified attribute-based access control model covering DAC, MAC and RBAC, in *IFIP Annual Conference on Data and Applications Security and Privacy* (Springer, Berlin, Heidelberg, 2012), pp. 41–55
4. Ashish Singh, Kakali Chatterjee, Trust based access control model for securing electronic healthcare system. J. Ambient Intell. Humaniz. Comput. **10**(11), 4547–4565 (2019)
5. E.J. Neystadt, D. Alon, D. Rose, E. Levy, Location Based Access Control. U.S. Patent Application 13/114,044, filed, 29 Nov 2012
6. A. Van Cleeff, W. Pieters, R. Wieringa, Benefits of location-based access control: A literature study, in *2010 IEEE/ACM Int'l Conference on Green Computing and Communications & Int'l Conference on Cyber, Physical and Social Computing* (IEEE, 2010), pp. 739–746
7. S.M. Chandran, J.B.D. Joshi, LoT-RBAC: a location and time-based RBAC model, in *International Conference on Web Information Systems Engineering* (Springer, Berlin, Heidelberg, 2005), pp. 361–375
8. R. Bhatti, M.L. Damiani, D.W. Bettis, E. Bertino, Policy mapper: administering location-based access-control policies. IEEE Internet Comput. **12**, 38–45 (2008)
9. M. Decker, Requirements for a location-based access control model, in *Proceedings of the 6th International Conference on Advances in Mobile Computing and Multimedia* (ACM, 2008), pp. 346–349
10. C.A. Ardagna, M. Cremonini, S.D.C di Vimercati, P. Samarati, Access control in location-based services, in *Privacy in Location-Based Applications* (Springer, Berlin, Heidelberg, 2009), pp. 106–126
11. J. Bravo, J.L. Crume, Expected location-Based Access Control. U.S. Patent 10,027,770, issued 17 July 2018
12. C.A. Ardagna, M. Cremonini, E. Damiani, S.D.C. di Vimercati, P. Samarati, Supporting location-based conditions in access control policies, in *Proceedings of the 2006 ACM Symposium on Information, computer and communications security* (ACM, 2006), pp. 212–222
13. Y. Xue, J. Hong, W. Li, K. Xue, P. Hong, LABAC: a location-aware attribute-based access control scheme for cloud storage, in *2016 IEEE Global Communications Conference (GLOBECOM)*, Washington, DC (IEEE, 2016), pp. 1–6
14. N. Ulltveit-Moe, V. Oleshchuk, Enforcing mobile security with location-aware role-based access control. Security Comm. Netw. **9**, 429–439 (2016). https://doi.org/10.1002/sec.879
15. N. Baracaldo, B. Palanisamy, J. Joshi, Geo-social-RBAC: a location-based socially aware access control framework, in *Network and System Security. NSS 2015*. Lecture Notes in

Computer Science, vol. 8792, ed. by M.H. Au, B. Carminati, C.C.J. Kuo (Springer, Cham, 2016), pp. 501–509
16. J. Rao, R. Pattewar, R. Chhallani, A privacy-preserving approach to secure location-based data, in *Intelligent Computing and Information and Communication. Advances in Intelligent Systems and Computing*, vol. 673, ed. by S. Bhalla, V. Bhateja, A. Chandavale, A. Hiwale, S. Satapathy (Springer, Singapore, 2018), pp. 48–55
17. Victor L. Yisa, Baba Meshach, Oluwafemi Osho, Anthony Sule, Application of geo-location-based access control in an enterprise environment. Int. J. Comput. Netw. Inf. Security **10**(1), 36–43 (2018)
18. D.-Y. Kim, D. Ko, S. Kim, Network access control for location-based mobile services in heterogeneous wireless networks. Mobile Inf. Syst. 1–11 (2017)
19. F.J. Martin-Vega, B. Soret, M.C. Aguayo-Torres, G. Gomez, I.Z. Kovacs, Analytical modeling of distributed location based access for vehicular ad hoc networks, in *2017 IEEE 85th Vehicular Technology Conference (VTC Spring)* (IEEE, 2017), pp. 1–5
20. M.I.H. Sukmana, K.A. Torkura, H. Graupner, A. Chauhan, F. Cheng, C. Meinel, Supporting internet-based location for location-based access control in enterprise cloud storage solution, in *International Conference on Advanced Information Networking and Applications* (Springer, Cham, 2019), pp. 1240–1253
21. F. Zhang, A. Kondoro, S. Muftic, Location-based authentication and authorization using smart phones, in *2012 IEEE 11th International Conference on Trust, Security and Privacy in Computing and Communications* (IEEE, 2012), pp. 1285–1292

Autonomous Electronic Guiding Stick Using IoT for the Visually Challenged

R. Krithiga and S. C. Prasanna

1 Introduction

Fundamentally, engineering focuses on the upliftment of the quality of well-being of people. There are lots of technological inventions and innovations which help in the well-being of physically and mentally challenged people in their day-to-day activities. Visually challenged people have the inability of visual perception due to various genetic, biological and neurological factors. Navigating from one place to other is one of the basic elementary actions performed in man's life. Visually challenged people face so many problems in navigation, and they are often dependent on traditional walking sticks, guide dogs or guides. It is a challenge for these people to navigate independently on a day-to-day basis. Situations are even worse in unfamiliar environments. Blind people do not know the surrounding information data regarding obstacles, potholes or other dangers and have very little knowledge regarding landmarks in a particular region.

With increasing human–machine interaction, there are lots of devices which help in navigation of the visually challenged in indoor and outdoor environments. The most preferred guiding solution by the blind is the walking cane. Guides and guide dogs can turn out to be an expensive option.

Therefore, we aim to design and implement a low-cost, real-time navigation system which is the guiding electronic stick that assists these visually challenged people in their navigation by audio instructions and can autonomously move which resembles the traditional walking cane. The main objectives of this work are:

1. To create a low-cost autonomous guiding system controlled by Raspberry Pi.
2. To create a smartphone app that receives input and sends navigation instructions to the user

R. Krithiga (✉) · S. C. Prasanna
Department of Electronics and Instrumentation Engineering, SRM Valliammai Engineering College, Kattankulathur, Chennai, India
e-mail: priyakrithi99@gmail.com

S. Patnaik et al. (eds.), *Advances in Machine Learning and Computational Intelligence*, Algorithms for Intelligent Systems, https://doi.org/10.1007/978-981-15-5243-4_59

3. To interface sensors that collect environmental information to Raspberry Pi.
4. To alert the user regarding forthcoming vehicles or obstacles in outdoor navigation.

The work has inspirations from various technological advancements for the upliftment of the visually challenged, and the remaining paper is structured as follows: Sect. 2 elaborates on the literature survey in a detailed manner, Sect. 3 discusses the research gap found in the base papers, and Sect. 4 explains the proposed methodology. Section 5 is about conclusions and future research work.

2 Literature Survey

The paper Krishnan et al. [1] proposes the idea of "Assistor" which is a smart walking stick that enables people with visual impairment to identify the impediments and assists them to reach their destination. The device works on the principles of echolocation. The device uses ultrasonic sensors to detect obstacles, and an image sensor is used to find and identify the real-time objects in front of the user. For navigation, a smartphone app is used using GPS and maps. The search algorithm implemented in this device is iterative deepening depth-first search.

The paper by Ganz et al. [2] presents an indoor navigation system for the visually challenged using an android smartphone that runs the Percept-II application. In this application, near-field communication (NFC) tags are deployed on specific landmarks on the environment. The users obtain audio navigation instructions from their phone when they touch these NFC tags. The building is denoted as a node–link structure. The tool then generates navigation routes based on the inputs and stores all the landmark information and navigation routes in a server database. The node–link structure has nodes that represent the landmarks, and the links represent the physical connection between the nodes. The latitude and the longitude of the landmark are stored in the database. The weights on each link denote the degree of toughness to travel via each link. The route with least weight is generated using the Dijkstra algorithm. The application was tested with sighted users to choose known locations such as entrance and restroom.

The paper by Kaminski et al. [3] proposes the idea of a prototype application that supports the street navigation. This system makes use of GIS database of geometric network of pedestrian paths in a city. It is capable of finding way from a specific source to destination. The prototype utilizes a spatial database of the pedestrian path network of the city. The system obtains the information from the user, and the direction towards the user is moving through the GPS receiver and the gyrocompass. The communication is carried by means of wireless keyboard and voice messages which regenerated by the voice synthesizer. The system consists of five main modules including the spatial database, system kernel, GPS unit, compass unit and user interface. The spatial database stores the geometric and logical network data of the pedestrian paths and the information about the user surroundings. The system kernel implements algorithms for path finding. The whole function is carried within the mobile device, and no Internet connection is necessary. The prototype of this system was developed and tested by using a notebook as a hardware platform.

The paper by Harsur et al. [4] has developed a navigation system that uses audio support to provide instructions for navigation to the user. Sphinx and Google API are used to convert the speech into text, and eSpeak software is used to convert text to speech. The navigation procedure is carried out through Raspberry Pi. This project uses embedded C to obtain the GPS data and Python to measure the distance of objects and also for obstacle detection. The text-to-speech synthesis has been implemented for English and Hindi languages and longitude information. For obstacle detection, ultrasonic sensors are used.

The paper by Owayjan et al. [5] focuses on low-cost navigation for the visually challenged. The prototype consists of an interface between Microsoft Kinect and MATLAB. An app which ensures a connection between the user and their respective guide is also developed. Microsoft Kinect is the primary sensor, and the laptop is the processing unit. Arduino Mega which accepts serial interface activates the mobile application. A portable battery supplies power to the system, and a vest holds all the components. The Arduino activates wireless communication between the mobile application and the MATLAB software in the laptop. This project was tested in three different locations based on navigation safety on daily basis. The results were distinguished on the basis of collisions, travel time and error. The total error percentage obtained in the test was of 15.6%.

This paper by Yang et al. [8] proposed an auditory navigation system and conducted an experiment which concluded significant results. The main objectives of this study were as follows: one was to verify the effect of the information detail on the navigation performance, and the other was to verify the effect of broadcasting distance. There were two independent variables considered which were completeness of information and broadcasting time. Completeness of information was divided onto two levels—complete and simple. Broadcasting time consisted of two levels—5 m and 7 m. A set of six visually impaired people participated in an experiment conducted with this system. The participants performed a way-finding test form one point to other where they were required to listen navigation information. If the instruction was missed, it is recorded as missed routes. Finally, a questionnaire was recorded. The results showed that broadcasting time significantly affected the number of information requests. Walking time, workload and missed routes were also influenced by it.

The paper by Yang et al. [9] proposes a system that integrates the concepts of communication satellite location and cell location. This integrated navigation system can provide continuous navigation service and also controls the influence of measurement outliers. The method is used to determine the location of users using distinct location model. To determine the integrated model, the "robust least square" algorithm has been implemented. The results clearly explain that filtrating estimation and accuracy can be improved by this algorithm. It can also control the influence of disturbances in the dynamic model.

3 Research Gap

The paper by Krishnan et al. [1] proposes the idea of Assistor which helps visually impaired people to detect obstacles and helps them to reach their destination. For navigation, a smartphone app is used using GPS and maps. The input ultrasonic and image sensors send information to the smartphone via Bluetooth. Servo motors are used for the mobility of the stick. A smartphone app performs all computation and calculations with the data given by the input sensors.

The paper by Anushree and Chitra [4] has developed a navigation system that uses audio support to provide instructions to the user. The navigation procedure is carried out through Raspberry Pi. The destination address is obtained using a microphone which is connected to Raspberry Pi.

This project what we develop is a combination of these two ideas which highlight the concept of autonomous walking stick implemented with Raspberry Pi. The transmission of data takes place wirelessly implying the concept of Internet of Things (IoT).

4 Proposed Methodology

The user holds on to the electronic stick in the same way as the traditional white walking cane using a strap. The system initially gets the input destination from the user via speech through a microphone. This speech information is converted into relevant text using Google API or Microsoft Azure or IBM Watson. The text information is acquired by an application in the user's smartphone in the system. The destination address is located with the help of Google Maps, and the navigation is started. As per the route, the user is directed to go left or right or to keep walking straight through audio instructions which are reflected on the headphones.

The Raspberry Pi receives information from the smartphone wirelessly via Internet in the mobile. As per the navigation route, the wheels in the underlying chassis of the stick move in the respective direction. These wheels are powered by motors that rotate according to the control signal received from Raspberry Pi. There are ultrasonic sensors in the four directions of the chassis (front, both sides and downwards) to indicate nearby obstacles, and the sensor in the bottom is used to indicate potholes. Pi cam, which is an image sensor, is mounted on top of the stick which captures the environment at equal intervals of time and sends to the smartphone app as navigation history. Any obstacles or heavy vehicles or unknown location captured by the image sensor alerts the user by suitable voice commands. Thus, autonomous function of the electronic guiding stick is achieved. The autonomous chassis movement along with navigation instructions communicated via IoT makes this project distinct from all other projects discussed above.

The simplified architecture of the system is shown in Fig. 1. The detailed component positioning and placement are shown in Fig. 2. This is the representation of the final prototype.

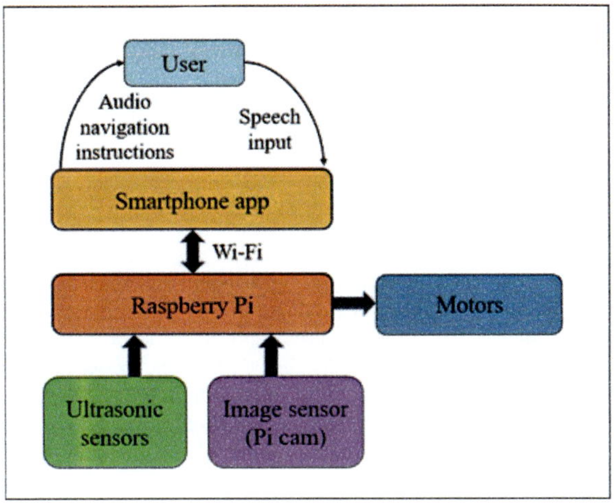

Fig. 1 System architecture

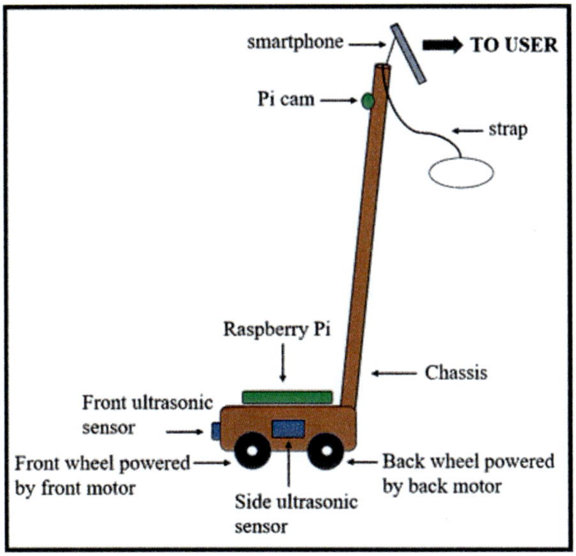

Fig. 2 Walking stick prototype representation

5 Conclusion and Future Work

The ground research work and the design of this project have been successfully completed. The construction of the stick and software development is already in progress. The project is estimated to be completed within a span of five months. We hope that this project motivates the visually challenged to navigate independently with self-confidence and pride.

References

1. A. Krishnan, G. Deepak raj, N. Nishanth, K.M. Anand Kumar, Autonomous walking stick for the blind using echolocation and image processing, in *2016 2nd International Conference on Contemporary Computing and Informatics* (ic3i)
2. A. Ganz, J.M. Schafer, Y. Tao, C. Wilson, M. Robertson, PERCEPT-II: smartphone based indoor navigation system for the blind, in *2014 36th Annual International Conference of the IEEE Engineering in Medicine and Biology Society*
3. L. Kaminski, R. Kowalik, Z. Lubniewski, A. Stepnowski, VOICE MAPS—portable, dedicated GIS for supporting the street navigation and self-dependent movement of the blind, in *Proceedings of the 2nd International Conference on Information Technology, ICIT 2010*, 28–30 June 2010, Gdansk, Poland
4. H. Anushree, M. Chitra, Voice based navigation system for blind people using ultrasonic sensor. Int. J. Recent Innov. Trends Comput. Commun. ISSN: 2321-8169 **3** (6)
5. M. Owayjan, A. Hayek, H. Nassrallah, M. Eldor, Smart assistive navigation systems for the blind and visually impaired individuals, in *2015 International Conference on Advances in Biomedical Engineering (ICABME)*
6. J. Ma, J. Zheng, High precision blind navigation system based on haptic and spatial cognition, in *2017 2nd International Conference on Image, Vision and Computing*
7. S. Kantawong, Road traffic signs detection and classification for blind man navigation system. Int. Conference on Control, Autom. Syst. 17–20 Oct 2007 in COEX, Seoul, Korea (2007)
8. C.-H. Yang, J. Wang, S.-L. Hwang, The design and evaluation of an auditory navigation system for blind and visually impaired, in *Proceedings of the 2014 IEEE 18th International Conference on Computer Supported Cooperative Work in Design*
9. X.Y. Yang, H. He, S. Wang, Communication satellite location/cell location Integrated navigation system for all blind war, in *2011 International Conference on Electric Information and Control Engineering*
10. R. Tapu, B. Mocanu, T. Zaharia, A computer vision system that ensure the autonomous navigation of blind people, in The 4th IEEE International Conference on E-Health and Bioengineering—EHB (2013)

Handling Heterogeneity in an IoT Infrastructure

B. M. Rashma, Suchitha Macherla, Achintya Jaiswal, and G. Poornima

1 Introduction

Smart-connected homes are one of the most useful advantages of IoT. They help users to control all the devices in their home with very less help from the users. But one of the most important challenges in building a smart-connected home is the use and control of multiple heterogeneous devices connected through Internet. Heterogeneity among devices can be based on the communication protocols they use to communicate. However, heterogeneity can also be based on the control commands used by various devices to control them. These control commands are basically the variables defined by the manufacturer and the devices are programmed using these variables, thus, one who wants to control the device need to change the values of these control commands.

Heterogeneity among the control commands of devices force the users to use different applications to control different devices. For example, in a room, we have all Philips Hue Lights which are controlled by the app provided by Philips, but if we replace one Philips Hue Light with another smart light such as Syska Smart Light, then the user needs to use another app provided by Syska to control this light alone. Thus, the user needs to reprogram the sensor every time he adds a new kind of light to the network or needs to use lights provided by one vendor only. Solution to the above problem is provided in this work in the form of prototype, which is exemplified on Contiki operating system using Cooja simulator.

B. M. Rashma (✉)
Department of CSE, PESU RR Campus, BMS College of Engineering, Bangalore, India
e-mail: rashmabm@pes.edu

S. Macherla · A. Jaiswal
PESIT Bangalore South Campus, Bangalore, India

G. Poornima
BMS College of Engineering, Bangalore, India
e-mail: gpoornima@bmsce.ac.in

S. Patnaik et al. (eds.), *Advances in Machine Learning and Computational Intelligence*, Algorithms for Intelligent Systems, https://doi.org/10.1007/978-981-15-5243-4_60

2 Literature Survey

IP-based sensor network solution is used in automating cargo container with accessible logistic processes. They emphasize the use of the CoAP protocol for the retrieval of sensor data during data transfer in any medium [1]. To address the challenge of heterogenity, the IEEE paper SPOT [2], provides a resolution by means of database to preserve the format of semantics of the. SPOT is reminiscent of integrated system as it consists of several open-device driver models scripted using XML. This is an approach by which user-driven device driver implementation and cohesive appliance control can be achieved.

An access-control protocol is premeditated for IoT to secure data collected by smart devices in the network. Reference [3] Illustrates a protocol, deployed into Contiki OS and evaluated using the power trace and few other tools. In article [4], they have discussed about the model of heterogeneous IoT data streams in order to prevail over the challenges of heterogeneity. Purpose of the VITAL platform is to facilitate rapid development of cross-platform and cross-context IoT-based applications for smart cities.

CoAP framework Californium provides a scalable and RESTful API to handle IoT devices. Thus, it helps us to form an open, interoperable, scalable, reliable, and low power WSN stack using these open tools and technologies [5]. Nathaniel et al. have a supporting system for heterogeneous IoT devices to avoid security vulnerability and the device compatibilities issues [6].

Bedhief et al. have proposed solution to manage heterogeneity by utilization of Dockers implemented on devices [7]. This work focuses on connecting heterogeneous devices beyond heterogeneous networks. Implementation results prove the feasibility of architecture with the use of a centralized SDN controller. On the other hand, [8] discusses another similar architecture to manage heterogeneous IoT devices in a home network.

Raj Jain et al. [9] have surveyed several standards by IEEE, IETF, and ITU that enable technologies enabling the rapid growth of IoT. These standards comprises communications, routing, network and session layer protocols that are being developed to meet IoT requirements.

Luca et al. [11] have proposed an IoT aware, smart architecture for automatic monitoring, and tracking of patients. Smart hospital system (SHS) relies on diverse, yet complementary, technologies, specifically RFID, WSN, and smart mobile, interoperating with each other through a CoAP/6LoWPAN/REST network infrastructure.

Sarkar et al. [12] proposed a distributed architecture for things (DIAT). It explicitly addresses scalability and heterogeneity of IoT devices, coupled with cognitive capabilities. Qin et al. [13] explained about deployments of IoT sub-networks. This system is capable of managing geographically distributed and heterogeneous networking infrastructures in dynamic environments. For the optimal use of IoT network, authors have used genetic algorithms. Blackstock and Lea [14] proposed IoT 'hubs' to aggregate things using Web protocols. They have introduced the HyperCat IoT catalog specification. This work explores the use of integration tools, to map existing IoT systems to diverse data sources.

Jin et al. [15] present a framework for the realization of smart cities through the Internet of things (IoT). The framework includes complete urban information system. They have designed a three-level approach, sensors at one level, networking support structure at next level, and finally data management and cloud-based integration of respective systems and services.

Cooja simulator [10] is a network simulator particularly designed for wireless sensor networks. Cooja simulator includes Contiki Mote. The system is controlled and analyzed by Cooja. Several different Contiki libraries can be compiled and loaded in the same Cooja simulation, representing diverse sensor nodes (heterogeneous networks).

3 Proposed Work

Problem statement: This work makes use of Database Node which has a record of the control commands for each distinct device in the network. Database Node can be a router or any edge device that is the first node to be deployed. When subsequent new device is installed in the network, it first registers itself to the Database Node by sending its ServiceID (which is a unique id given to distinct devices). Using control commands, it first normalizes and then sends the message. This normalized message is then denormalized by the receiver for further analysis. This work is a simulation model using Contiki OS and Cooja simulator.

Objective: Bring about hazel free communication among heterogeneous devices with same functionality (Philips bulb and Syska bulb) and heterogeneous devices with different functionality.

4 Implementation

For better illustration of hazel free communication of heterogeneous devices implementation is carried out in two scenarios, where scenario1 considers heterogeneous but similar kind of devices, like Syska smart bulb and Philips smart bulb. The other scenario considers two devices which are heterogeneous and dissimilar devices like bulbs and fans communicating with TV.

4.1 Database Node

The Database Node is expected to be the first node to be initialized in the network, because only the nodes (i.e. the sensors and smart devices) that are initialized after this node can register themselves to the database. Functionality of this node is to store the control commands of different smart devices in the network. In the format, "0|DeviceType|ServiceID|ControlCommands#".

If the first token after unpacking is "0", it states that a new device wants to register itself to the network. If the first token is equal to DATABASE_SERVICE_ID, it states that the source of the message wants to search for devices belonging to a "DeviceType".

4.2 Device Registration

When a smart device joins the network, to register itself to other nodes in the network, in the format: "0|DeviceType|ServiceID|ControlCommands#" then sends this message as a broadcast message (using UDP/IP), which is then received by the Database Node where the ControlCommands of the Device along with the ServiceID and the DeviceType are added to the database completing the registration process of the device.

4.3 Source and Destination Nodes (Normalizing and Denormalization)

For every 100 s source node senses the parameter required and communicates the sensed data to all destination nodes. Source node functions as follows, along with their ControlCommands, the message is normalized to retrieve ServiceID's and ControlCommands of all different destination nodes in the network. Using these service ID's, control commands and decided state of the destination node source node can take action. For each different destination node in the network, a normalized message is formed in the form, "ServiceID|variable1:value|variable2:value|…#". This message is then sent as a broadcast message (using UDP/IP) to all the nodes in the network where it is denormalized and the values are set to the respective variables to change the state of the device at the receiving end.

4.4 Scenarios Considered

For both the scenarios considered, Database Node (Node 1) should be within the radio medium range of all the devices. Database Node, Temperature and Humidity Sensor Node, Samsung AC Node, Crompton Greaves Fan Node, Havells Fan Node, LG AC Node, Light Sensor Node, Philips Bulb Node, Syska Bulb Node, Syska Bulb Node in scenario1. In scenario 2, additional to all the nodes in scenario 1 Sony TV is added. In scenario 1, the Temperature and Humidity Sensor (Node 2) and Light Sensor (Node 7) act as the source nodes. In scenario 2, Crompton Greaves Fan and Syska Bulb act as both Source Node and Destination Node.

5 Results and Analysis

Once all the Nodes in the network are placed, the simulation is started. On starting of simulation, all the nodes in the network are initialized, i.e., all the nodes are assigned a unique global IP Address.

Once the Database Node is initialized all the other nodes have to register themselves to the Database Node by sending their ServiceID, DeviceType, and ControlCommands as a broadcast message, which can be received by the Database Node where it adds the records to the database if they are not present already, as shown in Fig. 1.

On completion of registration of the devices, we witness that for every 100 s, the light sensor senses the luminosity value of the room, based on which the state and brightness of the bulbs is set which is then sent to all the bulbs in the network and the state and brightness of the bulbs (both Philips and Syska Bulbs) changes, as shown in Fig. 2. Similarly, the temperature and humidity sensed by the Temperature and Humidity Sensor (random numbers are generated for simulation) decides the fan speed, fan state, AC state, AC fan speed, and AC temperature.

Adding a new device, the Database Node checks if the device of same type (i.e., same ServiceID) is already present in the network or not, if fount it does not add a

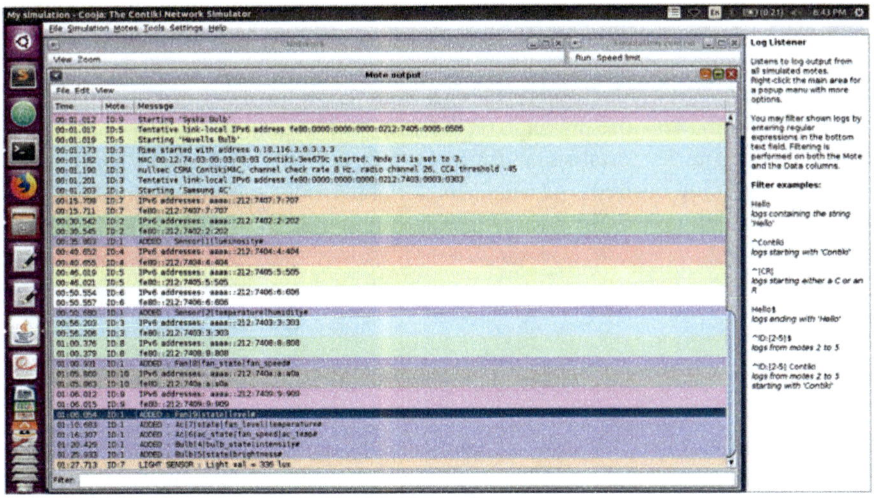

Fig. 1 Initialization of node in the network

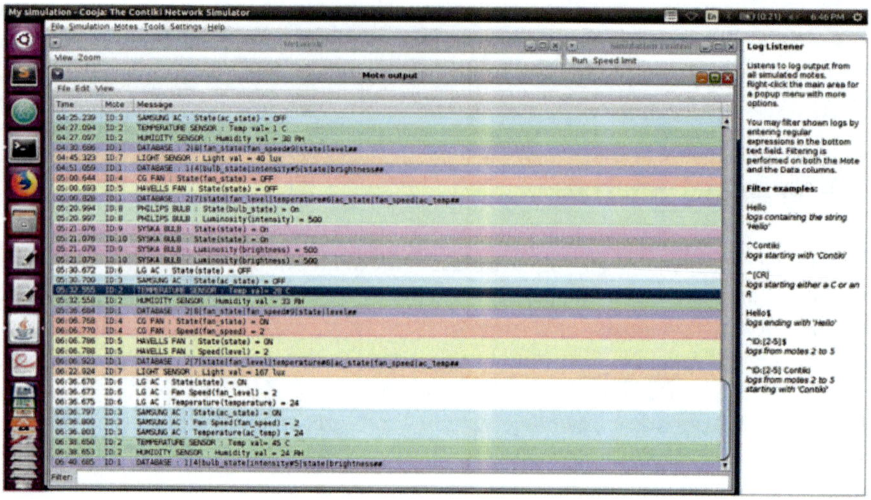

Fig. 2 Communication between sensors and smart devices

record to the database, since it already has the ControlCommands and to eliminate redundancy.

In scenario 2, after the bulb's, fans and AC's in the network update their state based on the control commands sent by sensors, the Syska Bulb and Crompton Greaves Fan then individually request the Database Node for the control commands of Sony TV, upon receiving which they normalize a message with their respective states and send a broadcast message. Registered nodes on retiraval of data from source, denormalises accordingly. for instance, when Sony receives messages it undertakes the process of denormalisation by considering fan's state control command sent by source node in order to update it's state.

This scenario shows that a node can act both as a source and destination node, where it de-normalizes the message received from one node for commands which are used to update its variables and also normalizes a message using the control commands of some desired nodes and its variables values and sends messages to other nodes which update their variable values based on the variable values of this node.

5.1 Analysis

Table 1 tabulation values that show evidently that the fan speed changes based on data sensed by temperature and humidity sensors. It is observed that when temperature is slightly higher and humidity is less, fan speeds up moderately. Fan reaches its highest speed when both temperature and humidity is high. Figure 3 is the graphical representation for the same readings.

Table 1 Auto increase of fan speed based on temperature and humidity sensor

Temperature in celsius	Humidity	Fan speed
20	<40	0
	40–60	0
	>60	0
23	<40	0
	40–60	0
	>60	1
26	<40	0
	40–60	1
	>60	3
>28	<40	4
	40–60	5
	>60	5

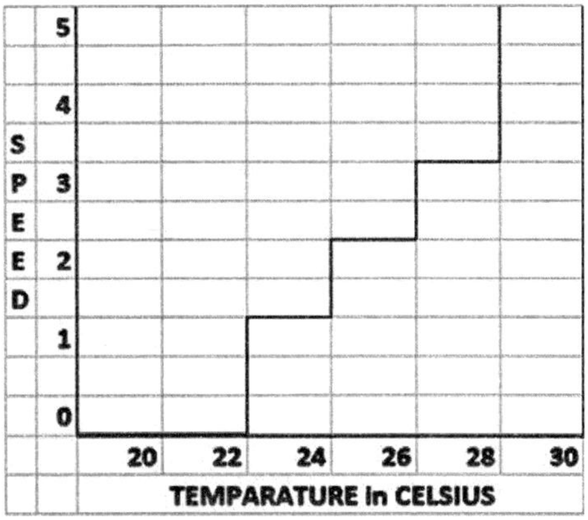

Fig. 3 Graphical representation of fan speed based on data in Table 1

Bulb senses the luminosity of the environment. Brightness of bulb is inversely proportional to the brightness of the room. This work measures luminosity flux in lumen (lm). Maximum luminosity is considered as 500 lm. Once the luminosity falls beyond 500 lm, bulb starts glowing. Consider an instance where luminosity of natural light falls to 300 lm, bulb starts glowing at 200 lm to make it up to 500 lm. This is graphically represented in Fig. 4.

6 Conclusions and Future Enhancement

Here, we proposed a way to handle heterogeneity among devices based on the control commands used by various devices to control them. Our solution makes use of a Database Node which is a repository of the control commands of different devices

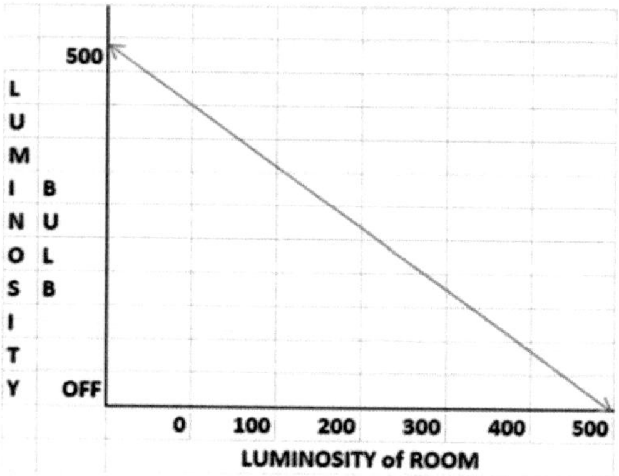

Fig. 4 Bulb luminosity variation based on luminosity of room

in the network. Our solution also makes it easy for the user to add a number of new devices to the network. One of the limitations of this solution is that the Database Node should be the first node to be initialized in the network.

Any device wanting to communicate with a bunch of devices having the same use (such as bulbs or fans) can request the database for the control commands of these heterogeneous devices in the network, using which they can communicate with the bunch of devices easily.

References

1. K. Kuladinithi, O. Bergmann, T. Pötsch, M. Becker, C. Görg, Implementation of CoAP and its application in transport logistics. Res. Gate, 3 May 2014. https://www.researchgate.net/publication/229057545
2. M.-M. Moazzami, G. Xing, D. Mashima, W.-P. Chen, U. Herberg, SPOT: a smartphone-based platform to tackle heterogeneity in smart-home IoT Systems, in *2016 IEEE 3rd World Forum on Internet of Things*, Accession Number: 16666885, 09 Feb 2017
3. X. Wu, R. Steinfeld, J. Liu, C. Rudolph, An Implementation of Access Control Protocol for IoT Home Scenario. 978-1-5090-5507-4/17 IEEE ICIS 2017, 24–26 May 2017
4. A. Kazmi, Z. Jan, A. Zappa, M. Serrano, Overcoming the heterogeneity in the internet of things for smart cities, in *International Workshop on Interoperability and Open-Source Solutions, Springer, InterOSS-IoT 2016: Interoperability and Open-Source Solutions for the Internet of Things*, pp. 20–35
5. S. Thombre, R.U. Islam, K. Andersson, M.S. Hossain, IP based wireless sensor networks: performance analysis using simulations and experiments. J. Wireless Mobile Netw. Ubiquitous Comput. Dependable Appl. **7** (3) (Sept 2016), pp. 53–76
6. N. Gyory, M. Chuah, IoT one: integrated platform for heterogeneous IoT devices, in *2017 International Conference on Computing, Networking and Communications (ICNC)*, Accession Number: 16725955, IEEE Xplore, 13 Mar 2017
7. I. Bedhief, M. Kassar, T. Aguili, SDN-based architecture challenging the IoT heterogeneity, in *2016 3rd Smart Cloud Networks & Systems (SCNS)*, Accession Number: 16724298, IEEE Xplore, 06 Mar 2017

8. C. Pham, Y. Lim, Y. Tan, Management architecture for heterogeneous IoT devices in home network, in *2016 IEEE 5th Global Conference on Consumer Electronics*, Accession Number: 16560555 IEEE Xplore: 29 Dec 2016
9. T. Salman, R. Jain, A survey of protocols and standards for internet of things, eprint Adv. Comput. Commun. **1** (1) ()2017
10. B. Sobhan babu, P. Lakshmi Padmaja, T. Ramanjaneyulu, I. Lakshmi Narayana, K. Srikanth, Role of COOJA simulator in IoT. Int. J. Emerg. Trends Technol. Comput. Sci. (IJETTCS) **6** (2) Mar–Apr 2017 ISSN 2278-6856
11. L. Catarinucci, D. De Donno, L. Mainetti, L. Palano, L. Patrono, M.L. Stefanizzi, L. Tarricone, An IoT-aware architecture for smart healthcare systems. IEEE Internet of Things J. **2**(6), 515–526 (2016)
12. C. Sarkar, S.N. Akshay Uttama, R.V. Abdur Rahim, R. Neisse, G. Baldini, DIAT: a scalable distributed architecture for IoT. IEEE Internet of Things J. **2** (3), 230–239 (2015)
13. Z. Qin, G. Denker§, C. Giannelli, P. Bellavista, N. Venkatasubramanian, A software defined networking architecture for the internet-of-things, in *IEEE Network Operations and Management Symposium (NOMS)*, pp. 1–9 (2014)
14. M. Blackstock, R. Lea, IoT interoperability: a hub-based approach, in *IEEE Internet of Things (IOT), International Conference*, pp. 79–84 (2014)
15. J. Jin, J. Gubbi, S. Marusic, M. Palaniswami, An information framework of creating a smart city through internet of things. IEEE Internet of Things J. **1** (2), 112–121 (2014)

Recent Trends in Internet of Medical Things: A Review

Ananya Bajaj, Meghna Bhatnagar, and Anamika Chauhan

1 Introduction

The Internet of things, or IoT as it is called, has been a buzzword in technology for a few years now. Along with more nascent technologies such as block chain, quantum computing, deep learning, and so on, it is finding its way into mainstream application domains. Internet of things (IoT) is a network of interconnected things; the "things" could be devices such as sensors and wearables, gateways, humans, and the Cloud, all linked together via the Internet. It involves real-time transmission of data, its aggregation, and analysis on that data.

IoT has a multitude of applications, with healthcare being a fast-growing sector with a very high potential. Out of all application domains of IoT, the healthcare sector is projected to grow the fastest, [1] at a Compound Annual Growth Rate (CAGR) of 18.9% and be valued at over USD 135 billion by 2025 [2] (Fig. 1). IoT will revolutionize the healthcare industry, by providing diagnostics at the click of a button from the comfort of one's homes. It provides several benefits over traditional healthcare systems, such as remote monitoring of patients, greater insight into medical data, management and tracking of patients, staff and inventory, reduced errors, and pre-emptive detection of diseases.

A. Bajaj (✉) · M. Bhatnagar · A. Chauhan
Delhi Technological University, New Delhi 600042, India
e-mail: ask4ananya@gmail.com

M. Bhatnagar
e-mail: meghna_bt2k16@dtu.ac.in

A. Chauhan
e-mail: anamika@dce.ac.in

S. Patnaik et al. (eds.), *Advances in Machine Learning and Computational Intelligence*, Algorithms for Intelligent Systems, https://doi.org/10.1007/978-981-15-5243-4_61

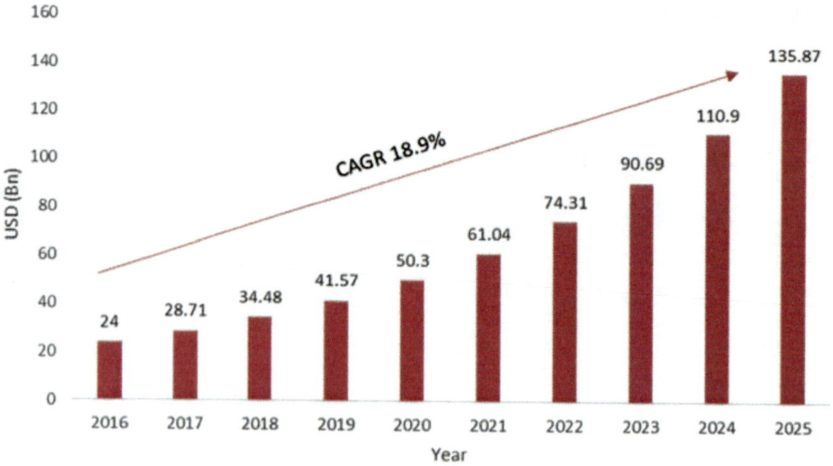

Fig. 1 Market size of IoT in healthcare (worldwide)

1.1 Use Cases

Following are the different use cases for IoT in healthcare (Fig. 2):

a. **Patients**: By using sensors and wearables, their data is transmitted to the doctors for monitoring their status.
b. **Physicians**: Doctors have access to patient data at the click of a button and can monitor, track, and give their analysis without physically meeting the patient.
c. **Hospitals**: Hospitals can keep track of their staff, equipment, inventory, drugs, and other devices. It can also be used for monitoring physical parameters such as temperature and humidity.

Fig. 2 Four different use cases of IoT in healthcare

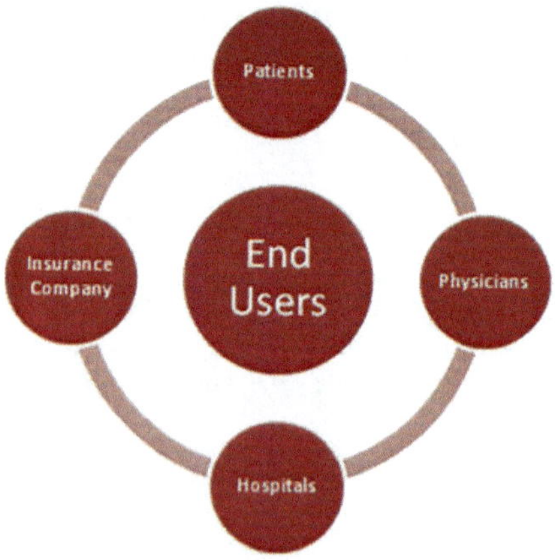

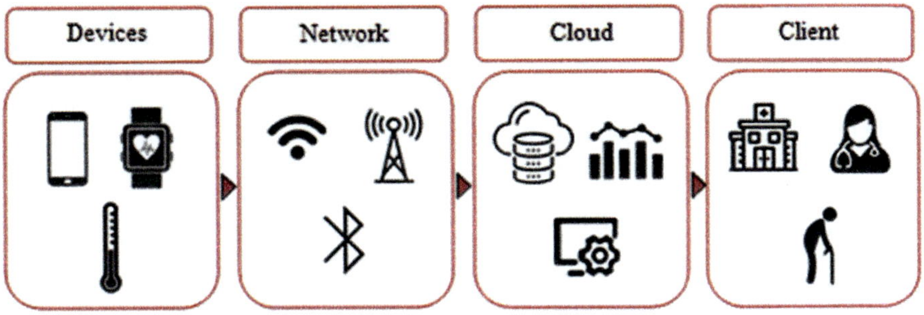

Fig. 3 General system architecture of IoMT

d. **Insurance Companies**: By using data through IoT-enabled healthcare devices, transparency will be ensured between customers and insurance companies, ease the process of risk assessment, and reduce the cases of fraud.

1.2 Internet of Medical Things (IoMT): General System Architecture

The general system design (see Fig. 3) of IoT solutions consists of the following components:

1. Devices: This consists of the interconnected devices such as sensors, actuators, and wearables that collect the data.
2. Network: This includes the network such as Wi-Fi and Bluetooth through which the collected data is transmitted.
3. Cloud: This is the cloud infrastructure where the data is stored, transformed, and analysed.
4. Client: This is the end user such as hospital, patient, physician that interacts with the rest of the system components.

1.3 Applications

Internet of things (IoT) has changed the face of the medical domain. The introduction of radical healthcare solutions, such as remote monitors and patient diagnostic machines, has bridged the gap between the patients and the doctors. IoT has a large number of applications in a plethora of fields; however, its contribution in the field of medicine has helped to unbridle the potential of the healthcare sector. Following are some applications of IoT in healthcare:

Telemedicine. The analysis carried out by Ramu et al. [3] sheds light on telemedicine, an up-and-coming concept that allows patients in rural areas to access expert doctors in urban hospitals. IoMT incorporates the use of sensors, wearables and actuators connected via the Internet to exchange medical information with the doctors. Furthermore, [4] demonstrate a comparative study between the three leading wireless technologies namely Bluetooth, GSM and Wi-Fi which allow doctors to view physiological data of patients from anywhere in real time, thereby making the healthcare sector more efficient and accessible.

Fitness Bands (Wearables). Combined with a variety of different sensors (e.g. global positioning system (GPS); optical heart rate monitor; galvanic skin response sensor; ultra violet sensor, etc.), fitness bands as per Kumbhar [5] contain small and powerful sensors connected via IoT, which help monitor activities such as heart rate, calories burnt, weight, oxygen level, etc. This enables patients to review their health. Additionally, this data is sent to the server using smartphones which allows doctors to analyse its patients' health status.

Ingestible Sensors and Cameras. Ingestible sensors set the bar high for modern healthcare facilities. Ingestible sensors or smart pill on daily ingestion measures a range of health metrics including non-adherence with medication. The system combines ingestible sensors embedded in tablets, small wearable sensor patches, a mobile application and a portal which helps doctors analyse and eliminate complications caused by medical non-adherence. The latest technologies have also brought forth ingestible cameras which could take snaps of the human body, thereby replacing the existing endoscopy procedures, per Hafezi et al. [6].

IoT-based Early Warning Score (EWS) Systems. In the era of a data-driven world, large volumes of medical data are generated by a wide range of bio-sensors. This makes it impossible to monitor every patient. In order to provide timely healthcare to patients, an IoT-aware early warning score system is utilized to effectively detect early signs of deterioration in a patient's health. The risk is analysed by mapping certain measured vital signs to a patient deterioration risk score. Azimi et al. [7] delve deep into personalized healthcare and propose a self-aware early warning score system. They address the need for long-term monitoring of patients with critical illnesses outside the hospital environment and hence introduce the concept of automated EWS.

M-Health Monitoring. Santosa et al. [8] discuss M-Health or the mobile healthcare system. IoT uses radio frequency identification (RFID) to create smart objects. Putting an identification label onto each object paves the way for a smart system and helps extract information in real time without any physical contact. RFID tags in medical context ensure a quick and accurate identification of each smart entity, enabling ubiquitous access to personal health records over an Internet of things.

2 Challenges

The development and implementation of any new technology come with its own roadblocks and pitfalls, and IoT in healthcare is no exception. Below are a few of the major hurdles this technology faces.

2.1 Security and Privacy Issues

It is forecasted that by 2025, healthcare will generate the most data than any other sector. Of this, a significant percentage will be through IoT devices. An increase in data transactions such as these makes the system highly prone to attacks by people with malicious intentions. The whole basis of IoT devices is the communication of data between sensors and the cloud, or any other storage database. But, sensors can be easily hacked, and networks can be breached. Moreover, since IoT devices generally run on the principle of automatic data collection and negligible verification and authentication, a security breach is highly likely, as per Alasmari et al. [9], the theft of medical data can be disastrous for both the patient as well as the medical infrastructure. To alleviate this issue, proper cryptographic algorithms must be deployed, and authentication and verification protocols must be in place as per Agrawal et al. [10].

2.2 Lack of Infrastructure

Despite the hype surrounding IoT, it is yet to be implemented in the field of healthcare with the same level of enthusiasm. Many hospitals and health monitoring systems lack the proper infrastructure that can be upgraded to accommodate IoT devices. This is especially the case for rural areas, where even traditional healthcare amenities are limited. Along with this, narrowband IoT (NB-IoT) devices play a huge role in healthcare IoT devices, due to advantages such as low power and bandwidth consumption. But, the services available at present day are unsuitable for real-time applications due to lack of control over delay. This can be solved by using efficient resource sharing algorithms, and by deploying protocols such as IPv6 over WPAN (6LoWPAN) and constrained application protocol (CoAP) as per Anand et al. [11].

2.3 Data Storage and Computation

The IoT sensors and actuators collect vast amounts of healthcare information, which is not always structured. The characteristics of this data closely resemble that of big data, in terms of volume generated, variety, and data generation frequency. This huge amount of data now needs to be structured and processed in order to make intelligent insights from it. But, IoT devices have limited processing capabilities and cannot perform complex computations on-site. Darwish et al. [12] propose Cloud technologies are used to resolve this issue. It is a powerful, convenient and cost-effective technology to handle bulk data sent by IoT devices, as well as in the data integration and aggregation. Along with that, it contains unlimited processing capabilities and an on-demand usage model.

2.4 Interoperability/Heterogeneity/Standardization

The IoT paradigm is heterogeneous in terms of the devices, platforms used, operating systems, and interfaces (device to device, human to device, and human to human). It aims to connect objects with different capabilities and complexity, different dates and release versions, which may be designed for a completely different function altogether. These issues, coupled with the lack of standards, are a big hindrance towards integration of IoT in healthcare. Therefore, protocols must be designed to integrate different devices and interfaces, to mitigate the issue of interoperability as per Mahmoud et al. [13].

2.5 Latency and Delay

Several applications of IoT in healthcare involve collection of data, and analysing it, they may not always be life threatening conditions. But in several life threatening scenarios, for example, real-time monitoring of a patient prone to cardiovascular diseases, a delay in the transmission of data can lead to a loss of life. This is often critical and time sensitive and hence cannot be relied upon the cloud capabilities to handle it, since they are usually located too far away to process the data and respond in a timely manner. Farahani et al. [14] propose this to be handled by fog computing technologies, which are present between the IoT sensors and the Cloud. Have the capability to perform short-term operations on the data when the data is time critical. Another method is using 5G Millimeter Wave (mmWave) Communication System with Software Defined Radio as proposed by Agrawal and Sharma [15].

2.6 Cost

While hardware costs have seen a significant decline with the advancement in technology, it would be unwise to discount the infrastructure and setup cost of the IoT healthcare system. The components involved are the sensors that collect data from patients, the network through which the data is transmitted to and from different nodes, middleware such as fog computing technologies, and the Cloud technology that stores, processes, and analyses the data. To integrate these technologies means an increase in costs as compared to traditional healthcare systems, but considering the return on investment in the long run, many researchers argue that it is worth the initial investment.

2.7 *Scalability*

A big challenge in any system design is the scalability of the system. In the healthcare scenario, we can have different levels at which to provide the IoT services, starting with individual patients and doctors, which goes up to span an entire hospital, which can then be scaled up to cover a city, and can finally encompass an entire country or even cross-continent. This will ensure a seamless healthcare experience and enhance the quality and turnaround time of delivery of medical services. But extensive infrastructure is needed to make this a reality as per Firouzi et al. [16].

3 Literature Review

Doukas and Maglogiannis [1] explore the use of Cloud computing with Internet of things in the healthcare domain. The immense amount of medical data collected by both wearable textile sensors and mobile sensors needs to be stored and managed efficiently. They propose a system in which a sensor gateway acquires the data from the sensors and forwards them to the cloud infrastructure using lightweight interfaces such as Representational State Transfer (REST) API. The cloud infrastructure deploys all the necessary resources for effective communication between sensors and other external systems. The communication channels have been secured using standard symmetric encryption methods.

Hassan et al. [17] investigate the performance of Narrowband IoT (NB-IoT) technology for healthcare applications. This technology consumes low data, costs less and has a longer battery life. NB-IoT supports only half-duplex mode in Frequency Division Duplex (FDD), i.e. users can either transmit or receive at one point of time. They compare and contrast two types of system architectures: Single sensor node design (SND) and multi-sensor node design (MND) and conclude that MND design is superior to SND in terms of throughput and number of patients per cell but comes at the cost of increased delay.

Moosavi et al. [18] analyse the performance of end-to-end security systems in IoT applications for healthcare. They propose to revamp the existing cryptographic key generation schemes since sensors in healthcare IoT systems have a low memory, processing power, and bandwidth. They use a three-layer architecture, which includes key generation by integrating interpulse interval (IPI) sequence of ECG signal with pseudorandom numbers; mutual authentication of devices and users using a smart gateway; and secure end-to-end communication between end points that use handover mechanisms.

Talpur et al. [19] design an energy optimized algorithm for a smartphone-based healthcare monitoring system for chronic heart disease. Body sensors send data to the smartphone using Bluetooth, which compiles and transfers data to the remote database servers. The types of data are grouped according to severity, and the energy consumption of smartphone is increased or decreased according to the priority of attack. High priority attacks are immediately sent to the database server for analysis and communication to concerned parties and may consume more energy.

Ali et al. [20] explore a less talked about topic, i.e. detecting depression symptoms and providing appropriate assistance using IoT. They use a Web of Objects (WoO) framework to virtualize the real-world objects and integrate heterogeneous devices that are linked with resources to develop IoT services and enable their deployment and operations. They propose a microservices model wherein each functionality of the system, such as Preference Learner and Recommendation Analyser, is modelled as a microservice. The model utilizes user feedback and preferences using the real-world knowledge (RWK) module that is updated by the user, to generate a recommendation. It also implements the semantic ontology module to perform overall integration and linking of virtual objects (VOs) and generates an ontology graph with the different objects as nodes and their characteristics as branches.

Romero et al. [21] examine the use of IoT in diagnosing and monitoring Parkinson's Disease, which is a neurodegenerative disorder that affects the elderly and causes involuntary tremors. They propose integrating body sensors to facilitate the clinical tests included in the Unified Parkinson's Disease Rating Scale (UPDRS) and help detect and quantify the tremors. This will benefit the patient as a short-term hospital visit may not reveal the true nature of the disease, but a prolonged collection and analysis of data from the comfort of the patient's home are where the IoT aspect comes into play.

Banerjee et al. [22] explore the cardiovascular research domain. They propose a system for heart attack detection, and the system keeps track of various parameters including heartbeat (BPM) and temperature. A certain protein called fatty acid binding protein (FABP3) is secreted in excess prior to a cardiac arrest, which is measured by the system. The information obtained from the sensor is interfaced with a microcontroller and is sent to the cloud via the Internet.

Hemalatha and Vidhyalakshmi [23] propose a novel system which uses a wide range of sensors namely ECG, chest belt, oximeter, accelometer, etc., to detect and compute chronic coughs. They arranged for a battery-powered MEMS vibration sensor to be placed on the neck. It is further connected to a smartphone, which transmits data to a cloud-based health platform which is monitored by medical personnel.

Matar et al. [24] propose a contemporary approach for posture monitoring, a subfield of remote medical monitoring, via body pressure distribution on the mattress. A pressure sensing mattress acquires body pressure distribution data, which is then processed by a computer workstation. The results are then analysed for diagnosis. It is used for various clinical purposes such as in sleep studies and anaesthetic surgical procedures. The proposed method uses a supervised learning support vector machine (SVM) algorithm.

Lamonaca et al. [25] cater to over a third of the adult population suffering from hypertension, which causes high BP. Their study depicts the rising trends of the wearables' market size, which is the prime application of Internet of Medical Things. The IoMT-based devices, capable of measuring blood pressure, can be used at anytime and anywhere. Their research addresses the absence of reliability and traceability of the BP measurements and sheds light on potential solutions which cater to this lack from a metrological point of view.

Magsi et al. [26] survey the effect of the state-of-the-art 5G technology when combined with IoT. The requirement of high data, high speed and long battery life with reliable connectivity is fulfilled by deploying 5G and provides support for diagnosis and treatment in the IoT healthcare domain. Apart from that, they shed light on the 5G-based sensor node architecture for health monitoring of patients and the long-term connectivity that 5G offers.

Saha et al. [27] explore the need for an intelligent smart health system. Due to overpopulation, India continues to neglect the need for timely provisions of healthcare systems. They propose to devise a system that will sense the patients' vital parameters and share the data with micro-controllers, which will further be sent to the Raspberry Pi. The Raspberry Pi is connected with the Internet or IoT cloud so that specialist doctors at health facilities can monitor it. If the patient is in critical condition, an ambulance will be allotted. In order to reach the patient on time, the ambulance uses Google maps along with an integrated hardware, made using Arduino UNO, ultrasonic sensors, buzzer, resistors, LEDs, buzzer, connecting wires and breadboard so as to avoid accidents and obstacles.

Fan et al. [28] investigate the Cloud-based RFID solution, a novel, cutting edge technology that caters to the healthcare sector. The problem that befalls this technology is the unreliable cloud server that deals with private medical data which is transmitted over a public wireless network, thereby exposing it to a risk of leakage. The proposed solution to deal with such privacy and security issues is a lightweight authentication scheme based on quadratic reSIDuals and a pseudo random number generator. The system maintains privacy by ensuring tag anonymity; resistance to tag-tracking attacks; mutual authentication; resistance to replay attacks, and resistance to de-synchronization.

Onasanya and Elshakankiri [29] put forth the implementation of an IoT-based medical system for enhanced treatment and diagnosis of cancer patients. The system is based on business analytics/cloud services. The use of WSN and smart connected devices has made it possible since WSN permits a number spatially distributed autonomous sensors to be linked to the network fabric based on geographical routing from source to destination, which facilitate data transmission/exchange. They exploit business analytics/cloud services for the generation of patient data stream in order to make informed decisions. The study augments the existing cancer treatment procedure by exploiting the potential of the IoT-WSN combination.

4 Conclusions and Future Scope

In this paper, we have evaluated how the IoT in healthcare paradigm has come to the forefront of research and is only expected to grow multifold in the near future. Major work done in this field and the applications and challenges regarding this have been discussed in depth. Thus, healthcare is a promising field for the implementation of IoT technology and one that has far reaching impacts on the lives of people.

While present day IoT in healthcare encompasses a wide range of integrated technologies including remote monitors, wearables, smart sensors, glucose monitors, smart mattresses, medication dispensers, etc., the investment and research in IoMT continue to ensue. The following are potential areas which can be exploited to provide improved healthcare:

- **Quantified Health**: Data has a huge impact on performance, i.e. IoT has the power to influence the outcomes in favour of the patient via improved object measurement and health evaluation.
- **Customized treatments**: IoT enables the provision of treatments, tailored to the patients' needs. Consistent monitoring can be done regardless of the doctor's presence. An IoT data stream can be used to transmit real-time data to a portal which can further be analysed by a physician. This can be further researched to provide befitting treatments.
- **A block chain future**: With the increase of investment in time and money for developing better and better medical solutions using IoT, and security concerns tend to grow as well. For this reason, the future of security lies in integrating IoT with block chain. Block chain can be used to ensure the integrity of medical data [30].
- **5G compliance**: Mobile network technology using 5G is rising in popularity. It is of utmost importance that new IoT solutions/systems are 5G compliant so as to have improved speed and battery life. The introduction of 5G is paving the way for enhanced IoMT systems [31].

References

1. C. Doukas, I. Maglogiannis, Bringing IoT and cloud computing towards pervasive healthcare, in *2012 Sixth International Conference on Innovative Mobile and Internet Services in Ubiquitous Computing* (2012)
2. S.R. Department, Statista, Dec 2016 [Online]. Available: https://www.statista.com/statistics/997959/worldwide-internet-of-things-in-healthcare-market-size/
3. R. Ramu, A. Sukesh Kumar, Real time diagnosis of rural cardiac patients through telemedicine, in *International Conference on Modelling and Simulation* (Springer, Cham, 2017)
4. W. Mohamed, M.M. Abdellatif, Telemedicine: an IoT Application For Healthcare systems, in *Proceedings of the 2019 8th International Conference on Software and Information Engineering* (ACM, 2019)
5. R. Kumbhar, Health monitoring using fitness band and IOT, in *Celal Bayar Üniversitesi Fen Bilimleri Dergisi* (2018)
6. H. Hooman, T.L. Robertson, G.D. Moon, K.-Y. Au-Yeung, M.J. Zdeblick, G.M. Savage, An ingestible sensor for measuring medication adherence. IEEE Trans. Biomed. Eng. **62** (1) (2014)
7. I. Azimi, A. Anzanpour, A.M. Rahmani, P. Liljeberg, H. Tenhunen, Self-aware early warning score system for IoT-based personalized healthcare, in *eHealth 360°* (Springer, Cham, 2017), pp. 49–55

8. A. Santos, J. Macedo, A. Costa, M. João Nicolau, Internet of things and smart objects for M-health monitoring and control. Proc. Technol. **16**, 1351–1360 (2014)
9. S. Alasmari, M. Anwar, Security & privacy challenges in IoT-based health cloud, in *2016 International Conference on Computational Science and Computational Intelligence (CSCI)* (2016)
10. S. Agrawal, K. Sharma, Software defined millimeter wave 5th generation communications system. Appl. Theory Computer Technol. **2** (2017)
11. S. Anand, S.K. Routray, Issues and challenges in healthcare narrowband IoT, in *International Conference on Inventive Communication and Computational Technologies (ICICCT)* (2017)
12. A. Darwish, A.E. Hassanien, M. Elhoseny, A.K Sangaiah, K. Muhammad, The impact of the hybrid platform of internet of things and cloud computing on healthcare systems: opportunities, challenges, and open problems. J. Ambient Intell. Humanized Comput. **10** (2019)
13. R. Mahmoud, T. Yousuf, F. Aloul, I. Zualkernan, Internet of things (IoT) security: current status, challenges and prospective measures, in *10th International Conference for Internet Technology and Secured Transactions (ICITST)* (2015)
14. B. Farahani, F. Firouzi, V. Chang, M. Badaroglu, N. Constant, K. Mankodiya, Towards fog-driven IoT eHealth: promises and challenges of IoT in medicine and healthcare. Future Gener. Comput. Syst. **78** (2018)
15. S. Agrawal, Sharma, K.: 5G millimeter wave (mmWave) communication system with software defined radio (SDR), in *Proceedings of the International Conference on Recent Trends in Engineering & Science (ICRTES-16)* (2016)
16. F. Firouzi, B. Farahani, M. Ibrahim, K. Chakrabarty, Keynote paper: from EDA to IoT eHealth: promises, challenges, and solutions, in *IEEE Transactions on Computer-Aided Design of Integrated Circuits and Systems* **37** (2018)
17. M. Hassan, M.M. Alam, Y.L. Mo, Narrow Band-IoT Performance Analysis for Healthcare Applications (IEEE, 2018)
18. S. R. Moosavi, E. Nigussie, T. N. Gia, A. M. Rahmani, S. Virtanen, H. Tenhunen and J. Isoaho, "End-to-end security scheme for mobility enabled healthcare Internet of Things," in Future Generation Computer Systems 64, 2016
19. M.S.H. Talpur, M.Z.A. Bhuiyan, G. Wang, Energy-efficient healthcare monitoring with smartphones and IoT technologies. Int. J. High Perform. Comput. Netw. **8** (2) (2015)
20. S. Ali, M. Golam Kibria, M. Aslam Jarwar, S. Kumar, I. Chong, Microservices model in WoO based IoT platform for depressive disorder assistance, in *2017 International Conference on Information and Communication Technology Convergence (ICTC)* (IEEE, 2017)
21. L.E. Romero, P. Chatterjee, R.L. Armentano, An IoT approach for integration of computational intelligence and wearable sensors for Parkinson's disease diagnosis and monitoring. Health Technol. **6** (3) (2016)
22. S. Banerjee, P. Souptik, R. Sharma, A. Brahma, Heartbeat monitoring using IoT, in *2018 IEEE 9th Annual Information Technology, Electronics and Mobile Communication Conference (IEMCON)* (2018)
23. P. Hemalatha, R. Vidhyalakshmi, A study on chronic cough detection using IoT and machine learning. Int. J. Res. Arts Sci. **5** (2019)
24. G. Matar, J.-M. Lina, J. Carrier, A. Riley, G. Kaddoum, Internet of things in sleep monitoring: an application for posture recognition using supervised learning, in *2016 IEEE 18th International Conference on e-Health Networking, Applications and Services (Healthcom)* (2016)
25. F. Lamonaca, D. Luca Carnì, F. Bonavolontà, V. Spagnuolo and A. Colaprico, "An Overview on Internet of Medical Things in Blood Pressure Monitoring," in 2019 IEEE International Symposium on Medical Measurements and Applications (MeMeA), 2019
26. H. Magsi, A. Hassan Sodhro, F. Akhtar Chachar, S. A. Khan Abro, G. Hassan Sodhro, S. Pirbhulal, Evolution of 5G in internet of medical things, in *2018 International Conference on Computing, Mathematics and Engineering Technologies (iCoMET)* (2018)
27. H.N. Saha, N. Firdaush Raun, M. Saha, Monitoring patient's health with smart ambulance system using Internet of Things (IOTs), in *2017 8th Annual Industrial Automation and Electromechanical Engineering Conference (IEMECON)* (2017)

28. K. Fan, Q. Luo, H. Li, Y. Yang, Cloud-based lightweight RFID mutual authentication protocol, in *2017 IEEE Second International Conference on Data Science in Cyberspace (DSC)* (2017)
29. A. Onasanya, M. Elshakankiri, IoT implementation for cancer care and business analytics/cloud services in healthcare systems, in *Proceedings of the 10th International Conference on Utility and Cloud Computing* (ACM, 2017)
30. M. Banerjee, J. Lee, K.K. Choo, A blockchain future for internet of things security: a position paper. Digital Commun.Netw. **4** (2018)
31. S. Agrawal, K. Sharma, 5G millimeter wave (mmWave) communications, in *3rd International Conference on Computing for Sustainable Global Development (INDIACom)* (IEEE, 2016)

Applications

Cryptanalysis of Modification in Hill Cipher for Cryptographic Application

K. Vishwa Nageshwar and N. Ravi Shankar

1 Introduction

The Internet has shrunk the globe into a tiny village. Communications and transactions over the Internet have emphasized the need for strong measures for securing the information. Cryptography is a tool employed for ensuring the confidentiality of Information. The algorithms that achieve confidentiality through data encryption are classified as public key and private key encryption algorithms based on key(s) used for encrypting the data and decrypting the data [2]. If same key is used for encryption and decryption, such algorithms are categorized as private key encryption algorithms [3]. If two different keys are used for encryption and decryption, such algorithms are termed as public key encryption algorithms. The encryption algorithms are also classified based on the unit of data encryption. If one byte/one character is encrypted at a time such algorithms are called stream ciphers [4]. If a group of characters or bytes are encrypted at a time such algorithms are known as block ciphers [5].

Hill cipher [6, 7] is the first block cipher reported in the literature of data encryption algorithms. This is a cipher which is based on algebraic properties of number theory and employs matrix algebra for data encryption and decryption. Hill cipher, though exhibits very good diffusion property, is vulnerable to the known plaintext attack. This cipher is governed by the relation

$$c = Kp \mod 26 \text{ and } p = K^{-1}c \mod 26$$

K. Vishwa Nageshwar
Computer Science and Engineering, Geethanjali College of Engineering and Technology (Autonomous), Hyderabad, India
e-mail: nag7524@gmail.com

N. Ravi Shankar (✉)
Computer Science and Engineering and Technology, Geethanjali College of Engineering and Technology (Autonomous), Hyderabad, India
e-mail: ravish00@yahoo.com

S. Patnaik et al. (eds.), *Advances in Machine Learning and Computational Intelligence*, Algorithms for Intelligent Systems, https://doi.org/10.1007/978-981-15-5243-4_62

where K denotes the key matrix, p the plaintext and c represents the ciphertext. The data encryption is done by a matrix multiplication operation. However, if one has the knowledge of p, one can easily obtain the inverse of p and get 'K'. A number of researchers have offered modifications to the classical Hill cipher to overcome this vulnerability against known plaintext attack. Sastri et al. devoted their time in studying the methods to make the Hill cipher stronger. In this process, they have used an iterated Hill cipher with interweaving [8, 9], permutation [10], interlacing [11, 12] and blending with generalized Playfair cipher [13] as additional operations in each of the iterations. In addition to multiplication with the key matrix, to make the relationship between the plaintext and the ciphertext as complex as possible. Several researchers have devoted their to strengthen the classical Hill cipher [14–16]. Further, J. R. Paragas et al. offered modifications to the Hill cipher through S-boxes [15, 17]. ElHabshy [18] experimented with augmented Hill cipher. Sharma et al. [19, 20] offered their modifications to the Hill cipher through non-invertible matrices. M. Attique ur Rehman et al. modified Hill cipher through non-square matrix to strengthen the cipher against known plaintext attack [21]. Recently, Rachmawati et al. [22] ported Hill cipher to securing messages on android.

Qazi et al. [1] have proposed a modification to the Hill cipher to improve the cipher performance. This cipher will henceforth be referred to as target algorithm. In their paper, they have used an orthogonal matrix [23] to be used as a key matrix. However, this scheme exhibits fatal weaknesses in the face of chosen plaintext/ciphertext attacks.

In Sect. 2 of this paper the target algorithm is described, which forms the motivation for this research. Section 3 details the cryptanalytic attack through which the cipher is broken. Section 4 draws conclusions and proposes ways of strengthening the target algorithm.

2 Description of the Target Algorithm

The researchers [1] have derived a key matrix through a set of complex maneuvers as described in [23]. However, the complexity induced in the generation of the key matrix do not enhance the strength of the cipher as the elements of the key matrix are as good as any randomly generated integers. The cipher functioning is almost similar to the classical Hill cipher with the exception of the addition of a few linear, highly predictable operations before the matrix multiplication.

The researchers have generated a key matrix

$$k = \begin{bmatrix} K_{11} & K_{12} & K_{13} & K_{14} \\ K_{21} & K_{22} & K_{23} & K_{24} \\ K_{31} & K_{32} & K_{33} & K_{34} \\ K_{41} & K_{42} & K_{43} & K_{44} \end{bmatrix} \quad (1)$$

The plaintext p is taken as

$$p = \begin{bmatrix} p1 \\ p2 \\ p3 \\ p4 \end{bmatrix} \tag{2}$$

This plaintext is inverted and taken as

$$p' = \begin{bmatrix} p_4 \\ p_3 \\ p_2 \\ p_1 \end{bmatrix} \tag{3}$$

Now a random number × (less than 26) is added to the elements of p and a modulo 26 is taken.

$$p'' = \begin{bmatrix} p_4 + x \\ p_3 + x \\ p_2 + x \\ p_1 + x \end{bmatrix} \bmod 26 \tag{4}$$

The elements of p'' are converted into their binary equivalent and a 4×8 matrix is formed.

$$p''' = \begin{bmatrix} \text{binary equivalent of } (p''_4) \\ \text{binary equivalent of } (p''_3) \\ \text{binary equivalent of } (p''_2) \\ \text{binary equivalent of } (p''_1) \end{bmatrix} \tag{5}$$

Now each row of p''' is given a one-bit right shift. In this process, all the bits in the least significant positions are stored in a vector. Zeros' are padded into the most significant positions of p'''. This new vector is denoted as p'''. Finally, the key matrix and the modified plaintext, p'''' are multiplied (modulo 26), to obtain the ciphertext.

The illustration of the target algorithm is given below:

Consider a plaintext:
Plaintext = SECURITY

The first four letters of the plaintext are taken as

$$p = \begin{bmatrix} S \\ E \\ C \\ U \end{bmatrix} = \begin{bmatrix} 18 \\ 4 \\ 2 \\ 20 \end{bmatrix} \tag{6}$$

Applying transposition

$$p' = \begin{bmatrix} U \\ C \\ E \\ S \end{bmatrix} = \begin{bmatrix} 20 \\ 2 \\ 4 \\ 18 \end{bmatrix} \quad (7)$$

Applying Caesar cipher substitution

$$p'' = \begin{bmatrix} 20+7 \\ 2+7 \\ 4+7 \\ 18+7 \end{bmatrix} = \begin{bmatrix} 27 \\ 9 \\ 11 \\ 25 \end{bmatrix} (\text{mod } 26) = \begin{bmatrix} 1 \\ 9 \\ 11 \\ 25 \end{bmatrix} \quad (8)$$

Right shifting

$$p''' = \begin{bmatrix} 1 \\ 9 \\ 11 \\ 25 \end{bmatrix} = \begin{bmatrix} 00000001 \\ 00001001 \\ 00001011 \\ 00011001 \end{bmatrix} = \begin{bmatrix} 00000000 \\ 00000100 \\ 00000101 \\ 00001100 \end{bmatrix} \quad (9)$$

Save the shifted bits in a vector named shift

$$\text{shift} = \begin{bmatrix} 1 \\ 1 \\ 1 \\ 1 \end{bmatrix} \quad (10)$$

Binary to decimal conversion of p''' gives

$$p'''' = \begin{bmatrix} 0 \\ 4 \\ 5 \\ 12 \end{bmatrix} \quad (11)$$

For encryption follow general method of hill cipher, so, will be as follow:

$$c = \begin{bmatrix} 25 & 12 & 10 & 22 \\ 12 & 20 & 5 & 24 \\ 10 & 5 & 16 & 20 \\ 22 & 24 & 20 & 19 \end{bmatrix} * \begin{bmatrix} 0 \\ 4 \\ 5 \\ 12 \end{bmatrix} = \begin{bmatrix} 362 \\ 393 \\ 340 \\ 424 \end{bmatrix} (\text{mod } 26) \quad (12)$$

$$c = \begin{bmatrix} 24 \\ 3 \\ 2 \\ 8 \end{bmatrix} \tag{13}$$

The decryption is the reverse process of encryption.

3 The Cryptanalytic Attack

This cipher can be broken with the help of one chosen ciphertext and four chosen plaintexts. This cipher exhibits fatal weaknesses when an all zero ciphertext is decrypted. It can be easily verified that an all zero plaintext, when deciphered, will give a plaintext whose contents equal to 26-x, where x is the constant added to the elements of the plaintext in the process of encryption. This calculation is demonstrated below:

We present a ciphertext of $[0000]^T$ to the encipher and ask for its corresponding plaintext.

$$c = \begin{bmatrix} 0 \\ 0 \\ 0 \\ 0 \end{bmatrix} \tag{14}$$

$$p'''' = \begin{bmatrix} 12 & 15 & 11 & 41 \\ 16 & 7 & 23 & 09 \\ 10 & 22 & 44 & 14 \\ 4 & 08 & 21 & 33 \end{bmatrix} * \begin{bmatrix} 0 \\ 0 \\ 0 \\ 0 \end{bmatrix} = \begin{bmatrix} 0 \\ 0 \\ 0 \\ 0 \end{bmatrix} \tag{15}$$

$$p''' = \begin{bmatrix} 0 \\ 0 \\ 0 \\ 0 \end{bmatrix} = \begin{bmatrix} 00000000 \\ 00000000 \\ 00000000 \\ 00000000 \end{bmatrix} = \begin{bmatrix} 0 \\ 0 \\ 0 \\ 0 \end{bmatrix} \tag{16}$$

Performing reverse of constant addition gives rise to

$$p'' = \begin{bmatrix} 0-7 \\ 0-7 \\ 0-7 \\ 0-7 \end{bmatrix} (\text{mod } 26) = \begin{bmatrix} 19 \\ 19 \\ 19 \\ 19 \end{bmatrix} \tag{17}$$

Here, Eq. (17) reveals that constant added in the encryption is $x = 26 - 19 = 7$. With this all zero ciphertext, the additive constant can be found.

If the elements of the key matrix are also found, the cryptanalytic attack is complete. If the p is chosen in such a way that $p''' = [1000]^T$, the ciphertext c, which is obtained by multiplying p''' with the key matrix K contains, first column of the key matrix K. This is illustrated below:

$$\begin{bmatrix} K_{11} & K_{12} & K_{13} & K_{14} \\ K_{21} & K_{22} & K_{23} & K_{24} \\ K_{31} & K_{32} & K_{33} & K_{34} \\ K_{41} & K_{42} & K_{43} & K_{44} \end{bmatrix} * \begin{bmatrix} 1 \\ 0 \\ 0 \\ 0 \end{bmatrix} = \begin{bmatrix} K_{11} \\ K_{21} \\ K_{31} \\ K_{41} \end{bmatrix} \tag{18}$$

Similarly,

$$\begin{bmatrix} K_{11} & K_{12} & K_{13} & K_{14} \\ K_{21} & K_{22} & K_{23} & K_{24} \\ K_{31} & K_{32} & K_{33} & K_{34} \\ K_{41} & K_{42} & K_{43} & K_{44} \end{bmatrix} * \begin{bmatrix} 0 \\ 1 \\ 0 \\ 0 \end{bmatrix} = \begin{bmatrix} K_{12} \\ K_{22} \\ K_{32} \\ K_{42} \end{bmatrix} \tag{19}$$

$$\begin{bmatrix} K_{11} & K_{12} & K_{13} & K_{14} \\ K_{21} & K_{22} & K_{23} & K_{24} \\ K_{31} & K_{32} & K_{33} & K_{34} \\ K_{41} & K_{42} & K_{43} & K_{44} \end{bmatrix} * \begin{bmatrix} 0 \\ 0 \\ 1 \\ 0 \end{bmatrix} = \begin{bmatrix} K_{13} \\ K_{23} \\ K_{33} \\ K_{43} \end{bmatrix} \tag{20}$$

$$\begin{bmatrix} K_{11} & K_{12} & K_{13} & K_{14} \\ K_{21} & K_{22} & K_{23} & K_{24} \\ K_{31} & K_{32} & K_{33} & K_{34} \\ K_{41} & K_{42} & K_{43} & K_{44} \end{bmatrix} * \begin{bmatrix} 0 \\ 0 \\ 0 \\ 1 \end{bmatrix} = \begin{bmatrix} K_{14} \\ K_{24} \\ K_{34} \\ K_{44} \end{bmatrix} \tag{21}$$

However, it is necessary to choose p such that $p''' = [1\,0\,0\,0]^T$ for this, p'' should be $[2\,0\,0\,0]^T$ since there is a one position right shift involved, and hence the p should be $[21\ 19\ 19\ 19]^T$. In this case, there is an additive constant 7 modulo 26 operation involved. Since it is known that the additive constant is 7 from (17), the p is taken as $[19\ 19\ 19\ 21]^T$. There is a transposition also involved here. Hence, p is taken in reverse. Transposition again reverses the p.

$$\begin{bmatrix} 21+7 \\ 19+7 \\ 19+7 \\ 19+7 \end{bmatrix} (\text{mod } 26) = \begin{bmatrix} 2 \\ 0 \\ 0 \\ 0 \end{bmatrix} \tag{22}$$

In order to satisfy this, p should be taken as $[19\,19\,19\,21]^T$. Thus, a first chosen plaintext $p = [19\,19\,19\,21]^T$, when encrypted using this algorithm, gives a ciphertext equal to $[K_{11}\ K_{21}\ K_{31}\ K_{41}]^T$.

A second chosen plaintext of $[19211919]^T$ reveals the second column of the key matrix K. Similarly, $[19192119]^T$ and $[19191921]^T$ give away third and fourth columns of the key matrix K.

Thus, the cipher is broken.

4 Conclusions

The target algorithm suffers from the lack of complex operations which make the relationship between the plaintext and the ciphertext as complex as possible. This is a basic requirement for any cipher. This algorithm can be marginally strengthened if there is an additive constant introduced after multiplication with the key matrix. Effectively, the equation for encryption in the original research is

$$c = K* < \$\%p\%\$ >,$$

where $\%p\%$ denotes transpositioning/reversing the plaintext vector,

$\$p\$$ represents adding a constant (modulo 26),

<> operation signifies converting the character into binary number, right shift by one position.

If this equation can be modified as

$c = \%K * < \$\% p \%\$ > \%$, where one more constant is added to c after multiplication with the key matrix, the all zero ciphertext do not expose the additive constant directly.

This research exposes the vulnerabilities in the target algorithm [1] and shows how the cipher can be broken with chosen ciphertext and chosen plaintext attacks. It also suggests means of strengthening the algorithm against the type of cryptanalytic attack carried in this paper.

References

1. F. Qazi, F.H. Khan, D. Agha, S. Ali Khan, S. ur Rehman, Modification in Hill cipher for cryptographic application. 3C Tecnología. Glosas de innovaciónaplicadas a la pyme(Special Issue 2), 240–257. http://dx.doi.org/10.17993/3ctecno.2019 (2019)
2. B. Schneier, Applied cryptography, in *Protocols, Algorithms, and Source Code in C*, 2nd edn. (Wiley, New York, NY, USA, 1995)
3. A.J. Menezes, S.A. Vanstone, P.C. Van Oorschot, *Handbook of Applied Cryptography*, 1st edn. (CRC Press Inc., Boca Raton, FL, USA, 1996)
4. C. Paar, J. Pelzl, *Understanding Cryptography: A Textbook for Students and Practitioners*. (Springer, 2010), p. 7. ISBN 978-364204100-6
5. W. Stallings, *Cryptography and Network Security: Principles and Practices*, 4th edn (Pearson Education, Asia, 2005)
6. L.S. Hill, Cryptography in an algebraic alphabet. The American Mathematical Monthly **36**(6), 306–312 (1929)

7. L.S. Hill, Concerning certain linear transformation apparatus of cryptography. The American Mathematical Monthly **38**(3), 135–154 (1931)
8. V.U.K. Sastry, N. Ravi Shankar, S. DurgaBhavani, A modified Hill cipher involving interweaving and iteration. Int. J. Netw. Secur. **11**(2), 51–56 (2010)
9. V.U.K. Sastry, S. DurgaBhavani, N. Ravi Shankar, A large block cipher involving interweaving and iteration, in *International Conference on Advances and Emerging Trends in Computing Technologies (ICAET10)* (Chennai, India, 2010), pp. 328–333
10. V.U.K. Sastry, N. Ravi Shankar, S. DurgaBhavani, A large block cipher involving key dependent permutation, interlacing and iteration. Cybern. Inf. Technol. **13**(3), 50–63 (2013)
11. V.U.K. Sastry, N. Ravi Shankar, Modified Hill cipher with interlacing and iteration. J. Comput. Sci. Sci. Publ. **3**(11), 854–859 (2007)
12. V.U.K. Sastry, N. Ravi Shankar, Modified Hill cipher for a large block of plaintext with interlacing and iteration. J. Comput. Sci. Sci. Publ. **4**(1), 15–20 (2008)
13. V.U.K. Sastry, N. Ravi Shankar, S. DurgaBhavani, A blending of a generalized Playfair cipher and a modified Hill cipher. Int. J. Netw. Mob. Technol. **2**(1), 35–43 (2011)
14. K. Man, A. BarakathBegam, Generation of Keymatrix for Hill cipher encryption using quadratic form. Int. J. Sci. Technol. Res. **8**(10), 964–968 (2019)
15. J.R. Paragas, A.M. Sison, R.P. Medina, A new variant of Hill cipher algorithm using modified S-box. Int. J. Sci. Technol. Res. **8**(10), 615–619 (2019)
16. S. Kumar, S. Kumar, G. Mittal, S. Narain, *A Vector Space Approach to Generate Dynamic Keys for Hill Cipher*. arXiv preprint arXiv:1909.06781 (2019)
17. J.R. Paragas, A.M. Sison, R.P. Medina, Hill cipher modification: a simplified approach, in *2019 IEEE 11th International Conference on Communication Software and Networks (ICCSN)* (Chongqing, China, 2019), pp. 821–825. https://doi.org/10.1109/iccsn.2019.8905360
18. A.A. ElHabshy, Augmented Hill cipher. Int. J. Netw. Secur. **21**(5), 812–818 (2019). https://doi.org/10.6633/IJNS.201909_21(5).13
19. P. Sharma, T. Dalal, Experimental analysis of modified Hill cipher to perform encryption and decryption for invertible and noninvertible matrices. Int. J. Enhanced Res. Manag. Comput. Appl. **7**(7) (2018)
20. P. Sharma, T. Dalal, Study of encryption and decryption for matrices using modified Hill cipher. Int. J. Enhanced Res. Manag. Comput. Appl. **7**(5) (2018)
21. M. Attiqueur Rehman, H. Raza, I. Akhter, Security enhancement of Hill cipher by using non-square matrix approach, in *4th International Conference on Knowledge and Innovation in Engineering, Science and Technology* (2018)
22. D. Rachmawati, A. Sharif, Erick, Hybrid cryptosystem combination algorithm of Hill cipher 3 × 3 and Elgamal to secure instant messaging for Android, in *Third International Conference on Computing and Applied Informatics* (2018)
23. F.H. Khan, R. Shams, F. Qazi, D. Agha, Hill Cipher Key generation algorithm by using orthogonal matrix. Proc. Int. J. Innovative Sci. Mod. Eng. **3**(3), 5–7 (2015)

Wearable Assistance Device for the Visually Impaired

Devashree Vaishnav, B. Rama Rao, and Dattatray Bade

1 Introduction

Even though machines surpass humans in precise task calibrations and provide accurate results for tasks which seem tedious for humans; the machines find it computationally extensive to implement and comprehend the world through sight in terms of visual perception and listening which is seemingly trivial for humans. The field of computer vision seeks to locate features from images or videos to aid in discriminating among objects. The field of artificial intelligence [1] has been flourishing since the past few years due to the boost in big data [2] availability paired with access to data science tools and the increment in computational capabilities of modern hardware. Large volumes of information can be hence processed within a short time span by combining artificial intelligence techniques with cognitive technologies. The quality of life of the visually impaired individuals can be greatly improved using the techniques of deep learning [3] and computer vision [4]. A wearable assistance device will aid the visually impaired to navigate through their surroundings by interpreting objects in the scenario through voice commands.

2 Related Work

Various combinations of existing technologies have been utilized to provide solution to the navigation and surrounding recognition problems faced by the people with visual disabilities. Early research for the visually impaired individuals is heavily dependent on embedded devices. It mostly involves electronic travel aid (ETA)

D. Vaishnav (✉) · B. Rama Rao · D. Bade
Vidyalankar Institute of Technology, Mumbai, India
e-mail: devashree.vaishnav@vit.edu.in

S. Patnaik et al. (eds.), *Advances in Machine Learning and Computational Intelligence*, Algorithms for Intelligent Systems, https://doi.org/10.1007/978-981-15-5243-4_63

devices for indoor navigation along with obstacle detection [5, 6] and distance calibration [7]. Further work included sensors for the utilization of IoT techniques with the use of GPS [8], RFID Tags [9], and Wi-Fi [10]. Certain researchers emphasized on robotic assistants for the visually impaired individuals [11]. Recent research tends more toward frameworks and software/application development. It involves obstacle detection [12], path detection [13], face detection [14], object tracking [15], gesture recognition [16], and interpretation through computer vision techniques [17].

Navigation guidance for the visually impaired individuals has evolved from walking sticks and guide dogs to smart sticks and guiding robots. Yet there is a scope for improvement when it comes to guidance devices.

3 Deep Learning Pre-requisites

Deep Learning is a subset of machine learning [18] which utilizes the concepts of machine learning and its techniques to implement cognitive decision-making algorithms in areas like vision, speech, etc.

3.1 Object Detection and Object Recognition

The terms object detection [19] and object recognition [20] are often used interchangeably leading to confusion, whereas both terms have distinct meaning. Object detection pertains to mere detection of the presence of an object in a particular scenario; while object recognition implies the identification of said object in the given image. Object recognition techniques utilize matching, learning, mapping, or pattern recognition algorithms.

3.2 Neural Networks

The synaptic linkage of the neurons in the human brain is emulated by a neural network by abstracting as a graph of nodes (neurons) connected by weighted edges (synapses) [21]. The weights affect the amount of forward propagation going through the network and can be re-adjusted during the backpropagation [22] process when the network enters learning stage. Network is trained by iteratively conducting these two processes for each piece of training information. Neural network forms the base of architectures on which the object detection models are designed.

3.3 Convolutional Neural Networks

Convolutional neural networks [23] can be thought of as multiple image filters working in a simultaneous parallel fashion on large volumes of image data for the purpose of image classification. The architectures mostly involve alternate convolution and pooling layers with activation functions such as ReLU [24] to activate the neurons with weights of higher probability, calibrating the loss function [25] to determine the underlying probability errors and prediction errors and further relaying the information to fully connected layers for decision making and performing the cumulative action of image classification. When it comes to object detection, classification, and recognition, convolutional neural networks have proven to have achieved superior performance pertaining to the effective feature mapping which tends to dramatically improve performance when compared to other deep learning techniques.

3.4 You Only Look Once (YOLO) Object Detection

Redmon et al. [26] proposed YOLO, a regression-based object detection technique with grid-based detections for entire image at once; unlike sliding window techniques that scan with arbitrary window size or region-based techniques which utilize regions of image and might confuse the object with the background; moreover, they require cross-checking the bounding boxes and re-scoring which is not required by YOLO.

The YOLO model tends to scan entire image only once by dividing it into a grid and determining center of object in each grid cell by scoring the class probabilities and drawing bounding boxes around objects. The model has been pretrained on the common objects in context (COCO) dataset containing 80 conditional class probabilities $P_r(\text{Class}_i \mid \text{Object})$ considering there exists an object given that the probability of predicted object is of $\text{Class}_{i.}$

4 Prototype and Implementation

4.1 Prototype Specifications

The device comprises a Raspberry Pi Zero W with 5MP Sony IMX Raspberry Pi Camera interface for image acquisition, with a portable 7800 mAH power bank for power supply of 5 V, 1 A and a set of Bluetooth enabled headphones for audible feedback to the user. The processor is accompanied by a MicroSD card containing the operating system, and YOLO object detection and recognition configuration files required for the algorithm. The Raspberry Pi Zero W is utilized due to the compact

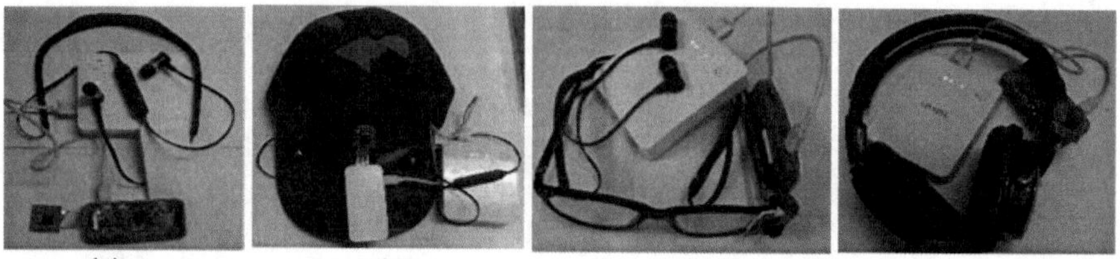

(a) (b) (c) (d)

Fig. 1 Prototype comprising Raspberry Pi Zero W, camera module, Bluetooth enabled headphones and a power bank with USB 2.0 (a) device attached to headphones (b) device attached on a cap, (c) device attached on a pair of glasses, (d) device attached to a pair of Bluetooth enabled headsets

size along with Wi-Fi connectivity for easier access to the cloud server. The ESpeak engine is used for test-to-speech (TTS) conversion along with OpenCV, a computer vision framework based on the Python scripting language has been used for aiding the prototype software development (Fig. 1).

4.2 Implementation

The proposed portable device is capable of object detection and recognition with the use of Internet as well as without the Internet. The prototype is implemented in two portions: Via cloud platform (over the Internet) and on the edge (in the absence of Internet) shown in Fig. 2.

The acquired images are uploaded to YOLO object detection model to perform a series of convolutions and pooling by comparing class probabilities and drawing bounding boxes and ultimately object classification with a label (text) generated to interpret the recognized object. For online process via cloud platform, the model has been deployed on IBM cloud platform with 256 MB RAM and 1 GB of disk memory. The YOLO model is accessed via Flask API through a POST request and the object labels are returned as a response to the processor which further converts text-to-speech and generates the audible feedback. The same model, along with the ESpeak engine is directly deployed on the Raspberry Pi Zero W processor for detections

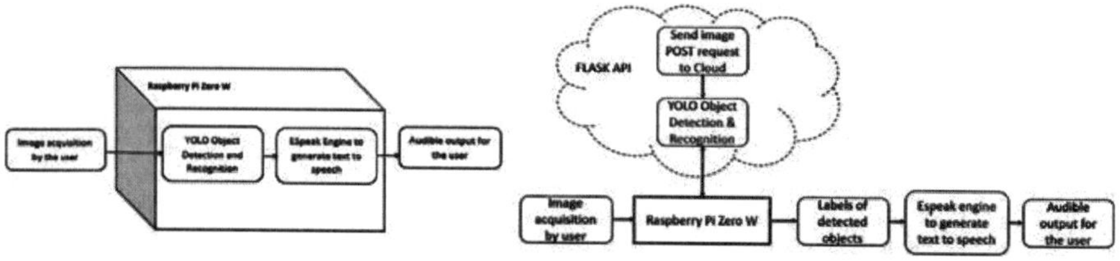

(a) (b)

Fig. 2 Block diagram of device **a** on the edge (offline) and **b** via cloud platform (online) on the edge (offline). The user controls the image acquisition process via a switch provided on the device and at the end of the entire process the user receives a voice message to inform that current detection is completed.

4.3 Algorithm

The algorithm flowchart for device on the edge is shown in Fig. 3a, while Fig. 3b shows the algorithm flowchart for device via cloud platform.

The algorithm initiates with extracting spatial dimensions of the image and initializing the YOLO network weights and configurations stored on the device memory. Object detection and recognition is performed by partitioning the image into a grid and detecting center(x, y) occurring in a grid cell. The detection is executed by calibrating the x, y, w, h co-ordinates and cross-checking with the confidence score threshold. On learning the presence of an object, the class ID and probability (confidence score) are determined. Non-maxima suppression (NMS) is applied to output only those classifications with maximal probability while suppressing the near-by probabilities which are non-maximal. When multiple detections are associated with the same object, a threshold is applied to highlight the bounding box with highest confidence score while darkening the boxes with high IOU and high overlap. until …one highlighted box is retained'. This process is repeated until all the darkened bounding boxes with high IOU are suppressed completely and only one highlighted box is retained (Fig. 4).

Each bounding box per object is determined by its center co-ordinates (x, y) along with the width (w) and height (h). The prediction error, i.e., separation of predicted value and actual value (ground-truth) is given by the intersection over union (IOU). The IOU is given by the union/(predicted-actual). Box confidence score is determined by $P_{r(\text{object})} * \text{IOU}_{(\text{truth pred})}$. It establishes the likelihood of existence of an object in that grid along with the accuracy of prediction by the model. Once the object is specified

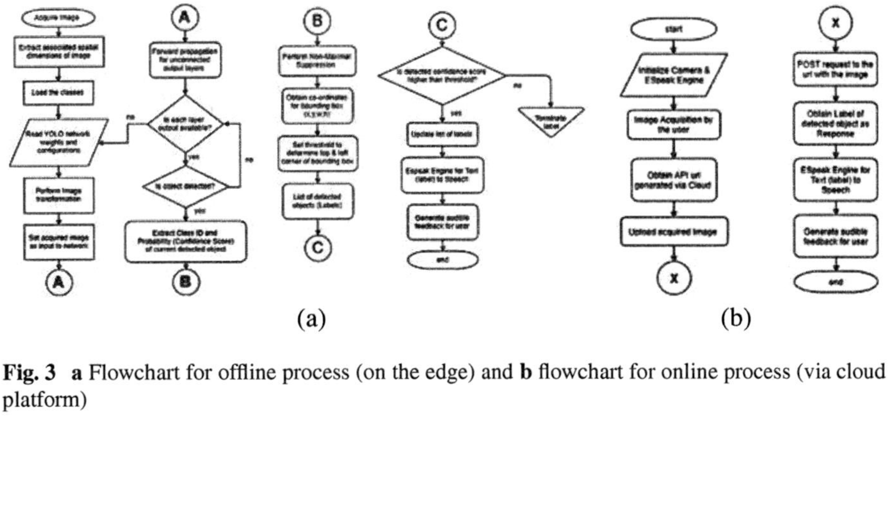

(a) (b)

Fig. 3 **a** Flowchart for offline process (on the edge) and **b** flowchart for online process (via cloud platform)

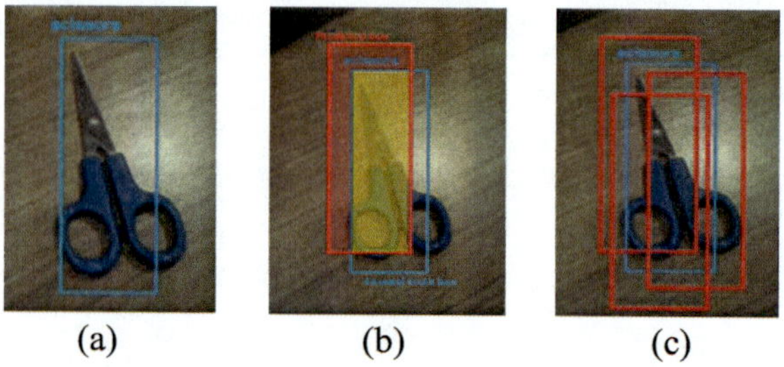

Fig. 4 **a** Center location of object along with height and width of bounding box, **b** IOU, **c** multiple bounding boxes detected by the model which can be removed using NMS

and confidence score is updated for single highlighted bounding box, the class label (object name) is generated which completes the task of object recognition. The label is provided to ESpeak text-to-speech engine for generation of audio and conveyed to the user through Bluetooth enabled headphones. Pertaining to the presence of multiple objects in the same image, a message conveying "Done" is ultimately given as a means to inform the user regarding the completion of task.

4.4 *Placement*

The device can be placed by the user on the glasses, cap, headphone wire, headset corner, backpack straps, over the front pocket, over trouser pocket, on the belt or on the walking cane; as deemed comfortable. Depending on the region to be covered by the camera, the device placement can be changed by the user.

5 Results and Discussion

The YOLO and SSD models were tested for object detection with respect to time taken to run the model and confidence score generated for each detected object. Figure 5 shows the images used to test for both models, in the presence and absence of salt and pepper noise. Figure 6 shows the graph of time and confidence scores

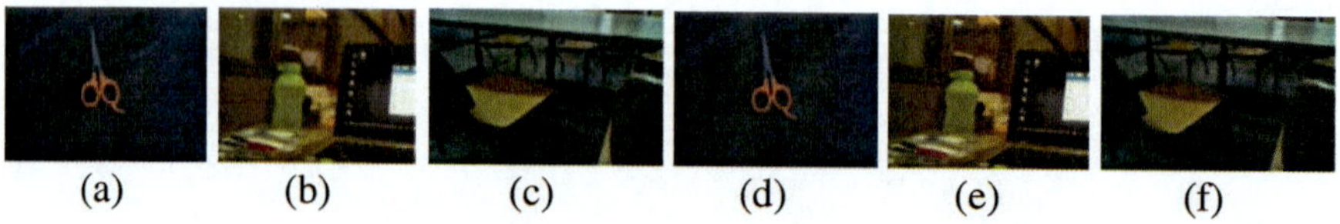

Fig. 5 Images tested for YOLO and SSD. **a** Scissor, **b** bottles and laptop, **c** chairs in absence of noise, respectively, **d** scissor, **e** bottles and laptop, **f** chairs in presence of noise

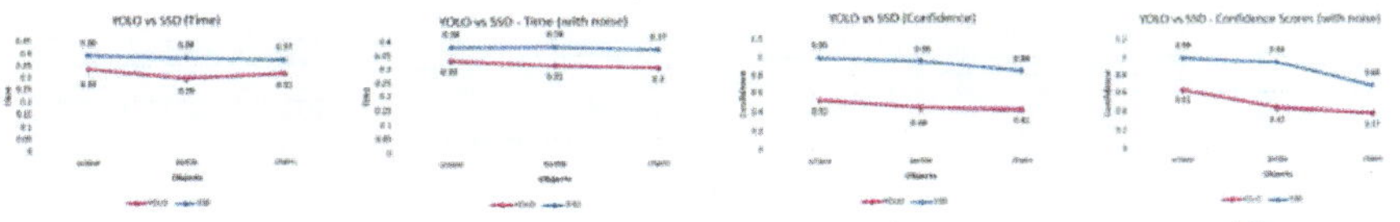

(a) (b) (c) (d)

Fig. 6 Graph of YOLO versus SSD. **a** Time taken to run model for image without noise, **b** time taken to run model for image in presence of noise, **c** confidence score for image without noise, **d** confidence score for image in presence of noise

Table 1 Performance mapping of YOLO and SSD based on time taken to run the model and generated confidence scores for images in presence and absence of noise

Objects		YOLO		SSD	
		Time	Confidence score	Time	Confidence score
Scissors	Without noise	0.33	0.52	0.39	0.99
	With noise	0.33	0.62	0.38	0.99
Bottle and laptop	Without noise	0.29	0.44	0.38	0.95
	With noise	0.31	0.43	0.38	0.94
Chairs	Without noise	0.31	0.41	0.37	0.84
	With noise	0.30	0.37	0.37	0.68

generated by YOLO and SSD interpreted based on Table 1. YOLO performs better in terms of time taken to run the model, while SSD requires more time to perform the same operation. Detection accuracy is determined by the confidence scores which show that SSD provides detections with a higher accuracy compared to YOLO. This is due to the ability of SSD to predict small objects occurring in closer proximity in an image, which YOLO is unable to identify easily and tends to skip similar objects occurring together. The effect of noise can be observed from Fig. 6b, d that the presence of noise affects the YOLO technique leading to poor accuracy in predicting the detected object, whereas the effect on SSD is negligible.

The experiment has been carried out on two versions of the Raspberry Pi family; Raspberry Pi B + v3 and Raspberry Pi Zero W for both online (via cloud) and offline (on edge) detections for objects occurring indoors. The output generated by the algorithm shows the bounding boxes drawn around each individual detected object in the acquired image as shown in Fig. 7 with respective graphs for Table 2. The output generated by the algorithm shows the bounding boxes drawn around each individual detected object in the acquired image as shown in Fig. 7 with the respective graphs for Table 2 in Fig. 8.

For online detections, the interface of both versions of the Raspberry Pi; B + v3 and Zero W performed relatively same considering the time required to process the entire algorithm with average processing time of 5 s. Considering offline detections, the Raspberry Pi Zero W tends to lag due to its lower processing frequency of 1 GHz compared to the 1.4 GHz frequency of Raspberry Pi B + v3. Irrespective of the

(a) (b) (c) (d)

Fig. 7 Results obtained by object detection and recognition performed by the YOLO network as **a** persons, **b** scissors, **c** bottle, **d** apple

Table 2 Comparison of speed (in seconds) for two versions of Raspberry Pi for online and offline detections

Objects	Raspberry Pi B + v3				Raspberry Pi Zero W			
	Via cloud (online)		On the edge (offline)		Via cloud (online)		On the edge (offline)	
	Algorithm	Model	Algorithm	Model	Algorithm	Model	Algorithm	Mode l
Person	2.90	2.51	6.76	6.34	6.72	5.86	30.2	25.2
Scissors	2.75	2.63	6.78	6.41	5.90	4.27	27.4	25.6
Bottle	2.77	2.64	6.69	6.40	6.18	5.29	27.5	25.0
Apple	2.83	2.66	6.58	6.35	6.54	3.22	28.1	25.2

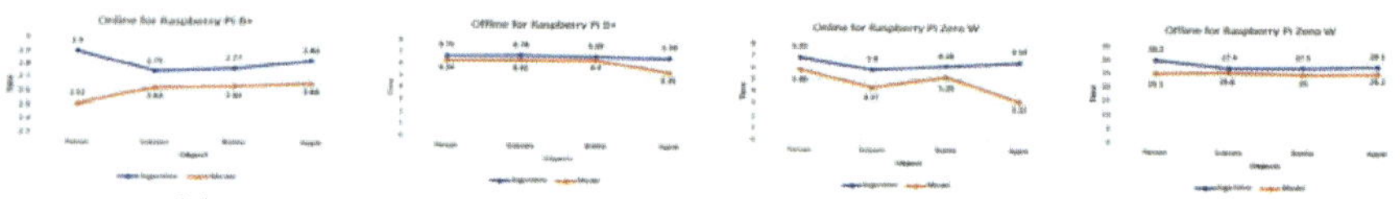

(a) (b) (c) (d)

Fig. 8 Graph showing speed considerations of Raspberry Pi B+ detections. **a** Online (via cloud platform), **b** offline (on the edge) and Raspberry Pi Zero W detections, **c** online (via cloud platform), **d** offline (on the edge)

object, the speed tends to remain constant for both processors individually implying that detection time is not affected by the number of objects present in the image. Trade-off between speed and size is likely depending upon user requirements.

6 Conclusion and Future Scope

Performance mapping of the algorithm is realized by comparing the time required to run the model and the generated prediction accuracy, i.e., confidence score of YOLO and SSD models where it has been noted that YOLO provides a higher speed but performs poor in the presence of noise while SSD provides higher accuracy but consumes more time in its operation. On comparing the performance of YOLO model in the offline and online modes on two versions of the Raspberry Pi; B+ and Zero W, it indicated that the Zero W performed drastically slow in the absence of Internet and low processing power while it performed better when the detections are performed via cloud platform.

The device can be utilized in both online and offline modes making it a viable option for people residing in rural areas, especially the remotely located individuals with minimal access to Internet facilities. Ease of use is of prime significance with flexibility of device placement. Though the prototype has been designed for indoor object detection and recognition for static images; further research shall extend the design for external navigation since the model has been pretrained for objects occurring in exterior scenarios. Applying CNNs to video frame classification is a possibility since there is relatively less work carried out on videos due to the existence of temporal dimension.

References

1. S. Russell, P. Norvig, *Artificial Intelligence: A Modern Approach* (Prentice Hall, Englewood Cliffs, NJ, 1995)
2. J. Yang, B. Jiang, B. Li, K. Tian, Z. Lv, A fast image retrieval method designed for network big data. IEEE Trans. Industr. Inf. **13**(5), 2350–2359 (2017)
3. I. Goodfellow, Y. Bengio, A. Courville, *Deep Learning* (MIT Press, 2016)
4. C. Szegedy, V. Vanhoucke, S. Ioffe, J. Shlens, Z. Wojna, Rethinking the inception architecture for computer vision, in *Proceedings of the IEEE Conference on Computer Vision and Pattern Recognition* (2016), pp. 2818–2826
5. S. Bharathi, A. Ramesh, S. Vivek, Effective navigation for visually impaired by wearable obstacle avoidance system, in *International Conference on Computing, Electronics and Electrical Technologies (ICCEET)* (2012)
6. F.B.H.A. Hanen Jabnoun, Object recognition for blind people based on features extraction, in *International Image Processing Applications and Systems Conference*, Sfax, Tunisia (2014)
7. J. Bai, S. Lian, Z. Liu, K. Wang, D. Liu, Virtual-blind-road following-based wearable navigation device for blind people. IEEE Trans. Consum. Electron. **64**(1), 136–143 (2018)
8. K.U.M. Naveen Kumar, Voice based guidance and location indication system for the blind using GSM, GPS and optical device indicator. Int. J. Eng. Trends Technol. (IJETT) **4**(7) (2013)
9. J. Na, The blind interactive guide system using RFID-based indoor positioning system, in *Lecture Notes in Computer Science*, vol. 4061 (2006), pp. 1298–1305
10. J. Bai, S. Lian, Z. Liu, K. Wang, D. Liu, Smart guiding glasses for visually impaired people in indoor environment. IEEE Trans. Consum. Electron. **63**(3), 258–266 (2017)
11. W. Elmannai, K. Elleithy, Sensor-based assistive devices for visually impaired people: current status, challenges, and future directions. Sensors **17**(3), 565–606 (2017)
12. M. Liang, X. Hu, Recurrent convolutional neural network for object recognition, in *2015 IEEE Conference on Computer Vision and Pattern Recognition (CVPR)* (2015)
13. V. Kuljukin, C. Gharpure, J. Nicholson, G. Osborne, Robot-assisted way finding for the visually impaired in structured indoor environments. Auton. Robots **21**(1), 29–41 (2006)
14. K. Imaizumi, V. Moshnyaga, *Network-Based Face Recognition on Mobile Devices* (IEEE ICCE, Berlin, 2013), pp. 406–409
15. D. Wang, H. Lu, M.-H. Yang, Online object tracking with sparse prototypes. IEEE Trans. Image Process. **22**(1) (2013)
16. S.S. Rautaray, A. Agrawal, Real time hand gesture recognition system for dynamic applications. Int. J. UbiComp (IJU) **3**(1) (2012)

17. M. Ravanbakhsh, M. Nabi, H. Mousavi, E. Sangineto, N. Sebe, Plug-and-play CNN for crowd motion analysis: an application in abnormal event detection. CoRR. abs/1610.00307 (2016)
18. C.M. Bishop, *Pattern Recognition and Machine Learning* (Springer, New York, 2006)
19. P. Viola, M. Jones, Rapid object detection using a boosted cascade of simple features, in *Proceedings of 2001 IEEE Computer Society Conference on Computer Vision and Pattern Recognition* (2001), pp. 8–14
20. K. He, X. Zhang, S. Ren, J. Sun, Deep residual learning for image recognition, in *Proceedings of 2016 IEEE Conference on Computer Vision and Pattern Recognition* (2016), pp. 770–778
21. G.E. Hinton, R.R. Salakhutdinov, Reducing the dimensionality of data with neural networks. Science **313**(5786), 504–507 (2006)
22. A. Krizhevsky, I. Sutskever, G.E. Hinton, Imagenet classification with deep convolutional neural networks, in *Advances in Neural Information Processing Systems 25: 26th Annual Conference on Neural Information Processing Systems* (2012), pp. 1106–1114
23. Y. Bengio, Learning deep architectures for AI. Found. Trends Mach. Learn. **2**(1), 1–127 (2009)
24. H. Ide, T. Kurita, Improvement of learning for CNN with ReLU activation by sparse regularization, in *IEEE International Joint Conference on Neural Networks*, USA (2017, May), pp. 2684–2691
25. D.E. Rumelhart, G.E. Hinton, R.J. Williams, *Learning Internal Representations by Error Propagation* (MIT Press, Cambridge, 1986)
26. J. Redmon, S. Divvala, R. Girshick, A. Farhadi, You only look once: unified, real-time object detection, in *Proceedings of the IEEE Conference on Computer Vision and Pattern Recognition* (2016), pp. 779–788

A Hybrid Approach Based on Lp1 Norm-Based Filters and Normalized Cut Segmentation for Salient Object Detection

Subhashree Abinash and Sabyasachi Pattnaik

1 Introduction

In today's world, with advancement of technology, most machine vision applications are for computer automation. For efficient results, researchers have been working on development of faster and powerful mathematical models which will reduce computer response time and fasten the operation. For example, in military applications while choosing and locking a visual target, we have to choose a model which must have very less time complexity. But in case of medical applications, the time is not a constraint but the accuracy to detect a tumor or any abnormality inside a living object. In real world, it is difficult for machines to identify the objects and backgrounds accurately as compared to humans. Salient object detection in computer vision is very difficult due to complexities present in natural scenes.

The efficiency of the algorithm also affects the identification process [1–3]. Salient object detection is based on efficient identification of a region of interest or particular object instead of segmentation of the scene. During last decade, research on salient object detection has gained momentum due to object tracking and object identification in a given video or scene or a photograph collage [4–8].

To extract foreground from background, different algorithms based on graph cut have been proposed [9–15]. These methods are based on finding the min-cut of undirected edge-weighted graphs, multiclass graph cut, gradient-based kernel selection with graph cut, but with increase in complexity of the mathematical models there is a

S. Abinash (✉)
Synergy Institute of Engineering and Technology, Dhenkanal, India
e-mail: abinash77@gmail.com
URL: https://www.synergyinstitute.net

S. Pattnaik
Fakir Mohan University, Balasore, India
e-mail: spattnaik40@yahoo.co.in
URL: https://www.fmuniversity.nic.in

S. Patnaik et al. (eds.), *Advances in Machine Learning and Computational Intelligence*, Algorithms for Intelligent Systems, https://doi.org/10.1007/978-981-15-5243-4_64

high chance of increase in time complexity. In real-time application, the algorithms should have less time complexity. In our work, we have tried to achieve low time complexity without increasing the complexity of the model.

In this paper, we have proposed two models based on graph cut for salient object detection in a given image. Previously, the algorithm proposed in [12] has been modified, in which we have proposed two new weight functions based on Lp1 norm to improve the efficacy of the model. Two standard algorithms are considered to prove the superiority of our model by considering validity measures like Dice and Jaccard coefficients as well as mean square error (MSE).

In this paper, Sect. 2 is consists of fundamentals of graph cut algorithm and its application on image segmentation. Section 3 explains the proposed graph cut algorithms. We have discussed different experimental results in Sect. 4 and concluded the work in Sect. 5.

2 Graph Cut

The standard image partitioning based on graph cut algorithm, proposed by Shi and Malik [18], is known as normalized cut. This algorithm segments a group of pixels where the pixels are mapped to a weighted undirected graph. Here, 'V' denotes a pixel or node and 'E' denotes the edge or the distance between two pixels. This is followed by the construction of a undirected weighted graph $G = (V, E)$, where a nonnegative weight W has been assigned to each weight.

The weight is defined [12] as follows.

$$W(i, j) = \mathrm{e}^{\frac{-\|F(i)-F(j)\|^2}{\sigma_I}} \times \begin{cases} \mathrm{e}^{\frac{-\|X(i)-X(j)\|^2}{\sigma_X}} & \text{if} \|X(i) - X(j)\| < r \\ 0 & \text{otherwise} \end{cases} \tag{1}$$

Here, $F(i)$ denotes the feature based on intensity and $X(i)$ denotes the spatial location of ith node. The radius of the circle is denoted by r, and the pixels are placed within this circle. If $S1$ and $S2$ are two disjoint sets of image pixels, then

$$S_1 \cup S_2 = V$$
$$S1 \cap S2 = \varphi$$

The cut, in graph cut, is defined as

$$\mathrm{cut}(S_1, S_1) = \sum_{i \in S_1, j \in S_2} W(i, j) \tag{2}$$

To efficiently partition the graph, the cut value in (2) should be minimized. The graph cut has been reformulated by Shi and Malik [18] and is given as follows

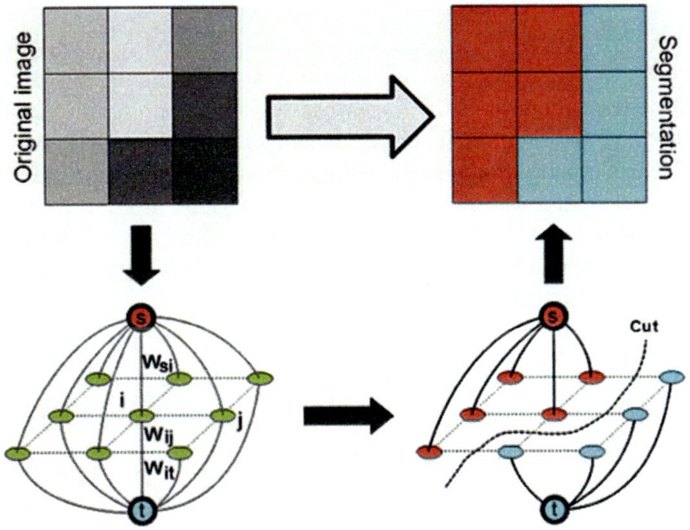

Fig. 1 Graph cut

$$\mathrm{Ncut}(S_1, S_2) = \frac{\mathrm{cut}(S_1, S_2)}{\mathrm{assoc}(S_1, V)} + \frac{\mathrm{cut}(S_2, S_1)}{\mathrm{assoc}(S_2, V)} \tag{3}$$

In a graph, $\mathrm{assoc}(S_1, V) = \sum_{i \in S_1, j \in V} W(i, j)$ is the total number of connection from nodes in $S1$ to all the available nodes (Fig. 1).

3 Proposed Models

According to many literatures [9, 16–23], efficacy of graph cut depends on the appropriate edge selection where the edges are assigned with a function of weights. This motivated us to propose two algorithms based on graph-cut-based algorithms for detection of salient objects in natural scene. Instead of gray images, we have considered color model based on RGB values of image pixels. Here, *R*, *G* and *B* channels are considered separately to estimate features and segmentation algorithm based on graph cut is applied on these features not the whole image to reduce the complexity of the model. To handle the outliers and noise might have corrupted the image; we have proposed weights that are function of neighborhood. Here, the high frequency components have been removed from individual color channels and the local properties of the scenes are retained to avoid loss of information.

3.1 Model I

In Model I, we have proposed a Lp1 norm-based Gaussian noise filter and this filter is applied on *R*, *G* and *B* color channels. The output of the filter is used to calculate the weight matrix not the whole image, which helps in reducing the time complexity of the proposed model. The high frequency components are removed from each channel with the help of Lp1 norm-based Gaussian filter. These further increase the efficiency of graph-cut-based segmentation algorithm. The new weight function based on Gaussian feature is redefined as

$$W(m,n) = \mathrm{e}^{\frac{-\|G(m)-G(n)\|^2}{\sigma_I}} \times \begin{cases} \mathrm{e}^{\frac{-\|X(m)-X(n)\|^2}{\sigma_X}} & \text{if} \|X(m) - X(n)\| < r \\ 0 & \text{otherwise} \end{cases} \quad (4)$$

where Gaussian feature is denoted by $G(m)$ and $X(n)$ denotes the *mth* node spatial location. The radius of the circle, within which the image pixels are located, is denoted by r.

$$\|G(m) - G(n)\| = |(G_r(m) - G_r(n))| + \left|(G_g(m) - G_g(n))\right| + |(G_b(m) - G_b(n))| \quad (5)$$

3.2 Model II

In Model II, we have proposed a Lp1 norm-based median filter which has been applied to each available color channels which can efficiently handle irregularities in an image as well as noisy pixels. The weight function has been redefined as

$$W(m,n) = \mathrm{e}^{\frac{-\|M(m)-M(n)\|^2}{\sigma_I}} \times \begin{cases} \mathrm{e}^{\frac{-\|X(m)-X(n)\|^2}{\sigma_X}} & \text{if} \|X(m) - X(n)\| < r \\ 0 & \text{otherwise} \end{cases} \quad (6)$$

where median feature is denoted by $M(m)$ and $X(n)$ denotes the mth node spatial location. The radius of the circle, within which the image pixels are located, is denoted by r.

$$\|M(m) - M(n)\| = |(M_r(m) - M_r(n))| + \left|(M_g(m) - M_g(n))\right| + |(M_b(m) - M_b(n))| \quad (7)$$

4 Results and Discussion

To prove the superiority of the proposed algorithm performance, we have considered Berkley image segmentation datasets and COREL image segmentation datasets with natural scenes. We have carried out different experiments on the images from each segmentation dataset. Otsu and C-means algorithms are considered to prove the efficiency of the proposed algorithms as compared to these existing standard algorithms. The validity indices like Dice coefficient (DC), Jaccard coefficient (JC) and mean square error (MSE) are considered for measuring the algorithm performance. Let $S1$ be the segmented image and $S2$ be the ground truth, and both are of size $M \times N$. Then, JC is defined as

$$\mathrm{JC}(S_1, S_2) = \frac{S_1 \cap S_2}{S_1 \cup S_2} \tag{8}$$

DC is defined as

$$\text{Dice Coefficient}(S_1, S_2) = \frac{2 \times |S_1 \cap S_2|}{|S_1| + |S_2|} \tag{9}$$

DC and JC values closure to unity prove the high efficiency of the model. The formula for MSE is

$$\mathrm{MSE} = \frac{1}{M \times N} \sum_{i=1}^{M \times N} |S_1(i) - S_2(i)|^2. \tag{10}$$

MSE value close to zero proves the high efficiency and close to unity shows low efficiency.

Figure 2 shows the salient objects detected in Berkley image segmentation dataset. Figure 2a and b shows the original images and ground truth images, respectively. The results for Otsu thresholding are shown in Fig. 2c, and results for C-means clustering are shown in Fig. 2d. The JC for the segmented images is given in Table 1. Table 1 shows that Otsu thresholding and C-means clustering are not able to identify the salient object from the scene. Due to this, the JC values are close to zero, which means less efficient algorithms. But JC of proposed algorithms is very high, i.e., close to 1, which means the efficacy of proposed graph cut methods is high.

Tables 2 and 3 show the DC and MSE values of images given in Fig. 2. Values in both the tables prove that the proposed graph cut methods are providing better results as compared to the existing ones.

The proposed graph cut methods have also been used on COREL image segmentation dataset. The original image from COREL image segmentation dataset is given in Fig. 3a and the corresponding ground truth in Fig. 3b. Otsu thresholding and C-means

Fig. 2 Segmented results for Berkley image segmentation dataset: **a**, **b** original image and ground truth, and **c**, **d** Otsu and C-means

clustering algorithm are also applied on these images, and the results are shown in Fig. 3c and d, respectively. As we can see in Fig. 3, the efficiency of both Otsu thresholding and C-means clustering is less than the proposed graph cut methods. The JC, DC and MSE values are shown in Tables 1, 2 and 3, respectively. The superiority of performance of the proposed graph-cut-based methods from the existing algorithms is shown in these tables.

Table 1 Jaccard coefficient

Image no.	Otsu thresholding	C-means	Model I	Model II
Berkley image segmentation dataset				
1	0.4001	0.2311	0.89	0.91
2	0.1911	0.2012	0.95	0.96
3	0.8821	0.8921	0.92	0.93
4	0.2221	0.4923	0.93	0.94
5	0.8311	0.7012	0.92	0.93
6	0.2912	0.3111	0.91	0.92
COREL image segmentation dataset				
1	0.3311	0.3322	0.89	0.90
2	0.6221	0.5231	0.93	0.91
3	0.7322	0.7331	0.93	0.92
4	0.4411	0.5722	0.92	0.91
5	0.3411	0.1711	0.92	0.93
6	0.3733	0.3821	0.91	0.92

Table 2 Dice coefficient

Image no.	Otsu thresholding	C-means	Model I	Model II
Berkley image segmentation dataset				
1	0.5721	0.3812	0.94	0.95
2	0.3111	0.3414	0.95	0.96
3	0.9122	0.9121	0.93	0.92
4	0.3631	0.6522	0.94	0.95
5	0.9133	0.8213	0.96	0.97
6	0.4521	0.4721	0.94	0.95
COREL image segmentation dataset				
1	0.5011	0.4921	0.91	0.92
2	0.7622	0.6831	0.97	0.96
3	0.8431	0.8511	0.93	0.91
4	0.6114	0.6414	0.95	0.93
5	0.5012	0.2924	0.93	0.92
6	0.5421	0.5544	0.91	0.90

Table 3 MSE

Image no.	Otsu thresholding	C-means	Model I	Model II
Berkley image segmentation dataset				
1	0.2921	0.4911	0.07	0.09
2	0.6042	0.5741	0.02	0.03
3	0.0944	0.0934	0.05	0.03
4	0.3013	0.0621	0.02	0.03
5	0.0422	0.1143	0.03	0.04
6	0.0411	0.36	0.04	0.05
COREL image segmentation dataset				
1	0.1912	0.2001	0.08	0.09
2	0.0631	0.1423	0.05	0.04
3	0.0440	0.0441	0.02	0.03
4	0.1304	0.1012	0.04	0.05
5	0.2302	0.4611	0.05	0.03
6	0.1703	0.1611	0.03	0.04

5 Conclusion

The performance of graph-cut-based segmentation algorithm depends upon the proper choice of weight function. In this work, we have formulated two different weight functions for efficient salient object detection. The proposed weight functions depend upon the color features instead of the whole image. We have considered three color channels, R, G and B, as our color features. The proposed models have been tested on two standard image segmentation datasets, i.e., Berkley image segmentation dataset and COREL image segmentation dataset. The validity indices like JC, DC and MSE are considered to measure the performance of proposed models. Results of all validity indices conclude the effectiveness of these models over the standard ones.

Fig. 3 Segmented results for COREL image segmentation dataset: **a**, **b** original image, ground truth, and **c**, **d** Otsu, C-means

References

1. T. Liu et al., Learning to detect a salient object. IEEE Trans. Pattern Anal. Mach. Intell. **33**(2), 353–367 (2011)
2. A. Borji, M. Cheng, H. Jiangand, J. Li, Salient object detection: a benchmark. IEEE Trans. Image Process. **24**(12), 5706–5722 (2015)
3. M. Donoser, M. Urschler, M. Hirzerand, H. Bischof, Saliency driven total variation segmentation, in *2009 IEEE 12th International Conference on Computer Vision*, Kyoto (2009), pp. 817–824
4. C. Kananand, G. Cottrell, Robust classification of objects, faces, and flowers using natural image statistics, in *2010 IEEE Computer Society Conference on Computer Vision and Pattern Recognition*, San Francisco, CA (2010), pp. 2472–2479

5. D. Waltherand, C. Koch, Modeling attention to salient proto-objects. Neural Netw. **19**(9), 1395–1407 (2006)
6. S. Goferman, A. Tal, L. Zelnik-Manor, Puzzle-like collage. Comput. Graph. Forum **29**, 459–468 (2010)
7. T. Liu, J. Wang, J. Sun, N. Zheng, X. Tang, H. Shum, Picture collage. IEEE Trans. Multimedia **11**(7), 1225–1239 (2009)
8. S. Abinash, Two novel graph theory based algorithms for salient object detection, in *2018 4th International Conference on Computing Communication and Automation (ICCCA)*, Greater Noida, India (2018), pp. 1–6
9. P. Zhao, J. Yu, H. Zhang, Z. Qin, C. Wang, How to securely outsource finding the min-cut of undirected edge-weighted graphs. IEEE Trans. Inf. Forensics Secur. **15**, 315–328 (2020)
10. J. Williams et al., 3D segmentation of trees through a flexible multiclass graph cut algorithm. IEEE Trans. Geosci. Remote Sens. **58**(2), 754–776 (2020)
11. J. Dogra, S. Jain, M. Sood, Gradient-based kernel selection technique for tumour detection and extraction of medical images using graph cut, in *IET Image Processing*, vol. 14, no. 1 (2020), pp. 84–93
12. S. Qu, Q. Li, M. Chen, Supervised image segmentation based on superpixel and improved normalised cuts, in *IET Image Processing*, vol. 13, no. 12, (2019) pp. 2204–2212
13. M. Wang, W. Jia, Q. Liu, F. Miao, Image spectral data classification using pixel-purity kernel graph cuts and support vector machines: a case study of vegetation identification in Indian pine experimental area, in *IGARSS 2019—2019 IEEE International Geoscience and Remote Sensing Symposium*, Yokohama, Japan (2019), pp. 3736–3739
14. K. Abiko, K. Uruma, M. Sugawara, S. Hangai, T. Hamamoto, Image segmentation based graph-cut approach to fast color image coding via graph Fourier transform, in *2019 IEEE Visual Communications and Image Processing (VCIP)*, Sydney, Australia (2019), pp. 1–4
15. T. Liu, J. Sun, N. Zheng, X. Tang and H. Shum, "Learning to Detect A Salient Object," 2007 IEEE Conference on Computer Vision and Pattern Recognition, Minneapolis, MN, 2007, pp. 1–8
16. Y. Zhai, M. Shah, Visual attention detection in video sequences using spatiotemporal cues, in *Proceedings of the 14th ACM International Conference on Multimedia (MM '06)*. ACM, New York (2006), pp. 815–824
17. M. Cheng, N.J. Mitra, X. Huang, P.H.S. Torrand, S. Hu, Global contrast based salient region detection, in *IEEE Transactions on Pattern Analysis and Machine Intelligence*, vol. 37, no. 3 (2015), pp. 569–582
18. J. Shi, J. Malik, Normalized cuts and image segmentation, in *IEEE Transactions on Pattern Analysis and Machine Intelligence*, vol. 22, no. 8 (2000), pp. 888–905
19. C. Yang, L. Zhang, H. Lu, X. Ruan, M. Yang, Saliency detection via graph—based manifold ranking, *2013 IEEE Conference on Computer Vision and Pattern Recognition*, Portland, OR (2013), pp. 3166–3173
20. B. Jiang, L. Zhang, H. Lu, C. Yang , M. Yang, Saliency detection via absorbing Markov Chain, *2013 IEEE International Conference on Computer Vision*, Sydney, NSW, pp. 1665–1672 (2013)
21. J. Dong, J. Xue, S. Jiang, K. Lu, A novel approach of multiple objects segmentation based on graph cut, *2018 IEEE Conference on Multimedia Information Processing and Retrieval (MIPR)*, Miami, FL (2018), pp. 328–333
22. O.M. Oluyide, J. Tapamo, S. Viriri, Automatic lung segmentation based on graph cut using a distance-constrained energy, in *IET Computer Vision*, vol. 12, no. 5 (2018), pp. 609–615
23. X. Chen, L. Pan, A survey of graph cuts/graph search based medical image segmentation. IEEE Rev. Biomed. Eng. **11**, 112–124 (2018)

Decentralizing AI Using Blockchain Technology for Secure Decision Making

Soumyashree S. Panda and Debasish Jena

1 Introduction

The fast evolution of Blockchain is paving the way for a system, where multiple parties who do not know each other or who do not trust each other can work together to achieve consensus. System designed using Blockchain works like a classical distributed system where all the entities have an equal privilege. Even though they are geographically scattered, they are connected via various types of networks. It was proposed by S. Nakamoto in 2008 [1]. Traditional transaction management systems face issues like privacy, single-point failure, cost because it depends upon a centralized trusted entity for certificate and key management [2]. Decentralization a key attribute of Blockchain can be practiced to address the above issues. The very first usage that manifested Blockchain is Bitcoin. But its use is no more restricted to the financial domain only. Indeed, it has achieved much acclaim in other domains like government, agriculture, manufacturing, enterprises, supply chain [3].

Another promising technology that is getting huge attraction is artificial intelligence (AI) that allows a machine to have cognitive functions to learn, theorize, and conform to the data it gathers. AI can deal with massive amount of data having a lot of information at a particular time, and even it can outperform human capabilities in certain areas. The huge amount of data produced by sensing systems, social networks, and applications has helped to the rise of AI [4]. But since the data, AI deal with is scattered over the Internet, there should be some mechanism which can securely share this data. Apart from this, since data may belong to a number of stakeholders who do not trust each other, so the privacy of the data must also be ensured. Moreover, Blockchain technology can be used to provide a decentralized and tamperproof

S. S. Panda (✉) · D. Jena
Information Security Laboratory, IIIT Bhubaneswar, Bhubaneswar, Odisha 751003, India
e-mail: C117011@iiit-bh.ac.in

D. Jena
e-mail: debasish@iiit-bh.ac.in

S. Patnaik et al. (eds.), *Advances in Machine Learning and Computational Intelligence*, Algorithms for Intelligent Systems, https://doi.org/10.1007/978-981-15-5243-4_65

environment that can facilitate data sharing in secure and trustful manner. Hence, AI can become more powerful and secure when integrated with Blockchain technology. So in this paper, we have studied how the integration of Blockchain technology with AI can provide better security and privacy for data [5]. It will not be surprising to see that using Blockchain-based smart contracts and enough data, AI can become one of the most powerful tools to enhance cyber-security. This is because it will be able to check massive amount of data in a very less time and reduce risks more quickly. Thus, it can provide more accurate prediction and decision [6].

The rest of the paper organized in this way. In Sect. 2, recent work on AI and Blockchain technology is described. In Sect. 3, Blockchain and its consensus mechanisms are explained. In Sect. 4, AI and its security and privacy requirements are discussed. The advantages of using Blockchain in AI system are presented in Sect. 5. Finally, Sect. 6 concludes the paper with the future work.

2 Related Work

A number of factors should be addressed in order to make AI secure. These factors include reducing bias, enhancing understandability of AI models, ensuring trust among different stakeholders and providing transparency. Providing the origin of data and its flow within the system, i.e. data provenance can be one of the aspects for providing trust among different stakeholders in AI. In [7], the authors have addressed the issue of data provenance in relational model. The authors in [8] expand provenance to decentralized/distributed systems. They have modeled the nodes which are kept in tables and codes as a set of declarative rules in [7]; the authors have proposed a method which contains an expressive provenance model that is capable of finding the origin of any data with different granularity. McDaniel et al. [9] have studied and analyzed the privacy and security aspect of a provenance system. Wylot et al. [10] analyzed provenance methods employed in the linked data model for application domains such as social networks.

The fast evolution of Blockchain technology offers a number of opportunities for decentralized/distributed trust computing. Though this is a very new technology, ample amount of work has been conducted in integrating this technology to provide decentralized and secure solutions to AI. Recently, in [11], a framework called ProvChain is presented and implemented which gathers and verifies origin of cloud data by storing this into Blockchain transactions. This framework provides a secure, reliable, trustful, and tamperproof features with low overhead. Lately, in [5], a model known as "SecNet" is proposed which combines Blockchain technology with AI to provide security to the entire system. This architecture ensures that the privacy and security of the data shared among different users of the system are maintained. The authors have also performed the security analysis of the architecture which proved that it resists DDOS attack. In [12], the authors have presented a secure and tamper-proof scientific data origin management system using Blockchain. They have used smart contracts, and an open provenance model (OPM) to store records that ensures the validity of data origin against unauthorized alterations.

3 Blockchain

Blockchain, the key element of Bitcoin, has been growing at an unbelievable pace over the last few years and its application is not limited to digital currency anymore. A Blockchain is basically a distributed database used by the interested network elements in a P2P network [13]. The distributed database records the transactions in the blocks of the Blockchain, and each block is connected to its previous block through a hash function to maintain the chain as shown in Fig. 1. The network elements/nodes will receive a pair of public key and private key upon registering to the Blockchain network which can be used for encryption and decryption purpose. In addition to this, the private key also helps to sign transactions in Blockchain network and public key works as a unique identifier for each element. Whenever a node wants to communicate with another node belonging to the same Blockchain network, the source node signs the transaction with its private key and broadcasts it to its peers at one-hop distance. Once the peers of the source node get the signed transactions, first they verify its authenticity. If it is valid, then only the peer nodes will re-transmit it to other peers. The transactions, which are received by all the nodes of the Blockchain network and are validated, are grouped into a timestamped block by few nodes designated as miners [14]. A consensus algorithm is used to select a block, among the number of blocks created by the miners, which will be added to the Blockchain network.

3.1 Consensus

Achieving consensus is crucial in a multiparty environment since all the parties take part in the decision-making process. The classical methods for attaining consensus will not work for Blockchain system. Consensus is required to attach a new block to the Blockchain in case of Bitcoin. Those entities that construct and propose a new block are known as minors. A Bitcoin network can have multiple miners, and all these minors may construct a block having different transaction [15].

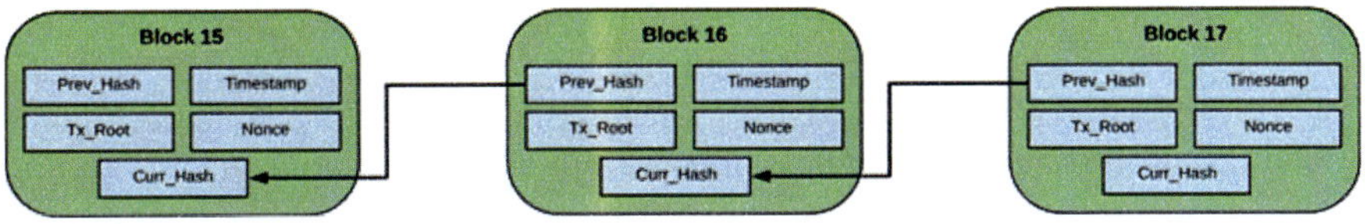

Fig. 1 Blocks in blockchain

Table 1 Consensus mechanisms

	Environment	Energy efficiency	Degree of de-centralization	Transaction throughput	Scalability (in terms of number of nodes)	Reward	Examples
PoW	Open	Very poor	High (Entirely Decentralized)	Very low	Unlimited	Yes	Bitcoin
PoS	Open	Poor (Better than POW)	High	Low	Unlimited	Yes	Peercoin
PBFT	Permissioned	Good	Low	Average	Limited	Mostly no	Hyperledger
Paxos	Permissioned	High	Low	Very high	Limited	Mostly no	Ethereum

In case of permissioned Blockchain, smart contracts are used to attain consensus. Smart contracts are self-executing piece of program which contains the terms and conditions of the network. Some of the relevant consensus mechanisms have been summarized in Table 1.

4 Artificial Intelligence

Artificial intelligence is the intelligence shown by the machines, contrary to the natural intelligence shown by humans.

In computer science, artificial intelligence (AI), sometimes called machine intelligence, is intelligence demonstrated by machines, in contrast to the natural intelligence displayed by humans. In other words, it is a computation tool that is capable of substituting for human intelligence in carrying out certain works [16]. Leading AI textbooks AI has defined as the field the reads about "intelligent agents." An intelligent agent is any device that understands its surrounding and takes necessary actions to achieve objective [1]. The fields of AI include machine learning, deep learning, natural language processing, robot, etc. Date is the key component in AI, and it can substantially increase the performance of AI if data can be networked systematically and properly merged. AI can be made more powerful and efficient by ensuring sharing of data which is scattered across different stakeholders [17].

AI can be divided into two categories, namely (a) weak AI and (b) strong AI.

- Weak AI: It is also known as narrow AI, and it generally used to perform narrow tasks or some predefined tasks. Machines under weak AI simulate human intelligence passively. An example of narrow intelligence can be "Siri."
- Strong AI: In strong AI, the system is capable to perform tasks like human; i.e., they will have human like high cognitive ability. The machines under strong AI are believed to have common sense, sentience, and self-awareness. Strong AI

is capable of simulating human intelligence actively and can perform different intelligent tasks.

At present, all the existing AI systems belong to weak AI type. It is believed that it would take more time to realize strong AI for humans [4]. But nowadays, AI is being used in almost all the fields of our day-to-day life-like speech text input, customized network shopping, various intelligent answering systems. Applications based on AI have helped and will continue to help in long term in different research domains [18].

4.1 Security and Privacy Issues in AI Applications

As already stated, AI is being applied in many areas and a number of people are getting benefited from it. However, AI can also be a way to steal the private data and can thus launch a number of security attacks by illegal users [12]. There exists a number or threats in different applications of AI which are described below.

- Technical Faults: since AI is a new technology and it is at the preliminary stage of development, it may not be secure due to technical faults as a huge amount of data is handled by AI systems. These data can be highly sensitive information which needs to be handled with great care.
- Data Alteration: The freshness of data set is crucial for an AI system during training phase. Malicious stakeholders can inject false samples into the data set to reduce the efficiency and performance of the AI system. Attackers use data with similar features but incorrect labels to distort the training phase.
- Impersonation Attack: Adversaries can impersonate valid users and can inject fakes samples of data to the data set to hamper the performance of the system.
- DDOS Attack: Since AI system works in a centralized manner, if an adversary tries to disturb the central system, then the entire AI system fails.
- Data Privacy and Anonymity: In an AI system, the data set can consists of highly sensitive information like healthcare information. These data are generally scattered among different systems in the Internet. Hence, utmost importance care should be taken while handling such sensitive data. In addition to it, sharing of the same should also do securely among the different stakeholders so that data integrity will be maintained. Apart from this, since in some cases like in health care, the patient's personal information should not be revealed without his or her consent while sharing his or her health records.
- Authentication: Only genuine and valid stakeholders should be allowed to access the data set. So that uniqueness of data will be ensured.

5 Integration of Blockchain with AI

This section describes how Blockchain technology can be used to mitigate the above-mentioned issues of an AI system and also presents the extra security benefits. As already stated, Blockchain technology can be used to provide a decentralized and tamperproof environment that can facilitate data sharing in secure and trustful manner. Hence, AI can become more powerful and secure when integrated with Blockchain technology. So in this paper, we have studied how the integration of Blockchain technology with AI can provide better security and privacy for data. It will not be surprising to see that using Blockchain-based smart contracts and enough data, and AI can become one of the most powerful tools to enhance cyber-security [5]. The features like openness, distributed, verifiable, and permanent makes Blockchain a suitable technology to rely upon for security and privacy. Using Blockchain technology a distributed architecture can be designed for AI system where multiple stakeholders who do not trust each other can come together and take part in computation. Since the Blockchain is an immutable database, malicious stakeholders cannot forget the system.

- Data Alteration: Blockchain technology eliminates the trust requirement in case of multiparty computations. This guarantees data integrity in an AI system. Apart from this, the distributed aspect of Blockchain completely eliminates single-point failure. Hence, increases the reliability of AI systems.
- DDOS Attack: The distributed aspect of Blockchain technology helps to resist DoS or DDoS attack in systems where it is utilized. Besides, Blockchain transactions are expensive so the adversary gets de-motivated to perform this attack.
- Data Privacy and Anonymity: Decentralization of data ensures that the sensitive data like personal information is not under control of any third party. Besides, the anonymity aspect of Blockchain can be exploited to provide privacy to the users in certain AI systems like health care and data sharing systems.
- Authentication: Since in Blockchain, every user has a unique address which acts like a certificate. Besides that, all transactions are signed using the private key (provided by the Blockchain system) which ensures the authentication of the messages as well as the sender.
- Impersonation Attack: Blockchain helps to resist impersonation attack as an entity can possess maximum one identity at a particular time. Hence, an adversary cannot fake identities to dominate the network.

6 Conclusion

This paper presents an in-depth survey on the Blockchain technology and AI with a special focus on how adoption of Blockchain technology in AI systems can empower AI by providing a secure and trustful environment. The paper started by defining the concept of Blockchain technology which is a distributed and tamperproof database. Blockchain technology accomplishes immutability by using the distributed consensus mechanisms. Hence, a trustless environment can be designed for communication using Blockchain. Next, a brief introduction about AI has been presented in the paper followed by the security and privacy requirements of it. Following the degree of decentralization that Blockchain has obtained in crypto-currency networks, Blockchain is considered as the possible solution to decentralizing the AI. A detailed discussion on how Blockchain can be used to provide the required security, and privacy to an AI system is also presented in the paper. As the future work, we plan to (a) explore more about AI and its applications and (b) design and implement security protocol using Blockchain technology for AI application area.

References

1. S. Nakamoto, *Bitcoin: A Peer-to-Peer Electronic Cash System* (2008)
2. A. Dorri, M. Steger, S.S. Kanhere, R. Jurdak, Blockchain: a distributed solution to automotive security and privacy. IEEE Commun. Mag. **55**(12), 119–125 (2017)
3. C. Shen, F. Pena-Mora, Blockchain for cities—a systematic literature review. IEEE Access **6**, 76787–76819 (2018)
4. J.H. Li, Cyber security meets artificial intelligence: a survey. Front. Inf. Technol. Electron. Eng. **19**(12), 1462–1474 (2018)
5. K. Wang, J. Dong, Y. Wang, H. Yin, Securing data with Blockchain and AI. IEEE Access **7**, 77981–77989 (2019)
6. A.B. Kurtulmus, K. Daniel, Trustless machine learning contracts; evaluating and exchanging machine learning models on the Ethereum blockchain. arXiv preprint arXiv:1802.10185 (2018)
7. A. Woodruff, M. Stonebraker, Supporting fine-grained data lineage in a database visualization environment, in *Proceedings 13th International Conference on Data Engineering* (IEEE, 1997, April), pp. 91–102)
8. B.T. Loo, T. Condie, M. Garofalakis, D.E. Gay, J.M. Hellerstein, P. Maniatis, et al., Declarative networking: language, execution and optimization, in *Proceedings of the 2006 ACM SIGMOD International Conference on Management of Data* (ACM, 2006, June), pp. 97–108
9. A.A. Jabal, E. Bertino, SimP: Secure interoperable multi-granular provenance framework, in *2016 IEEE 12th International Conference on e-Science (e-Science)* (IEEE, 2016, October), pp. 270–275
10. M. Wylot, P. Cudr-Mauroux, M. Hauswirth, P. Groth, Storing, tracking, and querying provenance in linked data. IEEE Trans. Knowl. Data Eng. **29**(8), 1751–1764 (2017)
11. A. Chen, Y. Wu, A. Haeberlen, B.T. Loo, W. Zhou, Data provenance at internet scale: architecture, experiences, and the road ahead, in *CIDR* (2017, January)
12. A. Ramachandran, M. Kantarcioglu, SmartProvenance: a distributed, blockchain based dataprovenance system, in *Proceedings of the Eighth ACM Conference on Data and Application Security and Privacy* (ACM, 2018, March), pp. 35–42
13. P.D. McDaniel, K.R. Butler, S.E. McLaughlin, R. Sion, E. Zadok, M. Winslett, Towards a secure and efficient system for end-to-end provenance, in *TaPP* (2010, February)
14. B.K. Mohanta, D. Jena, S.S. Panda, S. Sobhanayak, Blockchain technology: a survey on applications and security privacy challenges. Internet of Things, 100107 (2019)
15. T.M. Fernndez-Carams, P. Fraga-Lamas, A review on the use of blockchain for the Internet of Things. IEEE Access **6**, 32979–33001 (2018)
16. X. Liang, S. Shetty, D. Tosh, C. Kamhoua, K. Kwiat, L. Njilla, Provchain: a blockchain-based data provenance architecture in cloud environment with enhanced privacy and availability,

in *Proceedings of the 17th IEEE/ACM international symposium on cluster, cloud and grid computing* (IEEE Press, 2017, May), pp. 468–477
17. B.N. Harini, T. Rao, An extensive review on recent emerging applications of artificial intelligence. AsiaPacific J. Convergent Res. Interchange **5**(2), 79–88 (2019)
18. B.K. Mohanta, S.S. Panda, D. Jena, An overview of smart contract and use cases in blockchain technology, in *2018 9th International Conference on Computing, Communication and Networking Technologies (ICCCNT)* (IEEE, 2018, July), pp. 1–4

Speech Recognition Using Spectrogram-Based Visual Features

Vishal H. Shah and Mahesh Chandra

1 Introduction

The fundamental idea of audio-visual speech recognition is to obtain information from visual aids and use this data to complement acoustic recognition processes [1]. This visual aid is usually provided in parallel with the speaker's audio instances, in the form of either still images or video clips [2]. However, creation of effective databases and selection of ideal feature extraction techniques are critical to the performance of the whole process. Besides choosing effective features, combination of different types of features has also to be taken into consideration. In this paper, we have employed the early integration technique for generation of hybrid audio-visual features.

Speech recognition processes rely heavily on acoustic data. A crucial step in analysing this data is audio feature extraction [3]. Cepstral features are widely used for this purpose, and some of the techniques belonging to this family are the LPC features, the BFCC features, HPC features and the MFCC features. Mel frequency cepstral coefficients have been used for speech recognition as audio features only as well as along with visual features in research field [1, 4, 5].

Decades of research into speech recognition techniques have shown the capability of audio-visual speech recognition as an alternative approach. Machine learning approaches serve as efficient optimization techniques. While other classifying technique such as hidden Markov model is popular among researchers, support vector machines lead the domain of speech recognition [6–8] and have been implemented here as the classifying technique.

The visual segment of AVSR can be processed using different approaches [4, 5]. A common basic step in every approach is face detection. Isolating the speaker's face

V. H. Shah · M. Chandra (✉)
Birla Institute of Technology Mesra, Ranchi, India
e-mail: shrotriya69@gmail.com

V. H. Shah
e-mail: vishalhshah@bitmesra.ac.in

S. Patnaik et al. (eds.), *Advances in Machine Learning and Computational Intelligence*, Algorithms for Intelligent Systems, https://doi.org/10.1007/978-981-15-5243-4_66

from the rest of the image is important for ignoring other objects that might come in the frame. Several face detection techniques are available. However, since the database for most speech recognition models is large, probability of wrong prediction and false positives must be minimized. Here, the Viola–Jones technique has been used for face detection, as it has been shown to give better accuracy than others [5, 9]. Feature extraction techniques for visual media are again as varied as the images themselves. Key point-based feature extraction is one popular approach. Another perspective on image feature extraction is to transform the image into the frequency domain. Discrete cosine transform is suited to construct feature vectors from face images [10].

Spectrographic speech processing is a separate field which involves calculation and analysis of spectrograms. A spectrogram is a visual representation of the amplitude of a sound signal, plotted with respect to the frequencies comprising it and time or some other variable. It is very useful when recognizing distinctive patterns. The speech patterns of a word remain the same no matter who spoke it, and the spectrograms are an efficient way to define this underlying pattern.

A detailed description of the different feature extraction techniques implemented has been presented in the following section. This is followed by a comprehensive study of support vector machines (SVMs) and the kernel used for classification. The subsequent sections describe the implementation of our model, results derived from these experiments and finally a conclusion along with the applications of audio-visual speech recognition.

2 Feature Extraction Techniques

The human brain processes a tremendous of data in an infinitesimal amount of time while doing the most basic of tasks such as recognizing an object or understanding a word. The attributes of any image or audio instance can be translated into features for a machine to identify and classify them.

2.1 Mel Frequency Cepstral Coefficients (MFCC)

MFCC are the most commonly used acoustic features for speech and speaker recognition. An ideal feature is characterized by its ability to be stable across all environments. MFCC features do justice to this criterion by being effectively noise robust. A cepstrum is the inverse Fourier transform of the logarithm of a spectrum. The Mel scale is a cognitive scale of pitches equally spaced from each other. Conclusively, Mel frequency cepstral coefficients are features that collectively make up a Mel frequency cepstrum. The first step in this process is pre-processing, windowing and framing of the speech signal to enable the algorithm to operate on the signal frame by frame. Next, Fourier transform of the frame is computed. After this, the Mel frequency bank is calculated which is designed to simulate the filtering process of a human auditory system. It is linear up to 1 kHz and then turns logarithmic. Mathematically, this can be represented as:

$$M(f) = 1125 * \log_e\left(1 + \frac{f}{100}\right) \tag{1}$$

2.2 *Discrete Cosine Transform (DCT)*

The discrete cosine transform represents an image as a sum of sinusoids of varying frequencies. It breaks down an image into different subparts, each of which bears varying levels of importance to image. The DCT transforms the image from spatial domain to spectral domain which gives a quantifiable ability to it. The DCT features are widely used for image feature extraction and compression as they use cosine functions instead of sine functions, and fewer cosine functions are required to approximate a signal than sine. The general equation for a 2D DCT can be written as:

$$\mathrm{DCT}(i, j) = \frac{1}{\sqrt{2N}} C(i)C(j) \sum_{x=0}^{N-1} \sum_{y=0}^{N-1} \mathrm{pixel}(x, y) \cos\left[\frac{(2x+1)i\pi}{2N}\right] \cos\left[\frac{(2y+1)i\pi}{2N}\right] \tag{2}$$

where

$$C(x) = \frac{1}{\sqrt{2}} \quad \text{if} \quad x = 0, \quad \text{else} \quad 1 \quad \text{if} \quad x > 0$$

Together with efficient classification techniques such as SVM, the DCT can constitute an integral part of successful pattern recognition systems. The final DCT matrix is arranged such that along the rows frequency increases from left to right, while along the columns frequency increases from top to bottom. Since most of the information is contained in the lower-frequency components, a zigzag method of scanning is used to select the highest value components [9, 10].

3 Support Vector Machines (SVMs)

SVM shown in Fig. 1 [7, 8] is a supervised machine learning algorithm that can be applied for classification and regression goals. We plot each data item as a point in n-dimensional space where n is the number of features with the value of each

Fig. 1 SVM classifier with support vectors on the hyper-planes

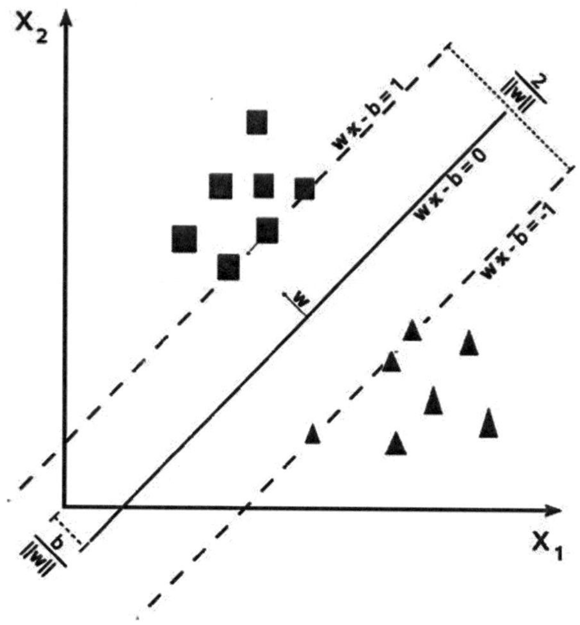

feature being the value of a particular coordinate. Support vectors are defined as the co-ordinates of individual observations. Thereafter, classification is performed by finding the hyper-plane that discerns the two classes by maximizing the distance between nearest data points (either class) and hyper-plane. This distance is known as the margin. The margin of separation is maximized, thereby minimizing the generalization error. The hyper-plane is known as maximum-margin hyper-plane, the classifier is known as maximum-margin classifier, and the process is that of linear classification. SVM is a frontier which best segregates the two classes. Nonlinear classification is also possible using the kernel trick. These are functions which take a low-dimensional input space and transform it to a higher-dimensional space, i.e. it converts non-separable problem to separable problem. Besides this, one versus one and one versus all techniques are used for multi-class classification.

Linear SVM: For a given set of training data $(\mathbf{x_i}, \mathbf{p_i})$, $\mathbf{x_i}$ is an n-dimensional real vector and $\mathbf{p_i}$ is either 0 or 1, depending on whether $\mathbf{x_i}$ belongs to a particular class. The objective is to find a maximum-margin hyper-plane. The hyper-plane can be written as the set of points $\mathbf{x}$, satisfying:

$$w.x - b = 0 \tag{3}$$

where w is the normal vector to the hyper-plane and $\frac{b}{w}$ is the offset from origin, as shown in Fig. 1.

For linearly separable data, we have two parallel hyper-planes, with the equations:

$$\begin{aligned} w.x - b &= 1 \\ w.x - b &= -1 \end{aligned} \tag{4}$$

The margin between the two is $\frac{2}{w}$. Hence for maximum margin, $||w||$ should be minimum. Also, the data points should not lie in the margin; therefore, adding this constraint produces an optimization problem, i.e. minimize $||w||$ subject to the condition:

$$p_i(w.x_i - b) \geq 1, \quad \text{for} \quad i = 1, \ldots, n \tag{5}$$

With the help of Lagrange's multiplier, this problem can be equated as:

$$\min(w, b), \max(\alpha \geq 0) \sum_{i=1}^{n} \alpha_i - \frac{1}{2} \sum_{i,j} \alpha_i \alpha_j p_i p_j x_i x_j \tag{6}$$

Nonlinear SVM: For times when the data set is not separable linearly in the given dimensions, a common practice is to project the data in a higher-dimensional space, such that it becomes linearly separable. This is done with the help of the kernel trick. However, a transformation in the higher-dimensional space might increase the generalization error. The feature vector x_i is transformed as $\varphi(x_i)$, and the kernel is given by:

$$k(x_i, x_j) = \varphi(x_i).\varphi(x_j) \tag{7}$$

Therefore, the optimization problem of linear SVM is transformed to:

$$\min(w, b), \max(\alpha \geq 0) \sum_{i=1}^{n} \alpha_i - \frac{1}{2} \sum_{i,j} \alpha_i \alpha_j p_i p_j k(x_i, x_j) \tag{8}$$

in the transformed feature space. In this manner, similarity between the training and testing data sets in the original feature space is obtained. Of all the kernel functions, radial basis function kernel is used in this experiment due to its infinite intricacy [11]. RBF kernel on two samples is defined as

$$K(x, x') = \exp^{-\left(\frac{||x-x'||^2}{2\sigma^2}\right)}, \tag{9}$$

where σ is a free parameter.

The feature space of this kernel has an infinite number of dimensions.

The value of RBF kernel ranges between zero and one and decreases with distance, and this measure can be interpreted as the degree of similarity.

4 Methodology

Methodology consists of following steps.

4.1 Database Preparation

For audio-visual speech recognition, a suitable database is prerequisite. The database used in this experiment is composed of videos of eighteen speakers, each speaking ten digits from "zero" to "nine", three times, making a total of 540 utterances. The background conditions have been maintained to be approximately same for each speaker. The video files are first split into audio-only and video-only files. The audio files are re-sampled at a frequency of 16 kHz. The video instances for each digit are split into frames. The audio is stored as .wav files, video is split digit-wise as .avi files, and their frames as .bmp. These file formats are readable in MATLAB, which has been used for this experiment.

4.2 Lip Tracking

Face Detection: The first step in lip tracking method is face detection in each frame of video. Viola–Jones algorithm has been employed in this experiment. This technique results in a square-face boundary box when applied to an image. Since the algorithm works on greyscale images only, the RGB images were first converted to greyscale.

Lip Localization: Since lips lie in the lower portion of the face, the region of interest has been confined to the lower one-third of the face-detected region. Colour intensity mapping has been used as lip localization approach. This method is based on colour mapping the lips by integrating the colour and intensity information. Experimentally, it is shown that a new colour space is defined as:

$$C = 0.2R - 0.6G + 0.3B. \tag{10}$$

Normalization is carried on the new colour space, and then, the colour and intensity information were combined to get an enhanced version of the lip colour map. To further enhance the separability between the lip and non-lip region, binarization has been performed. Trail-and-error method was applied to get the threshold for the location of lips as accurately as possible.

4.3 Feature Extraction

Acoustic and visual features have been extracted from the database corpus. For audio feature extraction, Mel frequency cepstral coefficients and a spectrogram-based technique have been employed. In MFCC feature extraction, first 24 features were extracted and quantized using LBG algorithm. Then, first 12 features were selected for each utterance. Also, spectrogram for each utterance was plotted, and DCT coefficients were found for each spectrogram. A total of 15 DCT coefficients were then selected using zigzag selection. For visual features, DCT has been employed on the binarized lip region, and again 15 coefficients were selected using zigzag method. The audio and visual features were integrated using early integration technique, yielding two sets of hybrid features consisting of 27 features (12 MFCC + 15 visual features) and 30 features (15 spectrogram features + 15 visual features).

4.4 Classification

In this paper, nu-SVC has been employed as a classifier, with radial basis function as the kernel. All the simulations were done in MATLAB, using an external library LIB-SVM for implementing SVM. The parameter gamma has been fixed at 0.07, and nu was varied from 0.01 to 0.95, in increments of 0.01. Gamma is a free parameter of Gaussian radial basis function whose value is inversely proportional to the variance. The experimental set-up is shown in Fig. 2. Out of 18 speakers, the utterances of last 17 speakers were used for training purpose and the testing was done using the data of the first speaker. Therefore, a total of 510 utterances were used for training, and 30 samples were used for testing. The values of parameters are varied, and accuracies are calculated. Many people have also used neural networks [11, 12] and many other classifiers for classification [13, 14].

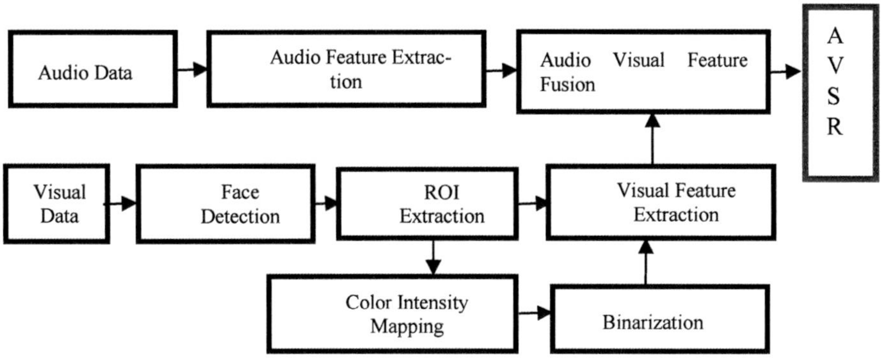

Fig. 2 Experimental setup

5 Experimental Results

Method of feature extraction for AVSRA is shown in Fig. 3, and the results are presented through Figs. 4, 5 and Table 1. It can be observed that the classification accuracy of MFCC and visual features is more than that of spectrogram DCT and visual features. Besides this, an AVSR system outperforms both audio-only and visual-only recognition techniques, as can be inferred from Fig. 4. The best accuracy was found to be 80%, in case of MFCC + visual features at a value of nu equal to 0.02 and gamma equal to 0.07. A comparison of the performances of visual + audio features and visual + spectrogram features shows their inverse dependency on the value of nu, as shown in Fig. 5.

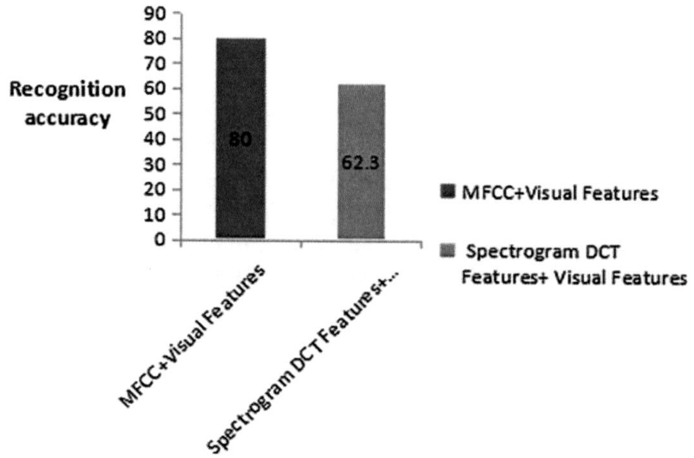

Fig. 3 % Recognition of fusion features

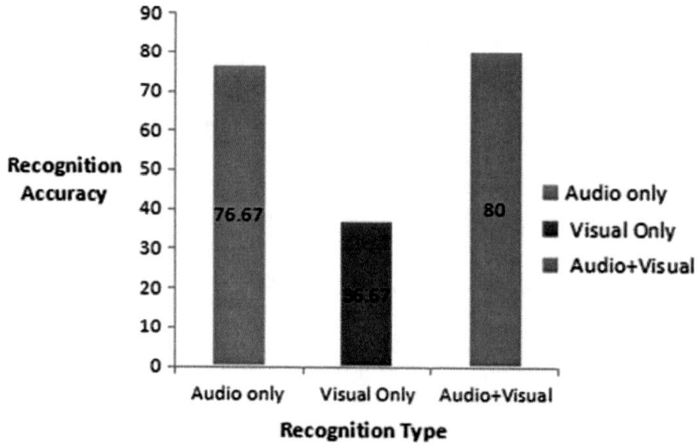

Fig. 4 % Recognition with audio-visual features

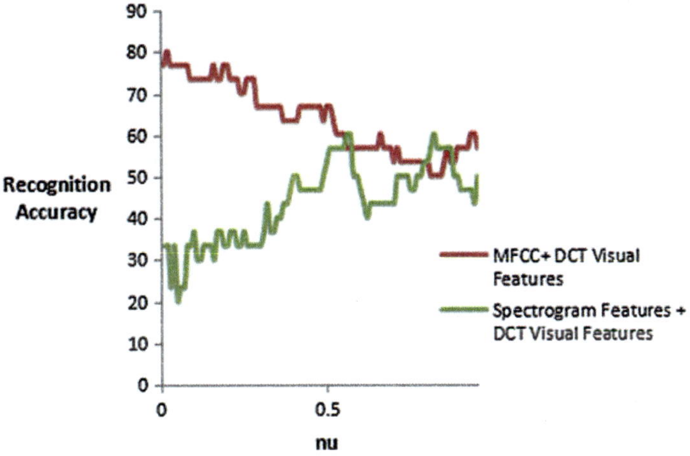

Fig. 5 Variation of recognition efficiency with nu

Table 1 Recognition accuracies of different feature extraction techniques

Recognition technique	Accuracy (%)	Value of nu
MFCC	76.67	0.07
Spectrogram	60	0.49
Visual (DCT)	36.67	0.16
MFCC + Visual (DCT)	80	0.02
Spectrogram + Visual (DCT)	62.3	0.57

6 Conclusion and Future Scope

Audio-visual speech recognition is always better than audio only for noisy environments. This results from the fact that AVSR has both these features, lip movement tracking and acoustic speech recognition. Also, spectrographic technique of speech recognition is outperformed by MFCC techniques. MFCC behaves as the most robust feature extraction technique for audio-only recognition as it resembles with human ear perception. However, the pattern in the variation of SVM parameters, especially nu, with accuracy cannot be established due to lack of evidences. With the growing penetration of autonomous technology in our everyday lives, the need for smart and robust speech recognition is now greater than ever. In future, AVSR can be developed in the field of forensic analysis of noisy speech data, air traffic control systems, for medical applications such as therapeutic and cognitive aides, smart gadgets for hearing impaired people. AVSR also holds great potential in the field of education where it can be employed for recognizing and teaching a fixed data with minor variations. Autonomous vehicles, audio passwords and automatic translators are some other interesting areas where AVSR can be implemented.

References

1. D. Stewart, R. Seymour, A. Pass, J. Ming, Robust audio-visual speech recognition under noisy audio-video conditions. IEEE Trans. Cybern. **44**(2), 175–184 (2014)
2. T. Chen, R. Rao, Audio-visual integration in multimodal communication. Proc. IEEE **86**(5), 837–852 (1998)
3. L. Rabiner, B.H. Juang, *Fundamentals of Speech Recognition*, 1st ed. (Prentice-Hall, Upper Saddle River, NJ, 1993)
4. A. Biswas, P.K. Sahu, M. Chandra, Multiple camera in car audio-visual speech recognition using phonetic and visemic information. Comput. Electr. Eng. **47**, 35–50 (2015)
5. A. Biswas, P.K. Sahu, M. Chandra, Multiple cameras audio visual speech recognition using active appearance model visual features in car environment. Int. J. Speech Technol. **19**(1), 159–171 (2016)
6. W.M. Campbell, J.P. Campbell, D.A. Reynolds, E. Singer, P.A. Torres-carrasquillo, Support vector machines for speaker and language recognition. Comput. Speech Lang. **20**(1), 210–229 (2006)
7. Chih-Chung Chang, Chih-Jen Lin, LIBSVM: a library for support vector machines. ACM Trans. Intell. Syst. Technol. **2**(2), 1–27 (2011)
8. B. Scholkopf, K.K. Sung, C.J. Burges, F. Girosi, P. Niyogi, T. Poggio, V. Vapnik, Comparing support vector machines with Gaussian kernels to radial basis function classifiers. IEEE Trans. Signal Process. **45**(11), 2758–2765 (1997)
9. P. Viola, M. Jones, Rapid object detection using a boosted cascade of simple features, in *Proceedings of the 2001 IEEE Computer Society Conference on Computer Vision and Pattern Recognition*, pp. I-511–518, USA, 2001
10. S. Dabbaghchian, M.P. Ghaemmaghami, A. Aghagolzadeh, Feature extraction using discrete cosine transform and discrimination power analysis with a face recognition technology. Pattern Recogn. **43**(4), 1431–1440 (2010)
11. C. Laurent, G. Pereyra, L.P. Brake et al., Batch normalized recurrent neural networks, in *IEEE International Conference on Acoustics, Speech and Signal Processing*, pp. 2657–2661, Shanghai, China, March 2016
12. Y. Wu, S. Zhang, Y. Zhang et al., On multiplicative integration with recurrent neural networks, pp. 2856–2864, December 2016
13. D. Bahdanau, J. Chorowski, D. Serdyuk et al., End-to-end attention-based large vocabulary speech recognition, in IEEE International Conference on Acoustics, Speech and Signal Processing, pp. 4945–4949, Shanghai, China, March 2016
14. K. Jokinen, T.N. Trong, V. Hautamäki, Variation in spoken north Sami language, in *Interspeech* 2016, pp. 3299–3303, 2016

Deployment of RFID Technology in Steel Manufacturing Industry—An Inventory Management Prospective

Rashmi Ranjan Panigrahi, Duryodhan Jena, and Arpita Jena

1 Introduction

Radio frequency identification (RFID) technology evolved from barcode and palette technology. RFID practices were introduced in the manufacturing firms for reducing means and variation of inventory/stock cycle time. This practice is used for reduction of production cost and improving firm's performance. In recent times, manufacturing firms are suffering from increasing cost of operation and try to find out ways of cost reduction and adoption of optimisation techniques in operations. In this context, RFID has considered to be the best wireless networks' techniques used as "intelligent monitoring systems" in manufacturing industry. Operational performance can be improved through the use of RFID, because it helps to reduce inventory misplacement [1]. Manufacturing firms enjoy the benefits of using RFID not only in identification of shrinkage and misplacement but also in providing economy and efficiency in operation [2]. Radio frequency practices had been proved as effective and most promising solutions to inventory misplacement and inventory shrinkage [3–5]. Main reasons for using RFID technologies are tracking of stocks/inventory with an RFID tag. The RFID helps to track sources. In automobile industry (Toyota), if tag of RFID used in the airbags, detailed information regarding source and date of production can be tracked [6]. RFID technology has provided extra millage towards operations in automotive part industry. It has brought comparability towards competition and efficiency in manufacturing firms, production function, distribution function and

R. R. Panigrahi (✉) · D. Jena · A. Jena
Institute of Business and Computer Studies (IBCS), Siksha 'O' Anusandhan (Deemed to be University), Khandagiri, Bhubaneswar, Odisha, India
e-mail: rashmiranjanpanigrahi@soa.ac.in

D. Jena
e-mail: duryodhanjena@soa.ac.in

A. Jena
e-mail: arpitajena@yahoo.co.in

S. Patnaik et al. (eds.), *Advances in Machine Learning and Computational Intelligence*, Algorithms for Intelligent Systems, https://doi.org/10.1007/978-981-15-5243-4_67

equipment installation process [7]. Radio frequencies have high degree of positive effect on the performance of SC in an apparel industry [8].

Different Automated Methods of Identifying Inventory in an Industry

For identification of inventory in industry, the following automated methods are being used (Table 1).

Table 1 Different types of automated method used

Technology	How it works	Brief description
Optical character reading (OCR)	Number, letters and characters are printed in a predetermined, standard character style, like barcode, image is illuminated and reflection is sensed and decoded	• Allows both for machine and human readability • Data density 10 characters per inch • Slow reading and higher error as compared to barcode
Machine vision	Camera takes the picture of objects, encodes it and sends it to a computer for further interpretation	• Very accurate under good light condition • Read at moderate speed • Expensive in nature
Magnetic stripe	It is like credit card or debit card and is encoded with information	• Proved technology • Readable through dirt and grease • High density of information—25 to 70 characters per inch • High-speed readings of contact details • Not readable by human
Surface acoustic wave (SAW)	Data are encoded on a chip that is encased in a tag Each tag is uniquely programmed. Wave is converted back to electronic signal and sent to again reader	• Used in highly hazardous environment • Read up to 6 feet away • Physically durable
Radio frequency tag	Data encoded with small chip, which is encased in a tag	• Tag can be easily programmable or permanently coded • Easily read up to 30 feet away • Physically durable—life span in excess of 10 years

Sources [9, p. 95]

2 Review of Literature

Inventory control and management in today's competitive are not so easy. All manufacturing firms try to control the inventory for increasing the firm's productivity through proper use of material management [10].

RFID as a technique used to address the issues related to material management in manufacturing firms. RFID as a concept was coined in the early 1980 and substantially invented in 1984. It was the first used in the Second World War to identify planes that flew over as sky territory came from and to prevent from false attack [11]. It was developed to replace barcode technology towards identification of automation flow [12]. In manufacturing industry, RFID used to check the status of inventory in different phases RM, WIP and FIG [13]. This technology is more useful for measuring work efficiency, reducing counting process, record process and above all majorly contributed in product arrangement and fast supply chain management with accuracy in relation to product [7]. According to him, RFID was applied to increase efficiency and effectiveness in industry operation. It has also considered to be the supportive tools for warehouse management WHM and minimisation of IT errors. This technology practice provides solution into manufacturing firms and can work in adverse business environment. RFID used as best practices as compared to barcode practices [14].

Most of the companies are adopting RFID technology as a practice in their manufacturing or service industry [15–18]. Among the above, Visich [15] discusses RFID technology for reducing stock-outs, improves inventory accuracy and increases sales in retail market. The study conducted by [17] analysed 630 odd papers on RFID from 2006 to 2009 and found that these are based on the different methods and benefits of RFID practices in retail industry. Papers are mainly focusing on better management and control of inventory, enhance operational efficiency and improve visibility and reducing cost. RFID practices improve the inventory accuracy as well as increase efficiency of supply chain which can improve the overall profitability of firms [19]. RFID can help to monitor and control time-sensitive product (perishable) through appropriate tagging [20]. Cost benefit of introducing RFID model investment in the organisation is being analysed [21]. Another article also focuses on cost and benefits analysis of RFID to control inventory shrinkage due to theft [22].

RFID addresses different issues of SCM and industry manufacturing process which is bit difficult on the part of barcode. It enhances visibility and security measures by adopting machine learning and artificial intelligence (AI) towards work monitoring activity. RFID tags are using for accurately recording inventory in store or stockyard [23].

RFID tags are very important in preventing: (a) low asset utilisation, (b) assets loss in a manufacturing unit and (c) limited visibility. RFID practice bring following benefits to the manufacturing units such as

1. Alert notification for assets movement/out of inventory,
2. regular information about assets in manufacturing process,

3. better visibility for work in process at different production processes for managers,
4. streamline the supply chain process of manufacturing firms [24].

2.1 Tags Used in RFID

In RFID, the main component is tag which is used to monitor the events. Tags are of different varieties and have different functions [12]. It is broadly divided into two classes: (a) active tags (used in industry to read and write longer range) and (b) passive tags (used in industry with shorter range).

2.2 RFID Model for Steel Manufacturing Firms

This is the RFID model used in the manufacturing firms (Fig. 1).

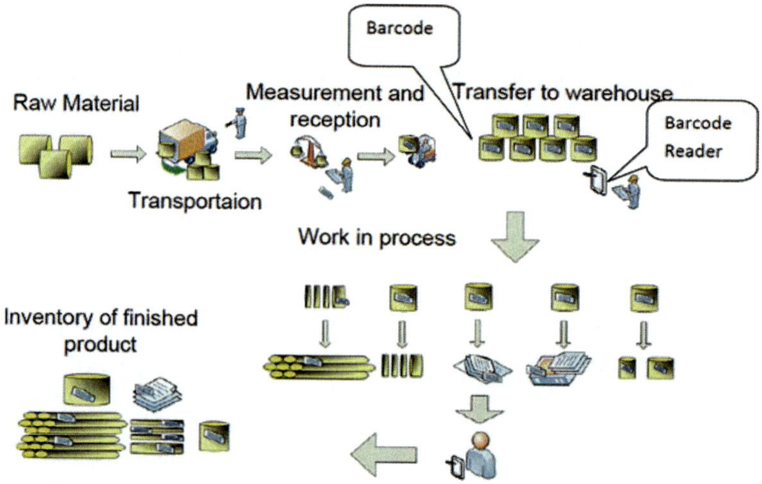

Fig. 1 Radio frequency identification model. *Source* [25]

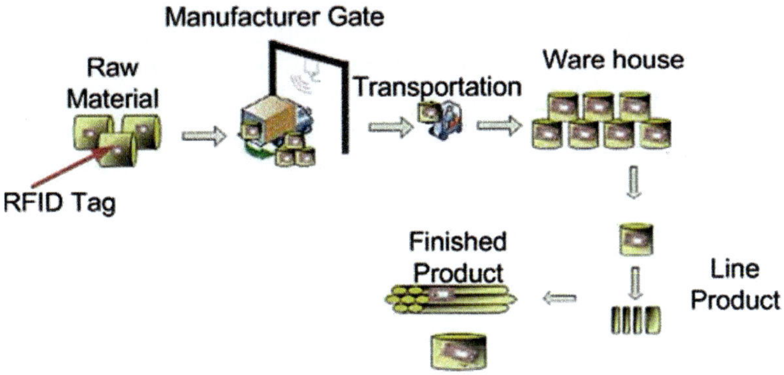

Fig. 2 Area of deploying radio frequency identification techniques. *Source* [25]

2.3 *Area of Deploying RFID Techniques*

2.4 *Different Function of RFID in a Manufacturing Industry*

- **RFID used for traceability of inventory**: By the use of RFID technology, status of raw material, WIP and finished products are visible to the manufacturing systems. It is useful for the manager to get the required data for decision making.
- **Equipment tracking and monitoring using RFID technology**: RFID practice used to track Equipment and monitoring its movements. The use of RFID practice, contributed towards minimisation of assets misplacement, proper recording of damaged assets etc. This practice gives automatic record of above facts in a systematic manner. It provides strength to production process.
- **RFID helps in production scheduling**: Production scheduling in manufacturing firms helps proper utilisation of materials and other assets by the adoption of RFID practices.
- **It helps monitor waste and shrinkage**: Tagging of component like raw material, WIP and finished goods helps industry to monitor the level of wastage, mishandling and shrinkage, internal theft, etc.
- **RFID helps to reduce cost**: RFID practices provide accurate data within no time which will save time and cost. This will also help to introduce just-in-time techniques in the systems.
- **RFID tracking and security**: Tracking different activities of staff and providing security measure for them will automatically track employee's productivity in manufacturing units.
- **RFID brings machinery integration**: Some of the large manufacturing firms depends upon the RFID practices to control and monitor the machinery activities. These types of manufacturing environment RFID were considered to be integral part of machine control system practices [26]. In supply chain systems, radio

frequency techniques provide error-free data with real-time efficiency and visibility in production process [27]. RFID techniques have a long future because it minimises the gap between time and location. Due to the higher connectivity, data can be available within no time at work station [28]. Smart manufacturing and warehousing systems are possible to implement in manufacturing firms with the help of barcode and RFID practices [29]. The main reason behind deploying the RFID practices in manufacturing firms is to get inventory accuracy and access of data at high speed, bringing visibility in minimum investment of time and cost [30].

3 Theoretical Contribution

3.1 Systems Theory of Management

System theory applied in every manufacturing firm because internal subsystems need continuous alignment with other units for collecting information regarding the present stock position, production process, purchase order, output in store, etc. [31]. As we know, business cannot run in vacuum. Every organisation has to depend on external environment for maintaining its socio-economic factors [32]. As per these papers, every organisation needs to accept input then transform it into output as per demand of the society, which shows basic input–output model. These models help to address different management issues in the areas of decision making. Source: Weihrich, [32]

Interdependency increases lots of burden on the part of key officials and manager to control the activities inside the organisation. Synchronisation in between the departments through systems helps to bring operation efficiency (OE) in a manufacturing firm. It is only possible through the systems theory of management. As per the study of [31], modern business organisation has to adopt systems approach for enhancing the efficiency, effectiveness and profitability in modern times. By using theoretical model of system theory as a basis, the paper proposed the contribution of RFID practices for improving the operational efficiency of manufacturing firms. The conceptual model is shown in Fig. 3.

Fig. 3 Conceptual model

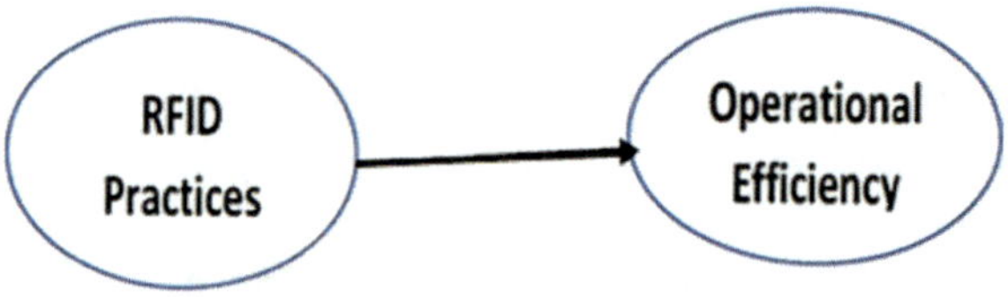

4 Our Article and Its Contribution

In this paper, an attempt has been made to study the feasibility of deploying radio frequency identification as a tool towards inventory management practice in steel manufacturing industry for reducing the cost of production and improving operational efficiency. In manufacturing firms, different techniques are being used to address the current issues related to effective maintenance of inventory and how these techniques help in increasing operational efficiencies by considering minimum cost of production. This study will contribute to the existing knowledge base as well as industry practitioners in the area of inventory management of steel manufacturing industry.

4.1 Hypothesis Formulation

Inventory management practices have positive impact on the operational performances of steel manufacturing firms [33]. According to Angeles [34], expected utility theory is one of the explanations of the associating RFID techniques for improving the firm's efficiency. Utility theory deals with proper utilisation of different tools and techniques to promote the efficiency of manufacturing firms, where RFID is one of them. Automatic identification and tracking facilities of RFID practices inventory level in a manufacturing firm can be reduced. These practices also minimise the number of lot or misplaced units through tracking, which facilitate in picking and shipping process [35]. From the above theoretical justification and empirical analysis, the following hypothesis is developed.

H_1: RFID practices positively impact on operational efficiencies (OEs) of steel manufacturing firms.

5 Methodological Foundation

In this study, the descriptive survey research design is being used to focus on larger steel manufacturing firms in the state of Odisha, India. ANOVA, correlation and regression analysis are used to measure the relationship between explanatory variables [36, 37] and explained variable [38]. The targeted respondents of this article are of all production manager, store cum warehouse manager, operation manager, quality control manager, sales manager, distribution and marketing manager, senior officers of production and operation department of steel manufacturing industry. In this study, three manufacturing industry has been surveyed, i.e. *SAIL, Rashmi Group, MESCO Steel.* A purposive sampling technique was used to select 100 sample respondents from three manufacturing firms in the state of Odisha, India, as mentioned above. By using descriptive statistics, the article tries to find out the relationship of radio frequency identification with operational efficiency of manufacturing firms. Eight variables represent RFID practices such as identification of numbers of items of stock, deals with stock replenishment, identification products and components in supply chain, real-time access of information regarding inventory, avoid product interruption and availability of the right product at the right place with zero defects and its contributed in planning process. On the opposite side, there are eight variables of operational efficiency (OE) such as enhancement of customer response, improvement in continuous production, reduced production cost, reduction of resources wastage, boosting moral employees, reduction of scrap and rejects, prevention of shortage and stock-out cost and lastly reduction of delivery lead time.

6 Analysis

RFID is one of the inventory automation practices used in the steel manufacturing firms in India. It is being used to measure the improved operational efficiency of firms. The respondent was asked to rate to what extent RFID practices in large steel manufacturing firms impact on operational efficiencies in a five-point Likert scale (5—very effective, 4—effective, 3—somewhat at, 2—not effective, 1—not at all effective).

Table 2 deals with different factors associated with the use of radio frequency identification (RFID) practices in steel manufacturing firms. It depicts highest mean value −3.75 and S.D. −1.158 of RFID 8 (improve planning processes). It tells us that the main reason of RFID was implemented in manufacturing firms to help in planning process of production unit. Followed by mean value −3.53 and S.D. −1.210 of RFID 6 deals with enabling the firm to avoid interrupted production, mean value −3.40 and S.D. −1.206 of RFID 3 deals with enabling the firm to identify different products and components in the process of supply chain. And mean value −3.31 and

Table 2 Descriptive statistics of RFID practices

	Mean	Standard deviation	Skewness	Kurtosis
	Statistic	Statistic	Statistic	Statistic
RFID 1	3.11	1.222	0.023	−1.222
RFID 2	3.31	1.285	−0.398	−1.204
RFID 3	3.40	1.206	−0.254	−0.941
RFID 4	3.23	1.213	−0.074	−1.269
RFID 5	3.25	1.313	0.043	−1.552
RFID 6	3.53	1.210	−0.281	−1.414
RFID 7	3.16	1.253	−0.058	−1.329
RFID 8	3.75	1.158	−0.650	−0.927

Source Author's Computation, 2019

Table 3 Descriptive statistics of operational efficiency (OE) in steel manufacturing firms

	Mean	Standard deviation	Skewness	Kurtosis
	Statistic	Statistic	Statistic	Statistic
OE1	3.24	1.224	−0.102	−1.198
OE2	3.23	1.213	−0.074	−1.269
OE3	3.25	1.313	0.043	−1.552
OE4	3.53	1.210	−0.281	−1.414
OE5	3.23	1.213	−0.074	−1.269
OE6	3.25	1.313	0.043	−1.552
OE7	3.53	1.210	−0.281	−1.414
OE8	3.16	1.253	−0.058	−1.329

S.D. −1.285 of RFID 2 help to recognise practices to handle stock replenishment throughout production process. According to the basic rule of skewness, less than −1/greater than 1 is highly skewed; all the eight factors of RFID fall under the categories of highly skewed which shows that data are normally distributed. Acceptance criteria for kurtosis are −2 to 2, and accordingly, all the data are under acceptable range. It represents data which are evenly distributed.

Table 3 shows the results which help to describe measure of operational efficiency. Descriptive Statistics represent the performance "OE" through mean score and standard deviation scores. 2 items of operational efficiencies i.e. OE 4 and OE 7 have highest impact on efficiency which deals with RFID helps in reduction of resource wastages and RFID practices helps to prevents shortages and stock-out costs (mean −3.53 and S.D. −1.210), OE 3 and OE 6 (mean −3.25 and S.D. −1.313) which deals with RFID Practices Reduces the production costs and RFID practices helps to minimises scrap and rejects followed by OE 1 deals with enhancing level of responsiveness towards customers' orders and enquiries (mean −3.24 and S.D. −1.224).

Table 4 represents the Pearson's correlation in between RFID and OE of manufacturing firms. From then, it was observed that RF7—availability right goods in the right place without any discrepancies and errors (0.803), RF4—Identification of components and products tracking throughout the supply chain (0.778), RF1—Identification to transmit the number of an item of stock to a reading device (0.777) and RF5—Identification to provide real-time information about inventory (0.757) are independently and highly positive relation with operational efficiency with significant level of 1%. Maximum positive correlation ($r = 0.803, p < 0.001$) is in between RF7 and OE. This analysis represents that the following 5–6 RF (radio frequency identification) factors have effect on operational efficiencies.

In regression analysis of RFID and OE, the adjusted *R* square (0.959) tells us coefficient of determination (which shows degree of variation of DV as with respect to change of PV). In the model, adjusted r square is 0.955, which says 95.5% of variation in operational efficiency. *R* stands to explain correlation coefficient in between RFID

Table 4 Pearson's correlation matrix of relationship in between RFID and operational efficiency

Correlations										
		OE	RF1	RF2	RF3	RF4	RF5	RF6	RF7	RF8
Pearson correlation	OE	**1.000**								
	RF1	**0.777**	**1.000**							
	RF2	−0.044	−0.099	**1.000**						
	RF3	−0.135	−0.092	0.108	**1.000**					
	RF4	**0.778**	0.855	−0.085	−0.167	**1.000**				
	RF5	**0.757**	0.581	−0.034	0.000	0.540	**1.000**			
	RF6	0.295	−0.142	0.121	−0.091	−0.146	−0.091	**1.000**		
	RF7	**0.803**	0.965	−0.125	−0.070	0.853	0.608	−0.116	**1.000**	
	RF8	0.063	−0.087	0.080	−0.166	0.013	−0.051	0.233	−0.077	**1.000**

Source Author's Computation, 2019
Test of Pearson's Correlation in Between RFID & Operational Efficiency

and OE, which shows positive relationship at a score of 0.979. This model comprises RF8, RF4, RF2, RF3, RF6, RF5, RF1 and RF7 and can able to explain adjusted R square (0.959)/95.9% of total variation on OE at level of significance of 5%. In regression analysis, all the observations should be independent. The independency of observation and no auto-correlation can be measured through the use of Durbin-Watson test. The value of D-W statistics should be in between 0-2 for certifing non-auto correlation between the variable. Calculated value of D-W statistics is 1.877 which implies that there is non-autocorrelation in the regression model.

ANOVA represents that F_{eal} is greater than the F_{tab} at 5% level of significance and shows that RFID practices have significant effect on operational efficiency of steel manufacturing firms.

7 Conclusion

Findings of the study are not exception, rather than it is in the line of [39] study which explains that RFID tag made up of a silicon chip, in which a unique identification number helps to carry information through a reading device. This practice provides extra millage to manufacturing firms regarding reduction of the production costs, reduction of resource wastages, minimisation of scrap and rejects, prevention of shortages and stock-out costs, reduction in delivery lead time and boosting employee work morale. These above factors are contributing to improving operational efficiency of steel manufacturing firms. The study provides certain benefits to employees and key officials in the area of stock replenishment, tracking of components and products throughout the supply chain, and enables the firm to avoid interrupted flow of production; lastly, it contributes to improved planning processes of an industry.

Due to the slowdown of economy, global steel price has been in decline stage since 2014–15 onwards, and the country like India, who is a good exporter of steel in the world market, is majorly affected. Our study has focused on how technological automation practices (RFID practices) help to provide economy in production. In coming times, this practice will be helpful in reducing cost of production and bringing production as well as operational efficiency in steel manufacturing firms. From the study, it has observed that still manufacturing firms are not fully equipped with automation technology and they are still following "rule of thumb" practices in production units.

Future research should be based on comparing more than two to three automation practices like MRP and VMI along with RFID to give more realistic view on impact of inventory automation practices on operational efficiency of steel manufacturing firms. The objective measurement will give more accurate information for better analysis. The same study can be applied to other sectors like textile industry, chemical industry and retail industry.

Table 5 Result on regression analysis of RFID and operational efficiency

Model summary										
Model	*R*	*R* Square	Adjusted *R* Square	Standard error of the estimate	Change Statistics					Durbin–Watson
					R Square change	*F* change	*df*1	*df*2	Sig. *F* change	
1	0.979[a]	0.959	0.955	0.15633	0.959	262.949	8	91	0.000	1.877

[a]Predictors: (Constant), RF8, RF4, RF2, RF3, RF6, RF5, RF1, RF7
Source Author's Computation, 2019

Table 6 Analysis of variances (ANOVA) of regression model

ANOVA[a]						
Model		Sum of squares	*df*	Mean square	*F*	Sig.
1	Regression	51.407	8	6.426	262.949	0.000[b]
	Residual	2.224	91	0.024		
	Total	53.631	99			

[a]Dependent variable: OE
[b]Predictors: (Constant), RF8, RF4, RF2, RF3, RF6, RF5, RF1, RF7
Source Author's Computation, 2019

Acknowledgements I want to convey my thanks to Dr. Duryodhan Jena, Associate Professor, Faculty of Management Science, IBCS, SOA Deemed to be University, for his valuable suggestion and guidance towards developing the concept and write this research paper. The contribution from Ms. Arpita Jena, research scholar of IBCS, SOA cannot ignore. Above all, I convey my thanks to Dean Prof. (Dr.) A. K. Samantaray, Faculty of Management Science, IBCS, SOA Deemed to be University, for providing me the opportunity to participate in international conference and complete the research paper successfully. At last, I convey my special thanks to my family members and almighty for motivational support.

References

1. F. Tao, T. Fan, K.K. Lai, L. Li, Impact of RFID technology on inventory control policy. J. Oper. Res. Soc. **68**(2), 207–220 (2017)
2. N. Denuwara, J. Maijala, M. Hakovirta, Sustainability benefits of RFID technology in the apparel industry. Sustainability **11**(22), 6477 (2019)
3. H. Dai, M.M. Tseng, The impacts of RFID implementation on reducing inventory inaccuracy in a multi-stage supply chain. Int. J. Prod. Econ. **139**(2), 634–641 (2012)
4. A.Z. Camdereli, J.M. Swaminathan, Misplaced inventory and radio-frequency identification (RFID) technology: Information and coordination. Prod. Oper. Manag. **19**(1), 1–18 (2010)
5. T.J. Fan, X.Y. Chang, C.H. Gu, J.J. Yi, S. Deng, Benefits of RFID technology for reducing inventory shrinkage. Int. J. Prod. Econ. **147**, 659–665 (2014)
6. P.J. Zelbst, K.W. Green, V.E. Sower, P.M. Reyes, Impact of RFID on manufacturing effectiveness and efficiency. Int. J. Oper. Prod. Manage. **32**(3), 329–350 (2012)
7. S. Kingsida, N. Jaturat, C., Kuntonbutr, RFID utilization on operational performance through supply chain management. Int. J. Appl. Comput. Technol. Inf. Syst. **7**(1) (2017)
8. A. Ali, M. Haseeb, Radio frequency identification (RFID) technology as a strategic tool towards higher performance of supply chain operations in textile and apparel industry of Malaysia. Uncertain Supply Chain Manage. **7**(2), 215–226 (2019)
9. M. Muller, *Essentials of Inventory Management*, 2nd edn (American Management Association, 2011), p. 9
10. R.R. Panigrahi, D. Jena, Inventory control for materials management functions—A conceptual study, in *New Paradigm in Decision Science and Management* (Springer, Singapore, 2020), pp. 187–193
11. P. Chiewnawin, *Utilization of Radio Frequency Identification (RFID) Technology for Supply Chain Management of Automotive Parts Industry in Thailand*. Doctoral dissertation, Mahidol University (2009)

12. Z. Asif, Integrating the supply chain with RFID: A technical and business analysis. Commun. Assoc. Inf. Syst. **15**(1), 24 (2005)
13. P.J. Zelbst, K.W. Green Jr., V.E. Sower, G. Baker, RFID utilization and information sharing: the impact on supply chain performance. J. Bus. Ind. Market. **25**(8), 582–589 (2010)
14. A. Akbari, S. Mirshahi, M. Hashemipour, Comparison of RFID system and barcode reader for manufacturing processes, in *2015 IEEE 28th Canadian Conference on Electrical and Computer Engineering (CCECE)* (IEEE, New York, 2015, May), pp. 502–506
15. J.K. Visich, S. Li, B.M. Khumawala, P.M. Reyes, Empirical evidence of RFID impacts on supply chain performance. Int. J. Oper. Prod. Manage. **29**(12), 1290–1315 (2009)
16. A. Sarac, N. Absi, S. Dauzère-Pérès, A literature review on the impact of RFID technologies on supply chain management. Int. J. Prod. Econ. **128**(1), 77–95 (2010)
17. M. Bhattacharya, Impact of RFID on the retail value chain: An exploratory study using a mixed method approach. J. Technol. Manage. Innov. **7**(4), 36–49 (2012)
18. X. Zhu, S.K. Mukhopadhyay, H. Kurata, A review of RFID technology and its managerial applications in different industries. J. Eng. Tech. Manage. **29**(1), 152–167 (2012)
19. S. Shin, B. Eksioglu, Effects of RFID technology on efficiency and profitability in retail supply chains. J. Appl. Bus. Res. (JABR) **30**(3), 633–646 (2014)
20. A. Chande, S. Dhekane, N. Hemachandra, N. Rangaraj, Perishable inventory management and dynamic pricing using RFID technology. Sadhana **30**(2–3), 445–462 (2005)
21. S. Baysan, A. Ustundag, The cost–benefit models for RFID investments, in *The Value of RFID* (Springer, London, 2013), pp. 13–22
22. A.G. De Kok, K.H. Van Donselaar, T. van Woensel, A break-even analysis of RFID technology for inventory sensitive to shrinkage. Int. J. Prod. Econ. **112**(2), 521–531 (2008)
23. B.C. Hardgrave, S. Langford, M. Waller, R. Miller, Measuring the impact of RFID on out of stocks at Wal-Mart. MIS Q. Executive **7**(4) (2008)
24. How to use RFID for Work in Process (WIP) Tracking in Manufacturing. Available online https://msmsolutions.com/how-to-use-rfid-for-work-in-process-wip-tracking-in-manufacturing/. Accessed on 25 Dec 2019
25. L.O. Kovavisaruch, P. Laochan, The study of deploying RFID into the steel industry, in *PICMET'09–2009 Portland International Conference on Management of Engineering & Technology* (IEEE, August, 2009), pp. 3391–3397
26. Using RFID in Manufacturing Operations. Available online: https://www.datexcorp.com/using-rfid-in-manufacturing-operations. Accessed on 25 Dec 2019
27. Zetes. RFID in Supply Chain. Available online: www.zetes.com/en/technologies-consumables/rfid-insupply-chain. Accessed on 29 Oct 2019
28. Advanced Mobile Group. 10 Companies, 10 Cost Effective RFID Solutions. Available online: www.advancedmobilegroup.com/blog/10-companies-10-rfid-solutions. Accessed on 26 July 2016
29. B. Wilkerson, The Top Benefits of Smart Manufacturing and Warehouse using IoT RFID Technology. Available online: https://msmsolutions.com/the-top-benefits-of-smart-manufacturing-and-warehouse-using-iot-rfid-technology/. Accessed on 2 May 2019
30. B. Wilkerson, RFID Manufacturing Process for the Automotive Industry. Available online: https://msmsolutions.com/rfid-manufacturing-automotiveindustry. Accessed on 21 Aug 2018
31. C.C. Chikere, J. Nwoka, The systems theory of management in modern day organizations—A study of Aldgate congress resort limited Port Harcourt. Int. J. Sci. Res. Publ. **5**(9), 1–7 (2015)
32. H. Weirich, M.V. Cannice, H. Koontz, *Management: A global and Entrepreneurial Perspective*. New Delhi (2008)
33. R.R. Panigrahi, J.R. Das, D. Jena, G. Tanty, Advance inventory management practices and its impact on production performance of manufacturing industry. Int. J. Recent Technol. Eng. (IJRTE) **8**(4), 3875–3880 (2019)
34. R. Angeles, An empirical study of the anticipated consumer response to RFID product item tagging. Ind. Manage. Data Syst. **107**(4), 461–483 (2007)
35. C.T. Stambaugh, F.W. Carpenter, RFID. Strategic Fin. **91**(6), 35–40 (2009)

36. Y. Dodge, D. Commenges (eds.), *The Oxford Dictionary of Statistical Terms* (Oxford University Press on Demand, 2006)
37. A. McBratney, *Everitt, BS, 2002. The Cambridge Dictionary of Statistics* (Cambridge University Press, Cambridge, UK, 2004), 410p. ISBN 0-521-81099. AU $75. (Geofisica Int. **121**, 155–156)
38. A.N. Sah, *Data Analysis Using Microsoft Excel* (Excel Books India, 2009)
39. F.N. Ngumi, Inventory *Management Practices and Productivity of Large Manufacturing Firms in Nairobi, Kenya* (2015)

Sparse Channel and Antenna Array Performance of Hybrid Precoding for Millimeter Wave Systems

Divya Singh and Aasheesh Shukla

1 Introduction

Due to the huge amount of spectrum and low cost of consumer devices, mmWave technology has become a hot topic of research in wireless communication. Millimeter wave (mmWave) communication is the latest technology for allowing much more high data rate and system capacity for the transmissions of indoor and outdoor networks [1–3]. The frequency range of this technology varies from 30 GHz to 300 GHz. To get the benefits of mmWave, hybrid precoding is used with reduced hardware complexity and less power consumption in mmWave communication systems [4, 5]. But it suffers from various technical issues like the use of high carrier frequency and high propagation loss. To mitigate the problem of attenuation, phased antenna arrays with extremely narrow beams are used to focus power along one direction. Low hardware complexity and less power consumption of mmWave communication are major factors of its popularity.

Recently, low cost and energy-efficient analog/digital hybrid precoding is the existing solution in which a smaller number of RF chains are used in the analog section and high number of antennas are used in the digital section. Low number of RF chains reduces the hardware complexity; thus, cost and high number of antennas increase spectral efficiency. In this paper, an analog structure is used to increase the spectral efficiency of hybrid precoder. The proposed structure can approach the performance of the optimal precoding technique.

Hybrid precoding can be categorized into two architectures. When all BS antennas are connected to each RF chain via PSs, then it is the fully connected architecture; and in the partially connected architecture, each RF chain is linked to only a subgroup

D. Singh (✉) · A. Shukla
Department of Electronics and Communication, GLA University, Mathura, India
e-mail: divya.singh@gla.ac.in

A. Shukla
e-mail: aasheesh.shukla@gla.ac.in

S. Patnaik et al. (eds.), *Advances in Machine Learning and Computational Intelligence*, Algorithms for Intelligent Systems, https://doi.org/10.1007/978-981-15-5243-4_68

of BS antennas via PSs. Besides, based on the number of users, hybrid precoding can also be categorized into one user and many users. In this paper, we will only discuss only the one-user mmWave system, where two hybrid precoding schemes, that is, the successive interference cancellation (SIC)-based hybrid precoding (sub-connected architecture) and the spatially sparse hybrid precoding (fully connected architecture) are instigated. In this paper, hybrid precoding is used for mmWave systems having large antenna arrays. OMP algorithm based on the principle of basis pursuit is used in this paper to approach accurately approximate optimal unconstrained precoder. Simulation results show the performance of the pursuit algorithm and also show that they allow mmWave systems to approach their unconstrained performance limits.

The mostly used mapping structure in previous works on hybrid precoding is the SPS implementation in fully connected architecture [4, 6–9]. However, this structure suffers from a drawback in the analog part. The use of a large number of phase shifters having variable high resolution increases the hardware cost and the power consumption of the millimeter wave system. Therefore, sub-connected mapping is used to reduce the number of phase shifters, which improves the hardware efficiency and reduces the power consumption [10–12]. A semidefinite relaxation-based alternating minimization (SDR-AltMin) [12] algorithm and successive interference cancellation (SIC) algorithm were proposed for sub-connected structure on hybrid precoding. The partially connected structure increases the hardware efficiency by sustaining too much performance degradation. On the other part, several hybrid precoding algorithms, e.g., orthogonal matching pursuit (OMP) [4], manifold optimization [13–15], and successive interference cancellation (SIC), have been proposed suspecting phase shifters with arbitrary precision.

In this paper, the remaining part is structured as follows. A system model of millimeter wave system followed by problem formulation is introduced in Sect. 2. Section 3 introduces the hybrid precoding algorithm. Simulation results to verify the work are shown in Sect. 4. Finally, the conclusion of the work is presented in Sect. 5.

The following notations are used throughout this paper: A is a matrix, "a" is a vector, $\mathcal{A}$ is a scalar, and $\mathcal{A}$ is a set. $|A|$ is the determinant of A, $\|A\|_F$ is its Frobenius norm, whereas A^{T}, A^*, and A^{-1} represents its transpose, conjugate transpose, and inverse, respectively.

2 System Model

Consider a system for mmWave communication having N_t antennas at the transmitter side and N_r antennas at the receiver side for a single user to transmit N_s data streams. Each user receives N_s data streams on each subcarrier from the base station with the help of N_r antennas. The signal received by the user at the receiver is given by

$$y = gH_x + n \tag{1}$$

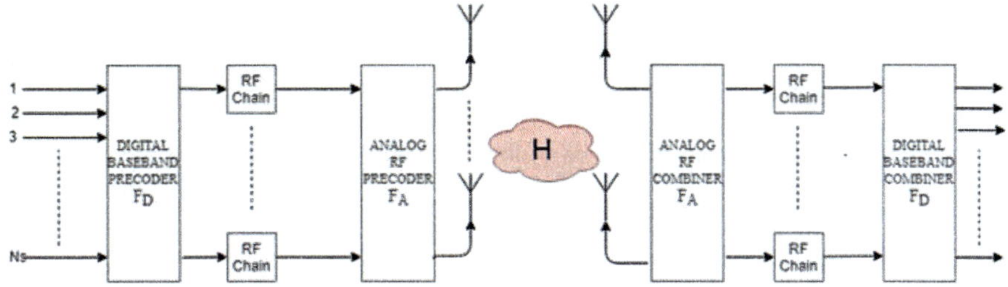

Fig. 1 Hybrid precoding mmWave system

where y is the received signal in complex vector form as $y \in \mathbb{C}\, N_r \times 1$. n represents the additive thermal noise modeled as the complex white Gaussian vector with variance ρ_n, i.e., $n \sim C\ (0,\ \rho_n)$. The average gain of the channel is represented by g. The transmit signal after precoding in the complex vector form such that $x \in \mathbb{C}\ N_t \times 1$ is given by:

$$x = F_A F_D s \tag{2}$$

The digital baseband precoder and analog RF precoder are denoted as F_D and F_A having the dimension $N_{\mathrm{RF}} \times N_s$ and $N_t \times N_{\mathrm{RF},}$ respectively. The original signal vector before precoding is denoted by s of size $N_s \times 1$ (Fig. 1).

The channel matrix H of size $N_r \times N_t$ having L propagation path using clustered channel model is modeled by the following equation:

$$H = \sqrt{\frac{N_t N_r}{L}} \sum_{l=1}^{L} \alpha_l a_r(\theta_{l,}^r \emptyset_l^r) a_t^H(\theta_l^t, \emptyset_l^t) \tag{3}$$

where α_l is the complex gain of the lth propagation path, whereas the vectors $a_r(\theta_{l,}^r \emptyset_l^r) a_t^H(\theta_l^t, \emptyset_l^t)$ represent the normalized array response vectors at an azimuth (elevation) angle of $(\theta_{l,}^r \emptyset_l^r)$ and $(\theta_l^t, \emptyset_l^t)$ of receiver and transmitter, respectively. These parameters are controlled by the structure of transmitter and receiver antenna array.

In matrix form, H is represented as,

$$H = \sqrt{\frac{N_t N_r}{L}} [a_r(\theta_1^r) a_r(\theta_2^r) \ldots a_r(\theta_l^r) * \begin{bmatrix} \alpha_1 & \cdots & 0 \\ \vdots & \ddots & \vdots \\ 0 & \cdots & \alpha_l \end{bmatrix} * \begin{bmatrix} a_t(\theta_1^t) \\ \cdot \\ \cdot \\ a_t(\theta_1^t) \end{bmatrix} \tag{4}$$

3 Problem Formulation

For the transmission of multistream data, the order of transmitting antenna, number of RF chain, and number of data stream must in such manner $N_s \le N_{\mathrm{RF}} \le N_t$.

To optimize the spectral efficiency R (F_A, F_D), we have to design (F_A, F_D) to achieve maximum mutual information by Gaussian signaling over the mmWave channel in such manner:

$$R(F_A, F_D) = \log_2\left(\left|I + \frac{\rho}{N_S\sigma_n^2} HF_AF_DF_D^HF_A^HH^H\right|\right) \tag{5}$$

Orthogonal matching pursuit (OMP) algorithm is very appropriate for large-scale multiple input multiple output (MIMO) systems due to its low complexity. First, we decompose the mmWave channel matrix H using the singular value decomposition (SVD) algorithm. Secondly, to get the analog precoding, we use the OMP algorithm by decomposing Fopt. In this section, we propose an OMP algorithm-based precoder for the considered quantized hybrid mmWave system to maximize the achievable rate (5), e.g., $R(F_A, F_D)$.

With the design of precoding parameters (F_A, F_D), the corresponding optimization problem of precoder can be represented by:

$$F_A^{\mathrm{opt}} F_D^{\mathrm{opt}} = \operatorname{Arg} \min_{F_A, F_D} ||F^{\mathrm{opt}} - F_AF_D||$$

$$\text{s.t.} \quad F_A \in \mathcal{F}, \tag{6}$$

$$||F_AF_D||_F^2 = N_s = \left|\left|F_D^HF_A^HF_AF_D\right|\right|_F \tag{7}$$

$$= \sqrt{Tr}(F_D^HF_A^HF_AF_DF_D^HF_A^HF_AF_D) \tag{8}$$

where $\mathcal{F}$ represents the set that consists of all feasible analog beamformers, that is, the set of $N_t \times N_{\mathrm{RF}}$ matrices having constant magnitude entries. Due to the non-convex nature of $F_A \in \mathcal{F}$, it is not possible to generate the solution of the above equation. Therefore, an approximation is used to solve the above equation in which achievable sum rate is transformed into the "distance" between the practical hybrid precoder F_A F_D and the optimal digital precoder F^{opt}.

4 Algorithm. OMP for Solving Problems

Input: 1: Assuming s is Gaussian distributed: Singular Value Decomposition:

2: H = U $\sum$V^H, F$_{opt}$ = V^H
OMP (orthogonal matching pursuit):
3: Fres = Fopt
4: For 1 ≤ I ≤ NRF
5:$\psi = A_t^H F_{res}$: Initially V^H
6: k = arg max $[\psi\psi^H]_{1,1}$, 1 ≤ l≤G
7: $F_{RF} = \left[F_{RF}|A_t^{(k)}\right]$: Add kth column of A$_t$
8: $F_{BB} = \left(F_{RF}^H F_{RF}\right)^{-1} F_{RF} F_{opt}$
9: $F_{res=} \frac{F_{opt} - F_{RF}F_{BB}}{||F_{opt} - F_{RF}F_{BB}||_F}$
10: End for

5 Simulation Results

In this section, the performance of the OMP-based hybrid precoder design is examined and compared with optimal precoder for single-user millimeter wave system. Assume that BS transmits $N_s = 8$ data streams to the user in a millimeter wave system with $N_t = N_r = 16$ and $N_t = N_r = 32$. The sparsity of the channel is taken as $l = 10$ and the angles of departure and arrival follow the uniform distribution from 0 to 2Π.

Figure 2 shows the average achievable rate of optimal precoder and hybrid precoder with the minimum number of RF chains, i.e., $N_{RF} = 8$. At 0 dB SNR, spectral efficiency is 13 bps/Hz with 16 * 16 transmitting/receiving antennas and 26 bps/Hz with 32 * 32 transmitting/receiving antennas. So, it is clear that the performance of precoder increases in increasing the number of antenna at the transmitter side as well as receiver side.

Figure 3 shows the performance of system with the variation of number of RF chains for the same configuration of antenna. When number of RF chains is kept fixed, e.g., 15, obtained spectral efficiency is 46 bps/Hz with 16 * 16 transmitting/receiving antennas and 68 bps/Hz with 32 * 32 transmitting/receiving antennas. So, it is clear

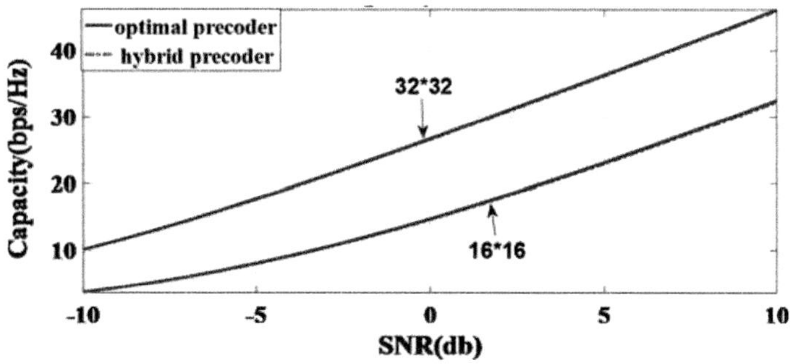

Fig. 2 Spectral efficiency for different combinations of the antenna when $l = 10$

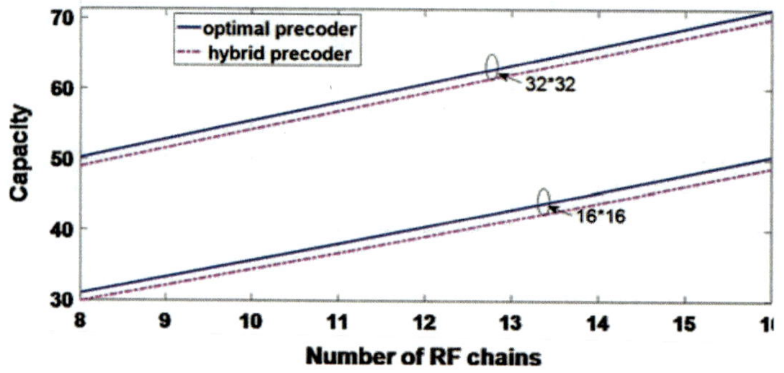

Fig. 3 Spectral efficiency for different RF chains at SNR = 0 dB

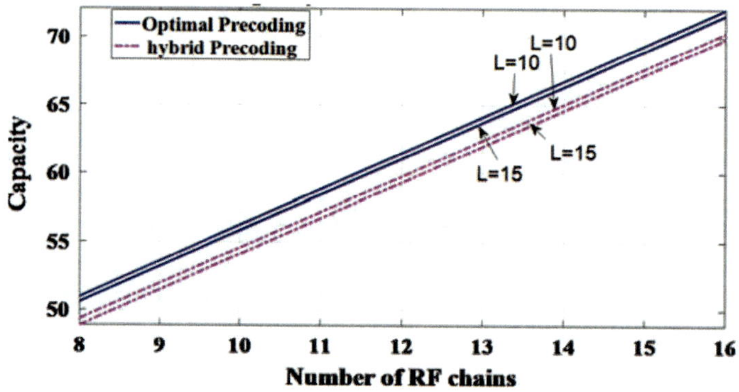

Fig. 4 Spectral efficiency for different RF chains at SNR = 0 dB

from the figure that capacity of the mmWave system improves with increase in the number of RF chains with condition that number of required RF chains should be greater than the number of data streams.

Figure 4 compares the performance of optimal precoder and hybrid precoder for different RF chain numbers at SNR 0 dB and different values of channel sparcity.

Figure 5 gives the performance comparison of precoder at different values of SNR and sparcity of the channel. At 0 dB SNR, capacity of the system is 26 bps/Hz at channel sparcity of 10 and 28 bps/Hz at channel sparcity of 15 when number of transmitting/receiving antennas are 32 * 32.

6 Conclusion

In this paper, the performance of hybrid precoder is examined using the OMP algorithm for single-user millimeter wave system. The performance of the precoder is measured in terms of the spectral efficiency. From the results, it is clear that spectral

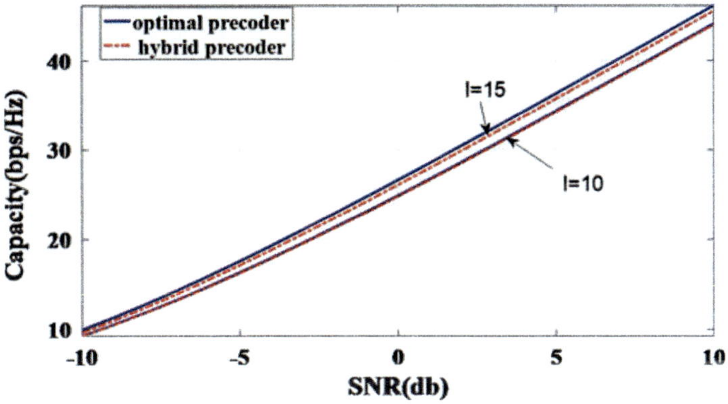

Fig. 5 Spectral efficiency achieved by optimal and hybrid precoder when $N_t = N_r = 32$

efficiency increases with increase in the number of antenna and number of chains in analog RF section. In future, hybrid precoding will be the widely used research area in millimeter wave systems.

References

1. H. Yuan, J. An, N. Yang, K. Yang, T.Q. Duong, Low complexity hybrid precoding for multiuser millimeter wave systems over frequency selective channels. IEEE Trans. Veh. Technol. **68**(1), 983–987 (2019)
2. X. Yu, J. Zhang, K.B. Letaief, Alternating minimization for hybrid precoding in multiuser OFDM mmWave systems, in *Conference Record—Asilomar Conference on Signals, Systems and Computers*, pp. 281–285 (2017)
3. J.C. Chen, Energy-efficient hybrid precoding design for millimeter-wave massive MIMO systems via coordinate update algorithms. IEEE Access **6**, 17361–17367 (2018)
4. X. Yu, J. Zhang, K.B. Letaief, Hybrid precoding in millimeter wave systems: how many phase shifters are needed? in *Proceedings of the IEEE Global Communications Conference (GLOBECOM)*, Singapore, Dec. 2017, pp. 1–6
5. T.S. Rappaport, S. Sun, R. Mayzus, H. Zhao, Y. Azar, K. Wang, G.N. Wong, J.K. Schulz, M. Samimi, F. Gutierrez, Millimeter wave mobile communications for 5G cellular: It will work! IEEE Access **1**, 335–349 (2013)
6. F. Sohrabi, W. Yu, Hybrid digital and analog beamforming design for large-scale antenna arrays. IEEE J. Sel. Topics Signal Process. **10**(3), 501–513 (2016)
7. T.E. Bogale, L.B. Le, A. Haghighat, L. Vandendorpe, On the number of RF chains and phase shifters, and scheduling design with hybrid analog-digital beamforming. IEEE Trans. Wireless Commun. **15**(5), 3311–3326 (2016)
8. J.C. Chen, Hybrid beamforming with discrete phase shifters for millimeter-wave massive mimo systems. IEEE Trans. Veh. Technol. **66**(8), 7604–7608 (2017)
9. R. Méndez-Rial, C. Rusu, N. González-Prelcic, A. Alkhateeb, R.W. Heath, Jr., Hybrid MIMO architectures for millimeter wave communications: Phase shifters or switches?. IEEE Access **4**, 247–267 (2016)
10. L. Liang, W. Xu, X. Dong, Low-complexity hybrid precoding in massive multiuser MIMO systems. IEEE Wireless Commun. Lett. **3**(6), 653–656 (2014)

11. S. Park, A. Alkhateeb, R.W. Heath Jr., Dynamic subarrays for hybrid precoding in wideband mmWave MIMO systems. IEEE Trans. Wireless Commun. **16**(5), 2907–2920 (2017)
12. A.M. Abbosh, Broadband fixed phase shifters. IEEE Microw. Compon. Lett. **21**(1), 22–24 (2011)
13. F. Sohrabi, W. Yu, Hybrid beamforming with finite-resolution phase shifters for large-scale MIMO systems, in *Proceedings of the IEEE International Workshop on Signal Processing Advances in Wireless Communications (SPAWC)*, Stockholm, Sweden, June 2015, pp. 136–140
14. X. Gao, L. Dai, S. Han, C.-L. I, R.W. Heath, Jr., Energy-efficient hybrid analog and digital precoding for mmwave MIMO systems with large antenna arrays. IEEE J. Sel. Areas Commun. **34**(4), 998–1009 (2016)
15. C.C. Chang, R.H. Lee, T.Y. Shih, Design of a beam switching/ steering Butler matrix for phased array system. IEEE Trans. Antennas Propag. **58**(2), 367–374 (2010)

Electromyography-Based Detection of Human Hand Movement Gestures

C. H. Shameem Sharmina and Rajesh Reghunadhan

1 Introduction

The most important parts of human body which help to do lot of day-to-day activities are hands. One of the ways by which the activities of hands can be conceptualized is by understanding the muscle movements. These conceptualizations are useful for the control of robotic hands and even for those who use artificial limbs [1]. The signals produced during the muscular movements (like contraction and distension) can be measured by electromyography (EMG) or surface EMG (sEMG) [2–7].

Many applications are there in the classification of EMG signals including robotic control [2, 8], pattern recognition, bio-signal processing [2], clinical diagnosis [9], robot-aided therapy [10], control in power prostheses [11], and control in prosthetic hand [12].

Many approaches have been introduced which deals with various gestures. Englehart et al. [9] and Akhmadeev et al. [13] have developed EMG classification of six hand movements. Al Omari et al. [14] developed methods for the classification of eight hand movements. Khushaba et al. [1] developed a method for classifying forearm orientations based on muscle contraction level, forearm orientation, and limb position. Al-Timemy et al. [15] proposed a method for the finger movement classification.

The features used for the classification of EMG signals in the literature include but not limited to the features in the time domain, frequency domain, wavelet transform domain, and empirical mode decomposition (EMD) [1, 2, 9, 13–16].

One of the most important and widely used features which measure the disorder/randomness of a signal is the entropy features. It seems surprising that

C. H. Shameem Sharmina (✉) · R. Reghunadhan
Department of Computer Science, Central University of Kerala, Periya, Kerala, India
e-mail: sharmina.ch@gmail.com

R. Reghunadhan
e-mail: kollamrajeshr@gmail.com

S. Patnaik et al. (eds.), *Advances in Machine Learning and Computational Intelligence*, Algorithms for Intelligent Systems, https://doi.org/10.1007/978-981-15-5243-4_69

no work has reported the use of entropy in the analysis of EMG signal. Hence, this paper focuses on the prediction of hand gestures using entropy feature and also in combination with other relevant features.

This paper is organized as follows. Section 2 provides information regarding the dataset and features. Experimental results and discussion are provided in Sect. 3. Section 4 concludes the paper.

2 Multi-feature Classification of Hand Movements

2.1 Dataset

Basic hand movements' dataset developed by Christos Sapsanis et al. is used for EMG classification [2]. The signals in the dataset were obtained from two channels at the sampling rate of 500 Hz and are already pre-processed with Butterworth band-pass filter with 15 and 500 Hz as lower and higher cutoff. Moreover, artifacts like interference are removed using Notch filter. Each hand movement in the dataset contains signals from two channels/sensors. Each signal in the dataset corresponds to a particular movement for a duration of six seconds. The six types of hand movements are performed for 30 times by each of the 5 subjects consisting of 2 males and 3 females. Figure 1 shows the six types of hand movements performed in the dataset.

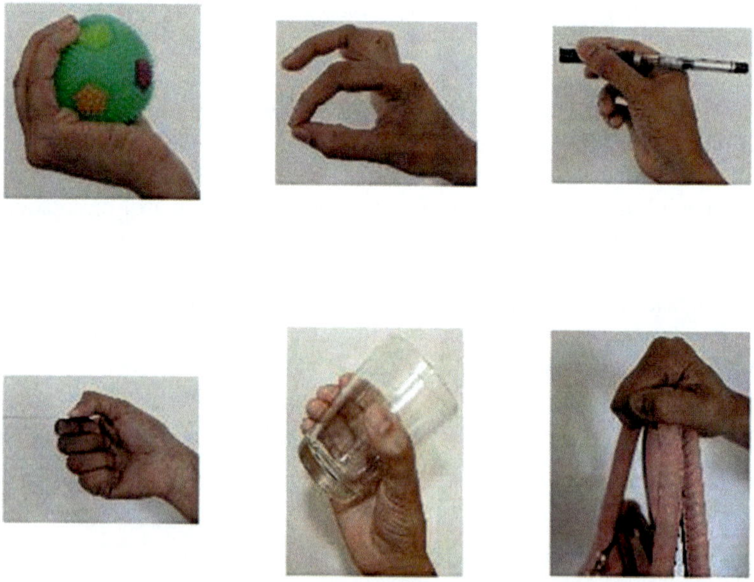

Fig. 1 Six hand gestures—spherical, tip, palmar, lateral, cylindrical, and hook

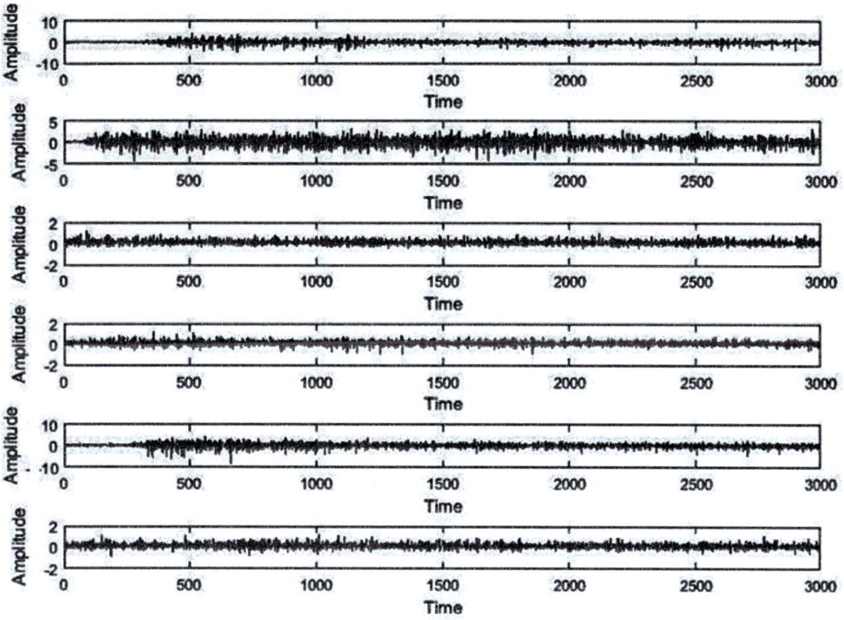

Fig. 2 EMG signal produced from a subject doing cylindrical, hook, lateral, palmar, spherical, and tip

Figure 2 shows the EMG signal produced from one of the channels of a subject doing cylindrical, hook, lateral, palmar, spherical, and tip.

2.2 *Features*

Various types of features can be extracted from raw EMG signal, namely from time domain (features like integrated EMG, mean, energy, variance, moments, log detector, waveform length, mean change in amplitude, zero-crossing, autoregressive coefficients, etc.), frequency domain (power, spectral moments, etc.), wavelet transform domain, empirical mode decomposition (EMD), etc. [1, 2, 9, 13–15, 17].

Only those features which used by Sapsanis et al. [2] and the features used in our study are shown in Table 1.

2.3 *Feature Extraction*

Each signal is segmented into segments of one-second duration (thus leading to 500 data points in each segment). Our proposed method examines the performance of a

Table 1 Features relevant to the proposed work and the features used by Christos Sapsanis

Feature	Formulae
Mean absolute value [17, 2, 10]	$\frac{1}{N}\sum_{i=1}^{N} w_i\|x_i\|$, where $w_i = 1$ for integrated EMG (IEMG) and for type 2 case $w_i = \begin{cases} 1, & \text{if } 0.25N \leq i \leq 0.75N \\ \frac{4i}{N}, & \text{if } i < 0.25N \\ \frac{4(i-N)}{N}, & \text{otherwise} \end{cases}$
Energy [17]	$\sum_{i=1}^{N} x_i^2$
Absolute third moment [17]	$\left\|\frac{1}{N}\sum_{i=1}^{N} x_i^3\right\|$
Absolute fourth moment [17]	$\left\|\frac{1}{N}\sum_{i=1}^{N} x_i^4\right\|$
Absolute fifth moment [17]	$\left\|\frac{1}{N}\sum_{i=1}^{N} x_i^4\right\|$
Log detector [17]	$e^{\frac{1}{N}\sum_{i=1}^{N}\log(\|x_i\|)}$
Length of waveform [17]	$\left\|\sum_{i=1}^{N-1}\|x_{i+1} - x_i\|\right\|$
Mean amplitude change [17]	$\frac{1}{N}\left\|\sum_{i=1}^{N-1}\|x_{i+1} - x_i\|\right\|$
Absolute mean [17]	$\frac{1}{N}\sum_{i=1}^{N}\|x_i\|$
Variance [17]	$\frac{1}{N}\sum_{i=1}^{N}(x_i - \mu)^2$, where μ is the mean
Willison amplitude [17]	$\frac{1}{N}\sum_{i=1}^{N} f(\|x_i - x_{i+1}\|)$ where $f(x) = \begin{cases} 1, & \text{if } x \geq \text{threshold} \\ 0, & \text{otherwise} \end{cases}$
Zero crossing [17]	$\sum_{i=1}^{N}\left[\text{sgn}(x_i \times x_{i+1}) \cap \|x_i - x_{i+1}\| \geq \text{threshold}\right]$ where $\text{sgn}(x) = \begin{cases} 1, & \text{if } x \geq \text{threshold} \\ 0, & \text{otherwise} \end{cases}$

feature called entropy, which gives the disorder information or defines how randomly signal is distributed. Different types of entropy measures are existing, and we used Shannon entropy for our experiments. Shannon entropy [18] is defined as

$$E(x) = \sum_{i=1}^{N} P_i \log_2 P_i \tag{1}$$

where N is the total number of samples in the window and P_i is the probability of $x_i \in X$. We use other popular features, namely mean absolute value (type 2), energy, absolute value of third, fourth, and fifth temporal moments, log detector, length of waveform, mean amplitude change, and zero crossing. The dataset contains signals from two channels; hence, these features are extracted from both the channels.

3 Experimental Results and Discussions

Support vector machine (with linear basis kernel) is used as the classifier. Two classification studies are performed. In the first classification study, each subject will act as its own control. Hence, 50% of the dataset of each subject is used for training and the remaining 50% is used for testing. The experiment is repeated by reversing the sets, and the average is taken. This study is conducted separately for each subjects, so as to compare the results with Sapsanis et al. [2]. Table 2 shows the results of comparison.

In the second study, all the subjects' data are taken together. 50% of each class/movement [(50% of 30 records) × 5 subjects = 75 records] is used for training, and the remaining 50% is used for testing. We repeat the experiment by reversing the dataset. Entropy in combination with other nine popular features provides an average classification accuracy of 99.66% (see Table 3), and it is promising when compared with the results of Sapsanis et al. [2].

Table 2 Performance comparison

Features + Classifier	Classification accuracy				
	Male 1	Male 2	Female 1	Female 2	Female 3
Raw EMG features + linear classifier [2]	86.92	92.38	85.24	83.88	84.82
First IMF features + linear classifier [17]	78.03	84.97	83.32	78.94	77.68
Raw EMG + first IMF + linear classifier [17]	90.42	94.80	87.25	88.05	85.53
Proposed: entropy + SVM (linear)	86.63	95.5	73.85	86.05	73.25
Proposed: nine features (see Sect. 2.3) + entropy + SVM (linear)	98.8	100	100	98.85	99.4

Table 3 Performance for all subjects taken together

Features + Classifier	Classification accuracy
Proposed: nine features (see Sect. 2.3) + entropy + SVM (linear)	99.96

4 Conclusion

This paper has focused on the entropy-based classification of human hand movement gestures by using EMG signals. From the results, it is clear that entropy is a strong feature and it increases the classification accuracy when combined with other features. The classification results are promising as compared with the results of Christos Sapsanis et al.

References

1. R.N. Khushba, A. Al-Timemy, S. Kodagoda, K. Nazarpour, Combined influence of forearm orientation and muscular contraction on EMG pattern recognition. Expert Syst. Appl. **61**, 154–161 (2016)
2. C. Sapsanis, G. Georgoulas, A. Tzes, D. Lymberopoulos, Improving EMG based Classification of basic hand movements using EMD, in *Proceedings of 35th Annual International Conference of the IEEE EMBS Osaka*, Japan, 3–7 July (2013)
3. D. Gamet, O. Fokapu, Electromyography. Laboratory of Biomechanics and Bioengineering, UMR CNRS 6600, Research Centre Royallieu, University of Technology of Compiegne, France (2008)
4. J.C. Navarro, F.L. Vangas, J.B. Perez, EMG—based system for basic hand movement recognition. Dyna **79**(171), 41–49 (2012)
5. H.P. Huang, C.-Y. Chiang, DSP-based controller for a multi-degree prosthetic hand, in *Proceedings of the IEEE International Conference on Robotics and Automation San Francisco, CA* (2000)
6. J. Tomaszewski, T.G. Amaral, O.P. Dias, A. Wołczowski, M. Kurzyński, EMG signal classification using neural network with AR model coefficients, in *IFAC Proceedings* vol. 42(13), 2009, pp. 318–325 (2009)
7. G.R. Naik, A. Al-Timemy, H.T. Nguyen, Transradial amputee gesture classification using an optimal number of sEMG sensors: An approach using ICA clustering. IEEE Trans. Neural Syst. Rehab. Eng. (2015)
8. K. Andrianesis, A. Tzes, Design of an anthropomorphic prosthetic hand driven by Shape Memory Alloy actuators. BioRob **2008**, 517–522 (2008)
9. K. Englehart, B. Hudgin, P.A. Parker, A wavelet—based continuous classification scheme for multifunction control. IEEE Trans. Biomed. Eng. **48**(3) (2001)
10. M. Jahan, M. Manas, B.B. Sharma, B.B. Gogoi, Feature extraction and pattern recognition of EMG based signal for hand movements, in *Proceedings of International Symposium on Advanced Computing and Communication (ISACC)* (2015)
11. K. Englehart, B. Hudgin, P.A. Parker, Myoelectric signal processing for control of power Limb prostheses. J. Electromyography Kinesiol. **16**(6), 541–548 (2006)
12. P. McCool, N. Chatlani, L. Petropoulakis, J.J. Soraghan, R. Menon, H. Lakany, Lower arm electromyography (EMG) activity detection using local binary patterns. IEEE Trans. Neural Syst. Rehab. Eng. **22**(5) (2014)
13. K. Akhmadeev, E. Rampone, T. Yu, Y. Aoustin, E. Le Carpentier, A testing system for a real-time gesture classification using surface EMG. IFAC Papers Online 50-1, 11498–11503 (2017)
14. F. Al Omari, G. Liu, Analysis of extracted forearm sEMG signal using LDA, QDA, K-NN classification algorithms. Open Autom. Control Syst. J. **6**, 108–116 (2014)
15. A.H. Al-Timemy, G. Bugmann, J. Escudero, N. Outram, Classification of finger movements for the dexterous hand prosthesis control with surface electromyography. IEEE J. Biomed. Health Inf. **17**(3) (2013)

16. A. Phinyomarka, F. Quainea, S. Charbonniera, C. Servierea, F. Tarpin-Bernardb, Y. Laurillau, Feature extraction of the first difference of EMG time series for EMG pattern recognition. Comput. Methods Programs Biomed. **117**, 247–256 (2014)
17. A. Phinyomark, P. Phukpattaranont, C. Limsakul, Feature reduction and selection for EMG signal classification. Expert Syst. Appl. **39**, 7420–7431 (2012)
18. M. Young, *The Technical Writer's Handbook* (University Science, Mill Valley, CA, 1989)
19. E.A. Biddiss, T.T. Chau, Upper limb prosthesis use and abandonment: A survey of the last 25 years. Prosthet. Orthot. Int. **31**, 236–257 (2007)

Bluetooth-Based Traffic Tracking System Using ESP32 Microcontroller

Alwaleed Khalid and Irfan Memon

1 Introduction

Bluetooth is a radio protocol that operates in 2.4 GHz radio band and used mostly for personal area networks (PANs). It was invented in Sweden by the phone company Ericsson. Bluetooth was based on an early wireless technology called MC-link which was intended to serve as a bridge to the worlds of PC and phone, providing a low power connection that is capable of handling audio and data [1].

Bluetooth low energy (or Bluetooth Smart) was developed by SIG as part of Bluetooth Core Specification Version 4.0 [2]. This newly developed device not only consumes low power but also complements classic Bluetooth. Bluetooth low energy (BLE) can be considered a different technology than classic Bluetooth, with different design goals and different market segment although it based a lot of technology on its parent [3].

Central devices are usually the devices with higher processing power such as smartphones and tablets, whereas peripheral devices are low power and small devices that connect to a central device; these are like a heart rate monitor or a BLE-enabled beacon. There are three main channels for BLE to broadcast of which (39, 38, and 37). These channels have been chosen as it does not crash with the Wi-Fi channels (11, 6, and 1). By this, we decreased the interference significantly. Figure 1 shows the channels arrays for both Bluetooth and Wi-Fi.

A. Khalid (✉)
The University of the West of England, Bristol, UK
e-mail: alwaleed@gcet.edu.om

I. Memon
Global Collage of Engineering and Technology, Muscat, Oman

S. Patnaik et al. (eds.), *Advances in Machine Learning and Computational Intelligence*, Algorithms for Intelligent Systems, https://doi.org/10.1007/978-981-15-5243-4_70

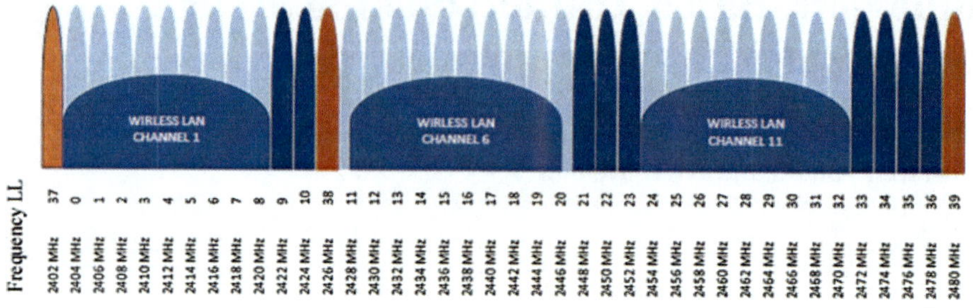

Fig. 1 Bluetooth low energy channel. GAP defines how BLE devices can interact with each other

2 Traffic Tracking

Global positioning system (GPS) is a radio wave receiver used to determine the correct position of an element in a space by providing the coordinates for that element. A constellation of orbiting satellites operates as a reference system for GPS and emits radio waves. In fact, if some aspects of the technology that the GPS is based on are exploited and the period of observation is maintained sufficiently, the technology could reach higher precision, even to less than a millimeter [4].

The triangulation of three visible satellites is implemented by the GPS to measure the distance between the receiver and each of these satellites. This is done with the calculation of time between each satellite and the GPS receiver using precise time signals emitted by the satellite [4]. Then, the distance can be calculated simply by multiplying the time by speed. Since the radio waves emitted by the satellites and light are both forms of electromagnetic radiation, the speed here, used in distance calculation, is the same as the speed of light. Given the distance, the receiver calculates the latitude and longitude of its position using triangulation, but the third coordinates (altitude) is determined by a fourth satellite giving the height the GPS receiver is at [4].

The accuracy of the position measured using GPS has some limitations as illustrated in Fig. 2. The conditions in the ionosphere (the region of upper atmosphere) can cause the radio waves to slow down slightly, resulting in a non-constant speed. This causes the signal to arrive at the GPS receiver at a slightly longer time, implying the satellite is at a distance further away than the actual distance.

The Kalman filter is one of the data fusion algorithms that are still being used commonly today. This is due to the small computational requirement of Kalman filter and its reputation as the optimal estimator for linear systems with Gaussian noise. In Kalman filter model, it is assumed that the previous state of the system at time ($k - 1$) determines the current state at time t according to the equation below.

$$X_k = AX_{k-1} + BX_{k-1} + w_k \tag{1}$$

where (X_k) is the state vector which contains the values of interest, while (U_k) is the control input vector. Moreover, (W_k) is the process noise vector. On the other hand,

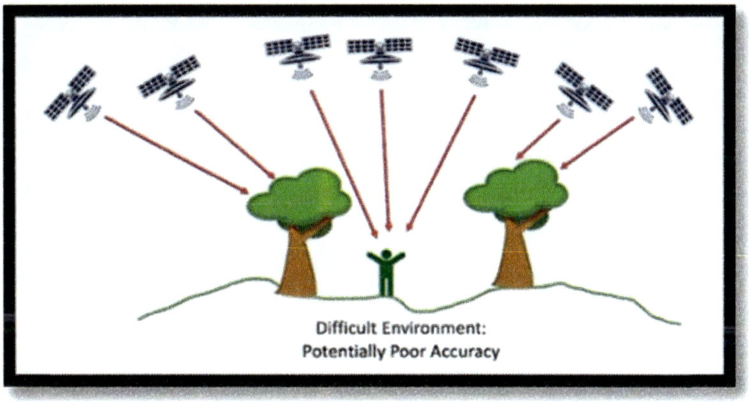

Fig. 2 GPS signal block by an object. The accuracy of positioning is really bad

(A) is the transition matrix, while (B) is the control input matrix. The measurement of the system is performed based on the equation below.

$$Z_k = HX_{k-1} + V_k \tag{2}$$

where (Z_k) is the measurement vector, while (H) is the transformation matrix. Moreover, (V_k) is the measurement noise. The measurement and process noise are independent of each other, assumed to be Gaussian with normal probability distribution. Kalman filter uses some form of feedback control to estimate a process.

3 Traffic Tracking System Design

The main function of the system is to determine the current position of a vehicle using multiple BLE-enabled modules located along the road and act as beacons that broadcast the position information to be received and interpreted by the microcontroller on the vehicle as shown in Fig. 3. The flowchart of the system is illustrated in Fig. 4.

Each beacon is based on the HC-05 Bluetooth low energy module, designed by Jinan Huamao Technology Company which, in turn, is based on the Texas Instruments CC2541 System-on-Chip. The module can use the advertisement principle which allows the module, as a BLE device, to periodically broadcast data (position, for this project) to nearby devices. The module will be programmed using AT commands that will be sent through using UART pins on the Arduino Nano (TX and RX).

The GPS receiver module will measure the current position of the tracking system before a signal from a beacon is detected to simulate outdoor tracking, and once a beacon is discovered, the vehicle will rely on position data received from the beacon

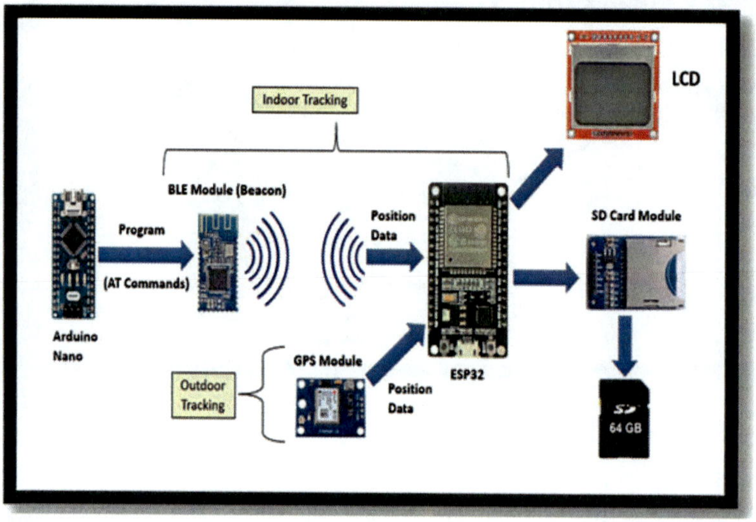

Fig. 3 Block diagram of the project, using ESP32 and Arduino microcontrollers

to perform indoor tracking. The position value recorded last from the GPS will be used as an initial value for Kalman filter. Although the vehicle will depend on the beacons for tracking, GPS will continue measuring the vehicle's position in the background, so the values will serve as a reference for comparison against the position data received from the beacons.

3.1 Beacon Design

Each beacon consists of a BLE module and battery holder that holds the 3 V battery that will power the module as shown in Fig. 5.

To get the location data from the beacon, the ESP32 microcontroller on the tracking system must receive the advertisement transmission from the beacon, identify the name of the beacon, and read the latitude and longitude values embedded in the advertisement as discussed in Fig. 6.

The universal unique identifier (UUID) is a 128-bit number stored in the Bluetooth devices used to identify a service. The UUID of the Bluetooth module is divided into four parts (IBE0, IBE1, IBE2, IBE3), and each one is 4-bytes long. These parts are configured individually through AT commands, and they will be used to store the necessary data that will be transmitted to the tracking system. The functionality of each part of the UUID is chosen as given in Table 1.

The IBE3 part of the UUID is configured as shown in Table 2.

Location Coordinates Conversion. The latitude ranges between 0° and 90°, whereas the longitude ranges between 0° and 180°. Those values are presented as

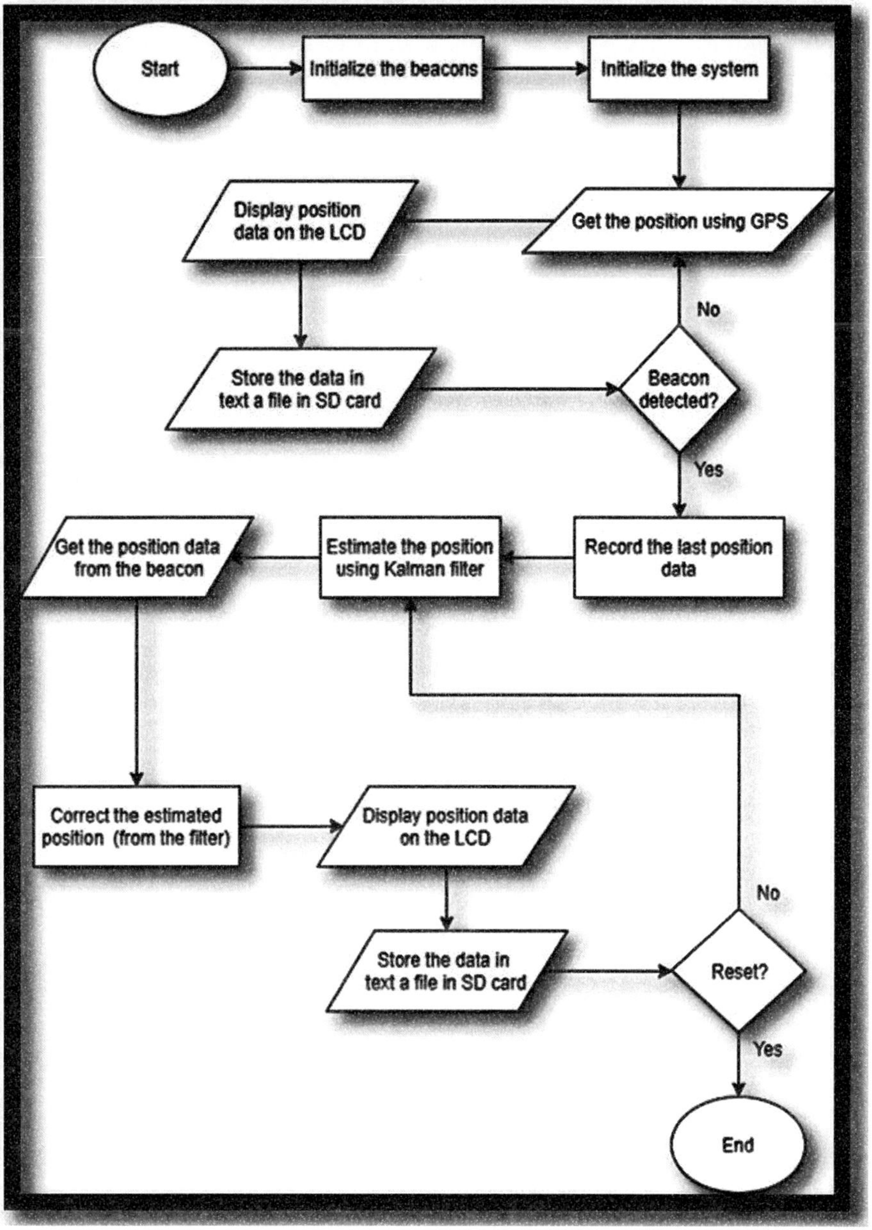

Fig. 4 Flowchart of planned processed for indoor traffic tracking

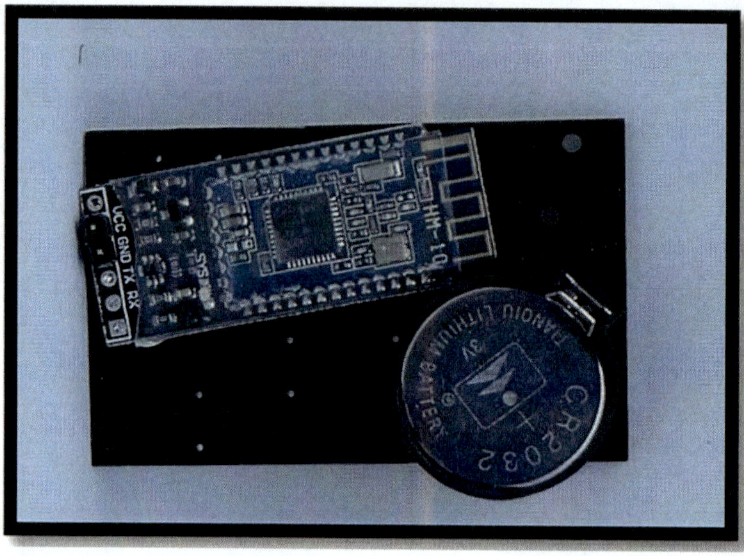

Fig. 5 Bluetooth low energy front design

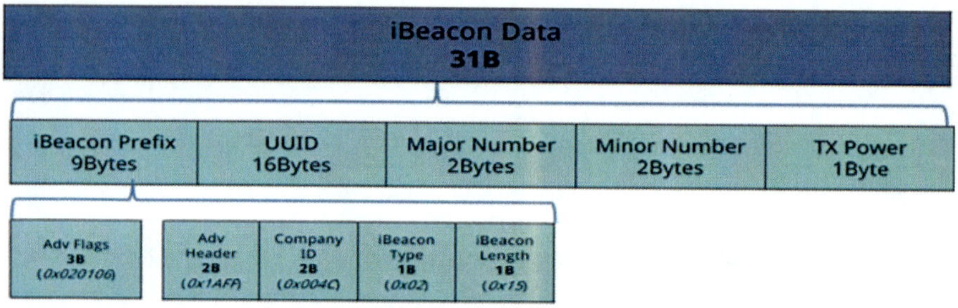

Fig. 6 Advertisement data broadcasted by beacons

Table 1 UUID functionality

UUID	Function
IBE0	A unique hexadecimal value that it used to differentiate the beacons from other Bluetooth devices. For this project, this value will be set to "**4EB00215**"
IBE1	Hexadecimal representation of the latitude
IBE2	Hexadecimal representation of the longitude
IBE3	Hexadecimal value that identifies the location with respect to the Equator and the Prime Meridian (North, South, East, and West)

Table 2 Configuration of IBE3

	Location	Hexadecimal value
IBE3	South, East	000F000F
	North, East	000E000E
	South, West	000D000D
	North, West	000C000C

either positive or negative where the positive latitude and longitude refer to the North and East, respectively, while the negative values indicate the South and West directions. Both IBE1 and IBE2 are 4 byte values which they can be used to represent integer values that range between 0 and (232-1). Therefore, the position data can be converted into integers using the formulas below.

$$\text{LtI} = \text{LtD} \times \frac{\left(2^{32} - 1\right)}{90} \tag{3}$$

$$\text{LnI} = \text{LnD} \times \frac{\left(2^{32} - 1\right)}{180} \tag{4}$$

This is where LtI is latitude as integer value, LtD is latitude in decimal degree, LnI longitude as an integer value and longitude in decimal degree. The integer values that represent both the latitude and longitude are converted into hexadecimal using Microsoft Excel and assigned to IBE1 and IBE2 parts of the UUID.

Configuring Beacons Using AT Commands. AT commands are simple instructions used to communicate with certain devices. The commands are used to communicate with the Bluetooth modules in order to configure them as beacons and assign the UUID to each beacon. Such configuration is done, first, by connecting TX and RX pins of the beacon to pins 8 and 9 of the Arduino Nano in order to establish UART communication as shown in the schematic in Fig. 7.

Tracking System Design. The system will function initially by tracking the location using the GPS module to simulate tracking in an outdoor environment. Then, the GPS signal will be deactivated by pressing a button to allow the system to turn to Bluetooth to track its location. The system will scan for nearby beacons and receive the location data from the nearest beacon. The main components of the system are ESP32 microcontroller, LCD module, SD card module, and GPS module. Connections between the components are shown in the schematic Fig. 8.

4 Testing and Discussion

When the tracking system discovers multiple beacons available nearby, it has to decide which beacon to receive the position data from. This is achieved by comparing the RSSI value of each discovered beacon. The RSSI value is negative and indicates

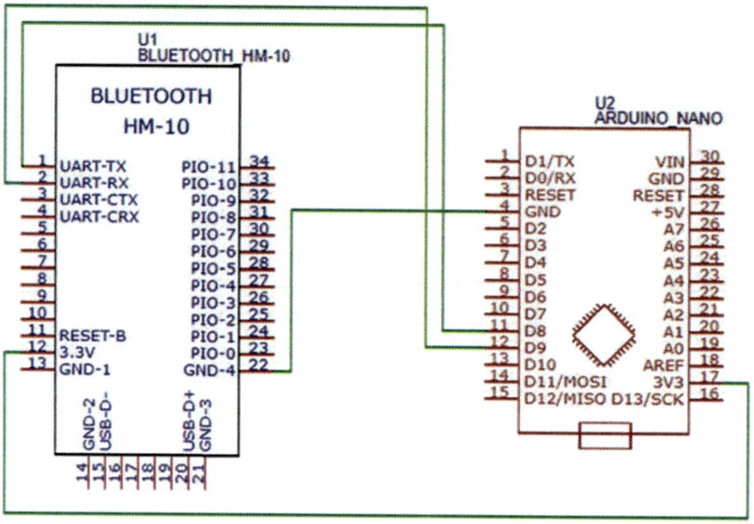

Fig. 7 Arduino UNO and Bluetooth module (iBeacon schematic diagram)

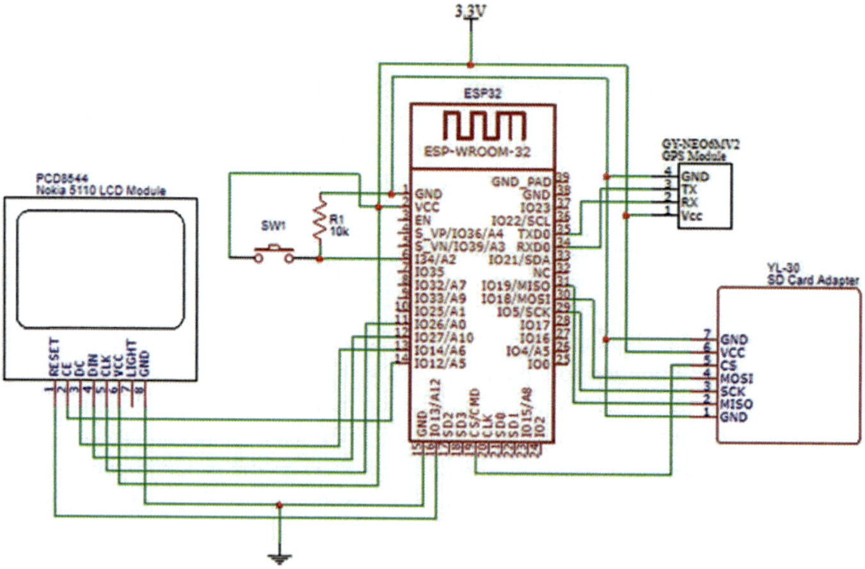

Fig. 8 Schematic diagram of the tracking system

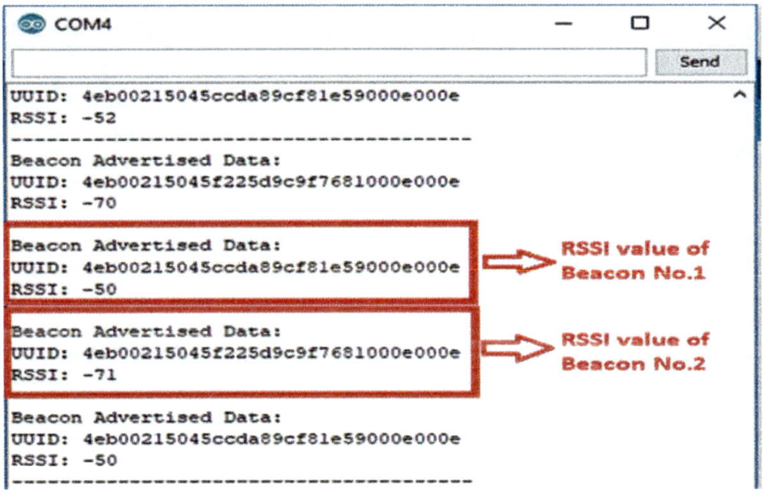

Fig. 9 Received RSSI values from the beacons (beacon no. 1 is closer to the system)

the power level of the received signal. The further the beacon is from the tracking system, the higher is the absolute value of the RSSI. Therefore, when the tracking system discovers multiple beacons nearby, it will choose to receive location data from the beacon with the lowest RSSI value. The resulting UUID of the beacons is demonstrated in the table below and both beacons will be configured with the given hexadecimal values using AT commands in Fig. 9.

To test the system, two beacons are placed away from the tracking system at different distances. Beacon no. 1 is placed nearer to the system than beacon no. 2, and the system is supposed to scan the area and receive the data advertised by beacon no. 1 since it is the one that is near to the system. Next, the position of the beacons will be reversed so that beacon no. 2 is beside the tracking system. The system will be forced to receive the location data from beacon no. 2 rather than no. 1. When beacon no. 1 is closer to the system than beacon no. 2, the serial terminal indicates the absolute value of RSSI of beacon no. 1. Figure 10 shows the received RSSI value.

5 Conclusion

Satellite navigation systems like the global positioning system have huge disadvantage and noticeable limitations when it comes to tracking within indoor settings such as tunnels. Therefore, this project is intended to provide an alternative tracking method when the GPS signal is weak or lost based on Bluetooth technology. The system was able to receive the advertised data from the beacon and convert the

Fig. 10 Position data received when beacon no. 1 is near the system

hexadecimal data into decimal values that can be interpreted as position data. Furthermore, the system's capability to receive the position data from the absolute nearest beacon was tested as well to ensure the accuracy of position of the system.

References

1. N. Hunn, *Essential of Short-Range Wireless* (Cambridge University Press, Cambridge, 2010), pp. 81–114
2. J. Gutierrez del Arroyo, J., Bindewald, S. Graham, Rice, Enabling Bluetooth low energy auditing through synchronized tracking of multiple connections. Int. J. Crit. Infrast. Prot. **18**(Supplement C), 58–70 (2017)
3. R. Haydon, What is bluetooth low energy? in *Bluetooth Low Energy: The Developer's Handbook*, Chap. 1 (Prentice Hall, Indiana, 2012), p. 3
4. M. Salvemini, Global positioning system A2, *in International Encyclopedia of the Social & Behavioral Sciences*, 2nd edn, ed. by J.D. Wright (Oxford, 2015), pp. 174–177

A Novel Solution for Stable and High-Quality Power for Power System with High Penetration of Renewable Energy Transmission by HVDC

C. John De Britto and S. Nagarajan

1 Introduction

The arrangement of energy sources and electricity accessible is to advance the proposed system. The power system will have superior permeability in the future generation of renewable energy. The total of power creation time might be equal to the power used in stability cases. The power invention of wind turbine or photovoltaic power differs from one place to another. In case of generating electricity from renewable energy provides reasonably slight portion of electricity in China. Most of the manufacturing and population are in the southeast and east part of the country. But the appropriate power in the northwest and north, long-distance program, is needed about 2000 km in many cases. The HVDC transmission lines are now in operation and recent days it is in the growth stage in china. After 2010 the operations were established by the transmission lines at ±500 KV and at ±1100 KV. The total magnitude of the lines exceeds more than 32,600 km. The progress of PV and wind and HVDC transmission is giving the control characteristics of the power grid.

Power electronics is generally is a way to enable the latest technologies for both HVDC transmission and renewable power generation. For the system adjustment, the characteristics of inertia and damping play a major role. The controlling of the wind turbine model can be easily limited. More synchronous generators are replaced by suitable power components, and then the level of damping and inertia is reduced in a way. The virtual synchronous generators control the renewable energy converters to operate the system level as stability like a synchronous generator. The limit of

C. John De Britto (✉) · S. Nagarajan
Department of EEE, Jerusalem College of Engineering, Chennai 600100, Tamilnadu, India
e-mail: johndebritto89@gmail.com

S. Nagarajan
e-mail: nagu.shola@gmail.com

C. John De Britto
Department of EEE, Jeppiaar Engineering College, Chennai 600119, Tamilnadu, India

S. Patnaik et al. (eds.), *Advances in Machine Learning and Computational Intelligence*, Algorithms for Intelligent Systems, https://doi.org/10.1007/978-981-15-5243-4_71

the converter and the merits of synchronous generators are discussed with motor-generator pair which is a new grid connection method to drive synchronous motor and to increase the permeability. We are going for this analysis in detail. In order to study the performance of MGP, including HVDC converters, DC circuit breakers are used. In a way for HVDC transmission, the gap between power electronics converter and grid system technologies gives a grid operation and stability of renewable energy. The nonlinearity and the complexity of the power electronics system structure vary in nature. The overall operation of the rotating equipment varies and affected by harmonics. The proposed system analyzes the transmission and harmonic equipment to justify the MGP circuit and improves the stability and increases the quality of power. Several schemes are implemented to improve the grid stability and to support the future expansion of renewable energy generation (Fig. 1).

Motor-Generator Pair. The figure shows the conversion of electrical and mechanical energy. The synchronous machines can provide damping level as well as inertia. Motor-generator pair is a new model of inertia circuit. By this proposed analysis, MGP set can produce efficiency level more than 94%. The production cost of synchronous generator will be lower than by comparing remaining methods to that of large power transmission.

High-quality harmonic response of MGP:

The harmonic input currents can be measured by the following way.

$iA = \varepsilon_{m=2} \infty$ Im Cosmωt

$iB = \varepsilon_{m=2} \infty$ Im Cosm$(\omega t - 2\,\pi/3)$

$i_C = \varepsilon_{m=2} \infty$ Im Cosm$(\omega t + 2\,\pi/3)$

The magnetic force can be generated by the motor side current rotates at synchronization. Motor-generator pair output is highly stable power in the order of 100 W. The

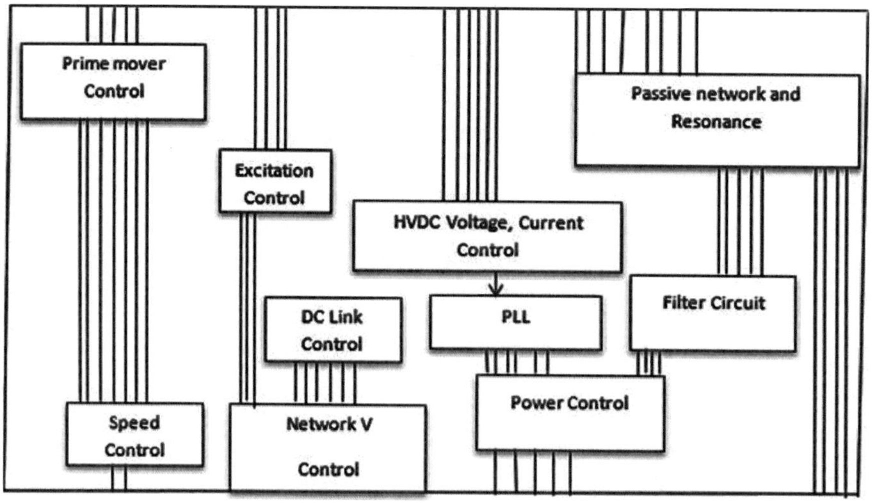

Fig. 1 Power system model for hybrid power transmission system

voltage level at different point is also measured by the fundamental wave frequency level of 50 Hz to get the normalized value.

2 Control and Dynamics

The advanced technology of modern power electronics circuits is designed by steady-state increasing operations with the improvement of advanced power semiconductor circuit topology. Switching frequency levels not only helps to increase control speed and advances the act of the system. The control system limits the use of inductors capacitors and transformers by reducing volume, cost, and weight. The frequency level in high-power high-voltage design is lowered by more frequency in the range of KHz which is possible and normal for PV and wind stations. DC-DC motor coupled generator pair acts as an inverter circuit. The bandwidth level of most control processors used for higher fundamental frequency as fast control complex dynamics in the synchronous level may be higher or lower, i.e., subsynchronous level and super-synchronous levels are used. This gives a proposed stability with suitable power quality and resonance problem (Figs. 2 and 3).

A modern DC-DC converters are designed for solar emergency applications by allowing soft switching of semiconductors and to improve the density of the converters. The popular type of resonance is used in the series coupling and with the highest possible usage of the passive network in the circuit model. The leakage inductance coupled in a isolation transformer is used as resonant inductors also as coupling capacitors of the voltage dynamic restorer circuit, and these circuits are generally applied in the output portion of the converter to reach maximum voltage gain at a given transformer turns ratio. The advanced model of semiconductor devices enables

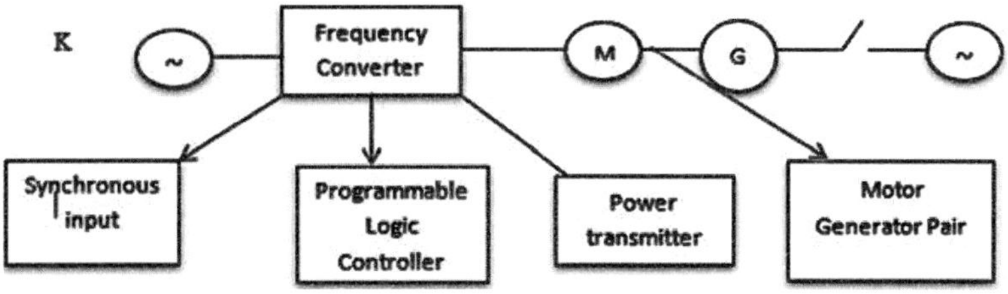

Fig. 2 Power transmission equipment system

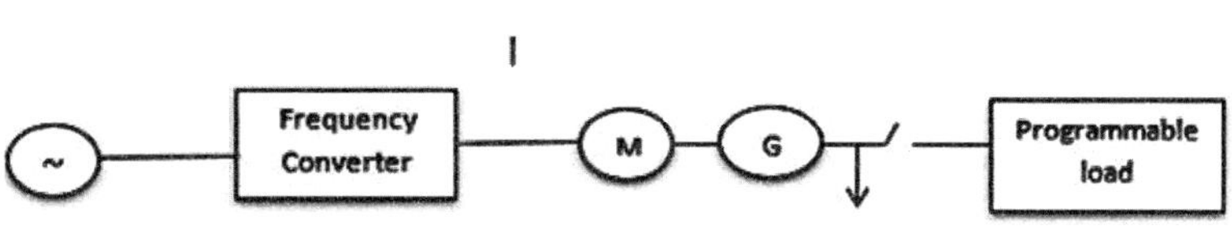

Fig. 3 Power transmission equipment system with programmable load

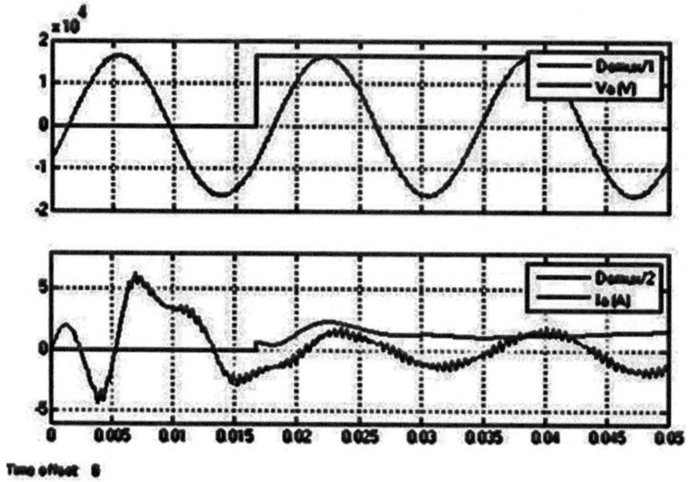

Fig. 4 Grid output power. Here, the first figure indicates the output power of 15 KW, X-axis indicates 0.005 s per cm and Y-axis indicates 1 cm and the output current is 0.5 A

the maximum efficiency to be 98%. The diodes in the VDR in addition with the output voltages over 400 V DC, by low recovery voltage. Practically in current mode of operation to improve the gain and performance of the converters and maximum power can be derived for advanced model.

3 Simulation Results

To extend the input voltage range, without affecting efficiency, the following Simulink results were analyzed by MATLAB Simulink processor (Figs. 4, 5, 6 and 7).

4 Hardware Implementation of Renewable Energy Systems

The proposed system is developed by means of 50 W Solarix and 12 V Nominal voltage. The hardware model is interfaced with Arduino software and the PV, wind battery voltages are displayed in a MATLAB software also it is executed in a PV–wind–battery hardware model (Fig. 8).

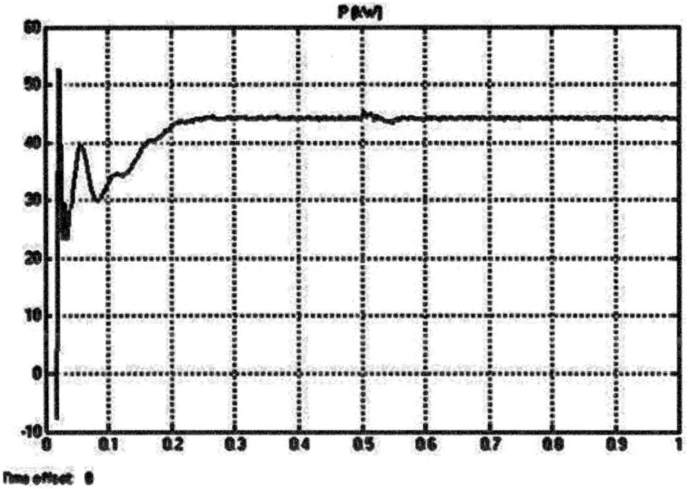

Fig. 5 Output power 45 KW. Here, *X*-axis indicates time 0.1 s per cm and *Y*-axis indicates 10 KW per cm

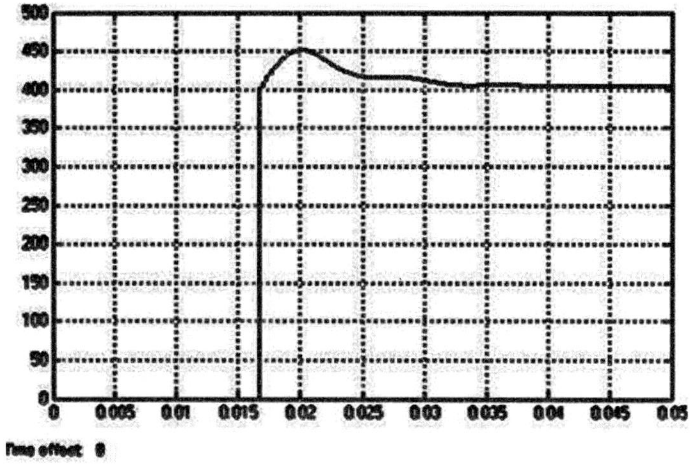

Fig. 6 Photovoltaic and wind energy voltage = 400 V. Here, *X*-axis indicates 0.005 s per cm and *Y*-axis indicates 50 V per cm

5 Conclusion

A novel technology is verified based on the theory of high-quality power which is good for growing level of permeability of renewable energy. The shaft model of motor-generator pair circuit topology is verified and designed with the stability

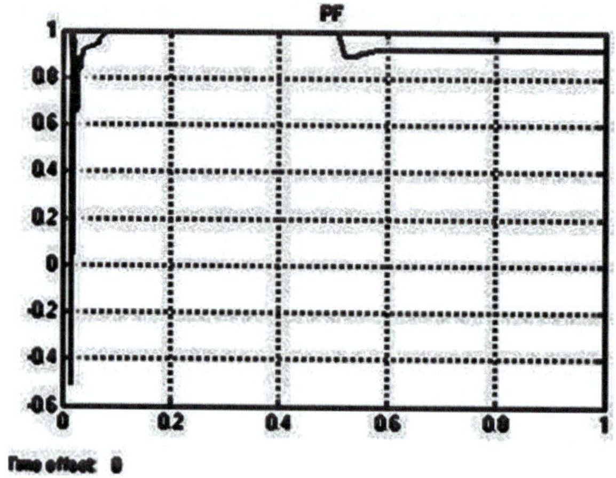

Fig. 7 The developed output power factor of the proposed system is unity. Here *X*-axis indicates 0.2 s per cm and *Y*-axis indicates 0.2 s per cm

Fig. 8 Implementation of PV-Wind Based Renewable energy Generation systems. Here the output power is achieved by means of discrete angular rotation. And the wind–solar-Battery voltages are displayed in a software module through arduino interface

analysis and it can be maintained at a constant level. The traditional power generation circuit proposes and the power generation unit gives a complementary power by two motors. More flexible damping control can be achieved by MGP control system. The developed renewable energy system with PV–wind model delivers a suitable output power for HVDC transmission. The proposed modeling of the total electromagnetic system and the output power of the motor-generator pair with high power quality is compared to the input power. New HVDC transmission lines are developed by the modeling of hybrid renewable systems. The review and analysis-based stability theory gives a high-frequency power electronics and fast control. New control methodologies will give a highly coupled converter-based HVDC systems and the theories were verified.

References

1. A. Hamadi, S. Rahmani, K. Addoweesh, K. Al-Haddad, A modeling and control of DFIG wind and PV solar energy source generation feeding four wire isolated load, in *IECON 2013—39th Annual Conference of the IEEE Industrial Electronics Society*, Vienna, pp. 7778–7783 (2013)
2. S.K. Tiwari, B. Singh, P.K. Goel, Design and control of autonomous wind-solar energy system with DFIG feeding 3-phase 4-wire network, in *2015 Annual IEEE India Conference (INDICON)*, New Delhi, pp. 1–6 (2015)
3. S.K. Tiwari, B. Singh, P.K. Goel, Design and control of micro-grid fed by renewable energy generating sources, in *2016 IEEE 6th International Conference on Power Systems (ICPS)*, New Delhi, pp. 1–6 (2016)
4. H. Polinder, F.F.A. van der Pijl, G.J. de Vilder, P. Tavner,"Comparison of direct-drive and geared generator concepts for wind turbines, in *IEEE International Conference on Electric Machines and Drives*, 2005, San Antonio, TX, pp. 543–550 (2005)
5. E.A. Bakirtzis, C. Demoulias, Control of a micro-grid supplied by renewable energy sources and storage batteries, in *XXth International Conference on Electrical Machines (ICEM)*, pp. 2053–2059, 2–5 Sept. 2012
6. S. Heier, *Grid Integration of Wind Energy Conversion Systems* (Wiley, Hoboken, NJ, 1998)
7. Z.M. Salameh, M.A. Casacca, W.A. Lynch, A mathematical model for lead-acid batteries. IEEE Trans. Energy Convers. **7**(1), 93–97 (1992)
8. A.B. Rey-Boué, R. García-Valverde, F. de A. Ruz-Vila, J.M. Torrelo-Ponce, An integrative approach to the design methodology for 3-phase power conditioners in photovoltaic grid-connected systems. Energy Convers. Manage. **56**, 80–95 (2011)
9. Z. Xuesong, S. Daichun, M. Youjie, C. Deshu, The simulation and design for MPPT of PV system based on incremental conductance method, in *2010 WASE International Conference on Information Engineering,* Aug 2010, pp. 314–317
10. S.K. Tiwari, B. Singh, P.K. Goel, Design and control of autonomous wind-solar hybrid system with DFIG feeding a 3-phase 4-wire system. IEEE Trans. Industry Appl. **99**, 1–1
11. W. Luo, B. Wang, H.S. Zhao, Modeling and simulation of non-linear dynamic process of the induction motor system with fluctuating potential loads. Sci. China Tech. Sci. **57**(9), 1729–1737 (2014)
12. W. Luo, H.S. Zhao, B. Wang, Y.L. Luo, Time-step FEM analysis of start and operational characteristics of beam pump motors. Proc. CSEE **34**(27), 4691–4699 (2014)
13. H. Zhao, Y. Wang, Y. Zhan, G. Xu, X. Cui, J. Wang, Practical model for energy consumption analysis of beam pumping motor systems and its energy saving applications, in Proceedings of the 2017 EEE Industry *Application Society Annual Meeting*, Cincinnati, OH, USA, pp. 1–9, https://doi.org/10.1109/ias.2017.8101721, Oct. 2016
14. S. Mallik, K. Mallik, A. Barman, D. Maiti, S.K. Biswas, N.K. Beb, S. Basu, Efficiency and cost optimized design of an induction motor using genetic algorithm. IEEE Trans. Ind. Electron. **64**(12), 9854–9863 (2017)
15. O.S. Ebrahim, M.A. Badr, A.S. Elgendy, P.K. Jain, ANN-based optimal energy control of induction motor drive in pumping applications. IEEE Trans. Energy Convers. **25**(3), 652–660 (2010)

Blockchain Technology: A Concise Survey on Its Use Cases and Applications

B. Suganya and I. S. Akila

1 Introduction

Blockchain technology is a new concept for a rapidly growing part of foundational technology like Internet or cloud computing. In earlier, Bitcoin was perceived as a cryptocurrency; however, major theories of blockchain architectures used today were first outlined and defined in the paper written by Satoshi Nakamoto in 2008. Blockchain is a shared, immutable ledger for simplifying the process of recorded transactions. The need for a blockchain is an efficient, cost-effective, consistent and secure system for accompanying and recording financial transactions. Transaction proportions have grown exponentially and will surely increase the complexities and costs of current transaction systems. In essence, a blockchain is a distributed ledger, a consensus protocol and also a membership protocol. A blockchain is a kind of distributed ledger and need not necessarily deploy blocks or chain the transactions. Although the term 'blockchain' is used more often than 'distributed ledger' in considerations, a blockchain is only one among many types of data structures which provide secure and valid achievement of distributed consensus. The Bitcoin blockchain, which uses 'Proof of Work Mining', for attaining distributed consensus. However, other forms of distributed ledger consensus exist such as Ethereum, Ripple, Hyperledger, Multichain, Eris, and other private enterprise solutions.

As shown in Fig. 1, the blockchain is a chain of chronological blocks. It is essential for a distributed database of records or public ledger of all transactions or digital events that have been executed and shared among participating parties. Consensus is achieved through verification and validation of each transaction in the public ledger.

B. Suganya (✉) · I. S. Akila
Department of ECE, Coimbatore Institute of Technology, Coimbatore, India
e-mail: suganyabcse88@gmail.com

I. S. Akila
e-mail: akila@cit.edu.in

S. Patnaik et al. (eds.), *Advances in Machine Learning and Computational Intelligence*, Algorithms for Intelligent Systems, https://doi.org/10.1007/978-981-15-5243-4_72

Once entered, information can never be erased. The blockchain contains a verifiable record of data for every single transaction. The fact is that, in this digital world everyone is relying on a third-party entity for the security and privacy of digital assets. Therefore, the third-party sources can be hacked, manipulated or compromised. It is able to reform the digital world by enabling a distributed consensus, and all transactions involving digital assets are verified and validated in future. It does this without compromising the secrecy of the digital assets and parties involved. The distributed consensus and anonymity are two important characteristics of blockchain technology.

2 Blockchain Technology: Working Principle

The concept of the blockchain explains how Bitcoin works since it is intrinsically linked to the Bitcoin. However, the blockchain technology is applicable to any digital asset transaction traded online. The working process of Blockchain technology incorporates sender node, receiver node, and the network for its execution process. The sending node records new data and broadcast into the network. Then the receiving node checks for the message, if the message was correct, then it will be stored to a block. All receiving node in the network has to execute Proof of Work (PoW) or Proof of Stake (PoS) algorithm for the block creation. Furthermore, the block will be warehoused into the chain after executing consensus algorithm; every node in the network confesses for the block that gets added newly and will continuously lengthen the chain based on the block which gets entered later. Proof of work (PoW) is a consensus strategy used in Bitcoin network (Nakamoto 2008). PoW involves a complex computational process in the authentication of a block.

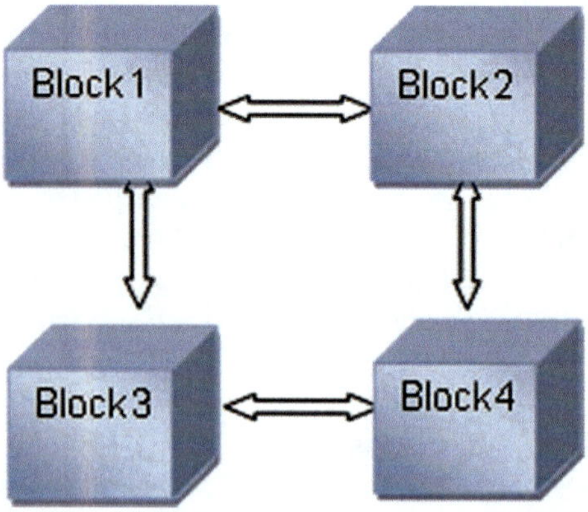

Fig. 1 Blockchain representation

3 Related Work

This concise survey focuses on Nakamoto blockchain where the literature survey described many aspects like security, food supply chain, agriculture and health care. Blockchain is a new type of database because the data in the database is saved and stored in a block. By linking one block to other block in a chain, thus a blockchain is created. In paper [1], they set out to provide a general exploration of blockchain. This paper focussed on the processing power to find a block that it is nearly impossible to alter the blockchain in future. On the other hand, there are limits to the possibilities and uses of blockchain. Moreover, the research indicates the utilization of energy factors that relies on the blockchain. In paper [2], the authors analysed the use of blockchain technologies in wireless networks. More specifically, the authors focused on possible solutions for the use of blockchain as a mechanism for anonymous access control in wireless networks and for the use of Proof of Work (PoW) as a replacement of the exponential backoff utilized for congestion avoidance in wireless access networks.

Moreover, the procedures executed for the Proof of Work are computing intensive which involves high energy consumption. The work in [3] described the privacy and security issues associated with the blockchain. The authors analysed the privacy threats in blockchain and discussed existing cryptographic defence mechanisms, i.e., anonymity and transaction privacy preservation. Furthermore, the authors summarized typical implementations of privacy preservation mechanisms using ring signature for crypto note and using non-interactive zero-knowledge proof for Zcash in blockchain. In paper [4], the author focused on the challenges in implementation on blockchain and its associated security. The research method followed by the architecture and the working principle of blockchain and various application areas like IoT and healthcare has been identified in this paper. In paper [5], the authors surveyed the application of blockchain technology for smart communities and focused on vital components of the blockchain applications. The authors also studied the various process models used in the execution of secure transactions. Specifically, the proposed work presented a detailed taxonomy on the applications, such as financial system, intelligent transportation system, and the Internet of things. The process models used are behavioural model, business model and governmental model as the blockchain is decentralized and distributed. The communication infrastructure relied on both wired and wireless networks for its telecommunications. In paper [6], the focus was given on authentication over blockchain technology-based networking. The authors provided the valuable and classified information which can enhance the understanding of various authentication systems that can be combined with blockchain technology. In addition to this, complications associated with blockchain technology and proposed solutions are analysed to fulfil the requirements of the network applications. Furthermore, this work emphasized the importance, abilities, motivations, and challenges of blockchain technology with discrete applications in various fields.

Blockchain is a promising concept in agriculture, particularly for its potential to deliver peer-to-peer (P2P) solutions to small-scale producers. Thus, in paper [7], six challenges have been identified including storage capacity, scalability, privacy leakage, high cost, regulation problem and throughput and latency issues. The proposed work contributed to the extent literature in the field of agri-food value chain management by determining the potential of blockchain technology and its inferences for agri-food value chain performance improvements such as food safety, food quality and food traceability for improving the performance. In paper [8], the authors surveyed some emerging applications that are being implemented including supply chains, smarter energy, and health care. Moreover, the paper outlined successful adoptions of blockchain for IoT. The IoT devices and its number have increased significantly and are accompanied by increasing in its processing power and 5G networking speeds. The authors studied three specific scenarios including of healthcare applications, supply chain applications, smart energy applications, and the author collaborated those applications between IoT devices and the Blockchain. In paper [9], the blockchain technologies which can hypothetically address the critical challenges arising from the IoT and hence outfitting the IoT applications are identified. Furthermore, the paper elaborated on the blockchain consensus protocols and data structures for the effective integration of blockchain into the IoT networks. In paper [10], the authors described how the convergence of blockchain and artificial intelligence (AI) in IoT has taken place. Blockchain and AI are the core technologies for IoT in modern-day's applications. As Blockchain is decentralized and distributed platform for IoT applications; on the other hand, AI is used for analysing and processing the data in IoT applications, and it offers intelligent and decision-making capabilities for machine to human. The authors also presented the future directions with block IoT intelligence architecture of converging blockchain and AI to achieve the goal of scalable and secure IoT with cloud intelligence, fog intelligence, edge intelligence and device intelligence.

4 Components of Blockchain Technology

4.1 Block

Block in a blockchain is the collection of legal transactions. Any node in a blockchain can start the transaction and broadcast it to other nodes in the network.

4.2 Blocks

Many block forming a linear structure like linked list is the blockchain.

4.3 Blockchain Technology

In technical terms, blockchain technology is a time-stamped series of immutable record of data that is managed by a cluster of computers not owned by any single entity. Every blocks of data (i.e. block) are secured and bound to each other using cryptographic principles (i.e. chain). The blockchain is transparent so anyone can track the data if they want to view the transactions.

4.4 Properties of Blockchain Technology

Decentralization. A single object cannot store all the information in a decentralized system. In detail, everyone in the network retains the information using Bitcoin technology.

Transparency. When the person's real identity is secure, all the transactions that were done by using their public address.

Immutability. The data in the blockchain cannot be tampered by intruders [3].

4.5 Key Concepts of Blockchain

Encryption. Public key is a user's address on the blockchain. Private key gives its holder access to their digital assets (Table 1).

Replication. All the nodes in a decentralized system have a replica of the blockchain. No centralized copy of data exists, and no user is 'trusted' more than any other. Transactions are broadcast to the network using the applications. Mining nodes validate all the transactions, add them to the block that has been created, and then broadcast the completed block to other nodes. Blockchain can be customized for various time-stamping schemes, such as Proof of Work and Proof of Stack.

Integrity. Peers keep the count of the database. Whenever a peer count increases, they extend or overwrite their own database and retransmit the improvement to their peers.

Table 1 Public blockchain versus Private blockchain

Permissionless ledger	Permissioned ledger
Public blockchain is also termed as permissionless blockchain as it allows anyone to send data to the ledger with digital identity	Private blockchain is also termed as permissioned blockchain as it is limited for the distributed digital identity of the ledger

Mining. Miners use the computational power to perform transactions. When miners try to compute a block, they prefer all transactions that they want to be added in the block, plus one coin base (generation) transaction. Valid transactions are needed for a block to be accepted by the network.

Consensus Mechanism. Consensus depends on the trust between nodes. Blockchain does not need centralized authority, and hence, the consensus may be determined algorithmically. The above-said technologies are the components of the blockchain technology.

5 Blockchain Architecture

Figure 2 describes the attributes such as Prev_Hash, Timestamp, Tx Root, and a Nonce (random string) which gets appended to the hash.Prev hash is hash value of the previous data. The term timestamp is the 'Network-adjusted time' which is meant for median of the timestamps returned by all nodes. Tx_Root is hash of the Merkle root. A block header hash is used to identify a block in a blockchain. The block header hash is considered by running the block header through the SHA256 algorithm twice. Then it is used to compute the validation of a block.

Fig. 2 Blockchain system and its attributes with Merkle tree structure

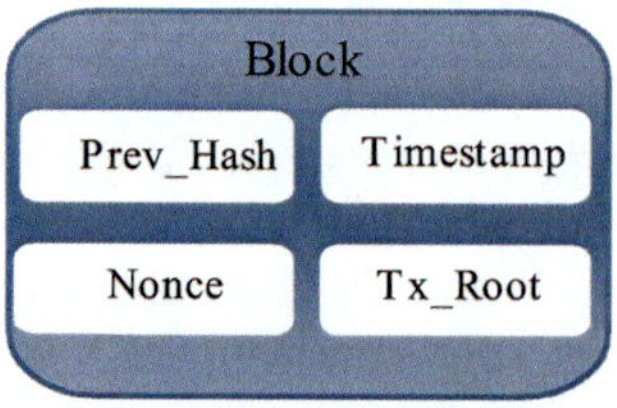

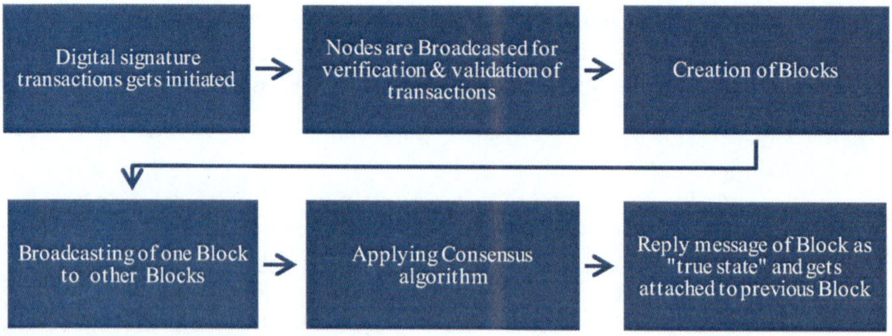

Fig. 3 Distributed ledger of blockchain

Figure 3 shows the distributed ledger architecture of blockchain. Basically, blockchain is an append-only database maintained by the nodes of a peer-to-peer (P2P) network. A blockchain is a technology that authorizes the transactions to be gathered into blocks of data and is recorded for future purposes. Later it can be modified into cryptographically chain of blocks in chronological order; and furthermore, it allows the resulting ledger to be accessed by different servers.

6 Applications of Blockchain

Figure 4 describes various applications of blockchain technology that are transforming society in recent days. In Paper [11] Blockchain financial services are redefining the existing barriers of our current financial market's infrastructure because of slow, expensive and several intermediaries.

The blockchain in insurance provides the automatic triggering of insurance policy for the customers and reduces the opportunity for insurance fraudulence.

In blockchain law applications, smart contracts are a digital contract verified on the blockchain, allowing for programmable, self-executable and self-enforcing contracts. The best example for smart contracts is Ethereum, which is an open-source blockchain platform.

Blockchain technology [also called distributed ledger technology (DLT)] helps to improve government services and nurture more transparent government–citizen relations. Governments can improve the way they deliver services, prevent tax fraud, by eliminating bureaucracy, and reducing the waste.

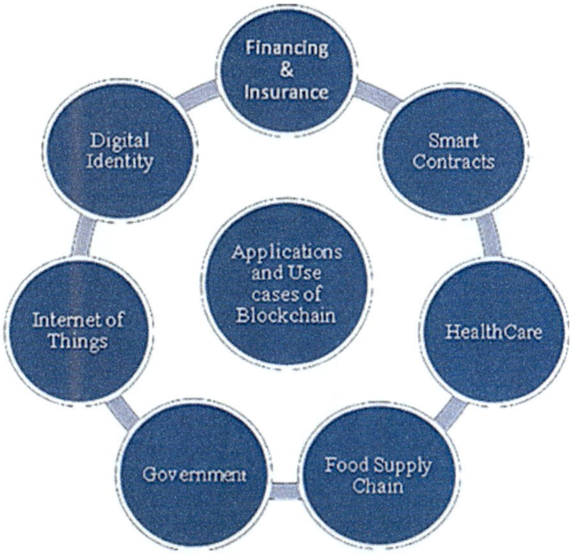

Fig. 4 Applications and use cases of blockchain

Blockchain technology has the latency to traverse the healthcare industry's centralized operations, for optimized business and service delivery. Moreover, the blockchain is the distributed ledger technology (DLT), and it gives a novelty fertile with the possibility of improved transparency, security and efficiency.

Blockchain technology offers the perfect solution to digital identities. Digital identities can be uniquely authenticated in an indisputable, immutable and secure manner. Blockchain-based identity systems also provide a better solution to the security issue with hardened cryptography and distributed ledgers.

Decentralized IoT will make device connectivity and data storage among the nodes that can operate without a centralized authority. The cryptographic algorithms used by blockchains would also help to make consumer data more private. If IoT and blockchain can resolve scalability issues over time, then the users will get improved global efficiency and connectivity.

Blockchain technology manages the modern, global, supply chain. A supply chain is how goods are sourced initially, manufactured and distributed to the end-user. The decisive goal for supply chain managers is producing effective goods for distribution and gets stuck to the budget for ensuring the customer satisfaction.

7 Conclusion

Blockchain technology looks promising still today in solving the existing centralized system in a decentralized way. The components such as consensus, architectural aspects and various applications like smart contracts, food supply chain were discussed in this paper. This concise survey of blockchain can be used by the researchers to have a glimpse over it. However, there is a scope for an extended version to explore the merits, limitations and features of blockchain technology as it is likely to be a worldwide ledger. The future direction is to resolve the issues in smart contracts such as agriculture and food supply chain.

References

1. S. Moin, A. Karim, Z. Safdar, K. Safdar, E. Ahmed, M. Imran, Securing IoT's in distributed Blockchain: analysis, requirements and open issues. Future Gener. Comput. Syst. **100**, 325–343 (2019)
2. A.A. Brincat, A. Lombardo, G. Morabito, S. Quattropanic., On the use of Blockchain technologies in WiFi networks. Comput. Networks **162**, 1–9 (2019)
3. Q. Feng, D. Hea, S. Zeadally, M.K. Khan, N. Kumar, A survey on privacy protection in Blockchain system. J. Network Comput. Appl. **126**, 45–58 (2019)
4. B.K. Mohanta, D. Jena, S.S. Pana, S. Sobhanayak, Blockchain technology: a survey on applications and security privacy challenges. Internet of Things (Elsevier) (2019)
5. S. Aggarwal, R. Chaudhary, G.S. Aujla, N. Kumar, K.K.R. Choo, A.Y. Zomayae, Blockchain for smart communities: applications, challenges andopportunities. J. Network Comput. Appl. **144**, 13–48 (2019)

6. A.H. Mohsin, A.A. Zaidan, B.B. Zaidan, O.S. Albahri, A.S. Albahri, M.A. Alsalem, K.I. Mohammed, Blockchain authentication of network applications: taxonomy, classification, capabilities, open challenges, motivations, recommendations and future directions. Comput. Standards Interfaces **64**, 41–60 (2019)
7. G. Zhaoa, S. Liu, C. Lopez, H. Lu, S. Elgueta, H. Chen, B.M. Boshkosk, Blockchain technology in agri-food value chain management—A synthesis of applications, challenges and future research directions. Comput. Industry **109**, 83–99 (2019)
8. A. Ravishankar Rao, D. Clarke, Perspectives on emerging directions in using IoT devices in Blockchain applications. Internet of Things (Elsevier) (2019)
9. X. Wang, X. Zha, W. Ni, R.P. Liu, Y. Jay Guo, X. Niu, K. Zheng, Survey on Blockchain for internet of things. Comput. Commun. **136**, 10–29 (2019)
10. S.K. Singh, S. Rathore, J.H. Park, Block IoT intelligence: A Blockchain-enabled intelligent IoT architecture with artificial intelligence. Future Gener. Comput. Syst. (2019)
11. K. Christidis, M. Devetsikiotis, Blockchains and smart contracts for the Internet of Things. IEEE Access **4**, 2292–2303 (2016)
12. M. Gupta, *Blockchain for Dummies* (IBM Limited Edition, 2017), pp. 1–51

Design and Implementation of Night Time Data Acquisition and Control System for Day and Night Glow Photometer

K. Miziya, P. Pradeep Kumar, Vineeth C., T. K. Pant, and T. G. Anumod

1 Introduction

The atmospheric airglows emissions are weak radiations originating because of de-excitation of atoms or molecules at different heights of the earth's atmosphere. Photochemical reactions happening during the course of a day are the reason for excitement of these molecules or atoms [1]. The role of optical airglow measurement in understanding the upper atmosphere/ionosphere process has been identified long back [2]. The simultaneous measurements of these emissions at different wavelengths are very significant as these emissions can act as a faultless tracer for the phenomenon occurring in atmospheric region where it originates. Hence, these measurements are widely used for the study of upper atmosphere/ionosphere, which is very important due to the role of ionosphere in radio communication. Measurement of dayglow is possible with the development of unique Multi-Wavelength Dayglow Photometer (MWDPM) [3]. A mask assembly for obtaining the faint dayglow from intense solar background is the novel part of MWDPM [4]. 'Day-Night Photometer' (DNP) is another instrument capable of measuring single wavelength of airglow during day and night time. The emerging DNPM is an enhanced version of its predecessors MWDPM and DNP.

Figure 1 shows the whole schematic depiction of the DNPM, including optical, opto-mechanical, and electronic components. One section of the device (shown at the left side of Fig. 1) is the daytime optics, which is used to measure the airglow emissions during daytime. The other section (shown at the right side of Fig. 1) is night time optics, which has a role in night glow measurements. The top side of

K. Miziya (✉)
Department of Electrical and Electronics, BITS Pilani Dubai Campus, International Academic City, UAE
e-mail: miziyak@gmail.com

P. Pradeep Kumar · C. Vineeth · T. K. Pant · T. G. Anumod
SPL, VSSC, Thiruvananthapuram 695022, India

S. Patnaik et al. (eds.), *Advances in Machine Learning and Computational Intelligence*, Algorithms for Intelligent Systems, https://doi.org/10.1007/978-981-15-5243-4_73

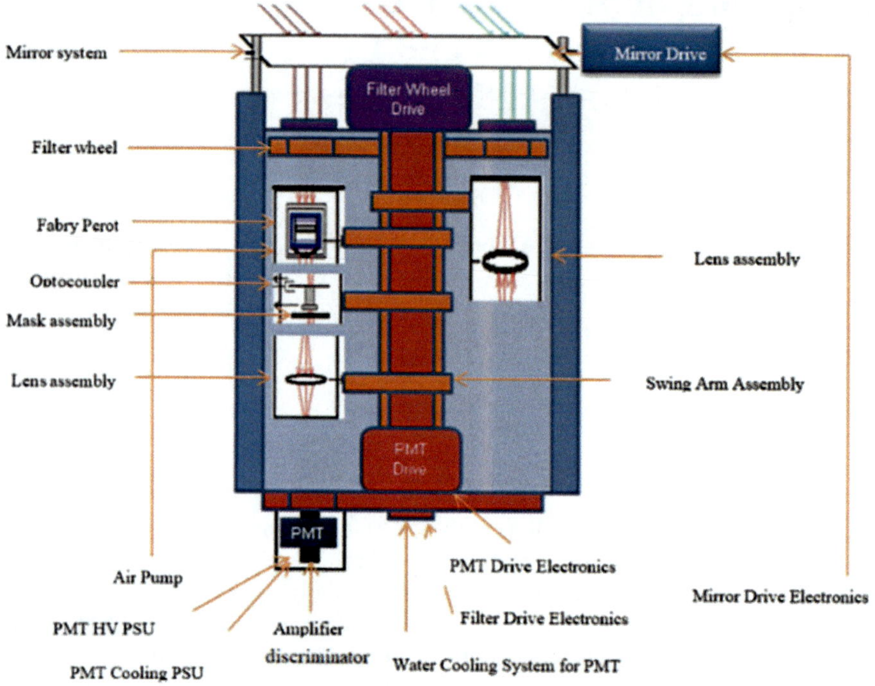

Fig. 1 A schematic diagram of the day and night glow photometer (DNPM)

the DNPM is filter wheel, which is a rotating platform containing three interference filters, arranged in such a way that only one of the filters becomes optically aligned with either night time optics or daytime optics. Rotating platform is provided with three holes, each hole near to separate filter. The position of the hole is kept such that it coincides with any of the three optocouplers present at the night time optics (or another three sets in daytime optics) during filter selection. These optocouplers present at day or night time optics gives the signal to identify the filter selected for extracting airglow. The position of the filter, if needed can be changed by the rotation of the filter wheel by a stepper motor connected to it. The stepper motor is controlled through a computer.

In daytime optics section, a Fabry-Perot (FP)etalon is kept under filter wheel drive, which produces circular fringes on the mask below to it only for the filtered airglow emission on the mask assembly and not for solar continuum. The mask assembly consisting of a rotor and stator (spiral) mask. With the rotation of rotor over spiral mask, optical spatial information from the three filters, in the form of fringes on the stator mask is converted into temporal information enabling simultaneous measurement of three wavelengths.

The lens assembly present beneath the mask assembly can focus this temporal information to a photon multiplier tube (PMT) for further processing. In DNPM, a PMT is also kept on a platform, which is connected to a stepper motor. The position of the PMT is changed 180° with this stepper motor controlled through a computer whenever a switching between daytime and night time is desired. This arrangement allows the usage of the same PMT module during night time and daytime. The PMT converts the optical temporal signal intensity to electrical signal. Further, this electrical signal is passed to amplifier discriminator, which remove the noise and amplify the signal. The difference between the counts of the amplified signals (both signal + background and background) gives the measure of airglow emission from a selected filter.

In night mode operation, below the filter section, only a lens assembly, which focusses the radiations to PMT is present. FP, mask assembly is not needed as there is no need of separation of airglow from solar background. For the proper night glow measurement, the proper placement of filter above the night time optics and PMT below lens assembly should be done before data acquisition of nightglow.

The scan mode operation and data acquisition of the predecessor MWDPM has been successfully performed by means of LabVIEW along with NIPCI 6602 DAQ module both at software and hardware level in [5] and [6], respectively. In this context, the automation of the emerging DNPM using LabVIEW along with NI PCI6602 DAC can help in integrating the operations of MWDPM and DNP.

2 Automation of DNPM Using LabVIEW

LabVIEW is a Windows-based Graphical Programming Language (GPU), suitable in making user-friendly interfaces for instruments so that it can be easily operated by computer users [7, 8]. The data acquisition module NI PCI 6602 provides eight up/down 32-bit counter/timer and up to 32 lines of TTL/CMOS compatible digital I/O [9]. To ensure the highest performance and user-friendliness, DNPM is automated using LabVIEW along with National Instruments (NI) DAQ PCI 6602.

2.1 Operational Requirements

As discoursed in introduction, DNPM is an enhanced version of its predecessors MWDPM and DNP. The operation mode of DNPM involves both daytime and night mode operations. In any case, the DNPM is capable of measuring near simultaneous emissions of three filters. The complexity involved in airglow measurements during night time is less compared to daytime both in design and control. The operational requirements for the night time airglow measurement, the DNPM, (1) Selects night time optics (2) Selects a filter (3) Obtain data from the selected filter for an integration time defined by user (4) Records and plots the acquired data and (5) Acquire data from all the three filters and record continuously until the device is functioned to stop.

System Design. As displays in Fig. 2, the modular design (including hardware and software) for the automated night time functioning of DNPM consists of (i) DNPM, which sends control input signals (NF1, NF2, NF3, P1, P2 and AD6 out) to NIDAQPCI 6602 module (ii) DAQ module, which receives control data from DNPM through the digital logic input (PFI lines 0.3, P0.4, P0.5, P0.6, P0.7 and source control input of CNTR 0) and sends control output signals (pulse, direction bit) to PMT drive assembly and stepper motor drive assembly from CNTR 6 and CNTR 7, respectively, (iii) PMT drive assembly, which changes the position of PMT by the rotation of a stepper motor connected to it (iv) Stepper motor assembly which alters the position of filter wheel by the rotational movement of another stepper motor (v) Computer. The virtual instrument (VI) developed with LabVIEW for the faultless functioning of DNPM is in the computer and it provides the user interface. The VI established for DNPM provides perfect control and co-ordination of the DAQ module and user-friendliness too. The following section discusses how each requirement is fulfilled with this robust design.

Night Time Optics Selection. The night time optic selection involves the proper positioning of PMT. Once the position of PMT is adjusted to acquire night time data, the position is maintained till the user demands to stop the data acquisition and to

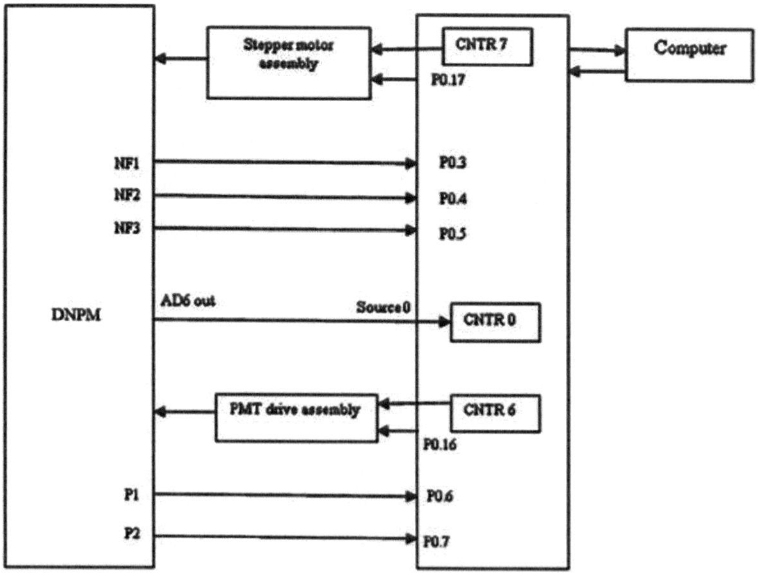

Fig. 2 Modular design of DNPM

switch to daytime optics. For the choice of night time optics, the VI detects the status of line PFI 6. The logical 'high' status of PFI 6 is recognized as the selection of night time optics. The 'low' status of PFI 6 is used for enabling the program used for controlling the PMT drive assembly to change position of PMT. Execution of the controlling program makes the counter (CNTR6) to send pulse and direction bit (PFI 16) to rotate the stepper motor till the status of PFI 6 becomes 'high' or in turn night time optics is selected. The pulse generation mode of counters of PCI 6602 is utilized in sending pulses to stepper motor.

Filter Selection. The selection of any filter or wavelength of operation involves the alignment of the holes on the filter wheel drive with any of the three optocouplers present at the night time optics. Once night mode operation is selected, the access of filters is in the order 1, 2, and 3 to acquire, record, and plot data. Status 'high' of the PFI lines 3,4,5 connected to night time optocouplers (NF1, NF2, and NF3) indicates the DNPM filter selection 1, 2, and 3, respectively. For the initial selection of the filter to start data acquisition in night time, VI senses the status of PFI 3 and rotates the filter wheel until the line corresponds to optocoupler becomes high. The VI rotates the filter wheel by giving control signals (pulse and direction bit) from counter 7 (CNTR 7) and PFI 7 of PCI 6602 to stepper motor assembly connected to filter wheel. A single pulse rotates the stepper motor by 1.8° and CNTR7 produces pulse till the desired filter is opted.

Data acquisition, plotting, and recording. Obtaining data during night time is simply performed by executing VI to operate CNTR0 in simple event counting mode. Figure 2 clearly shows that the output of amplifier discriminator is linked to the source input of CNTR0. Thus, CNTR 0 simply counts the AD6 out pulses which in turn is the airglow measurement of a particular wavelength selected by a filter during night time. These data or count values for the particular integration time are displayed in real time using a running display and archived automatically in the computer for further studies. Once the acquisition, displaying, and recording is performed for filter 1, the same steps are repeated for filter 2 and 3, respectively, which involves accessing next filter with the help of filter wheel assembly. The whole process repeats until the device is functioned to stop.

3 Results and Discussions

Data acquisition of DNPM during night time is realized. The front panel (FP) of the night mode operation VI of DNPM is shown in Fig. 3. For setting DNPM in data acquisition during night time, the tabs *Data Acquisition* followed by night *time* should be selected. The FP tab for night mode operation has one numeric control for setting integration time, two Boolean control ('Start' and 'OK'), and three radio buttons (to provide the stopping condition in three ways). The three radio buttons give the

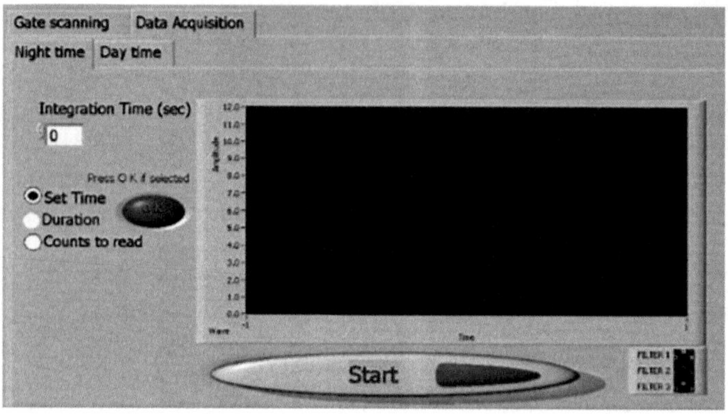

Fig. 3 Front panel of data acquisition of night mode operation of DNPM

options set time, duration, and counts to read as any one of the stopping conditions. Radio button ‘OK’ prevents the user from multiple selection of stopping condition. The stopping condition must be selected before pressing the ‘Start’ button. Once the stopping condition is selected, cluster or numeric control to set time, duration, or counts appears along with deactivation of ‘OK’ Boolean signal.

Once stopping condition is entered followed by the activation of ‘Start’ button, the system verifies the position of PMT under night time optics. The VI changes the position of PMT by giving control signals to stepper motor present in PMT drive assembly, if the position of PMT is inaccurate. During this time, the direction bit for PMT drive as well as the message displaying ‘PMT drive is under action’ is made visible in the front panel. Also the cluster input for the selected stopping condition would be displayed to enter the stopping condition by the user. The FP output of night mode operation of DNPM with stopping conditions ‘Set Time,’ is shown in Fig. 4a. The time entered in the cluster, which appears after selecting the radio button ‘Set Time’ is seen to be 4:27 PM. The graphical plot displays the count obtained as a result of the test data corresponds to night glow from all the three filters for an integration time of one second till the system clock time is 4.27 PM. Figure 4b shows the data obtained during night time operation of DNPM from different filter (1, 2, & 3) till the time 4:27 PM. From Fig. 4b, it can be clearly seen that data from three filters are saved in different folders within a separate folder named ‘nightglow’ (C:\nightglow\NG data F1).

Figure 5 shows the output, when the data acquired from the three filters for a duration of 3 min with one second integration time. It can be seen from the radio button that the stopping condition ‘Duration’ is selected. The acquired data from separate filters is saved automatically in separate files. Figure 6 shows the four sets of samples obtained from the three filters during night mode operation.

(a)

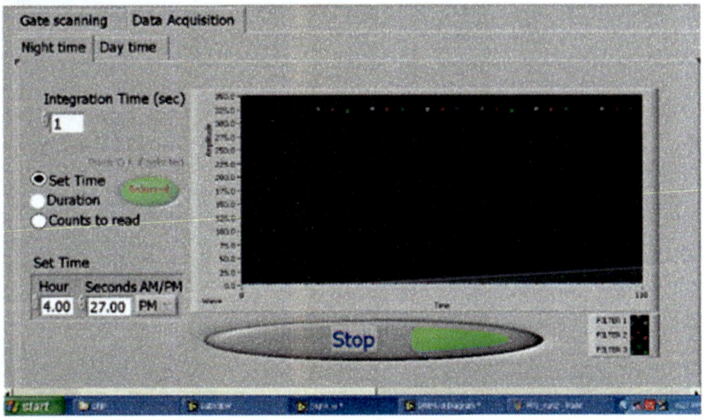

(b)

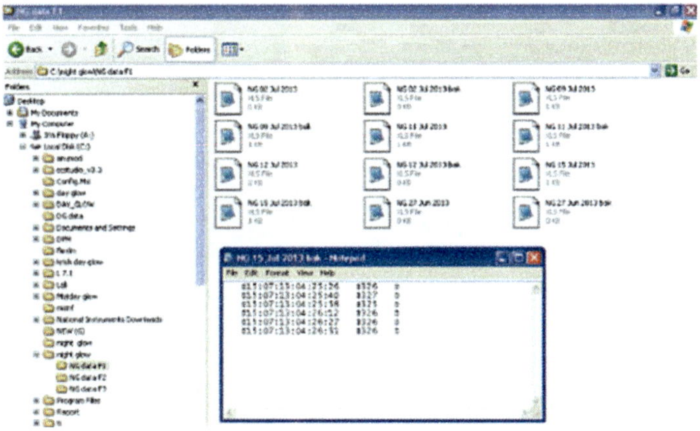

Fig. 4 **a** Data obtained from all the three filters till 4:27 PM when night time is selected. **b** Data obtained from filter 1 till 4:27 PM

4 Conclusion

The automation of Day and Night glow Photometer has been done efficiently using LabVIEW. The simulation results show that designed virtual instrument together with NI DAQ PCI 6602 is capable of performing data acquisition in night mode operation of DNPM successfully. Also, the automatic recording of acquired data in night mode operation is fruitfully performed and archived.

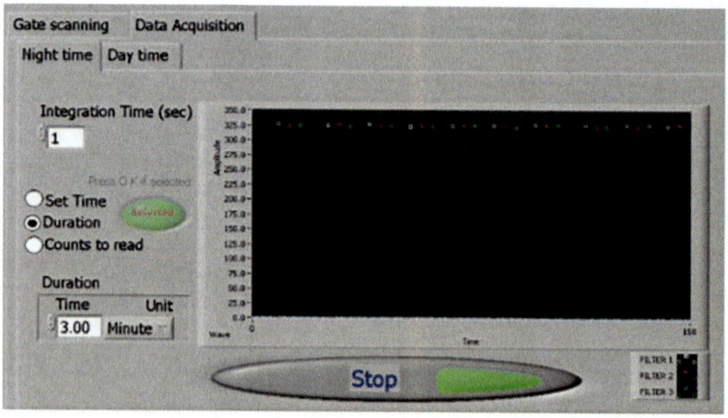

Fig. 5 Data obtained from all the three filters for duration of 3 min when night time is selected

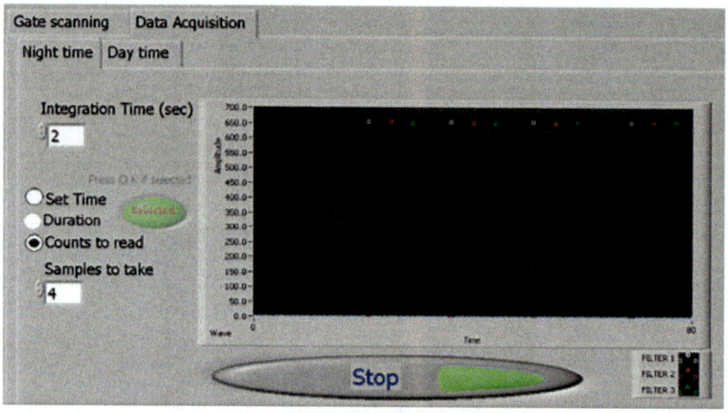

Fig. 6 Four sets of samples obtained from each filter when night time is selected

References

1. A.L. Broadfoot, K.R. Kendall, The airglow spectrum, 310010000°Ao. J. Geophys. Res. **73**(1), 426–428 (1968)
2. J.W. Chamberlain, *Physics of the Aurora and Airglow* (Academic Press, New York, 1961)
3. R. Sridharan, N.K. Modi, D. PallamRaju, R. Narayanan, T.K. Pant, A. Taori, D. Chakrabarty, Multiwavelength daytime photometer—a new tool for the investigation of atmospheric processes. Meas. Sci. Tech. **9**, 585–591 (1998)
4. R. Sridharan, R. Narayanan, N.K. Modi, D. PalalarRaju, Novelmask design for multiwavelength dayglow photometry. Appl. Opt. **32**, 4178–4180 (1993)
5. K. Miziya, K.G. Gopchandran, P. Kumar, T. Vineeth, T.K. Pant, T.G. Anumod, Automation of the gate scanning mechanism of the multi-wavelength dayglow photometer using LabVIEW, in *Fourth International Conference on Computing, Communications and Networking Technologies (ICCCNT)* (IEEE, Tiruchengode, 2013), pp. 36–43

6. K. Miziya, K.G. Gopchandran, P. Kumar, C. Vineeth, T.K. Pant, T.G. Anumod, Design and implementation of data acquisition and control system for multi-wavelength dayglow photometer, in *Fourth International Conference on Computing, Communications and Networking Technologies (ICCCNT)* (IEEE, Tiruchengode, 2013), pp. 7–13
7. A. M. Hassan, H.I. Ali, A.F. Lutfi, A Lab VIEW based data acquisition system for radial temperature distribution. IJCCCE **9**(1) (2009)
8. J. Kryszyn, D. Wanta, P. Kulpanowski, W.T. Smolik, LabVIEW based data acquisition system for electrical capacitance tomography, in *2018 International Interdisciplinary PhD Workshop (IIPhDW)*, Swinoujście, pp. 348–352 (2018)
9. Data Acquisition manual of National instrument, USA

Supplier's Strategic Bidding for Profit Maximization with Solar Power in a Day-Ahead Market

Satyendra Singh and Manoj Fozdar

1 Introduction

Deregulation in the electrical energy market has been begun since the 1980s to make competition at all stages, from production to the purchaser. This procedure has developed for improving the operational efficiencies and planning of the electrical power system. In this type of markets, different problems have been introduced for example oligopolistic behavior of the market, abuses of the market power, price-demand elasticity, strategic bidding behavior of power producers, etc. [1]. Theoretically, power suppliers maximize their profit by utilizing bid at, or very near to, their marginal production cost in a perfectly competitive power market. Additionally, suppliers, who are little enough to influence market costs with their offers, are price players, and their optimum strategy is to offer at the marginal production cost. However, in practice, the power markets are oligopolistic, and power providers may try to build their benefit by providing a price higher than the marginal cost of production. The power suppliers faces the problem when they try to develop the best optimum bid, due to the knowledge of their expenses, technical restrictions, market behavior and their expectation of rival. This is known as a strategic bidding problem.

The extensive studies on strategic bidding have been attempted by many researchers to how power producers dispatch their every unit's generation output for the market operator and how they maximize their profit by utilizing predicted market-clearing price. The literature survey on strategic bidding on competitive power markets for power producers has been introduced [2]. In [3–5], authors show that due to the uncertain bidding behavior of rival producers is mostly effect to the

S. Singh (✉)
School of Electrical Skills, Bhartiya Skill Development University, Jaipur, India
e-mail: satyendagur@gmail.com

M. Fozdar
Department of Electrical Engineering, Malaviya National Institute of Technology, Jaipur, India
e-mail: mfozdar.ee@mnit.ac.in

S. Patnaik et al. (eds.), *Advances in Machine Learning and Computational Intelligence*, Algorithms for Intelligent Systems, https://doi.org/10.1007/978-981-15-5243-4_74

process of strategic bidding decision model. Taking this reason as point of concern, the vast majority of the researchers have utilized a linear bid function and uniform market-clearing price model to construct the strategic bidding problem for profit maximization of the competitive power producers. In this type of strategic bidding, authors have modeled the rival's behavior using normal probability function and then solve the maximization of profit for competitive power producers utilizing heuristic algorithms [3–5]. However, these strategies only have been employed for traditional generators. Renewable energy in recent years is rapidly growing and further clean power productions with low carbon emission are incorporated into the electrical power system. Electrical power productions and percent of installed capacities of solar power plants go higher and will turn into the significant power generators soon. In this manner, considering the new round power foundation change and the expanding renewable power source in entire world, it is of incredible significance for the renewable power organizations to study the optimal bidding strategy of solar-based units taking an interest in the power market.

There have been several studies on the development of the joint strategic bid of renewable energy source (RES) with conventional generators. In this context, an optimal strategic bidding with solar photovoltaic (SPV) has also been considered in bidding strategy [6–10] that the uncertain output of solar power increases power imbalances and costs and also reduces solar power revenues. This mismatch between actual and forecasted power has been addressed by introducing a penalty. However, these works have not considered uncertainty associated with SPV. Thus, a method is required to model solar prediction in convincing way. In this work, a coordinated strategic bidding with the objectives of profit maximization is formulated with the amalgamation of traditional power suppliers and substantial solar power generation. This problem has been solved using novel heuristic technique [11], gravitational search algorithm (GSA), which is based on the law of gravity and interactions of masses with its application on power system problem is well established in [11–14]. Further, solar is used as a probabilistic manner to model the uncertainty and their prediction error is considered in cost function using underestimation and overestimation.

2 Modeling of Solar Power

It is necessary to handle the uncertainty associated with solar irradiation in order to deal with strategic bidding in the presence of solar power. Conversion of the solar irradiations is usually dependent upon the solar cell temperature, insolation of solar and technical properties of different PV modules. The output of solar power can be calculated by using solar irradiance and temperature. To model the solar power, a temperature and solar irradiance-based equations have been used and then a probabilistic beta distribution is utilized to model the uncertainty of solar power. For more detail, reader may refer [15].

2.1 Solar Power Scenarios Reduction

One thousand (1000) scenarios of solar power are generated. However, the probability of few scenarios might be very small and is some cases probabilities may be same. Subsequently, it is important to scrutinize the scenarios to obtain significant fewer scenarios while remaining lower and equal probability scenarios. The reduction should be such that the stochastic properties do not change. The amount of scenarios decreased depends on the type and nature of the problem to be optimized and must be reduced to or less than one-fourth of generated scenarios. In this, a scenario reduction technique known as Kantorovich distance matrix (KDM) has been employed [12]. It is based on the Euclidian distance between scenarios and their corresponding probabilities. This reduced the scenarios with closest and low probabilities.

2.2 Assessment of Scheduled Solar Power Amount for Bidding

The planned wind (W_g) and solar (S_g) power are obtained using KDM and the appropriate probabilities are calculated as follows

$$S_g = \sum_{i=1}^{\upsilon_i} S_{ai} \times \text{prob}_i \tag{1}$$

Here, prob_i is the probability of reduced ith generated scenario.

2.3 Solar Power Cost Evaluation

The imbalance cost of solar measure difference in forecasted and actual power which is the summation of underestimation and underestimation cost. It can be expressed as

$$\text{IMC}(Sg_n) = O_c(S_g) + U_c(S_g) \tag{2}$$

where $O_c(S_g)$ represents the overestimation cost, $U_c(S_g)$ represents the cost of underestimation for available solar energy. The underestimation and overestimation of the available solar energy is assessed as

$$U_c(S_g) = K_u * \int_{S_g}^{S_{\max}} (S_a - S_g) * f_{S_a}(S_a) * \mathrm{d}S_a \quad (3)$$

$$O_c(S_g) = K_o * \int_{0}^{S_g} (S_g - S_a) * f_{S_a}(S_a) * \mathrm{d}S_a \quad (4)$$

where K_u is the penalty for situational loss of profit per \$/kWh due to power underestimation. K_o is the penalty coefficient for overestimating power.

3 Problem Formulation

It is assumed that each power supplier (PS) has one generator to formulate the proposed strategic bidding problem. In addition, each generator bus has only one power supplier (PS). Any cost function of the power supplier can be formulated as follows

$$\mathrm{PC}_m(Pg_m) = a_m Pg_m + b_m Pg_m^2 \quad (5)$$

where the mth power supplier's cost parameters are a_m and b_m, and active power generation is Pg_m. It is known that all PS submit their bid to independent system operator (ISO) using linear supply function [3–5]. The function of the linear supply model is as follows:

$$CP_m(Pg_m) = \pi_m + \phi_m Pg_m \quad m = 1, 2, \ldots, \mathrm{CPS} \quad (6)$$

where π_m and ϕ_m are coefficients of bidding that must be non-negative. After receiving the offers from the PS, the ISO matches the output of power with the system's total demand and then minimizes the purchase costs. It is to be noted that Eqs. (6)–(9) should be satisfied when considering the constraints of power balance (8) and the constraints of power inequality (9).

$$\pi_m + \phi_m Pg_m = R \quad (7)$$

$$\sum_{m=1}^{\mathrm{CPS}} Pg_m + \sum_{n=1}^{sg} sg_n = Q(R) \quad (8)$$

$$Pg_{\min,m} \le Pg_m \le Pg_{\max,m} \quad (9)$$

where the market-clearing price is R, the market operator's forecast load is $Q(R)$. Let us suppose that

$$Q(R) = L_c - k * R \tag{10}$$

where L_c is constant and $k = 0$ is non-negative load price elasticity. Solution for Eqs. (7) and (8) when ignoring (9)

$$\left(R = \frac{L_c - \sum_{n=1}^{sg} Sg_n + \sum_{m=1}^{cps} \frac{\pi_m}{\phi_m}}{k + \sum_{m=1}^{\mathrm{CPS}} \frac{1}{\phi_m}} \right) \tag{11}$$

$$Pg_m = \frac{R - \pi_m}{\phi_m} \tag{12}$$

If in Eq. (12), the solution of Pg_m exceeds its maximum limits, it will be set in accordance with (9).

It is possible to express the optimal strategic bid by conventional power suppliers (CPS) with renewable sources to maximize profit.

Maximize

$$F(\pi_m, \phi_m) = R \times Pg_m - \mathrm{PC}_m(Pg_m) + R \times Sg_n - \mathrm{IMC}(Sg_n) \tag{13}$$

Subject to: Eqs. (1), (2), (5), (11) and (12).

Information about the next bidding period is hidden in a competitive energy market with a sealed bid. Members do not have information about the bid data of other members in this way. Be that as it might be, information of previous bidding data is available; the estimation of market-clearing price (MCP) is possible in the light of this information. Each member is endeavoring to evaluate various members that bid coefficients, but this is troublesome. Therefore, the coefficient of bidding follows a normal joint distribution with the following probability density function (pdf)

$$\mathrm{pdf}(\pi_m, \phi_m) = \frac{1}{2\pi\sigma_m^{(\pi)}\sigma_m^{(\phi)}\sqrt{1-\rho_m^2}} \times \exp\left\{ -\frac{1}{2(1-\rho_m^2)} \left[\left(\frac{\pi_m - \mu_m^{(\pi)}}{\sigma_m^{(\pi)}} \right)^2 + \left(\frac{\phi_m - \mu_m^{(\phi)}}{\sigma_m^{(\phi)}} \right) - \frac{2\rho_m\left(\pi_m - \mu_m^{(\pi)}\right)\left(\phi_m - \mu_m^{(\phi)}\right)}{\sigma_m^{(\pi)}\sigma_m^{(\phi)}} \right] \right\} \tag{14}$$

where the combined distribution parameters are $\mu_m^{(\pi)}$, $\mu_m^{(\phi)}$, $\sigma_m^{(\pi)}$ and $\sigma_m^{(\phi)}$, the coefficient of correlation between π_m and ϕ_m is ρ_n. Considering the mean values $\mu_m^{(\pi)}$ and $\mu_m^{(\phi)}$, and standard deviations values $\sigma_m^{(\pi)}$ and $\sigma_m^{(\phi)}$ of π_m and ϕ_m, marginal distribution for both is normal. Build on the last hour bidding; these can be calculated [3]. Joint distribution of π_m and ϕ_m is characterized by the function of distribution of probabilities with the objective function (13) subject to Eqs. (1), (2), (5), (11) and (12) is a stochastic optimization problem.

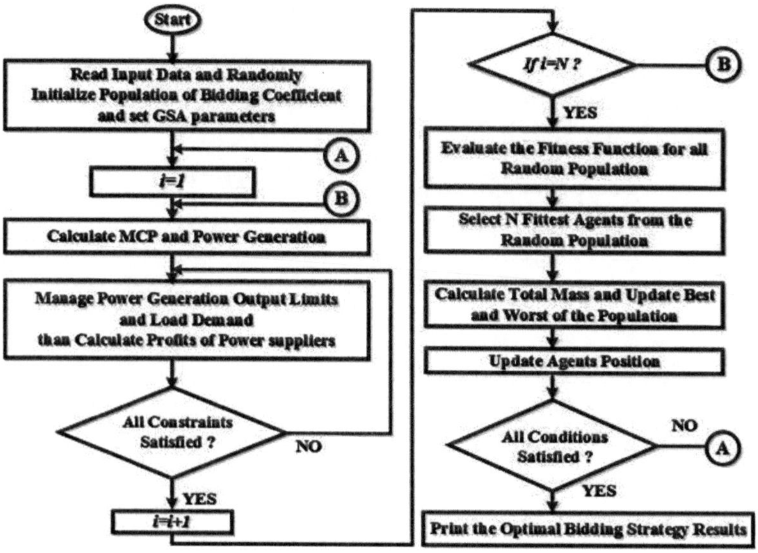

Fig. 1 Solution approach as a flowchart

4 Gravitational Search Algorithm

In [11], authors proposed a GSA to solve the problems of non-differentiable and nonlinear optimization. The flowchart solution procedure is given in Fig. 1.

5 Result and Discussion

In this segment, optimal strategic bidding model is considered for IEEE 30-bus system. The considered framework is utilized to obtain the maximum profit for power suppliers. The system data are taken from [12]. First, the test is conducted and analyzed on this system. Further, the considered model is modified to accommodate one solar power supplier to extent the influence of solar source. One solar supplier of 200 MW rated capacity is assumed in this work. The suggested formulation is solved on a 3.20 GHz, i5 processor, 4 GB RAM PC using the gravitational search algorithm (GSA) in MATLAB R2014a. For the solar power estimation, single hour wind speed data from January 1 to December 31, 2013 of Barnstable city, Massachusetts, USA is taken as study [16]. Solar irradiation is converted into solar power by using PV module specifications are taken from [15]. The solar irradiation data is fitted into various probability distributions are shown in Fig. 2. The log-likelihood, mean and variance values are calculated using various distributions and are presented in Table 1. It should be noted that the log-likelihood value of beta distribution is better than others, indicating the best fitting of the data in the distribution.

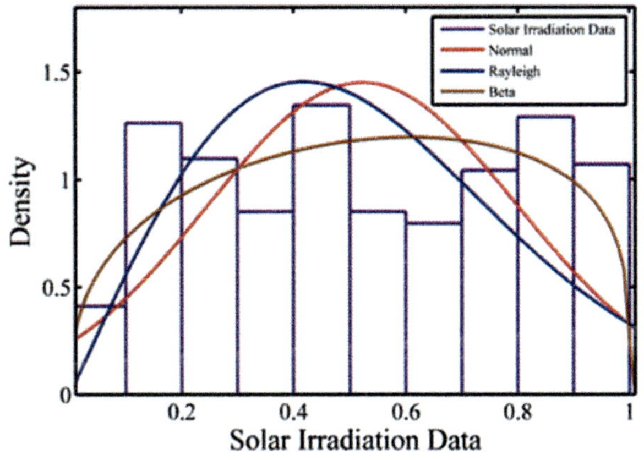

Fig. 2 Historical solar irradiation data for different distribution

Table 1 Mean, log-likelihood and variance values for historical solar irradiation data distribution

	Normal fit	Rayleigh fit	Beta fit
Log-likelihood value	−47.2392	−34.9839	10.1446
Mean	0.52266	0.523565	0.526305
Variance	0.0760555	0.0749005	0.06844

The values of beta distribution parameters are 1.3909 and 1.2518, respectively, for historical solar irradiation data. Then, a thousand solar irradiation scenarios are generated and convert into power scenarios using PV module specifications. The each generated scenario assigned a probability of normalization obtained using beta distribution to make their summation equal to unity. The beta and normalize the density function of probabilities shown in Fig. 3 for generated power scenarios. Since, the large number of scenarios predicts the uncertainty of solar power. However, there are few scenarios exhibit the same assessment. Therefore, KDM [12] method is employed to eliminate such scenario for better modeling of solar power. Here, 10 reduced scenarios are generated using 1000 scenarios for solar. Based on the final obtained value of solar power outputs and their corresponding probabilities, the expected values of solar power is 73.29 MW. Thereafter, the proposed optimal bidding strategies are investigated without solar and with solar using GSA. In this bidding process, the interdependency of bidding coefficients is contemplated by fixing one value and value of other coefficient is determined using an optimization method [3]. Therefore, the value of coefficient π_m is fixed in this work and the GSA is used to determine the optimum value of coefficient ϕ_m from the interval of $[b_m \; M * b_m]$. The value of M is set to be 10. The value of joint parameters of normal distribution in Eq. (14) is taken from [3]. The optimum value for coefficients of bidding of different CPS with and without solar and wind power using GSA is given in Table 2. The suggested optimal strategic bidding to clear the market-clearing price (MCP) for

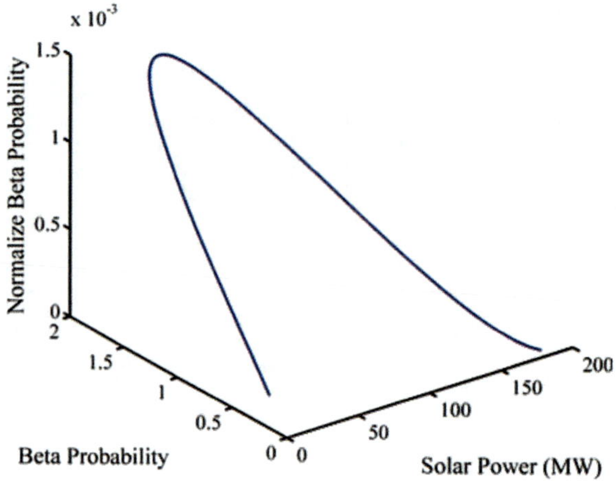

Fig. 3 Beta and normalize the density of beta distribution for generated solar power scenarios

Table 2 Optimum bidding results for standard IEEE 30-bus system with and without solar

PSs	π_m	Standard IEEE 30-bus system without solar			Standard IEEE 30-bus system with solar		
		ϕ_m	Pg (MW)	Profit ($)	ϕ_m	Pg (MW)	Profit ($)
1	2.0	0.049231	160	1815.32	0.049575	160	1495.4
2	1.75	0.224134	77.45	839.65	0.215113	55.53	512.23
3	1.0	0.722945	40.95	425.33	0.453362	32.27	288.17
4	3.25	0.097653	100	986.18	0.104385	91.44	725.46
5	3.0	0.289934	60.80	573.06	0.251243	43.74	343.46
6	3.0	0.289934	60.80	573.06	0.251243	43.74	343.46
MCP		13.9458			11.95		
Total profit for TPS ($)		5212.6			3708.22		
Total generation for TPS (MW)		500			426.71		
Sg (MW)		00			73.2897		
$O_c(S_g)$($)		00			114.7551		
$U_c(S_g)$($)		00			248.4213		
IMC(Sg_n)($)		00			363.1764		
Profit for SPS ($)		00			512.6355		

the standard IEEE 30-bus system is evaluated with the help of bidding coefficients predicated by the GSA and their corresponding profits and individual dispatch of generators are measured.

Effects of renewable power sources are successively considered on IEEE 30-bus system. For bidding strategy of renewable power, the system operator is allowed to modifying the existing demand, which means actual demand excluding renewable power generation, and then updates the bidding coefficients in accordance with the changing demand [12]. Based on this approach, solar power is considered to determine the new MCP. First, the solar power generator is considered and new value of MCP is calculated by updating the bidding coefficients at modified demand. In this analysis, the consideration of operating cost for solar power source has not been taken into account. However, due to the associated intermittency of these renewable sources it is acceptable to consider their imbalance cost. This cost is determined in terms of overestimation and underestimation of generation from solar. And the effect of this cost is reflected on total profit obtained by renewable suppliers in terms of revenue minus the imbalance cost. Also, the penalty coefficient and reserve coefficient linked with underestimation and overestimation separately are considered as 50% of MCP and equivalent to MCP, respectively [12].

The results of the proposed bidding strategy on considered system, considered system with solar by using GSA are presented in Table 2. From Table 2, it is observed that the market is cleared at MCP value of 13.94 $/MW, total generation of CPS is 500 MW and net profit is $5212.6 with standard CPS. In the second case, i.e., only solar power with CPS, the net profit value, overestimation and underestimation cost is 512.6355, $114.7551, and $248.4213, respectively. For this case, the MCP value is 11.95 $/MW with total generation of CPS 426.71 MW which is lower than conventional due to significant power generation from the solar. From Tables 2, it can be observed that all the purchase bids would satisfy by the lower MCP value. Due to the presence of solar supplier in the process of dispatch, there will be fewer CPS requirements in power system operation. Further, the overestimation estimate is very small compared to the underestimation of solar power uncertainties. Therefore, applying KDM in reduction of scenarios is better in modeling uncertainty. This will encourage to the solar power suppliers for bidding the extra power into the real-time market if the underestimation is positive.

6 Conclusion

This paper investigates optimal bidding strategies to maximize the profit of power producers with the amalgamation of renewable energy sources. Integration of solar is considered in the probabilistic approach using beta distribution and transform into power variable. Further, this power incorporated in cost model as overestimation and underestimation terms in order to consider the variability of power output. The proposed method considers the rival's behavior using normal probability distribution function to minimize the dynamics of competitor in the power market. Incorporation of solar power affects the bidding such as it reduces the CPS generation and provides lowered value of MCP which would deliver sufficient electricity from accepted sales bids to satisfy all the accepted purchase bids. Therefore, the proposed method is capable of satisfactory results with the consideration of uncertainty model of renewable sources.

References

1. K. Bhattacharya, M.H. Bollen, J.E. Daalder, *Operation of Restructured Power Systems* (Springer Science & Business Media, Berlin, 2012)
2. A.K. David, Competitive bidding in electricity supply, in *IEE Proceedings C-Generation, Transmission and Distribution*, vol. 140 (IET, 1993), pp. 421–426
3. F. Wen, A.K. David, Optimal bidding strategies and modeling of imperfect information among competitive generators. IEEE Trans. Power Syst. **16**(1), 15–21 (2001)
4. J.V. Kumar, D.V. Kumar, Generation bidding strategy in a pool based electricity market using shuffled frog leaping algorithm. Appl. Soft Comput. **21**, 407–414 (2014)
5. S. Singh, M. Fozdar, Generation bidding strategy in a pool-based electricity market using oppositional gravitational search algorithm, in *2017 14th IEEE India Council International Conference (INDICON)* (IEEE, New York, 2017), pp. 1–6
6. H.M.I. Pousinho, J. Contreras, P. Pinson, V.M.F. Mendes, Robust optimisation for self-scheduling and bidding strategies of hybrid CSP–fossil power plants. Electr. Power Energy Syst. **67**, 639–650 (2015). https://doi.org/10.1016/j.ijepes.2014.12.052
7. G.Q. He, Chen C. Kang, Q. Xia, Optimal offering strategy for concentrating solar power plants in joint energy, reserve and regulation markets. IEEE Trans. Sustain. Energy **7**, 1245–1254 (2016). https://doi.org/10.1109/TSTE.2016.2533637
8. I.L.R. Gomes, H.M.I. Pousinho, R. Melíco, V.M.F. Mendes, Bidding and optimization strategies for wind-PV systems in electricity markets assisted by CPS. Energy Proc. **106**, 111–121 (2016). https://doi.org/10.1016/j.egypro.2016.12.109
9. J. Martinek, J. Jorgenson, M. Mehos, P. Denholm, A comparison of price-taker and production cost models for determining system value, revenue, and scheduling of concentrating solar power plants. Appl. Energy **231**, 854–865 (2018). https://doi.org/10.1016/j.apenergy.2018.09.136
10. O. Abedinia, M. Zareinejad, M.H. Doranehgard, G. Fathi, N. Ghadimi, Optimal offering and bidding strategies of renewable energy based large consumer using a novel hybrid robust-stochastic approach. J. Cleaner Prod. **215**, 878–889 (2019). https://doi.org/10.1016/j.jclepro.2019.01.085
11. E. Rashedi, H. Nezamabadi-pour, S. Saryazdi, GSA: A gravitational search algorithm. Inf. Sci. **179**, 2232–2248 (2009). https://doi.org/10.1016/j.ins.2009.03.004
12. S. Singh, M. Fozdar, Optimal bidding strategy with inclusion of wind power supplier in an emerging power market. IET Gener. Transm. Distrib. (2019). https://doi.org/10.1049/iet-gtd.2019.0118
13. D. Serhat, G. Ugur, S. Yusuf, Y. Nuran, Optimal power flow using gravitational search algorithm. Energy Convers. Manage. **59**, 86–95 (2012). https://doi.org/10.1016/j.enconman.2012.02.024
14. R. Provas Kumar, Solution of unit commitment problem using gravitational search algorithm. Electr. Power Energy Sysy. **53**, 85–94 (2013). https://doi.org/10.1016/j.ijepes.2013.04.001
15. V.K. Jadoun, V.C. Pandey, N. Gupta, K.R. Niazi, A. Swarnkar, Integration of renewable energy sources in dynamic economic load dispatch problem using an improved fireworks algorithm. IET Renwab. Power Gener. **12**, 1004–1011 (2018). https://doi.org/10.1049/iet-rpg.2017.0744
16. Solar anywhere [Online]. Available: https://data.solaranywhere.com/Public/Tutorial.aspx

Police FIR Registration and Tracking Using Consortium Blockchain

Vikas Hassija, Aarya Patel, and Vinay Chamola

1 Introduction

We have proposed this system keeping in mind the difficulties that people face during the registration of an FIR or a complaint at the police station [1]. In the conventional system, the people have to physically visit the police station multiple times, which is very time-consuming. The same also consumes a whole lot of money and energy. The other disadvantages include the fear of getting abused or harmed by people against whom FIR is lodged. Filing FIR against a highly reputed person is sometimes a hard task. It is a common issue that people are often refused an FIR registration. The possible reasons could be that the police official genuinely does not believe the informant, or it could be that his refusal stems from the influence upon him of powerful vested personnel, who have managed to approach him before the informant. The Indian Legal System provides some options that one can exercise in such cases, but it is often seen that people do not have the required information, time, energy, and money to exercise the same. By allowing people to file their complaints directly, this system bypasses the police officers who are often reluctant to register the FIRs, mainly in kidnapping and ransom cases. This would also help eliminate corruption.

In recent years, blockchain technology [2, 3] has attracted increased interests worldwide in various domains such as finance, insurance, energy, logistics, and mobility. It has the potential to revolutionize the way applications will be built, operated, consumed, and marketed in the near future. In this context, consortium blockchains [4] emerged as an interesting architecture concept that has some advantages of both the private and the public blockchains. These consortium blockchains can also be described as being semi-decentralized. They possess the security features

V. Hassija · A. Patel
Department of CSE and IT, JIIT, Noida, India

V. Chamola (✉)
Department of Electrical and Electronics Engineering, BITS, Pilani Campus, Pilani, India
e-mail: vinay.chamola@pilani.bits-pilani.ac.in

S. Patnaik et al. (eds.), *Advances in Machine Learning and Computational Intelligence*, Algorithms for Intelligent Systems, https://doi.org/10.1007/978-981-15-5243-4_75

that are inherent in public blockchains, whilst also allowing for a greater degree of control over the network [5, 6]. Proper maintenance of police station records is a prerequisite for the smooth functioning of a police station. However, nowadays, these records are highly vulnerable and are exposed to the risk of being breached or forged. Our proposed system would ensure transparency, security, and privacy of the records stored [7]. The verification of the transaction information would require a consensus mechanism. Current consensus mechanisms designed for blockchains are slow due to the significant amount of time and energy consumed for block production and safety. Therefore, our objective is to design a consensus algorithm with high performance to be used in consortium blockchain [8]. In this paper, we propose a proof of vote consensus protocol based on voting mechanism and consortium blockchain.

2 Proposed Solution

We propose to develop a system, wherein the victim can lodge their FIR using a mobile application. As shown in Fig. 1, the victim would fill up the FIR form on his mobile application. He would provide the proofs and details related to the complaint in the application. The user can upload images, audio files, and video files as records. These details would then be converted into a transaction with the digital signature of the complainant and a smart contract [9] would also be associated with it. The transaction will now float in the network of consortium. The commissioner (the head of a police station) of the respective police station can go through all the FIRs registered at his police station and assign police inspectors (miner candidates) in-charge of the floating FIRs by directly messaging them through the app. All the police inspectors will have a score according to the work they perform. If the inspector in-charge fails to provide updates to the complainants, then the smart contract, which

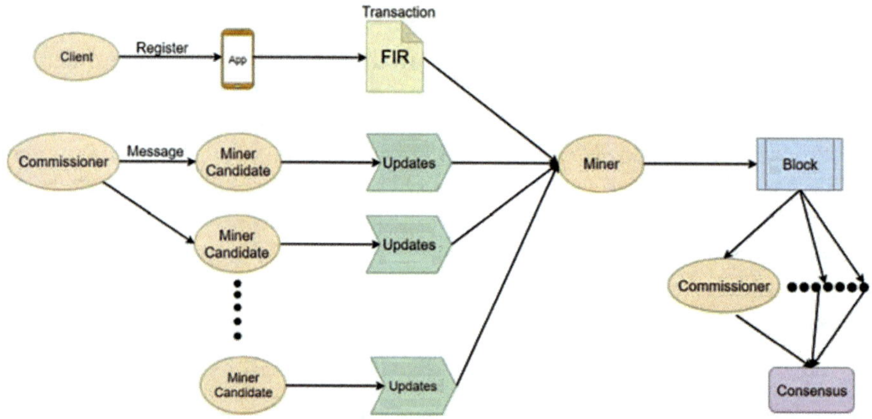

Fig. 1 Flow of the proposed solution

has a certain time limit by design, would notify the commissioner about the failure of the inspector in-charge. The commissioner now will decrease the score of that police inspector. Based on this score, the commissioner would select a group of miners whose work would be to collect all the floating transactions and combine them into a block and send it to all the commissioners in the network for verification. The scoring system would ensure honest behaviour on the part of the police officials. The commissioner would now validate the transactions in the block. If any transaction is found invalid, then the police inspector who formed this transaction would be held responsible, and the commissioner would take actions against him accordingly. Figure 1 shows the flowchart of the process to be followed to FIR registration and tracking in the proposed model.

3 Network Architecture

A consortium blockchain is a specialized blockchain with authorized nodes to maintain distributed shared databases [10, 11]. As our system is a consortium blockchain, it becomes very important to keep it distributed so that a singular entity does not possess all the power. To solve this problem, we have proof of vote as the consensus mechanism [12]. Consensus participants of a consortium blockchain are likely to be a group of pre-approved nodes on the network. Thus, consortium blockchain possesses the security [13, 14] features that are absent in public blockchain while allowing the required degree of control over the network [15, 16].

3.1 Consortium Blockchain Model

In our blockchain system, we have established different roles for the network participants that have been divided on the basis of their functionality. Figure 2 shows the format of a single transaction added in the blockchain for an FIR. The four roles are:

Commissioner. Just like in the real world, the commissioner in our consortium blockchain will have the right to recommend and vote miners to form a blockchain. They will have the power to choose a miner from all the miner nodes to form a block from the floating transactions. With that, they are also given the power to evaluate the transactions (FIR transaction from the client or case update from miner candidate) inside the block made by the miners. The blocks made by the miners will be considered valid and will get added to the longest chain of blockchain only when it will receive more than half the votes in their favour from these commissioners [16].

Miner. These nodes will be responsible for adding blocks to the blockchain. The commissioner will choose them on the basis of their reputation (based on their previous work). They will form a block out of all the floating transactions and sign the block. If the block is validated successfully and no malicious activity is discovered,

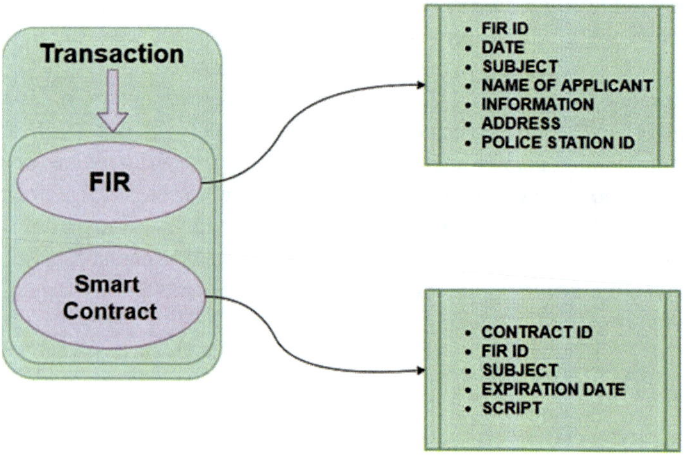

Fig. 2 Format of transactions on the blockchain

this node will be given points, and its reputation will be increased by the commissioners. Becoming a miner will be a two-step process that involves becoming a miner candidate and then winning the election for the miner. Re-election will take place after the expiration of their term of office.

Miner candidate. As the number of miner nodes is fixed, the other registered nodes act as miner candidates that are given special work. As mentioned before, the miner candidates will be given authority to investigate the FIR transactions, which will be given to them by the commissioner nodes via elections, and hence they will provide the client with updates about the case. The transaction will be signed by their digital signatures so that if they fail to fulfil all the clauses mentioned in the smart contract, they are identified, and their reputation is decreased. Also, if this happens, the information will be transferred to another miner candidate. They are eligible to be a miner only if elected by the commissioner nodes.

Client. All of the four roles use the digital signature to authenticate their identities. They need to sign the messages they send so that their actions be verified easily. Ordinary users can join or leave the blockchain network anytime without being authorized, and their behaviour can be arbitrary. They will need to have an account on the mobile application in order to add a FIRs in their name. They will only participate in the public part of the blockchain and can only take part in the process of making transactions of their FIR and mentioning additional clauses of smart contracts if any.

4 Proof of Vote

4.1 Consensus Process

The algorithm that we will use for consensus is proof of voting. The assumptions in this algorithm require that there be N_c total commissioners, N_m miners, N_{mc} miner candidates, and N_{cl} clients. Let the total number of nodes be N_{all}. Where $N_{all} = N_c + N_m + N_{mc} + N_{cl}$. For our algorithm, we consider one cycle to be composed of several rounds of consensus, each producing one block. The cycle gets over when a certain number of blocks are mined by the miners, say *B*, and one extra block pertaining to the elections, its results and the server information of the newly elected miners (as part of the proof of vote mechanism) get approved and added in the blockchain. The complete cycle thus generates a total of $B + 1$ blocks at the end of it and is termed as a tenure cycle. The duration of the tenure cycle is T_t. This is also the duration of completion of a miner's tenure. Let there be a random number assigned to each miner, from 0 to N_{m-1} and let the time period allotted to each miner within which he should mine a block be T_b. The end of each round of consensus produces a valid block, signed and approved by at least $N_{c/2+1}$, i.e., 51% of the commissioners [17]. After generating a block, if the block gets validated, the miner calls for a function that generates a random number *R* between $(0, N_m)$, which is matched with another miner having the same number. Now, this miner is the one who performs the next round of consensus and takes the procedure forward. If the block does not get validated, it must have happened because of one of the following two reasons.

The miner might have exceeded the time within which he was required to mine the block, i.e., T_b.

In such situations, the task of mining the block gets passed onto the subsequent miner, i.e., one with the number $R = R + 1$. If this pattern follows and *R* exceeds N_m, the procedure starts from $R = 0$. It finally reaches consensus when, at least and at most, one block gets validated. Such a procedure of consensus reaches consensus finality.

The miner candidates that are producing the updates and floating transactions can act maliciously and make wrong updates that will cause the commissioner to invalidate the block.

In this situation, the commissioner will delete the invalid transaction of the block. Decrypting the digital signature and finding out the exact person who has floated the malicious transaction, the commissioner can reduce the rewards and scores associated with that police officer's name in their vote-list and appoint some other police officer (miner candidate) to perform further updates on the case. The task of mining, however, is allotted to the same miner again, i.e., a miner with the number *R*.

After generating the required *B* number of blocks, the last round of consensus for that tenure runs to produce the special block. In this round, the miner candidates who were till now responsible only for floating transactions with updates will now be competing in elections to be a part of the next lot of miners. Each commissioner will give a vote-list of their own, containing scores of the miner candidates. In the end, topmost N_m contenders will win the election and form the group for carrying out the new round of tenure. The last block will contain the result of the election and related information. This ends one tenure cycle at the completion of which $B + 1$ rounds of consensus have already taken place, increasing the blockchain by $B + 1$ blocks.

4.2 The Generation of a Valid Block

One round of consensus may take C cycles before a block actually gets validated and added to the chain. Let C_1 be the number of cycles that fail due to the maliciously added transactions and C_2 be the number of cycles that fail due to the inability of a miner to generate the block within the given time restraint (T_b). Thus, the total no of cycles $C = C_1 + C_2$. This means there have been $C - 1$ invalid blocks that have already been rejected. Therefore, the total time for one round of consensus comes out to be $T_r = C * T_b$, where C is equal to $C_1 + C_2$ and ($1 \leq C_2 \leq N_m$). (C_2 causes the authority of mining to be passed onto the next miner, whereas C_1 reauthorizes the same miner for the mining task).

1. In our model, the transactions that will be floating are the FIR made by the users attached with a smart contract and the updates made by the policeman (miner candidates) corresponding to some FIR number. Anybody floating a transaction should have their digital signature attached to it.
2. The unconfirmed FIR gets stored in the transaction pool.
3. As soon as a transaction containing an FIR floats from a user, the commissioner on that network (the network of the police station where FIR was filed) appoints a miner candidate who will handle the case by sending him a message on the app.
4. The transactions containing the FIR will be picked up by the appointed police officer (miner candidates) who will work on the case and float an update within the time specified in the smart contract (T_{sc}) failing which will affect their rewards and reputation.
5. The transactions containing the updates will be picked up by a miner numbered i where $i = R$ and R is the random number of the previous block. For the addition of the genesis block, the value of R is zero by default.
6. These transactions are packed into one block and sent to all the commissioners. If a commissioner verifies a block, he signs the block header and sends it back.
7. The cut-off time for this block is T_{cutoff} = time for the previous block to get confirmed + T_b.
8. If the miner receives at least $N_{c/2+1}$ signed blocks back within T_{cutoff}, he can add the random number of the block and his own signature and finally add it to the blockchain.

9. After T_{cutoff} though, the block becomes invalid and $R = R + 1$ as discussed in Sect. 4.1.
10. If the block gets rejected due to some malicious actions, follow the procedure described in Sect. 4.1.
11. In either case, however, the time between the addition of two blocks, i.e., passing of an update by the police station to the client, should not exceed T_{sc} as per the terms of the contract. If this happens, the miner or the miner candidate whosoever is responsible shall have their reputation on the vote-list at stake.
12. If $C \geq N_{\text{b}}$, this will imply that none of the miner presents could actually mine a valid block, and now C will equate to 1 again. If this continues to happen, the network may fall into a dead circle.

5 Numerical Analysis

In this section, we perform some simulations to test, verify, and compare the effectiveness of the proposed model with the current system. The simulations were perfomed on a blockchain data structure created using various Node js libraries for cryptographic functions. The proof of vote consensus algorithm was written on top of the blockchain data structure. For the purpose of simulation we have taken the average values of the incentives and time taken per case for the current model and have compared them with the average values of proposed model. A total of 200 cases are considered to compare the incentive model and 400 cases are considered to compare the change in time taken to solve a single case. We have used the National Crime Records Bureau data set for analyzing the changes in the time taken to solve a single case with the increasing number of cases and have intutively compared the growth of values with the proposed model [18]. Figure 3 shows a plot between incentive ($ per case) and the number of cases solved. In the current system, there is no per case incentive or a bonus for the officers. Therefore, a lack of motivation is observed among the officers to pick more cases. In the proposed model, the officers are allowed to become the miners based on the efficiency they show in solving the cases. Therefore, the officers are more motivated to pick more cases and solve them in as little time as possible. Therefore, the figure shows exponential growth in terms of incentive earned by the officers in the proposed model. The growth in incentive is very slow and linear in the current system, as shown in the Fig 3. Figure 4 shows a plot between the average time at solving each case (months) and the number of FIRs registered. In the current system, as more FIR gets registered, it will increase the time to solve the individual case. Whereas in our proposed model, police officers will get incentive when they successfully solve the case. This will motivate the police officers to solve the cases correctly and quickly. Therefore, the average time will decrease exponentially as the number of FIRs increases. It can be observed in Figs. 3 and 4 that initially till 100 cases the proposed model is showing low incentive values and more time per case. This is so because of the lack of participation in the model. However, gradually

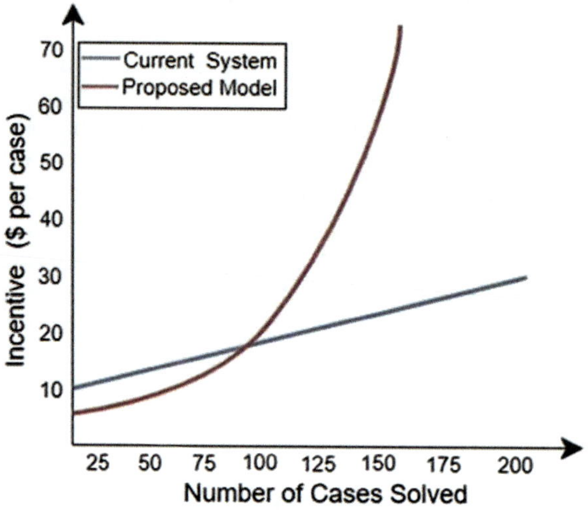

Fig. 3 Comparison of growth of incentive for the officers solving the FIR cases

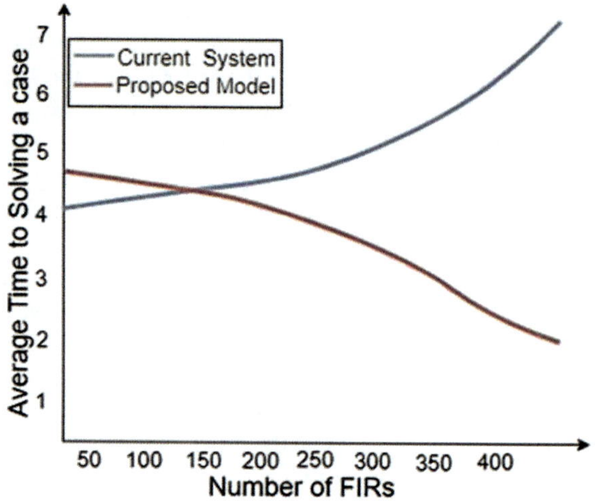

Fig. 4 Comparison of average time to solve a case (months) with the increasing number of FIRs

as more and more officers start participating and start competing for becoming the miners the values of incentive starts increasing and the time taken to solve per case starts decreasing.

6 Conclusion

The conventional method is prone to delays and inefficiency. This paper proposes to simplify and speed up the process of FIR registration and tracking through the use of consortium blockchain technology. We presented the complete consensus process using the proof of vote protocol. We designed four identities for network participants based on the key idea of the voting mechanism. This guarantees the separation of voting and execution rights that enhance the independence of miner's role, so does the internal control system within the consortium. With the advancement of information and communication technology, our proposed method will definitely boost up the FIR proceedings. Therefore, this paper aims to help the citizens and the police officials alike. The proposed system would guarantee the acceptance and response of the FIRs from the police department to the complainants. Thus, the ease of access, registry, and tracking will encourage a more judicial and lawful society.

References

1. W. Zhao, Dubai Plans to 'Disrupt' Its Own Legal System with Blockchain. https://www.coindesk.com/dubai-plans-to-disrupt-its-own-legalsystem-with-blockchain. Accessed 11 March 2019
2. Y. He, H. Li, X. Cheng, Y. Liu, C. Yang, L. Sun, A blockchain based truthful incentive mechanism for distributed p2p applications. IEEE Access **6**, pp. 27,324–27,335 (2018)
3. S. Yao, J. Chen, K. He, R. Du, T. Zhu, X. Chen, Pbcert: Privacy-preserving blockchain-based certificate status validation toward mass storage management. IEEE Access **7**, 6117–6128 (2019)
4. Z. Li, J. Kang, R. Yu, D. Ye, Q. Deng, Y. Zhang, Consortium blockchain for secure energy trading in industrial internet of things. IEEE Trans. Industr. Inf. **14**(8), 3690–3700 (2018)
5. T. Alladi, V. Chamola, K. Choo, Consumer IoT: Security vulnerability case studies and solutions. IEEE Consumer Electronics (Sep 2019) (2019)
6. T. Alladi, V. Chamola, J. Rodrigues, *Blockchain in smart grids: A review on different use cases* (Sensors, MDPI, 2019)
7. N. Fabiano, Internet of things and blockchain: legal issues and privacy. the challenge for a privacy standard, in *2017 IEEE International Conference on Internet of Things (iThings) and IEEE Green Computing and Communications (GreenCom) and IEEE Cyber, Physical and Social Computing (CPSCom) and IEEE Smart Data (SmartData)* (IEEE, New York, 2017), pp. 727–734
8. G. Bansal, V. Hassija, V. Chamola, N. Kumar, M. Guizani, Smart stock exchange market: A secure predictive decentralised model, in *IEEE Globecom*, Waikoloa, USA, Dec 2019, pp. 1–6
9. C. Sillaber, B. Waltl, Life cycle of smart contracts in blockchain ecosystems. Datenschutz und Datensicherheit-DuD **41**(8), 497–500 (2017)
10. V. Hassija, V. Chamola, V. Saxena, D. Jain, P. Goyal, B. Sikdar, A survey on IoT security: application areas, security threats, and solution architectures. IEEE Access (2019)
11. K. Christidis, M. Devetsikiotis, Blockchains and smart contracts for the internet of things. IEEE Access **4**, 2292–2303 (2016)
12. C. Cachin, M. Vukolić, Blockchain consensus protocols in the wild. arXiv preprint arXiv:1707.01873 (2017)
13. Z. Zheng, S. Xie, H.-N. Dai, H. Wang, Blockchain challenges and opportunities: A survey, in *Work* Paper–2016 (2016)
14. I.-C. Lin, T.-C. Liao, A survey of blockchain security issues and challenges. Int. J. Network Security **19**(5), 653–659 (2017)
15. V. Hassija, G. Bansal, V. Chamola, V. Saxena, B. Sikdar, Blockcom: A blockchain based commerce model for smart communities using auction mechanism, in *2019 IEEE International Conference on Communications Workshops (ICC Workshops)*, May 2019, pp. 1–6

16. V. Hassija, V. Saxena, V. Chamola, Scheduling drone charging for multi-drone network based on consensus time-stamp and game theory. Comput. Commun. (2019)
17. V. Hassija, M. Zaid, G. Singh, A. Srivastava, V. Saxena, Cryptober: A blockchain-based secure and cost-optimal car rental platform
18. NCRB, Crimes in India statistics 2018. Accessed 18 May https://ncrb.gov.in/crime-india-2018. (2020)

KYC as a Service (KASE)—A Blockchain Approach

Dhiren Patel, Hrishikesh Suslade, Jayant Rane, Pratik Prabhu, Sanjeet Saluja, and Yann Busnel

1 Introduction and Background

1.1 Know-Your-Customer (KYC)

Know-your-customer (KYC) refers to the steps taken by a business to establish customer identity, understand the nature of a customer's activities and to assess risks (if any) involved with the customer. It is a legal requirement for the financial institutions for on-boarding a customer. As shown in Fig. 1, KYC requires the submission of the identity documents by the customer to the businesses or organizations on which they wish to onboard. Individual verification of the documents is done and thus establishing the identity of the customer independently.

Know-your-customer (KYC) is used for customer management and identity verification. This document is submitted by the customer to an organization for authentication and verification purposes.

D. Patel · H. Suslade (✉) · J. Rane · P. Prabhu · S. Saluja
Veermata Jijabai Technological Institute, Mumbai, India
e-mail: hrishikesh.suslade1@gmail.com

D. Patel
e-mail: dhiren29p@gmail.com

J. Rane
e-mail: jayantrane811@gmail.com

P. Prabhu
e-mail: pratikprabhu27@gmail.com

S. Saluja
e-mail: sanjeetsaluja99@gmail.com

Y. Busnel
IMT Atlantique, Rennes, France
e-mail: yann-busnel@imt-atlantique.fr

S. Patnaik et al. (eds.), *Advances in Machine Learning and Computational Intelligence*, Algorithms for Intelligent Systems, https://doi.org/10.1007/978-981-15-5243-4_76

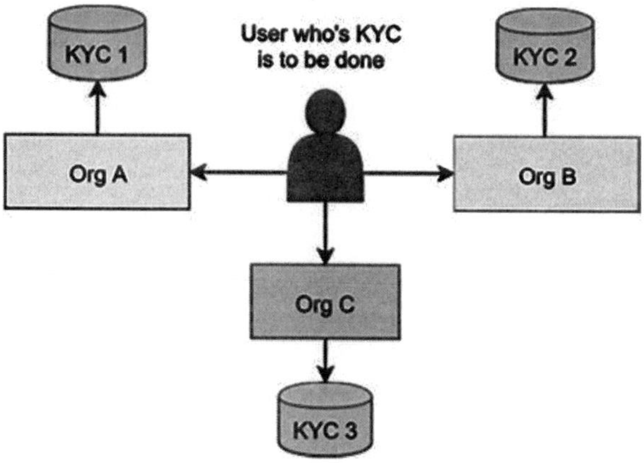

Fig. 1 Current KYC implementation

Currently, KYC is done individually by every business and the same data is provided by the users to multiple businesses and independently verified by each of them. It would be far more efficient if the KYC could be shared securely among the organizations and hence give a better quality of experience (QoE) to the customer. However, due to the lack of trust between organizations this data is not shared between them, hence, requiring a solution that can guarantee trust and reliability.

1.2 Blockchain

Blockchain is a distributed ledger technology that is used to ensure trust and reliability since the data and transactions are committed into the blockchain only after a consensus is reached among the participants [1]. There are various consensus mechanisms that have been implemented to ensure a reliable distributed consensus.

Interplanetary file system is a distributed file system that stores files in a decentralized and distributed manner [2]. Blockchain has many applications [3] and some of the use case are:

1. Industry and IoT: Major use case under this topic include supply chain management, healthcare—patient data management and smart power grids [4], Agriculture—agriculture food traceability and manufacturing industries [5].
2. Others: Creation of middleman free services such as blockchain-based auction mechanism [6] and blockchain-based death wills.

Until now, no centralized KYC verification system exists due to the lack of trust between institutions requiring individual and separate KYC processes and systems followed in each of them internally. Therefore, using a decentralized open technology such as blockchain would help ensure trust and integrity [7] from the ground-up and help in the open acceptance of this system.

Rest of the paper is organized as follows: Sect. 2 provides the design rationale for KASE with the detailed architecture of KASE is discussed in Sect. 3. Section 4 gives prototype implementation of KASE using solidity smart contracts with in-progress validation. Paper is finally concluded in Sect. 5 with references at the end.

2 KASE Design Proposal

The system proposed in this paper performs KYC as a service—a service which acts as a one-stop solution for all of a customers' and business's KYC needs. The customer provides the data to the service, where the service verifies the data using machine learning techniques and stores its encrypted format on a distributed file system and stores every transaction of the KYC data on a blockchain [8]. A blockchain is a decentralized data structure in which transactions are conducted only after certain consensus is reached through consensus mechanisms. The proposed system uses an Ethereum Blockchain [9, 10] with a proof-of-work consensus mechanism. This mechanism allows the blockchain to enforce "smart contracts" such that the transactions are only committed to the blockchain after certain conditions are satisfied.

The system initially asks the user to register on the service and provide his information, including identity proofs to the service voluntarily. The next time the customer wants to get onboarded onto a business he/she uses the service for the KYC process. This information is stored encrypted by the user's secret key on the distributed file system and the transaction is stored on the blockchain to ensure transparency.

If a customer wants to onboard to a business, he/she can register to the business using the service and provide basic details which would be given to the business and verified by the service.

The service first asks the customer for confirming and validating the KYC request in accordance with GDPR and then after receiving the permission verifies the customer's identity to the business. The request transaction is also pushed onto the blockchain to ensure transparency of the data flow and credibility of the transfer.

The service also provides businesses the feature of verification of any KYC documents they may request based on their internal policies and uses machine learning approaches to verify those documents and to confirm the identity of the individual.

The architecture proposed in this paper (Fig. 2) for the blockchain-based KYC system uses a decentralized database (IPFS). We also use machine learning-based image processing and data extraction for legacy KYC processes.

3 KASE—Detailed Architecture

As shown in Fig. 3, at the time of on-boarding onto the system (KASE), the user will have to provide his identity proofs he/she has. The user will have to fill in all the

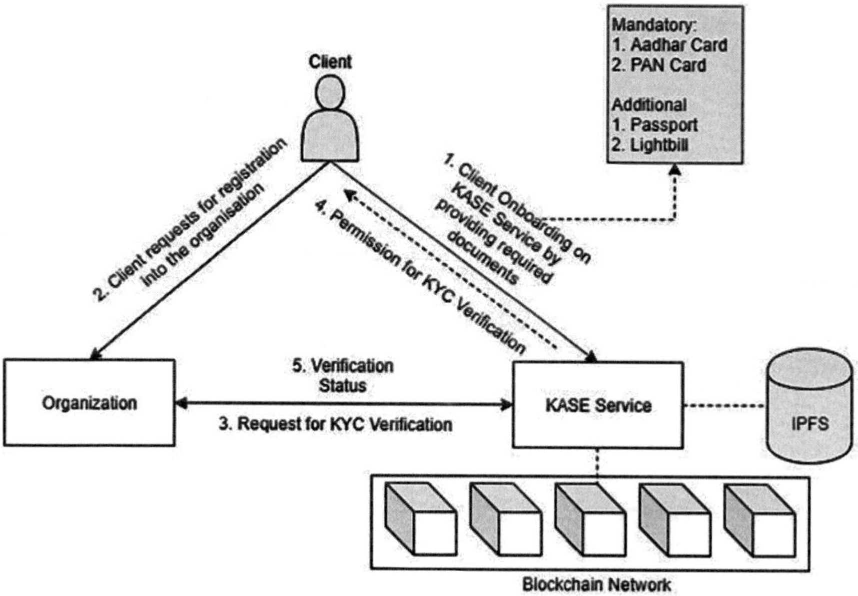

Fig. 2 High-level architecture

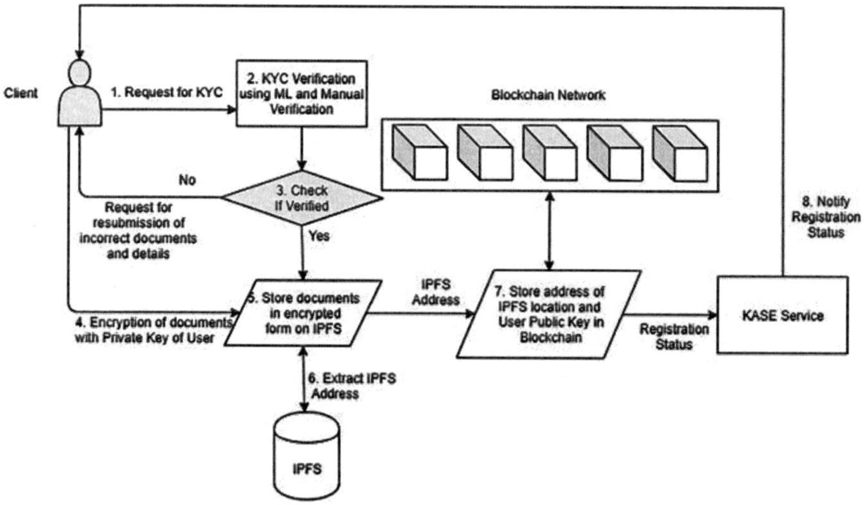

Fig. 3 User on-boarding on KASE

details manually also, which will be converted to a JSON object. The data entered by the user and data extracted from the documents uploaded will cross-checked for any irregularities using machine learning, and an extra layer of verification can be added by comparing the images of the user on various IDs and their image taken digitally.

Once all the checks are complete and all the data is verified, a public–private key pair is generated on the user's system. For security purposes, the private key will be a key file which could be stored on an isolated storage device such as a USB drive or a Gemalto Token.

The data stored in JSON object will be stringified so that it can be stored in IPFS along with the various ID documents, which are encrypted using the user's public key and stored on IPFS. All the documents that will be uploaded will have a different hash. The JSON file will also have a different hash.

All these set of hashes along with the username will be stored in Ethereum blockchain as a "KYC on-boarding request". The Ethereum wallet [9] address generated will be of 42 characters which are impossible to remember. A mapping functionality provided by solidity can be used to map username with the wallet address. At the time of data retrieval, the user has to only provide his unique username to access his/her details. From username, the wallet address can be accessed and through that one can get their stored data [11, 12].

If a business wants to do the KYC of a customer, it can use the proposed service in two ways:

1. It can either request directly for verification
2. It can request the customer for documents and get those verified with the service.

As per Fig. 4, when the business wants to KYC a customer, the business sends a request to the customer to allow the KYC to be processed by the business. KASE sends a notification to the user that the business is requesting KYC and the customer has to authenticate and allow the service to use the user's information to verify it to business. The system gets the address of the user's encrypted information from the blockchain and uses it to verify the customer details provided. After completing the request, the system pushes a "request transaction" to the blockchain.

In case of an event of a change in a user's KYC documents or details, the user has to provide the changed identity proofs to the system where the system verifies and ensures their credibility. The system finds the location of the previous documents using blockchain and pushes the new documents into that location. After completion of this request, this "update documents request" is pushed on the blockchain.

If the business requests documents from the customer, KASE sends a request to the customer to authenticate and allow the request, and the service then retrieves the location of the encrypted documents provided at the time of on-boarding and uses machine learning to verify the currently submitted documents to ensure credibility. After completing the request, the system pushes a "request transaction" to the blockchain.

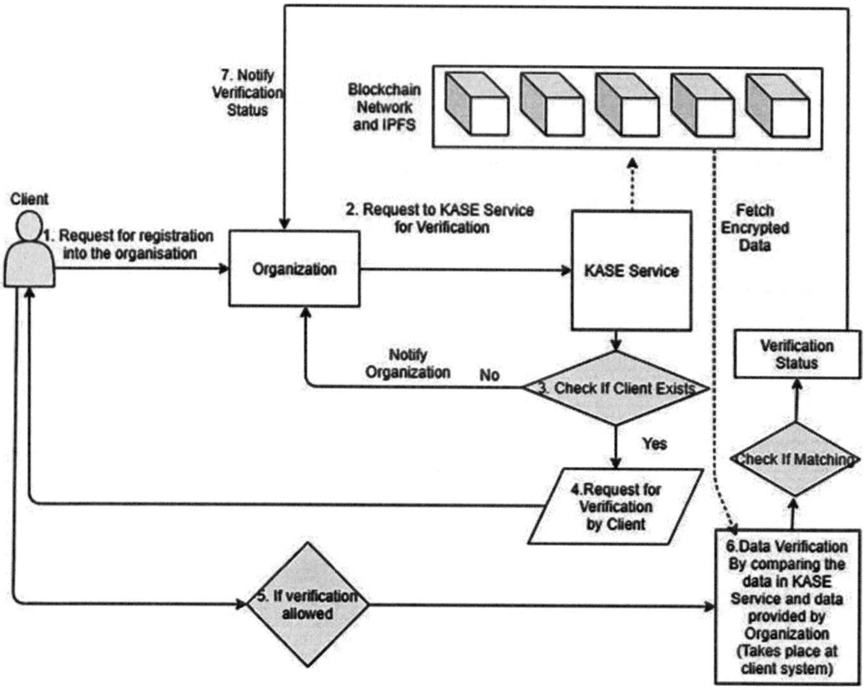

Fig. 4 KYC process

4 Prototype Implementation and Validation

Smart contracts are written in solidity which will be deployed on Ethereum blockchain. The solidity version used is 0.4.21. There are two main smart contracts used.

4.1 Customer Contract

This contract is invoked when the customer is successfully verified and can be onboarded on the KASE system. It has a total of six functions.

1. To add data to blockchain
2. To get customer name
3. To get customer data link (IPFS)
4. To get customer aadhaar image link (IPFS)
5. To get customer PAN card image link (IPFS)
6. To get customer passport image link (IPFS).

4.2 *Organization Contract*

This contract handles the functionalities of on-boarding or adding an organization to the KASE system. This contract has three functions.

1. To add an organization
2. To get organization name
3. To get organization details link.

4.3 *Validation*

The proposed solution leverages the core features of blockchain to ensure trust between the users, businesses and services to ensure the usage of the service. The service provides the following key advantages:

1. Cost-effective solution for a business's KYC needs.
2. It saves valuable time of the business to ensure a customer's reliability.
3. No single point of failure of the system due to the usage of inherently decentralized components. It also achieves immutability of user data.
4. Ensuring openness, trust and reliability through blockchain.
5. Follows GDPR guidelines [13] by ensuring that the user data is not used without permission of the user.

5 Conclusion and Future Scope

Blockchain is one of the latest technologies in the field of cybersecurity and ensures trust in trustless environments. The proposed blockchain-based KYC system that uses a decentralized database (IPFS), machine learning-based image processing and data extraction for legacy KYC processes. As shown in Fig. 5, through blockchain, KASE ensures that the parties using the service can trust the service and its reliability, and will use it over other solutions. The solution further uses a decentralized file store to ensure complete decentralization of data and reduce any single points of failure. Our prototype implementation through solidity smart contracts gives encouraging results.

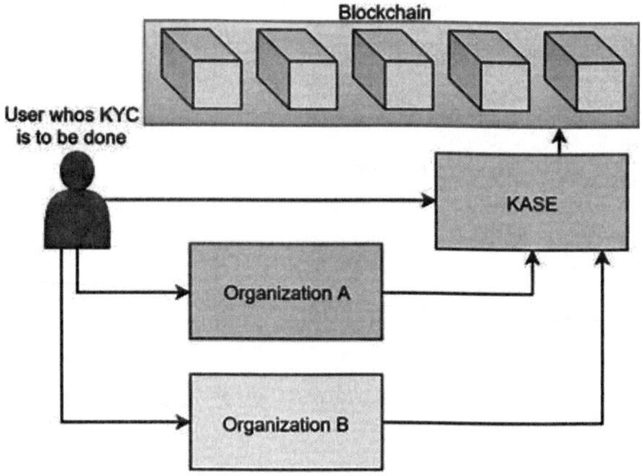

Fig. 5 KYC as a service

KASE service can be used as a one-stop solution of all KYC needs. By leveraging the power of ML, AI and explainable AI, we can make the system free of manual verification.

References

1. S. Nakamoto, *Bitcoin: A Peer-to-Peer Electronic Cash System* (2009)
2. IPFS-Content Addressed, Versioned, P2P File System. 14 July 2014. https://arxiv.org/abs/1407.3561
3. A. Ali, S. Latif, J. Qadir, S. Kanhere, J. Singh, J. Crowcroft, *Blockchain and The Future of the Internet: A Comprehensive Review* (2019). arXiv preprint arXiv:1904.00733
4. T. Alladi, V. Chamola, J.J. Rodrigues, S.A. Kozlov, Blockchain in smart grids: a review on different use cases. Sens. MDPI **19**(22), 4862 (2019)
5. T. Alladi, V. Chamola, R. Parizi, K.K.R. Choo, Blockchain applications for industry 4.0 and industrial IoT: a review. IEEE Access, Nov 2019
6. V. Hassija, G. Bansal, V. Chamola, V. Saxena, B. Sikdar, *BlockCom: A Blockchain Based Commerce Model for Smart Communities Using Auction Mechanism*. In: IEEE ICC, Shanghai, China, May 2019
7. *Blockchain and The Future of the Internet: A Comprehensive*, 23 Feb 2019. https://arxiv.org/abs/1904.00733
8. TCS Whitepaper—Reimagining KYC with Blockchain Technology. Last accessed, 8 Nov 2019
9. Gavin Wood, Ethereum: a secure decentralised generalised transaction ledger. Ethereum Project Yellow Paper **151**, 1–32 (2014)
10. V. Buterin, *The Meaning of Decentralization*, Feb 2017. Reterived from https://medium.com/@VitalikButerin/the-meaning-of-decentralization-a0c92b76a274

11. P. Sinha, A. Kaul, Decentralized KYC System (2018) https://www.irjet.net/archives/V5/i8/IRJET-V5I8204.pdf
12. J. Parra Moyano, O. Ross, Bus. Inf. Syst. Eng. **59**, 411 (2017). https://doi.org/10.1007/s12599-017-0504-2
13. GDPR European Union Guidelines https://gdpr-info.eu/. Last accessed 8 Nov 2019

Enhancing Image Caption Quality with Pre-post Image Injections

T. Adithya Praveen and J. Angel Arul Jothi

1 Introduction

The generation of semantically relevant captions for any given image has seen quite a surge in interest with the advent of deep learning methods. Initial mentions of deep learning architectures for image captioning dating back to 2014 have been referenced in [1–3]. Although [1] does not directly mention image captioning, the novel idea of an RNN-based encoder–decoder architecture introduced in the paper definitely served as the impetus for later papers that focused on image captioning.

The idea of machine translation was explored in [1] where both the encoder and decoder were recurrent neural networks (RNNs). Image caption generation is but merely an extension of this notion, with the encoder RNN replaced by an encoder CNN. Reference [2] delves into this idea involving multimodal neural language models. The contribution of the paper being the introduction of language models that can be conditioned on other modalities including images, audio, and so on. When the modality that the language model is conditioned on lies in the domain of images, the problem turns into that image caption generation. The implementation of architectures in [2, 3] was a radical improvement over syntactic trees, templates, and structured models, which were prevalent back then.

The huge spike in interest in this subfield is largely due to its useful applications involving accurate image retrieval given an image query and generating accurate descriptions for automated image annotation for web sites. These ideas are briefly mentioned in [2] but are dealt with great detail in [4]. Retrieval of images, in a way analogous to human perception of images, is detailed in [4]. The paper uses dense

T. Adithya Praveen · J. Angel Arul Jothi (✉)
Department of Computer Science, Birla Institute of Technology & Science, Pilani, Dubai Campus, Dubai International Academic City, Dubai 345055, UAE
e-mail: angeljothi@dubai.bits-pilani.ac.in

T. Adithya Praveen
e-mail: f20170199@dubai.bits-pilani.ac.in

S. Patnaik et al. (eds.), *Advances in Machine Learning and Computational Intelligence*, Algorithms for Intelligent Systems, https://doi.org/10.1007/978-981-15-5243-4_77

captions in unison with scene graph generation to understand various relations in an image, thereby facilitating higher quality image retrieval results. The idea of dense captions that was used to experiment with image retrieval in [4] has its roots in [5]. This paper talks about using fully convolutional localization networks (FCLN) to tackle the task of efficient localization and description tasks. This was made possible through the novel "dense localization layer" introduced in the paper.

Due to the bottleneck of the encoder–decoder layer in traditional image captioning architectures, the question of where the image must be injected for generating decent captions has long been debated. This notion is thoroughly explored in [6] and looks at various ways of injecting an image by using the proposed "merge" architecture. The paper empirically shows the modality that is used to condition the language model, is better introduced in a subsequent, "delayed" stage.

The quest to improve language models conditioned on images, has seen numerous advances. One of the most significant advances in this direction has been from attention-based image captioning networks. This approach inches closer toward the human way of describing images, by looking at different parts of the image at incremental stages, to get a more coherent understanding of the image being input. They are delineated more formally in [7–9]. These papers also visualize how the neural network decides to output words of a description for an image, utilizing soft attention or hard attention.

This paper is organized as follows: Sect. 2 details the current approach, and how the proposed approach modifies the former to get better captions, for no additional cost. Section 3 explains the dataset used while Sect. 4 details the evaluation metrics used to quantify performances of both approaches. Sections 5 and 6 delineate the experiments conducted and the subsequent results of those experiments.

2 Method: Pre-post Injections to Improve Caption Quality of Merge Architecture Networks

2.1 *Encoding the Image*

It is the encoding of the image that gets injected before and after the encoding of the caption by the LSTM. This is because the encoding would ideally be a lower dimensional dense representation of the image. Such a compact representation would mean the training would not be computationally expensive. This job is accomplished by using a pre-trained convolutional neural network (CNN). The CNN accomplishes the task of being a feature extractor, or in other words, the encoder of the image. The feature extractor used is the Inception V3 network.

$$o = W[\mathrm{CNN}_{\mathrm{theta}}(I)] + b \quad (1)$$

Here, in (1), the function CNN (I) defined using the parameters "theta" transforms the input pixels I into L-dimensional activations. This output of CNN (I) is then converted to the global visual information denoted by $[o]^D$ in (1) using weights $[W]^{(D \times L)}$ & bias $[b]^D$. The superscripts denote the dimensions of the respective vectors/tensors. The Inception V3 CNN used in this paper is the third iteration of the inception series and takes form through the introduction of factorization ideas expressed in [10].

2.2 The Post-merge Architecture

Flowcharts for both architectures described in this section and Sect. 2.2 are shown in Fig. 1. The merge architecture usually encodes the image in a single branch using the encoder described in Sect. 2.1 and encodes the text in a separate branch using an LSTM. Usually, the image gets injected after encoding the text, thereby receiving the context of the image only at a later stage of the network's forward pass. Figure 1 (left) illustrates the typical "post-inject and merge" architecture.

The partial caption described in Fig. 1 is a part of the caption corresponding to the training image. So, if the image has the following caption: "A big red apple," there would be three training data points for the same image. The data points correspond to the partial captions, "A", "A big," "A big red." The target word for each of these captions being the word that follows immediately after "big," "red," and "apple," respectively.

2.3 The Pre-post-Merge Architecture

Although the existing post architecture shows laudable performance, it makes sense to improve the caption quality by additionally introducing the image context at an earlier stage in the pipeline. This is exactly what the pre-post architecture does. It appends the image encoding generated from the Inception V3 network to the partial caption embedding after the embedding layer. The output from the append layer is then passed into the LSTM, thereby giving a better conceptually embedded encoding as the output. The flowchart for this network is depicted in Fig. 1 (right).

It makes sense to include this encoding in this manner to get more bang for your buck, since the LSTM would now have a hidden state that is conditioned not only on the partial caption of each training data point but also the image corresponding to it.

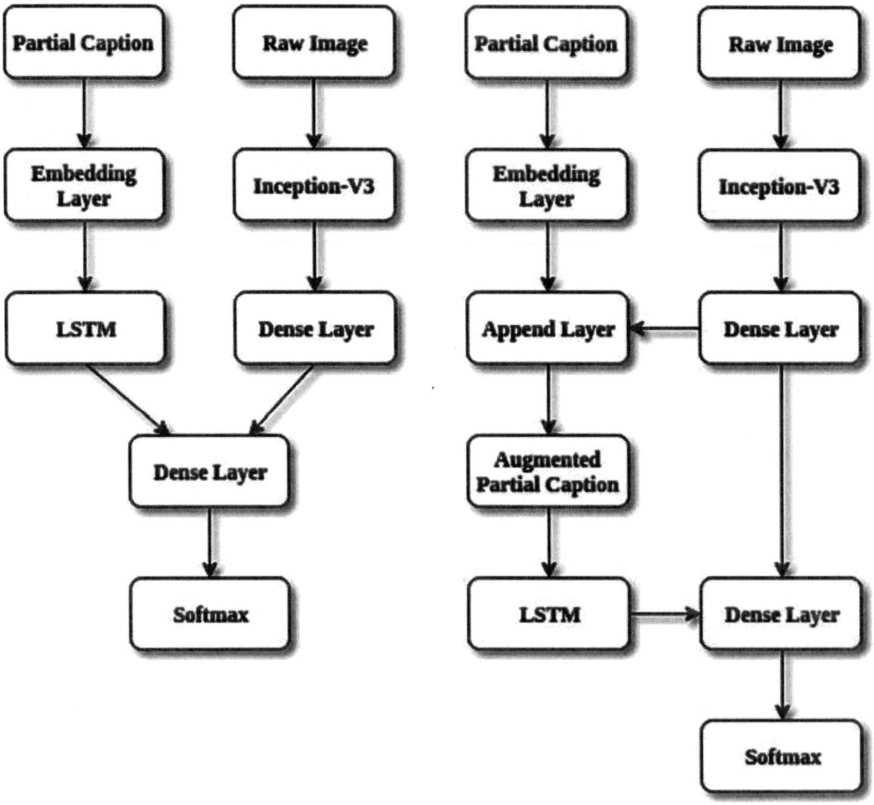

Fig. 1 The existing "post" architecture (left) and the proposed "pre-post" architecture (right). The post architecture never introduces the image until the last dense layer. However, in the proposed architecture, the image encoding is injected via the append layer before being introduced to the LSTM. The forward pass in the LSTM now accounts for the image data as well

3 Dataset

All the experiments conducted for this paper utilized the Flickr-8 k dataset. One of the earliest references to the dataset can be found in [11]. In terms of a description for the dataset, it contains various images depicting common-place scenarios. Each of the 8000 images in the dataset contains five manually labeled captions, forming a grand total of 40,000 captions. Each of the five captions for each image was labeled by five different human annotators. So, there is no machine-assisted captioning, unlike the larger scale datasets such as Google's Conceptual Captions [12]; the annotations are a hundred percent manual. Multiple captions per image help capture a good portion of the salient features present in the image. The images in the dataset were also chosen to keep specific locations and people at a bare minimum. The reasoning is simple—a model cannot be expected to discern the location or person from an image alone, without the help of added context.

4 Evaluation Metrics

The Bilingual Evaluation Understudy (BLEU) metric was used to quantify the quality of sentences produced by the currently available model and the proposed model and the math formulation for this evaluation metric are described in (2)–(4).

Unigram precision (2) finds the fraction of words in the predicted caption that are also present in the ground truth (1-word phrase matches). So, an n-gram precision quantifies n-word phrase matches. Here, in (2), "m" denotes the number of words in the predicted caption that is also present in the ground truth caption. w_t denotes the number of words in the reference caption that the prediction is being compared against (since the prediction can be compared against 5 separate reference captions). The brevity penalty in (3) penalizes the neural network for longer caption predictions. In addition, "r" and "c" in (3) are the respective effective and total lengths of the reference and translation corpora. P_n is computed using N, which is the length of the n-grams, where P_n is the average (geometric) of the modified definition of precision (4). The BLEU score essentially weights the n-gram precisions while accounting for the brevity penalty at the same time.

$$\text{Unigram precision } P = \frac{m}{w_t} \tag{2}$$

$$\text{Brevity penalty } p = \begin{cases} 1 & \text{if } c > r \\ e^{\left(1-\frac{r}{c}\right)} & \text{if } c \leq r \end{cases} \tag{3}$$

$$\text{BLEU} = p \cdot e^{\sum_{n=1}^{N}\left(\frac{1}{N} \times \log Pn\right)} \tag{4}$$

5 Experiments Conducted

Vocabulary threshold is the number of times a word occurs in the entirety of the caption corpus, which qualifies it to be included in the training process. So, if the threshold is 3, any word is included in the vocabulary while training only if it occurs for a minimum of 3 times among all the captions combined. The post-inject and the proposed pre-post-inject model where compared at different vocabulary thresholds ranging from 2 all the way up to 10. Using a vocabulary threshold is necessary in

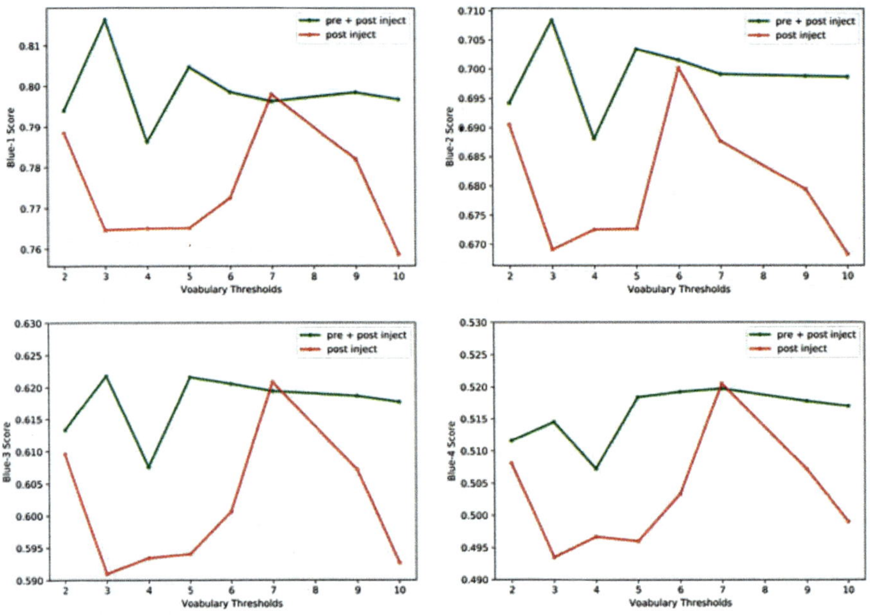

Fig. 2 BLEU- 1, 2, 3, and 4 scores for various vocabulary thresholds

order to remove the noise of misspelled words and the noise of words that are rarely used when describing captions. Both networks were trained for 50 epochs with a batch size of 128 "image-caption" pairs.

The BLEU 1, BLEU 2, BLEU 3, and BLEU 4 scores were computed and plotted for each model. The different BLEU scores differ in terms of the weights given to n-gram precisions. For example, BLEU 1 gives a weight of 1 to 1-gram matches and 0–2, 3, and 4 g matches. BLEU 2 gives a weight of 0.5 to 1-gram and 2-gram matches, and 0 for the rest, and so on. Plots for each of the above metrics, compare both the "post-inject" architecture and the "pre + post" inject architectures caption qualities (see Fig. 2).

6 Results and Discussions

6.1 Quantifying Caption Quality Using Caption Metrics

The BLEU scores for both architectures depicted in Fig. 1 have been tabulated in Table 1. Figure 2 shows plots of BLEU scores for the respective architectures.

We have quantified the caption quality using the Bilingual Evaluation Understudy (BLEU) score. The analysis has been done for both architectures depicted in Fig. 1. The caption metric values correspond to various "vocabulary thresholds" applied on

Table 1 Quantified BLEU scores of the existing post architecture and proposed pre + post architecture

Threshold	Post architecture				Pre + post architecture			
	BLEU1	BLEU2	BLEU3	BLEU4	BLEU1	BLEU2	BLEU3	BLEU4
2	0.7884	0.6905	0.6096	0.5081	0.7940	0.6942	0.6133	0.5115
3	0.7645	0.6691	0.5909	0.4935	0.8163	0.7084	0.6217	0.5145
4	0.7650	0.6725	0.5934	0.4966	0.7862	0.6881	0.6076	0.5072
5	0.7651	0.6726	0.5940	0.4959	0.8046	0.7034	0.6215	0.5184
6	0.7725	0.7002	0.6006	0.5033	0.7985	0.7015	0.6205	0.5192
8	0.7980	0.6877	0.6208	0.5205	0.7962	0.6991	0.6194	0.5197
9	0.7820	0.6794	0.6072	0.5071	0.7984	0.6988	0.8186	0.5177
10	0.7586	0.6683	0.5927	0.4989	0.7967	0.6987	0.6177	0.5169

The proposed architecture has a higher BLEU score in general

the caption data. As explained in Sect. 5, the vocabulary threshold of x retains only those words that have occurred for a bare minimum of "x" times in the corpus comprising of all the captions in the dataset. The plots consider vocabulary thresholds ranging from 2 occurrences all the way up to 10 occurrences of words. In general, the proposed "pre + post" architecture image captions have better accuracy, and consequently, better quality of captions as well. The BLEU scores for both architectures have been tabulated in Table 1.

7 Conclusion

This paper quantifies the benefits of injecting an image encoding along with caption embedding tensors, before being encoded by an LSTM in the merge architecture for generating image captions. Simply put, this is a pre-injection of the image encoding to produce a better-quality encoding from the LSTM. Subsequently, this is coupled with the post-injection of the image embedding vector, and the remainder of the merge architecture remains relatively the same. The pre-injection provides a better context of the image for the LSTM encoder, in relation to its corresponding caption. Consequently, the captions are enhanced while not having to train extra weights to achieve this added quality. The corroboration of this idea is quantified in the caption metrics resulting from the benchmark tests on the Flickr-8k dataset. This paper shows that the additional pre-injection provides the foundation for producing captions of better quality, rather than simply introducing the context of the image through a single post-injection; this comes without the overhead of training more weights than the latter model.

References

1. K. Cho, B. Van Merriënboer, Ç. Gùlċehre, D. Bahdanau, F. Bougares, H. Schwenk, Y. Bengio, *Learning Phrase Representations Using RNN Encoder-Decoder For Statistical Machine Translation*, in Conference on Empirical Methods on Natural Language Processing (2014), pp. 1724–1734
2. R. Kiros, R. Salakhutdinov, R. Zemel, *Multimodal Neural Language Models*, in Proceedings of the 31st International Conference on Machine Learning (ICML-14) (2014), pp. 595–603
3. D. Bahdanau, K. Cho, Y. Bengio, *Neural Machine Translation by Jointly Learning to Align and Translate*. CoRR, abs/1409.0473
4. X. Wei, Y. Qi, J. Liu, F. Liu, Image retrieval by dense caption reasoning, in *IEEE Visual Communications and Image Processing* (VCIP). St. Petersburg, FL (2017), pp. 1–4. https://doi.org/10.1109/vcip.2017.8305157
5. J. Johnson, A. Karpathy, L. Fei-Fei, *Densecap: Fully Convolutional Localization Networks for Dense Captioning*, in IEEE Conference on Computer Vision and Pattern Recognition (2016), pp. 4565–4574
6. M. Tanti, A. Gatt, K. Camilleri, Where to put the image in an image caption generator. Nat. Language Eng. **24**. https://doi.org/10.1017/S1351324918000098
7. A. Borji, L. Itti, State-of-the-art in visual attention modeling. IEEE Trans. Pattern Anal. Mach. Intell. **35**(1), 185–207 (2013)
8. Q. You, H. Jin, Z. Wang, C. Fang, J. Luo, Image captioning with semantic attention, in Proceedings of the IEEE Conference on Computer Vision and Pattern Recognition (2016), pp. 4651–4659
9. M. Khademi, O. Schulte, *Image Caption Generation with Hierarchical Contextual Visual Spatial Attention*, in 2018 IEEE/CVF Conference on Computer Vision and Pattern Recognition Workshops (CVPRW) (2018). https://doi.org/10.1109/cvprw.2018.00260
10. C. Szegedy, V. Vanhoucke, S. Ioffe, et al., *Rethinking the Inception Architecture for Computer Vision*, in 2016 IEEE Conference on Computer Vision and Pattern Recognition (CVPR) (2016). https://doi.org/10.1109/cvpr.2016.308
11. M. Hodosh, P. Young, J. Hockenmaier, Framing image description as a ranking task: data, models and evaluation metrics. J. Artif. Intell. Res. **47**, 853–899 (2013)
12. P. Sharma, N. Ding, S. Goodman, R. Soricut, *Conceptual Captions: A Cleaned, Hypernymed, Image Alt-text Dataset For Automatic Image Captioning*, in Proceedings of the 56th Annual Meeting of the Association for Computational Linguistics, vol. 1 (2018). https://doi.org/10.18653/v1/p18-1238

Integral Sliding Mode for Nonlinear System: A Control-Lyapunov Function Approach

Ankit Sachan, Herman Al Ayubi, and Mohit Kumar Garg

1 Introduction

Today, in the modern control application, there were enormous techniques to be developed in the literature to tackle the uncertainties of the system. Out of these, our attention goes to sliding mode control [1] due to its ability to deal with robustness of the nonlinear system. The intriguing aspects of sliding mode control are complete compensation of the so-called matched uncertainties (i.e., vanishing/non-vanishing uncertainties entering to the system through input channel) when the system is in sliding phase. Generally, sliding mode control becomes insensitive to the matched uncertainties and this insensitivity provides a control chattering in nonlinear system during movement of system trajectories toward the origin.

Prior too, many algorithms were proposed itself under the domain of sliding mode control. Thus, we interested in integral sliding mode control which was first proposed in [2]. The key feature of integral sliding mode control is that its system trajectory always stands in the sliding manifold. So, there is no reaching phase problem occurring while controlling the system. In this approach, the basic sliding mode controller collaborates with other existing controller like adaptive control [3], model predictive control [4], and so on. The other controller which is of full order is used to stabilize the nominal system within sliding manifold. The nominal controller is always acting on the original system. Whenever, the externally appeared unknown

A. Sachan (✉) · M. K. Garg
Department of Electrical Engineering, Indian Institute of Technology (BHU), Varanasi, India
e-mail: ankits.rs.eee15@itbhu.ac.in

M. K. Garg
e-mail: mohitkumargarg.eee18@itbhu.ac.in

H. Al Ayubi
School of Information and Communication Technology, Gautam Buddha University, Greater Noida, India
e-mail: herman7sda@gmail.com

S. Patnaik et al. (eds.), *Advances in Machine Learning and Computational Intelligence*, Algorithms for Intelligent Systems, https://doi.org/10.1007/978-981-15-5243-4_78

disturbance with known upper bound is the system, sliding mode control shows its existence and start compensating the external disturbance at very beginning while the nominal control tries to hold the system trajectories within the predefined manifold such that its states remains always in the predefined manifold.

This technique is expressed for nonlinear system to overcome both matched and unmatched type of disturbances in [5]. The central idea of ISMC is to maintain the system trajectories within sliding manifold. Defoot et al. [6] tried similar approach to track the mobile robot by keeping its trajectory in sliding surface with integral sliding mode control. The compatibility of this control with model predictive control is discussed in [4]. The discrete approach for integral sliding mode control is proposed in [7] to control the piezo-motor driven for linear motion. The analysis of integral sliding mode for different aspects under unmatched perturbation is driven in [8] where integral sliding mode approach for minimization of perturbation term for partial derivative of constant input matrix with a state dependent nonlinear drift term is discussed. The design of controller to compensate the matched or unmatched disturbance acting on nonlinear drift for constant input matrix in the presence of non-holomaic constraints via integral sliding mode is proposed by [9]. With several promising features of ISMC, there exists a drawback of chattering effect. Thus, super-twisting control (STC) [10] is standout amongst several other techniques to attenuate chattering. Firstly, the intriguing aspect of STC is introduced in the ISMC in [11]. Recently, a new work on ISMC is discussed in [12] for the analysis of 2-DOF helicopter setup where the STC is utilized for the continuous response of control signal while disturbance rejection and the stability of nominal system is ensured by composite nonlinear function.

The nominal control working for unperturbed system is to regulate its state variables to track the predefined manifold ($S = 0$). It can recast the states of system to bring to zero states with the help of control input. The order of nominal control is full such that each initial state is to stabilize the closed loop system. The order of nominal control is full such that each initial state having some control input to brought the states asymptotically at origin. The Lyapunov function was allowed as feedback to obtain a stabilizable state for closed-loop system [13] by allowing a relaxed control based on algebraic sum of lie-derivatives. The concept of control-Lyapunov function is extended in [14] to provide an analytical feedback dynamic of first order $k(x)$ by explicit recourse of derivative Lyapunov function. Further, the theory of control-Lyapunov function for stabilization of time-varying system is discussed in [15]. This concept was extended to non-affine nonlinear system [16] to maximize the angle between two vector function. Along this, the Eigen angle for stabilization of non-affine nonlinear system is also discussed.

An integral sliding mode control is presented in this technical note, which consists the combination of two control parts, i.e., (i) nominal control based on control-Lyapunov function approach that resembles the solution of nonlinear regulation theory of output feedback stabilization to recast the states of original system to predefined manifold. (ii) continuous control based on SMC to alleviate the matched Lipschitz disturbance completely. The generalized super-twisting [17] consists of a standard super-twisting control with extra linear correction term that renders faster convergence for a larger class of robustness as well as overcomes with demerit of a basic sliding mode, i.e., chattering effect. Thus, the system trajectory remains in predefined manifold and asymptotically approaches toward origin showing uniformity with time for global results.

2 Problem Statement

The description of nonlinear affine-type system for perturbed model as

$$\dot{\Theta} = \xi(t, \Theta) + \eta(t, \Theta)(v + d(t, \Theta)) \tag{1}$$

where state $\Theta \in \mathbb{R}^n$ having condition $\Theta(t_0) = \Theta_0$ with scaler control input for nonlinear system. $\xi(t, \Theta)$ and $\eta(t, \Theta)$ are known to be real-valued function of nth dimension for state $\Theta(t)$. While $d(\Theta(t))$ is a bounded unknown quantity showing matched uncertainties and disturbances entering through input channel.

The design of an appropriate controller for the nonlinear plant such that the system trajectory approaches to single equilibrium point globally with certain assumption to be satisfied as

Assumption 1 For the state variable $\Theta \in \mathbb{R}^n$, the pair $(\xi(t, \Theta), \eta(t, \Theta))$ is stabilizable.

Assumption 2 The bounded quantity of disturbance dynamics $d(t, \Theta)$ is as.

$$||d(t, \Theta)|| \leq q(t, \Theta)||\Theta|| + p(t) \tag{2}$$

where $q(t, \Theta)$ and $p(t)$ are positive functions for vanishing and non-vanishing disturbances, respectively

3 Problem Statement

To investigate the stability of perturbed model, we need to design a controller which renders all the state variable to a single equilibrium point globally as well as to withstand with unwanted disturbance signals. Therefore, we suggest integral sliding mode control to reach the desirable objective. The proposed technique for integral sliding mode control is the combination of two well-known classical approaches, i.e., control-Lyapunov function approach, which uses the estimated Lyapunov function as a feedback for closed-loop system to force the state trajectories to steers toward the origin asymptotically for disturbance-free model and a basic sliding mode approach as a generalized super-twisting controller act as disturbance observer with continuous control action to alleviate the disturbance signals to a predefined manifold ($S = 0$) as well as minimized chattering. Thus, the overall plant approaches to origin asymptotically with the concept of integral sliding mode control.

3.1 Control-Lyapunov Function Based Control Law

The main objective of this controller is to recast the states of original system for a predefined manifold ($S = 0$) via analytic feedback law on the basis of Lyapunov function for closed-loop system. The description of nonlinear affine system for disturbance-free system as

$$\dot{\Theta} = \xi(t, \Theta) + \eta(t, \Theta)u_{clf} \tag{3}$$

where u_{clf} is the nominal control allowed to perform to a disturbance-free model.

Let us consider a strictly increasing function α: $\mathbb{R}_{\geq 0} \to \mathbb{R}_{\geq 0}$ with zero at origin and intended to show the lower and upper boundaries for a C^1 Lyapunov function V: $[0, \infty) \times \mathbb{R}^n \to \mathbb{R}$ holds Lipchitz property as

$$\alpha_1(\|\Theta\|) \leq V(t, \Theta) \leq \alpha_2(\|\Theta\|) \quad \forall \Theta \in \mathbb{R}^n, \forall t \geq 0 \tag{4}$$

As the Lyapunov function approaches to infinity in (4), which shows radially unbounded property for $\alpha_1(||\Theta||)$ with $.||\Theta|| \to \infty$.

To indeed the applicability of C^1 Lyapunov function, we compute the derivative of Lyapunov function along the system trajectory such that

$$\dot{V}(t, \Theta) = \frac{\partial V}{\partial t} + \frac{\partial V}{\partial \Theta}\frac{\mathrm{d}\Theta}{\mathrm{d}t} \leq -W(\Theta) \tag{5}$$

where $W(\Theta)$ is a positive definite function.

Based on inequality (5), $V(t, \Theta)$ is a non-increasing function showing that the solution starting from Θ_0 stays in some invariant set $\Omega \subseteq \mathbb{R}^n$ for all future time. Since $\dot{V}(t, \Theta)$ is re-arranged as

$$\int_{t_0}^{t} W(\Theta)\mathrm{d}\tau \leq -\int_{t_0}^{t} V(t, \Theta)\mathrm{d}\tau = V(t_0, \Theta_0) - V(t, \Theta) \leq V(t_0, \Theta_0) \tag{6}$$

As the common approach to show stability is that energy of Lyapunov function cannot remain same throughout and similarly no trajectory stays identical except at the origin, i.e., showing asymptotic nature with invariance principle [18]. For the generalization of above inequality, a Lyapunov function is recommended as output feedback regulator which often provides a satisfactory control according to Artstein's theorem [13] for closed-loop system.

$$\begin{aligned}\dot{V}(t,\Theta) &= \frac{\partial V}{\partial t} + \frac{\partial V}{\partial \Theta}\xi(t,\Theta) + \frac{\partial V}{\partial \Theta}\eta(t,\Theta)u_{clf} \\ &= a(t,\Theta) + b(t,\Theta)u_{clf} \leq -\alpha_3(\|\Theta\|) \quad \forall \xi \in \mathbb{R}^n,\ \forall t \geq 0 \end{aligned} \tag{7}$$

Therefore, Lie derivative of first order is suggested to derive the control-Lyapunov function-based approach.

Assumption 3 For the state variable $\Theta \in \mathbb{R}^n$, the pair $(a(t,\Theta), b(t,\Theta))$ is stabilizable.

According to Assumption 3, the control-Lyapunov function satisfies a small control property, i.e., for any $\varepsilon > 0$, there exists a $\delta > 0$ such that, whenever $0 < |\varepsilon| < \delta$ then control input $|u| < \varepsilon$ is active showing a feedback law $u_{clf} = K(a(t,\Theta), b(t,\Theta))$ is almost smooth for closed-loop system with following condition.

$$\sup_{\xi\in\mathbb{R}^n} \inf_{u\in\mathbb{R}} \left(a(t,\Theta) + b(t,\Theta)u_{clf}\right) < 0, \quad \forall \Theta \neq 0$$

Using below Theorem, the nominal control is designed such that the state trajectories are always lie in a predefined manifold.

Theorem 1 Design of control-Lyapunov function for nonlinear system (3) as

$$u_{clf} = -m[\gamma V(t,\Theta) + a(t,\Theta)] \tag{8}$$

where $m = (b(t,\Theta))^{-1}$

Then, the system trajectories are infinitesimally decreasing to attain a zero state for any initial state showing uniform global asymptotic stability.

Proof The derivative of C^1 Lyapunov function along the system trajectory is written as

$$\begin{aligned}\dot{V}(\Theta) &= \frac{\partial V}{\partial t} + \frac{\partial V}{\partial \Theta}\frac{\mathrm{d}\Theta}{\mathrm{d}t} \leq -\alpha_3(\|\Theta\|) \\ &\leq -\alpha_3\left(-\alpha_2^{-1}(V)\right) \leq -\rho(V(t,\Theta))\end{aligned} \tag{9}$$

By comparison principle approach [18], we compute the right-hand derivative $\left(D^+\right)$ for boundedness of solution as

$$\dot{V}(t,\Theta) = -\gamma V(t,\Theta), \quad \forall \Theta \in \mathbb{R}^n, \quad \forall t \geq 0 \tag{10}$$

where γ is the tuning parameter which renders a suitable constant gain to ensure the system feedback stabilizable.

For design of nominal control, we going to re-look the derivative of Lyapunov function (4) while comparing (9) and (10) as

$$\frac{\partial V}{\partial t}+\frac{\partial V}{\partial \Theta}\xi(t,\Theta)+\frac{\partial V}{\partial \Theta}\eta(t,\Theta)u_{cLf}=-\gamma V\ (t,\Theta)$$

$$u_{cLf}\leq-\left(\frac{\partial V}{\partial \Theta}\eta(t,\Theta)\right)^{-1}\left[-\gamma V(t,\Theta)+\frac{\partial V}{\partial t}+\frac{\partial V}{\partial \Theta}\xi(t,\Theta)\right]=m\left[\gamma V(t,\Theta)+a(t,\Theta)\right]$$

Thus, the design of nominal control for disturbance-free nonlinear affine system to drive the system trajectories to predefined manifold and remains there in with continuous applicability of control action.

Special case. Assume for the disturbance-free nonlinear affine system (3) has the following canonical form:

$$\dot{\Theta}_1=\Theta_2,\dot{\Theta}_2=\Theta_3,\ldots\quad \dot{\Theta}_n=\xi(t,\Theta)+\eta(t,\Theta)u_{clf} \tag{11}$$

where ξ,η: $\mathbb{R}_{\geq 0}\times\mathbb{R}^n\rightarrow\mathbb{R}^n$ are known smooth functions with an assumption that $\eta(t,\Theta)\neq 0$. Let $z_1=\Theta_1,\psi_1=-C_1z_1$ for some $C_1>0$. Moreover, for $k=2,3,\ldots,n$, let $z_k=\Theta_k-\psi_{k-1}$ and for $k=2,3,\ldots,n-1$, let

$$\psi_k=-z_{k-1}-C_kz_k+\sum_{l=1}^{k-1}\frac{\partial\psi_{k-1}}{\partial\Theta_l}\Theta_{i+1}$$

where $C_k>0$. Then, the Lyapunov function $V=\frac{1}{2}\sum_{k=1}^{n}z_k^2$ is positive definite and radially unbounded. Moreover, it can satisfy a small control property. For that

$$A(t,\Theta)=\frac{\partial V}{\partial t}+\frac{\partial V}{\partial \Theta}\xi(t,\Theta)=-\sum_{k=1}^{n-1}C_kz_k^2+z_k\left(z_{n-1}+\xi(t,\Theta)\right)-\sum_{i=1}^{n-1}\frac{\partial\psi_{n-1}}{\partial x_l}\Theta_{i+1}$$

$$B(t,\Theta)=\frac{\partial V}{\partial t}+\frac{\partial V}{\partial \Theta}\eta(t,\Theta)=z_n\eta(t,\Theta)$$

where $B(t,\Theta)\neq 0$ and the controller design based on control-Lyapunov function is defined as.

Thus, from the Theorem 1, a state feedback controller is designed for the nonlinear system to be stabilizable.

$$u_{cLf}=-(B(t,\Theta))^{-1}\left[\gamma V(t,\Theta)+A(t,\Theta)\right] \tag{12}$$

3.2 Generalized Integral Sliding Mode-Based Control Law

A generalized integral sliding mode control consists of two control parts, i.e., (i)nominal control $\left(u_{clf}\right)$ as discussed in previous subsection and (ii) a generalized super-twisting control (u_{gstc}). The design of both controllers is exclusively independent to each other. The nominal control is designed for an output feedback controller to ensure asymptotic stability for an unperturbed model. Assume that, system trajectories always lie on a predefined manifold with continuous activeness of nominal control. Therefore, to alleviate the matched disturbance throughout, a generalized super-twisting control u_{gstc} is added with the nominal control u_{cLf}. So that the system trajectories universally lie in the predefined manifold.

Assumption 4 Some priori-known positive constant $\mathcal{L}$ such that

$$\mathcal{L} > \sup_{\theta \geq 0}(\|d(t, \Theta)\|)$$

The generalized super-twisting control (u_{gstc}) [17] as described by differential inclusion as

$$u_{gstc} = -\varepsilon_1 \Omega_1(\mathcal{S}) + q, \quad q = -\varepsilon_2 \Omega_2(\mathcal{S}) \tag{13}$$

where $\varepsilon_1 = 1.5\sqrt{\mathcal{L}}$ and $\varepsilon_2 = 1.1\mathcal{L}$ are the desired positive gain and nonlinear terms

$$\begin{aligned} \Omega_1(\mathcal{S}) &= \mu_1 |\mathcal{S}|^{\frac{1}{2}} \operatorname{sgn}(\mathcal{S}) + \mu_2 \mathcal{S}, \quad \mu_1, \mu_2 \geq 0 \\ \Omega_2(\mathcal{S}) &= \frac{\mu_1^2}{2} \operatorname{sgn}(\mathcal{S}) + \frac{3}{2} \mu_1 \mu_2 |\mathcal{S}|^{\frac{1}{2}} \operatorname{sgn}(\mathcal{S}) + \mu_2^2 \mathcal{S} \end{aligned} \tag{14}$$

which renders the summation of standard super-twisting control with the linear version of its algorithm to provide an extra linear correction term for larger class of robustness and faster convergence. Without loss of generality, we consider the values $\mu_1 = 1$ and $\mu_2 = 0$ to get the standard super-twisting control. In the similar way, if we set the values $\mu_1 = 0$ and $\mu_2 > 0$ the system reduces to classical PI controller.

Thus, predefined manifold is designed as a difference of nonlinear perturbed system with the nominal plant on the basis of integral sliding manifold proposed in [2] as

$$S = h(\Theta) - h_{clf}(\Theta) \tag{15}$$

where $h(\Theta)$: $\mathbb{R}^n \rightarrow \mathbb{R}$ for nonlinear system with disturbance acting toward the input channel and the derivative of function, i.e., $\dot{h}(\Theta)$ as

$$h(\Theta) = G\dot{\Theta} = G[\xi(t, \Theta) + \eta(t, \Theta)(v + d(t, \Theta))] \tag{16}$$

And $h_{cLf}(\Theta) : \mathbb{R}^n \rightarrow \mathbb{R}$ for nonlinear system nominal control input as

$$h_{cLf}(\Theta) = G\left[\Theta_0 + \int_{t_0}^{t} \left[\xi(t,\Theta) + \eta(t,\Theta)u_{cLf}\right]\mathrm{d}\tau\right]$$

while the derivative of function $h_{cLf}(\Theta)$ is defined as

$$\begin{aligned}\dot{h}_{cLf}(\Theta) &= G\left[\dot{\Theta}_0 + \int_{t_0}^{t} \frac{\mathrm{d}}{\mathrm{d}t}\left[\xi(t,\Theta) + \eta(t,\Theta)u_{cLf}\right]\mathrm{d}\tau\right] \\ &= G\left[\xi(t,\Theta) + \eta(t,\Theta)u_{cLf}\right]\end{aligned} \tag{17}$$

Here, we were comparing the one-dimensional sliding manifold to the nth-dimensional function. So, we introduce as vector quantity $G \in \mathbb{R}^{1\times n}$ to match the sliding manifold.

System trajectory of nonlinear system is always lying on the sliding manifold, i.e., $S = 0$. So there is no reaching phase for integral sliding mode control and system dimension is always equal to state space as the order of system will never reduce.

To define the state equation for sliding manifold. The difference between nominal and perturbed equation of state space is derivative of sliding manifold as zero after determining the value of equivalent control. The derivative of sliding manifold as

$$\begin{aligned} S &= G[\xi(t,\Theta) + \eta(t,\Theta)(v + d(t,\Theta))] - G\left[\xi(t,\Theta) + \eta(t,\Theta)u_{clf}\right] \\ &= G\eta(t,\Theta)\left(v + d(t,\Theta) - u_{clf}\right) \\ &= G\eta(t,\Theta)\left(u_{clf} + u_{gstc} + d(t,\Theta) - u_{clf}\right) \\ &= G\eta(t,\Theta)u_{gstc} + G\eta(t,\Theta)d(t,\Theta) \end{aligned} \tag{18}$$

In design, a scaling factor $(G\eta(t,\Theta))^{-1}$ should be used in control u_{gstc} for showing unit vector representation form as

$$\dot{S} = u_{gstc} + d(t,\Theta) \tag{19}$$

substituting (13) into (19), we get

$$\dot{S} = -\varepsilon_1\Omega_1(S) + \delta, \quad 4\dot{\delta} = -\varepsilon_2\Omega_2(S) + \dot{d}(t,\Theta) \tag{20}$$

where $\mathcal{L} > |d(t,\Theta)|$ and $\delta = q + d(t,\Theta)$.

Therefore,

$$\dot{\delta} = \dot{q} + \dot{d}(t,\Theta)$$

proves that $S, \xi = 0$ in finite time and disturbance is completely observed with generalized super-twisting control as $\xi = q + d(t,\Theta) \Rightarrow q = -d(t,\Theta)$.

From (19), we get

$$u_{gstc} = -d(t, \Theta) \tag{21}$$

Finally, we can say that matched disturbance present in the actuator is completely alleviated with the generalized super-twisting control that provides continuous control instead of discontinuous to avoid chattering and the system trajectory always remains in sliding manifold throughout.

4 Illustrative Example

Consider an illustrative example for effectiveness of the proposed algorithm as:

$$\begin{bmatrix} \dot{x}_1 \\ \dot{x}_2 \end{bmatrix} = 0.6 \times \begin{bmatrix} -x_1 + x_1^3 \\ x_2 - 3x_1^3 \end{bmatrix} + 0.2 \times \begin{bmatrix} x_1 - 5x_2 + 2x_1x_2 \\ -8x_2 - x_1^2 \end{bmatrix} + \begin{bmatrix} 2 - 3x_2 \\ -3 + 1.5x_1 \end{bmatrix}(u + d) \tag{22}$$

where $d = 1 + 2\sin(t)$. For the nonlinear nominal system of (22), select control-Lyapunov function $V = \frac{1}{2}x_1^2 + x_2^2$, $\gamma = 1.5$ and initial condition $x(0) = [11]$. Assume:

$$\begin{aligned} a(x) &= \frac{\partial V(x)}{\partial x}\left[-0.2x_1 - 2x_2 + 0.8x_1x_2 + 0.6x_1^3 - 2.6x_2 - 1.8x_1^3 - 0.4x_1^2\right] \\ &= -0.2x_1^2 - 2x_1x_2 + 0.8x_1^2x_2 + 0.6x_1^4 - 5.2x_2^2 - 3.6x_1^3x_2 - 0.8x_1^2x_2 \\ b(x) &= \frac{\partial V(x)}{\partial x}\begin{bmatrix} 2 - 3x_2 \\ -3 + 1.5x_1 \end{bmatrix} = 2x_1 - 3x_1x_2 - 6x_2 + 3x_1x_2 = 2x_1 - 6x_2 \end{aligned}$$

Therefore, from the Theorem 1, nominal control u_{cLf} asymptotically stabilizes the nominal system of (22) about the origin. For the nonlinear perturbed system (22), generalized super-twisting controller is used with the constant $\varepsilon_1 = 3$ and $\varepsilon_2 = 2.5$ to make system asymptotically stable about the origin. Therefore, combination of both controllers is u_{overall} which asymptotically stabilize the nonlinear perturbed system (22) about the origin. Figure 1 shows the convergence of state trajectory $x_1(t)$ and $x_2(t)$. Figure 2. describes the nominal control (u_{cLf}), the generalized super-twisting control (u_{gstc}) and the combination of both the control (u_{overall}).

5 Concluding Remarks

This paper addresses the design of a sophisticated control for nonlinear system with certain parametric uncertainty. Here, an integral sliding mode control allows the design of two continuous control algorithms which are externally independent to each

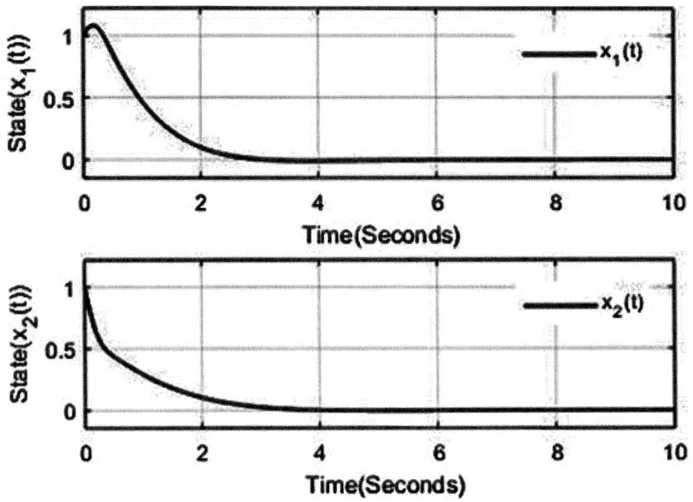

Fig. 1 Evolution of State trajectories

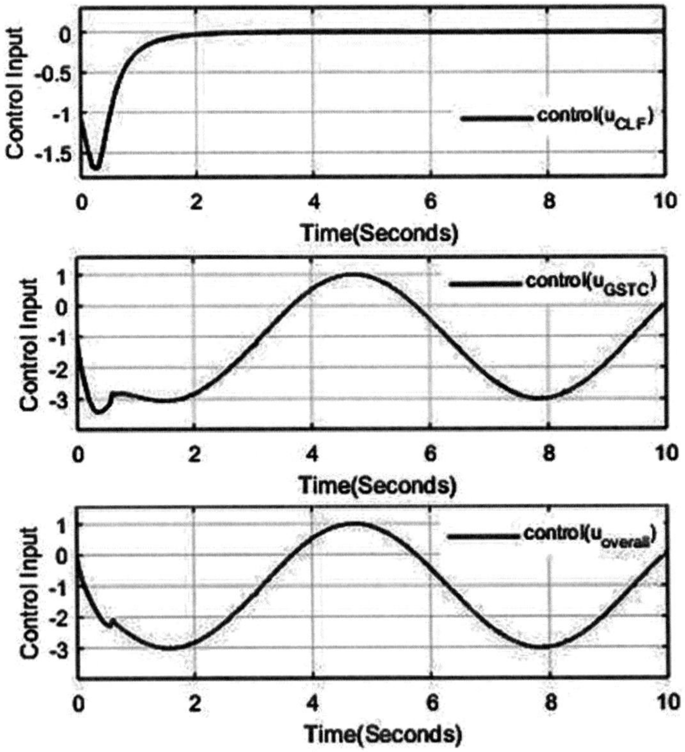

Fig. 2 Evolution of Control signals

other. The control-Lyapunov function existing in the literature is slightly modified to bring the states of nominal system to an equilibrium point. While, the uncertainty term is completely alleviated by a generalized super-twisting control and state trajectories will remain throughout in a predefined manifold. The proposed control law is finally illustrated by a numerical system.

References

1. V. Utkin,, J. Guldner, J. Shi, *Sliding Mode Control in Electromechanical Systems*. Boca Raton: CRC Press (2009)
2. V. Utkin, J. Shi, *Integral Sliding Mode in Systems Operating Under Uncertainty Conditions*, in Proceedings of the 35th IEEE Conference on Decision and Control, vol. 4. IEEE (1996), pp. 4591–4596
3. B. Yao, M. Tomizuka, Smooth robust adaptive sliding mode control of manipulators with guaranteed transient performance. J. Dyn. Syst. Meas. Contr. **118**(4), 764–775 (1996)
4. R. Matteo, D.M. Raimondo, A. Ferrara, L. Magni, Robust model predictive control with integral sliding mode in continuous-time sampled-data nonlinear systems. IEEE Trans. Autom. Control **56**(3), 556–570 (2011)
5. W.J. Cao, J.X. Xu, Nonlinear integral-type sliding surface for both matched and unmatched uncertain systems. IEEE Trans. Autom. Control **49**(8), 1355–1360 (2004)
6. M. Defoort, T. Floquet, A. Kokosy, W. Perruquetti, Integral sliding mode control for trajectory tracking of a unicycle type mobile robot. Integrated Comput. Aided Eng. **13**(3), 277–288 (2006)
7. J.X. Xu, K. Abidi, Discrete-time output integral sliding-mode control for apiezomotor-driven linear motion stage. IEEE Trans. Ind. Electron. **55**(11), 3917–3926 (2008)
8. F. Castanos, L. Fridman, Analysis and design of integral sliding manifoldsfor systems with unmatched perturbations. IEEE Trans. Autom. Control **51**(5), 853–858 (2006)
9. M. Rubagotti, A. Estrada, F. Castanos, A. Ferrara, L. Fridman. Integral sliding mode control for nonlinear systems with matched and unmatched perturbations. IEEE Trans. Autom. Control **56**(11), 2699–2704 (2011)
10. A. Levant, Sliding order and sliding accuracy in sliding modecontrol. Int. J. Control, **58**(6), 1247–1263 (1993)
11. A. Chalanga, S. Kamal, B. Bandyopadhyay, *Continuous Integral Slidingmode Control: A Chattering Free approach*, in IEEE International Symposium on Industrial Electronics (2013), pp. 1–6
12. S.P. Sadala, B.M. Patre, A new continuous sliding mode control approachwith actuator saturation for control of 2-DOF helicopter system. ISA Trans. **74**, 165–174 (2018)
13. Z. Artstein, Stabilization with relaxed controls. Nonlinear Anal. Theory, Methods Appl. **7**(11), pp 1163–1173 (1983)
14. E.D. Sontag, A universal construction of Artstein's theorem on nonlinear stabilization. Syst. Control Lett. **13**(2), 117–123 (1989)
15. Z.P. Jiang, Y. Lin, Y. Wang, Stabilization of Time-Varying Nonlinear Systems: A Control-Lyapunov Function Approach, in IEEE International Conference on Control and Automation (2007), pp. 404–409
16. A. Shahmansoorian, B. Moshiri, A.K. Sedigh, S. Mohammadi, A new stabilizing control law with respect to a control-Lyapunov function and construction of control-Lyapunov function for particular nonaffine nonlinear systems. J. Dynam. Control Syst. **13**(4), 563–576 (2007)
17. J.A. Moreno, *A Linear Framework for the Robust Stability Analysis of a Generalized Super-Twisting Algorithm*, in 6th International Conference on Electrical Engineering, Computing Science and Automatic Control (2009), pp. 1–6
18. G. Bitsoris, E. Gravalou, Comparison principle, positive invariance and constrained regulation of nonlinear systems. Automatica **31**(2), 217–222 (1995)

FileShare: A Blockchain and IPFS Framework for Secure File Sharing and Data Provenance

Shreya Khatal, Jayant Rane, Dhiren Patel, Pearl Patel, and Yann Busnel

1 Introduction

Blockchain is seen as a distributed ledger that can be accessed globally by anyone to verify stored data with high integrity, resilience, credibility, and traceability. Distributed ledger technology can be used to write smart contracts which are self-executing contracts, and can be replicated, shared, and supervised by a network of computers that run on blockchain. Smart contracts avoid middleman by automatically defining and enforcing rules and obligations made by the parties in the ledger. Blockchain, however, is an expensive medium for data storage. For efficient storage of digital content, we can use InterPlanetary File System (IPFS) which is a peer-to-peer hypermedia protocol and distributed file system. Since IPFS is distributed, it has no single point of failure.

This paper presents a framework where digital content is shared in a secure, tamper-proof environment and achieves provenance of read, modify and share operations on data. Our application is based on IPFS and smart contracts of Ethereum blockchain. Blockchain technology is utilized for access control of digital content

S. Khatal (✉) · J. Rane · D. Patel
Veermata Jijabai Technological Institute, Mumbai, India
e-mail: khatalshreya@gmail.com

J. Rane
e-mail: jayantrane811@gmail.com

D. Patel
e-mail: dhiren29p@gmail.com

P. Patel
Vidyalankar Institute of Technology, Mumbai, India
e-mail: pearl207@gmail.com

Y. Busnel
IMT Atlantique, Rennes, France
e-mail: yann.busnel@imt-atlantique.fr

S. Patnaik et al. (eds.), *Advances in Machine Learning and Computational Intelligence*, Algorithms for Intelligent Systems, https://doi.org/10.1007/978-981-15-5243-4_79

and storage of provenance data. The proposed FileShare application ensures that the digital content would only be accessible in the application and will not be available in the end-users' operating system.

Rest of the paper is organized as follows: Sect. 2 briefly explains underlying concepts and summarizes related work. Section 3 describes the proposed application's workflow. Section 4 discusses validation and analysis of FileShare application. Section 5 concludes the paper.

2 Background and Related Work

Morgan [1] presents the idea of using blockchain technology to prove the existence of a document using the timestamping concept. The legitimacy of the document can be verified, but this system is not focused about the authority of the owner on his/her document. IPFS [2] is the distributed and versioned file system which can connect many computing nodes with the same system of files and manage them by tracking their versions over time. Rajalakshmi et al. [3] propose a model of academic research record keeping with access control methods. Nizamuddin et al. [4] propose a solution that is based on using IPFS and smart contracts of Ethereum blockchain. Both of the papers mentioned above do not restrict duplication and piracy of the shared content as once the document is downloaded, it can be replicated and any other user can claim the ownership. Further, any changes made to the document in users' operating system in offline mode are not logged and hence ownership as well as originality is threatened.

Smart contracts provide an easy way to access the Ethereum blockchain written in a high-level coding language called Solidity [5]. To develop Ethereum smart contracts, Remix IDE [6] can be used, which is a browser-based IDE. Another one is the Truffle framework [7], which supports built-in smart contract compilation, linking, deployment, and binary management. In order to run Ethereum decentralized apps in the browser itself, without running a full node, MetaMask [8] can be used. The above tools can be combined for an effective Ethereum decentralized application development.

Data provenance is very critical as it provides the history of the origins of all changes to a data object, list of components that have either forwarded or processed the object and users who have viewed and/or modified the object. Hasan et al. [9] proposed SPROVE that protects provenance data confidentiality and integrity using encryption and digital signature. However, SPROVE does not possess provenance data querying capability. Ko and Will [10] proposed Progger which is a kernel-level logging tool which can provide log tamper-evidence at the expense of user privacy. Liang et al.'s [11] proposal ProvChain is a distributed, cloud-based data provenance architecture, which creates tamper-proof record of events by embedding the provenance records into the blockchain as transactions. All of the above methods have no restriction on piracy and hence possess a breach to integrity.

3 Proposed Application's Workflow

Figure 1 shows components of the proposed architecture. Users first register to the FileShare application. The registration details of the user are added to the Ethereum blockchain by the application. After creation of the file in the application's inbuilt text editor, user decides if the file has to be made shared or public. If the file is supposed to be shared, the file owner provides the public key of recipients with whom the file has to be shared with.

The application then deploys a smart contract which stores the file metadata. It then encrypts the document and adds to IPFS in an encrypted format. In order to access the files, users are required to use the file sharing application editor as the file would be decrypted only in the application editor. The application uses the file smart contract to access the file metadata, fetches the file from IPFS, decrypts the file, and opens in the inbuilt editor. In order to collect provenance data, calls to functions of smart contract are logged as file operations performed in the editor. After an operation is performed, the record is generated, which will be uploaded to the blockchain network and stored in the provenance blockchain.

The framework proposed here can be divided into four main phases: user registration and authentication, file creation and storage, file retrieval, and provenance data collection and storage.

3.1 User Registration and Authentication

Users are required to register to the system in order to have a unique identity. We propose to create a smart contract for every user, which will act as a unique identity for them. 'AllUsersMetadata' is a smart contract which acts as a factory to generate a smart contract for every user after their registration. During the registration process as shown in Fig. 2, user provides registration key in the form of a string as an input to the application. Using this registration key and current timestamp, application generates public–private key pair using ECDSA algorithm. Now, AllUsersMetadata deploys a smart contract of the registered user and obtains address of the deployed

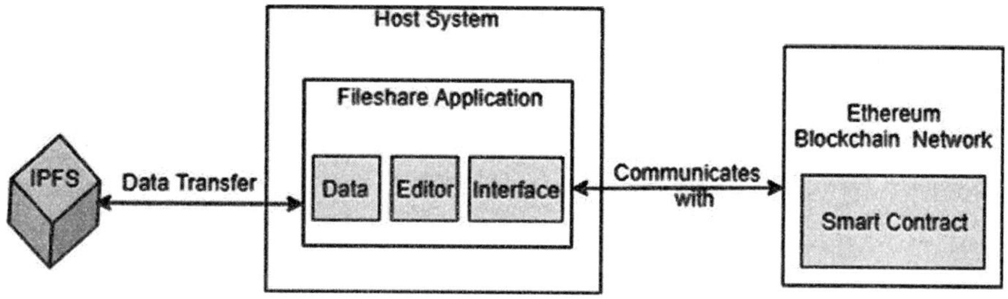

Fig. 1 Components of the proposed architecture

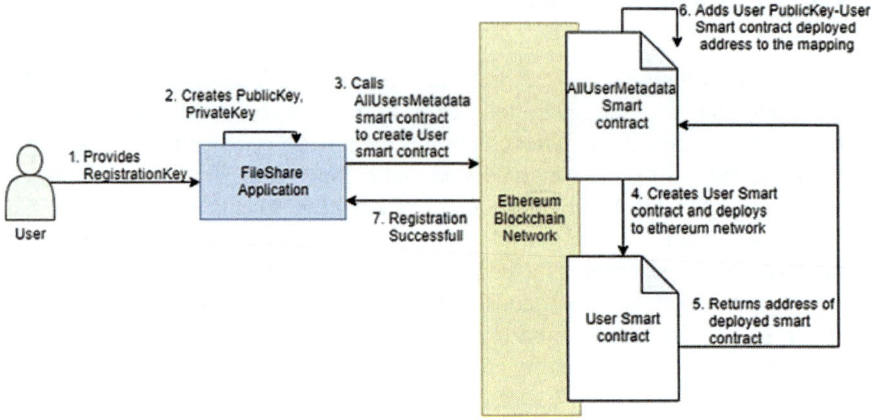

Fig. 2 FileShare system interaction for user registration and authentication

smart contract. The deployed user's smart contract contains user's metadata which includes user's public key, registration key, an array of information details regarding the files which have been shared with the user.

AllUsersMetadata also contains a mapping of every registered user's public key to the address of their deployed smart contract. After the deployment of the user's smart contract, the received deployed address of the user's smart contract is added to the mapping in AllUsersMetadata smart contract. Public key generated during the registration process will be used by file owner while specifying recipient to whom the file must be shared with, while registration key and private key will be used to validate the user authenticity during the login process of the FileShare application.

For authentication, user will provide registration, public as well as their private keys as an input to the FileShare application. Initially, registration key will get encrypted using private key, and generated string will be the 'EncryptedRegistrationKey'. Using the received public key as an input, user's smart contract deployed address will be fetched from the AllUsersMetadata's mapping. As the user's smart contract is fetched from the obtained address, to validate the user, the application will send the EncryptedRegistrationKey to validation function of the user's smart contract. Now, the EncryptedRegistrationKey will be decrypted using public key of the user, and if resulting string is same as registration key of the user's smart contract, then user will be validated; otherwise, the authentication would fail.

3.2 File Creation and Storage

Owner creates a file in FileShare application editor and requests for sharing this file on the FileShare application. FileShare now creates a random key, to encrypt the file using AES-256 symmetric encryption technique. This random key will be

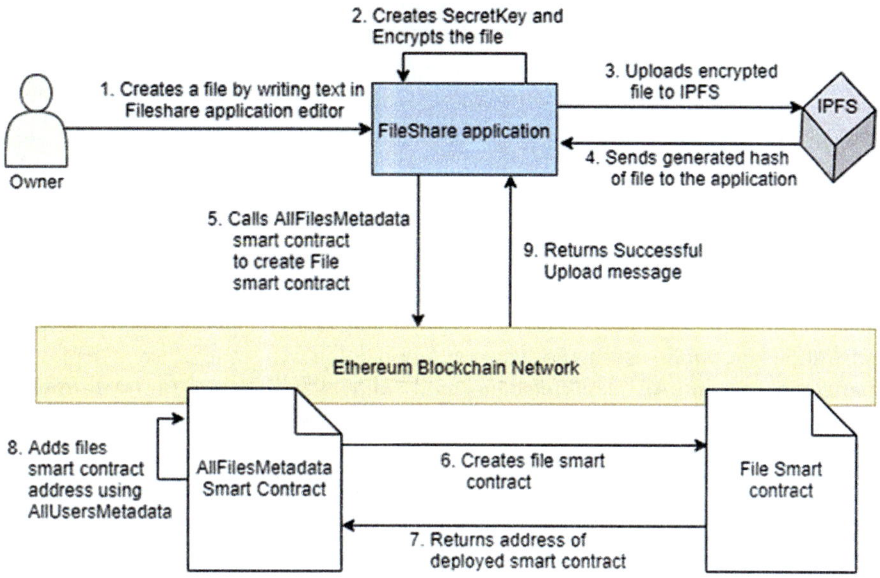

Fig. 3 FileShare system interaction for sharing files

'SecretKey' for given file which will only reside in owner's FileShare application. FileShare encrypts the file with the SecretKey. This encrypted file is added to the IPFS network. IPFS network returns hash of the uploaded file. As shown in Fig. 3, we propose to create a smart contract for every deployed file on IPFS. 'AllFilesMetadata' smart contract acts as a factory to generate smart contract for every file shared on the application. File's smart contract contains metadata which includes filename, IPFS address of the encrypted file and owner's public key. After deployment of the smart contract, FileShare application will receive deployed file smart contract's address. Now, owner can specify following types of access control for the specified file.

Shared. In this access control, the owner can share the file to other users by using the public key of the user they want to share the file with. After giving this public key to the FileShare application as an input, the application will encrypt 'SecretKey' of the file with 'PublicKey' of the user with whom the file has to be shared with to create an 'EncryptionKey'. This is asymmetric encryption, whereas 'EncryptionKey' can only be decrypted by the user who has corresponding PrivateKey. Smart contract of the file, for shared mode, contains a mapping of the receiver's public key to the EncryptionKey of the file. This mapping will be added to the shared file's smart contract. AllFilesMetadata will access AllUsersMetadata to obtain deployed address of receiver's smart contract. The shared files' smart contract address will be added to the receiver's smart contract. Thus, the receiver's smart contract will contain an array of deployed address of all the files which are shared with them.

Public. In this access control, the owner can share the file to every user who is registered on the FileShare application. Owner will specify the SecretKey in the file's smart contract. Also, owner will send their public key along with deployed file smart contract's address, to the AllFilesMetadata smart contract.

After these specifications are set to file's smart contract, other users will be able to access it if they are authorized of the FileShare application.

3.3 File Retrieval

On the FileShare application interface, after giving user's logging details such as registration key, public key, and private key, application will retrieve user's deployed smart contract using AllUsersMetadata smart contract as shown in Fig. 4. If user is validated, then FileShare application will now access the user's smart contract using the AllUsersMetadata. The user's smart contract contains the address of deployed smart contracts of all files shared with them. These files will appear on the application interface as 'Shared with me'. Application interface will also retrieve all files which are publicly shared using AllFilesMetadata smart contract. The following mechanisms are performed for the proposed access control types:

Shared. Using the file's deployed smart contract address, FileShare application will retrieve key available in the mapping of publicKeyToEncryptedKey using user's own public key. Received key will then be decrypted by user's private key in the application, and generated key will be used to decrypt the accessed files by the FileShare application.

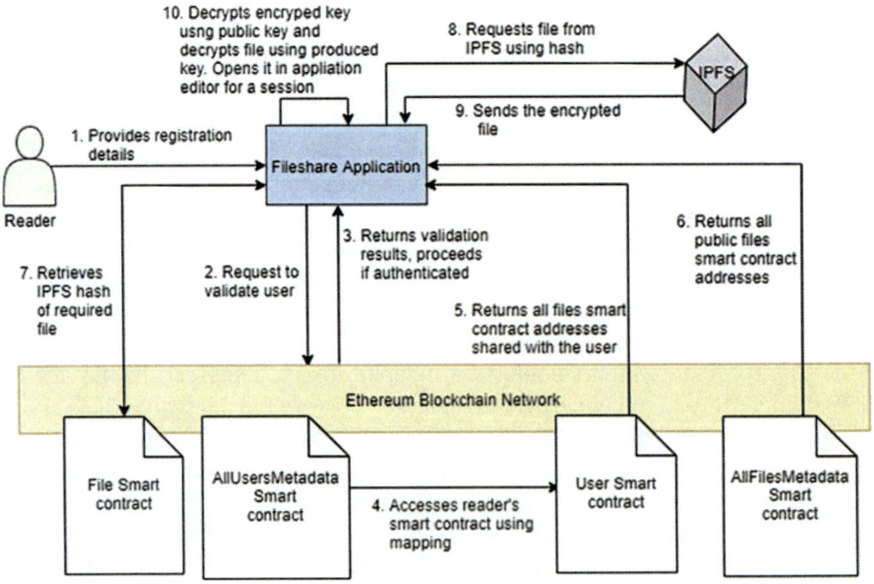

Fig. 4 FileShare system interaction for accessing files

Public. Using file's deployed smart contract address, FileShare application will request decryption key of file from corresponding file's smart contract deployed on blockchain. This key will be internally sent to the application, and FileShare will decrypt the file and open in the application editor.

File accessed will be available to read for a session where session time would be a defined parameter. Also, file can be modified in FileShare application, which will be redeployed in application along with original owner's public key attached to it. The uploaded content can only be accessed by using the application editor. The content cannot be downloaded nor be copied to clipboard of operating system, from the editor.

3.4 Provenance Data Collection and Storage

Every time a user performs operations such as read and sharing files, it needs decryption key of the file. This key is available only in corresponding deployed smart contract. Whenever user requests this key, smart contract logs these events in blockchain. The provenance data will contain the unique id of the user who has accessed the content, corresponding file's deployed smart contract address, time of access, and type of operation accessed by user. For publishing data records to blockchain network, we adopt Chainpoint standard [12].

4 Validation and Analysis of FileShare Application

Implementation Details

Figure 5 gives an overview of the FileShare application architecture. The Ethereum's Testnet/Ropsten blockchain is used to store the user details as well as the audit logs. IPFS gets free hosting forever in a decentralized platform. React.js with webpack is used for the front end. Solidity 0.4.11 is used for developing smart contracts. web3.js is used to interact with Ethereum node, using a HTTP connection. We used

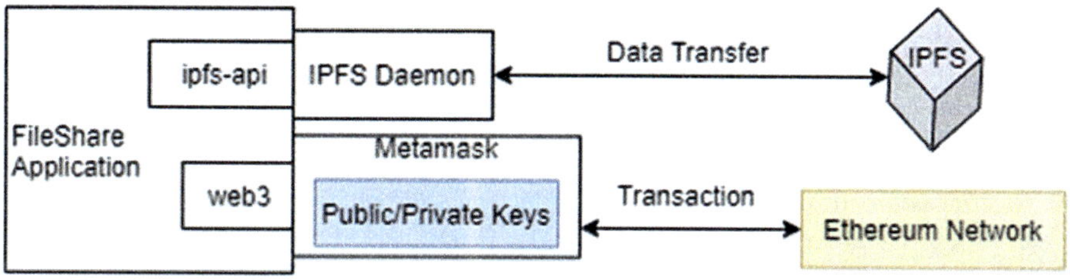

Fig. 5 FileShare application architecture

MetaMask to use the final application like the end-user would. As our solution does not store any unencrypted data on blockchain, it is not prone to Ethereum blockchain hacks. The FileShare application achieves the following five objectives:

No duplication of shared files. As any shared file may it be in public or private mode can only be decrypted using the FileShare application, it cannot be downloaded in any end-users' operating system. Thus, no copies of the file exist.

Real-time data provenance. The audit logs obtained include user information and the operations performed on the file may it be view, modify or share and are then added to the blockchain network making it tamper-proof.

Tamper-proof Environment. Data provenance record is collected and then published to the blockchain network which protects the provenance records.

End-to-End traceability. Users can access the provenance data to know the owner of the file, number of people who viewed the file, and the edit operations on the file. Thus, no ownership problem occurs.

Provenance Data Validation. Data provenance record is published globally on the blockchain network. Thus, provenance data is validated by the blockchain nodes.

5 Conclusion

The FileShare application provides a secure, tamper-proof model for sharing files in a distributed file system. The provenance data can be used to obtain analytical information. The file owners who made the file public can obtain analytical information about the number of people who viewed the file. The private file owners can keep accessing the modification operations performed by the users with whom the file has been shared. The owner of the files can be traced easily avoiding the ownership problems. Further, as the provenance data is stored in the blockchain, it creates an immutable record, and any malicious modifications to the provenance data can be prevented.

References

1. P. Morgan, *Using Blockchain Technology to Prove Existence of a Document*. Last accessed 20 Feb 2018
2. J. Benet. *IPFS-Content Addressed, Versioned, P2P Filesystem* (2014). arXiv preprint arXiv:1407.3561
3. A. Rajalakshmi, K.V. Lakshmy, P.P. Amritha, A blockchain and IPFS based framework for secure Research record keeping. Int. J. Pure Appl. Math. **119**, 1437–1442 (2018)
4. N. Nizamuddin, H. Hasan, K. Salah, *IPFS-Blockchain-Based Authenticity of Online Publications* (2018). https://doi.org/10.1007/978-3-319-94478-4_14
5. Solidity—Solidity 0.4.23 Documentation, in Solidity.readthedocs.io (2018). [Online]. Available: http://solidity.readthedocs.io/en/v0.4.23/
6. Remix—Solidity IDE, in Remix.ethereum.org (2018). [Online]. Available: https://remix.ethereum.org/

7. Truffle Suite—Your Ethereum Swiss Army Knife, Truffle Suite (2018). [Online]. Available: http://truffleframework.com/
8. MetaMask, in Metamask.io (2018). [Online]. Available: https://metamask.io/
9. R. Hasan, R. Sion, M. Winslett, Sprov 2.0: A *Highlyconfigurable Platform-Independent Library for Secure Provenance*, in ACM, CCS, Chicago, IL, USA (2009)
10. R.K. Ko, M.A. Will, *Progger: An Efficient, Tamperevident Kernel-Space Logger for Cloud Data Provenance Tracking*, in 2014 IEEE 7th International Conference on Cloud Computing. IEEE (2014), pp. 881–889
11. X. Liang, S. Shetty, D. Tosh, C. Kamhoua, K. Kwiat, L. Njilla, *Provchain: A blockchain-Based Data Provenance Architecture in Cloud Environment with Enhanced Privacy and Availability*, in Proceedings of the 17th IEEE/ACM International Symposium on Cluster, Cloud and Grid Computing. IEEE Press (2017), pp. 468–477
12. *Chainpoint: A Scalable Protocol for Anchoring Data in the Blockchain and Generating Blockchain Receipts*. http://www.chainpoint.org/. Last accessed 8 Nov 2019

NFC-Based Smart Insulin Pump, Integrated with Stabilizing Technology for Hand Tremor and Parkinson's Victims

Advait Brahme, Shaunak Choudhary, Manasi Agrawal, Atharva Kukade, and Bharati Dixit

1 Introduction

In the twenty-first century, the people suffering from various diseases and disorders are increasing proportionately with the number of healthcare devices coming up to counter these problems. Most of these problems being interrelated to each other, the need of the hour is to come up with solutions that can cater to multiple disorders. Our proposed methodology can be a solution that can benefit patients of three diseases/disorders namely diabetes, Parkinson's, and hand tremors. Diabetes is one of the most widespread diseases in the world. Parkinson's and hand tremors are mainly prevalent among elderly people. There is a sizeable chunk of population suffering from diabetes as well as one of the above two disorders. The proposed model will take into consideration these two disorders and provides one solution for all. This paper explores the idea of a device, which if developed, can cater to the needs of this population.

De Pablo-Fernandez et al. [1] talk about the interrelation between diabetes and Parkinson's disorder, and how one disease has an impact on the other and vice versa. Chung et al. [2] investigate the negative effect of Type 2 diabetes on patients of Parkinson's disorder. D'Attellis et al. [3] propose an intelligent insulin pump, which

A. Brahme · S. Choudhary · M. Agrawal · A. Kukade · B. Dixit (✉)
MIT College of Engineering, Pune, India
e-mail: bharati.dixit@mitcoe.edu.in

A. Brahme
e-mail: brahmeadvait@gmail.com

S. Choudhary
e-mail: shaunakc2013@gmail.com

M. Agrawal
e-mail: manasi1211@gmail.com

A. Kukade
e-mail: kukadeatharva9922@gmail.com

S. Patnaik et al. (eds.), *Advances in Machine Learning and Computational Intelligence*, Algorithms for Intelligent Systems, https://doi.org/10.1007/978-981-15-5243-4_80

can automate the process of insulin infusion into the body using control algorithms. Sivraj et al. [4] also talk about developing an insulin pump that makes the process of insulin infusion simpler. Gupta et al. [5] propose an assisting device for Parkinson's patients using Internet of things and machine learning. Chamraz et al. [6] deal with measurement of small accelerations using feedback sensor.

1.1 Diabetes Mellitus and Its Rise Through the Years

When glucose is absorbed from a recently consumed meal, the glucose gets released in the blood stream and pancreas are triggered by high blood glucose levels which result in the secretion of insulin. The insulin strikes the balance between hyperglycemia and hypoglycemia [3]. A person suffering from diabetes does not have enough insulin to maintain the blood sugar levels in the body. The pancreas not producing enough amount of insulin is the primary reason for the insufficiency of insulin. According to the reports, approximately 425 million people were affected by diabetes in the year 2017 worldwide and by 2045, this will rise to 629 million [7].

1.2 Parkinson's and Its Rise

Labeled as one of the most common age-related disease, Parkinson's is a neurodegenerative disorder. According to the recent reports by the Parkinson's News Today, approximately 7–10 million people are suffering from Parkinson's in the year 2019 [8]. The disease directly affects the midbrain of the victim in which dopamine producing cells die off in the substantia nigra of the brain which results in body tremors. According to a recent study carried out, the merged relative risk of developing Parkinson's disease is increased if the person is suffering from diabetes mellitus. The degree of risk was found to be more in younger individuals by which genetic factors may comparatively exert added effect [1].

1.3 Near-Field Communication (NFC)

NFC is a short-range contactless technology used for intentional communication. The technology used in NFC is based on radio frequency identification which uses the set of radio frequencies to transmit and read information. Because of its ability to be a passive component, they do not need their power supply. Instead, they can be powered by an electromagnetic field produced by an active NFC component when it comes into its range. Active devices can both send and receive data and can communicate with each other. Smartphones are the most common example of active devices. Average proximity is 4cms and maximum proximity at which data can be

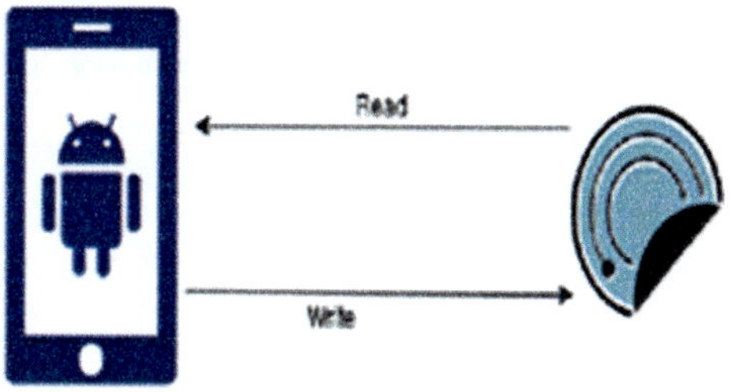

Fig. 1 NFC counter

transmitted and read is 10 cm [9]. The transmitted frequency for data across NFC is 13.56 MHz and data can be sent at either 106, 212, or 424 KB/s [9]. There are three modes of operations [10]:

- Peer-to-peer mode
- Card emulation mode
- Read Write Mode: We are proposing use of this mode of operation in our idea. In this mode, the active NFC device reads any other NFC embedded device. Further it has the option to write, delete, or modify on the tag or the device. NFC advertisements tags use this mode (Fig. 1).

2 Proposed Methodology

NFC-based smart insulin pump is an integrated module of multiple devices listed below. Devices work together to produce the result of modulating body glucose levels, and the stabilizer will be used to stabilize hand tremors for Parkinson's victims while affixing the sensor. The components used the proposed idea: Glucose sensor, stabilizer, insulin pump, and mobile application.

2.1 Components

Insulin pump. Insulin pumps are small, computerized devices which are used by people suffering from diabetes. This pump can substitute the need for multiple injections daily and substitute the need for long-acting insulin. Insulin pumps act as artificial pancreas by delivering short doses of insulin.

The contents of Table 1 are a reference for the insulin pump to calculate the amount of insulin to inject in the body. Table 2 is used to show the modulation required according to the weight of the user (Fig. 2) [11].

The readings from sensor will be transferred to the pump via mobile application/reader device using NFC technology. This eliminates the need to program the amount of insulin to be pumped at various times manually.

Table 1 Function F(GL) is the glucose level, calculating insulin units (IU) to be injected per day in function of glucose level and patient weight

Glucose level (GL, mg/dl)	Number of IU (IU/kg/day)
< 150	0.3
150–200	0.4
>200	0.5

Table 2 Additional insulin units to be injected in function of glucose level and patient weight (part of corrector factor)

Glucose level	<60 kg	60–90 kg	>90 kg
<80 mg/dl	−1	−1	−2
80–129	0	0	0
130–149	0	+1	+1
150–199	+1	+1	+2
200–249	+2	+3	+4
250–299	+3	+5	+7
300–349	+4	+7	+10
>349	+5	+8	+12

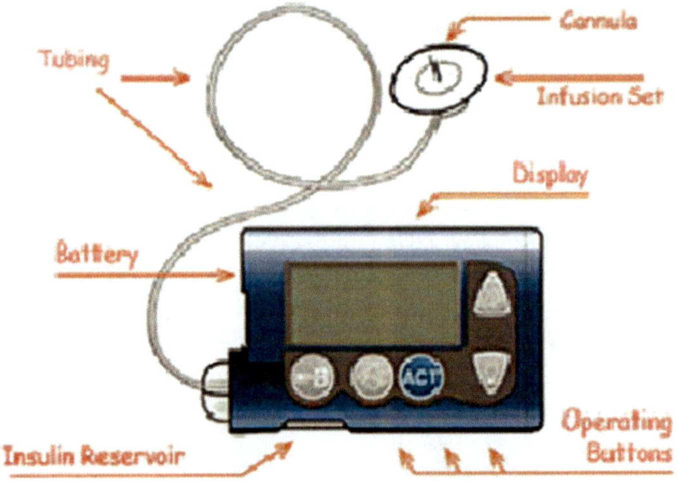

Fig. 2 Insulin pump

Glucose Sensor. The sensor is inserted into the subcutaneous tissue of the body. The sensor takes readings in periodic intervals. These readings are converted into user understandable data like values and graphs. The readings are to be stored in the sensor itself until the mobile application/reader device is used to read from the sensor using NFC technology [12]. The user can check the sensor reports spot patterns, review glucose patterns, identify areas to target with a medical professional and track the progress.

Fig. 3 Glucose sensor

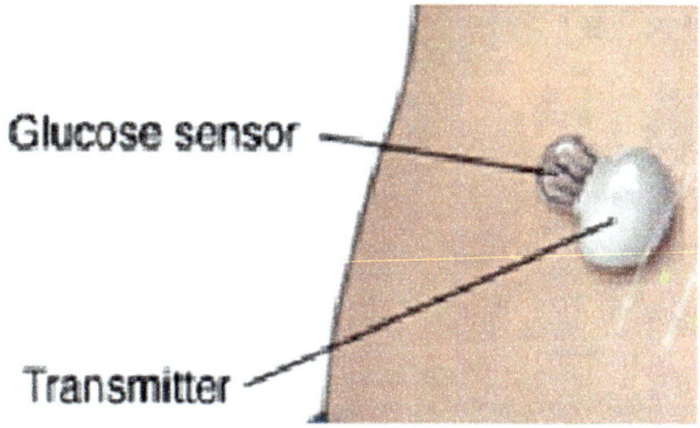

For automatic sensor calibrations, some important factors [12] are:

- Reduce sensitivity variations between the sensor
- Maintain sensor sensitivity over the assigned shelf life.
- Maintain sensor sensitivity over the wear duration
- Consistent blood/tissue relationship (Fig. 3).

Stabilizer. An applicator will be used to affix the glucose sensor to the body. This applicator will be equipped with a tremor stabilizer, for aiding patients of Parkinson's and hand tremors. This stabilizing technology consists of an accelerometer, closed-loop system, and a microcontroller and is like the technology used in the cameras for image and video stabilization. The accelerometer consisting of motion sensors detect the vibrations and measures the acceleration of the vibrations of the victim. When the handle will tilt to one side because of the motion, the accelerometer will identify the degree of the tilt and the acceleration and pass this information forward to the microprocessor. The microprocessor actively listens to these signals and differentiates between the victim's intended motion and unintended motion. Based on the detected input frequencies and direction of the tremor, the processor will then produce outputs to the actuator [13]. Further, the actuator initiates the vibrations in the direction opposite to the one to which the applicator has moved, and at the same angle on the opposite side.

Thus, it will be a big boon for patients of Parkinson's and hand tremors, as they would feel that the applicator is stable and not have to be reliant on anyone else for adhering to the glucose sensor. This stabilizing technology has been implemented in spoons, forks, and sporks by a company, Verily (Figs. 4 and 5).

2.2 System Architecture and Working

The process commences with affixing the glucose sensor beneath the bicep. This process of implanting the sensor is very difficult for Parkinson's victims without any medical aid. This is when the handle equipped with a stabilizer will come into

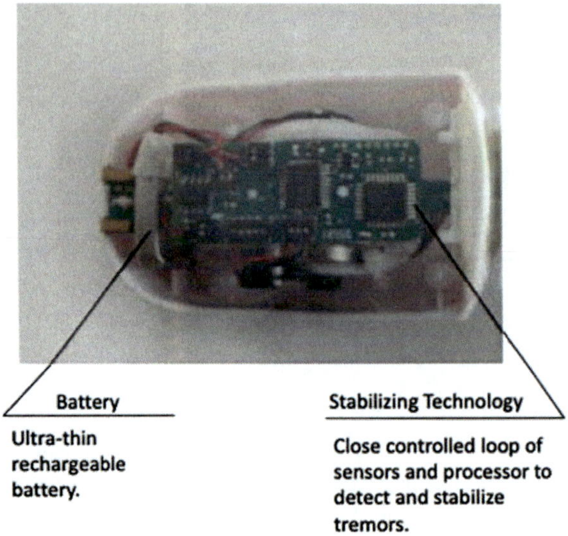

Fig. 4 Stabilizer

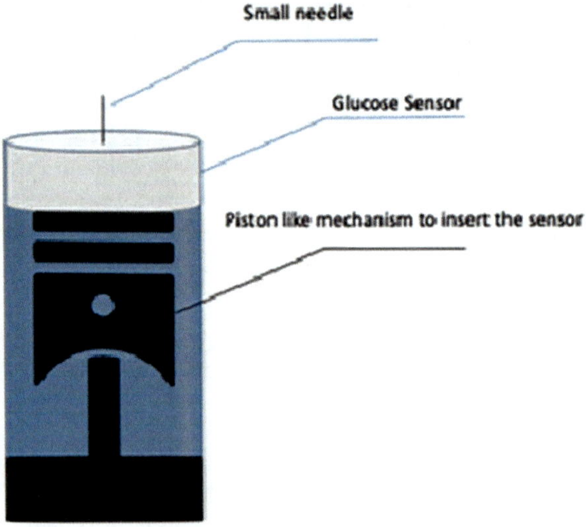

Fig. 5 Mechanism of stabilizer and sensor

action. The handle will contain the sensor to be affixed. While affixing the sensor, the accelerometer within the stabilizer will estimate the acceleration produced due to the hand tremors and the direction of tremors. The processor will then generate vibrations to counter the movements of the hand. This will assist the patients to inject the sensor precisely. The glucose sensor will take in the blood sample through the small needle/strip and identify the glucose level in the blood sample. It will store this data for further transfer. This process of assessing the glucose levels will take place after short, regular intervals.

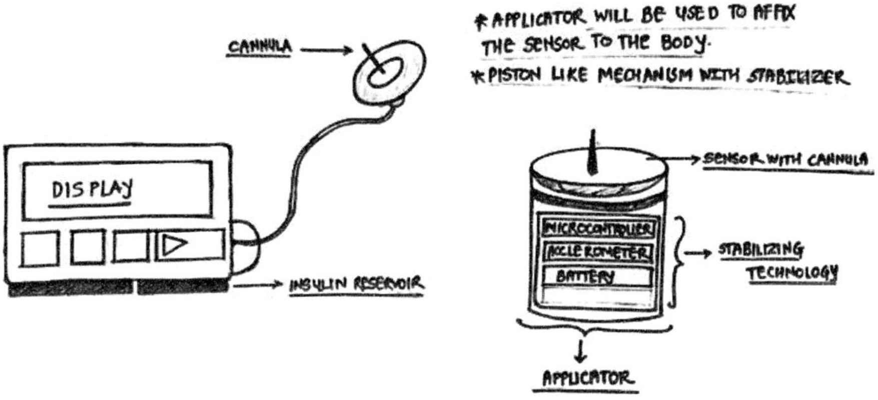

Fig. 6 Flow diagram of proposed idea

Further, this data will be transferred to the insulin pump with the use of NFC technology. The user must tap the glucose sensor for the data to be transferred from the sensor to the NFC-enabled mobile device. Nowadays, almost every cell phone is equipped with NFC technology. The mobile application will represent all the gathered data in graphical format. This will help the user and the doctor to understand the data better and analyze the patient's current health. For modulating the glucose levels, the proposed model will use an insulin pump. The insulin pump with the right data will infuse a certain amount of insulin to balance the glucose levels in the body. The insulin pump will contain an infusion set which will be injected into the user's abdomen. The pumped insulin will pass through a thin catheter into the infusion set. To set up the insulin pump, the patient will have to first inject the needle. Same as the glucose sensor, the handle will be equipped with a stabilizer to stabilize hand movements for better precision. Once the insulin pump setup is done, by just tapping the cell phone, all the required data for insulin to be pumped is transferred via onboard NFC chip. Using these parameters, it will regulate the glucose levels (Fig. 6).

3 Conclusion

The use of technology has enabled medical science to become really advanced in nature. These advancements in medical science have made it possible to design different equipment to cater to the needs in healthcare domain. It has generated a possibility to create efficient and generalized devices which can be used by people suffering from more than one disease. Our proposed methodology looks to eliminate the difficulties faced by patients of Parkinson's as well as diabetes while administering insulin doses. Automating the entire process not only makes it reliable but also proves to be a game-changer in the field of health care, providing one stop solution to the diasporas suffering from both diseases, to overcome a major hurdle in their everyday life. The proposed concept is new by itself, and no prototype has been developed yet. But this idea has immense potential to be developed into a device.

References

1. E. De Pablo-Fernandez, R. Goldacre, J. Pakpoor, A.J. Noyce, T.T. Warner, Association between diabetes and and subsequent Parkinson's disease. AAN (2018)
2. S.J. Chung, S. Jeon, H.S. Yoo, G. Kim, J.S. Oh, J.S. Kim, A.C. Evans, Y.H. Sohn, P.H. Lee, Detrimental Effect of Type 2 Diabetes Mellitus in a Large Case Series of Parkinson's Disease, vol. 9. Elsevier, Amsterdam (2019), pp. 54–59
3. G. Cocha, J. Rapallini, O. Rodriguez, C. Amorena, H. Mazzeo, C.E. D'Attellis, *Intelligent Insulin Pump Design*, in IEEE (2018)
4. T.T. Thomas, S. Nithin, P. Sivraj, K. Guruvayurappan, *Design of an Embedded Controller for Next Generation Low Cost Insulin Pump*, in IEEE (2019)
5. C.J. Baby, A. Mazumdar, H. Sood, Y. Gupta, A. Panda, R. Poon Kuzhali, *Parkinson's Disease Assist Device Using Machine Learning and Internet of Things*, in IEEE (2018)
6. S. Chamraz, R. Balogh, *Analysis of Capacitive MEMS Sensor for Small Accelerations*, in Research Gate Conference (2018)
7. Diabetes Facts and Figures. https://www.idf.org/aboutdiabetes/what-is-diabetes/factsfigures.html. Last accessed 15 Dec 2019
8. Parkinsons's Disease Statistics. https://parkinsonsnewstoday.com/parkinsons-disease-statistics/. Last accessed 15 Dec 2019
9. What is NFC and How Does it Work. https://www.androidauthority.com/what-is-nfc-270730/. Last accessed 15 Dec 2019
10. NFC Forum. https://nfc-forum.org/. Last accessed 15 Dec 2019
11. A.J. Jara, A. Skarmeta, M.A. Zamora-Izquierdo, An internet of things based personal device for diabetes therapy management in ambient assisted living (AAL). Pers. Ubiquit. Comput. **15**, 431–440 (2011)
12. U. Hoss, E.S. Budiman, Factory-calibrated continuous glucose sensors: the science behind the technology. Diabetics Technol. Therapeut. **19**(Suppl. 2) (2017)
13. D. Behera, M. Mohanty, Design of an assistive device for older age people suffering from essential tremor (2015)

Digital Forensics: Essential Competencies of Cyber-Forensics Practitioners

Chamundeswari Arumugam and Saraswathi Shunmuganathan

1 Introduction

The common cyber-forensics investigation that exists in today's society is identity theft, child exploitation, counter terrorism, dating, financial fraud, smuggling, etc. At present, the crime is encircled with many technological resources like smartphone, computer, cloud, network, IoT, wearable devices, etc. So, a sound technical skill and exposure about these resources are essential, to recover the evidence data by cyber-forensics practitioners. Therefore, the forensics practitioners should be trained to tackle the various cyber-investigations by equipping with the current technological advances. Many government sectors and private organization have training centers to train the practitioners for investigation.

Practitioners should acquire the supporting equipments to investigate evidence professionally and methodically to create strong legal case. Basic equipment for forensics laboratory should contain hardware devices, software applications, evidence collection accessories, evidence preservation devices, digital data investigation kits, data transmission cables, and connectors [1]. Portable forensics laboratory has a carrying case that holds the different connectivity drives, memory cards, etc., to capture evidence data at high speed from multiple types of digital media. Investigator offers services in wide range of spectrum. Some of the important areas where they offer services include data recovery, evidence captured from mobile phones, iphones, VMware, bitcoin, matrimonial digital case forensics, cell site analysis, etc.

Software organization provides enormous forensics tool [2–5] to recover the evidence data. Extensive free digital investigation tools are available for recovering

C. Arumugam (✉) · S. Shunmuganathan
Department of Computer Science and Engineering, SSN College of Engineering, Kalavakkam, Chennai, Tamil Nadu, India
e-mail: chamundeswaria@ssn.edu.in

S. Shunmuganathan
e-mail: saraswathis@ssn.edu.in

S. Patnaik et al. (eds.), *Advances in Machine Learning and Computational Intelligence*, Algorithms for Intelligent Systems, https://doi.org/10.1007/978-981-15-5243-4_81

the evidence data from memory [6], hard disk, smartphones, cloud, and network. Open-source forensics tools that aid to recover data are HexEditor [7], FTK [8], Sleuthkit [9], Linux "dd," etc. Many commercial tools that aid digital forensics are Encase [3], Oxygen [2], Autospy [9], etc. Some forensics tools [10–14] were also used to investigate the user sensitive data from instant messaging applications like Whatsapp, WeChat, Viber, etc. The practitioner should be certified to use and apply forensics tools for investigation.

The main objective of this work is to analyze the cyber-forensics practitioner's competencies in cyberspace in resolving the investigation. There may be many ongoing investigations in a forensics laboratory. How well the practitioner is effective in resolving the investigation is measured, by considering the project risk and process risk involved in various investigations. The role of multi-agent is incorporated in this proposed work to collect the practitioner's competencies in various forensics investigations.

The organization of this paper proceeds as follows. Section 2 discusses the related work, while Sect. 3 elaborates on the investigation approach. Section 4 details the practitioner's perspective, and Sect. 5 details the conclusion and future work.

2 Literature Survey

The expansion of the Internet and explosion of connectivity among various devices like computer, mobile, cloud, IOT, etc., led to the growth of cyber-criminal activities. Mobile criminal activity is vital as mobile phone usage has increased drastically and majority of all Internet traffic is generated by mobile phones [15]. The advances, challenges, and opportunities in mobile forensics were discussed in this work. Feng et al. [16] discussed Android mobile device's data migration challenges. Continuous upgradation of new version has an impact on forensics acquisition and migration. The existing migration methods cannot be adopted to new version. Also, the deployment of security technologies added the challenges like unlocking of screen-lock and the opening of USB debug mode. Quick et al. [17] discussed the challenges related to volume, variety, velocity, and veracity of the data. An investigation strategy to reduce the volume of data as data association increases was followed in this work. Further, a semi-automated scanning process to extract data from variety of devices like hard drive, mobile device, cloud, IOT, and other portable storage was proposed.

Quick et al. [18] discussed the investigator challenge of collecting relevant data from wide range of mobile phone using various forensics tools. The collection relevant to data stored in a single mobile device like contact details, sms, mms, video, audio, image, file, Internet history, location details, temperature, subscription detail, cloud storage, and even health details was discussed. In addition, the collection relevant to data stored in multiple devices like terrorist activity was also discussed. Case et al. [19] made a survey on the art of acquisition and analysis in the area of mobile forensic.

Casey et al. [20] focused on the challenges in the standardized form of representation and exchange of the information between tools, organizations, investigators, and jurisdiction. Also, cyber-investigation analysis standard expression, a specification language that supports combination and validation of information at less time during exchange was developed. Cheng et al. [21] developed a lightweight live memory forensic framework based on hardware virtualization. This framework enables the operating system migration to virtual machine without termination for data acquisition and analysis.

Thus, in this work, the view of forensics practitioner's competencies in analyzing an investigation is measured by accounting the risk as the parameter.

3 Investigation Approach

Practitioners are equipped to offer services to their corporate clients to improve business values and strengthen confidence. Apart from corporate clients, they offer services to sectors such as software organization, attorneys, law enforcement, individuals, military, automobile industry, and finance

Computer forensic practitioners can investigate the incidents by following the procedures, namely evidence discovery, recovery, preservation, verification, analysis, reconstruction, and reporting. Evidence discovery primarily focuses on collection of active data and metadata from digital media. In certain cases where data cannot be collected through evidence discovery, evidence recovery need to be applied. Appropriate forensic tools should be used to recover the damaged or inaccessible data. After collection, the evidence data are preserved without alteration for further investigation. Further, the integrity of collected evidence data is validated for submission in court. A detailed investigation is done on the evidence data to facilitate, detect, and prove the crime incident. The final investigation document is submitted to court as report.

Mobile [10, 13–15, 22] practitioners extract evidence from cellular phones, smartphones, tablets, PDAs, GPS units, etc., irrespective of carrier technologies, wireless service providers, and operating system. A routine procedure is followed in normal data recovery and extraction of unlocked mobile from SD card, micro-SD, and memory chip. But, a specialized training is essential for recovery and extraction of deleted data from locked, corrupted, encrypted, and broken mobile phones. After collection, data is integrated, preserved, analyzed, and submitted to court.

IoT [23] has a wide range of applications like transportation, health care, embedded systems, smart home, and smart cities, and it is quite challengeable to handle forensics incidents related to IoT due to vast data format. Accessing the evidence data from cloud storage services [24] can be done through mobile phones or Web, etc. But the procedure to collect the data is really challenging than traditional forensics. In this incident, the same procedure like of preservation, analyzing, reporting is followed as traditional forensics, but the method to collect the evidence data differs.

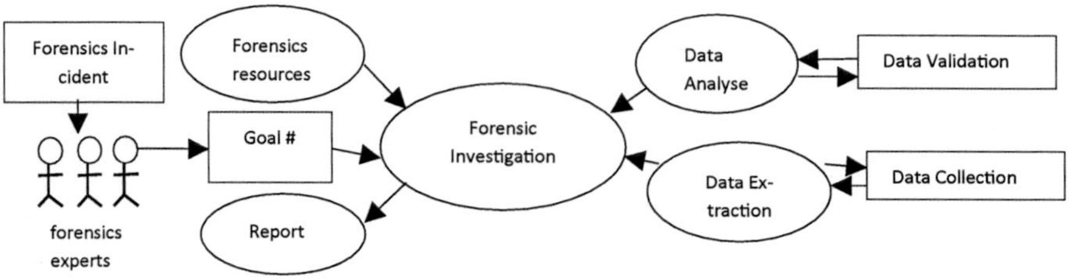

Fig. 1 A process flow of forensics practitioner's approach toward an investigation

Figure 1 represents the process flow of forensics practitioner's approach toward an investigation, like (1) sexual harassment, (2) illegal data transfer, (3) theft of business secrets, (4) matrimonial, (5) financial tampering, (6) corporate crime, etc. The above procedure will be applied by practitioners to investigate the incidents and submit the evidence to court. Thus, the practitioners should be trained to handle the various incidents irrespective of the challenges in it.

Digital forensics tool comes in three categories, namely freeware, open source, and commercial for the different types of user needs that includes extraction, recovery, analyzes. The most popular digital forensics tools that are commonly used by forensics community are sleuth kit [9], AccessData's FTK [8], EnCase [3], HexEdit [7], and Autospy [9]. Some popular email forensics tools that are vastly used by experts are eMailTrackerPro [25], Aid4Mail forensics [26], AbusePipe [27], FINALeMAIL [28], PlainSight [29], and so on. Memory forensics tools Registry Recon [30], Volatility [31], Bulk Extractor [32], and ProDiscover [33] are majorly used for analysis and extraction. Some popularly used forensics tools for extraction and analysis using mobile devices are XRY [4], Cellebrite [5], Oxygen [2], MOBILedit [34], etc.

4 Practitioner's Perspective

In education sector, a lot of scenarios arise due to the student community interaction in the Internet, namely two such scenarios are taken up here for discussion. First, online interview, and second is online money lenders. The various cases of evidence collection are discussed below.

Connection 1. In college, the various PC/laptop/mobile systems are connected via proxy server using wired or wireless LAN during the two transactions (Fig. 2).

In this case, the forensics practitioners can use the server squid access log data and find the culprit in case of fake online interview or money lenders. Mostly, the student would have used any one of the IP addresses of wired or wireless LAN connection for transactions and visited unauthenticated http Web site with unsecured port such as 8080, etc. As most authenticated Web site starts with https using secured port such as 443, etc.

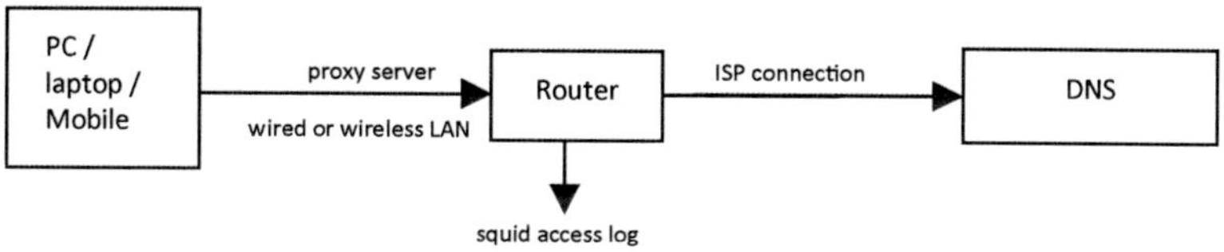

Fig. 2 Wired or wireless LAN connection

In this investigation, the practitioner can collect student IP address through which he established the connection for online transaction. The online transaction site can be connected to private or public Internet. In private Internet, location identification is really challengeable than public. But in both cases, after tracking location, the evidence data can be collected by seizing the device with proper legal procedure. If the victim device used PC/laptop, then the FTK tool [8], otherwise in case of mobile device, MOBILedit tool [34] can be used to collect and analyze the evidence data.

Connection 2. In college, the various laptop/mobile systems are connected via mobile service provider during the two transactions (Fig. 3).

In this case, the forensics practitioner needs to collect the log data to find the culprit in case of fake online interview or money lenders. Getting the log data from service provider is a difficult task. Otherwise, the student visited Web site data should be used to retrieve the suspect location. Then, by following proper legal procedure, the device is seized. Based on the device, an appropriate forensic tool can be used to collect the evidence data. The greatest challenge is the seizure of the device from the suspect. Once seized, the relevant data can be collected.

Thereby, the practitioners have many challenges from one investigation to another. For this purpose, multi-agent can be used to track the various forensics investigations. A model discussed in this work is represented in Fig. 4. The state agent and incident agent are involved to collect the information related to the various investigations in the forensics laboratory for various forensics practitioners. A stage agent can be deployed to collect information related to process risk concerning the process of an investigation, while an incident agent can be deployed to collect the information related to project risk concerning the various investigations that is happening in the forensics lab.

The process risk plays a predominant role to accomplish the completion of a forensics incident case. The mitigation of process risk may reduce the running time for completion. The four stages in which process risk can be identified include forensics

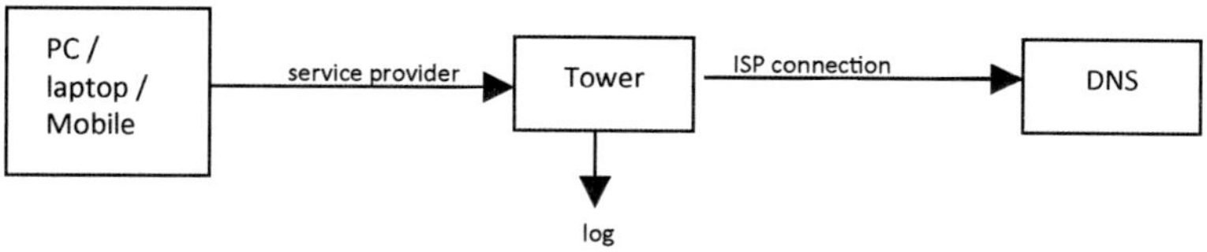

Fig. 3 Mobile service provider connection

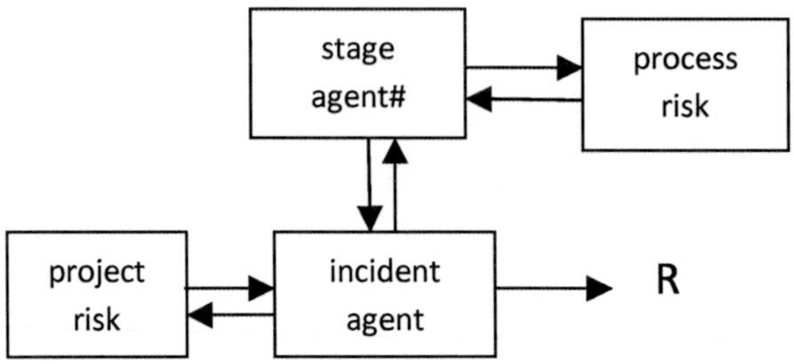

Fig. 4 Proposed model

Table 1 Risk measurement in forensics laboratory on various periods for an investigation

Risk name	Agent	Category	*R*		
Process	Stage	Forensics resources	0.19	0.04	0.50
		Data extraction	0.45	0.58	0.20
		Data analysis	0.30	0.30	0.05
		Report	0.30	0.11	0.58
Project	Incident	–	0.29	0.20	0.36

resources, data extraction, data analysis, and report. The identified risk includes access log, http Web site, unsecure port, private Internet, public Internet, tracking location, seizing the device, forensic software tool, forensic hardware tool, collection, preservation, recovery, integration risk, etc. A stage agent can be deployed to collect the risk information related to process risk for the success of the process completion.

The incident risk plays a predominant role to accomplish the completion of all the incident cases. This measures the total running time of the various project risks that is done in forensics laboratory. By following the risk measurement as specified in Arumugam et al. [35], the risk is calculated. A sample data related to a forensics laboratory on a particular day is tabulated in Table 1. The total risk, R, and risk probability are represented in Eqs. 1 and 2, respectively.

$$\text{Risk } (R) = \text{risk probability} \times \text{risk impact} \tag{1}$$

$$\text{Risk probability} = \text{No. of risks pending/Total no. of risks} \tag{2}$$

Let us consider the risk impact value as 4, 2, and 1 for high, low, very low, respectively. Table 1 represents the total risk on various periods for an investigation. For a period 1, consider no. of risk = 6, pending = 2, mitigation = 4, the probability is calculated as 2/6 = 0.33. Then, the impact is calculated as follows. Let two "Very low = 1" and two of "low = 2" risk has already been mitigated, and the remaining two pending risks have "High = 4" risk level. Impact is calculated as (4 × 1 + 4 ×

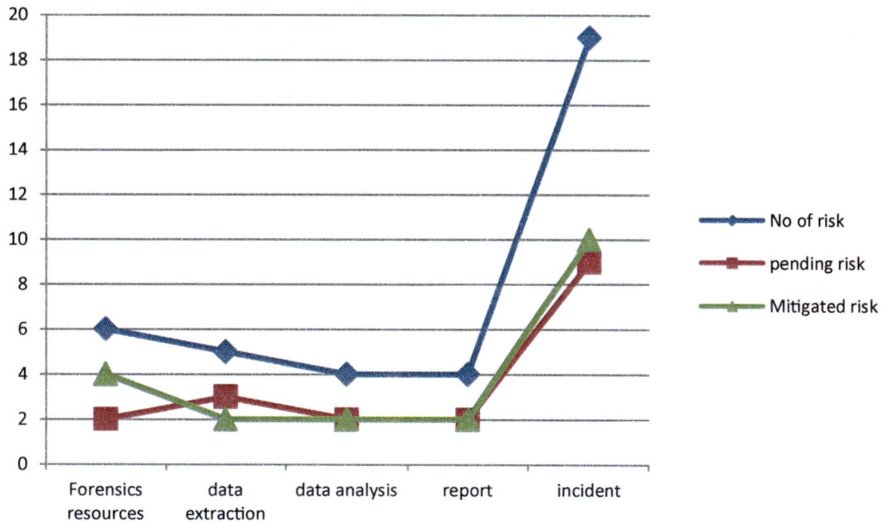

Fig. 5 Risk chart

1)/(4 × 1 + 4 × 1 + 1 × 2 + 1 × 2 + 1 × 1 + 1 × 1) = 8/14 = 0.57. Thus, the R = 0.33 × 0.47 = 0.19.

Table 1 summarizes the risk of the three incidents in forensics laboratory for project and process risk considering the risk mitigated, pending, and total risks. The first and second incident risks are less when compared to third one. Thus, the practitioners working on the third one have to be competent, to mitigate the risks and complete it. Figure 5 shows the risk chart for an incident that is pending, mitigated, and total number of risk. From this, the risks at a point of time for an incident can be tracked. The ideal case would be total number of risk and mitigated risk should be equal. In this sample graph, the mitigated are below the total number of risk and it concludes that a special attention is required by the practitioner to close an incident.

The various contributions in this work are as follows:

1. To view the forensics laboratory progress in various investigations.
2. To list the risk involved in solving the incident.
3. The practitioner's competencies can be made visible.
4. The various upcoming process risks can be identified and mitigated.
5. The various upcoming project risks can be identified from the process risk.

5 Conclusion

This paper takes a view at the process and project risk of a forensics laboratory that provides a solution to clients for the various investigations. An investigation is viewed as an incident and the risk pertaining to solve an incident is identified as project risk. Each project risk has four process risks that include forensics resources, data extraction, data analysis, and report. The number of risk is identified for the four process risks and based on the mitigation and pending risk the practitioner's competencies can be exposed. Multi-agent takes a control to identify the various risk status in an incident. From this, a practitioner's competency to complete an incident can be analyzed. In this paper, for a sample incident, how the practitioner competency solve the incident is measured by a project and process risk. The less gap between the mitigated risk and total number of risk indicate that an incident is getting ready to completion. This work can be implemented by taking the actual forensics laboratory data and study the various risks that arise in an investigation to measure the forensics practitioner's competencies.

References

1. www.forensicsware.com
2. https://www.oxygen-forensic.com/en/
3. encase - https://www.guidancesoftware.com/
4. XRY - https://www.msab.com/
5. https://www.cellebrite.com/en/home/
6. wireshark - https://resources.infosecinstitute.com/wireshark-open-source-forensic-tool/#gref
7. Hexeditor: https://www.hhdsoftware.com/free-hex-editor
8. FTK: https://accessdata.com/product-download
9. Sleuthkit: https://www.sleuthkit.org/sleuthkit/download.php
10. H. Zhang, L. Chen, Q. Liu, *Digital Forensic Analysis of Instant Messaging Applications on Android Smartphones*, in International Conference on Computing, Networking and Communications (2018), pp. 647–651
11. C. Anglano, M. Canonico, M. Guazzone, Forensic analysis of telegram messenger on android smartphones. Digital Invest. **23**, 31–49 (2017)
12. L. Zhang, F. Yu, Q. Ji, *The Forensic Analysis of WeChat Message.* In 2016 Sixth International Conference on Instrumentation & Measurement, Computer, Communication and Control (IMCCC) (2016), pp. 500–503
13. A.H. Lone, F.A. Badroo, K.R. Chudhary, A. Khalique, Implementation of forensic analysis procedures for WhatsApp and Viber android applications. Int. J. Comput. Appl. **128**(12), 26–33 (2015)
14. C. Anglano, M. Canonico, M. Guazzone, Forensic analysis of the chat secure instant messaging application on android smartphones. Digital Invest. **19**, 44–59 (2016)
15. M. Chernyshev, S. Zeadally, Z. Baig, A. Woodward, Mobile forensics: advances, challenges, and research opportunities. IEEE Secur. Priv. **15**(6), 42–51 (2017)
16. P. Feng, Q. Li, P. Zhang, Z. Chen, Logical acquisition method based on data migration for android mobile devices. Digital Invest. **26**, 55–62 (2018)
17. D. Quick, K.K.R. Choo, Digital forensic intelligence: data subsets and open source intelligence (DFINT + OSINT): a timely and cohesive mix. Fut. Gener. Comput. Syst. **78**, 558–567 (2018)
18. D. Quick, K.K.R. Choo, Pervasive social networking forensics: intelligence and evidence from mobile device extracts. J. Network Comput. Appl. **86**, 24–33 (2017)
19. A. Case, G.G. Richard III, Memory forensics: the path forward. Digital Invest. **20**, 23–33 (2017)
20. E. Casey, S. Barnum, R. Griffith, J. Snyder, H. van Beek, A. Nelson, Advancing coordinated cyber-investigations and tool interoperability using a community developed specification language. Digital Invest. **22**, 14–45 (2017)

21. Y. Cheng, X. Fu, X. Du, B. Luo, M. Guizani, A lightweight live memory forensic approach based on hardware virtualization. Inf. Sci. **379**, 23–41 (2017)
22. C. Rajchada, V. Wantanee, R.C. Kim-Kwang, Forensic analysis and security assessment of Android m-banking apps. Austr. J. Forens. Sci. **50**(1), 3–19 (2018)
23. Y. Ibrar, I.A.T. Hashem, A. Ahmed, S.M. Ahsan Kazmi, C.S. Hong, Internet of things forensics: Recent advances, taxonomy, requirements, and open challenges. Fut. Gener. Comput. Syst. **92**, 265–275 (2019)
24. C.-T. Huang, H.-J. Ko, Z.-W. Zhuang, P.-C. Shih, S.-J. Wang, *Mobile Forensics for Cloud Storage Service on iOS Systems*, In ISITA2018, Singapore, 28–31 Oct 2018
25. http://www.emailtrackerpro.com/
26. https://www.aid4mail.com/ediscovery-forensics-trial
27. https://www.datamystic.com/abusepipe
28. http://finalemail.findmysoft.com/
29. http://www.plainsight.info/
30. http://arsenalrecon.com/apps/recon/
31. http://code.google.com/p/volatility/
32. http://digitalcorpora.org/downloads/bulk_extractor/
33. http://prodiscover-basic.freedownloadscenter.com/windows/
34. https://www.mobiledit.com/forensic-solutions
35. C. Arumugam, S. Kameswaran, B. Kaliamourthy, Risk assessment framework: ADRIM process model for global software development, in *Towards Extensible and Adaptable Methods in Computing*, ed. by S. Chakraverty, A. Goel, S. Misra (Springer, Singapore, 2018)

Hybrid Pixel-Based Method for Multimodal Medical Image Fusion Based on Integration of Pulse-Coupled Neural Network (PCNN) and Genetic Algorithm (GA)

R. Indhumathi, S. Nagarajan, and K. P. Indira

1 Introduction

Medical imaging plays a vital role in medical diagnosis and treatment. However, distinct imaging modality yields information only in limited domain. Studies are done for analysis information collected from distinct modalities of same patient. This led to the introduction of image fusion in the field of medicine and the progression of image fusion techniques. Image fusion is characterized as the amalgamation of significant data from numerous images and their incorporation into seldom images, generally a solitary one. This fused image will be more instructive and precise than the individual source images that have been utilized, and the resultant fused image comprises paramount information. The main objective of image fusion is to incorporate all the essential data from source images which would be pertinent and comprehensible for human and machine recognition. Image fusion is the strategy of combining images from distinct modalities into a single image [1]. The resultant image is utilized in variety of applications such as medical diagnosis, identification of tumor and surgery treatment [2]. Before fusing images from two distinct modalities, it is essential to preserve the features so that the fused image is free from inconsistencies or artifacts in the output.

Medical images can be obtained from distinct modalities such as computed tomography (CT), magnetic resonance imaging (MRI), positron emission tomography (PET), single-photon emission tomography (SPECT) and X-ray. For instant, X-ray

R. Indhumathi (✉) · S. Nagarajan
EEE Department, Jerusalem College of Engineering, Chennai 600100, India
e-mail: indhuraja.phd@gmail.com

S. Nagarajan
e-mail: nagu_shola@yahoo.com

K. P. Indira
ECE Department, BVC Engineering College, Allavaram, India
e-mail: kpindiraphd@gmail.com

S. Patnaik et al. (eds.), *Advances in Machine Learning and Computational Intelligence*, Algorithms for Intelligent Systems, https://doi.org/10.1007/978-981-15-5243-4_82

and computed tomography (CT) are used to provide information about dense structures like bones, whereas magnetic resonance imaging (MRI) is used to provide information about soft tissues, while positron emission tomography (PET) provides information about metabolic activity taking place within the body. Hence, it necessitates to integrate information from distinct images into a single image for a comprehensive view. Multimodal medical image fusion helps to diagnose the disease and also reduces the cost of storage by amalgamating numerous source images into a single image.

Image fusion can be accomplished at three levels—pixel level, feature level and decision level [3]. Pixel-level image fusion is usually performed by combining pixel values from individual source images. Feature-level image fusion performs fusion only after segmenting the source image into numerous features such as pixel intensity and edges. Decision-level image fusion is a high-level fusion strategy which utilizes fuzzy rule, heuristic algorithms, etc. In this paper, image fusion has been performed on pixel level owing to their advantages such as easy implementation and improved efficiency.

Image fusion techniques are partitioned into two major categories—spatial-domain and frequency-domain techniques [4, 5]. Spatial-domain techniques are further categorized into average, maximum fusion algorithms and principle component analysis. Frequency-domain techniques are further categorized into pyramid-based decomposition and wavelet transforms.

Averaging method is a simple image fusion strategy where the output is determined by calculating the average value of each pixel [6]. Though easier to understand, averaging method yields output with low contrast and washout appearance. Choosing maximum fusion rule selects the maximum pixel value from the source images as the output. This in turn yields a highly focused output [7]. PCA is a vector space transformation methodology which reduces multi-dimensional datasets to lower dimensions which is a powerful tool to analyze graphical data since elucidation is tedious for data of higher dimensions. Frequency-domain strategies are further categorized into pyramidal method and wavelet-based transforms. Pyramidal method consists of bands of source images. Each level will be usually smaller compared to its predecessor. This results in higher levels concentrated on higher spatial frequencies. Pyramidal method is further classified into [8] Gaussian pyramid method [9], Laplacian pyramid method, ratio of low-pass pyramid method and morphological pyramid method. Gaussian pyramid method is a low pass filtered version of its predecessor. LP is a bandpass filtered version of its predecessor. In ratio of low-pass pyramid, the image at each level is the ratio between two successive levels in Gaussian pyramid method. Morphological pyramid strategy used morphological filters to extract the details of an image without causing any adverse effects.

Wavelet image fusion techniques are further classified into [10] discrete wavelet transform (DWT) [11], Stationary wavelet transform (SWT) [12], non-subsampled contourlet transform (NSCT). DWT technique decomposes an image into low- and high-frequency components. Low-frequency components provide approximation information, while high-frequency components provide detailed information contained within an image. DWT strategy suffers from a major drawback called shift variance. In order to provide a shift-invariant output, Stationary wavelet transform (SWT) was introduced. This technique accomplishes upsampling strategy which results in the decomposed image which has the same size as the input image. Though these wavelet transforms perform well, they failed to perform well along edges. To provide information along edges, non-subsampled contourlet transform (NSCT) was proposed which is a geometric analysis procedure which increases localization, shift invariance, etc. In spite of providing various advantages, NSCT technique requires proper filter tuning and proper reconstruction filters for any particular application [13]. To overcome the above disadvantages, neural technique strategies were introduced. Pulse-coupled neural network (PCNN) is a neural network technique evolved by utilizing the synchronous pulse emergence from cerebral visual cortex of some mammals.

Pulse-coupled neural network (PCNN) is a most commonly used image fusion strategy for medical diagnosis and treatment [14]. Initially, manual adjustment was done to tune PCNN variables. Lu et al. [14] have utilized a distinct strategy where pulse-coupled neural network (PCNN) was optimized by utilizing multi-swarm fruit fly optimization algorithm (MFOA). Quality assessment was utilized as hybrid fitness function which enhances the performance of MFOA. Experimental results illustrate the effectiveness of the proposed strategy.

Gai et al. [15] have put forward a novel image fusion technique which utilized pulse-coupled neural network (PCNN) and non-subsampled shearlet transform (NSST). Initially, the images were decomposed by utilizing non-subsampled shearlet transform (NSST) strategy. This decomposes the image into low- and high-frequency components. Low-frequency components have been fused by utilizing improved sparse representation method. This technique eliminates the detailed information by using Sobel operator, while information preservation has been done by using guided filter. High-frequency components have been fused by utilizing pulse-coupled neural network (PCNN) strategy. Finally, inverse transform is done to yield the fused output. The effectiveness of the proposed strategy has been validated against seven different fusion methods. The author has also fused information from three distinct modalities to justify the superiority of the proposed method. Subjective and objective analyses illustrate the effectiveness of the proposed method.

A multimodal fusion approach which utilizes non-subsampled shearlet transform (NSST) and simplified pulse-coupled neural network model (S-PCNN) was put forward by Hajer et al. [16]. Initially, the images were transformed into YIQ components. The images were initially disintegrated into low- and high-frequency components using NSST strategy. The low-frequency components were fused using weight region standard deviation (SD) and local energy, and high-frequency components are fused by utilizing S-PCNN strategy and finally, inverse NSST and inverse YIQ technique. The final discussion illustrates that the proposed strategy outperforms quantitatively in terms of performance measures such as mutual information, entropy, SD, fusion quality and spatial frequency.

Jia et al. [17] have put forward a novel framework which utilized improved adaptive PCNN. PCNN is a technique that emerged from the visual cortex of mammals and has proved to be very suitable in the field of image fusion. The source images were initially fed to the parallel PCNN, and the gray value of the image was utilized to trigger PCNN. Meanwhile, sum-modified Laplacian was chosen as the evaluation function, and the linking strength of neuron which corresponds to PCNN was evaluated. The ignition map was generated after ignition of PCNN. The clearer part of the images was chosen to yield the fused image. Quantitative and qualitative analyses illustrated that the proposed strategy outperformed than the existing strategies.

In this paper, Wang et al. [18] have put forward an image fusion technique which utilized discrete wavelet transform (DWT) and dual-channel pulse-coupled neural network (PCNN). For fusing low-frequency coefficients, choosing maximum fusion rule has been utilized, while spatial frequency of high-frequency components has been chosen to motivate dual-channel PCNN. Finally, inverse DWT has been utilized to yield the fused image. Visual and quantitative analyses illustrated the superiority of the proposed approach than other image fusion strategies.

Arif et al. [19] proposed an existing image fusion strategies that lacked the capability to produce a fused image which could preserve the complete information content from individual source images which utilized combination of curvelet transform and genetic algorithm (GA) to yield a fused image. Curvelet transform helped in preserving the information along the edges, and genetic algorithm helped to acquire the fine details from the source images. Quantitative analysis demonstrated that the proposed strategy outperformed than the existing baseline strategies.

Fu et al. [20] have put forward a novel image fusion approach which utilized non-subsampled contourlet transform (NSCT) and pulse-coupled neural network (PCNN) jointly in image fusion algorithms. High- and low-frequency coefficients have been processed using modified PCNN. Determining the degree of matching between input images is utilized in fusion rules. Finally, inverse NSCT has been employed to reconstruct the fused image. Experimental analysis illustrated that the proposed strategy outperformed wavelet, contourlet and traditional PCNN methods in terms of higher mutual information content. Also, the proposed strategy preserved edge as well as texture information, thereby including more information content in the fused image. The author concluded by stating that research about selection of parameters for image fusion should be performed deeply.

Image fusion is the strategy in which the input from multiple images is combined to yield an efficient fused image. Lacewell et al. [21] have put forward a strategy which utilized combination of discrete wavelet transform and genetic algorithm to produce an efficient fused image. DWT has been utilized to extract features, while genetic algorithm (GA) has been utilized to yield an enhanced output. Quantitative and comparison analyses illustrated that the proposed strategy produced superior results in terms of mutual information and root mean square error.

Wang et al. [22] have put forward a novel image fusion approach which utilizes pulse-coupled neural network and wavelet-based contourlet transform. In order to motivate PCNN, spatial high frequency in WBCT has been utilized. High-frequency coefficients can be selected by utilizing weighted method of firing times. Wavelet transform strategies perform better at isolated discontinuities but not along curved edges especially for 3D images. In order to overcome the above drawbacks, PCNN has been utilized which performs better for higher-dimensional images. Experimental analysis illustrated that WBCT-PCNN performed better from both subjective and objective analyses.

From the literature survey, it is inferred that combination of two distinct image fusion techniques provides better results in terms of both quality and quantity. Though the solution may be obtained by utilizing any image fusion strategy, the optimal solution can be obtained only by utilizing genetic algorithm (GA). Hence, an attempt has been made to integrate the advantages of both PCNN and GA to yield an output image from both quality and quantitative analyses.

The rest of the paper is organized as follows: Sect. 2 provides a detailed explanation about pulse-coupled neural network (PCNN), Sect. 3 illustrates about the proposed methodology, the proposed algorithm is provided in Sect. 4, qualitative and quantitative analyses have been provided in Sect. 5 and conclusion has been provided in Sect. 6.

2 Pulse-Coupled Neural Network (PCNN)

A new type of neural network, distinct from traditional neural network strategies, is pulse-coupled neural network (PCNN) [23]. PCNN is developed by utilizing the synchronous pulse emergence from cerebral visual cortex of some mammals. A gathering of neurons is usually associated to frame PCNN. Each neuron correlates to a pixel value whose intensity is contemplated as external stimulant. Every neuron interfaces with another neuron in such a manner that a single-layer two-dimensional cluster of PCNN is constituted. When linking coefficient beta is zero, every neuron pulses naturally due to external stimulant. When beta is nonzero, the neurons are associated mutually. At a point, when the neuron fires, its yield subscribes the other adjacent neurons leading them to pulse before the natural period. The yield of captured neurons influences the other neurons associated with them to change the internal activity and the outputs. When iteration terminates, the output of every stage is added to get an aggregate yield which is known as the firing map.

There are two primary issues in the existing image fusion strategies [24]. Firstly, pyramid and wavelet transform strategies treat each pixel in an individual manner rather than considering the relationship between them. Further, the images should be entirely registered before fusion.

In order to overcome the above drawbacks, PCNN has been proposed [25]. Fusion strategy usually takes place in two major ways—either by choosing the better pixel value or by outlining a major–minor network for different networks, thereby choosing the yield of the first PCNN as the fusion result.

The pulse-coupled neural network comprises three compartments: receptive field, linking part or modulation and pulse generator.

Receptive field is an essential part which receives input signals from neighboring neurons and external sources. It consists of two internal channels—feeding compartment (F) and linking compartment (L). Compared to feeding compartment, the linking inputs have quicker characteristic response time constant. In order to generate the total internal activity (U), the biased and multiplied linking inputs are multiplied with the feeding inputs. The net result constitutes the linking/modulation part. At last, pulse generator comprises step function generator and a threshold signal generator.

The ability of neurons in the network to respond to external stimulant is known as firing which enables the internal activity of neuron to exceed a certain threshold value. Initially, the yield of neuron is set to 1. The threshold value starts rotting till the next internal activity of the neuron. The output generated is then iteratively nourished back with a delay of single iteration. As soon as the threshold exceeds the internal activity (U), the output will be reset to zero. A temporal series of pulse outputs are generated by PCNN after n number of iterations which carries the data about the input images. Input stimulus which corresponds to pixel's color intensity is given to feeding compartment. The pulse output of PCNN helps to make a decision on the content of the image.

Initially, double-precision operation is performed on the acquired input CT and PET images. In order to reduce the memory requirements of an image, unsigned integers (unit 8 or unit 16) can be utilized. An image whose information lattice has unit 8 and unit 16 class is known as 8-bit image and 16-bit image, respectively.

Though the measure of colors emitted cannot be differentiated in a grayscale image, the aggregate sum of radiated light for each pixel can be partitioned since a small measure of light is considered as dark pixels, while more amount of light is considered as bright pixels. On conversion from RGB to grayscale image, the RGB value of each pixel is ought to be taken and made as the yield of a solitary value which reflects the brightness of the pixel.

Normalization (norm) converts the scale of pixel intensity values and is known as contrast stretching or histogram stretching. Normalization can be determined by using the following formula

$$\text{I norm} = (\text{I abs} - \text{I min})/(\text{I max} - \text{I min})$$

where

abs represents absolute value
min represents minimum value
max represents maximum value

Each neuron in firing pulse model comprises receptive field, modulation field and pulse generator.

Two important features are necessary to fire a pulse generator—spiking cortex model (SCM) and synaptic weight matrix. SCM has been demonstrated in compliance with Weber–Fechner law, since it has higher sensitivity for low-intensity stimulant and lower sensitivity for high-intensity stimulant. In order to improve the performance and make the output reachable, synaptic weight matrix is applied to linking field and sigmoid function is applied in firing pulse. PCNN comprises neuron capture property which causes any neuron's firing to make the nearby neurons whose luminance is similar to be fired. This property makes the information couple and transmission to be automatically acknowledged which makes PCNN satisfactory for image fusion.

In this model, one of the original images is chosen as input to the main PCNN network randomly and another image as input to the subsidiary network. The firing information about the subsidiary network is transferred to the main PCNN network with the help of information couple and transmission properties. By doing so, image fusion can be figured out. When a neuron is fired, the firing information about the subsidiary network is communicated to the adjacent neurons and neurons of the main PCNN network. The capturing property of PCNN makes it suitable for image fusion. Eventually, the output obtained is transformed to unit 8 format, and finally, the fused image is obtained.

3 Genetic Algorithm

Genetic algorithm (GA) is a heuristic search algorithm used to solve the problems of optimization [26]. The essence of GA is to simulate the evolution of nature, i.e., a process in which a species experiences the selective evolution and genetic inheritance. At first, a random group is formed, and then, with mutual competition and genetic development, the group goes through the following operation process: selection, crossover, mutation, etc. The subgroup who has better fitness will survive and form a new generation. The process cycles continuously until the fittest subgroups are formed. The surviving groups are supposed to well adapt to the living conditions. Thus, the genetic algorithm is actually a type of random search algorithm. Moreover, it is nonlinear and parallelizable, so it has great advantages when compared with the traditional optimization algorithm.

Four entities that help to define a GA problem are the representation of the candidate solutions, the fitness function, the genetic operators to assist in finding the optimal or near optimal solution and specific knowledge of the problem such as variables [27]. GA utilizes the simplest representation, reproduction and diversity strategy. Optimization with GA is performed through natural exchange of genetic material between parents. Offspring are formed from parent genes. Also, fitness of offspring is evaluated. The best-fitting individuals are only allowed to breed.

Image fusion based on GA consists of three types of genetic operations—crossover, mutation and replication [28]. The procedure is as follows:

1. Encode the unknown image weight and define the objective function $f(xi)$. *SML* is used as the fitness function using GA.
2. N stands for the initial population size of fusion weight. Pm represents the probability of mutation and Pc the probability of crossover.
3. Generate randomly a feature array whose length is L to form an initial fusion weight group.

4. Follow the steps below and conduct iterative operation until termination condition is achieved.

   (a) Calculate the adaptability of an individual in the group.
   (b) On the basis of adaptability, *Pc* and *Pm*, operate crossover, mutation and replication.

5. The best-fitting individuals in the surviving subgroup are elected as the result. So, the optimal fusion weight is obtained.

From the steps above, the optimal fusion weights of the images to be fused can be obtained after several iterations. However, since this method does not take into account the relationship between the images to be fused, a lot of time is wasted to search for the fusion weights in the clear focal regions, which leads to low accuracy of the fusion.

4 Optimization of Image Fusion Using PCNN aND GA

A new image fusion algorithm using pulse-coupled neural network (PCNN) with genetic algorithm (GA) optimization has been proposed which uses the firing frequency of neurons to process the image fusion in PCNN. Aiming at the result of image fusion being affected by neuron parameters, this algorithm is dependent on the parameters image gradient and independent of other parameters. The PCNN is built in each high-frequency sub-band to simulate the biological activity of human visual system. On comparing with traditional algorithms where the linking strength of each neuron is set constant or continuously changed according to features of each pixel, here, the linking strength as well as the linking range is determined by the prominence of corresponding low-frequency coefficients, which not only reduces the calculation of parameters but also flexibly makes good use of global features of images.

The registration has been conducted to the images to be fused, and the image focusing is different; the fusion coefficients (Q) of the two pixels in the same position have a certain hidden relationship [29]. Besides, the constraint condition of the fusion coefficient is $Q1 + Q2 = 1$, $Q1 > 0$, $Q2 > 0$, so modeling for either coefficient is enough. Therefore, in the process of wavelet transformation, the PCNN for fusion weight coefficient (Q) of each pixel of each layer in a hierarchical order from high to low has been built. Q stands for fusion coefficient and S stands for hidden state.

In this model, S is defined as three states, i.e., $S \in \{1, 2, 3\}$. When the pixel is in the focused region of image 1, or when located much nearer the focal plane of image 1 than that of image 2, S is set as 1. When the pixel is in the focused region of image 2, or if it is located much nearer the focal plane of image 2 than that of image 1, S is set as 3. If the pixel is not in the focused region of image 1 or image 2 and there is no obvious difference between the distances from focal plane of image 1 and that of image 2, S is set as 2.

In addition, if the fusion coefficient in the neighboring region is greater than 0.9 or less than 0.1, then S is set to be 1 or 3. The state transfer matrixes from the parent node Si to the sub-node $Si + 1$ are defined as follows.

According to the fact that details of the low frequency in the clear area are more than that in unclear area, this tries to get fusion weights from high scale to low scale by using GA after the process of wavelet transformation. Meanwhile, the author constructs a PCNN for each fusion weight in each layer and figures out its hidden states layer by layer with the help of the fusion weights calculated by GA. PCNN is a single-layered, two-dimensional, laterally connected neural network of pulse coupled neurons where the inputs to the neuron are given by feeding and linking inputs. Feeding input is the primary input from the neurons receptive area. The neuron receptive area consists of the neighboring pixels of corresponding pixel in the input image. Linking input is the secondary input of lateral connections with neighboring neurons. The difference between these inputs is that the feeding connections have a slower characteristic response time constant than the linking connections. Guided by the hidden states of the fusion weights in the upper layer, the author gets the values directly from the clear area of the next layer without GA. In this way, the population size in the process of GA is reduced, which contributes a lot to improving the precision of calculation in the same amount of time.

5 Algorithm

Step 1: Apply PCNN to N layers of the image and then figure out the fusion coefficient of Layer N by using GA. Set variate $i = 0$.

Step 2: Search the neighboring region of Layer $N - i$ for the pixels, whose fusion coefficients are greater than 0.9 or less than 0.1. Set $Si = 1$ or 3.

Step 3: Search Layer $N - (i + 1)$ for the pixels whose parent node are $Si = 1$ or 3, and then, the Qs of these pixels are set to be 1 or 0, and accordingly, $Si + 1$ are set to be 1 or 3.

Step 4: Search Layer $N - (i + 1)$ and find out the pixels whose parent node are $Si = 2$ and then apply GA to work out the fusion coefficients of these pixels. Set their $Si + 1$ to be 2, and set variate $i = i + 1$. After that, go back to Step 2 and circulate the operation.

Step 5: Circulate Step 2, Step 3 and Step 4. Work out the fusion coefficients until the last layer.

6 Results and Discussion

6.1 *Quantitative Analysis*

Percentage Residual Difference (PRD): Percentage residual difference reflects the degree of deviation between source image and the fused image [24]. Lower the value of PRD, higher is the quality of the image. On comparing PRD values of PCNN and GA with PCNN, it is inferred that PCNN and GA offer better results for all 16 datasets (Table 1).

Table 1 Objective analysis of PCNN and hybridization of PCNN and GA

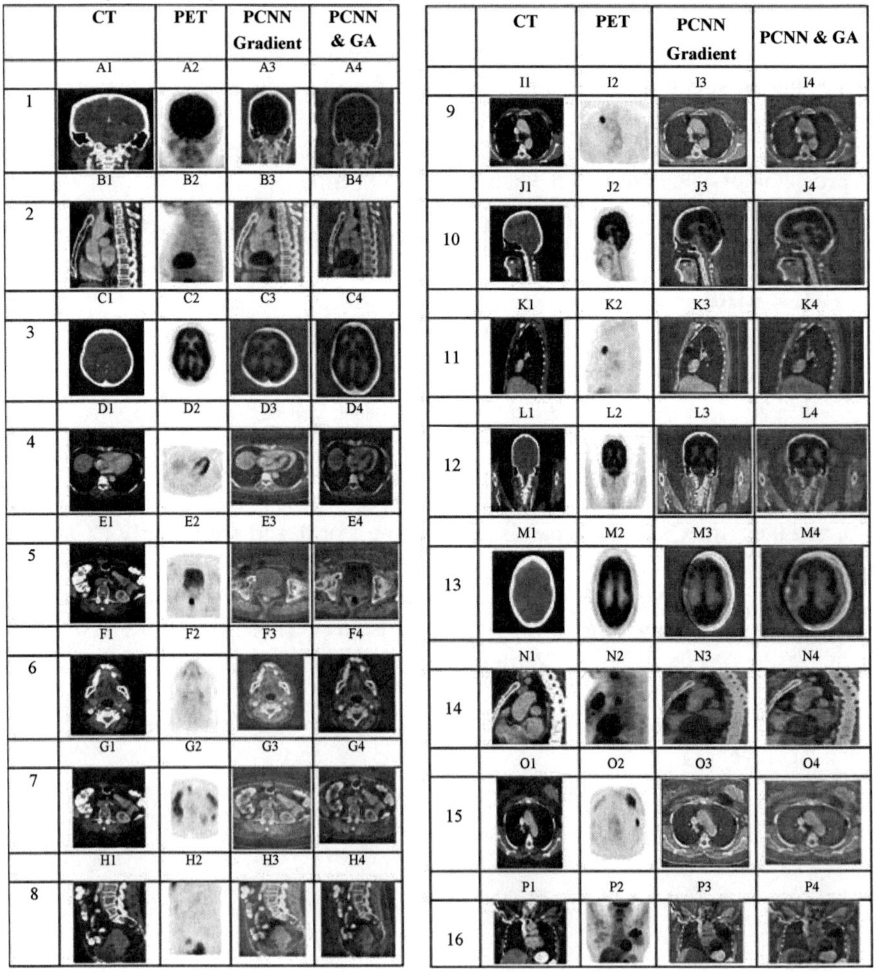

	CT	PET	PCNN Gradient	PCNN & GA		CT	PET	PCNN Gradient	PCNN & GA
1	A1	A2	A3	A4	9	I1	I2	I3	I4
2	B1	B2	B3	B4	10	J1	J2	J3	J4
3	C1	C2	C3	C4	11	K1	K2	K3	K4
4	D1	D2	D3	D4	12	L1	L2	L3	L4
5	E1	E2	E3	E4	13	M1	M2	M3	M4
6	F1	F2	F3	F4	14	N1	N2	N3	N4
7	G1	G2	G3	G4	15	O1	O2	O3	O4
8	H1	H2	H3	H4	16	P1	P2	P3	P4

Root Mean Square Error (RMSE): RMSE is a measure of difference between predicted value and the actual value [25]. Lower the value of RMSE, higher is the quality of the image. On comparing RMSE values of PCNN and GA with PCNN, it is inferred that PCNN and GA offer better results for all 16 datasets (Table 2).

Peak Signal-to-Noise Ratio (PSNR): PSNR is the ratio between maximum possible power of a signal and the power of corrupting noise affecting the image. The quality of an image will be better if the value of PSNR is high. On comparing PSNR values of PCNN and GA with PCNN, it is inferred that PCNN and GA offer better results for all 16 datasets (Table 3).

Entropy: Entropy reflects the amount of information content which is available in the fused image. Higher the value of entropy, higher is the quality of fused image. On comparing entropy values of PCNN and GA with PCNN, it is inferred that PCNN and GA offer better results for almost all datasets.

7 Conclusion and Future Scope

A new image fusion algorithm using pulse-coupled neural network (PCNN) with genetic algorithm (GA) optimization has been proposed which uses the firing frequency of neurons to process the image fusion in PCNN. The performance of the proposed algorithm has been evaluated using sixteen sets of computed tomography (CT) and positron emission tomography (PET) images obtained from Bharat Scans. Qualitative and quantitative analyses demonstrate that "optimization of image fusion using pulse-coupled neural network (PCNN) and genetic algorithm (GA)" outperforms PCNN technique.

The proposed strategy can be extended to merge color images since color carries remarkable information and our eyes can observe even minute variations in color. With the emerging advances in remote airborne sensors, ample and assorted information is accessible in the fields of resource investigation, environmental monitoring and disaster prevention. The existing strategies discussed in the literature survey introduce distortion in color. The algorithm proposed can be extended to fuse remote sensing images obtained from optical, thermal, multispectral and hyperspectral sensors without any color distortion.

Multimodal medical image fusion has been implemented with static images in the proposed work. At present, fusion of multimodal video sequences generated by a network of multimodal sources is turning out to be progressively essential for surveillance purposes, navigation and object tracking applications. The integral data provided by these sensors should be merged to yield a precise gauge so as to serve more efficiently in distinct tasks such as detection, recognition and tracking. From the fused output, it is conceivable to produce a precise representation of the recognized scene which in turn finds its use in variety of applications.

Table 2 Comparison analysis of PRD and RMSE for PCNN and hybridization of PCNN and GA

Percentage residual error (PRD)		
Datasets	PCNN (gradient)	Hybridization of PCNN and GA
1	0.4273	3.3080e−008
2	0.3893	4.8480e−008
3	0.4878	4.8667e−008
4	0.4283	1.5920e−008
5	0.3807	5.0838e−008
6	0.4216	7.4327e−008
7	0.4041	7.2718e−008
8	0.3904	8.0547e−008
9	0.1121	6.9992e−008
10	46.6390	5.1606e−008
11	0.1795	6.5642e−008
12	6.9132	4.8696e−008
13	0.1654	4.4340e−008
14	1.1723	5.2005e−008
15	0.1393	6.8369e−008
16	1.5822e−004	7.0856e−008
Root mean square error (RMSE)		
Datasets	**PCNN (gradient)**	Hybridization of PCNN and GA
1	0.0057	5.4255e−013
2	0.0054	6.9413e−013
3	0.0085	6.7392e−013
4	0.0094	2.2869e−013
5	0.0069	8.2856e−013
6	0.0091	8.7955e−013
7	0.0076	8.5858e−013
8	0.0070	8.2650e−013
9	0.0089	9.0324e−013
10	0.0098	7.6707e−013
11	0.0110	9.1334e−013
12	0.0085	8.2101e−013
13	0.0088	6.7181e−013
14	3.2809e−004	4.552e−013
15	0.0077	8.7183e−013
16	0.0049	5.7605e−013

Table 3 Comparison analysis of PSNR and entropy for PCNN and hybridization of PCNN and GA

Peak signal-to-noise ratio(PSNR)		
Datasets	PCNN (gradient)	Hybridization of PCNN and GA
1	54.5123	55.7234
2	55.0001	57.2346
3	53.4523	54.8976
4	56.1234	57.1042
5	55.6321	57.1234
6	56.1235	57.8432
7	55.4567	56.7046
8	55.1732	56.5460
9	57.8432	58.4389
10	57.2341	58.8975
11	57.6574	59.1004
12	55.6978	57.9874
13	54.2054	55.5512
14	56.1800	58.1254
15	57.8358	58.2657
16	55.8526	57.2116
Entropy		
Datasets	PCNN (gradient)	Hybridization of PCNN and GA
1	7.3120	8.1423
2	7.4250	8.4146
3	7.3690	8.0799
4	7.9854	8.3523
5	7.8453	8.0733
6	8.0001	8.0452
7	7.7785	8.3483
8	7.4567	8.4209
9	7.3001	8.2272
10	7.5254	7.6642
11	7.3001	7.3740
12	7.9784	8.1282
13	7.8546	8.1151
14	7.8945	8.2251
15	7.9000	8.0205
16	8.1234	8.6886

References

1. P. Hill, M. Ebrahim Al-Mualla, D. Bull, Perceptual image fusion using wavelets. IEEE Trans. Image Process. **26**(3), 1076–1088
2. N. Mittal, et al., *Decomposition & Reconstruction of Medical Images in Matlab Using Different Wavelet Parameters*, in 1st International Conference on Futuristic Trend In Computational Analysis and Knowledge Management. IEEE (2015). ISSN 978-1-4799-8433-6/15
3. K.P. Indira, et al., Impact *of Co-efficient Selection Rules on the Performance of Dwt Based Fusion on Medical Images*, in International Conference On Robotics, Automation, Control and Embedded Systems-Race. IEEE (2015)
4. Y. Yang, M. Ding, S. Huang, Y. Que, W. Wan, M. Yang, J. Sun, *Multi-Focus Image Fusion Via Clustering PCA Based Joint Dictionary Learning*, vol. 5, pp.16985–16997, Sept 2017
5. A. Ellmauthaler, C.L. Pagliari, et al., Image fusion using the undecimated wavelet transform with spectral factorization and non orthogonal filter banks. IEEE Trans. Image Process. **22**(3), 1005–1017 (2013)
6. V. Bhateja, H. Patel, A. Krishn, A. Sahu, Multimodal medical image sensor fusion framework using cascade of wavelet and contourlet transform domains. IEEE Sens. J. **15**(12), 6783–6790 (2015)
7. B. Erol, M. Amin, *Generalized PCA Fusion for Improved Radar Human Motion Recognition*, in IEEE Radar Conference (RadarConf), Boston, MA, USA (2019), pp. 1–5
8. V.S. Petrovic, C.S. Xydeas, Gradient based multi resolution image fusion. IEEE Trans. Image Process. **13**(2), 228–237 (2004)
9. P.J. Burt, E.H. Adelson, The Laplacian pyramid as a compact image code. IEEE Trans. Commun. **31**(4), 532–540 (1983)
10. J. Tian, L. Chen, Adaptive multi-focus image fusion using a waveletbased statistical sharpness measure. Signal Process. **92**(9), 2137–2146 (2012)
11. M.D. Nandeesh, M. Meenakshi, *A Novel Technique of Medical Image Fusion Using Stationary Wavelet Transform and Principal Component Analysis*, in 2015 International Conference on Smart Sensors and Systems (IC-SSS), Bangalore (2015), pp. 1–5
12. Q.M. Gaurav Bhatnagar, W. Jonathan, Z. Liu, Directive contrast based multimodal Medical image fusion in NSCT domain. IEEE Trans. Multimedia **15**(5), 1014–1024 (2013)
13. S. Das. M.K. Kundu, A neuro-fuzzy approach for medical image fusion. IEEE Trans. Biomed. Eng. **60**(12), 3347–3353 (2013)
14. T. Lu, C. Tian, X. Kai, Exploiting quality-guided adaptive optimization for fusing multimodal medical images. IEEE Access **7**, 96048–96059 (2019)
15. D. Gai, X. Shen, H. Cheng, H. Chen, Medical image fusion via PCNN based on edge preservation and improved sparse representation in NSST domain. IEEE Access **7**, 85413–85429
16. O. Hajer, O. Mourali, E. Zagrouba, Non-subsampled shearlet transform based MRI and PET brain image fusion using simplified pulse coupled neural network and weight local features in YIQ colour space. IET Image Proc. **12**(10), 1873–1880 (2018)
17. Y. Jia, C. Rong, Y. Wang, Y. Zhu, Y. Yang, *A Multi-Focus Image Fusion Algorithm Using Modified Adaptive PCNN Model*, in 12th International Conference on Natural Computation, Fuzzy Systems and Knowledge Discovery (ICNC-FSKD). IEEE (2016), pp. 612–617
18. N. Wang, W. Wang, *An Image Fusion Method Based on Wavelet and Dual-Channel Pulse Coupled Neural Network*, in 2015 IEEE International Conference on Progress in Informatics and Computing (PIC) (2015), pp. 270–274
19. M. Arif, N. Aniza Abdullah, S. Kumara Phalianakote, N. Ramli, M. Elahi, *Maximizing Information of Multimodality Brain Image Fusion using Curvelet Transform with Genetic Algorithm*, in IEEE 2014 International Conference on Computer Assisted System in Health (CASH) (2014), pp. 45–51
20. L. Fu, L. Yifan, L. Xin, *Image Fusion Based on Nonsubsampled Contourlet Transform and Pulse Coupled Neural Networks*, in IEEE Fourth International Conference on Intelligent Computation Technology and Automation, vol. 2 (2011), pp. 180–183

21. C.W Lacewell, M. Gebril, R. Buaba, A. Homaifar, *Optimization of Image Fusion using Genetic Algorithm and Discrete Wavelet Transform*, in Proceedings of the IEEE 2010 National Aerospace and Electronics Conference (NAECON) (2010), pp. 116–121
22. X. Wang, L. Chen, *Image Fusion Algorithm Based on Spatial Frequency-Motivated Pulse Coupled Neural Networks in Wavelet Based Contourlet Transform Domain*, in 2nd Conference on Environmental Science and Information Application Technology, vol. 2. IEEE (2010), pp. 411–414
23. Y. Yang, J. Dang, Y. Wang, *Medical Image Fusion Method Based on Lifting Wavelet Transform and Dual-channel PCNN*, in 9th IEEE Conference on Industrial Electronics and Applications (2014), pp. 1179–1182
24. Y. Wang, J. Dang, Q. Li, S. Li, Multimodal Medical Image Fusion Using Fuzzy Radial Basis Function Neural Networks, in *IEEE, Proceedings of the 2007 International Conference on Wavelet Analysis and Pattern Recognition*, vol. 2 (2007), pp. 778–782
25. T. Li, Y. Wang, Multi scaled combination of MR and SPECT images in neuroimaging: a simplex method based variable-weight fusion. Comput. Method Programs Biomed. **105**:35–39
26. C.W Lacewell, M. Gebril, R. Buaba, A., *Optimization of Image Fusion using Genetic Algorithm and Discrete Wavelet Transform*, in Proceedings of the IEEE 2010 National Aerospace and Electronics Conference (NAECON) (2010), pp. 116–121
27. R. Gupta, D. Awasthi, *Wave-Packet Image Fusion Technique Based on Genetic Algorithm*, in IEEE, 5th International Conference on Confluence The Next Generation Information Technology Summit (2014), pp. 280–285
28. A. Krishn, V. Bhateja, Himanshi, A. Sahu, *Medical Image Fusion Using Combination of PCA and Wavelet Analysis*, in IEEE International Conference on Advances in Computing, Communications and Informatics (2014), pp. 986–991
29. A. Sahu, V. Bhateja, A. Krishn, Himanshi, *Medical Image Fusion with Laplacian Pyramids*, in IEEE, 2014 International Conference on Medical Imaging, m-Health and Emerging Communication Systems (2014), pp. 448–453

Advanced Colored Image Encryption Method in Using Evolution Function

Shiba Charan Barik, Sharmilla Mohapatra, Bandita Das, Mausimi Acharaya, and Bunil Kumar Balabantaray

1 Introduction

Media communication is one of the important aspects that affects innovations and economies of the society. The methods for the present media transmission significantly rely on general society arrange. Today, we traded huge measure of information in different configuration over general society organize. Furthermore, as we probably am aware people in general system consistently have the issues of security. In such situation, we cannot depend just on the security arrangements of the system. Aside from this, our framework must have fit to encode the information before they transmit it over general society organize. Pictures are likewise a productive method to speak to the information, so to present a strong picture encryption system is the primary goal of this paper. Since the most recent decades, different encryption procedures have been proposed to scramble a picture. Henon's chaotic system-based image encryption method is suggested in [1]. Using chaotic permutation, block-based image encryption is proposed in [2]. For an entire shading picture, DWT-based lossless confused encryption in recurrence area is suggested in [3]. Lorenz system-based encryption

S. C. Barik (✉) · S. Mohapatra · M. Acharaya
Department of Computer Science & Engineering, DRIEMS, Cuttack, Odisha, India
e-mail: barikshiba@gmail.com

S. Mohapatra
e-mail: sharmilla.cs@gmail.com

M. Acharaya
e-mail: acharya.mousumi@gmail.com

B. Das
Department of CS, RD Women's University, Bhubaneswar, India
e-mail: bandita.gunu@gmail.com

B. K. Balabantaray
Department of Computer Science and Engineering, NIT Meghalaya, Shillong, India
e-mail: bunil@nitm.ac.in

S. Patnaik et al. (eds.), *Advances in Machine Learning and Computational Intelligence*, Algorithms for Intelligent Systems, https://doi.org/10.1007/978-981-15-5243-4_83

is also proposed in [4]. In spatial domain, utilizing Henon map, logistic Map, cubic map, and quadratic Map is suggested in [4].

Image encryption strategies that are used in the spatial domain have confounded the area of the pixel. The image encryption method used [5–9] in frequencies is either puzzled the area of frequencies or diffused the recurrence's worth utilizing different riotous maps, and/or with appropriate adjustment or change in tumultuous maps. Picture encryption acted in either area has its own advantages and disadvantages. The focal points that are accessible in one area are commonly not accessible in other. Our proposed calculation evokes the highlights of both encryption for spatial area monitored in recurrence space. Our proposed picture encryption calculation firstly diffuses the original image utilizing the logistic chaotic map in spatial area, at that point confounded the frequencie estimations of diffused picture in the recurrence space utilizing the Arnold's cat map and discrete cosine transformation. The investigation of trial result communicated that how much our proposed picture encryption system is vigorous to oppose any kinds of assaults.

2 Applied Methodology

2.1 Logistic Map

A series of complex random numbers are generated through a nonlinear chaotic map which is an evolution function. This series can be utilized either to confuse or to diffuse the pixels. The evolution function as logistic chaotic map(LCM) is a 1-D chaotic map, defined as:

$$Y_{n+1} = \mu Y_n(1 - Y)$$

where

μ chaotic factor
n number of iterative operations, and

where $\mu \in [0,3]$ and $Y \in [0,1]$ in which, the chaotic characteristic is observed when $\mu \in [2.54,3]$.

2.2 Arnold's Cat Map

The proposed technique is using Arnold's cat map (ACM), a two-dimensional map that converts the image $f(x, y)$ of size $N \times N$ to $f'(x', y')$. The ACM is defined as:

$$\begin{bmatrix} x_{i+1} \\ y_{i+1} \end{bmatrix} = \begin{bmatrix} 1 & W_b \\ W_a & W_b W_b + 1 \end{bmatrix} \begin{bmatrix} x_i \\ y_i \end{bmatrix} \bmod (N) \tag{1}$$

where (x_{i+1}, y_{i+1}) is repositioned at (x_i, y_i) of the original image. The W_a and W_b parameters are any positive integers.

The inverse of ACM is

$$\begin{bmatrix} x_i \\ y_i \end{bmatrix} = \begin{bmatrix} 1 & W_b \\ W_a & W_b W_b + 1 \end{bmatrix} \begin{bmatrix} x_{i+1} \\ y_{i+1} \end{bmatrix} \bmod (N) \tag{2}$$

2.3 Discrete Cosine Transformation

The two-dimensional DCT of an M by N matrix is defined as follows:

$$f(u, v) = \alpha_u \alpha_v \sum_{y=0}^{N-1} I(x, y) \cos \frac{\pi(2x+1)u}{2M} \cos \frac{\pi(2y+1)v}{2N} \tag{3}$$

and inverse DCT (IDCT) is as follows

$$I(x, y) = \alpha_u \alpha_v \sum_{x=0}^{M-1} \sum_{y=0}^{N-1} C(u, v) \cos \frac{\pi(2x+1)u}{2M} \cos \frac{\pi(2y+1)v}{2N} \tag{4}$$

where $\alpha_u = \begin{cases} \frac{1}{\sqrt{M}}, & \text{for } u = 0 \\ \sqrt{\frac{2}{M}}, & \text{for } 1 \le u \le M-1 \end{cases}$ And $\alpha_v = \begin{cases} \frac{1}{\sqrt{N}}, & \text{for } v = 0 \\ \sqrt{\frac{2}{N}}, & \text{for } 1 \le v \le N-1 \end{cases}$

$f(u, v)$ is the DCT coefficients of image I, whereas the top extreme left component is termed as DC coefficient and all other components are termed as AC coefficients. However, DCT is used for transformation of the entire image [2].

3 The Proposed Method

Consider, input as the original image, and initial values for the two parameters W_a and W_b are of μ and x_0, respectively. Then, I_R, I_G, and I_B (I_R: Red, I_G: Green, and I_B: Blue) are used for storing the color components of each pixels. In the next step, the diffusion operation using LCM is performed on each matrices I_R, I_G, and I_B. The LCM algorithm yields a stream of keys, and XOR operation is applied on the keys along with every element of matrices I_R, I_G, and I_B. This results that matrices I_R, I_G, and I_B now contain diffused values of color components of pixel of original image. Next, the three matrices are divided into the block size of 8 × 8. Then, DCT is performed on each block of the matrices. As a result, we get I_{R}, I_{G}, and I_{B} with the modified values as DCT coefficients. The topmost left coefficient with the size of 8 × 8 blocks of I_{R}, I_{G}, and I_{B} are considered as DC coefficient and others are AC coefficients. Now, the diffused image is converted from the spatial domain to the frequency domain. This will not degrade the image as the DC component keeps the necessary data of the image. Then, ACM transformation is used along with keys W_a and W_b to add confusion in every blocks. Lastly, inverse DCT is used to acquire the encrypted image.

Encryption Algorithm

The matrices I_{R}, I_{G}, and I_{B} are collectively represented as I_{RGB} for making the algorithm simple.

STEP 1. Consider original image as $I(x,y)$ with keys W_a and W_b with the preliminary values of μ and x_0.
STEP 2. For $x = 0$ to $I.width$
  For $y = 0$ to $I.height$
    $I_{RGB_{x,y}} = I.getRGB(y, x)$
STEP 3. For $x = 0$ to $I.width$
  For $y = 0$ to $I.height$
    $L_{KEY_{x,y}} = logisticMAP(\mu, x_0)$
STEP 4. For $x = 0$ to $I.width$
  For $y = 0$ to $I.height$
    $I_{RGB_{x,y}} = I_{RGB_{x,y}}$ XOR $L_{KEY_{x,y}}$
STEP 5. For $x = 0$ to $I.width/8$
  For $y = 0$ to $I.height/8$
    I_{RGB}.dctT(8,8)
STEP 6. For $x = 0$ to $I.width/8$
  For $y = 0$ to $I.height/8$
    I_{RGB} .arnoldMAP()
STEP 7. For $x = 0$ to $I.width$
  For $y = 0$ to $I.height$
    I_{RGB} .idctT(8,8)
STEP 8. For $x = 0$ to $I.width$
  For $y = 0$ to $I.height$
    $drawImage(x, y, I_{RGB})$
STEP 9. End.

Excepting the logistic map operation, the decryption steps are in the reverse order.

4 Results and Investigation

In our experiment, three different images such as Lena, peppers, and cat are used for investigation of the proposed method. Different standard image encryption techniques are also performed along with the proposed method on these images. A comparative calculation is also presented in the tables. The initial values of parameters for all these images are as follows: $\mu = 3.9600$; $x_0 = 0.1234$; $W_a = 94$; and $W_b = 50$. The result of the proposed method is shown in Fig. 1 which shows our method

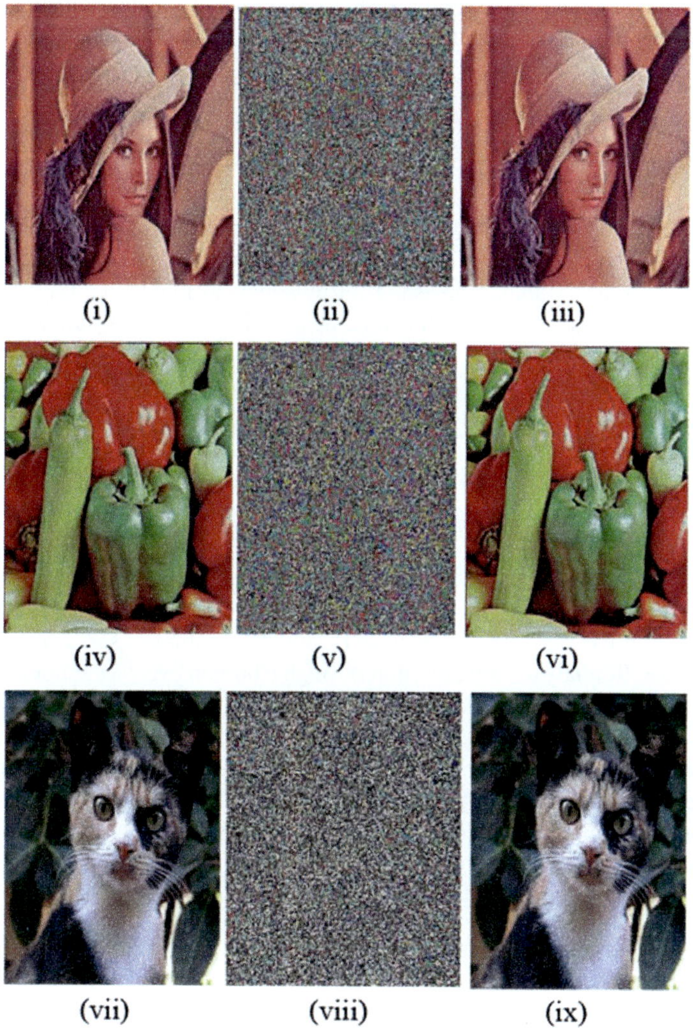

Fig. 1 **i**, **iv**, and **vii** are original image, **ii**, **v**, and **viii** are encryption result image, and **iii**, **vi**, and **ix** are images after decryption

has outperformed the others. Experiments are performed on the MATLAB R2015a platform.

4.1 Observation Through Histogram

The intensity distribution of the color components of images is shown in histogram of the images. The histograms of the encrypted images must have uniform distribution which indicate that the method used for encryption is robust to any statistical attacks. In Fig. 2, histogram of original along with encrypted image of Lena is shown.

4.2 Entropy Observation

The degree of randomness of the pixel intensity is described by the entropy analysis of an image. If the encrypted image is having more randomness, then it is secured from statistical attack. Table 1 shows the entropy analysis in which it is observed that the encrypted images are having the value near 8. This indicates that the proposed method is more robust attack.

Table 2 represents the entropy analysis of encrypted image using the existing method and the proposed method. It is observed that the proposed method outperforms the existing one as it is nearly equal to 8.

4.3 Peak Signal-to-Noise Ratio Analysis

Peak signal-to-noise ratio (PSNR) is used to study the nature of scrambleness in the image. The PSNR analysis indicates that through a human eye we cannot distinguish the two images. The greater PSNR value concerning the original and its separate unscrambled image affirms the less twisting in the first plain picture. Table 3 has demonstrated the PSNR analysis of images that is used in the experiment and it is observed that it is constantly over the normal value. In this way, our proposed calculation is sufficiently skilled to recuperate the decoded picture with more visual appealing.

4.4 Correlation Coefficient

The association of any two contiguous pixels of a picture is a hint for assailants to pick up the measurable data about the picture. The estimation of the connection coefficient close to zero of a picture has less important measurable data of any two nearby pixels.

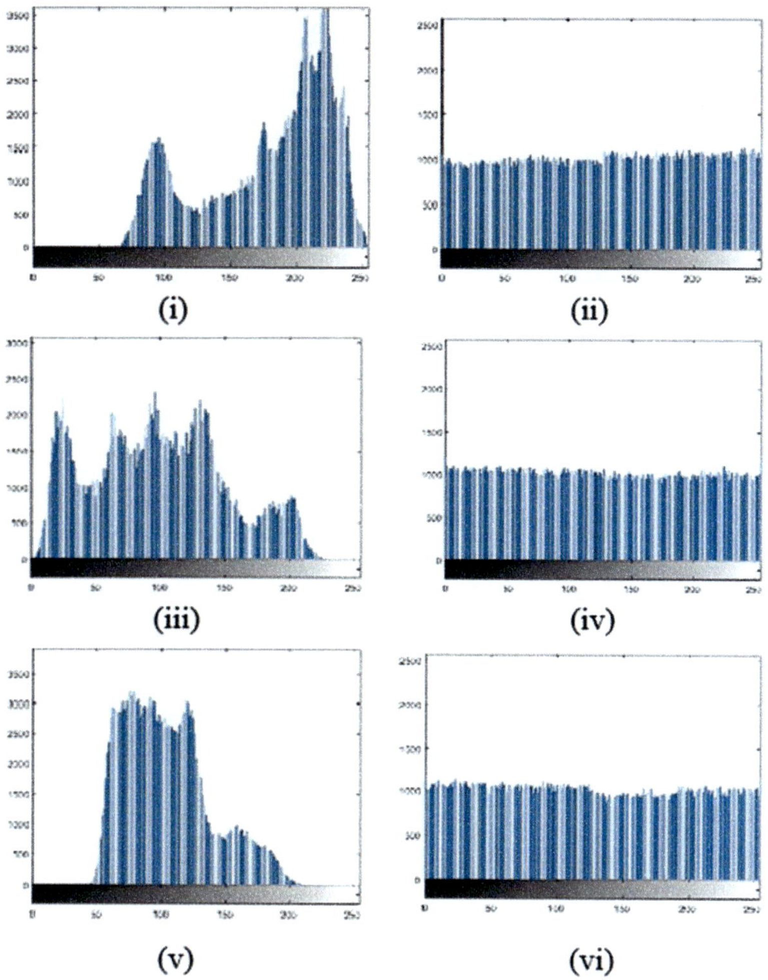

Fig. 2 Histogram for colored component of **i** red of original image Lena, **ii** red of the encrypted Lena image, **iii** green of original Lena image, **iv** green of the encrypted image, **v** blue of the original Lena image, and **vi** blue of encrypted image

Table 1 Entropy calculation of the original and the corresponding encrypted image

	Original			Encrypted		
	Red	Green	Blue	Red	Green	Blue
Lena.jpeg	7.2451	7.6831	6.7562	7.8899	7.8798	7.8979
Peppers.jpeg	7.4321	7.4818	7.3162	7.8897	7.7881	7.8897
Cat.jpeg	7.5761	7.8275	7.8591	7.8978	7.8879	7.8897

Table 2 Entropy analysis for Lena.jpeg using existing method [3] and proposed method

Existing method [3]			Proposed algorithm		
R	G	B	R	G	B
7.3306	7.5387	7.3163	7.8988	7.8998	7.8899

Table 3 PSNR values between original and its decrypted image

Decrypted image	PSNR value
Lena	33.3545
Peppers	33.5621
Cat	32.782

Along these lines, we take all sets of two adjoining pixels first in on a level plane and afterward vertically from the plain picture and scrambled picture. Moreover, we haphazardly take 2000 sets of the slantingly nearby pixels from the plain pictures and encoded pictures. Table 4 has indicated the connection coefficient of each plain picture and relationship coefficient of the separate scrambled pictures that utilized in this examination. The relationship coefficients of scrambled pictures are exceptionally near zero that legitimizes the safe highlights of the proposed calculation against any factual assaults.

Table 5 contains the correlation coefficient of the proposed encryption method and the existing one applied to Lena image. The proposed technique has a correlation coefficient of encrypted Lena image which is superior than coefficient in [3], and nearly to zero.

Table 4 Correlation values of the two neighboring pixels in different directions in both original and encrypted image (Direction: H—horizontal, V—vertical, and D—diagonal)

Image	Correlation values of the original versus encrypted						
	Directional value	Original			Encrypted		
		R	G	B	R	G	B
Lena	H	0.9897	0.9979	0.9987	0.0023	0.0062	0.0011
	V	0.9988	0.9978	0.9867	0.0231	0.0035	0.0151
	D	0.9888	0.9694	0.9986	0.0207	0.0212	0.0021
Peppers	H	0.9864	0.9920	0.8899	0.0066	0.0027	0.0231
	V	0.9875	0.8975	0.8989	0.0134	0.0033	0.0041
	D	0.8987	0.8949	0.9762	0.0065	0.0177	0.0046
Cat	H	0.8876	0.9976	0.9932	0.0202	0.0031	0.0607
	V	0.8990	0.8977	0.9679	0.0211	0.0235	0.0032
	D	0.8390	0.8990	0.8981	0.0122	0.0210	0.0075

Table 5 Correlation coefficients of neighboring pixels in the decrypted image using existing and proposed method

Directional values	Method in Ref. [3]			Proposed algorithm		
	Encrypted			Encrypted		
	R	G	B	R	G	B
H	0.3715	0.4634	0.4432	0.0023	0.0062	0.0011
V	0.4321	0.4764	0.4765	0.0231	0.0035	0.0151
D	0.3762	0.3672	0.4334	0.0207	0.0212	0.0021

4.5 Centainty Analysis

In the proposed method, the underlying estimation of μ and x_0, and assessment of W_a and W_b are considered to be the important factors of calculation. In this work, it is tried to decode an encoded images of peppers, cat, and Lena along with same initial values of W_a and W_b, and x_0 with a minor distinction in the beginning estimation of μ. The following values are considered:

$$\mu = 3.8500; \quad x_0 = 0.1123; \quad W_a = 90; \quad W_b = 60;$$

Now, the same values of W_a and W_b, and x_0 are used with a minor change in the value of μ from $\mu = 3.8500$ to $\mu = 3.8501$. Figure 3 shows the three decrypted images.

The outcome has demonstrated that the least conceivable change in the key produces a picture that is absolutely good for nothing and even not to be near the first plain picture. In this way, it is explained that how much our proposed calculation is delicate about its key.

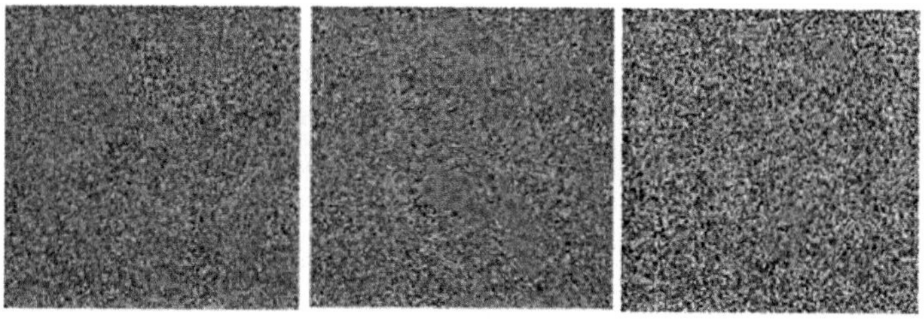

(a) (b) (c)

Fig. 3 **a** Decrypted Lena image, **b** decrypted peppers image, and **c** decrypted cat image

5 Conclusion

A colored image encryption method is proposed by using a turbulent change in spatial domain with a recurrence area. The investigation result has indicated that the scrambled images which come are robust against every attack. As it is a block-based encryption method, it can be used widely for JPEG images as well as other formats.

References

1. C. Wei-Bin, Z. Xin, *Image Encryption Algorithm Based on Henon Chaotic System*, in *2009 International Conference on Image Analysis and Signal Processing*. IEEE (2009), pp. 94–97
2. R. Munir, *A Block-Based Image Encryption Algorithm in Frequency Domain Using Chaotic Permutation*, in 2014 8th International Conference on Telecommunication Systems Services and Applications (TSSA). IEEE (2014), pp. 1–5
3. X. Wu, Z. Wang, *A New DWT-Based Lossless Chaotic Encryption Scheme for Color Images*, in 2015 International Conference on Computer and Computational Sciences (ICCCS). IEEE (2015), pp. 211–216
4. X. Chai, K. Yang, Z. Gan, A new chaos-based image encryption algorithm with dynamic key selection mechanisms. Multimedia Tools Appl. **76**(7), 9907–9927 (2017)
5. L. Xu, X. Gou, Z. Li, J. Li, A novel chaotic image encryption algorithm using block scrambling and dynamic index based diffusion. Opt. Lasers Eng. **91**, 41–52 (2017)
6. X. Chai, Z. Gan, M. Zhang, A fast chaos-based image encryption scheme with a novel plain image-related swapping block permutation and block diffusion. Multimedia Tools Appl. **76**(14), 15561–15585 (2017)
7. Y. Li, C. Wang, H. Chen, A hyper-chaos-based image encryption algorithm using pixel-level permutation and bit-level permutation. Opt. Lasers Eng. **90**, 238–246 (2017)
8. C. Guanrong, M. Yaobin, K. Chui Charles, A symmetric image encryption scheme based on 3D chaotic cat maps. Chaos Solitons Fractals **21**(3), 749–761 (2004)
9. Y. Wang, K.W. Wong, X. Liao, G. Chen, A new chaos-based fast image encryption algorithm. Appl. Soft Comput. **11**(1), 514–522 (2011)